2024 台达杯国际太阳能建筑设计竞

Awarded Works from International Solar Building Design Competition 2024

阳光 · 乡村韧性
SUNSHINE & RURAL RESILIENCE

中国建筑设计研究院有限公司　编
Edited by China Architecture Design & Research Group
执行主编：张　磊　鞠晓磊　张星儿
Chief Editor: Zhang Lei　Ju Xiaolei　Zhang Xing'er

中国建筑工业出版社

图书在版编目（CIP）数据

阳光·乡村韧性：2024台达杯国际太阳能建筑设计
竞赛获奖作品集 = SUNSHINE & RURAL RESILIENCE：
Awarded Works from International Solar Building
Design Competition 2024 / 中国建筑设计研究院有限公
司编；张磊，鞠晓磊，张星儿执行主编 . -- 北京：中
国建筑工业出版社，2025. 4. -- ISBN 978-7-112-31090-
6

Ⅰ. TU18
中国国家版本馆 CIP 数据核字第 20250GC548 号

　　本次作品集以"阳光·乡村韧性"为赛题，贯彻落实《"十四五"国家综合防灾减灾规划》精神，围绕如何构建具有韧性的村镇基础设施、提升农村地区的抗灾减灾能力展开，重点关注村镇的规划布局、建筑结构、公共设施、生态环境、绿色能源技术等方面。覆盖了新农村建设、美丽乡村、城市住宅，有建筑改造、阳光小学、适老建筑、幼儿园、科学考察站等多种类型，反映了社会民生的实际需求，共分为张家口市怀安县二堡子村建设项目和保定市清苑区李八庄村建设项目。该作品集收录了参赛项目中的一、二、三等奖和优秀奖作品，旨在提高农村的防灾减灾能力，推动可再生能源在建筑上的应用，以期打造出安全、可持续、富有活力的农村社区。本书适用于从事建筑学各分支领域研究的研究人员、工程技术人员、科技管理人员和高等院校师生阅读参考。

责任编辑：唐　旭　吴　绫　张　华
责任校对：王　烨

阳光·乡村韧性
SUNSHINE & RURAL RESILIENCE

2024 台达杯国际太阳能建筑设计竞赛获奖作品集
Awarded Works from International Solar Building Design Competition 2024
中国建筑设计研究院有限公司　编
Edited by China Architecture Design & Research Group
执行主编：张　磊　鞠晓磊　张星儿
Chief Editor: Zhang Lei　Ju Xiaolei　Zhang Xing'er
　＊
中国建筑工业出版社出版、发行（北京海淀三里河路9号）
各地新华书店、建筑书店经销
北京雅盈中佳图文设计公司制版
临西县阅读时光印刷有限公司印刷
　＊
开本：787毫米×1092毫米　1 / 12　印张：25²⁄₃　插页：1　字数：330千字
2025 年 4 月第一版　2025 年 4 月第一次印刷
定价：**178.00**元
<u>ISBN 978-7-112-31090-6</u>
　　　　　（44792）

随着气候变化带来的极端天气频发，乡村地区面临的生态压力日益凸显，提升防灾韧性、发展可再生能源成为乡村振兴的重要课题。党的二十大报告强调，要统筹乡村基础设施和公共服务布局，建设宜居宜业的和美乡村。本届竞赛以"阳光·乡村韧性"为主题，设置张家口市怀安县二堡子村防灾社区建设和保定市清苑区李八庄村可再生能源应用示范两项赛题，探索以光伏为主的可再生能源技术与韧性设计在乡村建设应用的创新融合。参赛者以专业智慧回应时代需求，通过作品展现了太阳能建筑与防灾体系的协同设计潜力，为构建安全、可持续、充满活力的未来乡村提供了宝贵实践。

感谢台达集团对2024台达杯国际太阳能建筑设计竞赛的全程支持！

谨以本书致敬所有投身于乡村绿色发展、生态建设的探索者与践行者！

As climate change intensifies and extreme weather events become more frequent, the ecological challenges confronting rural areas have grown increasingly significant. Strengthening disaster resilience and promoting renewable energy development have emerged as critical priorities for rural revitalization. The report of the 20th National Congress of China underscores the importance of integrating rural infrastructure and public service planning to create harmonious, livable villages.

The 2024 International Solar Architecture Design Competition, themed "SUNSHINE & RURAL RESILIENCE", featured two key tasks: constructing a disaster-resilient community in Erpuzi Village, Huai'an County, Zhangjiakou City; and demonstrating renewable energy applications in Libazhuang Village, Qingyuan District, Baoding City. These initiatives explored the innovative integration of photovoltaic and other renewable energy technologies with resilient design principles in rural contexts. Participants leveraged their professional expertise to address contemporary challenges, showcasing through their designs the potential synergy between solar architecture and disaster prevention systems. Their contributions provide valuable insights into building safe, sustainable, and vibrant rural communities for the future.

Many thanks are extended to Delta Electronics for its comprehensive support of the 2024 International Solar Building Design Competition.

This book is dedicated to all pioneers and practitioners committed to advancing green development and ecological construction in rural areas.

目　录
CONTENTS

综合奖·优秀奖·保定市清苑区李八庄村建设项目
Comprehensive Awards – Honorable Mention – Libazhuang Village Project, Qingyuan District, Baoding City

综合奖·入围奖·保定市清苑区李八庄村建设项目
Comprehensive Awards – Nomination – Libazhuang Village Project, Qingyuan District, Baoding City

单项奖·设计创意奖·张家口市怀安县二堡子村建设项目
Individual Award – Design Creativity Prize – Erpuzi Village Project, Huai'an County, Zhangjiakou City

单项奖·技术专项奖·张家口市怀安县二堡子村建设项目
Individual Awards – Technical Excellence Prize – Erpuzi Village Project, Huai'an County, Zhangjiakou City

有效作品参赛团队名单
Name List of All Participants Submitting Valid Works

2024台达杯国际太阳能建筑设计竞赛办法
Brief for International Solar Building Design Competition 2024

2024台达杯国际太阳能建筑设计竞赛回顾
Review of International Solar Building Design Competition 2024

主题：阳光·乡村韧性

Theme: Sunshine & Rural Resilience

一、赛题设置

为乡村振兴注入新活力、乡村基础设施建设提供新思路，2024 台达杯国际太阳能建筑设计竞赛以"阳光·乡村韧性"为主题，设置张家口市怀安县二堡子村建设项目和保定市清苑区李八庄村建设项目两个赛题，旨在提高农村的防灾减灾能力，推动可再生能源在建筑上的应用，以期打造出安全、可持续、富有活力的农村社区。

二、竞赛启动

2024 年 4 月 12 日，2024 年台达杯国际太阳能建筑设计竞赛在天津大学启动。中国工程院院士、全国工程勘察设计大师、中国建设科技集团股份有限公司首席科学家、台达杯国际太阳能建筑设计竞赛评审专家组组长崔愷，中国建设科技集团股份有限公司副总裁刘志鸿，天津大学党委常委、常务副校长胡文平，天津大学建筑学院党委书记、常务副院长宋昆，中国建筑设计研究院有限公司副总经理景泉，台

I. Competition Projects

In our efforts to infuse new vitality into rural revitalization and explore innovative approaches for rural infrastructure development, International Solar Building Design Competition 2024 has chosen "SUNSHINE & RURAL RESILIENCE" as its theme and initiated two target projects as follows: i) Erpuzi Village Project (located in Huai'an County, Zhangjiakou City, Hebei Province); and ii) Libazhuang Village Project (located in Qingyuan District, Baoding City, Hebei Province). This year's competition aims to enhance disaster prevention and mitigation capabilities in rural areas, with a focus on the integration of renewable energy in construction projects. Our goal is to contribute to the creation of safe, sustainable, and vibrant rural communities.

组委会前往竞赛场地调研
A Site Visit to the Competition Venue by the Organizing Committee

达公益事业部总监陈奕祥等领导与竞赛评审专家、各高校师生团队及媒体一同出席启动会，共同开启本届竞赛。

三、校园宣讲

自2005年第一届竞赛举办以来，组委会已先后前往清华大学、天津大学、东南大学、重庆大学、山东建筑大学等近80所高校开展竞赛校园宣讲活动，受到了高校师生的积极响应和好评。

2024年4月至7月，组委会前往中国矿业大学、厦门大学、福州大学、浙江理工大学、北京建筑大学、合肥工业大学、内蒙古工业大学、新疆大学、昆明理工大学等高校举办了宣讲会，吸引了千余名师生现场参与。宣讲会上，在竞赛介绍和赛题讲解的同时，组委会也邀请了可再生能源领域专家与师生们共同围绕太阳能建筑进行交流，并针对在"双碳"目标下建筑行业发展及建筑相关专业毕业生工作方向等问题进行深入讨论，为学生们对建筑行业未来发展方向打开了新思路。

同时，竞赛联合"迈向产能建筑"专家团队，精心策划多次线上讲堂。其内容涵盖了太阳能建筑技术应用现状与趋势、天正绿色能碳系列计算分析软件讲解、历届竞赛获奖作品分析和本届竞赛答疑等方面。

II. Commencement of the Competition

The 2024 Competition was officially launched on April 12th, 2024, at Tianjin University. Esteemed experts and VIPs took part in the Kick-off Ceremony along with college students, faculty members and media professionals. Experts include Mr Cui Kai, Academician of Chinese Academy of Engineering, Master of National Engineering Survey and Design of China, Chief Scientist of China Construction Technology Consulting Co., Ltd. (CCTC), and Head of the Jury Panel of the Competition; Mr. Liu Zhihong, Deputy President of CCTC; Mr Hu Wenping, Executive Vice President of Tianjin University; Mr Song Kun, Executive Vice Dean of the School of Architecture of Tianjin University; Mr Jing Quan, Deputy General Manager of China Architecture Design & Research Group (CADG); and Mr Chen Yixiang, Director of the Public Welfare Division of Delta Electronics.

中国工程院院士崔愷（右三）、中国建设科技集团股份有限公司副总裁刘志鸿（左三）、中国建筑设计研究院有限公司副总经理景泉（左一）、中国建筑设计研究院有限公司总工程师仲继寿（右二）、台达公益事业部总监陈奕祥（左二）、国家住宅工程中心主任张磊（右一）共同启动2024台达杯国际太阳能建筑设计竞赛
Academician Cui Kai of the Chinese Academy of Engineering, Vice President Liu Zhihong of China Construction Technology Consulting Group (CCTC), Vice President Jing Quan of China Architecture Design & Research Group (CADG), Chief Engineer Zhong Jishou of CADG, Director Chen Yixiang of Delta's Public Welfare Division, and Director Zhang Lei of the China National Engineering Research Center for Human Settlements Jointly Inaugurated the 2024 International Solar Architecture Design Competition. Their Positions are Noted for Reference: Academician Cui Kai (Third from the Right), Vice President Liu Zhihong (Third from the Left), Vice President Jing Quan (First from the Left), Chief Engineer Zhong Jishou (Second from the Right), Director Chen Yixiang (Second from the Left), and Director Zhang Lei (First from the Right)

来自高校、设计院等机构近两百名嘉宾和师生参加竞赛启动会
Nearly 200 Participants, Including Guests, Faculty Members, and Students from Universities, Design Institutes, and Other Relevant Organizations, Attended the Official Launch Meeting of the Competition

通过线上讲堂，竞赛进一步扩大了影响力，并拉近了与观众的距离。通过主旨汇报、现场答疑等环节，让观众深入了解低碳建筑技术的理念与应用实践，更透彻地理解竞赛赛题，进而激发了参赛团队在设计思维上的创意与灵感，催生出诸多创新性的思考与实践方案，使竞赛宣讲会成为太阳能建筑领域行业知识、技术、理念的重要科普和交流平台。

竞赛组委会与福州大学师生合影
A Group Photo with Faculty Members and Students from Fuzhou University and the Organizing Committee

竞赛组委会与合肥工业大学师生合影
A Group Photo with Faculty Members and Students from Hefei University of Technology and the Organizing Committee

III. Campus Roadshow

Since the inaugural of the Competition in 2005, the Organizing Committee has conducted campus roadshows at nearly 80 colleges and universities, including Tsinghua University, Tianjin University, Southeast University, Chongqing University, Shandong Jianzhu University. These promotional activities have received positive responses from both faculty members and students.

From April to July, 2024, the Organizing Committee conducted roadshow sessions in various academic institutions across China, including China University of Mining and Technology, Xiamen University, Zhejiang Sci-Tech University, Beijing University of Civil Engineering and Architecture, Hefei University of Technology, Inner Mongolia University of Technology, Xinjiang University, and Kunming University of Science and Technology. These sessions were introduced to present the Competition and explain related issues. In addition, the organizers successfully invited experts, educators, and students from the renewable energy field to engage in in-depth discussions on the topic of solar buildings. They shared insightful views on the development of the construction industry and career prospects for architecture graduates, particularly in response to the national ″dual-carbon″ objective. These discussions provided students with valuable insights into the future of the construction industry.

Concurrently, the Competition collaborated with experts from the ″Towards Productive Buildings″ team to conduct a series of online lectures covering various topics such as the current industrial landscape, trends in solar building technology applications, demonstrations of Tangent ″*Green Energy and Carbon Series*″ software for calculating building energy consumption and carbon emissions, analysis of winning works from previous competitions, and Q & A sessions for 2024 Competition.

Through online lectures, the Competition's influence has been further expanded, effectively bridging the gap between us and our audience. Through keynote reports, on-site Q & A sessions, and other interactive segments, the

竞赛组委会与内蒙古工业大学师生合影
A Group Photo with Faculty Members and Students from Inner Mongolia University of Technology and the Organizing Committee

竞赛组委会与新疆大学师生合影
A Group Photo with Faculty Members and Students from Xinjiang University and the Organizing Committee

四、研学活动

竞赛已走过近 20 年历程，始终秉承着"让梦想照进现实"的理念，不断推动太阳能与建筑集成应用的技术进步，让绿色、环保的可持续发展理念深入人心。截至目前，共有六项作品实地建成并投入使用，一项作品即将竣工，这些作品不仅展示了太阳能技术在建筑领域的应用成果，更让大众切实感受到了清洁能源为生活带来的美好变化。

2024 年 7 月 19 日至 21 日，竞赛首次开展实地建设研学活动，邀请建筑领域专家、高校教师、竞赛参与者与往届获奖团队成员组成的研学团队一同前往 2015 年台达杯竞赛的建设项目——位于青海省西宁市湟源县兔尔干村的"日月山下二十四个庄廓"民宿聚落。该项目建设因地制宜，充分利用丰富的太阳能资源，设置被动式太阳能供暖技术、光伏发电技术、太阳能热水技术、智能微电网系统等技术，实现产能量大于用能需求的全年零能耗目标，被列为青海省"科技促进新农村建设计划"项目，实现科技创新支撑新农村建设的示范推广。研学团队深入庄廓及

audience gained a profound understanding of the principles and practical applications of low-carbon building technology, as well as a comprehensive grasp of the competition theme. This has further stimulated the creativity and inspiration of the participating teams in design thinking, giving rise to numerous innovative ideas and practical solutions. As a result, these lectures have become a vital platform for popularizing and exchanging industry knowledge, technologies, and concepts within the realm of solar buildings.

IV. Academic Study Tours

The Competition has been ongoing for nearly 20 years, consistently adhering to the concept of "making dreams come true", continuously promoting technological advancements in the integrated application of solar energy and buildings, and deeply ingraining the green and environmentally friendly sustainable development philosophy in people's minds. To date, a total of six projects have been constructed on-site and put into use, with one project nearing completion. These projects not only showcase the application achievements of solar energy technology in architecture but

竞赛线上宣讲会海报
Poster for Online Briefing Session of the Competition

竞赛海报（中、英文版）
Competition Poster (Chinese and English)

also enable the public to truly experience the positive changes that clean energy brings to life.

From July 19th to 21st, 2024, the Competition held its first academic study tour for on-site construction research. A team composed of experts in the field of architecture, university teachers, competition participants, and members of previous winning teams visited the *Twenty-Four Zhuangkuo under the Sun and Moon Mountain* (translator's note: Zhuangkuo is a traditional quadrangle courtyard composed of tall earthen walls and heavy gates for defence, popular in the vast rural areas of Qinghai Province, characterized by the plateau continental climate and historically affected by the scourge of war and banditry since the Tang and Song Dynasties). It is a guesthouse cluster located in Tu'er Gan Village, Huangyuan County, Xining City, Qinghai Province, which was the construction project of the 2015 Competition. The project was built in accordance with local conditions, making full use of the abundant solar resources and

"日月山下24个庄廓"是2015年台达杯竞赛一、二、三等奖和优秀奖获奖作品设计，为竞赛获奖作品首次成组示范建设

First, Second, and Third Prize-Winning Entries, as well as the Excellence Awards from the 2015 Competition, were Successfully Incorporated into the *Twenty-four Zhuangkuo under the Sun and Moon Mountain Project*, Representing the Inaugural Group Demonstration Construction of Award-winning Designs from the Competition

incorporating passive solar heating technology, photovoltaic power generation technology, solar water heating technology, and intelligent micro-grid systems, achieving the goal of zero energy consumption throughout the year with energy production exceeding energy demand. It was listed as a project of the "Science and Technology Promotion of New Rural Construction Plan" in Qinghai Province, serving as a demonstration of how scientific and technological innovation can support new rural construction. The tour delved into the Zhuangkuo and both the new and old sites of Tu'er Gan Village, focusing on the practical application of solar architectural designs. Simultaneously, a symposium themed *Integration of Traditional Regional Farmhouses and Renewable Energy* was held to discuss the latest green building technology design concepts, aiming to promote the application of renewable energy in buildings and bring benefits to people's lives. This event also received extensive coverage from media outlets such as CCTV, People's Daily Online, China News Service, *Beijing Daily*, and *Economic Daily*.

兔尔干新村与旧址，关注太阳能建筑设计的实践应用。同步举办以"传统地域性农宅与可再生能源集成应用"为主题的研讨会，探讨前沿绿色建筑技术设计理念，以期推动可再生能源在建筑上的应用，让科技走进生活、惠及民生。此项活动还获得了中央电视台、人民网、中国新闻网、北京日报、经济日报等媒体的广泛报道。

V. Media and Publicity

Since the launch of 2024 Competition, the Organizing Committee has conducted publicity efforts through various media forms, including: real-time bilingual progress reports on the official website; simultaneous online and offline promotion of solar buildings; keyword search in Baidu for easy public access to the official website; coverage on over 50 domestic and international websites such as Xinhuanet.com, Tencent, and Sina.com, with direct links to the official website; liaisons with various domestic and foreign university-level media outlets for releasing competition information and updates; real-time news through the Wechat public account, Weibo and other media channels, providing downloadable information resources and case analyses. These endeavors effectively enhance both the influence of the Competition and the technical capabilities of the participating teams.

五、媒体宣传

自竞赛启动以来，组委会通过多种媒体形式开展竞赛宣传工作，包括：竞赛官方网站（双语）实时报道竞赛进展情况；线上线下同步开展太阳能建筑的科普宣传；在百度设置关键字搜索，方便大众查询，从而更快捷地登录竞赛网站；在新华网、腾讯网、新浪网等50余家国内外网站上报道竞赛相关信息并链接到竞赛官方网站；与多所国内外校级媒体取得联系，并发布竞赛信息与动态；通过微信公众号、微博等媒体平台实时发布竞赛相关动态，提供竞赛相关资料下载与案例分析等，有效提高竞赛的影响力及参赛团队的技术能力。

六、竞赛注册及提交情况

竞赛的注册时间为2024年4月12日至2024年8月15日，共786组团队通过竞赛官网完成注册，其中有来自我国港澳台及美国、英国、荷兰、越南等注册团队7组。截至2024年9月30日24时，组委会收到提交作品204件。

七、竞赛评审

竞赛评审由形式审查、竞赛初评、中评、终评四个阶段组成，其中终评阶段分为线上评审和现场评审会两个部分组成。

10月1日至18日，组委会组织专家对通过形式审查的有效作品开展初评工作。根据竞赛办法中的评比标准，专家们逐一针对每件作品进行审查。历经合规性审查、竞赛初评及评审分数汇总、排名等环节后，共有80件作品进入中评。

青海省科技厅原副厅长周卫星（左三）、中国建筑设计研究院有限公司总工程师仲继寿（左二）、台达首席品牌官郭珊珊（左四）、台达首席可持续发展官暨发言人周志宏（左一）、国家住宅与居住环境工程技术研究中心主任张磊（左五）、陕西省建筑设计研究院有限公司西宁办事处商选平（左六）共同参与研学活动启动会
Mr Zhou Weixing, Former Deputy Director of the Department of Science and Technology of Qinghai Province (Third from the Left), Mr Zhong Jishou, Chief Engineer of CADG (Second from the Left), Ms Guo Shanshan, Chief Brand Officer of Delta Electronics (Fourth from the Left), Mr Zhou Zhihong, Chief Sustainability Officer and Spokesperson of Delta Electronics (First from the Left), Ms Zhang Lei, Director of the China National Engineering Research Center for Human Settlements (Fifth from the Left), and Mr Shang Xuanping, Director of the Qinghai Branch of Shaanxi Provincial Architectural Design and Research Institute (Sixth from the Left), Collectively Attended the Opening Ceremony of the Research and Study Program

VI. Registration & Submission

The registration period of the Competition lasted from April 12th to August 15th, 2024. A total of 786 teams successfully completed registrations through the official website, including 7 teams from Hongkong, Macao, Taiwan of China, and other countries such as the United States, UK, the Netherlands, and Vietnam. As of midnight (24：00) on September 30th, 2024, the Organizing Committee has received a total of 204 submitted works.

VII. Entries Appraisal

The Entries Appraisal comprises four stages: formal examination, preliminary appraisal, mid-term appraisal, and final appraisal. The final appraisal consists of both online appraisal and on-site appraisal sections.

From October 1st to 18th, the Organizing Committee arranged for experts to conduct a preliminary appraisal for valid works that had passed the formal examination. Following the appraisal criteria outlined in the Competition Guide,

建筑专家、高校教师、竞赛参与者与往届获奖成员及媒体共同参与青海研学行
A Diverse Group Comprising Architectural Experts, University Faculty Members, Competition Participants, Past Award Recipients, and Representatives from the Media Collectively Took Part in the Qinghai Study Tour

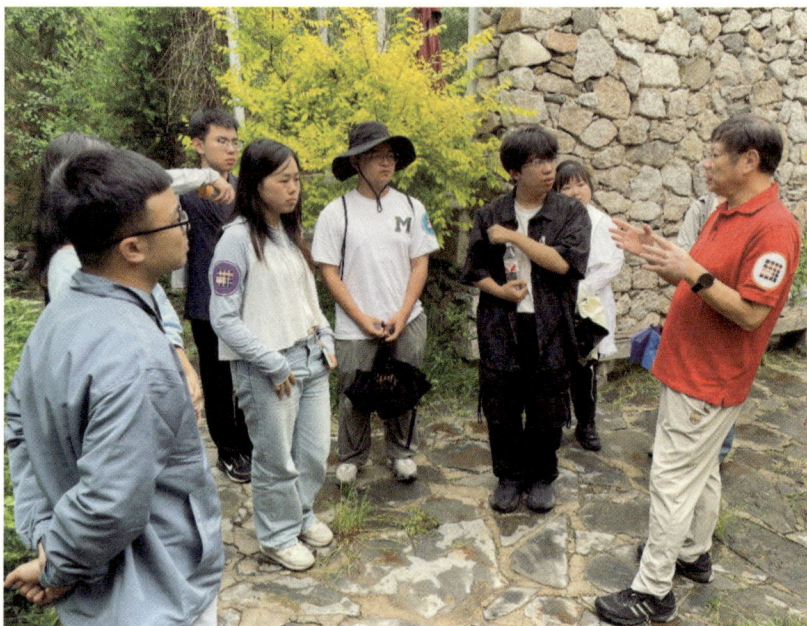

研学团队深入兔尔干新村以及老村调研，分析当地自然、人文环境
The Research and Study Team Conducted in-depth Investigations in Both the New and Old Villages of Tuergan, Analyzing the Local Natural and Cultural Environment

the experts evaluated each work through compliance inspection, preliminary assessment, and scoring. A total of 80 works proceed to the mid-term appraisal procedure.

From October 21st to 28th, the Organizing Committee invited nine esteemed experts to assess the entries for mid-term appraisal. Through strict assessment by the experts, a selection of 30 works advanced to the final appraisal procedure.

The Jury Panel for the final appraisal was composed of nine distinguished experts, including：Mr Cui Kai, Academician of Chinese Academy of Engineering, Master of National Engineering Survey and Design of China; Mr King Mun Yeang, Founder of T. R. Hamzah & Yeang Sdn. Bhd. of Malaysia, Winner of the Liang Sicheng Architecture Prize 2016; Mr Qian Feng, Master of National Engineer Survey and Design of China, Professor of College of Architecture and Urban Planning of Tongji University; Mr Prof. Peter Luscuere, professor of the Department of Architecture, Delft University of Technology, Netherlands; Mr Takashi Hokoiwa, Chairman of AXS Satow Inc. Japan; Mr Huang Qiuping, Chief Architect of East China Architectural Design &

中央电视台对竞赛研学活动进行报道
CCTV Report on the Competition and Study Tour Activities

现场评审专家组与马来西亚汉沙杨建筑师事务所创始人杨经文（右下）、荷兰代尔夫特理工大学教授Peter Luscuere（左上）等国际专家连线讨论
The On-site Review Expert Panel Conducted an Online Dialogue with Prominent International experts, Including Mr Kenneth King Mun Yeang, Founder of T.R.Hamzah & Yeang Sdn. Bhd., Malaysia (Lower Right), and Mr Peter Luscuere, a Professor at Delft University of Technology in the Netherlands (Upper Left)

在 10 月 21 日至 28 日的中评阶段，组委会邀请了行业内 9 位知名专家对竞赛作品进行评审。经由专家组严谨且细致的评审后，评分位居前 30 名的作品进入终期评审环节。

终期评审团队由中国工程院院士、全国工程勘察设计大师崔愷；马来西亚汉沙杨建筑师事务所创始人、2016 年梁思成建筑奖获得者杨经文；全国工程勘察设计大师、同济大学建筑与城市规划学院教授钱锋；荷兰代尔夫特理工大学建筑学院教授 Peter Luscuere；日本株式会社佐藤综合计画代表取缔役社长鉾岩崇；华东建筑设计研究总院有限公司总建筑师黄秋平；中国建筑设计研究院有限公司总工程师仲继寿；中国建筑科学研究院有限公司建筑设计院首席总建筑师薛明；清华大学建筑学院长聘教授宋晔皓 9 位专家组成。

终评专家组首先复核了中评结果、确定入围作品并评审，择取成绩最高的 15 组入围现场评审会。

Research Institute Co., Ltd.(ECADI); Mr Zhong Jishou, Chief Engineer of China Architectural Design and Research Co., Ltd.(CADG); Mr Zhong Jishou, Chief Engineer of China Architecture Design & Research Group (CADG); Mr Xue Ming, Chief Architect of the Architecture Design Institute of China Academy of Building Research Co., Ltd. (CABR); and Tenured Mr Song Yehao, Tenured Professor of School of Architecture of Tsinghua University.

The Jury Panel first rechecked the intermediate review results, determined the shortlisted works, and then selected the top 15 works with the highest scores to enter the on-site review procedure.

On December 6th, 2024, the on-site review session was held in Beijing in the form of oral defense and panel discussions, providing an excellent opportunity for participating teams to effectively showcase their works from multiple angles, particularly in terms of design thinking, innovative highlights, and applicability. This also allowed the Jury Panel to evaluate the works with a

（合影由左至右）中国建筑科学研究院有限公司建筑设计院首席总建筑师薛明；日本株式会社佐藤综合计画代表取缔役社长鉾岩崇（两位专家首次参与评审）；与华东建筑设计研究总院总建筑师黄秋平；中国工程院院士、全国工程勘察设计大师崔愷；全国工程勘察设计大师、同济大学建筑与城市规划学院教授钱锋；中国建筑设计研究院有限公司总工程师仲继寿；清华大学建筑学院长聘教授宋晔皓教授
(From Left to Right in the Group Photo) Mr Xue Ming, Chief Principal Architect of the Architectural Design Institute, China Academy of Building Research Co., Ltd.; Mr Hokoiwa Takashi, President of AXS SATOW Inc., Japan and an Expert from the Company Participated in the Review for the First Time; Mr Huang Qiuping, Chief Architect of East China Architectural Design and Research Institute (ECADI); Mr Cui Kai, Academician of the Chinese Academy of Engineering and National Engineering Design Master; Mr Qian Feng, National Engineering Design Master and Professor at the School of Architecture and Urban Planning, Tongji University; Mr Zhong Jishou, Chief Engineer of CADG; and Mr Song Yahao, Tenured Professor at the School of Architecture, Tsinghua University

中国建筑设计研究院有限公司总工程师仲继寿（左一）与现场专家讨论评审
Mr Zhong Jishou, Chief Engineer of CADG (Second from the Left), Engaged in a detailed Discussion and Review Session with the On-site Review Experts

2024 年 12 月 6 日，现场评审会在北京举行。本次会议采用现场答辩这一极具互动性与直观性的形式，为入围团队搭建了一个绝佳的展示平台，使其能够全方位、多角度地展示作品的设计思路、创新亮点以及可实施性等内容，使评审专家综合评价竞赛作品。进入决赛的 15 组团队通过图像、文字、模型与视频等多元方式展示阐述作品，并回答评审专家提问。最终，经过视频会议的方式，由 9 位评审专家，历经 3 轮评选和讨论，选出本届竞赛的综合奖、设计创意及技术单项奖作品共计 66 项。

more comprehensive understanding. The top 15 teams successively presented their works using various methods such as images, text, physical models, and videos while addressing questions raised by experts. Through video conference, after three rounds of evaluation and extensive panel discussions, a total of 66 exceptional works successfully won various types of prizes, including comprehensive prizes, design creation prizes, and individual awards.

竞赛评审专家与现场答辩小组师生合影
A Formal Group Photo of the Onsite-review Experts and the Faculty Members and Students of the Defense Teams

2024台达杯国际太阳能建筑设计竞赛评审专家介绍
Introduction to Jury Members of International Solar Building Design Competition 2024

评审专家
Jury Members

杨经文，马来西亚汉沙杨建筑师事务所创始人、**2016 年梁思成建筑奖获得者**
Mr King Mun Yeang: Founder of T. R. Hamzah & Yeang Sdn. Bhd of Malaysia, Winner of Liang Sicheng Architecture Award 2016

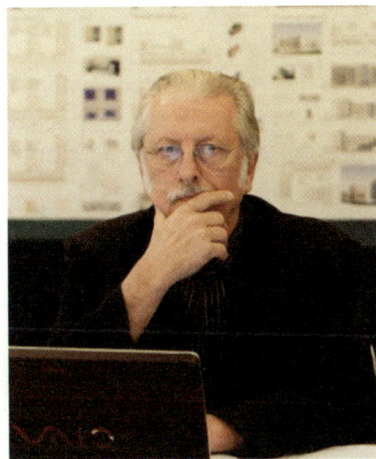

Peter Luscuere，荷兰代尔伏特理工大学建筑系教授
Mr Peter Luscuere: Professor of the Department of Architecture, Delft University of Technology, Netherlands

崔愷，中国工程院院士、全国工程勘察设计大师、中国建筑设计研究院有限公司总建筑师
Mr Cui Kai: Academician of China Academy of Engineering, Master of National Engineering Survey and Design of China, and Chief Architect of China Architecture Design and Research Group Co., Ltd. (CADG)

钱锋，全国工程勘察设计大师、同济大学建筑与城市规划学院教授、博士生导师
Mr Qian Feng: Master of National Engineering Survey and Design of China, Professor and Doctoral Supervisor of College of Architecture and Urban Planning of Tongji University

鉾岩崇，日本建筑家协会登录建筑家、日本一级注册建筑师、株式会社佐藤综合计画代表取缔役社长
Mr Takashi Hokoiwa, Registered Architect of the Japan Institute of Architects (JIA), Grade I Registered Architect of Japan, and Chairman of AXS Satow Inc. Japan

薛明，教授级高级建筑师、国家一级注册建筑师、中国建筑科学研究院有限公司建筑设计院首席总建筑师
Mr Ming Xue, Professor-level Senior Architect, Grade I Registered Architect of China, and Chief Architect of the Architecture Design Institute of China Academy of Building Research Co., Ltd. (CABR)

黄秋平，华东建筑设计研究院有限公司总院总建筑师
Mr Huang Qiuping: Chief Architect of East China Architectural Design & Research Institute Co., Ltd. (ECADI)

冯雅，中国建筑西南设计研究院顾问总工程师
Mr Feng Ya: Chief Consulting Engineer of China Southwest Architectural Design and Research Institute Co., Ltd.

仲继寿，中国建筑设计研究院有限公司总工程师、中国建筑学会主动式建筑专业委员会主任委员
Mr Zhong Jishou: Chief Engineer of China Architecture Design and Research Group Co., Ltd. (CADG); Chairman of Committee of Active House of the Architectural Society of China (ASC)

杨明，华东建筑设计研究院有限公司首席总建筑师、教授级高级工程师
Mr Yang Ming, Chief Architect and Professor-level Senior Engineer of East China Architectural Design and Research Institute Co., Ltd. (ECADI)

袁烽，同济大学建筑与城市规划学院教授、博士生导师、副院长
Mr Yuan Feng: Professor, Doctoral Supervisor and Deputy Dean of College of Architecture and Urban Planning of Tongji University

张宏，东南大学建筑学院教授、博士生导师
Mr Zhang Hong: Professor and Doctoral Supervisor of School of Architecture of Southeast University

彭晋卿，湖南大学土木工程学院教授、博士生导师
Mr Peng Jinqing: Professor, Doctoral Supervisor of School of Civil Engineering, Hunan University

刘恒，中国建筑设计研究院有限公司副总建筑师、绿色建筑设计研究院院长
Mr Liu Heng: Deputy Chief Architect of China Architecture Design and Research Group Co., Ltd. (CADG); Director of Green Architecture Design and Research Institute of CADG

宋晔皓，清华大学建筑学院长聘教授、博士生导师、副系主任，清华大学建筑与技术研究所所长，清华大学建筑设计研究院副总建筑师
Mr Song Yehao: Tenured Professor, Doctoral Supervisor and Deputy Dean of School of Architecture of Tsinghua University; Director of Architecture and Technology Institute of Tsinghua University; Deputy Chief Architect of Architectural Design and Research Institute of Tsinghua University (THAD)

钟辉智，工学博士，中国建筑西南设计研究院有限公司双碳中心总工程师
Mr Zhong Huizhi, Doctor of Engineering, and Chief engineer of the "Double Carbon" Center of the Southwest Institute of China State Construction Engineering Corporation (CSCEC)

获奖作品

Prize Awarded Works

99-110004

综合奖·一等奖·张家口市
怀安县二堡子村建设项目
Comprehensive Awards －
First Prize － Erpuzi Village
Project, Huai'an County,
Zhangjiakou City

注册号：110004
Register Number：110004
项目名称：互生叶序
Entry Title：Alternate Phyllotaxy
作者：周琪云、张焕悦
Authors：Zhou Qiyun, Zhang Huanyue
作者单位：西安建筑科技大学
Authors from：Xi'an University of Archi-
　　　　　　tecture and Technology
指导教师：王芳、陈敬
Tutors：Wang Fang, Chen Jing
指导教师单位：西安建筑科技大学
Tutors from：Xi'an University of Architec-
　　　　　　ture and Technology

Alternate Phyllotaxy

Sunshine · Rural Resilience
International Solar Building Design Competition

Design Note

Design is based on the theme of "Alternate leaf Sequence", creating Erpuzi Village as a rural community that combines resilience and sustainability. Through the corridor as a "stem", the adjacent buildings alternately grow like leaves, realizing the low-carbon and green design of the complex. Here, solar energy is widely used, through photovoltaic thermal panels, not only to provide energy for the building, but also as an emergency energy source in case of emergency. The landscape and spatial design emphasize the blending of public and private Spaces and the mixing of functions, while reducing energy consumption through passive design and utilizing natural wind and light. In addition, the layout and structure of the building ensure resistance to natural disasters such as earthquakes. We also introduce smart energy management systems for efficient energy use and revenue generation. Through these innovations, the goal is to create a safe and vibrant rural community that is not only resilient to natural disasters, but also contributes to sustainable economic and social development.

设计以"互生叶序"为主题，将二堡子村打造为一个集韧性和可持续性于一身的农村社区。通过长廊作为"茎"，相邻建筑像叶片一样交替生长，实现建筑群的低碳、绿色设计。这里，太阳能被广泛利用，通过光伏光热板，不仅为建筑提供能源，还能在紧急情况下作为备急能源。景观和空间设计强调公共与私密空间的融合和功能组合，同时，通过被动式设计减少能源消耗，利用自然风和光线。此外，建筑布局和结构确保了对地震等自然灾害的抵抗力。我们还引入智慧能源管理系统，实现能源的高效利用和创收。通过这些创新，目标是创造一个既安全又富有生机的农村社区，不仅能够抵御自然灾害，还能促进经济和社会的可持续发展。

Climatic Analysis

Temperature

Average temperature: 9.6℃.
Average annual maximum temperature: 15℃.
Average annual minimum temperature: 3℃.
Record high temperature:39 ℃ (2002)
Record low temperature:-26 ℃ (1958)
Average annual rainfall: 405 mm

Zhangjiakou city is a temperate continental monsoon climate. The seasons are distinct, and the winters are cold and long. The spring is dry and windy. Summer hot short precipitation concentration. Autumn is sunny and warm. Rain and heat in the same season, the growing season: climate cool; There are few hot days with high temperature and high humidity.

Sunshine

Through this analysis, Zhangjiakou City can see that the overall light resources are rich, and the direct radiation is very high throughout the year, with the highest value around 12.6% Indirect radiation is mainly higher at noon in summer. Suitable for the installation of photovoltaic system, the use of light resources for power generation and heating.

Optimal inclination annual irradiation:
1824.4 kWh / ㎡. year
Optimum azimuth angle: 1°

In this site, the building faces north and south, especially the corridor part, and the daily sunshine on the roof part is more than 5h. The lighting of the south-facing building is normal, and the lighting of the north-facing building is normal through the east-west direction, daylight and patio can also ensure the lighting of more than 5h on the winter solstice.

Ventilate

In Zhangjiakou City, northwest wind prevails in winter and southeast wind prevails in summer. Spring is dry and windy. Need to pay attention to winter wind protection.

During the design process, several iterations were made so that there was no calm wind zone inside the building. At the same time, the east-west corridor uses the southwest part of the pool for cooling in summer, and can block the wind in winter.

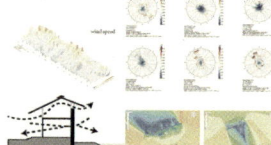

Humidity

The humidity characteristics of Zhangjiakou city are mainly affected by climate. Because Zhangjiakou city is located in the inland area, and affected by the topography and terrain, the overall performance is dry climatic characteristics. Winter is cold and dry, the air humidity is low, especially in partly sunny climates, humidity is often very low. The relative humidity is high in summer, but it is still dry and hot on the whole. Spring and autumn are relatively dry and humid. In general, the humidity characteristics of Zhangjiakou City show obvious arid climate characteristics, which has a certain impact on the local agricultural production and life.

The precipitation in Zhangjiakou City is mainly concentrated in summer and less in winter. Summer (June to August) is the main precipitation season to Zhangjiakou City, with a large amount of precipitation and even heavy rain in the local area.

In winter (December to February of the following year), there is relatively little precipitation, and the precipitation is significantly reduced, and even the characteristics of drought. The precipitation in spring and autumn is relatively moderate, and the overall distribution is relatively uniform. Due to the influence of precipitation characteristics, Zhangjiakou city needs to prepare and plan for the seasonal characteristics of precipitation in agricultural production and domestic water use. At the same time, more precipitation in summer also provides certain conditions for the growth of local vegetation and ecological environment.

Site Plan 1：600

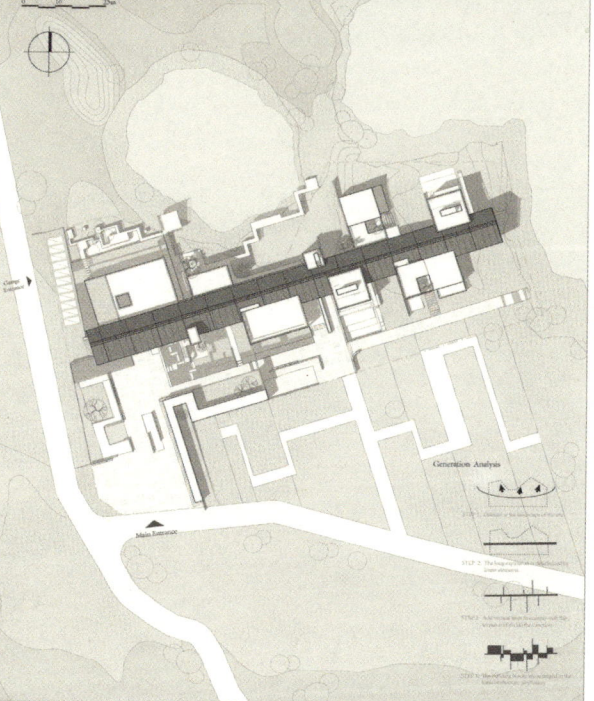

Site Analysis

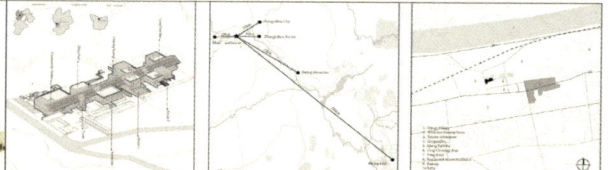

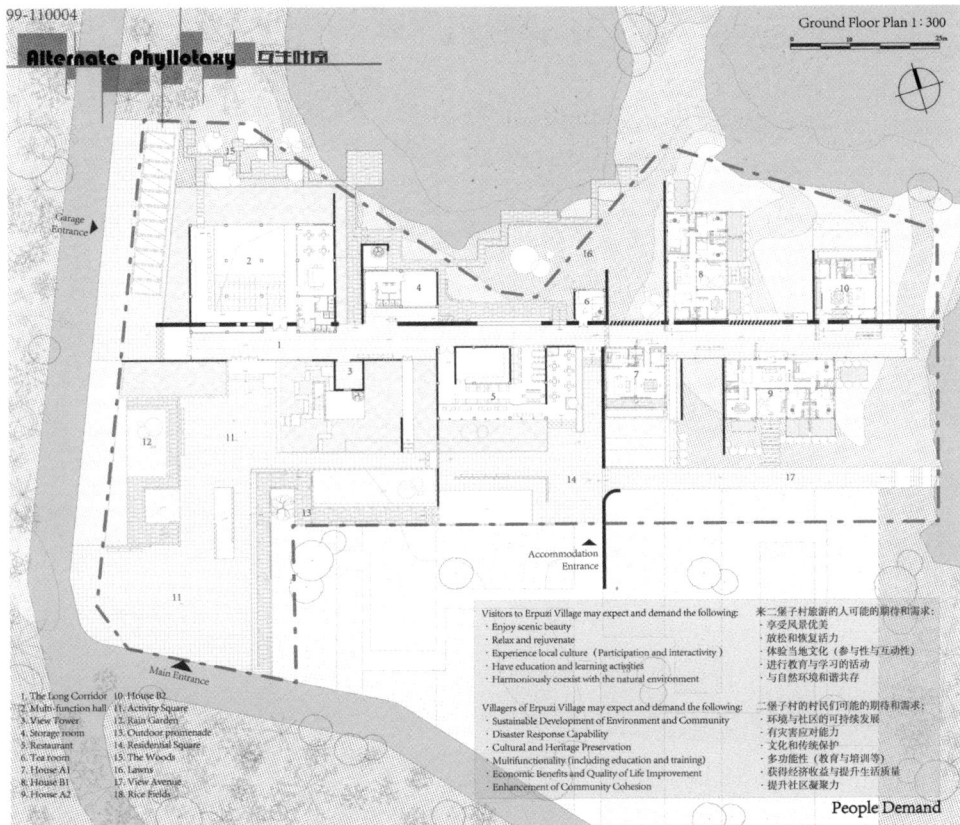

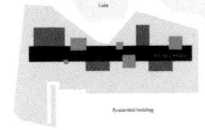

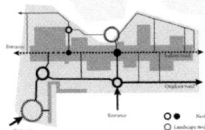

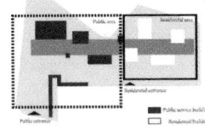

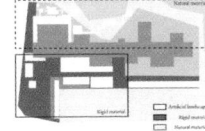

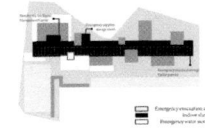

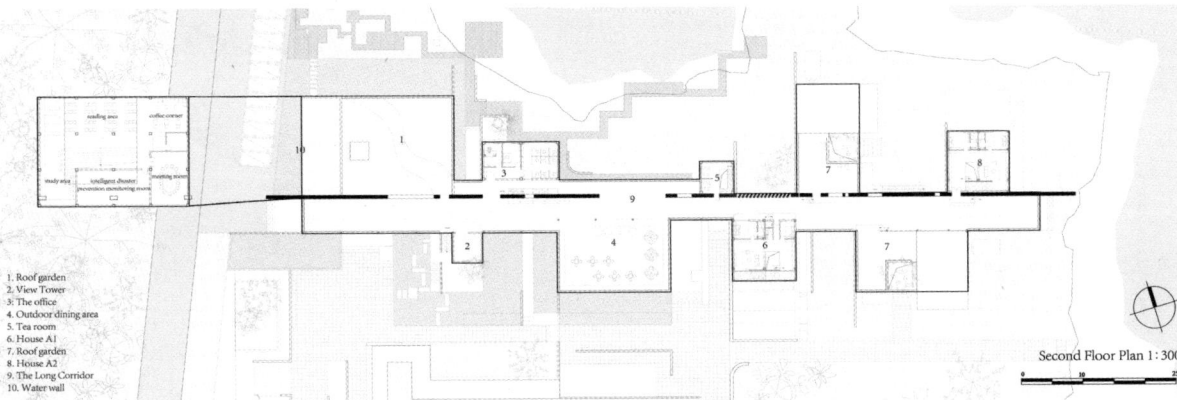

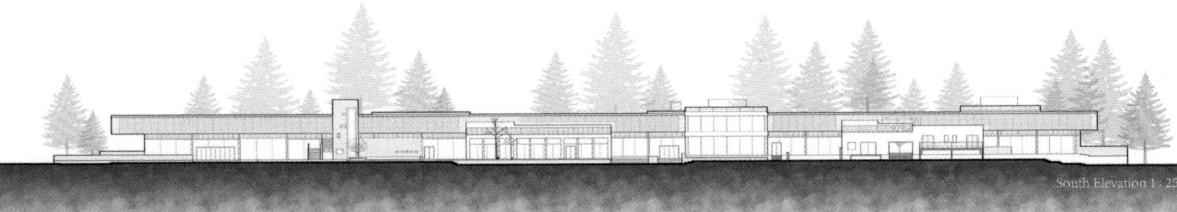

Alternate Phyllotaxy 互生叶序

Ground Floor Plan 1:300

The Relationship Between Picture and Background
Lake
Residential building

Road Network Analysis

Building Function Analysis
Public service building
Residential building

Landscape Design Analysis
Artificial landscape
Rigid material
Natural material

Calamity Prevention Plan
Emergency evacuation area
Indoor shelter
Emergency water storage

Garage Entrance

Main Entrance

Accommodation Entrance

People Demand

Visitors to Erpuzi Village may expect and demand the following:
· Enjoy scenic beauty
· Relax and rejuvenate
· Experience local culture (Participation and interactivity)
· Have education and learning activities
· Harmoniously coexist with the natural environment

Villagers of Erpuzi Village may expect and demand the following:
· Sustainable Development of Environment and Community
· Disaster Response Capability
· Cultural and Heritage Preservation
· Multifunctionality (including education and training)
· Economic Benefits and Quality of Life Improvement
· Enhancement of Community Cohesion

来二堡子村旅游的人可能的期待和需求：
· 享受风景优美
· 放松和恢复活力
· 体验当地文化（参与性与互动性）
· 进行教育与学习的活动
· 与自然环境和谐共存

二堡子村的村民们可能的期待和需求：
· 环境与社区的可持续发展
· 有灾害应对能力
· 文化和传统保护
· 多功能性（教育与培训等）
· 获得经济收益与提升生活质量
· 提升社区凝聚力

1. The Long Corridor
2. Multi-function hall
3. View Tower
4. Storage room
5. Restaurant
6. Tea room
7. House A1
8. House B1
9. House A2
10. House B2
11. Activity Square
12. Rain Garden
13. Outdoor promenade
14. Residential Square
15. The Woods
16. Lawns
17. View Avenue
18. Rice Fields

Second Floor Plan 1:300

1. Roof garden
2. View Tower
3. The office
4. Outdoor dining area
5. Tea room
6. House A1
7. Roof garden
8. House A2
9. The Long Corridor
10. Water wall

Forest

Section Cutaway Diagram

Lake

South Elevation 1:250

专家点评：

该方案在总体布局方面有特色，长廊的设计有利于公共交流空间的形成，适合旅游及村民日常活动，也展现出多样性和现代感。架空地板系统的设计不仅有助于自然通风，还能使结构具备更大的灵活性来增强抗震能力。建筑的形态与体量经过精心设计，以适应未来的扩建和延伸，确保能适应不断变化的功能需求。长廊的布局虽富有创意，但也造成了布局与北面景观呼应不够，对北侧两组住宅日照和通风不利。

Experts' Commentary：

The scheme is distinctive in terms of its overall layout. The design of the corridor facilitates the formation of public communication spaces, making it suitable for both tourism and the daily activities of villagers while also showcasing diversity and modernity. The raised floor system not only promotes natural ventilation, but also enhances structural flexibility, thereby increasing earthquake resistance. The form and volume of the buildings have been carefully designed to accommodate future extensions, ensuring adaptability to changing functional requirements. Although the corridor's layout is creative, it results in insufficient correspondence between the layout and the northen landscape, which negatively impacts the sunshine and ventilation of the two residential groups on the north side.

Resilience Strategy

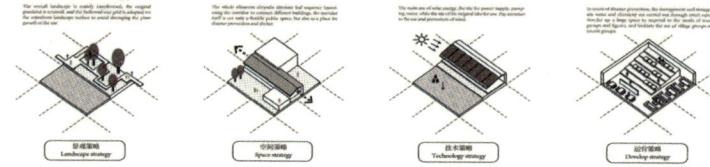

Residential Technology and Simulation

Exploded Axonometric

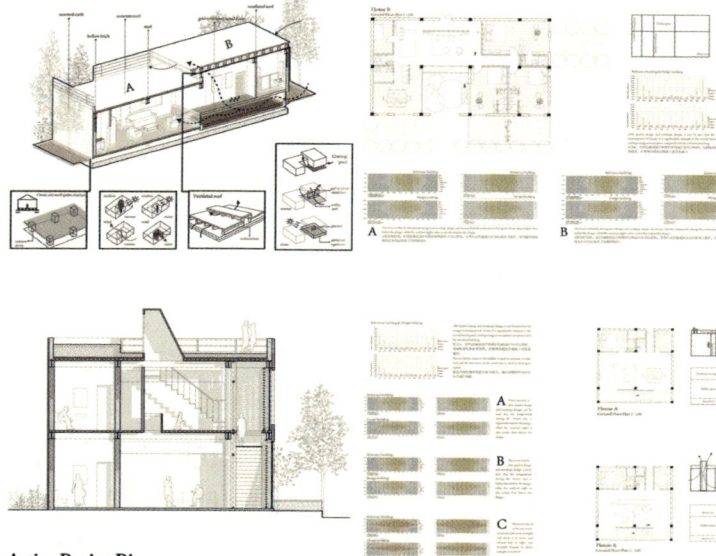
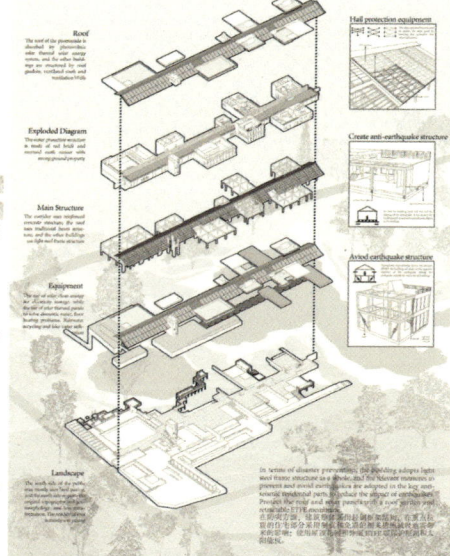

Active Device Diagram

Carbon Emission Analysis

Wind Analysis

STEP 1 Wind simulation using PHOENICS software. The design originally a solid wall, and you can see that the ventilation in summer is not good. In the summer, the north part cannot be ventilated and needs to be reoriented.

STEP 2 The design has added some holes to the solid wall on the north side of the corridor, hoping to provide ventilation while allowing views. Through the simulation of summer and winter, it can be seen that the ventilation has been improved to a certain extent, and detailed adjustments need to be made.

STEP 3 The adjustment of ventilation in the details has been improved, some windows have been expanded, and the existing art brick wall can be closed in winter, so that there is no quiet-wind area on the site. Next, the building details and internal ventilation were improved.

STEP 4 Finally, the wind simulation of building interior was carried out. After the simulation of plane and section, it was found that the ventilation could not be ventilated. HOUSE A was effectively ventilated by increasing the opening of ventilation shaft, and HOUSE B was ventilated by building ventilation roof.

Solar Energy Analysis

Site Condition

Generating Capacity

Equipment

Solar Radiation

Electricity Consumption

*Avoid emitting 61 tons of equivalent amounts of carbon dioxide

Through the above calculation, the electricity and heat generated by using 271 solar panels can cover the use of the building.

Comparison of winter air conditioning and floor heating

It can be calculated from the above table that the economy of air conditioning heating is not as good as floor heating, and the comfort is lower. So the use of ground floor heating for heating.

Energy Usage Analysis

Public Buildings

Residential Buildings

作品编号：110048

风曦·舟上村舍 01
Wind Sunshine - Rural dwellings on boat

综合奖·一等奖·保定市清苑区李八庄村建设项目
Comprehensive Awards – First Prize – Libazhuang Village Project，Qingyuan District，Baoding City

注册号：110048
Register Number：110048

项目名称：风曦·舟上村舍
Entry Title：Wind Sunshine—Rural Dwellings on Boat

作者：沈鑫、柴振国、李云龙、赵一凡、赵雅雯
Authors：Shen Xin，Chai Zhenguo，Li Yunlong，Zhao Yifan，and Zhao Yawen

作者单位：内蒙古工业大学
Authors from：Inner Mongolia University of Technology

指导教师：伊若勒泰、许国强
Tutors：Yiruo Letai，Xu Guoqiang

指导教师单位：内蒙古工业大学
Tutors from：Inner Mongolia University of Technology

Location Analysis

Libazhuangcun

Qingyuanqu

Climatic Simulation

Direct Normal Radiation

Diffuse Horizontal Radiation

Relative Humidity

Dry Bulb Temperature

Spring　　Summer

Autumn　　Winter

Spring

Summer

Autumn

Winter

Design Specification

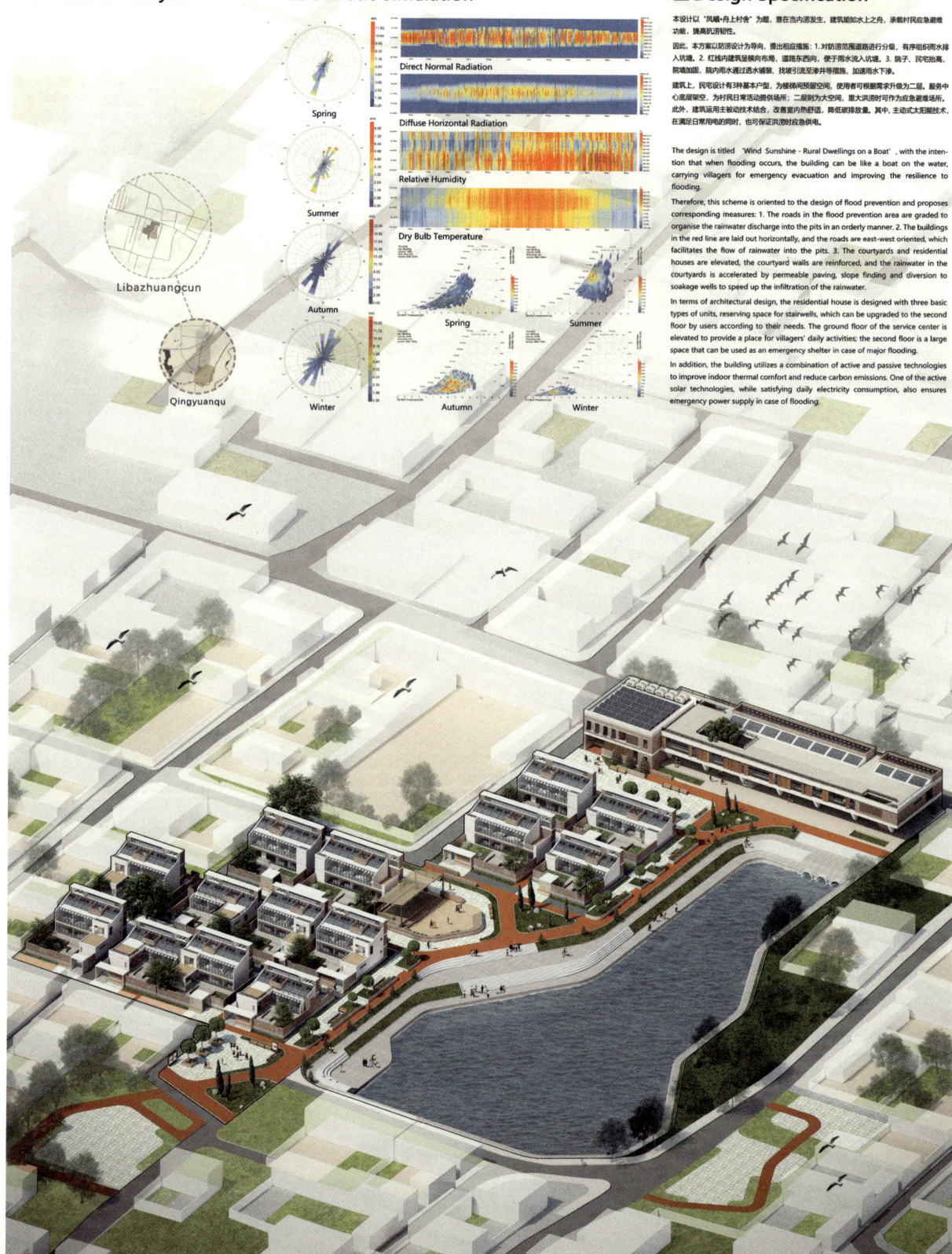

本设计以"风曦·舟上村舍"为题，意在当内涝发生，建筑能如水上之舟，承载村民应急避难功能，提高抗涝韧性。

因此，本方案以防涝设计为导向，提出相应措施：1. 对防涝范围道路进行分级，有序组织雨水排入坑塘。2. 红线内建筑呈横向布局，道路东西向，便于雨水流入坑塘。3. 院子、民宅抬高，防墙加固，院内雨水通过透水铺装、找坡引流至渗井等措施，加速雨水下渗。

建筑上，民宅设计有3种基本户型，为楼梯间预留空间，使用者可根据需求升级为二层，服务中心底层架空，为村民日常活动提供场所；二层则为大空间，重大洪涝时可作为应急避难场所。此外，建筑运用主被动技术结合，改善室内热舒适，降低碳排放量。其中，主动式太阳能技术，在满足日常用电的同时，也可保证洪涝时应急供电。

The design is titled "Wind Sunshine - Rural Dwellings on a Boat"，with the intention that when flooding occurs, the building can be like a boat on the water, carrying villagers for emergency evacuation and improving the resilience to flooding.

Therefore, this scheme is oriented to the design of flood prevention and proposes corresponding measures: 1. The roads in the flood prevention area are graded to organise the rainwater discharge into the pits in an orderly manner. 2. The buildings in the red line are laid out horizontally, and the roads are east-west oriented, which facilitates the flow of rainwater into the pits. 3. The courtyards and residential houses are elevated, the courtyard walls are reinforced, and the rainwater in the courtyards is accelerated by permeable paving, slope finding and diversion to soakage wells to speed up the infiltration of the rainwater.

In terms of architectural design, the residential house is designed with three basic types of units, reserving space for stairwells, which can be upgraded to the second floor by users according to their needs. The ground floor of the service center is elevated to provide a place for villagers' daily activities; the second floor is a large space that can be used as an emergency shelter in case of major flooding.

In addition, the building utilizes a combination of active and passive technologies to improve indoor thermal comfort and reduce carbon emissions. One of the active solar technologies, while satisfying daily electricity consumption, also ensures emergency power supply in case of flooding.

■ Site Plan 1:500

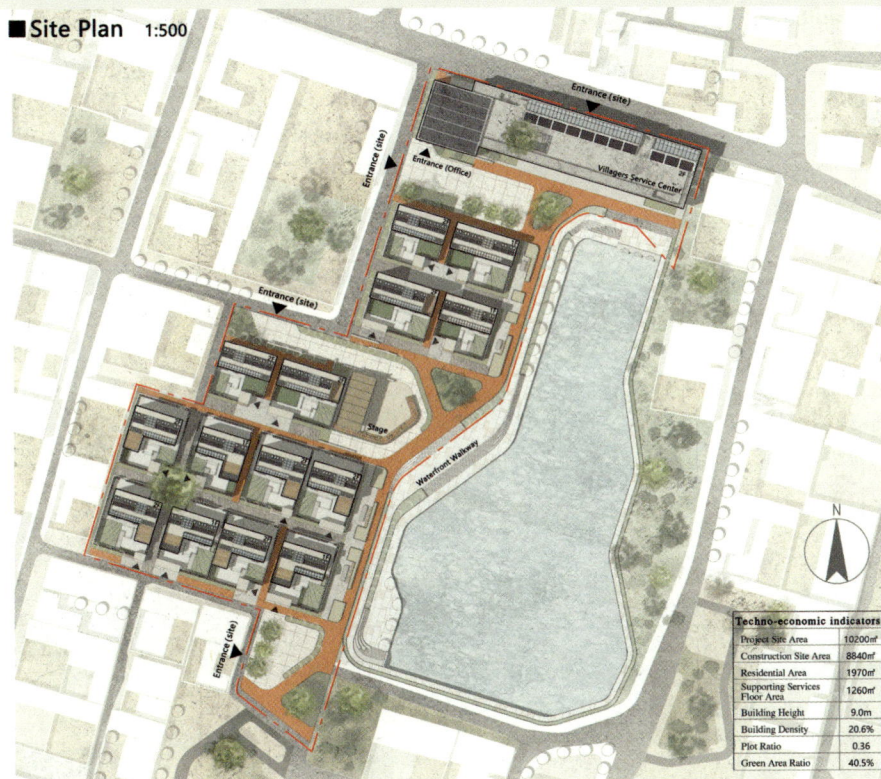

Techno-economic indicators	
Project Site Area	10200㎡
Construction Site Area	8840㎡
Residential Area	1970㎡
Supporting Services Floor Area	1260㎡
Building Height	9.0m
Building Density	20.6%
Plot Ratio	0.36
Green Area Ratio	40.5%

■ Building Plan 1:300

■House Type Ⅰ ■House Type Ⅱ
■House Type Ⅰ (plus) ■House Type Ⅱ (plus)
■House Type Ⅲ ■House Type Ⅲ (Plus)

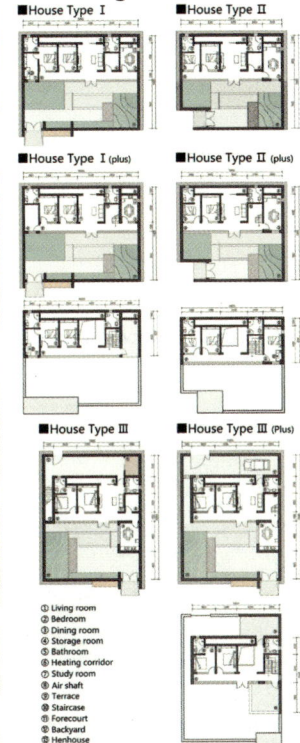

① Living room ② Bedroom ③ Dining room ④ Storage room ⑤ Bathroom ⑥ Heating corridor ⑦ Study room ⑧ Air shaft ⑨ Terrace ⑩ Staircase ⑪ Forecourt ⑫ Backyard ⑬ Henhouse

■ Site Design

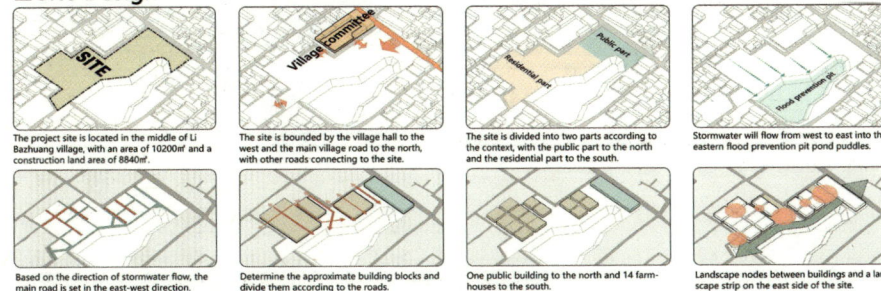

The project site is located in the middle of Li Bazhuang village, with an area of 10200㎡ and a construction land area of 8840㎡.

The site is bounded by the village hall to the west and the main village road to the north, with other roads connecting to the site.

The site is divided into two parts according to the context, with the public part to the north and the residential part to the south.

Stormwater will flow from west to east into the eastern flood prevention pit pond puddles.

Based on the direction of stormwater flow, the main road is set in the east-west direction.

Determine the approximate building blocks and divide them according to the roads.

One public building to the north and 14 farm-houses to the south.

Landscape nodes between buildings and a landscape strip on the east side of the site.

■ Creation of Small Spaces —— in Line with Living Rural Habits

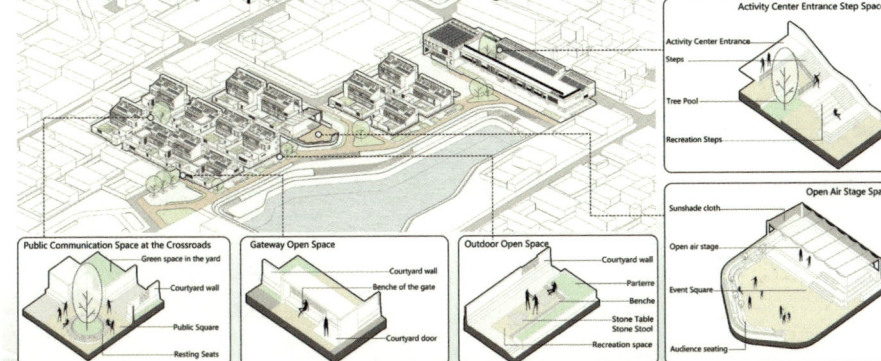

■ Residential Upgrading

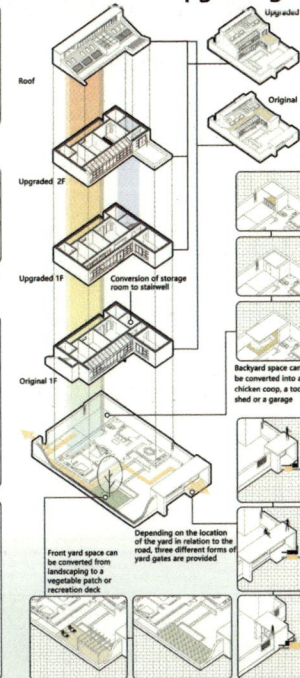

■ Elevation 1:300

专家点评：

该方案总体分区和布局合理，用地平衡，公建架空提升公共性，有利于平常开放式应用和内涝时预防水淹；建筑设计考虑了被动技术的应用，建筑单体造型地域特点体现略有不足。排洪设计考虑到村塘排洪，沿坑洼设生物洼地和绿化带的方案，能增强吸水性、促进排水以减轻洪灾影响，不过成本偏高；若能明确施工类型，完善太阳能利用规划，优化住宅通风井与排洪成本问题，将更为科学可行。

Experts' Commentary：

The overall zoning and layout of the scheme are reasonable. The planning of land use is well-balanced. The overhead design of the public area enhances its sharing feature, facilitating daily use in ordinary times and aiding in flood prevention when waterlogging occurs. The architectural design incorporates the application of passive technology, however, the building shape does not adequately reflect its regional characteristics. The flood drainage design takes into account the idea of village pond drainage, bioswales, and green belts along the depression areas, which enhances water absorption and promotes drainage to reduce the impact of flooding. However, the cost is relatively high. It would be more scientific and feasible to specify the construction types, improve the solar energy utilization plan, and optimize the costs associated with residential ventilation shafts and flood drainage.

风曦·舟上村舍 03
Wind Sunshine - Rural dwellings on boat

Drainage Analysis at the Planning Level

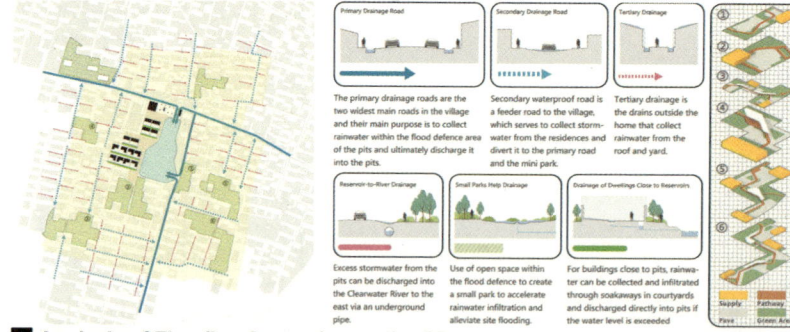

Primary Drainage Road

Secondary Drainage Road

Tertiary Drainage

The primary drainage roads are the two widest main roads in the village and their main purpose is to collect rainwater within the flood defence area of the pits and ultimately discharge it into the pits.

Secondary waterproof road is a feeder road to the village, which serves to collect stormwater from the residences and divert it to the primary road and the mini park.

Tertiary drainage is the drains outside the home that collect rainwater from the roof and yard.

Reservoir-to-River Drainage

Small Parks Help Drainage

Drainage of Dwellings Close to Reservoirs

Excess stormwater from the pits can be discharged into the Clearwater River to the east via an underground pipe.

Use of open space within the flood defence to create a small park to accelerate rainwater infiltration and alleviate site flooding.

For buildings close to pits, rainwater can be collected and infiltrated through soakaways in courtyards and discharged directly into pits if the water level is exceeded

Supply   Pathway
Pave   Green Area

Node Effect Diagram

Analysis of Flooding Scenarios on the Site

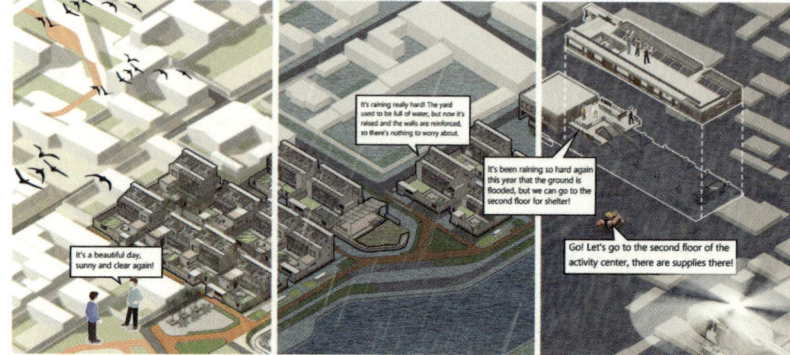

It's raining really hard! The yard used to be full of water, but now it's raised and the walls are reinforced, so there's nothing to worry about.

It's been raining so hard again this year that the ground is flooded, but we can go to the second floor for shelter!

Go! Let's go to the second floor of the activity center, there are supplies there!

It's a beautiful day, sunny and clear again!

Analysis of Village Service Centres

Medical Space   Meeting Room   Office   Public Toilet   Emergency water tank   Solar photovoltaic panel   Ground floor elevated (market)   Second Floor Activity Space   Roof platforms

Floor Plan of the Village Service Centre 1:300

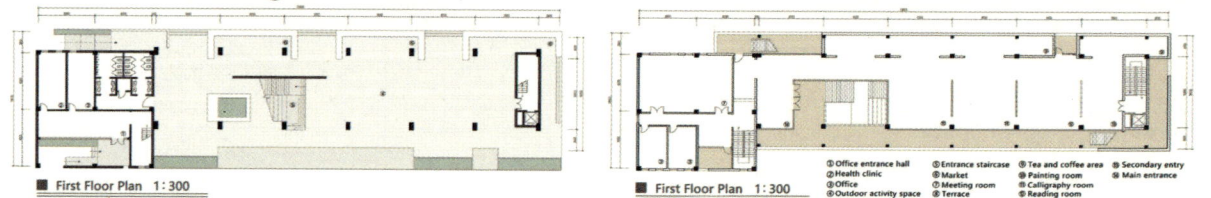

First Floor Plan 1:300

First Floor Plan 1:300

①Office entrance hall   ④Entrance staircase   ⑦Tea and coffee area   ⑩Secondary entry
②Health clinic   ⑤Market   ⑧Painting room   ⑪Main entrance
③Office   ⑥Meeting room   ⑨Calligraphy room
④Outdoor activity space   ⑥Terrace   ⑨Reading room

Drainage in Residential Courtyards

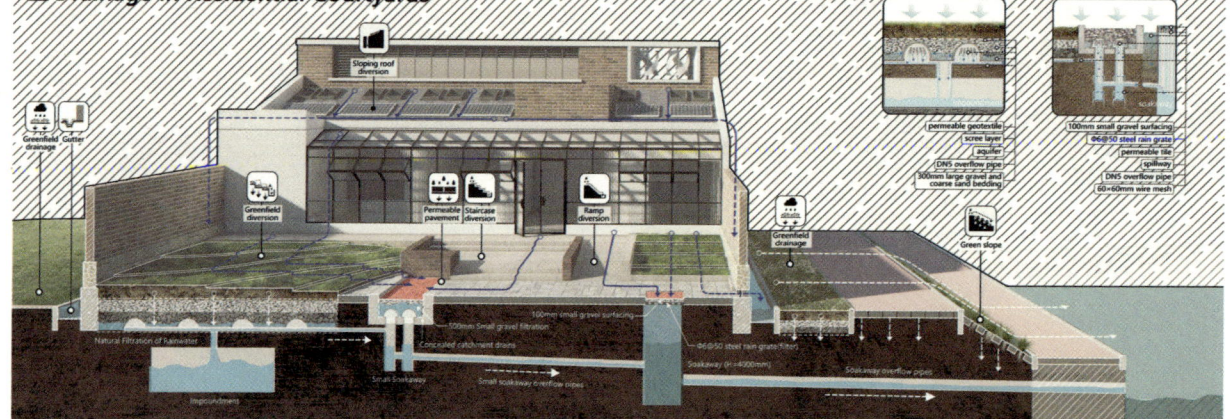

Sloping roof diversion

permeable geotextile
scree layer
aquifer
DN5 overflow pipe
300mm large gravel and coarse sand bedding

100mm small gravel surfacing
φ60/50 steel rain grate
permeable tile
spillway
DN5 overflow pipe
60×60mm wire mesh

Greenfield drainage   Gutter

Greenfield diversion

Permeable pavement

Staircase diversion

Ramp diversion

Greenfield drainage

Green slope

Natural Filtration of Rainwater

Concealed catchment drains

100mm Small gravel filtration

100mm small gravel surfacing

φ60/50 steel rain grate (filter)

Small Soakaway

Small soakaway overflow pipes

Soakaway (H=400mm)

Soakaway overflow pipes

作品编号：110048

风曦・舟上村舍 04
Wind Sunshine - Rural dwellings on boat

Section Perspective View

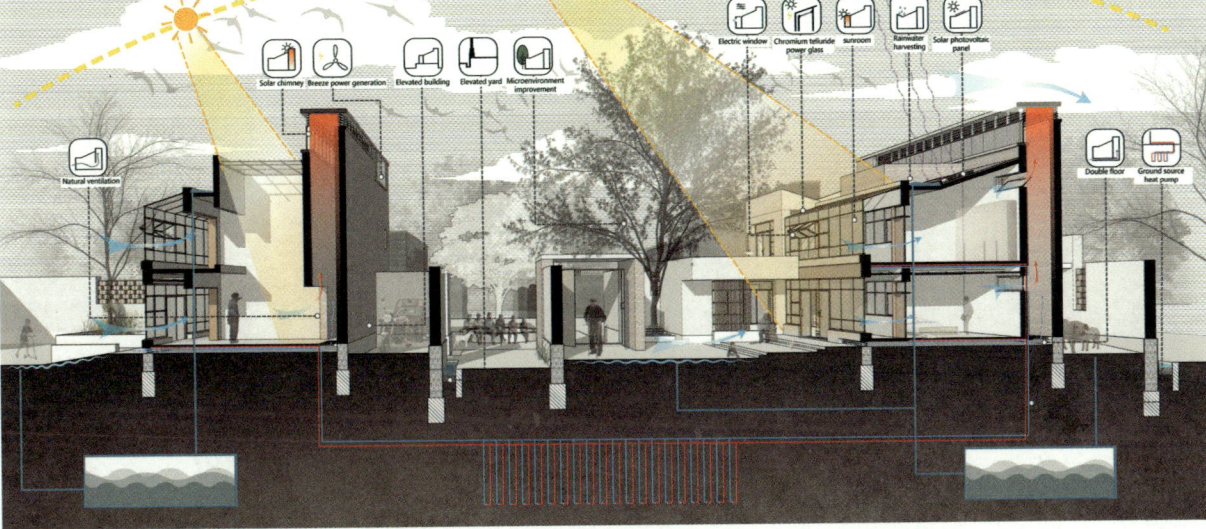

Natural ventilation

Solar chimney | Breeze power generation | Elevated building | Elevated yard | Microenvironment improvement

Electric window | Chromium telluride power glass | sunroom | Rainwater harvesting | Solar photovoltaic panel

Double floor | Ground source heat pump

Building Passive Technology

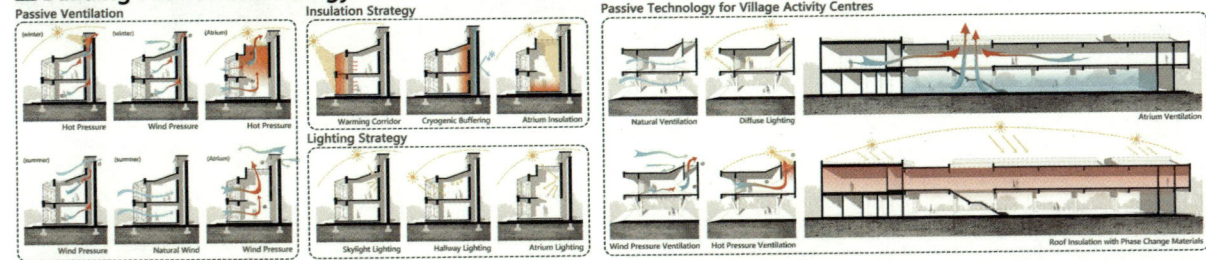

Passive Ventilation

(winter) | (winter) | (Atrium)
Hot Pressure | Wind Pressure | Hot Pressure

(summer) | (summer) | (Atrium)
Wind Pressure | Natural Wind | Wind Pressure

Insulation Strategy

Warming Corridor | Cryogenic Buffering | Atrium Insulation

Lighting Strategy

Skylight Lighting | Hallway Lighting | Atrium Lighting

Passive Technology for Village Activity Centres

Natural Ventilation | Diffuse Lighting | Atrium Ventilation

Wind Pressure Ventilation | Hot Pressure Ventilation | Roof Insulation with Phase Change Materials

Green Building Simulation

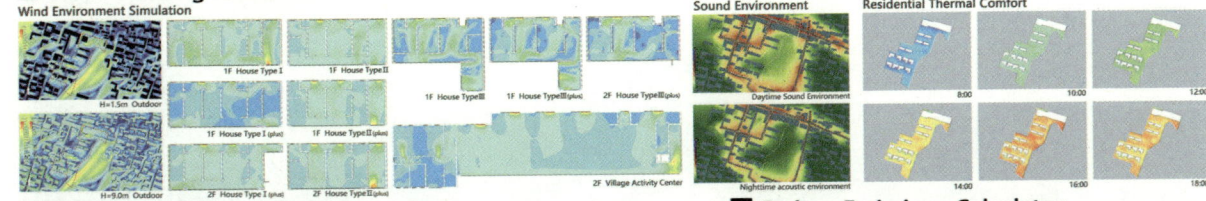

Wind Environment Simulation

H=1.5m Outdoor

1F House Type I | 1F House Type II
1F House Type I (plus) | 1F House Type II (plus)
2F House Type I (plus) | 2F House Type II (plus)

H=9.0m Outdoor

1F House Type III | 1F House Type III(plus) | 2F House Type III(plus)

2F Village Activity Center

Sound Environment

Daytime Sound Environment

Nighttime acoustic environment

Residential Thermal Comfort

8:00 | 10:00 | 12:00
14:00 | 16:00 | 18:00

Utilisation of Active Solar Technologies

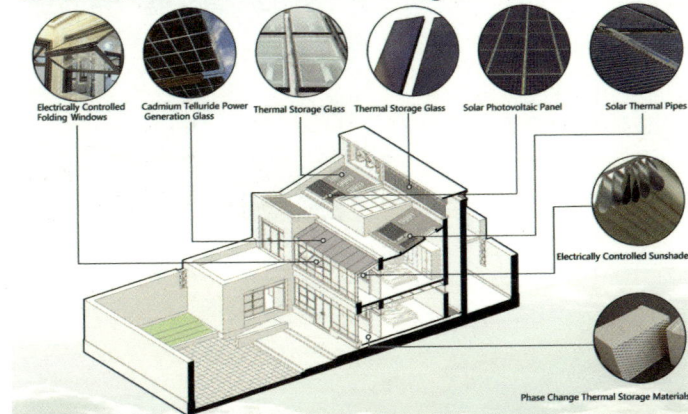

Electrically Controlled Folding Windows | Cadmium Telluride Power Generation Glass | Thermal Storage Glass | Thermal Storage Glass | Solar Photovoltaic Panel | Solar Thermal Pipes

Electrically Controlled Sunshade

Phase Change Thermal Storage Materials

Carbon Emissions Calculator

Electric energy	Category	Power consumption (kW・h/m²)	Carbon emission factor (kgCO₂/kW・h)	Carbon emissions (tCO₂)
Electric power	space cooling	289.45		1078.009
	heating	0		0.000
	air-conditioning fan	86.053	0.8843	320.49
	illumination	517.455		1927.176
	socket equipment	756.794		2818.555
	miscellaneous	955.804		3559.735
Fossil fuel	Category	Heat consumption (kW・h/m²)	Carbon emission factor (tCO₂/TJ)	Carbon emissions (tCO₂)
Coal	Heating: Heat source	1778.023	89	2399.269
There is no	Heating: Municipal	0.000	0	0.000
There is no	Domestic hot water	0.000	0	0.000
Gas	Cooking	-(m³/m²)	55.54	-
Refrigerant	Cooling	0		0.000
Renewable	Category	Power supply (kW・h/m²)	Carbon emission factor (kgCO₂/kW・h)	Carbon emission
Renewable energy	Solar power	3295.572	0.8843	12273.816
	Wind force	62.324		230.908
Total carbon emissions from building operations				1996.571
Building carbon sink				2155.000
Total project carbon emissions				-158.429

Section View 1:300

综合奖·二等奖·张家口市怀安县二堡子村建设项目
Comprehensive Awards – Second Prize – Erpuzi Village Project，Huai'an County，Zhangjiakou City

注册号：110003
Register Number：110003
项目名称：织光麦社
Entry Title：Weaving Sunshine—Wheat Field Village Community
作者：董春朝、吴梦雨、朱嘉宇、张吉森、房彦娜
Authors：Dong Chunchao, Wu Mengyu, Zhu Jiayu, Zhang Jisen, and Fang Yanna
作者单位：西北工业大学
Authors from：Northwestern Polytechnical University
指导教师：刘煜、黄姗、杜昱民、宋戈、邵腾
Tutors：Liu Yu, Huang Shan, Du Yumin, Song Ge, and Shao Teng
指导教师单位：西北工业大学
Tutors from：Northwestern Polytechnical University

CODE:110003-99

织光麦社 ——乡村会客厅和民宿民居设计 I
Weaving Sunshine—Wheat Field Village Community
——the Village reception and Homestay Hotel Design

Volume Analysis

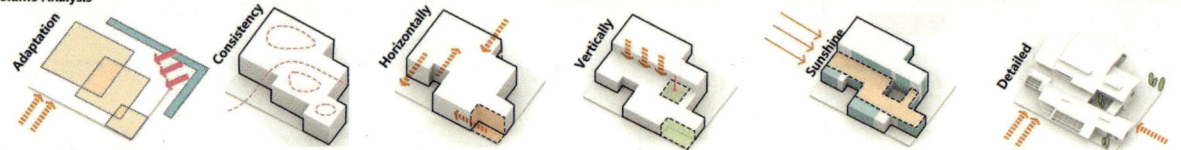

Adaptation | Consistency | Horizontally | Vertically | Sunshine | Detailed

General Plan 1:500

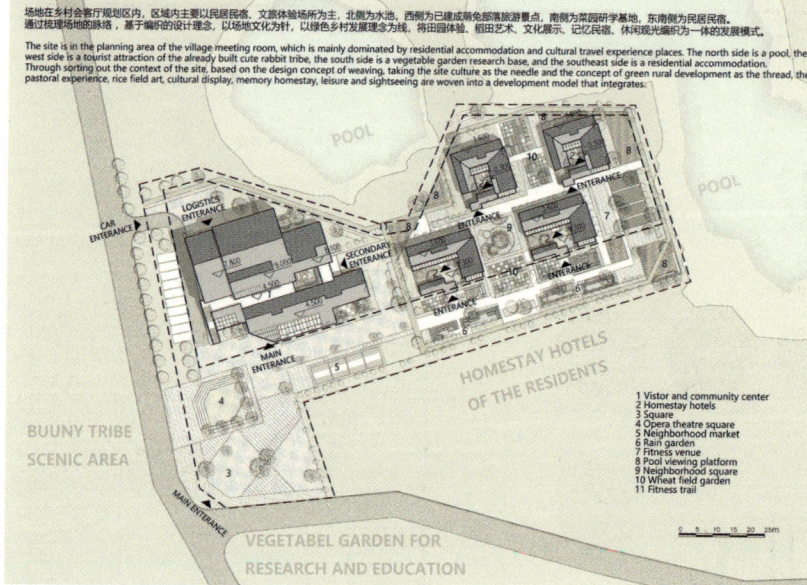

场地在乡村会客厅规划区内，区域内主要以民居民宿、文旅体验场所为主，北侧为水池，西侧为已建成萌兔部落旅游景点，南侧为菜园研学基地，东南侧为民居民宿。通过梳理场地的地脉络，基于编织的设计理念，以场地文化为针，以绿色乡村发展理念为线，将田园体验、稻田艺术、文化展示、记忆民宿、体闲观光编织为一体的发展模式。

The site is in the planning area of the village meeting room, which is mainly dominated by residential accommodation and cultural travel experience places. The north side is a pool, the west side is a tourist attraction of the already built cute rabbit tribe, the south side is a vegetable garden research base, and the southeast side is a residential accommodation. Through sorting out the context of the site, based on the design concept of weaving, taking the site culture as the needle and the concept of green rural development as the thread, the pastoral experience, rice field art, cultural display, memory homestay, leisure and sightseeing are woven into a development model that integrates:

POOL
POOL

CAR ENTRANCE
LOGISTICS ENTRANCE
SECONDARY ENTRANCE
ENTRANCE
MAIN ENTRANCE

BUUNY TRIBE SCENIC AREA

HOMESTAY HOTELS OF THE RESIDENTS

MAIN ENTRANCE

VEGETABEL GARDEN FOR RESEARCH AND EDUCATION

1 Vistor and community center
2 Homestay hotels
3 Square
4 Opera theatre square
5 Neighborhood market
6 Rain garden
7 Fitness venue
8 Pool viewing platform
9 Neighborhood square
10 Wheat field garden
11 Fitness trail

Design Specification

Yard | Quite | Noisy | Link | Tolit | sunroom | Rest

Crowd activity analysis

Aborigines | Shutterbug | Literary youth | Tourists

Cultural context

Combine unique agricultural culture, promote the integration of agriculture and tourism for tourists, and deeply develop all kinds of agricultural experience tourism activities

Profile Map 1:150

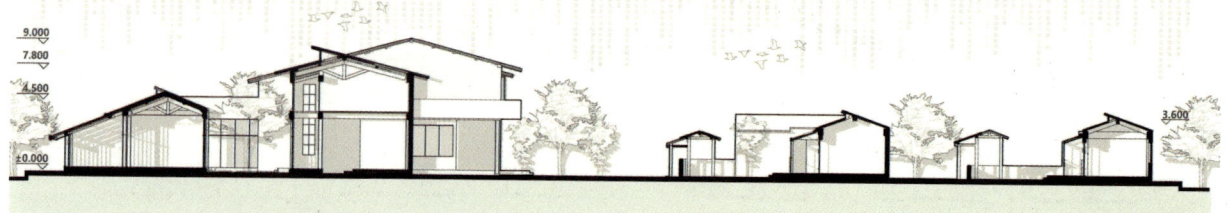

9.000
7.800
4.500
±0.000
3.600

CODE:110003-99

织光麦社——乡村会客厅和民宿民居设计 II

Weaving Sunshine-Wheat Field Village Community
—— The Villagereception and Homestay Hotel Design

TECHNICAL AND ECONOMIC INDEX	
LAND AREA	8260㎡
CONSTRUCTION LAND AREA	4275㎡
FIRST FLOOR BUILDING AREA	1461.18㎡
FLOOR AREA	1946.48㎡
PLOT RATIO	0.24
BUILDING DENSITY	0.18
RATIO OF GREEN SPACE	44.78%
PARKING SPACE	9
NUMBER OF FLOORS	2
BUILDING HEIGHT	9m

Plane Graph 1:300

1 Reception Hall
2 Service counter
3 Sand table & exhibition area
4 Lounge
5 Water Bar
6 Washroom
7 Weaving workshop
8 Leisure activity space
9 cildren's activity space
10 Reading space
11 Common stores
12 Souvenir shop
13 Restaurant
14 Kitchen
15 Restaurant outboard
16 Fire service control room
17 Equipment room
18 Courtyard
19 Livingroom
20 Study
21 Bedroom
22 Dining room
23 Sun room
24 Tea-stall
25 Rain garden
26 Square
27 Opera theatre square
28 Neighborhood market
29 Fitness venue
30 Pool viewing platform
31 Neighborhood square
32 Wheat field garden
33 Fitness trail

1st Floor Layout

1 Village activity hall
2 Village activity hall lobby
3 Village innovation and entrepreneurship commune
4 Office
5 Second floor platform

2nd Floor Laylot

0 5 10 15 20 25m

N

BUUNY TRIBE SCENIC AREA

HOMESTAY HOTELS OF THE RESIDENTS

VEGETABEL GARDEN FOR RESEARCH AND EDUCATION

Traffic Flow Line Analysis

North Elevation 1:400

South Elevation 1:400

专家点评：

该方案总体布局充分考虑了地方文化，重视面向水面的景观视野，营造出了独特的空间体验。建筑风格具有一定的地域特色，住宅部分的总体设计和单体设计比较理想。方案设计综合全面，建筑坡屋面结合冬季融雪和光伏。被动式设计表现优异，从被动设计到先进能源系统，混合运用多种模式，积极整合先进能源系统运营策略，全力降低建筑系统的碳足迹与运营能耗。同时，对空气源热泵供暖在严寒地区的适用性展开探讨，但设计方案中民宿与服务中心连接性有待加强，建筑单体坡屋面角度不足。

Experts' Commentary：

The overall layout of the scheme takes into account the local culture and emphasizes the view of the waterfront landscape to create a unique spatial experience. The architectural style, characterized by distinct regional features, embodies comprehensive design principles. In addition, both the overall design and unit design of the residential buildings are well-conceived. The sloping roofs incorporate snow-melting and photovoltaic functions. Passive design is excellent, successfully combining passive strategies with advanced energy systems. It integrates hybrid models and incorporates operational strategies of advanced energy systems to significantly reduce the carbon footprint and operational energy consumption. The design also discusses the applicability of ASHP（Air Source Heat Pump）heating in cold areas. However, the scheme could be further improved by strengthening the connection between the guest house and the service center. Additionally, the angle of the sloping roof of the unit building is inadequate to support its intended function.

CODE:1100003005

织光麦社——乡村会客厅和民宿民居设计 Ⅲ

Weaving Sunshine: Wheat Field Village Community
—— The Villagereception and Homestay Hotel Design

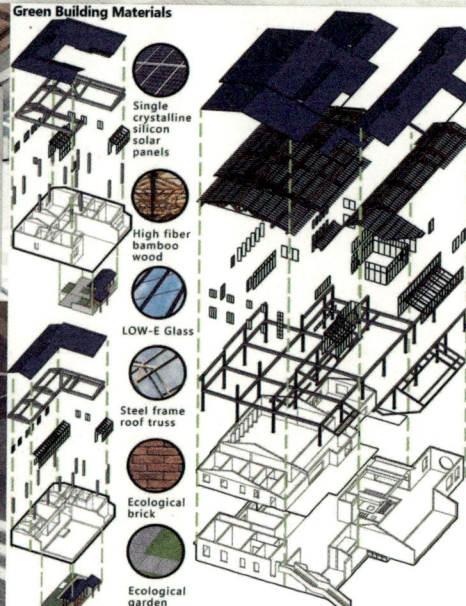

Green Building Materials

- Single crystalline silicon solar panels
- High fiber bamboo wood
- LOW-E Glass
- Steel frame roof truss
- Ecological brick
- Ecological garden

Active & Passive Energy Saving Strategy System

In the design, we have chosen appropriate technologies and strategies for green building design based on the local climate and environment. The design adopts a combination of passive and active design to improve building comfort while reducing building energy consumption. Passive design includes architectural form and spatial design, passive structure, and energy-saving building materials. Active design includes solar energy, ground source heat pumps, rainwater circulation, waste recycling, and garbage recycling, biomass energy utilization, while utilizing intelligent monitoring systems to monitor building energy operation.

BIPV-Building Integrated Photovoltaic

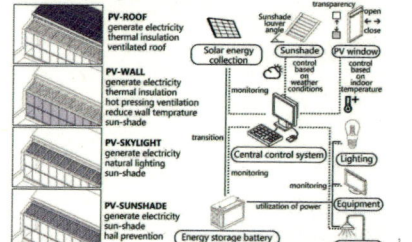

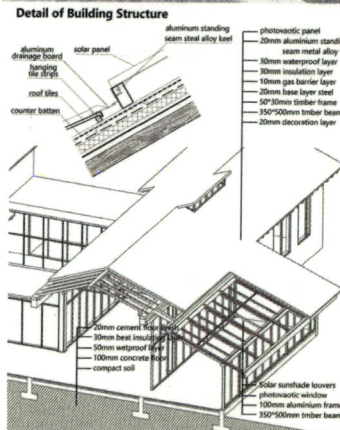

PV-ROOF
generate electricity
thermal insulation
ventilated roof

PV-WALL
generate electricity
thermal insulation
hot pressing ventilation
reduce wall temprature
sun-shade

PV-SKYLIGHT
generate electricity
natural lighting
sun-shade

PV-SUNSHADE
generate electricity
sun-shade
hail prevention

Detail of Building Structure

aluminium standing seam steal alloy keel
aluminum drainage board
solar panel
hanging tile strips
roof tiles
counter batten

photovaotic panel
20mm aluminium standing seam metal alloy keel
30mm waterproof layer
30mm insulation layer
10mm gas barrier layer
20mm base layer steel
50*30mm timber frame
350/500mm timber beam
20mm decoration layer

20mm cement floor board
30mm heat insulation
50mm wetproof liner
100mm concrete floor
compact soil

Solar sunshade louvers
photovoltaic window
100mm aluminium frame
350*500mm timber beam

CODE:110003-99

织光麦社——乡村会客厅和民宿民居设计 IV

Weaving Sunshine—Wheat Field Village Community —— The Villagereception and Homestay Hotel Design

Indoor Environment Analysis

Summer daytime

adjust the sunshade louvers. Open the windows and skylights to create air pressure and ventilation indoors.

Summer night

the building uses wind pressure for ventilation.

Winter daytime

close the doors, Windows and skylights, the sun room and the indoor travel heat circulation, thermal storage floor and wall start to store energy

Winter night

thermal storage floor wall for heat release

Hail-proof Design Strategies

Sunlight Room

The pitched roof protects against hail accumulation
The roof material is solar cell/elastomer composite material

OPEN | CLOSED

Wind — Elastic Material

tempered laminated bulletproof

Corners using PVC, rubber or anti-collision stickers or polyurethane of anti-collision strips

Intelligent air conditioning system

Adjust — air conditioning system

central control system

Temperature sensor

passive

The central system adjusts the passive structure based on the indoor temperature hrough sensors, and then adjusts the air conditioning and heating systems when the passive cannot meet the requirements.

Outdoor Comfort

According to the image above, people in the area feel more comfortable between march-may and september-october.

Wind speed chart

The wind speed map shows that the city from January to June wind speed is larger, up to 14 m/s, 10-12 months wind speed is smaller.

Dry ball temperature chart

The temperature in the city is high from June to September, reaching 35.6 °C.

Global radiation levels

The figure shows that radiation levels are highest at 12 noon and can reach 949.00 Wh/m2.

Relative humidity chart

The map shows the highest relative humidity between June and September.

Sky radiation model

The map shows that the area of greater sunshine time, the buildings are less sheltered sunshine time, the impact is greater.

Annual enthalpy-humidity chart

According to the strategies provided by the optimal combination of enthalpy-humidity diagram, 3-5 months have a relatively large proportion of comfortable time, with moderate humidity; 6-8 months have the largest proportion of comfortable time, with slightly dry conditions; 9-11 months have a relatively small proportion of comfortable time, with slightly humid conditions; and 12-2 months have no comfortable time.

Annual wind rose chart

According to the wind rose chart, the most frequent winds in the city are from the northwest in January to April, from the northwest and southeast in May to August, and from the northwest in September to December.

Spacial Scene

Eucommia ulmoides
Fraxinus
Albizia julibrissin
Populus
Ligustrum × vicaryi
Berberis thunbergii
Rosa chinensis
Sophora japonica
Platanus orientalis Linn.
Sabina chinensis
Lagerstroemia indica
Sabina chinensis
Forsythia suspensa
Cedrus deodara
Amygdalus
Salix babylonica
Koelreuteria

Sponge Green

Rain Garden
Cultural Living Room
Leisure Square
Parking Lot
Farm Produce Markets
Stage
Park Signage

Crop
Outdoor Exercises
Gathering Plaza
Rest Corner
Canopy Walking
Crop
Rest Space
Homestay Sign

Rain Collection
Infiltration
Runoff
Cropland Irrigation
Sponge Exhibition

Solar Roof
Sunken Green
Permeable Paving
Metal Channel
Drainage to Muncipal Pipe

Sponge City Concept

Primary Purification
Secondary Purification
Tertiary Purification
Stabilization Pool
Landscape Pool

Photovoltaic Power
Rain Water Green Cycle

Subsurface runoff

Perspective Section

8.100

4.200

Intelligent monitoring system

4.500

BIPV

energy-saving doors and windows

sun room

Village acticvity hall

hot water

Restaurant

power

Leisure activity space

weaving work shop

Reception hall

Heat pump unit

Garbage recycling

Ground source heating system

floor heating system

综合奖·二等奖·张家口市
怀安县二堡子村建设项目
Comprehensive Awards －
Second Prize － Erpuzi
Village Project，Huai'an
County，Zhangjiakou City

注册号：110031
Register Number：110031
项目名称：织宇向阳
Entry Title：Interwoven Roofing for Solar
Energy Harvesting
作者：殷梓芸、崔璨、贾龙庆、陈雄宇、
杜钰玮、李啸跃
Authors：Yin Ziyun, Cui Can, Jia Longqing,
Chen Xiongyu, Du Yuwei, and
Li Xiaoyue
作者单位：山东建筑大学
Authors from：Shandong Jianzhu University
指导教师：侯世荣、何文晶、李晓东
Tutors：Hou Shirong, He Wenjing, and Li
Xiaodong
指导教师单位：山东建筑大学
Tutors from：Shandong Jianzhu University

作品编号：110031

织宇向阳 I
Interwoven Roofing for Solar Energy Harvesting

Site Activity

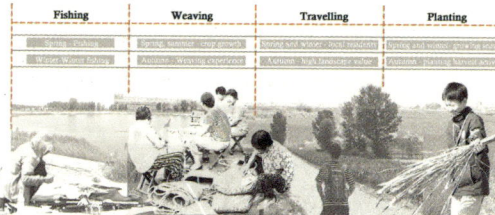

Fishing	Weaving	Travelling	Planting
Spring - Fishing	Spring, summer - crop growth	Spring and winter - pool resident	Spring and winter - grass has sprout
Winter-Waters fishing	Autumn - Weaving experience	Autumn - high landscape value	Autumn - Planting harvest activity

Description of Design

1.在方案构思上，整体采用模块化的手法，解决乡村生产、生活、生态问题，并且在场地中设计可变构筑物，以便在特殊情景下作为应急避难所，从而提高抗风险能力，体现"韧性"主题。
2.在方案特点上，结合太阳能光伏板设计南向单坡屋顶，同时穿插北向单坡屋顶实现热压通风，从而调节室内环境，南向北向屋顶交织，形成特色建筑形态。
3.在方案的结构与选材上，通过实地调研选用更加融入当地环境氛围的木结构作为主要抗震支撑结构，就地取材选用稻草砖进行保温处理以减少整个生命周期的碳排放量。

1.In the concept of the program, the overall use of modular approach, the design of variable construction in the site .To act as an emergency shelter in special situations, thereby enhancing the ability to resist risks.
2.In terms of the characteristics of the scheme, a south-facing single-slope roof is designed in combination with solar photovoltaic panels, and a north-facing single-slope roof is interspersed to achieve thermal pressure ventilation to adjust the indoor climate.
3.In the structure and material selection of the scheme, the wood structure which is more integrated into the local environmental atmosphere is selected as the main support structure through field investigation, and the straw brick is selected for heat preservation treatment to reduce the carbon emission of the whole life cycle.

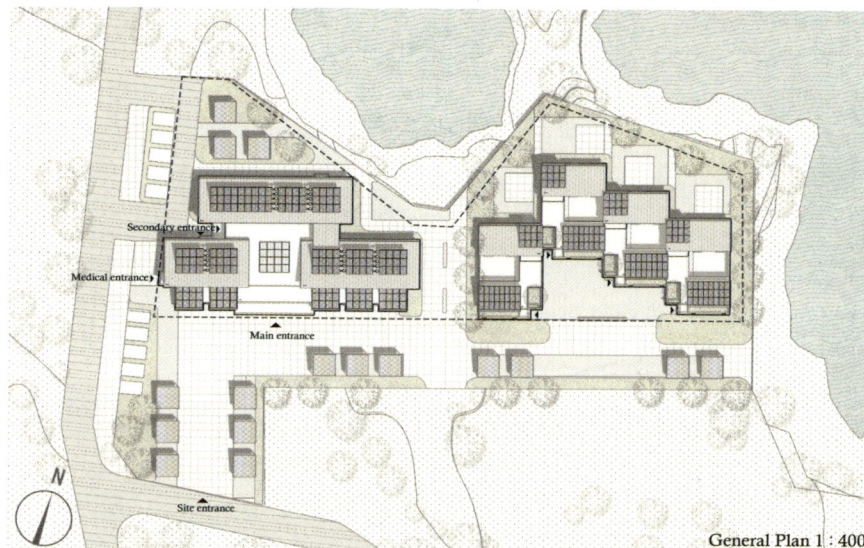

Analysis of Block Generation

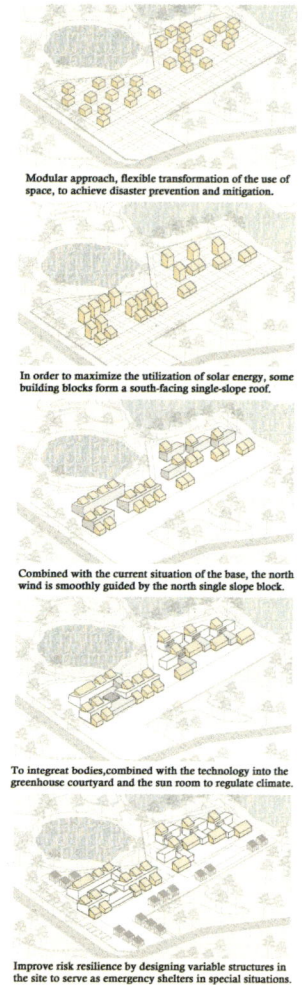

Modular approach, flexible transformation of the use of space, to achieve disaster prevention and mitigation.

In order to maximize the utilization of solar energy, some building blocks form a south-facing single-slope roof.

Combined with the current situation of the base, the north wind is smoothly guided by the north single slope block.

To integreat bodies,combined with the technology into the greenhouse courtyard and the sun room to regulate climate.

Improve risk resilience by designing variable structures in the site to serve as emergency shelters in special situations.

Secondary entrance
Medical entrance
Main entrance
Site entrance
N

General Plan 1：400

作品编号:110031

织宇向阳 II
Interwoven Roofing for Solar Energy Harvesting

1 Clinic　　　　　　5 Showroom
2 Consultation room　6 Sell along the street
3 Infusion room　　　7 Open reading area
4 Pharmacy　　　　　8 Coffeehouse
　　　　　　　　　　9 Fishing gear storage area

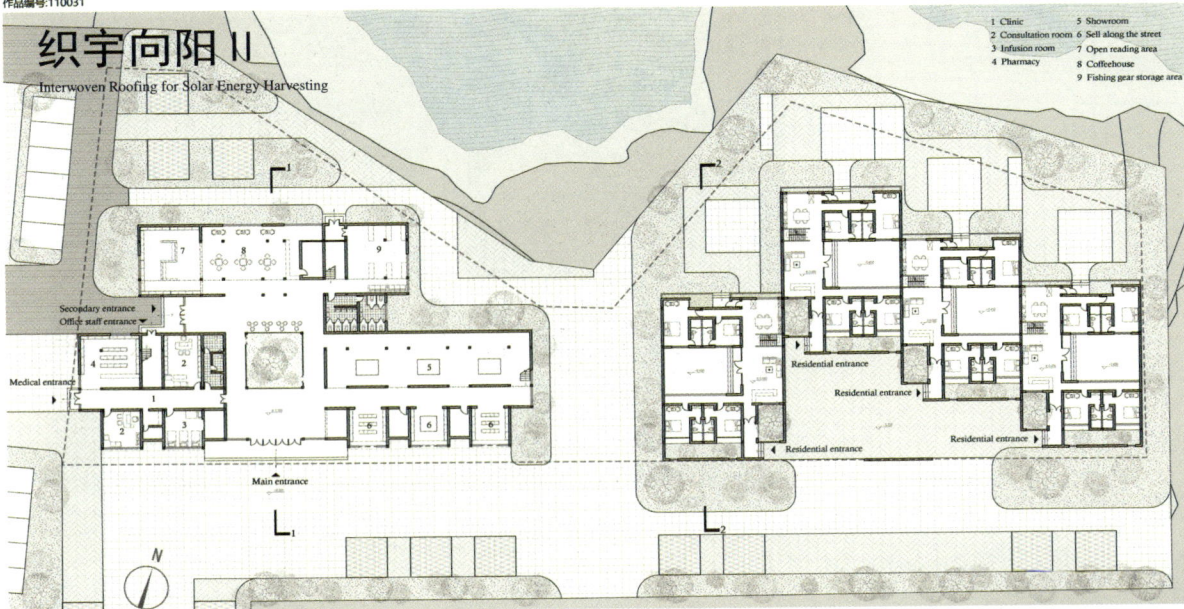

Secondary entrance
Office staff entrance

Medical entrance

Main entrance

Residential entrance
Residential entrance
Residential entrance
Residential entrance

N

Second Floor Plan　1：300

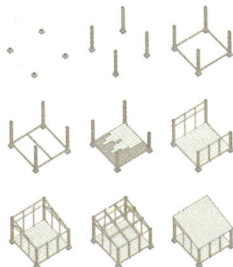

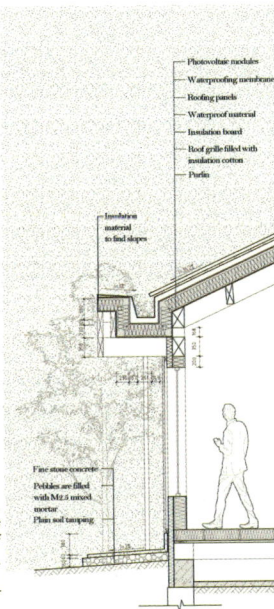

Photovoltaic modules
Waterproofing membrane
Roofing panels
Waterproof material
Insulation board
Roof grille filled with insulation cotton
Purlin

Insulation material to find slopes

Fine stone concrete
Pebbles are filled with M2.5 mixed mortar
Plain soil tamping

Usually or Disaster combination

The structures placed on the south side of the site are used as a garden for leisure, while structures on the north side are used as an interactive experience area for local weaving activities. When the disaster comes, the structures on the south side are quickly assembled into a centralized resettlement space surrounded by four sides, while the north side is mainly used as a transportation point and storage space for materials.

Garden　　Storage　　Weaving　　Emplacement

Sectional View 1-1　1：300

Sectional View 2-2　1：300

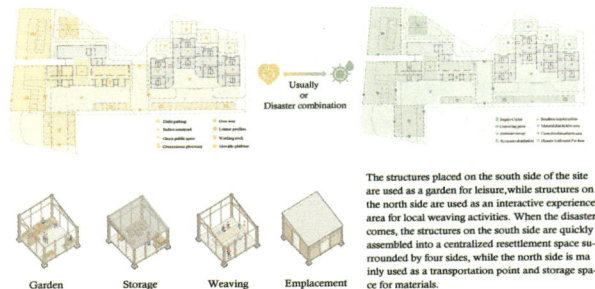

North Elevation View　1：300

专家点评：

该方案总体分区合理，民宿用地均衡性好。民宿和服务中心各自设置开放空间的构思巧妙。建筑设计上，每栋住宅楼配备三个庭院，有利于促进微观社区互动。建筑形式与北方建筑风格契合，地域文化与实用性兼备。坡顶与光伏排列结合较好，既满足能源利用需求，又增添建筑美感。空间剖面设计出色，优化了空间的层次感与功能性，但民宿与服务中心连接薄弱，开放式空间规划对多样化社区活动支持有限，可能会对可达性和融合性造成阻碍。另外，北侧景观与空间的互动存在缺陷。

Experts' Commentary：

The overall zoning of the scheme is reasonable, and the usage of residential land is well-balanced. The guest house and service center feature independently designed open spaces that are thoughtfully conceived. Architecturally, each residential building is equipped with three courtyards, which facilitates micro-community interaction. The architectural form conforms to the northern architectural style, integrating regional culture and practicality. The combination of sloping roofs and photovoltaic arrangements is skillfully executed, not only meeting energy utilization requirements but also enhancing the aesthetic appeal of the buildings. The spatial profile design is exemplary, optimizing the sense of hierarchy and functionality of the space. However, the connection between the guest house and the service center is weak, limiting support for diverse community activities in open spaces. This detracts from the scheme's accessibility and integration. Additionally, there are deficiencies in the spatial interaction design in the northern landscape areas.

作品编号:110031

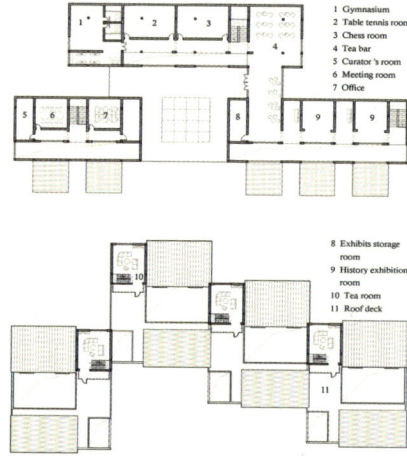

1 Gymnasium
2 Table tennis room
3 Chess room
4 Tea bar
5 Curator's room
6 Meeting room
7 Office

8 Exhibits storage room
9 History exhibition room
10 Tea room
11 Roof deck

Second Floor Plan 1:300

织宇向阳 III
Interwoven Roofing for Solar Energy Harvesting

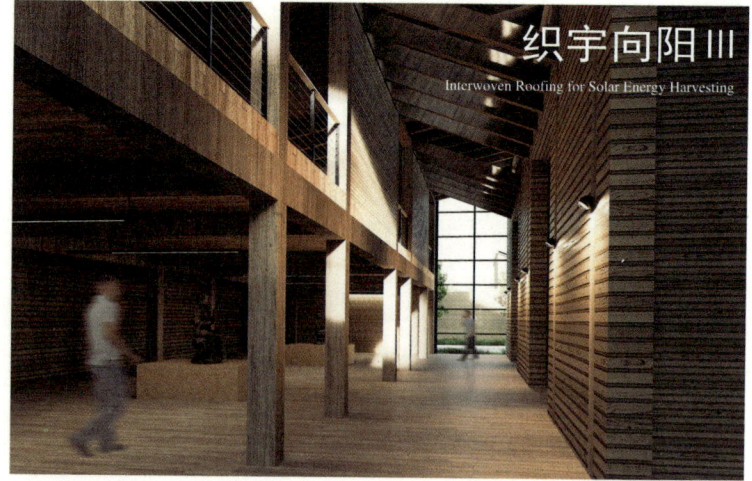

Explosive View

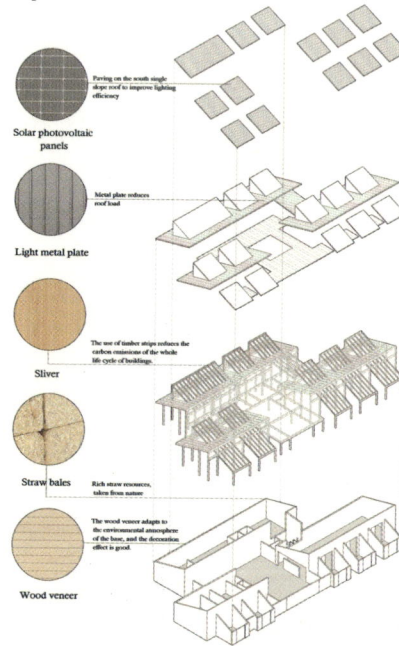

Solar photovoltaic panels
Paving on the south single slope roof to improve lighting efficiency

Light metal plate
Metal plate reduces roof load

Sliver
The use of timber strips reduces the carbon emissions of the whole life cycle of buildings.

Straw bales
Rich straw resources, taken from nature

Wood veneer
The wood veneer adapts to the environmental atmosphere of the base, and the decoration effect is good.

Structural Unit Analysis Diagram

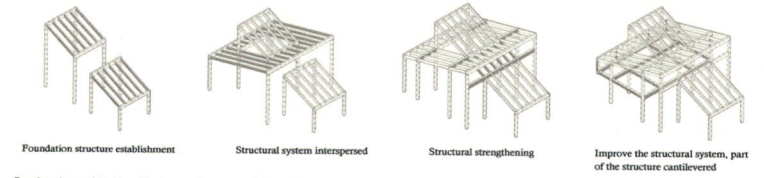

Foundation structure establishment

Structural system interspersed

Structural strengthening

Improve the structural system, part of the structure cantilevered

Based on the consideration of the base environment and low carbon emission reduction, the designer chooses the timber structure system, and focuses on the consideration of stability when organizing the structure, so as to increase the risk resistance ability of the building.

Functional Analysis of Public Space

Street sale area

Fishing gear storage space

Straw weaving space

Combined with field research to understand the local combination of natural resources to develop rural tourism, combined with summer and winter fishing activities, rich straw resources and open rabbit park for open space function layout, into the environment and drive to stimulate the vitality of the site.

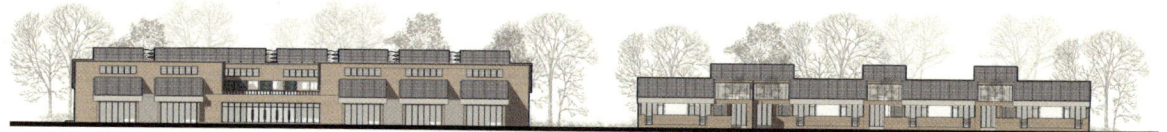

South Elevation 1:250

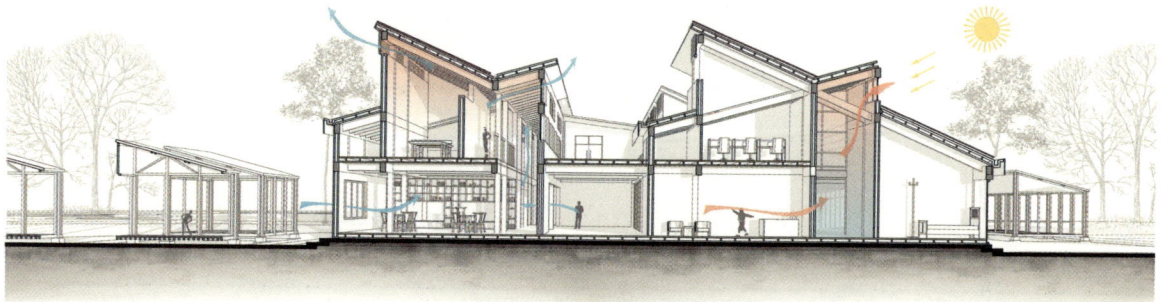

织宇向阳Ⅳ

Interwoven Roofing for Solar Energy Harvesting

Calculation results of carbon emissions	Public service buildings	Local dwelling houses
Heat consumption index of heating	19920.97 kW·h	1025.679 kW·h
Cooling heat consumption index	4639.13 kW·h	238.53 kW·h
Lighting heat consumption index	3451.38 kW·h	1550.62 kW·h
Electricity carbon emission factor	0.8843e/kW·h	
Total	27257.6632Kg	

Calculation of carbonemission reduction	Public service buildings	Local dwelling houses
Solar photovoltaic power generation	53785.70 kW·h	33923.88 kW·h
Wind turbine power generation	623.56 kW·h	0 kW·h
Electricity carbon emission factor	0.8843e/kW·h	
Total	-78112.58 kg CO2	

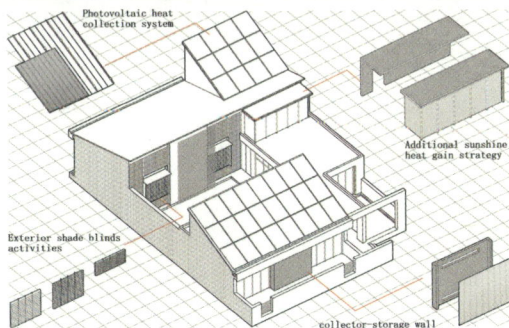

Photovoltaic heat collection system

Additional sunshine heat gain strategy

Exterior shade blinds activities

collector-storage wall

Wind Environment Simulation

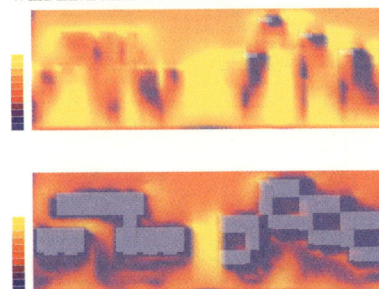

Ventilation Strategy

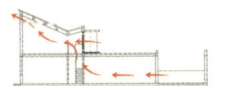

The building opens doors and windows at opposite positions on the north and south sides to ensure natural ventilation, and opens side windows locally on the high side to enhance ventilation.

Courtyard Strategy

Courtyard hot pressing ventilation

Under the irradiation of the sun, the air temperature near the ground increases, so the hot air rises, forming hot pressure ventilation.

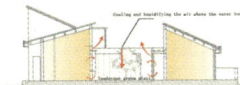

Courtyard greenery

The indoor thermal comfort can be improved by cooling the water.

Attached Solar Space

Collect heat during the day

The air in the sunroom and the public wall are heated directly through conduction and radiation to heat the room.

night heat preservation

Close the doors and windows on the public wall, and the sunlight room becomes a thermal buffer zone to slow down the heat loss in the room.

Collector-storage Wall

Ventilation during the day in summer

The flow of air in the interlayer prevents indoor overheating and takes away part of the indoor waste heat.

Winter gets hot during the day

The solar radiation is radiated to the heat storage wall through the glass cover.

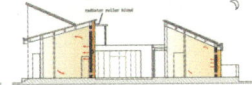

Summer night ventilation

The accumulated heat in the wall at night heats the air in the cavity to meet the thermal pressure ventilation.

Keep warm at night in winter

The heat stored by the regenerator continues to heat the room at night in a radiant and convective manner

Sunlight Analysis

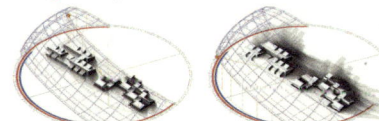

Summer solstice shadow range   Winter Solstice Sun Shadow Range

Solar Radiation

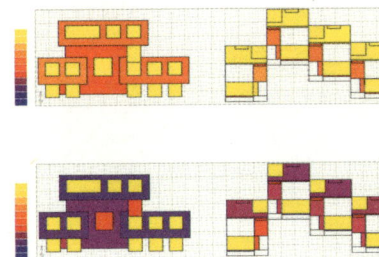

Analysis of Daylight Shading

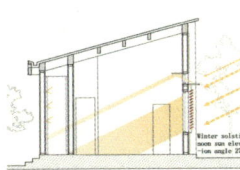

Heat-gaining in Winter

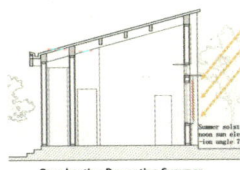

Over-heating Prevention Summer

Technical Overview

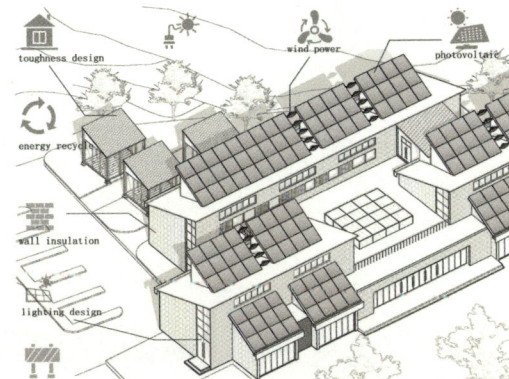

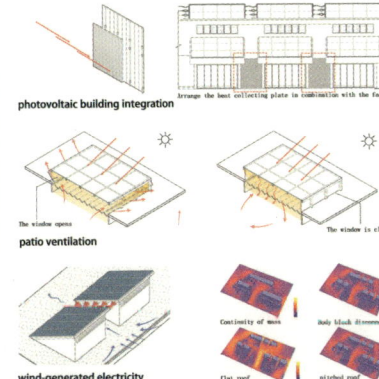

toughness design

wind power

photovoltaic

energy recycle

wall insulation

lighting design

photovoltaic building integration

Arrange the heat collecting plate in combination with the facade

patio ventilation

The window opens   The window is closed

wind-generated electricity

Continuity of roars   Body block dismantled

flat roof   pitched roof

自然之道
Tao of Nature

综合奖・二等奖・张家口市怀安县二堡子村建设项目
Comprehensive Awards — Second Prize — Erpuzi Village Project，Huai'an County，Zhangjiakou City

注册号：110666
Register Number：110666
项目名称：自然之道
Entry Title：Tao of Nature
作者：刘志波、刘洋、刘晓莹、侯玉淑、张畅、朱怡萱、高绮绮、张佳旭、柳洋
Authors：Liu Zhibo, Liu Yang, Liu Xiaoying, Hou Yushu, Zhang Chang, Zhu Yixuan, Gao Qiqi, Zhang Jiaxu, and Liu Yang
作者单位：中建八局发展建设有限公司设计研究院
Authors from：The Design Research Institute of China Construction Eighth Engineering Division Corp. Ltd. of CSCEC (China State Construction Engineering Corporation)
指导老师：王剑涛
Tutor：Wang Jiantao
指导老师单位：中建八局发展建设有限公司设计研究院
Tutor from：The Design Research Institute of China Construction Eighth Engineering Division Corp. Ltd. of CSCEC (China State Construction Engineering Corporation)

CONCEPT

In Huai'an County, Hebei Province, a land full of history and natural beauty, our architectural design philosophy is deeply rooted in "nature-based solutions", a core concept that aims to explore a new realm of harmonious symbiosis between architecture and the natural environment. The architecture we advocate is not only a simple division of space, but also an in-depth integration of the wisdom of traditional villages in Hebei and modern ecological concepts. Through carefully designed architectural forms, we skillfully echo the rustic appearance and layout wisdom of local traditional villages, preserving the cultural roots and giving them aesthetic value for the new era.

Hebei traditional village · Hebei traditional village · Site condition

GENERATING

Positive north-south orientation to fully utilize the sunlight conditions.

The building is in the form of a low-rise courtyard to enhance the comfort and convenience of use.

The long slope to the south adopts an angle of 14°, providing a suitable angle and sufficient roof area for photovoltaic.

The short slopes on the north side allow more low angle sunlight to enter the room in winter and optimize indoor ventilation.

MASTERPLAN

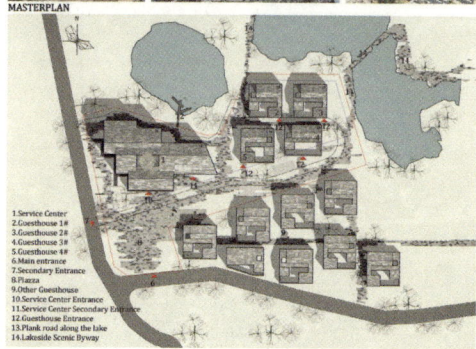

1.Service Center
2.Guesthouse 1#
3.Guesthouse 2#
4.Guesthouse 3#
5.Guesthouse 4#
6.Main entrance
7.Secondary Entrance
8.Plaza
9.Other Guesthouse
10.Service Center Entrance
11.Service Center Secondary Entrance
12.Guesthouse Entrance
13.Plank road along the lake
14.Lakeside Scenic Byway

ENERGY-SAVING EQUIPMENT

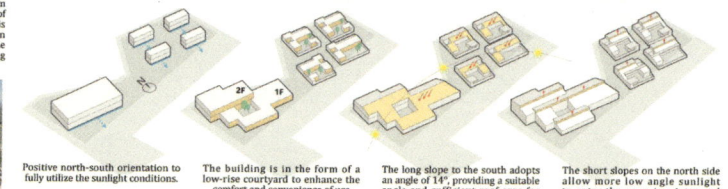

Solar Power Heat Generation · Ground source heat pump · Rainwater recycling

ELEVATION

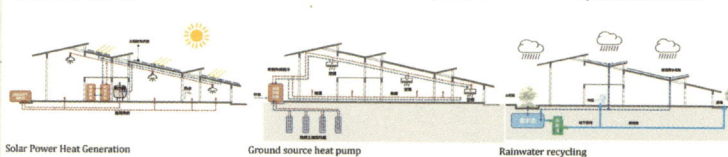

service center elevation · Guesthouse elevation

NAMAL FLOW

EMERGENCY FLOW

自然之道
The Sense of Nature

DINING ROOM RENDERING

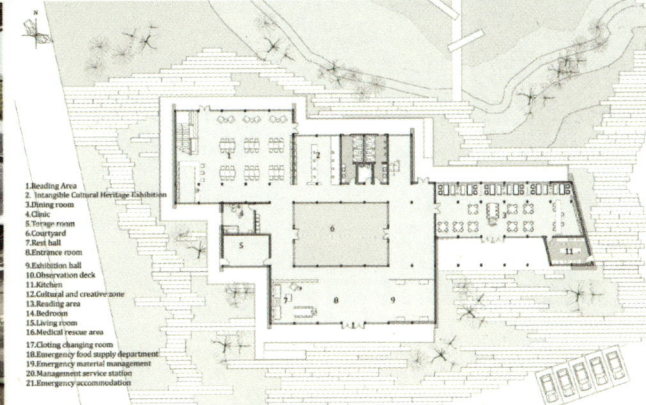

HALL RENDERING

HALL RENDERING

SERVICE CENTER SECTIONAL PERSPECTIVE

专家点评：

该方案在规划上总体布局合理，为后续的功能实现与空间利用提供了良好的框架。建筑设计从空间、形态到材质展现出独特的地域风格。建筑形态经过精心构思，可适应未来的扩建和延伸，确保能适应不断变化的需求并培养长期的功能灵活性，有力保障了建筑的可持续性。但民宿平面对节能存在不利因素。

Experts' Commentary：

The overall layout of the scheme is reasonable, providing a sound framework for subsequent function realization and space utilization. The architectural design showcases a unique regional style through its spatial configuration, formal composition, and material selection. The building form has been meticulously planned to facilitate future extensions, ensuring adaptability to changing needs and fostering long-term functional flexibility, which strongly supports the sustainability of the building. However, some factors in the architectural plan of the guest house negatively impact energy efficiency.

NORMAL PLAN(service center 1F)

NORMAL PLAN(Guesthouse 1#2#3#)

NORMAL PLAN(Guesthouse 4#)

PLAN OF EMERGENCY (Guesthouse 1#2#3#)

1.Reading Area
2. Intangible Cultural Heritage Exhibition
3.Dining room
4.Clinic
5.Storage room
6.Courtyard
7.Rest hall
8.Entrance room
9.Exhibition hall
10.Observation deck
11.Kitchen
12.Cultural and creative zone
13.Reading area
14.Bedroom
15.Living room
16.Medical rescue area
17.Cloting changing room
18.Emergency food supply department
19.Emergency material management
20.Management service station
21.Emergency accommodation

NORMAL PLAN(service center 2F) | PLAN OF EMERGENCY(service center 1F) | PLAN OF EMERGENCY(service center 2F) | PLAN OF EMERGENCY(Guesthouse 4#)

GUESTHOUSE SECTION

Roof warming promotes airflow — Natural building ventilation

Blocking of sunlight in summer (solar altitude angle 73°26') — Eaves create a physical environment that is warm in winter and cool in summer

Low angle sunlight in winter (solar altitude angle 26°34')

PVT for energy extraction, containerized unit for energy storage and power supply — Solar Photovoltaic System

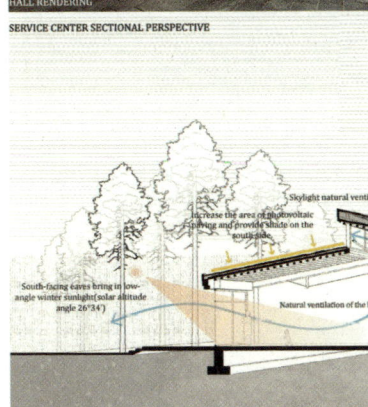

Heating tanks store rainwater for watering, indoor cleaning and dirt cleaning — Solar Photovoltaic System

South-facing eaves bring in low-angle winter sunlight (solar altitude angle 26°34')

Skylight natural ventilation

North facing eaves block sunlight creating a pleasantly cool outdoor space in summer (solar altitude angle 73°26')

increase the area of photovoltaic paving and provide shade on the south side.

increase the area of photovoltaic paving and provide shade on the south side.

Skylight natural ventilation

South-facing eaves bring in low-angle winter sunlight(solar altitude angle 26°34')

Natural ventilation of the building

Natural ventilation through internal courtyards

Natural ventilation of the building

自然之道
Tao of Nature

EXHIBITION RENDERING

GUESTHOUSE AREA RENDERING

CULTURAL AND CREATIVE ZONE RENDERING

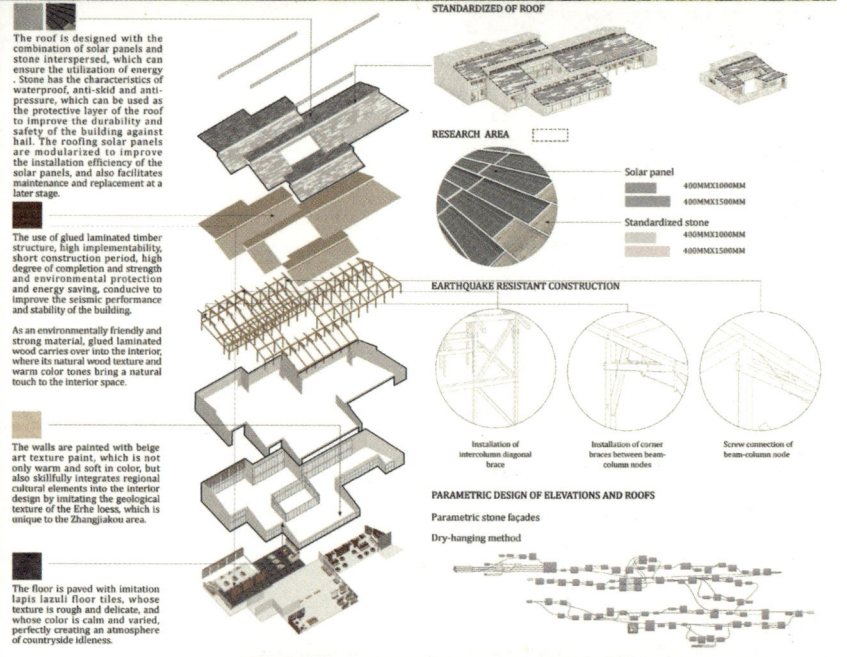

The roof is designed with the combination of solar panels and stone interspersed, which can ensure the utilization of energy. Stone has the characteristics of waterproof, anti-skid and anti-pressure, which can be used as the protective layer of the roof to improve the durability and safety of the building against hail. The roofing solar panels are modularized to improve the installation efficiency of the solar panels, and also facilitates maintenance and replacement at a later stage.

The use of glued laminated timber structure, high implementability, short construction period, high degree of completion and strength and environmental protection and energy saving, conducive to improve the seismic performance and stability of the building.

As an environmentally friendly and strong material, glued laminated wood carries over into the interior, where its natural wood texture and warm color tones bring a natural touch to the interior space.

The walls are painted with beige art texture paint, which is not only warm and soft in color, but also skillfully integrates regional cultural elements into the interior design by imitating the geological texture of the Erhe loess, which is unique to the Zhangjiakou area.

The floor is paved with imitation lapis lazuli floor tiles, whose texture is rough and delicate, and whose color is calm and varied, perfectly creating an atmosphere of countryside idleness.

STANDARDIZED OF ROOF

RESEARCH AREA

Solar panel
400MMX1000MM
400MMX1500MM

Standardized stone
400MMX1000MM
400MMX1500MM

EARTHQUAKE RESISTANT CONSTRUCTION

Installation of intercolumn diagonal brace

Installation of corner braces between beam-column nodes

Screw connection of beam-column node

PARAMETRIC DESIGN OF ELEVATIONS AND ROOFS

Parametric stone façades

Dry-hanging method

ENERGY-SAVING MATERIALS

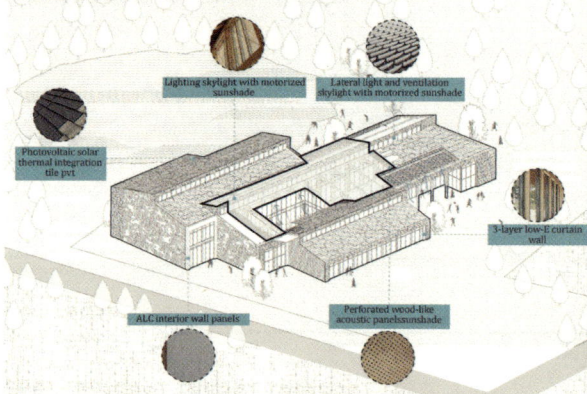

Lighting skylight with motorized sunshade

Lateral light and ventilation skylight with motorized sunshade

Photovoltaic solar thermal integration tile pvt

3-layer low-E curtain wall

ALC interior wall panels

Perforated wood-like acoustic panelssunshade

自然之道
Tao of Nature

CLIMATE ANALYSIS

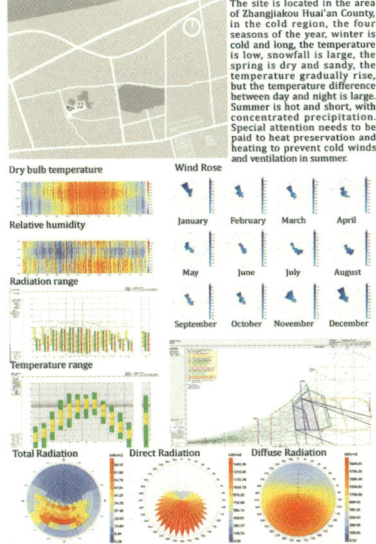

The site is located in the area of Zhangjiakou Huai'an County, in the cold region, the four seasons of the year, winter is cold and long, the temperature is low, snowfall is large, the spring is dry and sandy, the temperature gradually rise, but the temperature difference between day and night is large. Summer is hot and short, with concentrated precipitation. Special attention needs to be paid to heat preservation and heating to prevent cold winds and ventilation in summer.

Dry bulb temperature

Relative humidity

Radiation range

Temperature range

Wind Rose

January | February | March | April
May | June | July | August
September | October | November | December

Total Radiation | **Direct Radiation** | **Diffuse Radiation**

WIND ANALYSIS

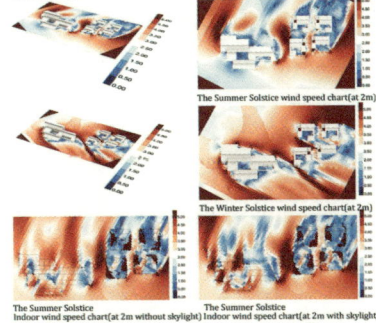

The Summer Solstice wind speed chart(at 2m)

The Winter Solstice wind speed chart(at 2m)

The Summer Solstice Indoor wind speed chart(at 2m without skylight) | The Summer Solstice Indoor wind speed chart(at 2m with skylight)

Skylights allow for better air circulation in the summer than rooms without skylights.

In this architectural design, we uphold the concept of respecting nature and integrating into the environment, and plan a set of design solutions integrating active and passive energy-saving technologies, aiming to create a positive energy-saving building that meets the local ecological needs as well as highlights the local cultural characteristics.

Passive design as the basis of the design strategy, through the building form, structure, special construction, material selection and other means, according to the local sunshine trajectory and wind direction characteristics, we designed a reasonable building orientation and form, to ensure that in winter can fully receive sunlight, and in summer can effectively block direct sunlight to reduce heat radiation.At the same time, the building's smooth form diverts the cold winds from the northwest, improves the efficiency of ventilation in summer,and reduces the energy consumption of air conditioning. In terms of structure, the use of local materials, the use of steel and wood structure increases the seismic resistance of the building, the use of local stone for façade design, the use of parametric façade texture standardized design, reduce transportation and replacement costs, and improve the adaptability and resilience of the building.The skylight design is made of three-glass, two-chamber Low-E glass, which not only efficiently insulates against external heat, but also substantially improves thermal insulation performance, effectively reducing reliance on artificial lighting during the day.

Passive technologies employ a range of advanced systems, including Photovoltaic Thermal (PVT) systems, ground-source heat pumps, and water recycling and purification techniques, to enhance energy efficiency and sustainability.

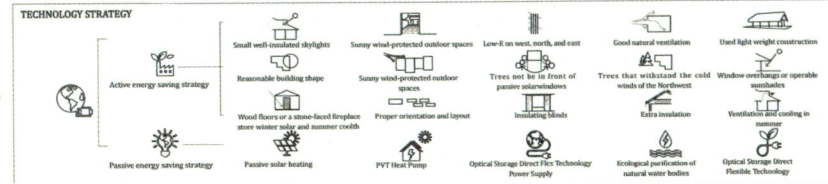

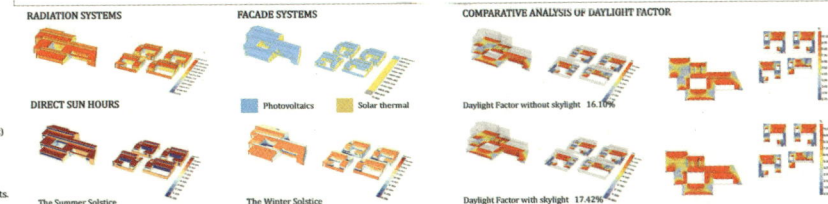

TECHNOLOGY STRATEGY

Active energy saving strategy

- Small well-insulated skylights
- Sunny wind-protected outdoor spaces
- Low-E on west, north, and east
- Good natural ventilation
- Used light weight construction
- Reasonable building shape
- Sunny wind-protected outdoor spaces
- Trees not be in front of passive solarwindows
- Trees that withstand the cold winds of the Northwest
- Window overhangs or operable sunshades
- Wood floors or a stone-faced fireplace store winter solar and summer coolth
- Proper orientation and layout
- Insulating blinds
- Extra insulation
- Ventilation and cooling in summer

Passive energy saving strategy

- Passive solar heating
- PVT Heat Pump
- Optical Storage Direct Flex Technology Power Supply
- Ecological purification of natural water bodies
- Optical Storage Direct Flexible Technology

RADIATION SYSTEMS

DIRECT SUN HOURS

The Summer Solstice | The Winter Solstice

The region is endowed with abundant light resources, high intensity of solar radiation. Combined with the solar radiation analysis, the building's roof and south elevation are the preferred areas for installing solar panels to maximize the use of solar energy due to the abundant sunlight.

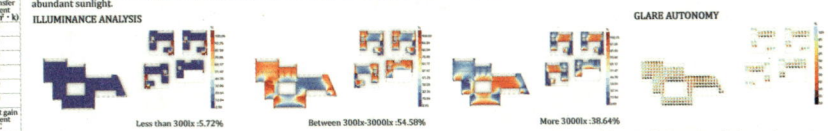

FACADE SYSTEMS

Photovoltaics | Solar thermal

COMPARATIVE ANALYSIS OF DAYLIGHT FACTOR

Daylight Factor without skylight 16.10%

Daylight Factor with skylight 17.42%

Skylights are designed to ensure that light is evenly distributed in the room and shadow areas are reduced. By comparison, the lighting situation is effectively improved.

CARBON EMISSION

Building Maintenance Structure Information

Building Envelope	Architectural Design Area(m²)	Heat Transfer Coefficient Value W/(m²·k)	Additional Heat Transfer Coefficient Value W/(m²·k)	Baseline Building Area(m²)	Heat Transfer Coefficient Value W/(m²·k)	Additional Heat Transfer Coefficient Value W/(m²·k)
East Exterior Wall	302.22	0.40	0.00	302.22	0.50	0.00
South Exterior Wall	611.63	0.40	0.00	611.63	0.50	0.00
West Exterior Wall	300.33	0.40	0.00	300.33	0.50	0.00
North Exterior Wall	611.63	0.40	0.00	611.63	0.50	0.00
Roof Surface	989.98	0.38	0.00	989.98	0.45	0.00
Floor Surface	952.51	0.44	-	952.51	0.37	0.00

External Window	Window-to-wall area ratio	Heat Transfer Coefficient W/(m²·k)	Solar heat gain coefficient SHGC	Window-to-wall area ratio	Heat Transfer Coefficient W/(m²·k)	Solar heat gain coefficient SHGC
East External Window	0.00	0.00	0.00	0.40	2.40	0.48
South External Window	0.42	1.40	0.44	0.40	2.40	0.48
West External Window	0.01	1.40	0.44	0.40	2.40	0.48
North External Window	0.57	1.40	0.44	0.40	2.40	0.41
Skylights(courtyard)	0.00	0.00	0.00	0.20	2.40	0.44

Airtightness and ventilation system

Project	Architectural Design	Baseline Building
Air infiltration rate of the exterior envelope at 50Pa Building (m³/hm²)	4.40	11.20
Natural ventilation	Yes	No

Type of HVAC (Heating, Ventilation, and Air Conditioning) system

		Design of HVAC system for buildings			
Number	Zone number	Type of cooling system terminals	Name of cold source	Type of heating system terminals	Name of heat source
1	1112-1114 1127 1129-1137	Single duct system	Ground source heat pump	Radiator terminal	Ground source heat pump

Carbon emission calculation results from green carbon sequestration during the operational phase of building design

Number	Name of green vegetation	Carbon sequestration from green vegetation kgCO₂
1	Evergreen trees	632700.00
2	Deciduous trees	533100.00
3	Evergreen shrubs	513300.00
4	Deciduous shrubs	294900.00
5	Groundcover herbs	170600.00
6	Flowers	88800.00
7	Climbing plants	27000.00
8	Aquatic plants	10000.00
	Total	2270400.00

Review of Carbon Emission Intensity Indicators during the Operational Phase of Buildings

Item	Value
Reduction in Carbon Emission Intensity (kg CO₂/m²·a)	124.96
Percentage Reduction in Carbon Emission Intensity	196.20%

ILLUMINANCE ANALYSIS

Less than 300lx :5.72% | Between 300lx-3000lx :54.58% | More 3000lx :38.64%

GLARE AUTONOMY

Regarding the international illuminance standard, the universally recognized appropriate range is set at 300-3000 lux, which is widely applicable to a variety of indoor environments, aiming to guarantee good visual conditions and comfort. In-depth analysis of our illuminance analysis illustration, you can clearly see that in the space evaluated, the proportion of areas that meet or even exceed this standard range is extremely significant, up to 90% or more, and this data fully demonstrates the science and effectiveness of lighting design.

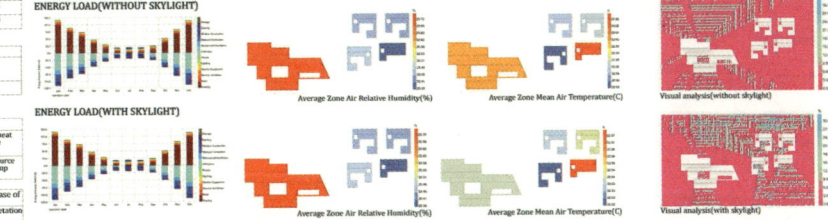

ENERGY LOAD(WITHOUT SKYLIGHT)

Average Zone Air Relative Humidity(%) | Average Zone Mean Air Temperature(C) | Visual analysis(without skylight)

ENERGY LOAD(WITH SKYLIGHT)

Average Zone Air Relative Humidity(%) | Average Zone Mean Air Temperature(C) | Visual analysis(with skylight)

The chart without skylight shows a full year cooling load of 84kW-h/m², heating load of 66kwh/m² and lighting load of 4.93 k·Wh/m². The higher energy consumption is due to the lack of effective energy-saving measures compared to the group with skylights. Skylights help reduce energy use by promoting natural ventilation in the summer, which lowers cooling demands, and by allowing solar heat gain in the winter, which reduces heating requirements. Additionally, skylights provide natural daylight, reducing the need for artificial lighting.

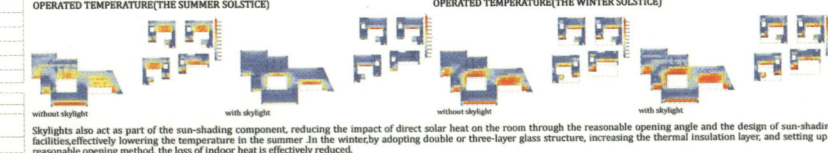

OPERATED TEMPERATURE(THE SUMMER SOLSTICE)

without skylight | with skylight

OPERATED TEMPERATURE(THE WINTER SOLSTICE)

without skylight | with skylight

Skylights also act as part of the sun-shading component, reducing the impact of direct solar heat on the room through the reasonable opening angle and the design of sun-shading facilities,effectively lowering the temperature in the summer .In the winter,by adopting double or three-layer glass structure, increasing the thermal insulation layer, and setting up a reasonable opening method, the loss of indoor heat is effectively reduced.

100-110030

光·韧 I

综合奖·二等奖·保定市清苑区李八庄村建设项目
Comprehensive Awards – Second Prize – Libazhuang Village Project, Qingyuan District, Baoding City

注册号：110030
Register Number：110030

项目名称：光·韧
Entry Title：Light – Resilience

作者：吴疏、朱书慧、严卓洋、龚友敏、
李京、李梦楚
Authors：Wu Shu, Zhu Shuhui,
Yan Zhuoyang, Gong Youmin,
Li Jing, and Li Mengchu

作者单位：山东建筑大学
Authors from：Shandong Jianzhu University

指导教师：侯世荣
Tutor：Hou Shirong

指导教师单位：山东建筑大学
Tutor from：Shandong Jianzhu University

Design Specfication

设计旨在回应场地的气候、环境和功能。通过对当地气候和功能需求的分析，我们确定了建筑的总体布局，并通过适应性的体形和缓冲空间回应气候的问题，同时结合场地的特殊环境，营造出丰富、错落有致的滨水空间。在防涝策略上，我们以韧性、海绵城市为理念，采取景观干预的方式从应急避险和居民自给自足、邻里交往三个层面进行设计并建立三道防洪线。在节能策略上，以空间调节为理念，采取被动优先，主动优化的方式从形体、表皮以及构造三个层面进行设计与应对。为了充分运用太阳辐射来解决冬季热舒适问题，通过建筑形态控制来获得更多有利的朝向同时采取单元化的建筑形式从形体成不同的热工分区。材料和施工上则主要采用轻钢结构和装配式施工方式以便于减少整个建筑生命周期的碳排放。在主动式应用方面，通过计算出最佳的太阳能光伏板角度，并将其合理地与建筑形态相结合，确保了其功能和美学方面的统一。

The design is designed to respond to the climate, environment, and function of the site. Through the analysis of the local climate and functional requirements, we determined the overall layout of the building, and responded to the climate through the adaptive shape and buffer space, while combining the special environment of the site to create a rich, scattered waterfront space. In terms of flood prevention strategy, we take the concept of resilience and sponge city as the concept, and adopt landscape intervention to design and establish three flood prevention lines from three levels: emergency avoidance, residents' self-sufficiency and neighborhood communication.In terms of energy saving strategy, the concept of space adjustment, passive priority, active optimization from the form, skin and structure of three levels of design and response. In order to make full use of solar radiation to solve the problem of thermal comfort in winter, more favorable orientation is obtained by building form control and different thermal zones are formed by building forms of units. In terms of materials and construction, light steel structure and prefabricated construction are mainly used to reduce carbon emissions throughout the building life cycle. In terms of active applications, the optimal solar photovoltaic panel Angle is calculated and reasonably combined with the building form to ensure the unity of its function and aesthetics.

Form Analysis

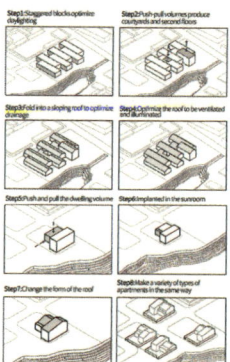

Vertical Design for Waterlogging Prevention

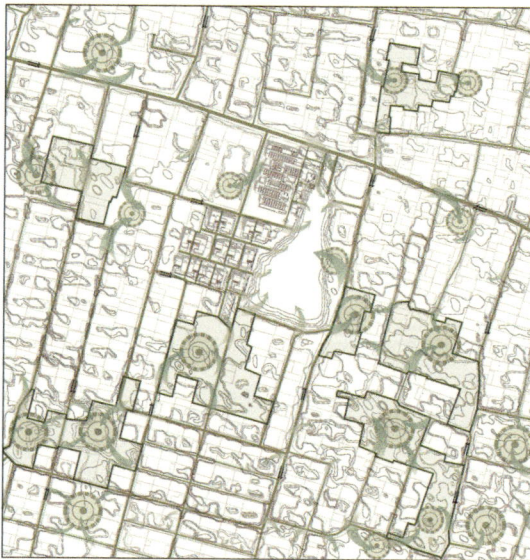

Annual runoff total control rate calculation table

Space Utilization in The Period of Abundant and Low Water

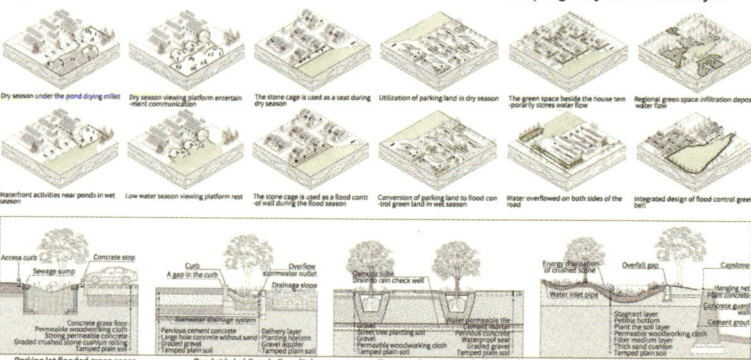

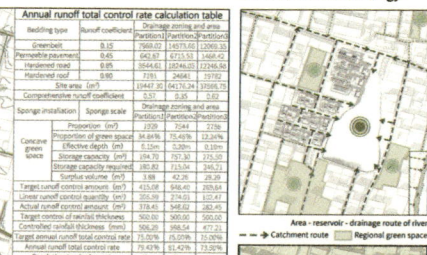

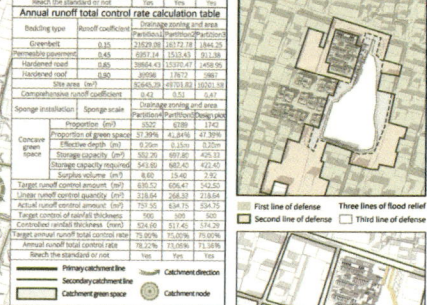

Sponge City Green Belt Analysis

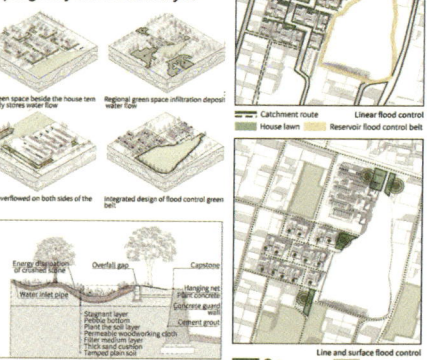

Flood Control Strategy

Site Plan

光·韧 Ⅱ

Economic and Technical Indicator Table		
Project	Unit	Quantity
Land area	hm²	0.96
Total building area	m²	2380
Village Supporting Facilities	m²	1300
vernacular architecture	m²	1810
Parking space		18
Plot ratio	%	24
Building density	%	22
Ratio of green space	%	18

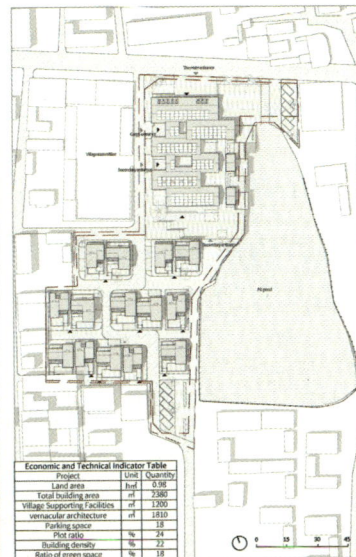

Ground Floor Plan **Second Floor Plan**

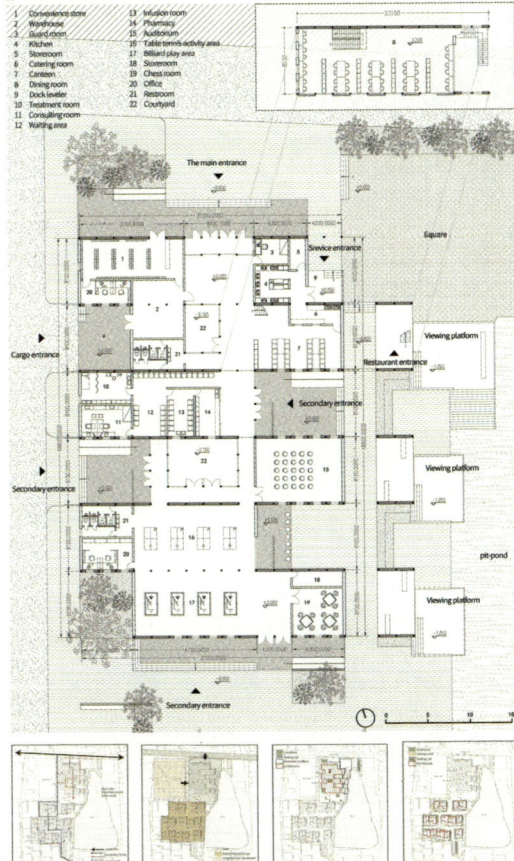

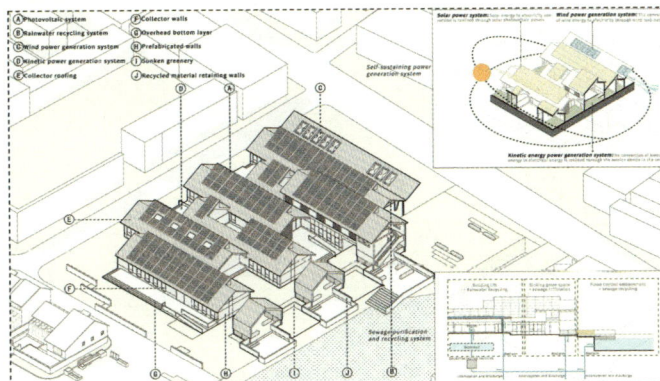

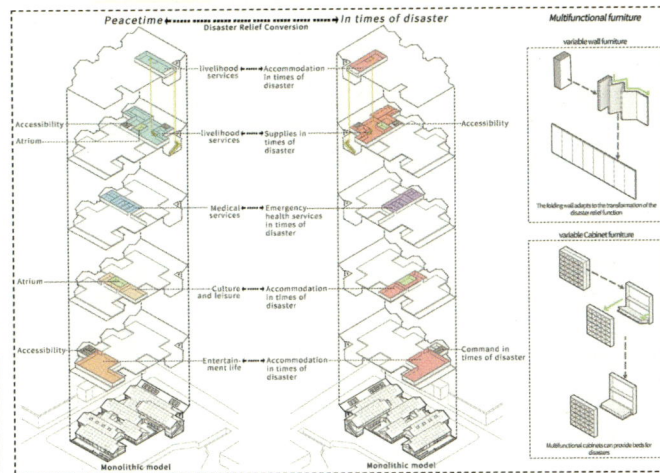

1. Convenience store
2. Warehouse
3. Guard room
4. Kitchen
5. Storeroom
6. Catering room
7. Canteen
8. Dining room
9. Dock leveler
10. Treatment room
11. Consulting room
12. Waiting room
13. Infusion room
14. Pharmacy
15. Auditorium
16. Table tennis activity area
17. Billiard play area
18. Storeroom
19. Chess room
20. Office
21. Restroom
22. Courtyard

The main entrance
Service entrance
Cargo entrance
Secondary entrance
Secondary entrance
Secondary entrance
Square
Restaurant entrance
Viewing platform
Viewing platform
Viewing platform
pit-pond

(A) Photovoltaic system
(B) Rainwater recycling system
(C) Wind power generation system
(D) Kinetic power generation system
(E) Collector roofing
(F) Collector walls
(G) Overhead bottom layer
(H) Prefabricated walls
(I) Sunken greenery
(J) Recycled material retaining walls

Solar power system
Wind power generation system
Self-sustaining power generation system
Kinetic energy power generation system
Sewage purification and recycling system

Peacetime — Disaster Relief Conversion — In times of disaster

livelihood services — Accommodation in times of disaster
livelihood services — Supplies in times of disaster — Accessibility
Medical services — Emergency health services in times of disaster
Culture and leisure — Accommodation in times of disaster
Entertainment life — Accommodation in times of disaster — Command in times of disaster

Accessibility
Atrium
Atrium

Monolithic model   Monolithic model

Multifunctional furniture
variable wall furniture
The folding wall adapts to the transformation of the disaster relief function
variable Cabinet furniture
Multifunctional cabinets can provide beds for disaster

Southeast Elevation   **Northeast Elevation**

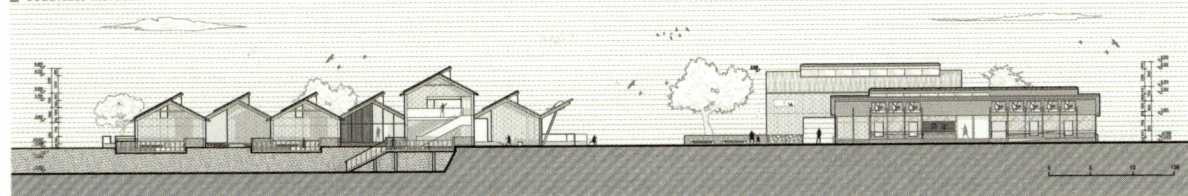

专家点评：

该方案在总体布局分区明确，为项目整体架构奠定了良好的基础，但住宅用地部分较为拥挤。建筑单体形式具有地域特色。住宅部分平面功能考虑较为完善，被动式与主动式技术应用较好，展现了一定的技术融合能力。防内涝策略上，以海绵城市技术为理念，采取景观干预的方式，分三个层级缓解内涝问题，但设计中水池岸边与水面高差变化缺乏呼应，道路两侧存在水溢出的风险，可能损坏道路基础设施。

Experts' Commentary：

The scheme features a well-defined zoning design in the overall layout, establishing a solid foundation for the project's overall framework. However, the residential area appears congested. The building designs reflect regional characteristics, with some residential floor plans showcasing well-considered and effective integration of passive and active technologies. For waterlogging prevention strategies, sponge city technology and landscape interventions are employed to address waterlogging issues at three levels. However, the design does not adequately address the elevation difference between the pond bank and the water surface. Moreover, there is a risk of overflow on either side of the road, which could potentially damage road infrastructure.

光·韧 Ⅲ

Ground Floor Plan

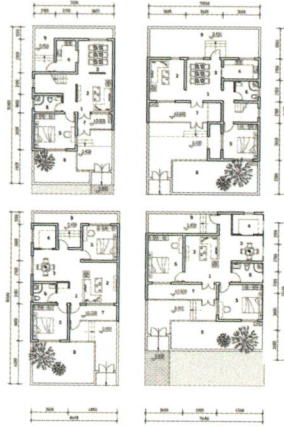

Second Floor Plan

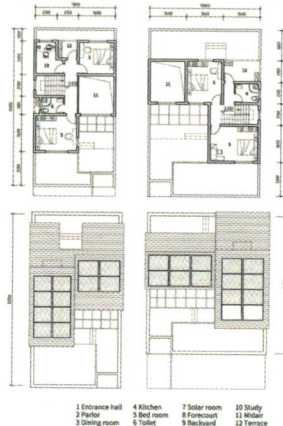

1 Entrance hall   4 Kitchen   7 Solar room   10 Study
2 Parlor   5 Bed room   8 Forecourt   11 Midair
3 Dining room   6 Toilet   9 Backyard   12 Terrace

Constructing Nodes and Techniques

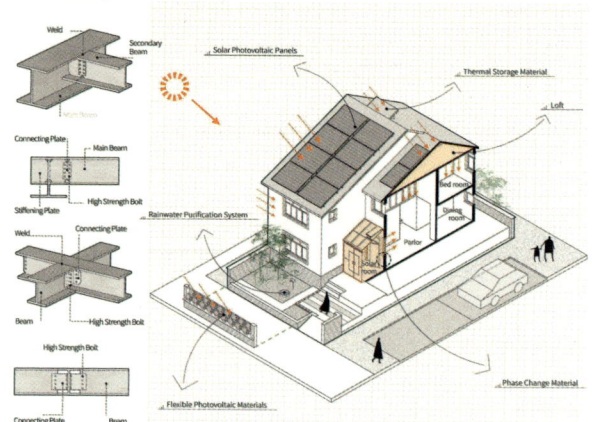

Solar Photovoltaic Panels
Thermal Storage Material
Loft
Rainwater Purification System
Flexible Photovoltaic Materials
Phase Change Material

Ventilation Analysis

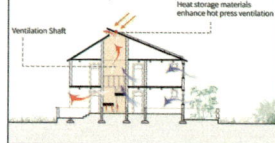

Heat storage materials enhance hot press ventilation
Ventilation Shaft

Loft & Solar Room

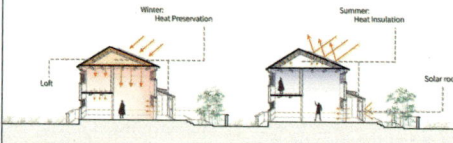

Winter: Heat Preservation
Summer: Heat Insulation
Loft
Solar room

Rainwater Purification & Sewage Disposal

Roof Rainwater
Three cylinder sewage treatment tank
Grassland Garden
Wastewater from showers, laundry, and kitchens
Rainwater purification pool
Reservoir

Quickly Build

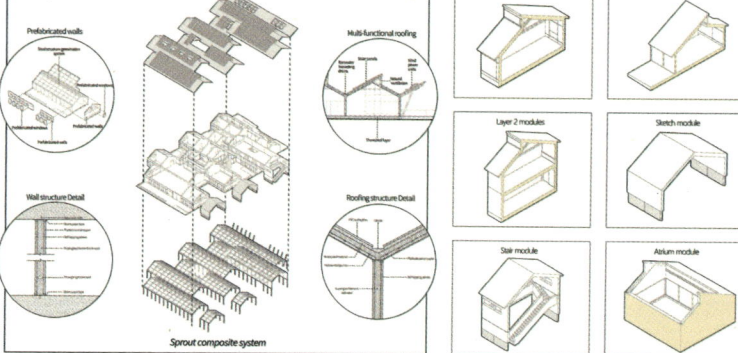

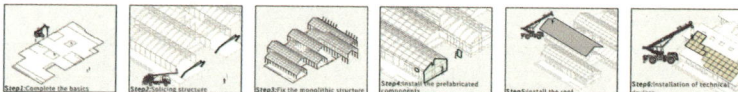

Prefabricated walls
Multi-functional roofing
Wall structure Detail
Roofing structure Detail
Sprout composite system

Basic modules   Entrance module
Layer 2 modules   Sketch module
Stair module   Atrium module

Step1:Complete the basics   Step2:Splicing structure   Step3:Fix the monolithic structure   Step4:Install the prefabricated components   Step5:Install the roof   Step6:Installation of technical devices

100-110030

光·韧 IV

■ Energy Saving Overall Instructions

在节能策略上，以空间调节为理念，采取被动优先、主动优化的方式从形体、表皮以及构造三个层面进行设计与应对，为了充分运用太阳辐射来解决季节性热舒适问题，通过建筑形态控制来获得更多有利的朝向面并采取单元化的建筑形式从而形成不同的热工分区。材料和施工上则主要采用轻钢结构和装配式施工方式以便于减少整个建筑生命周期的碳排放。在主动应用方面，通过计算得出最佳的太阳能光伏板角度，并将其合理地与建筑形态相结合，确保了其功能和美学方面的统一。

In terms otenergy saving strategy, the concept of space adjustment, passive priority,active optimization from theform, skin and structure of three levels of design andresponse.In order to make full use of solar radiatiorto solve the problem of thermal comfort in wintermore favorable orientation is obtained by buildingform control and different thermal zones are formedby building forms of units. In terms of materials andconstruction,light steel structure and prefabricatedconstruction are mainly used to reduce carbon emissions throughout the building life cycle. In terms ofactive applications,the optimal solar photovoltaicpanel Angle is calculated and reasonably combinedwith the building form to ensure the unity of its function and aesthetics.

■ Active Architectural Techonology Analysis

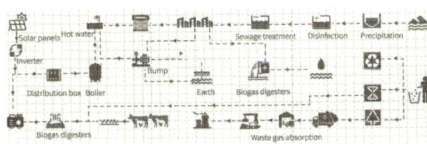

■ Energy Graded Utilization

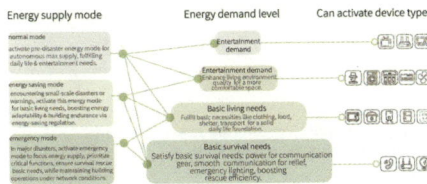

Energy supply mode   Energy demand level   Can activate device types

■ Carbon Emission Analysis

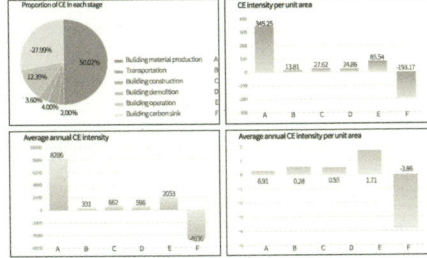

Proportion of CE in each stage

CE intensity per unit area

Average annual CE intensity

Average annual CE intensity per unit area

Building carbon sequestration

Vegetation types	CO₂ fixed amount kgCO₂e/(m²·a)	Green Area m²	Service life a	Carbon sink quantity tCO₂e
Grass	0.34	400	50	-6.8
Large and small trees	22.5	200	50	-225

CE Evaluation Indicators for Civil Buildings

Type	Name	Carbon emission values	Company
Total amount indicator	Overall carbon emissions of buildings	364.69	tCO₂e
	Carbon emissions from building operation	102.65	tCO₂e
	Building carbon emissions	464.01	tCO₂e
	Building external carbon emissions	493.84	tCO₂e
Strength index	Carbon emissions per unit area	303.91	kgCO₂e/m³
	Annual average carbon emissions	7293.8	kgCO₂e/a
	Annual average carbon emissions per unit area	6.08	kgCO₂e/(m²·a)
	Unit area physical and chemical carbon emissions	386.68	kgCO₂e/m³

Summary of Carbon Emissions Throughout the Life Cycle of Buildings

Activity Stage	Carbon emission sources	Total CE tCO₂e	Annual average CE kgCO₂e/a	Carbon intensity kgCO₂e/(m²·a)	Proportion of CE %
Building and	Stage total	493.84	9876.8	8.23	135.41
	Production	414.3	8286	6.91	113.6
	Transportation	16.57	331.4	0.28	4.54
Demolition	Construction	33.14	662.8	0.55	9.09
	Demolition	29.83	596.6	0.5	8.18
Building operation		102.65	2053	1.71	28.15
Carbon sink		-231.8	-4636	-3.86	-63.56
total		364.69	7293.8	6.08	100

■ Passive Design Strategy of Climate

In summer, the design can use a hybrid ventilation strategy combining wind pressure and hot pressure to remove heat and moisture from the room.In winter, use the Trunborough wall to heat indoor air, thereby relaxing and heating the core space.

Wind pressure ventilation in summer

Heat press ventilation in summe

use the Trunborough wall in winter

■ Daylighting Analysis

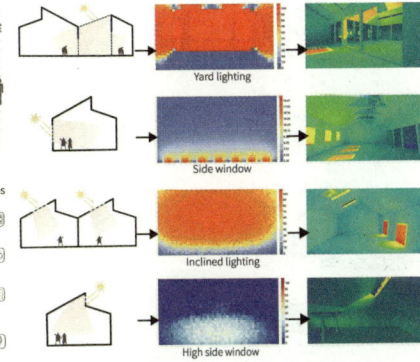

Yard lighting

Side window

Inclined lighting

High side window

■ System Diagram of "Light Storage Straight Flexible"

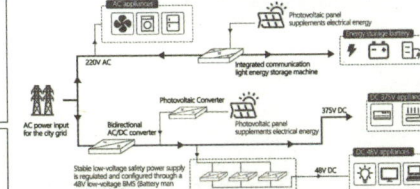

■ Analysis of Solar energy

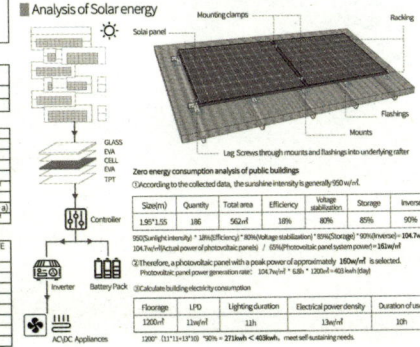

Zero energy consumption analysis of public buildings

①According to the collected data, the sunshine intensity is generally 950 w/㎡.

Size(m)	Quantity	Total area	Efficiency	Voltage stabilization	Storage	inverse
1.95*1.05	186	562㎡	80%	80%	85%	90%

950(Sunlight intensity) * 18%(Efficiency) * 80%(Voltage stabilization) * 85%(Storage) *90%(inverse)= 104.7w/㎡
104.7w/㎡(Actual power of photovoltaic panels) * 1.56㎡(Photovoltaic panel system power)= 161w/㎡

②Therefore, a photovoltaic panel with a peak power of approximately 160w/㎡ is selected.
Photovoltaic panel power generation rate: 104.7w/㎡ * 6.8h * 1300㎡= 403 kwh (day)

③Calculate building electricity consumption

Floorage	LPD	Lighting duration	Electrical power density	Duration of use
1200㎡	13w/㎡	11h	13w/㎡	10h

1200㎡ (11*11*13*10) *90% = 271kwh < 403kwh, meet self sustaining needs.

■ Climate Analysis

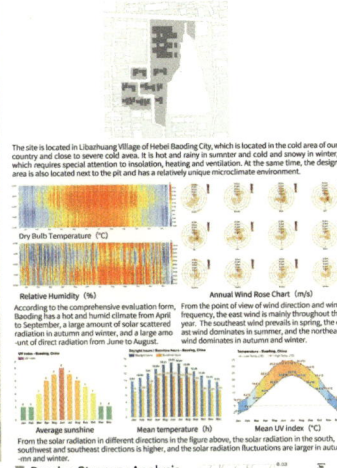

The site is located in Libazhuang Village of Hebei Baoding City, which is located in the cold area of our country and close to severe cold avea. It is hot and rainy in sumnter and cold and snowy in winter, which requires special attention to insolation, heating and ventilation. At the same time, the design area is also located next to the pit and has a relatively unique microclimate environment.

Dry Bulb Temperature (℃)

Relative Humidity (%)

According to the comprehensive evaluation form, Baoding has a hot and humid climate from April to September, a large amount of solar scattered radiation in autumn and winter, and a large amount of direct radiation from June to August.

Annual Wind Rose Chart (m/s)

From the point of view of wind direction and wind frequency, the east wind is mainly throughout the year. The southeast wind prevails in spring, the east wind dominates in summer, and the northeast wind dominates in autumn and winter.

Average sunshine   Mean temperature (h)   Mean UV index (℃)

From the solar radiation in different directions in the figure above, the solar radiation in the south, southwest and southeast directions is higher, and the solar radiation fluctuations are larger in autu -mn and winter.

■ Passive Straregy Analysis

Psychrometric Chart

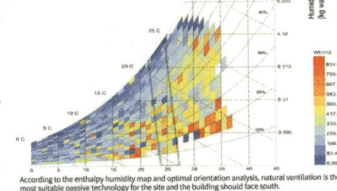

According to the enthalpy humidity map and optimal orientation analysis, natural ventilation is the most suitable passive technology for the site and the building should face south.

■ Radiation Strategy

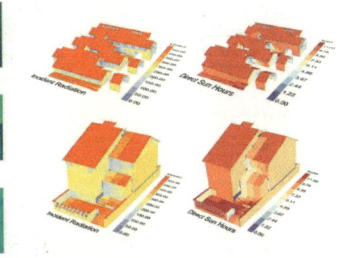

Since the site is in the northern hemisphere, the height Angle of the sun changes with the seasons, and the direct sunshine hours of the site show a downward trend from south to north. In the design, when installing solar panels, we should choose the direction of the south as far as possible, so as to obtain the maximum exposure of solar energy.

■ Wind Simulation

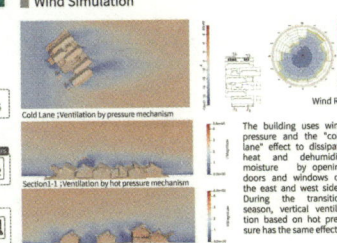

Cold Lane ;Ventilation by pressure mechanism

Section1-1 ;Ventilation by hot pressure mechanism

Section2-2 ;Ventilation by hot pressure mechanism

Wind Rose

The building uses wind pressure and the "cold lane" effect to dissipate heat and dehumidify moisture by opening doors and windows on the east and west sides. During the transition season, vertical ventilation based on hot pressure has the same effect

■ Active Energy Saving Strategy

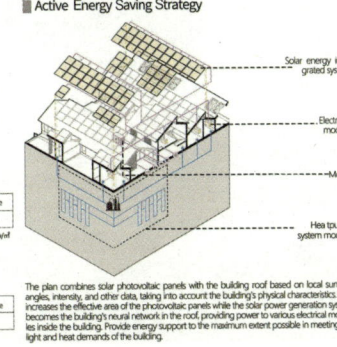

Solar energy integrated system
Electricity module
Motor
Hea tpump system module

The plan combines solar photovoltaic panels with the building roof based on local sunlight angles, intensity, and other data, taking into account the building's physical characteristics. This increases the effective area of the photovoltaic panels while the solar power generation system becomes the building's neural network in the roof, providing power to various electrical modules inside the building. Provide energy support to the maximum extent possible in meeting the light and heat demands of the building.

综合奖·三等奖·张家口市怀安县二堡子村建设项目
Comprehensive Awards – Third Prize – Erpuzi Village Project，Huai'an County, Zhangjiakou City

注册号：110411
Register Number：110411

项目名称：荫庇乡村 韧性环院
Entry Title：Village Shelter & Community Restorer

作者：杨洋、彭影、刘洋源、赵志鹏
Authors：Yang Yang, Peng Ying, Liu Yangyuan, and Zhao Zhipeng

作者单位：去边建筑设计（广州）有限公司
Authors from：Qubian Architectural Design (Guangzhou) Co.，Ltd.

指导老师：杨洋
Tutor：Yang Yang

指导老师单位：去边建筑设计（广州）有限公司
Tutor from：Qubian Architectural Design (Guangzhou) Co.，Ltd.

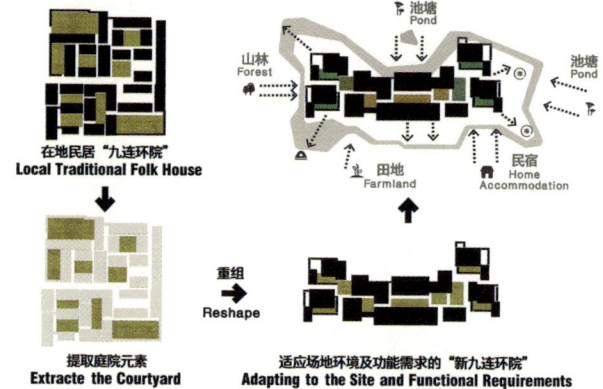

110411

乡村会客厅入口广场
Activity Center Entrance Square

乡村会客厅入口广场
Activity Center Entrance Square

市集摊位活动区
Market Stall Activity Area

乡村会客厅院落
Activity Center Courtyard

在地民居"九连环院"
Local Traditional Folk House

提取庭院元素
Extracte the Courtyard

重组
Reshape

适应场地环境及功能需求的"新九连环院"
Adapting to the Site and Functional Requirements

池塘
Pond

山林
Forest

池塘
Pond

田地
Farmland

民宿
Home Accommodation

打造乡村的庇护所，成为社区的修复者
Village Shelter & Community Restorer

竹境景观构筑
Bamboo Landscape Construction

民居民宿组团
Residential Home Accommodation Group

乡村会客厅组团
Multi Functional Activity Center

民居民宿组团
Residential Home Accommodation Group

民居民宿组团
Residential Home Accommodation Group

南侧鸟瞰效果 | 南入口集散广场
Aerial View | South Entrance Square

110411

经 济 技 术 指 标　Economic and Technical Indicators
用地面积：8260m²　Land Area: 8260 square meters
建筑面积：1800m²　Building Area: 1800 square meters
容 积 率：0.21　Floor Area Ratio (FAR): 0.21
密 度：0.15　Building Density: 0.15
绿地率：35%　Green Rate: 35%
停 车 位：10 辆　Parking Spaces: 10
建筑高度：9 m　Building Height: 9 meters
建筑层数：2 层　Number of Floors: 2F

N
总平面图 MASTERPLAN

1 乡村会客厅组团　1 Multi functional Activity Center Group
（应急抗灾应护）　(Emergency Disaster Resistance Shelter)
2 民居民宿组团　2 Residential Home Accommodation Group
3 露营区景观构筑　3 Landscape Structure in Camping Area
（应急临时应护）　(Emergency Temporary Shelter)
4 露营区休闲平台　4 Leisure Platform in Camping Area
5 露营区帐篷组团　5 Tent Group in Camping Area
6 中心篝火　6 Central Campfire
7 亲水休闲活动　7 Waterfront Leisure Activities
8 户外剧场　8 Outdoor Theater
9 院落(共9个)　9 Courtyard（Nine in total）
10 市集摊位　10 Market Stalls
11 市集休闲区　11 Market Leisure Area
12 入口广场　12 Entrance Square
13 活动广场　13 Activity Square
（应急临时集散）　(Emergency Temporary Evacuation Square)
14 停车场(10位)　14 Parking Lot(10 car parking space)

▲ 场地主要出入口　Site Main Entrance
▲ 停车场流线出入口　Parking Lot Entrance
◆ 露营区出入口　Camping Area Entrance
▲ 民宿出入口　Residential Entrance
△ 会客厅出入口　Activity Center Entrance

乡村会客厅二层挑台
Activity Center 2F-Platform

民居民宿街道
Street Scene

1-1剖面 SECTION 1:300

2-2剖面 SECTION 1:300

南立面 SOUTH ELEVATION 1:300

北立面 NORTH ELEVATION 1:300

设计说明：
在地民居"九连环院"由九个相连的院子组成，以此为设计概念起源，**将现代技术与中国传统民居智慧相结合**。抽取其图底及院落关系进行再设计，院落使得建筑外表面积增加，与周围环境进行能量交换，同时形成垂直的风道带走热量。庭院与天窗形成拔风效应，将室内热空气抽出，实现自然通风。

技术应用层面通过**减少各阶段的碳排放来实现建筑的近零能耗**：建筑运行阶段的直接碳排放，热电使用的间接碳排放，建筑材料制造运输的隐含碳排放。

The local residential building **"Nine Linked Courtyard"** consists of nine connected courtyards, which is the origin of our design concept, **combining modern technology with the wisdom of traditional Chinese folk house.** Extract the traditional overall layout pattern and redesign,including side and center court relationships which increases buildings' external surface area, making energy exchange between the buildings and its surroundings, also forming some vertical air ducts to assist in taking away the heat. At the same time,we creat some skylights act as chimneys to draw out hot air indoor, improving natural ventilation.

It would be an energy-saving building by reducing carbon emissions at every phase of its existence: direct carbon emissions from the operation of the buildings, indirect carbon emissions from heat and electricity usage, embodied carbon emission during the manufacturing of building materials.

北侧低点效果 | 亲水休闲活动场景
Rendering | Waterfront Activity Scene

民居民宿组团 Residential Home Accommodation Group　乡村会客厅组团 Multi Functional Activity Center　乡村会客厅北入口 North Entrance Of Activity Center　多功能户外剧场 Multi Functional Outdoor Theater　民居民宿组团 Residential Home Accommodation Group

专家点评：
该方案以庭院为中心的布局灵活有趣，9 个庭院承载着传统文脉的意义。建筑设计方面，造型有特色，完整封闭的屋顶在形态上有特点。太阳能主被动技术设计较好，但大屋顶影响采光，剖面设计也过于简单；社区空间较为分散，导致社区活动缺乏凝聚力和连通性。用桥梁连接公共服务建筑与住宅建筑的做法令人担忧，在地震多发地区，可能危及结构稳定性与安全性，需重新评估并调整抗震设计方案。

Experts' Commentary：
The scheme's layout, centered around the courtyard, exhibits a flexible and engaging design, with nine courtyards encapsulating the essence of traditional elements. While the architectural design possesses distinctive characteristics, the fully enclosed roof lacks variety in its design. The integration of active and passive technology is commendable; however, the expansive roof negatively affects natural lighting, and the sectional design appears overly simplistic. Community spaces are too dispersed to maintain cohesion and connectivity among community activities. Furthermore, the use of bridges to connect public service buildings to residential buildings raises concerns about structural stability and safety, particularly in seismically active regions, necessitating a reassessment and modification of the earthquake-resistant design strategies.

110411

竹编景观构筑
Bamboo landscape Construction

乡村会客厅组团
Multi functional Activity Center

民居民宿组团
Residential Home Accommodation Group

南侧低点效果丨南侧入口
Rendering丨South Entrance

乡村会客厅院落
Activity Center Courtyard

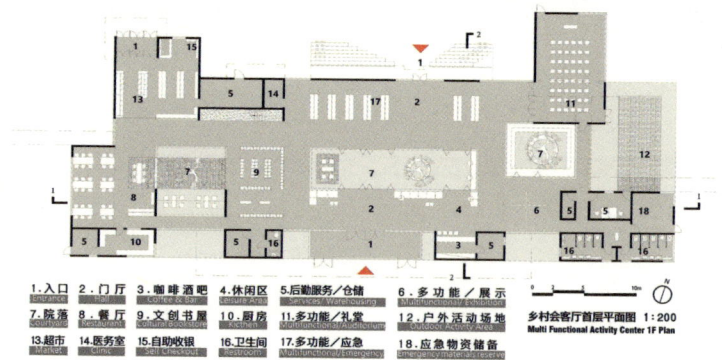

1.入口 Entrance   2.门厅 Hall   3.咖啡酒吧 Coffee & Bar   4.休闲区 Leisure Area   5.后勤服务／仓储 Services, Warehousing   6.多功能／展示 Multifunctional Exhibition
7.院落 Courtyard   8.餐厅 Restaurant   9.文创书店 Culture Bookstore   10.厨房 Kitchen   11.多功能／礼堂 Multifunctional Auditorium   12.户外活动场地 Outdoor Activity Area
13.超市 Market   14.医务室 Clinic   15.自助收银 Self Checkout   16.卫生间 Restroom   17.多功能／应急 Multifunctional Emergency   18.应急物资储备 Emergency materials reserve

乡村会客厅首层平面图 1：200
Multi Functional Activity Center 1F Plan

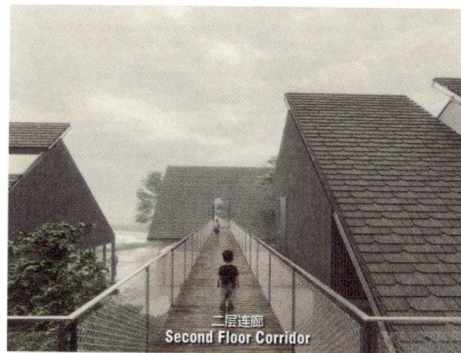

二层连廊
Second Floor Corridor

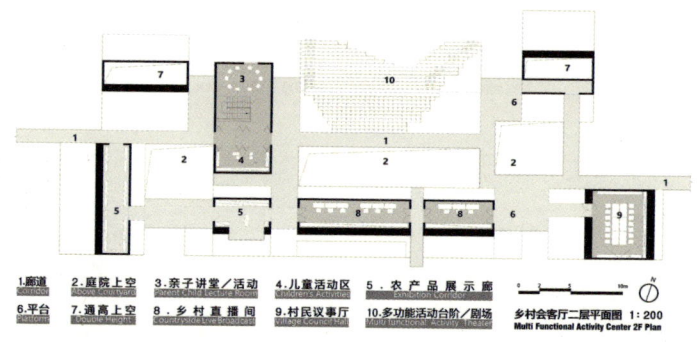

1.廊道 Corridor   2.庭院上空 Yard Overhead   3.亲子讲堂／活动 Parent & Child Lecture Hall   4.儿童活动区 Children's Activities   5.农产品展示廊 Agricultural Products
6.平台 Platform   7.通高上空 Double Height   8.乡村直播间 Country Live Broadcast   9.村民议事厅 Cottage Council Hall   10.多功能活动台阶／剧场 Multifunctional Activity Theater

乡村会客厅二层平面图 1：200
Multi Functional Activity Center 2F Plan

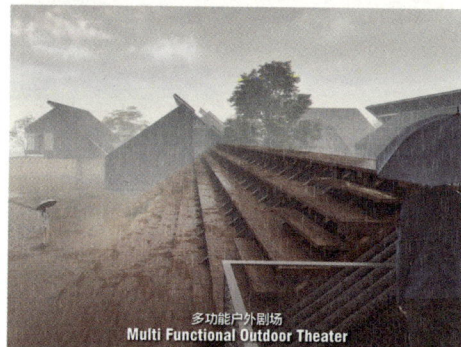

多功能户外剧场
Multi Functional Outdoor Theater

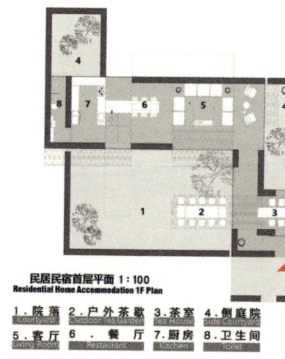

民居民宿首层平面 1：100
Residential Home Accommodation 1F Plan

1.院落 Courtyard   2.户外茶歇 Outdoor Tea Break   3.茶室 Tea House   4.侧庭院 Side Courtyard
5.客厅 Living Room   6.餐厅 Restaurant   7.厨房 Kitchen   8.卫生间 Toilet

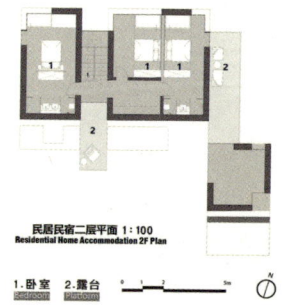

民居民宿二层平面 1：100
Residential Home Accommodation 2F Plan

1.卧室 Bedroom   2.露台 Platform

110411

柳篱遮阳休闲区
Wickerwork Sunshade Leisure Area

市集摊位模块装置
Market Stall Modular Devices

东侧鸟瞰效果 | 市集摊位活动场景
Aerial View | Market Stall Activity Scene

模数化&可拓展
Modular&Extensibility

简易结构搭建+轻质半透阳光板
Simple Structure Construction With Lightweight Material (PP Sheet)

灵活开合系统
Retractable Structures

建筑坡屋面大样图
Detail Drawing Of Building Slope Roof

建筑墙身大样
Detailed Drawing Of Building Wall

·太阳能光伏+光热屋瓦
南向45°屋面能更大程度接收收阳光，太阳能光伏和光热的利用为全屋提供电能、热水、地暖。
· Solar Tiles
Southward 45° roofs can better receive sunlight. Solar energy photothermal and photoelectricity provide electricity- hot water&underfloor heating for buildings.

·蒸压加气混凝土板 AAC Panel
材料生产过程无污染：材料轻质便于运输；工厂预制模块可降低建设过程碳排放，有效缩短工期。更优的保温隔热性能可减少暖气空调的使用，达到节能效果。
· Autoclaved Aerated Concrete Panel
The material production process is pollution-free; it's lightweight so that easy to transport; Prefabricated modules in factories not only can reduce carbon emissions during the construction process but also effectively shorten the construction period. Better insulation performance can reduce the use of heating and air conditioning.

·双层中空LOW-E玻璃
优化窗墙比且使用双层中空Low-E玻璃，更好地实现保温隔热。
· 6+12+6 Low-E Glass
Optimize the area ratio of window to wall and use double-layer low-e glass for better insulation and heat preservation.

·朴门永续景观绿化
景观绿化种植经济类作物，抗灾极端情况下作为应急果蔬补给。
· Sustainable Landscape Greening
Planting economic crops for landscape , serving as emergency fruit and vegetable supplies in extreme disaster situations.

·水循环处理系统
雨水及中水回收至初期弃流装置，再流入蓄水桶用于浇灌、洗车、马桶等生活用水；或经矿化装置深度处理，达到饮用水标准。
· Water Circulation System
Rainwater and reclaimed water are collected in the initial filtration device and then flow into the storage tank,using for irrigation, car washing, toilets and domestic water, while the other is recycled through mineralization treatment devices to meet the standards of drinking water.

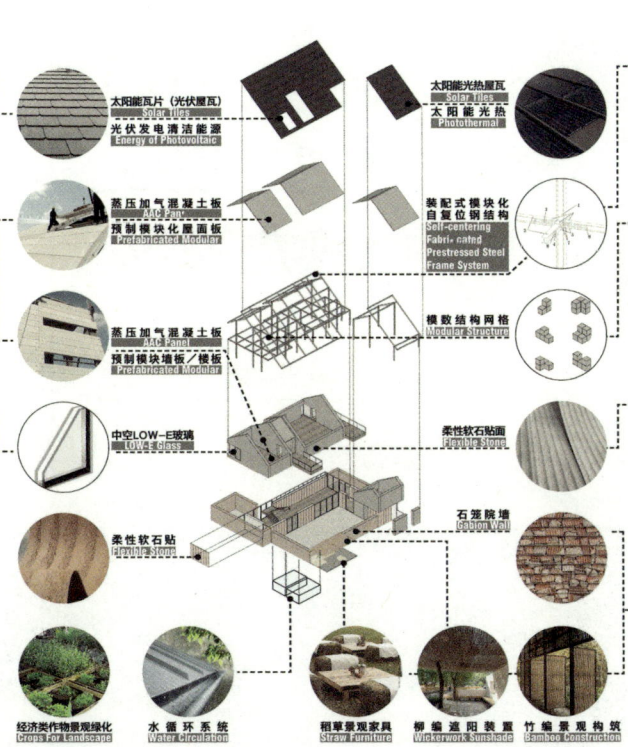

太阳能瓦片（光伏屋瓦）
Solar Tiles
光伏发电清洁能源
Energy of Photovoltaic

蒸压加气混凝土板
AAC Panel
预制模块化屋面板
Prefabricated Modular

蒸压加气混凝土板
AAC Panel
预制模块化墙板/楼板
Prefabricated Modular

中空LOW-E玻璃
LOW-E glass

柔性软石贴
Flexible Stone

经济类作物景观绿化
Crops For Landscape

水循环系统
Water Circulation

太阳能光热瓦
Solar Tiles
太阳能光热
Photothermal

装配式模块化自复位钢结构
Self-centering Fabricated Prestressed Steel Frame System

模数结构网格
Modular Structure

柔性软石贴面
Flexible Stone

石笼院墙
Gabion Wall

稻草景观家具
Straw Furniture

柳编遮阳装置
Wickerwork Sunshade

竹编景观构筑
Bamboo Construction

·装配式模块化自复位钢结构
工厂预制模块，施工便捷，减少建设垃圾、噪声、碳排放；该结构系统可在震后实现自复位。
· Self-centering Fabricated Prestressed Steel Frame System
standardized production of modular units in the factory, leading to fast construction , reduced construction waste and carbon emissions.The system reduces residual deformation of frame joints after earthquake and achieve self-resetting.

·模数化结构网格
建筑结构与空间模数化，使空间具拓展性，构件重复率最大化。
· Modular Structural Grid And Spatial Modularization
Ensuring both the expandability of functional spaces and the maximization of component repetition rates.

·柔性软石贴
以废建筑材料、石粉等无机物为原料，具良好耐候性和耐冻性。材质轻薄、抗裂防裂，运输方便。施工便捷且无粉尘、噪声等污染，可循环可完全再生，新能源生产，无三废排放。
· Flexible stone
It's an eco-friendly material made from construction waste and stone powder, recyclable and fully renewable. It has good weather fastness and freezing tolerance. It's an lightweight and seismic and crack resistant material, easy to transport, easy to install, and no pollution during material production and construction.

·石笼院墙、稻草景观家具、柳编遮阳装置、竹编景观构筑
在地资源利用、可回收循环、低碳可持续。石笼景观院墙，当地石、竹、木、稻杆等均可作为填充物料。景观构筑及家具利用柳编、竹编、稻草墩子、木墩等乡村物料，节省人力物力成本。
· Utilization of local resources
Filler of gabion wall can be replaced according to local rural resources,such as stone, bamboo, wood, straw poles .Landscape construction and furniture can also utilize willow or bamboo weaving, straw and wooden bales.

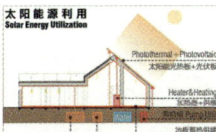

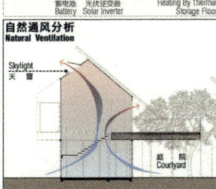

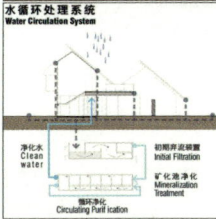

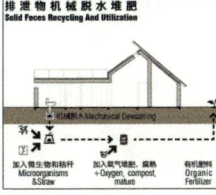

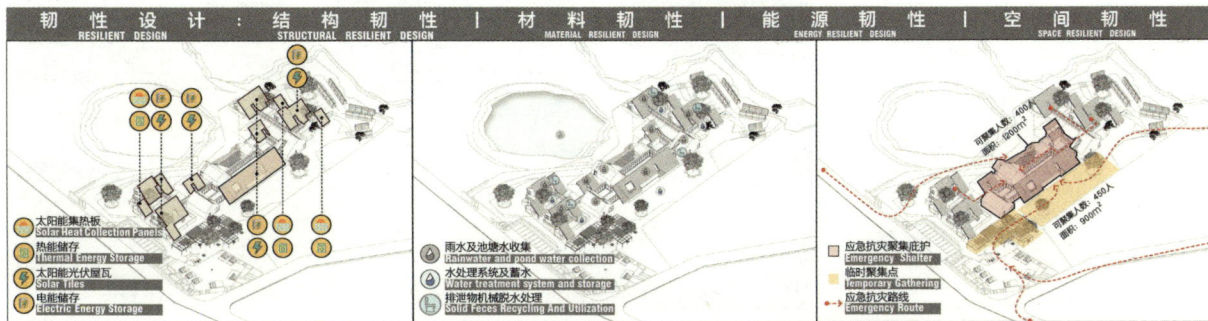

太阳能源利用 / Solar Energy Utilization
Photothermal + Photovoltaic
光热+光伏
Heater&Heating
采热&供暖
Battery   Solar Inverter
蓄电池   光伏逆变器
Heating By Thermal Storage Floor
地板蓄热层供暖

自然通风分析 / Natural Ventilation
Skylight
天窗
Courtyard
庭院

水循环处理系统 / Water Circulation System
Clean water
净化清水
Initial Filtration
初期弃流装置
Mineralization Treatment
矿化池净化
Circulating Purification
循环净化

排泄物机械脱水堆肥 / Solid Feces Recycling And Utilization
Mechanical Dewatering
机械脱水
Microorganisms &Straw
加入微生物秸秆+稻草
Oxygen, compost, mature
加入氧气、堆肥、腐熟
Organic Fertilizer
有机肥

韧性设计：结构韧性 | 材料韧性 | 能源韧性 | 空间韧性
RESILIENT DESIGN : STRUCTURAL RESILIENT DESIGN | MATERIAL RESILIENT DESIGN | ENERGY RESILIENT DESIGN | SPACE RESILIENT DESIGN

太阳能集热板
Solar Heat Collection Panels
热能储存
Thermal Energy Storage
太阳能光伏屋瓦
Solar Tiles
电能储存
Electric Energy Storage

雨水及池塘水收集
Rainwater and pond water collection
水处理系统及蓄水
Water treatment system and storage
排泄物机械脱水处理
Solid Feces Recycling And Utilization

应急抗灾聚集庇护
Emergency Shelter
临时聚集点
Temporary Gathering
应急抗灾路线
Emergency Route

综合奖·三等奖·张家口市
怀安县二堡子村建设项目
Comprehensive Awards –
Third Prize – Erpuzi Village
Project, Huai'an County,
Zhangjiakou City

注册号：110582
Register Number：110582

项目名称：沐阳稻舍
Entry Title：Sunbathed Rice Cottage

作者：曾雅清、李嘉欣、李思静、郝泽厚
Authors：Zeng Yaqing, Li Jiaxin,
        Li Sijing, and Hao Zehou

作者单位：厦门大学、北方工程设计研
        究院有限公司
Authors from：Xiamen University, Northern
        International Co., Ltd.

指导教师：李立新
Tutor：Li Lixin

指导教师单位：厦门大学
Tutor from：Xiamen University

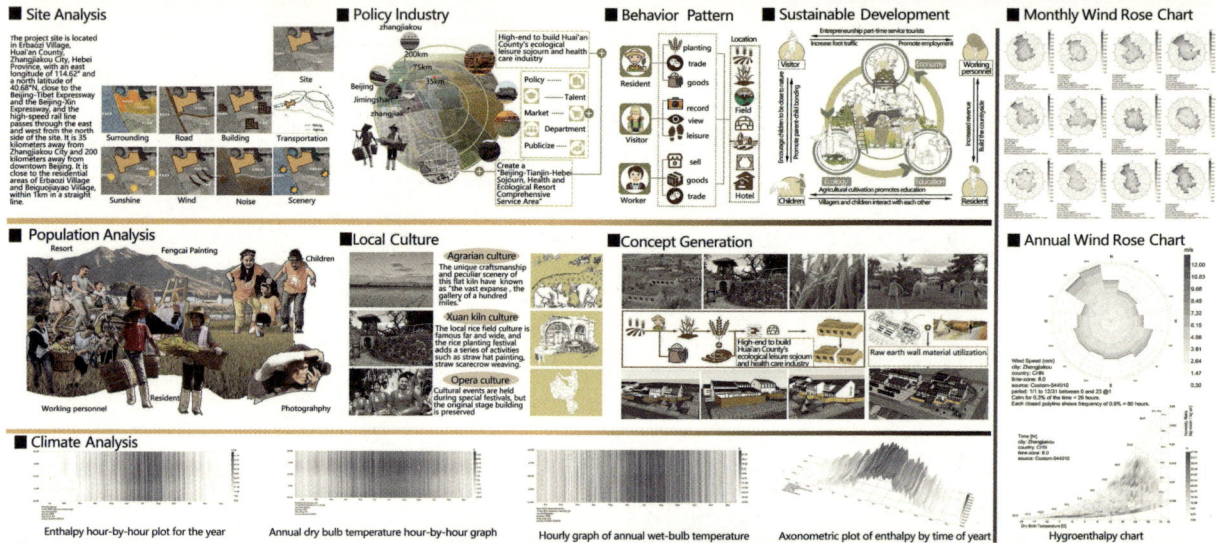

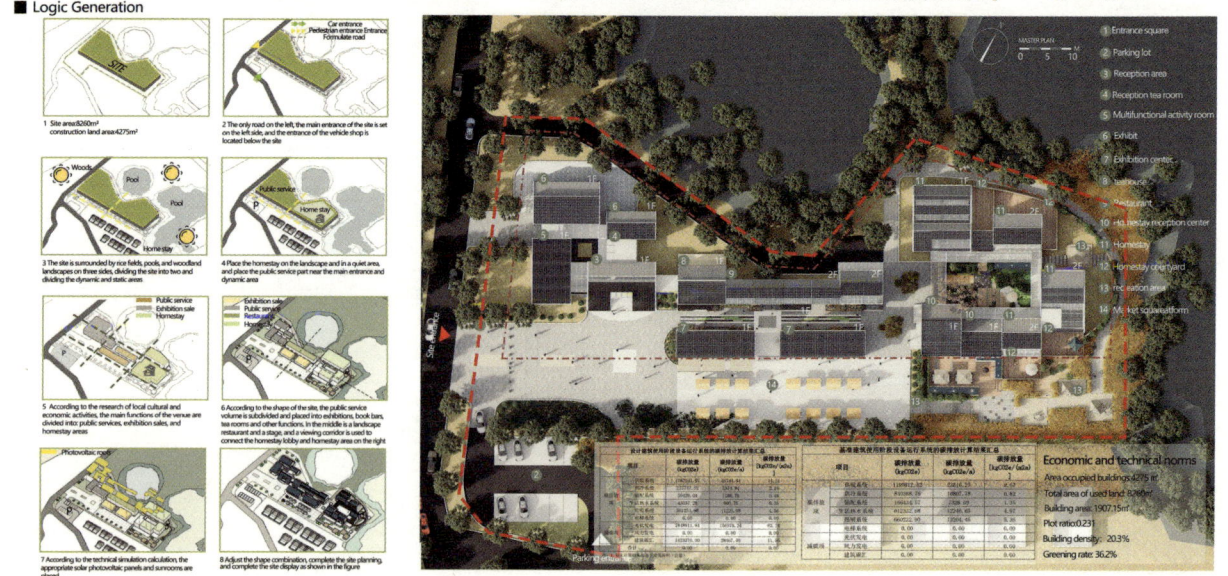

Solar photovoltaic panel
Solar collector system
Heat storage roof
Collector-Storage Wall
Ventilation
Solar chimney
Phase change heatstorage material
Rounding layout
Rainwater harvesting and reuse
Attached sunroom
Solar streetlights
The main wind direction entrance avoidings
Attached sunroom
Pervious brick

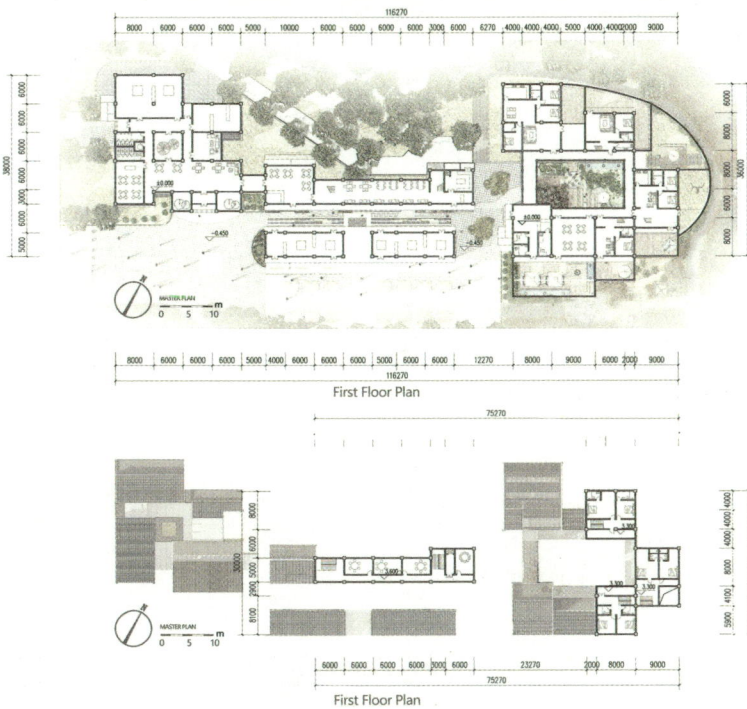

116270
First Floor Plan

MASTER PLAN
0 5 10 m

75270
First Floor Plan

MASTER PLAN
0 5 10 m

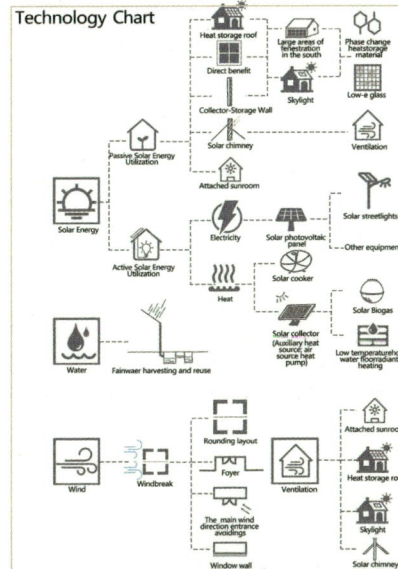

Multi functional particle heating furnace
biomass pellet fuel
Biomass clean heating
Agricultural and forestry wastes
Rainwater Storage
Air source heat pump heating system

Technology Chart

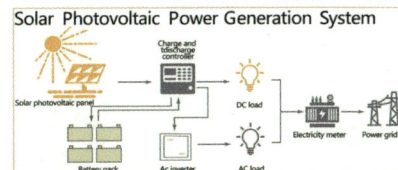

Heat storage roof
Large areas of fenestration in the south
Phase change heatstorage material
Direct benefit
Low-e glass
Skylight
Collector-Storage Wall
Solar chimney
Passive Solar Energy Utilization
Ventilation
Attached sunroom
Solar Energy
Solar streetlights
Electricity
Solar photovoltaic panel
Other equipments
Active Solar Energy Utilization
Heat
Solar cooker
Solar Biogas
Water
Solar collector (Auxiliary heat source: air source heat pump)
Low temperature hot water floor radiant heating
Rainwater harvesting and reuse
Attached sunroom
Rounding layout
Foyer
Heat storage roof
Wind
Windbreak
Ventilation
Skylight
The main wind direction entrance avoidings
Window wall area ratio
Solar chimney

Solar Photovoltaic Power Generation System

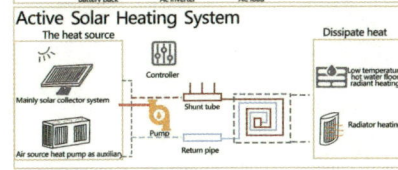

Charge and discharge controller
Solar photovoltaic panel
DC load
Electricity meter
Power grid
Battery pack
Ac inverter
AC load

Active Solar Heating System

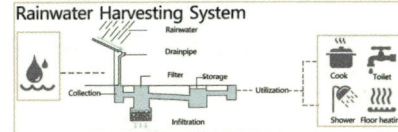

The heat source
Dissipate heat
Controller
Low temperature hot water floor radiant heating
Mainly solar collector system
Shunt tube
Radiator heating
Air source heat pump as auxiliary
Pump
Return pipe

Rainwater Harvesting System

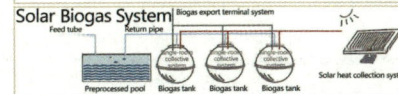

Rainwater
Drainpipe
Cook
Toilet
Filter
Storage
Collection
Utilization
Shower
Floor heating
Infiltration

Solar Biogas System

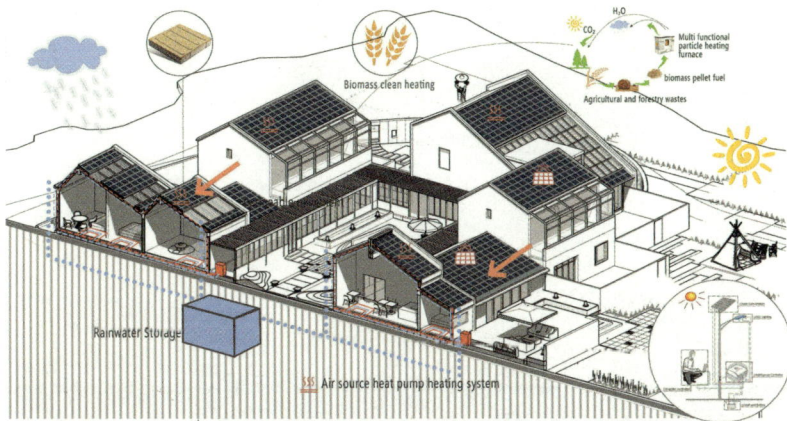

Biogas export terminal system
Feed tube
Return pipe
Preprocessed pool
Biogas tank
Biogas tank
Biogas tank
Solar heat collection system

专家点评：

该方案在设计上，以内庭院和开放空间为介质将建筑群有机结合，设计理念清晰易懂，有助于营造舒适且富有层次的空间环境。设计方案考虑了与周边环境的协调性，确保建筑形象能够自然融入周边环境，维护了区域的整体美感与和谐氛围，但方案在公共服务部分的位置设置、住宅布局与平面设计、对严寒气候和地域性的适应性、广场规模以及抵御灾害能力的策略展示等方面还有待改进。

Experts' Commentary：

In terms of design, the inner courtyards and open spaces serve as a medium to organically integrate the architectural ensembles. The design concept is clear and easy to apprehend, helping to create a comfortable and well-structured spatial environment. By considering coordination with the surrounding environment, the design ensures that the architectural image naturally blends into its surroundings, preserving the overall aesthetic appeal and harmonious ambiance of the area. However, improvements are necessary in several aspects of planning, including the positioning of public services, the layout and graphic design of residential buildings, adaptability to cold climates, regional characteristics, plaza dimensions, and disaster resistance strategies.

■ Disaster Resistance Design Strategy

Pre-disaster defense

Risk Mitigation During Disasters

Post-disaster recovery

Enhancement building Disaster resistance / Scientific Planning / Disaster monitoring / Emergency shelter / Solar energy storage system / Scientific Planning

Height control / Materials Structure / Evacuation Routes / Information release / Emergency evacuation communication / Security Recycling of building materials

Form ratio control / Seismic design Hail-resistant design / Refuge building / Emergency infrastructure / Emergency supplies / Water storage system / Multi-party cooperation

■ Enhance Disaster Resilience

building setback distance

Rescue road width

Control of building height and form ratio

■ Function Layout before Disaster

Normal time

Public service buildings multi-functional spaces

Rescue road

■ Disaster Site Response

After disaster

Material reserve

Emergency shelter and resource distribution center

Energy storage systems ensure communication

Solar energy storage systems provide power

■ Information Sharing

Establish an information dissemination system to ensure timely communication of safety information and guidance during disasters.

Resource integration
Collaborate with local governments and organizations to integrate resources and enhance the overall emergency response capability of the community.

■ Section

Solar photovoltaic panel / Solar collector panel / Heat storage roof / Collector-Storage Wall / Ventilation / Solar photovoltaic panel

■ Analysis of Construction Strategies   ■ Wall Structure

Orientation design / Functional partitioning

Body size factor / Window-to-wall ratio

Ventilation through the window / Wind and snow roofs

■ Sunroom Summer Design

Solar radiation / Solar radiation / Thermal buffer space / Ventilation at night takes away the heat

Phase change heat storage / Phase change heat storage / Phase change heat storage / Phase change heat storage

Sunroom

■ Sunroom Winter Design

Solar radiation / Solar radiation / Thermal buffer space / Ventilation at night takes away the heat

Phase change heat storage / Phase change heat storage / Phase change heat storage / Phase change heat storage

Sunroom

■ Roof Analysis

solar panel / heat gain / dissipate heat
disperse/collecting heat panel (roof) / heat dissipation ventilation effect
heat transfer
disperse/collecting heat panel (floor) / dissipate heat
heat panel (underground) / dissipate heat

SUMMER DAYTIME   SUMMER NIGHT

dissipate heat / heat gain / store heat / heat transfer / store heat / dissipate heat

WINTER DAYTIME   WINTER NIGHT

■ t Design

outside inside / outside inside / outside inside / outside inside

Winter / Summer and Transition Season / Heat Release in Daytime / Heat Release in Night

Braun wall

■ 45° Wall Design

The 45-degree angle is in line with the optimal angle of sunlight

The building shape coefficient is small and the thermal insulation is good

In winter, snow does not accumulate, effectively reducing the load on the house

The building door bucket is very good for thermal insulation

45 °wall

■ Integrated Wall Design

Shout down blinds in winter and transition season / Open blinds to ventilate in summer / Heat Release in Daytime / Heat Release in Daytime

Phase Change Material

■ North Facade

■ Exploded View

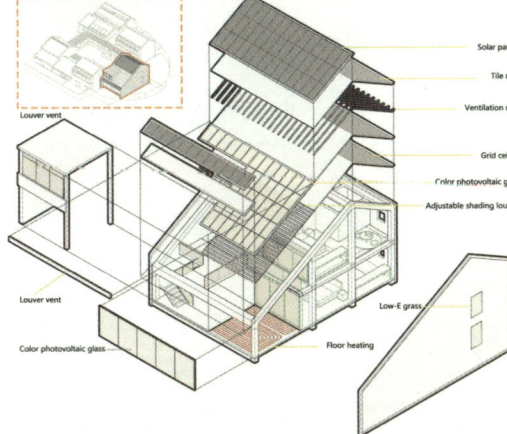

Louver vent
Louver vent
Color photovoltaic glass

Solar panels
Tile roof
Ventilation roof
Grid ceiling
Color photovoltaic glass
Adjustable shading louvers
Low-E grass
Floor heating

■ Technology

Solar cel construction

Solar collector

Low temperature hot water floar radiant heating

Thermal insulation construction

■ Concept Generation

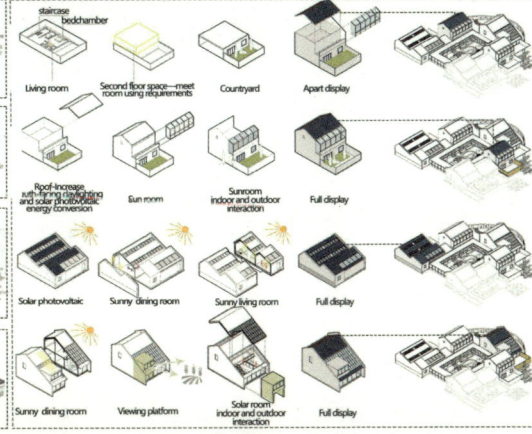

staircase
bedchamber

Living room | Second floor space—meet room using requirements | Countryard | Apart display

Roof-increase suth-firing day lighting and solar photovoltaic energy conversion | Sun room | Sunroom indoor and outdoor interaction | Full display

Solar photovoltaic | Sunny dining room | Sunny living room | Full display

Sunny dining room | Viewing platform | Solar room indoor and outdoor interaction | Full display

■ Material Analysis

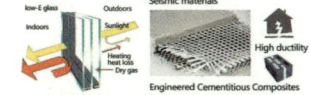

low-E glass
Indoors Outdoors Seismic materials
Sunlight
Heating heat loss Dry gas
High ductility
Engineered Cementitious Composites

■ Sunlight Analysis

■ Year-round Radiation Analysis

Incident Radiation

■ Outdoor Wind Environment Analysis

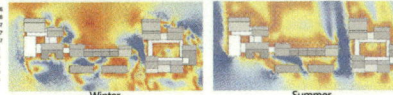

Winter   Summer

■ Operation Mode and Strategy

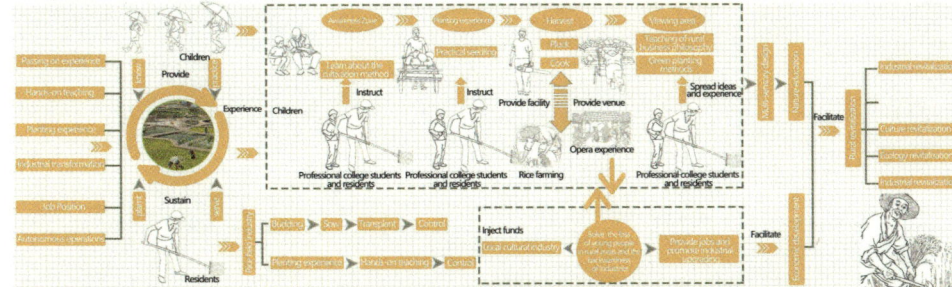

■ Activity Streamline Design

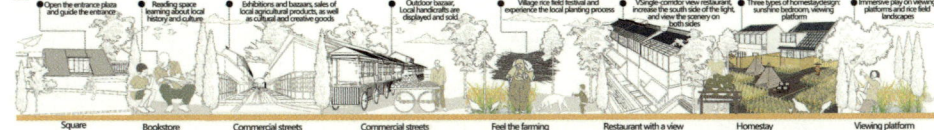

Square | Bookstore | Commercial streets | Commercial streets | Feel the farming | Restaurant with a view | Homestay | Viewing platform

综合奖·三等奖·保定市清苑区李八庄村建设项目
Comprehensive Awards – Third Prize – Libazhuang Village Project，Qingyuan District，Baoding City

注册号：110093
Register Number：110093
项目名称：向阳而居·依塘治水
Entry Title：Facing the Sun，Living by the Water
作者：白炳骏、蔡严宇、沈梓程、赵文清
Authors：Bai Bingjun, Cai Yanyu, Shen Zicheng, and Zhao Wenqing
作者单位：南京工业大学
Authors from：Nanjing Tech University
指导教师：董凌
Tutor：Dong Ling
指导教师单位：南京工业大学
Tutor from：Nanjing Tech University

01

Street Penetration: Continuing the surrounding streets and alleys, dividing the residential layout.

Functional zoning: The area along the street is public, with a layout of village service centers, squares, and parking lots. The interior is private, and residential areas are arranged

Guiding wind into residential areas: Residential areas are arranged parallel to the direction of the summer monsoon, allowing summer wind energy to pass through ponds and enter residential areas and streets.

Planting trees to prevent wind: Winter winds blow from the northwest, bringing sandstorms and cold air currents, so windproof trees are planted in the west and north of residential areas.

Site texture: The residential buildings are arranged in a clustered layout in the form of courtyards, surrounded by large areas of farmland. The drainage inside the mass is solved by pits and ponds.

Typical residential buildings in the surrounding area: mostly consisting of two courtyard, three courtyard, and quadrangle courtyards, with staggered entrances arranged.

Traditional quadrangle courtyard / Changfa Manor Master's Courtyard / Guangfazeng Manor / Hongzhonfa rear garden

A typical "two swinging sleeves" plane: the "swinging sleeves" space is in a "concave" shape, forming a platform gray space in the middle.

A B C D

Winter breeze enters the courtyard house

Courtyard houses are prone to theft

Large surface area and poor insulation

Adaptive evolution

In recent years, the "two swinging sleeves" have developed into the **form of a shed courtyard** on the basis of tradition, with the purpose of winter cold insulation, reducing heat dissipation area, but with poor aesthetics. If evolved into a buffer corridor form, it can combine functionality and aesthetics.

Site plan

向阳而居·依塘治水
——河北保定市清苑区李八庄村民居设计

设计说明：

设计基于对基地周边建筑肌理，两种户型以二合院及三合院的形式对周边民居形式做出适应性回应。

"两甩袖"是河北的一种典型合院形式，由传统四合院发展而来，其凹型内弯的"甩袖"空间是由于当地冬季的寒冷气候而形成的演变，功能上便成为气候缓冲空间。

我们以"甩袖"空间及其前部形成的月台灰空间为原型，设计阳光房及阳光廊，被动式利用太阳能集热收；建筑采用生土材料及木结构实现半装配式建造，便于就地取材，同时利用夯土墙良好的蓄热特性，使室内能做达到冬暖夏凉的预期；功能布局上将服务空间布置于被服务空间周围作为缓冲空间以增加室温；建筑屋顶形式为斜坡顶，倾角30°以达到最高的光伏发电效率，部分屋顶材料为半透明的聚碳酸酯板，可透入阳光以加热其中的缓冲空间；设计阁楼，作为隔温层缓和温度；设计高窗或天窗，利用烟囱效应达到通风换风的作用。服务中心多采用可移动隔墙，发生灾害时可成为大空间布置临时床位以做应急避灾。

防涝设计上，采用水渠、渗水绿地、生物滞留池、雨水花园等方式组织雨水排入池内，并在防涝设计范围内设计渗水绿地、增加小型坑塘以应对洪涝灾害。

Design Description:

The design is based on the surrounding architectural texture of the site, and two types of layouts are adapted to the surrounding residential forms in the form of two courtyard and three courtyard.

"Two swinging sleeves "is a typical courtyard form in Hebei Province, developed from traditional quadrangle courtyards. Its concave shaped" swinging sleeves "space evolved due to the cold winter climate in the local area, and functionally became a climate buffering space.

We designed a sunroom and a sunroom based on the "swinging sleeve" space and the platform gray space formed in front of it, passively utilizing solar energy for heat collection;The building adopts rammed earth materials and wooden structures to achieve semi assembled construction, which facilitate the use of local materials. At the same time, the good heat storage characteristics of rammed earth walls are utilized to achieve the expected warmth in winter and coolness in summer indoors; Arrange the service space around the serviced area as a buffer space in the functional layout to ease the room temperature; The roof form of the building is a sloping roof with a tilt angle of 30 ° to achieve the highest photovoltaic power generation efficiency. Some roof materials are semi transparent polycarbonate panels that can penetrate sunlight to heat the buffer space inside; Design an attic as a thermal insulation layer to ease room temperature; Design high windows or skylights to utilize the chimney effect for ventilation and air extraction. Service centers often use movable partition walls, which can serve as temporary beds in large spaces for emergency shelter in case of disasters.

In terms of flood prevention design, water channels, permeable green spaces, biological retention ponds, rainwater gardens, and other methods are used to organize drainage into ponds. Within the scope of flood prevention design, permeable green spaces are designed and small ponds are added to cope with flood disasters.

Technical and Economic Indicators

Land area:	10200㎡
Construction land area:	8840㎡
Building area:	3345㎡
Second generation residential area（10）：	133㎡
Third generation residential area（4）：	194㎡
Village center area:	1239㎡
Parking lot number:	20
Greening rate:	27.9%
Building density:	20.4%
Floor area ratio:	0.328
Height:	9m

向阳而居·依塘治水
——河北保定市清苑区李八庄村民居设计

Two Generation Residential South Facade

Two Generation Residential East Facade

Three Generation Residential South Facade

Three Generation Residential East Facade

专家点评：

该方案总体布局合理，空间布局展现出功能性的聚居方式，对居住建筑平面有深入研究，能够较好地结合地方"两甩袖"的特色元素，为建筑增添独特魅力。建筑形式整体与光伏结合良好，采用模块尺寸提高了施工的灵活性与效率，但方案在水池岸边的处理不够充分，没有考虑水面高度变化的关系，"两甩袖"对院子通风存在不利影响，北坡屋顶使用玻璃产生光污染，且过于关注造型而影响了功能，设计中选用的层压木材在易受洪水和高湿度影响的区域容易变质。

Experts' Commentary：

The overall layout of the scheme is well conceived, and the spatial arrangement effectively demonstrates a functional settlement approach. The designers have conducted an in-depth study of residential buildings' floor plans and successfully incorporated local "two swinging sleeves" characteristic elements into the architectural design, thereby enhancing its unique charm. The integration of photovoltaic technology into the building's form, along with the use of module sizes, significantly enhances construction flexibility and efficiency. However, the treatment of the pond bank is inadequate, as it fails to address the relationship with fluctuating water levels. Moreover, the design of the "two swinging sleeves" adversely affects yard ventilation, while the use of glass on the north-facing slope roof prioritizes aesthetics over functionality, leading to potential light pollution. Finally, the GLT (Glued Laminated Timber) selected in the design is prone to deterioration in flood-prone and high-humidity areas.

Village Center First Floor Plan

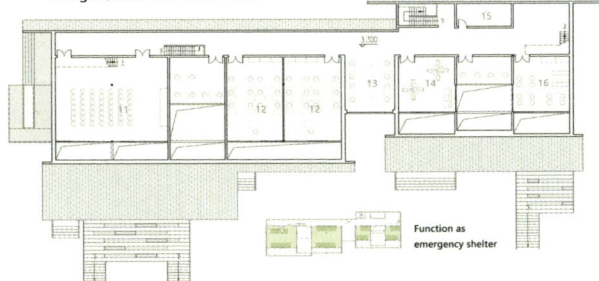

Function as emergency shelter

Village Center Second Floor Plan

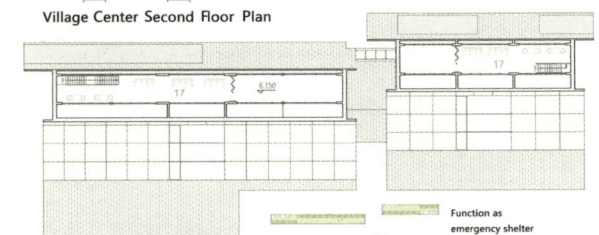

Function as emergency shelter

Village Center Attic Floor Plan

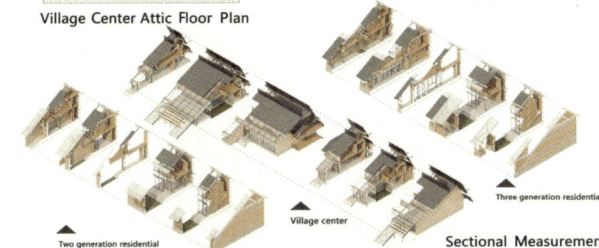

Function as emergency shelter

Two generation residential

Village center

Three generation residential

Sectional Measurement

Three Generation Residential Plan

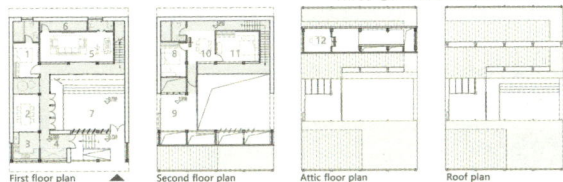

First floor plan   Second floor plan   Attic floor plan   Roof plan

Two Generation Residential Plan

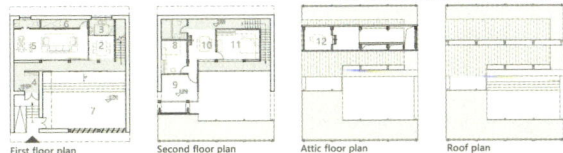

First floor plan   Second floor plan   Attic floor plan   Roof plan

Residential functions
1.Elderly room   7.Yard
2.Dinning room   8.Second bedroom
3.Kitchen   9.Balcony
4.Porch   10.Living room
5.Living room   11.Master bedroom
6.Store room   12.Attic

Village center functions
1.Lobby   11.Multi-Function Hall
2.Children's Education   12.Classroom
3.Digital Reading Room   13.Teahouse
4.Reading room   14.Card room
5.Pharmacy   15.Storeroom
6.Convalesce   16.Painting room
7.Medical room   17.Lounge
8.Kitchen
9.Restaurant
10.Cafe

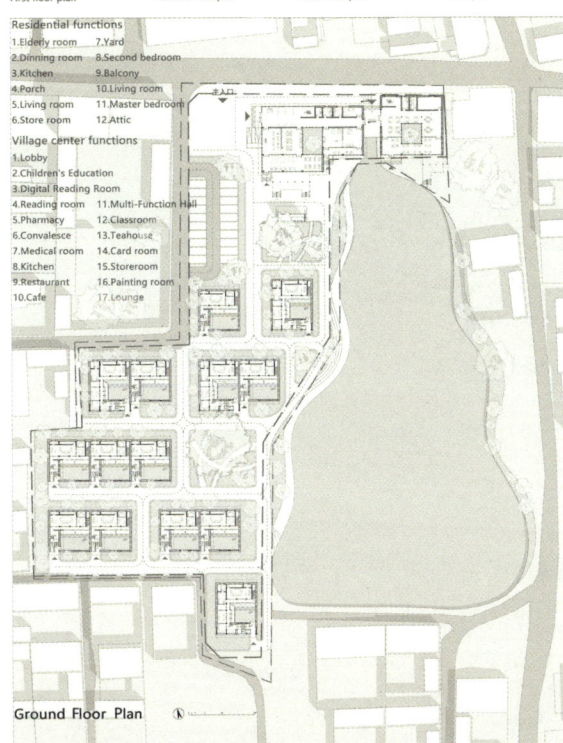

Ground Floor Plan

03

向阳而居·依塘治水
——河北保定市清苑区宋八庄村民居设计

Residential living room

Sunroom of service center

Sunroom of three residential

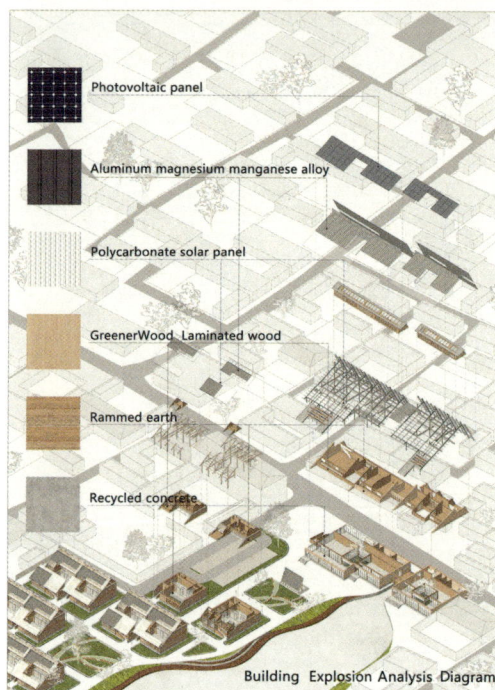

Photovoltaic panel

Aluminum magnesium manganese alloy

Polycarbonate solar panel

GreenerWood Laminated wood

Rammed earth

Recycled concrete

Building Explosion Analysis Diagram

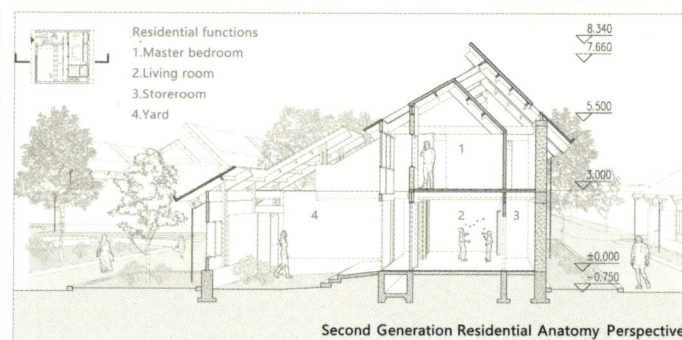

Residential functions
1.Master bedroom
2.Living room
3.Storeroom
4.Yard

8.340
7.660
5.500
3.000
±0.000
-0.750

Second Generation Residential Anatomy Perspective

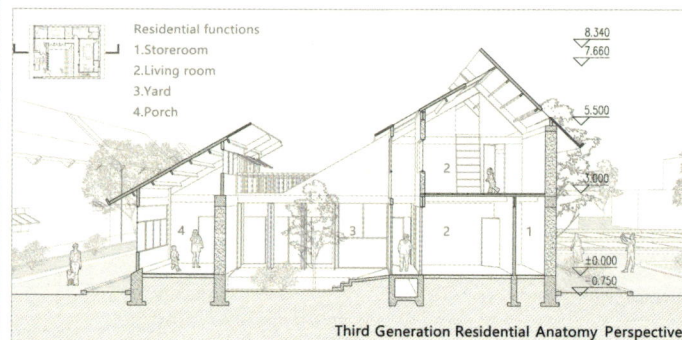

Residential functions
1.Storeroom
2.Living room
3.Yard
4.Porch

8.340
7.660
5.500
3.000
±0.000
-0.750

Third Generation Residential Anatomy Perspective

Village center functions
1.Lounge
2.Classroom
3.Convalesce
4.Pharmacy

8.970
7.600
6.300
3.400
±0.000
-0.750

Village Center Residential Anatomy Perspective

向阳而居·依塘治水
——河北保定市清苑区李八庄村民居设计

Wind pressure ventilation and reflected lighting in summer + Rainwater collection+Analysis of hot air during winter daytime

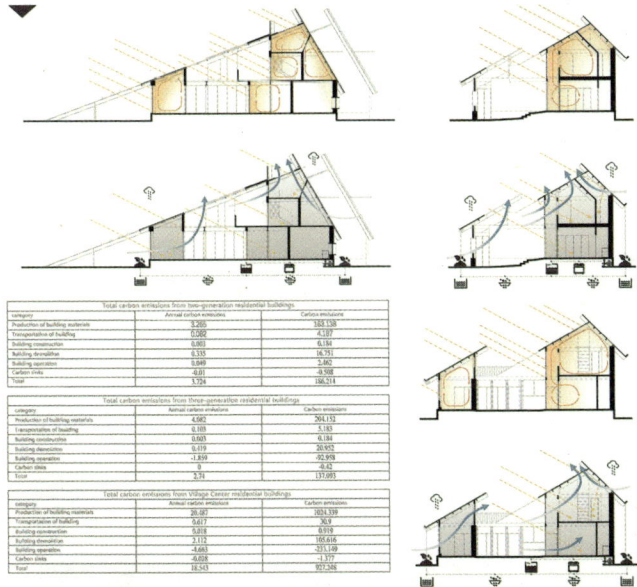

Simulation Analysis

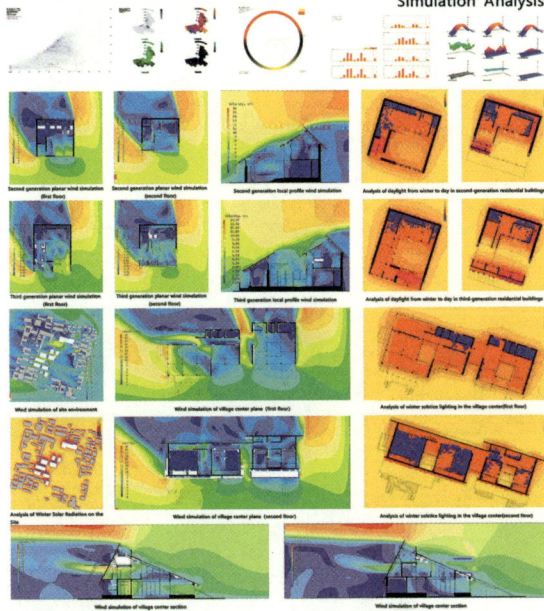

Flood Prevention Design Within the Region

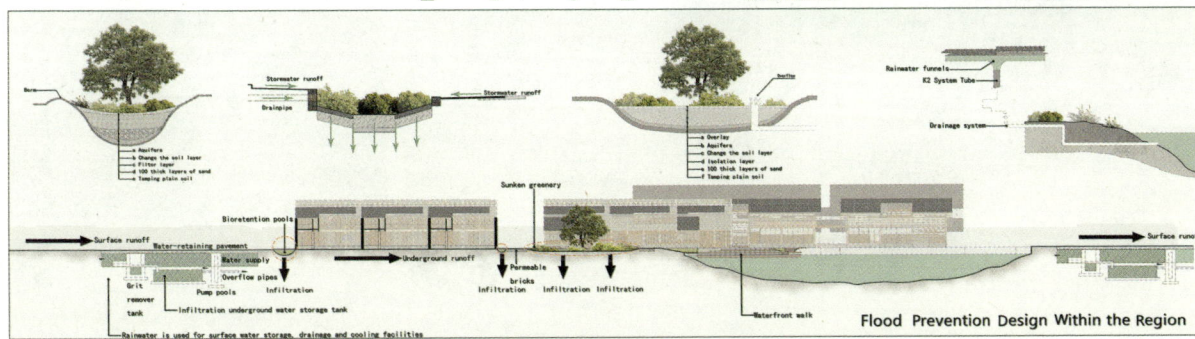

Site Issues in the Whole Year

Oct to Jan: Population loss and lack of festive

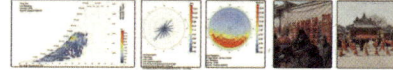

Feb to May: The area for grain drying is insufficient

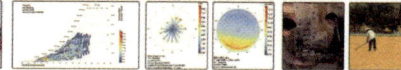

Jun to Sep: Floods are frequent

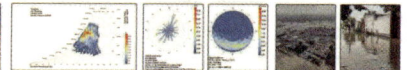

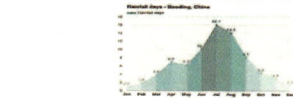

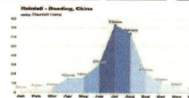

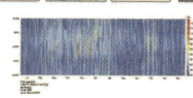

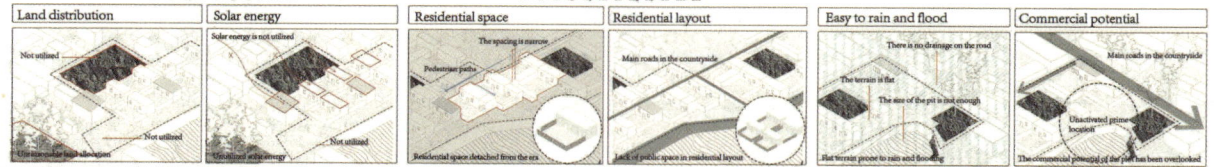

Land distribution	Solar energy	Residential space	Residential layout	Easy to rain and flood	Commercial potential
Not utilized	Solar energy is not utilized	Pedestrian paths	Main roads in the countryside	There is no drainage on the road	Main roads in the countryside
Not utilized	Unutilized water energy	Residential space detached from the era	Lack of public space in residential layout	The terrain is flat / The size of the pit is not enough / Flat terrain prone to rain and flooding	Unactivated primer location / The commercial potential of the pier has been overlooked

综合奖·三等奖·保定市清苑区李八庄村建设项目
Comprehensive Awards – Third Prize – Libazhuang Village Project, Qingyuan District, Baoding City

注册号：110160
Register Number：110160
项目名称：水过炊烟
Entry Title：Flood Sweeping Cross the Village Life
作者：舒心、陶禹竹、崔一然、何若珩
Authors：Shu Xin, Tao Yuzhu, Cui Yiran, and He Ruoheng
作者单位：沈阳建筑大学
Authors from：Shenyang Jianzhu University
指导教师：付瑶、张龙巍
Tutors：Fu Yao, Zhang Longwei
指导教师单位：沈阳建筑大学
Tutors from：Shenyang Jianzhu University

Toughness Design Concept Steps

设计理念 Design Concept
通过计算降雨量，引入调蓄层的理论，为抗内涝设计相应体积的坑塘和道路的排水管网。额外设计的坑塘在旱时作活动使用，涝时被水淹没扩大蓄水面率。将民居、公共建筑、坑塘放置在不同的高差上，之间用廊道连接，提供灵活的应对水患的策略。

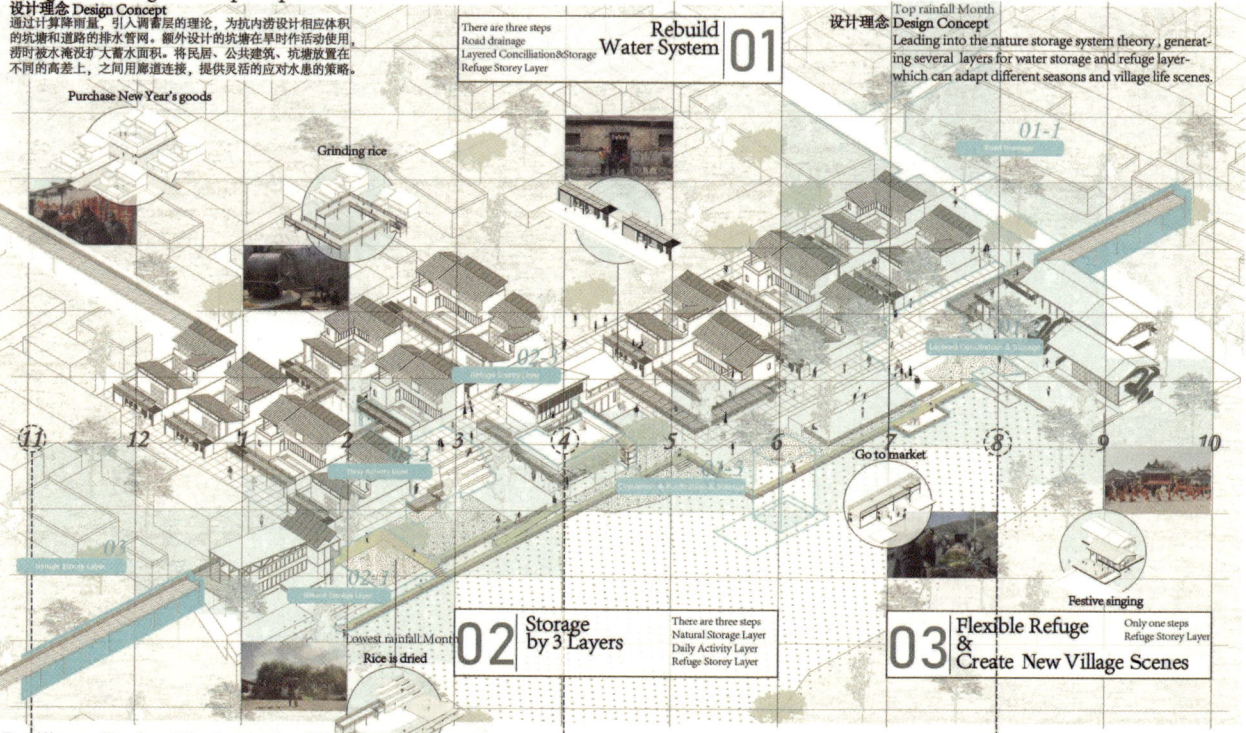

Rebuild Water System 01
There are three steps
Road drainage
Layered Conciliation & Storage
Refuge Storey Layer

设计理念 Design Concept
Leading into the nature storage system theory, generating several layers for water storage and refuge layer which can adapt different seasons and village life scenes.

Purchase New Year's goods

Grinding rice

Go to market

Festive singing

Rice is dried

02 Storage by 3 Layers
There are three steps
Natural Storage Layer
Daily Activity Layer
Refuge Storey Layer

03 Flexible Refuge & Create New Village Scenes
Only one steps
Refuge Storey Layer

Lowest rainfall Month

Resilience Design Display in the Whole Year

Oct to Jan Spring Festival Market

Feb to May Harvest wheat

Jun to Sep Rain and flood shelter

The holding of the Spring Festival market has added a lively atmosphere to the winter in rural areas, and can also drive tourism and increase residents' income

The autumn harvested rice is dried in the courtyard, which not only activates the public space but also enhances neighborhood interaction

When a flood disaster strikes, people can quickly evacuate through the shelter corridor to ensure safety and reduce the losses caused by the flood disaster

Rebuild Water System

Grass planting ditches and rainwater pipe networks are set up along the roads in the village, and three drainage ditches are set up within the red line according to the rainfall data and waterlogging area, and the domestic drainage system and septic tank are added at the same time, and biogas is used as a clean energy.

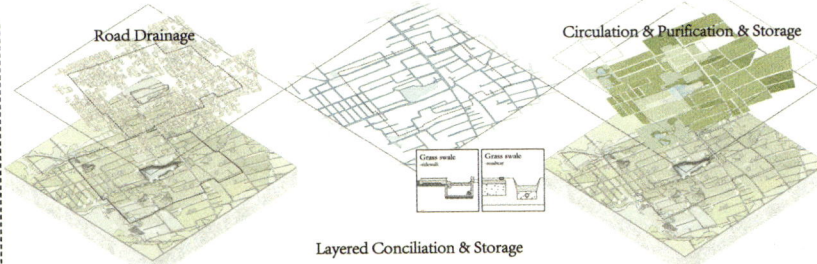

Road Drainage

Circulation & Purification & Storage

Layered Conciliation & Storage

Idea Extraction

Use Natural Storage System natural mode to extracted concept

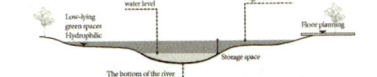

Collecting rainwater during floods and using it during droughts can not only avoid rainwater peaks, but also improve the utilization rate of rainwater, and also provide residents with activity venues during drought periods.

Rainfall Calculations

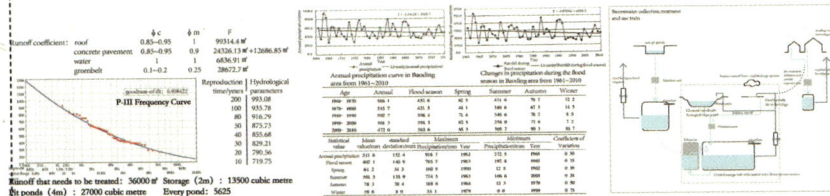

Hierarchical Storage System

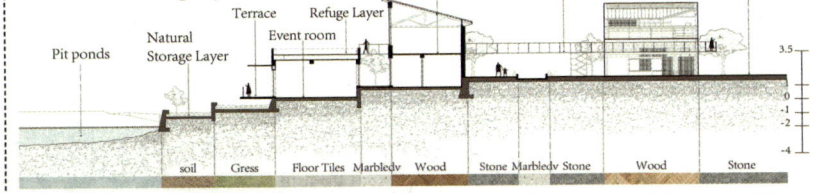

Master Plan 1:500

1.Village Event Room
2.Village Book House
3.Hospital
4.Stone Mill Cloister
5.Waterfront Staircase
6.Waterfront Experience Area
7.Pit Ponds
8.Refuge Square
9.Village

Scenario Generation

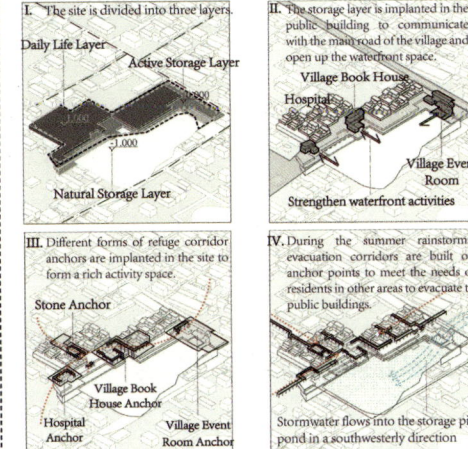

I. The site is divided into three layers.

II. The storage layer is implanted in the public building to communicate with the main road of the village and open up the waterfront space.

III. Different forms of refuge corridor anchors are implanted in the site to form a rich activity space.

IV. During the summer rainstorms, evacuation corridors are built on anchor points to meet the needs of residents in other areas to evacuate to public buildings.

Stormwater flows into the storage pit pond in a southwesterly direction

Caculation Carbon Emissions

East Elevation 1:250

专家点评：

该方案对不同季节自然天气的影响采取不同的处理方式。将服务中心按功能分解为三部分，利用屋檐下的连廊进行连接，在面临灾害条件下公共服务功能分开不利于高效协调和资源分配。村庄抗内涝防洪设计体现了对自然灾害的前瞻性思考。建筑节能措施不足，民居设计在功能和空间上存在不足，需要进一步优化。

Experts' Commentary：

The scheme takes different approaches to mitigate weather impacts in various seasons. It segments the service center into three functional areas, interconnected by a corridor under the eaves. However, this separation of public service functions impedes efficient coordination of facilities and resource allocation under disaster conditions. The village's anti-waterlogging and flood control design reflects forward-thinking in addressing natural disasters. Nonetheless, the buildings' energy-saving measures are insufficient, and residential designs, which have functional and spatial limitations, require further optimization.

Hebei Residential

Hebei residential standard unit size analysis

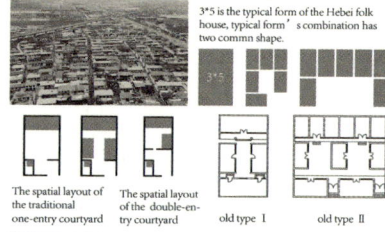

3*5 is the typical form of the Hebei folk house, typical form's combination has two commn shape.

The spatial layout of the traditional one-entry courtyard

The spatial layout of the double-entry courtyard

old type I

old type II

Green Construction Technology

Implant locally appropriate solar energy/thermal insulation /rainwater collection system technology

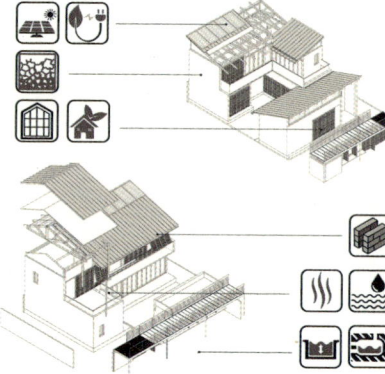

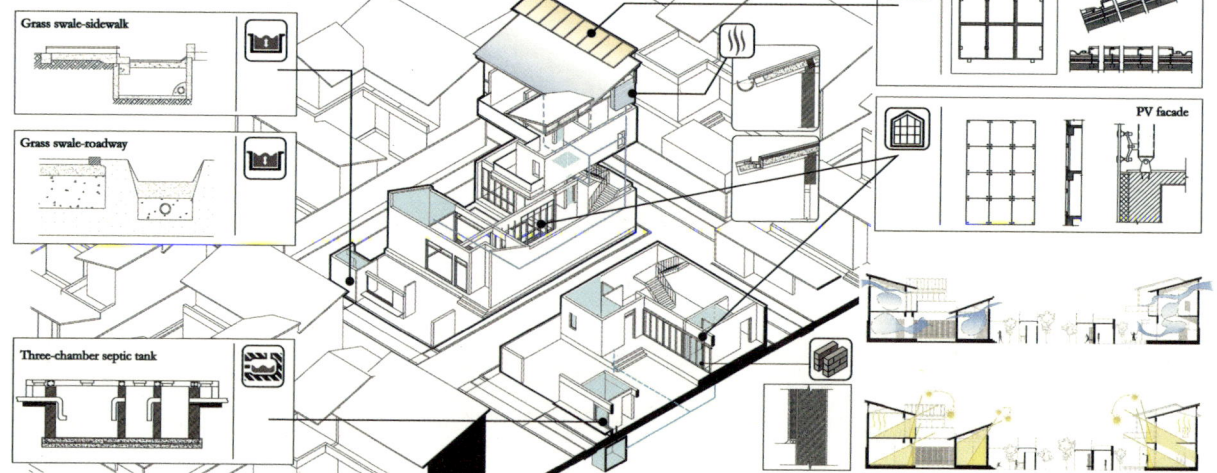

Plan Generation

01 — Change the size of the unit space to make it more suitable for the daily living of residents

02 — Re-divide space to reduce wasted space

03 — Re-plan the courtyard to enrich the activity space

04 — Reorganize rooms and clarify functional zoning

Type I Folk House

Ground level plan 1 : 200

Second level plan 1 : 200

1 Foyer
2 Dining room
3 Ketchen
4 Living room
5 Bedroom
6 Yard
7 Storeroom

Type II Folk House

Ground level plan 1 : 200

Second level plan 1 : 200

1 Foyer
2 Dining room
3 Ketchen
4 Living room
5 Bedroom
6 Yard
7 Storeroom

Green Construction Technology Apply

Implant locally appropriate solar energy/thermal insulation /rainwater collection system technology

Grass swale-sidewalk

Grass swale-roadway

Three-chamber septic tank

EL panel solar

PV facade

Shelter Corridor's Change
Shelter Corridor in Residential
Shelter Corridor in Public

dry time / flood time

Overall Connection Effect

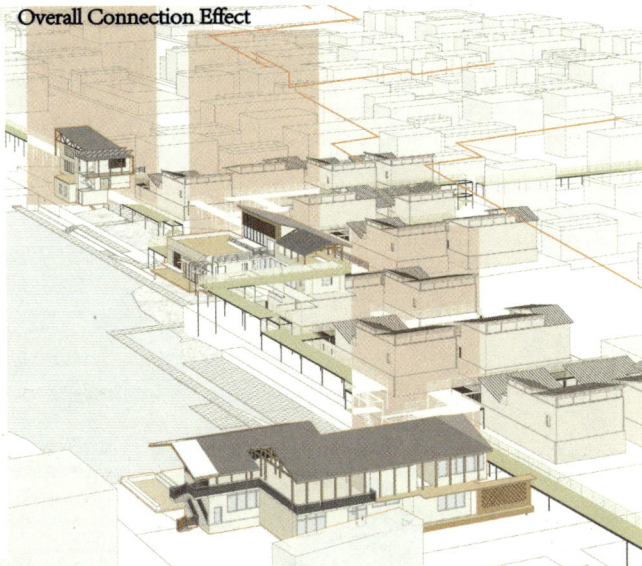

Shelter Corridor's Change

Activity / Shelter
dry time / flood time

Village Hospital

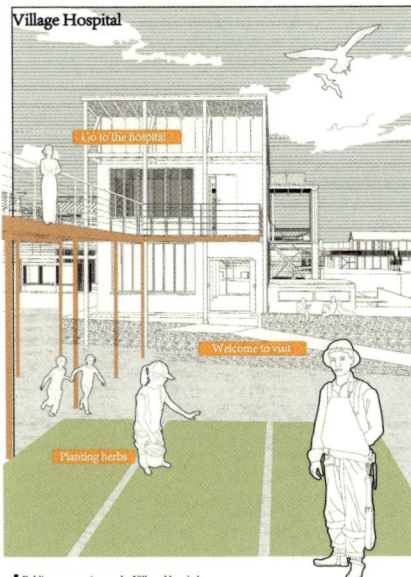

Go to the hospital
Welcome to visit
Planting herbs

Village Book House

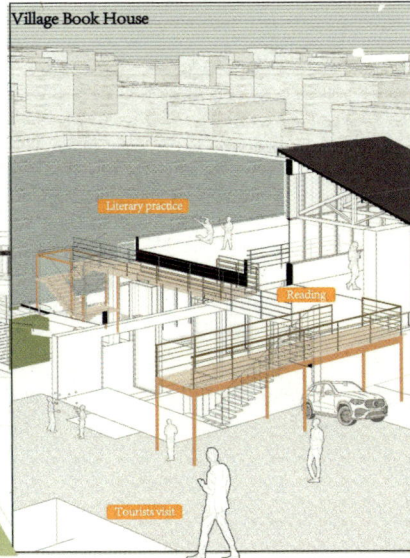

Literary practice
Reading
Tourists visit

Village Event Room

Visit and inspect
Rehearsing opera
Mediate disputes

■ Public construction node: Village Hospital

Introduction to Public Building Nodes:
The village hospital has both daily medical care and emergency rescue function in case of flood disasters, providing safety guarantees for villagers.
Ensuring the safety of village can enhance their sense of happiness.

1 X-ray room
2 Accessible restrooms
3 consultation rooms
4 guide stations

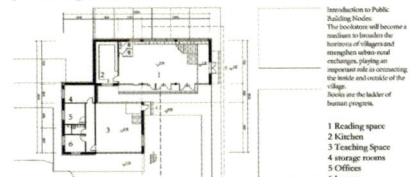

■ Public Building Node:Bookstore

Introduction to Public Building Nodes:
The bookstore will become a medium to broaden the horizons of villagers and strengthen urban-rural exchanges, playing an important role in connecting the inside and outside of the village.
Books are the ladder of human progress.

1 Reading space
2 Kitchen
3 Teaching Space
4 storage rooms
5 Offices
6 Lounge

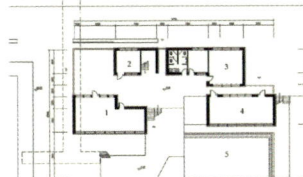

■ Public construction node: Village Committee

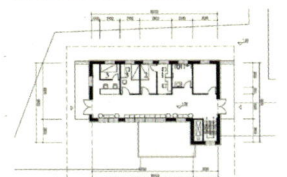

Introduction to Public Building Nodes:
The village committee combines with local characteristic theaters to provide villagers with village services while providing them with diverse life experiences.The village committee also serves as a village history museum, and the village history corridor can help tourists quickly understand the history and culture of the village.

1 Village Activity Room
2 Village History Museum
3 Mediation Room
4 Stage Reserve Room
5 stages

Sunlight Analysis

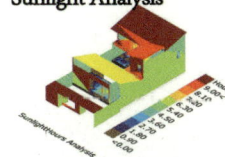

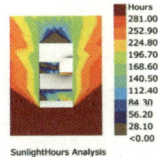

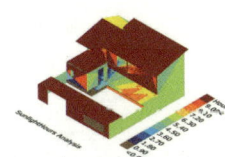

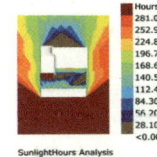

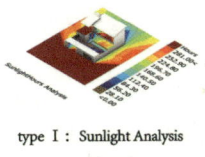

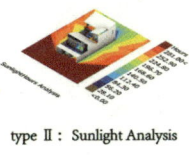

Hours
281.00<
252.90
224.80
196.70
168.60
140.50
112.40
84.30
56.20
28.10
<0.00

SunlightHours Analysis

type Ⅰ：Sunlight Analysis   type Ⅱ：Sunlight Analysis

综合奖·三等奖·保定市清苑区李八庄村建设项目
Comprehensive Awards – Third Prize – Libazhuang Village Project，Qingyuan District，Baoding City

注册号：110383
Register Number：110383
项目名称：林荫光坊
Entry Title：The Shaded Light Pavilion
作者：邹旭、包益槟、原建、邱伊雯
Authors：Zou Xu，Bao Yibin，Yuan Jian, and Qiu Yiwen
作者单位：南京工业大学
Authors from：Nanjing Tech University
指导教师：罗靖
Tutor：Luo Jing
指导教师单位：南京工业大学
Tutor from：Nanjing Tech University

110383

林 荫 光 坊 |
The Shaded Light Pavilion

设计说明

本方案旨在因应场地的环境特性与韧性乡居需求。在呼应场地肌理的基础上引入林荫景观和市集街坊，串联不同代际的民居，营造出丰富的交往空间。提升了居民的日常生活品质，同时利用场地高差变化，缓解了村落在雨季中的内涝问题。节能策略方面，以"光"为主题，优先采取被动式太阳能技术，辅以主动式太阳能技术的优化。通过转译传统片空间为阳光房，结合墙体保温提高室内舒适性。通过增加天井和天窗创造更多采光和自然通风，同时在林荫也成为居民的精神场所，因地选择木材等可预制构件，合理运用太阳能储能等主动式技术，提高了能源应益、优化了村落应灾与自维持能力。

Design Description

This plan aims to respond to the site's environmental characteristics and the resilient rural living needs. Based on the site's texture, it introduces shaded landscapes and marketplace neighborhoods, connecting residential areas across different generations, creating rich social interaction spaces. This enhances the quality of daily life for residents, while utilizing the site's elevation changes to alleviate flooding issues during the rainy season.

In terms of energy-saving strategies, the theme focuses on "light." Passive solar technology is prioritized, complemented by the optimization of active solar technology. Traditional veranda spaces are reinterpreted as sunrooms, combined with wall insulation to improve indoor comfort. The addition of courtyards and skylights increases natural lighting and ventilation, while the shaded areas also serve as spiritual spaces for residents. Locally suitable materials, such as prefabricated wood components, are chosen, and active technologies like solar energy storage are efficiently utilized, increasing energy benefits and optimizing the village's disaster resilience and self-sustainability.

Demographic Analysis

Main family structures and derived problems

First-Generation Solitary Elders | Second-Generation Left-Behind Families | Three-Generation Migrant Worker Families

Street Generation Logic

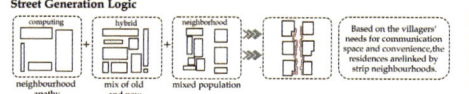

Climate Analysis

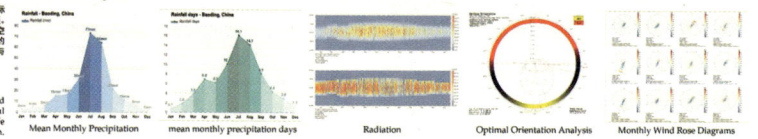

Mean Monthly Precipitation | mean monthly precipitation days | Radiation | Optimal Orientation Analysis | Monthly Wind Rose Diagrams

Libazhuang is located in Qingyuan District, Hebei Province, characterized by hot and rainy summers. The average monthly rainfall during summer can reach around 60 mm, with approximately 12 rainy days each month. Intense precipitation can easily trigger internal flooding disasters.

Flood Defense Strategy

We divide the flood control planning into three levels based on area scale:
First-level: Comprehensive flood control and drainage for the entire village.
Second-level: Flood control for internal flooding within the village.
third-level: Flood control measures for specific sites.

First Level Flood Control Strategy

Libazhuang, in Qingyuan District, Baoding City, has a terrain that slopes from west to east. The eastern side features flood control riverbanks and the Qingshui River to manage flooding. Main roads are designated as drainage outlets, directing water to the western riverbank.

Second Level Flood Control Strategy

We designed several flood control nodes in the second-level area, such as sunken rainwater squares and parking lots, to address internal flooding. These nodes provide localized protection during floods. Additionally, the village pond is designated as the drainage hub, using elevation differences to channel floodwaters to this central point.

Evacuation Routes and Shelter Distribution

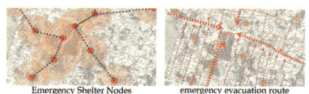

Emergency Shelter Nodes | emergency evacuation route

Technology System

Passive design	Active design	Sponge city technology
Sun room	Photovoltaic power	Impounding reservoir
Insulation wall	Solar energy storage	Rainwater garden
Thermal pressure ventilation	Water SourceHeat Pump	Permeable pavement
Air pressure ventilation		

Site Generation Logic

Site The site is located in the center of Libazhuang village in Qingyuan District, Hebei Province, 25 kilometers from downtown Baoding, and adjacent to a pond.

Function Based on the needs of local residents, we have divided the site into residential areas, public building areas, and landscape areas.

Main road Based on the planning of the main roads in the village, a north-south primary road has been designed to connect the site.

Minor Road Based on the planning strategy of the main roads and functional zoning, an east-west landscape road has been designed to connect the site.

Drainage Generation Logic

STEP 1 We designed a slope in the site from west to east, directing water from the west side into a pond on the east side.

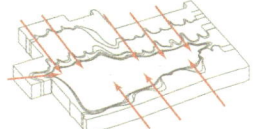

STEP 2 The water flows along the slope, moving along the east-west landscape path and exiting from the center of the building into the pond.

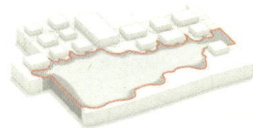

STEP 3 After the water flows out, it drains into the pond from the outlets of the various landscape paths. Flood embankments are established along the shoreline, along with a scenic walkway.

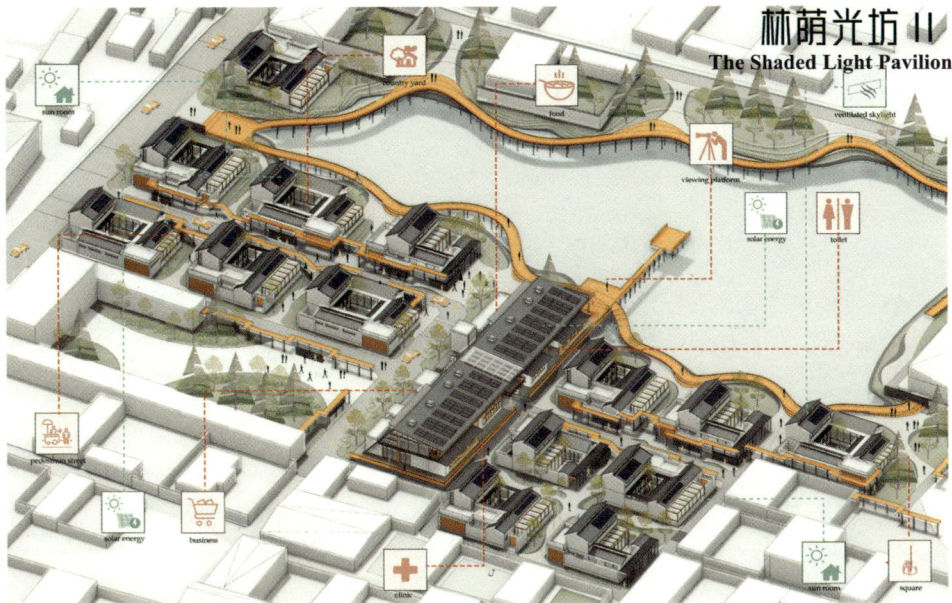

Sunlight and Wind

Winter Wind Speed | Summer Wind Speed | Summer Solstice Sunshine Hours | Winter Solstice Sunshine Hours | Summer Sunshine Hours | Winter Sunshine Hours

In the area, the prevailing wind in summer is from the southwest, while in winter it comes from the northeast. Therefore, the site is oriented in a southwest-northeast direction to allow the wind to flow freely within the area.

The area enjoys abundant sunlight, so we maintain a distance of over 6 meters between buildings to ensure that each residence receives ample light.

Flood Control Nodes

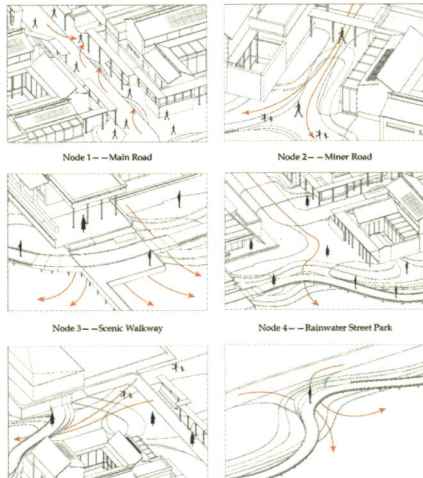

Node 1——Main Road

Node 2——Miner Road

Node 3——Scenic Walkway

Node 4——Rainwater Street Park

Node 5——Sunken Rainwater Plaza

Node 6——Water Walkway

Site Plan 1:1000

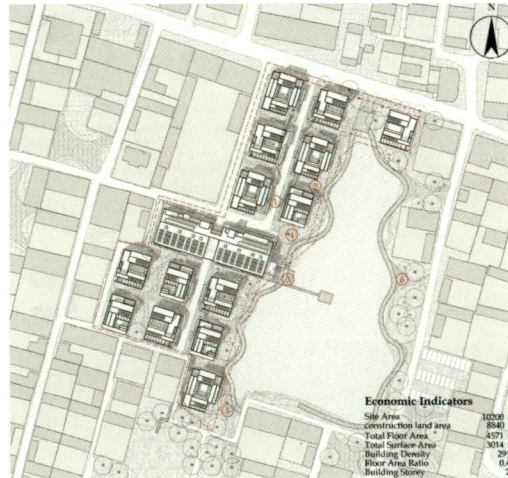

Economic Indicators
Site Area ... 10200 m²
construction land area ... 8840 m²
Total Floor Area ... 4571 m²
Total Surface Area ... 3014 m²
Building Density ... 29%
Floor Area Ratio ... 0.44
Building Storey ... 2F

Arnebia | Locust | Winter Jasmine | Chinese bellFlower | Cedarwood

Greenfield infiltration in courtyards
Greenfield infiltration in courtyards
Landscape pathway paving infiltration

Cobblestone | Plant Cover | Grit | Grass-planking

专家点评：

该方案将服务中心置于中间位置，方便村民使用各类服务设施，提高了使用效率和便利性，有效促进了村民之间的交流互动，但通往居民区中心公共建筑的主路可能导致交通拥堵，需要进一步优化交通流线。高架平台强调自然通风，有助于提升环境舒适度。阳光房等被动设计的运用体现了对节能和环保的考虑，但住宅房间进深太小会影响室内空间的使用效率。沿水塘设计环廊，未充分考虑水位高度变化对环廊使用和安全的影响。胶合层积木材成本高昂，经济性较差。

Experts' Commentary：

The design positions the service center in a central location to facilitate villagers' access to various service facilities, thereby improving usage efficiency and convenience while fostering communication and interaction among residents. However, main roads leading to public buildings within the residential core may cause traffic congestion, necessitating further optimization in circulation design. The design of elevated platform emphasizes natural ventilation, enhancing environmental comfort. The use of passive designs such as a sunroom reflects considerations for energy saving and environmental protection. However, the shallow depth of residential rooms may compromise the efficient utilization of indoor space. Additionally, the design of the ring corridor along the pond has not fully considered the impact of fluctuating water levels on its functionality and safety. Moreover, GLT (Glued Laminated Timber) is expensive, making it economically unfavorable.

110383

林荫光坊 III
The Shaded Light Pavilion

Eave Detail Construction

Insulated Wall Detail Construction

Ventilated Raised Floor Detail Construction

Hybrid home PV + Energy storage system

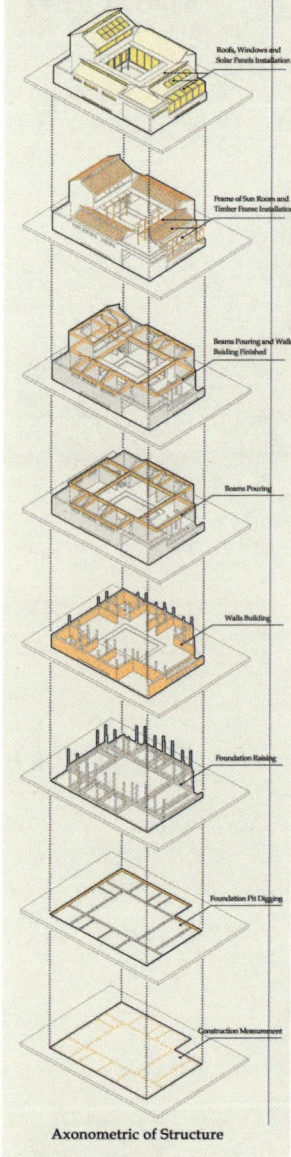

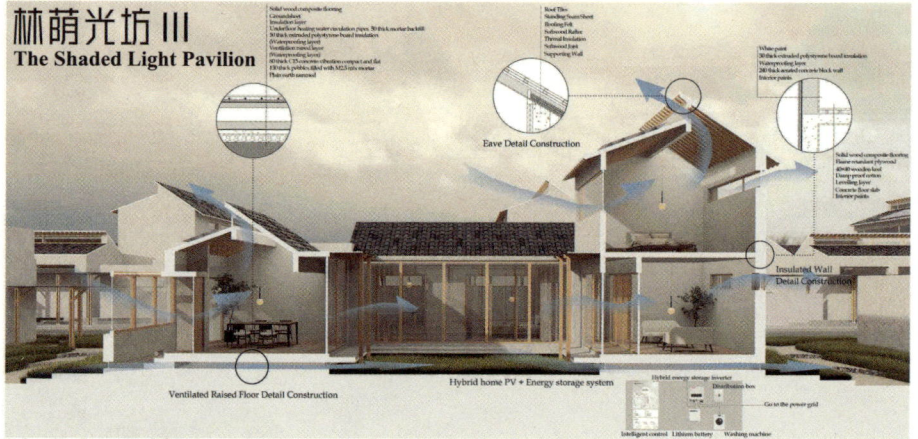

Indoor Environment Analysis

Summer | Winter | Day | Night

Section 1 : 200

Section 1-1 1 : 200 (First Generation) | Section 2-2 1 : 200 (Second Generation) | Section 3-3 1 : 200 (Third Generation)

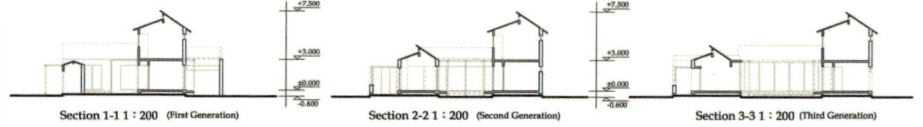

Concept Generation

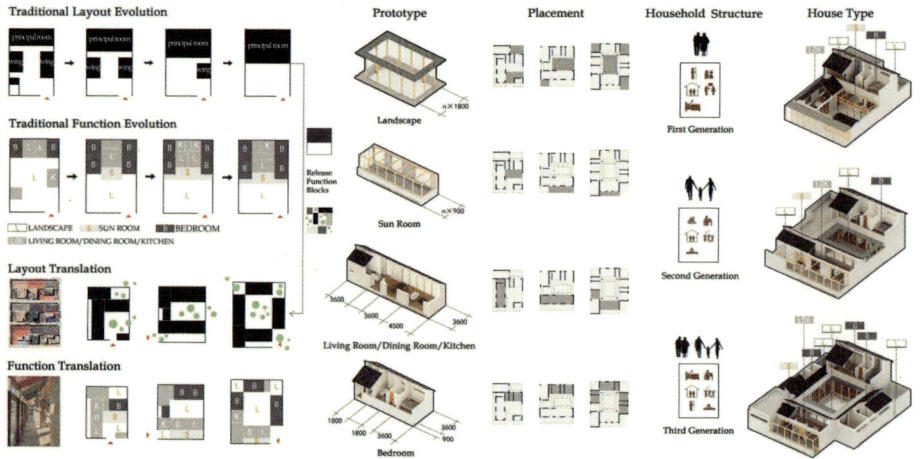

Traditional Layout Evolution

Traditional Function Evolution

Layout Translation

Function Translation

Prototype | Placement | Household Structure | House Type

Landscape

Sun Room

Living Room/Dining Room/Kitchen

Bedroom

First Generation

Second Generation

Third Generation

Axonometric of Structure

Roofs, Windows and Solar Panels Installation

Frame of Sun Room and Timber Frame Installation

Beams Pouring and Walls Building Finished

Beams Pouring

Walls Building

Foundation Raising

Foundation Pit Digging

Construction Measurement

Floor Plan and Elevation 1 : 200

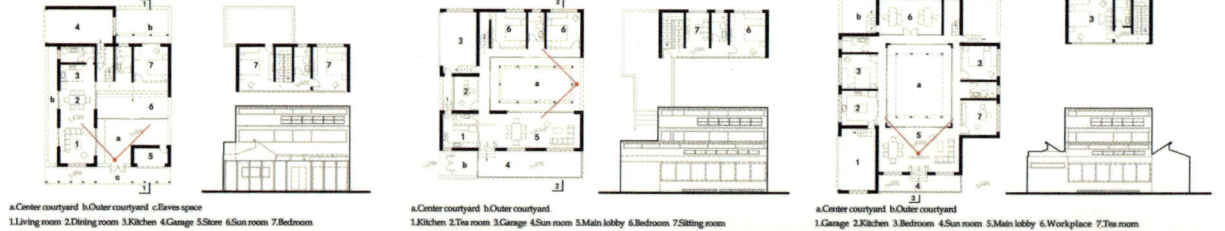

a.Center courtyard b.Outer courtyard c.Eaves space
1.Living room 2.Dining room 3.Kitchen 4.Garage 5.Store 6.Sun room 7.Bedroom

a.Center courtyard b.Outer courtyard
1.Kitchen 2.Tea room 3.Garage 4.Sun room 5.Main lobby 6.Bedroom 7.Sitting room

a.Center courtyard b.Outer courtyard
1.Garage 2.Kitchen 3.Bedroom 4.Sun room 5.Main lobby 6.Workplace 7.Tea room

First Generation House Type | Second Generation House Type | Third Generation House Type

林萌光坊 IV
The Shaded Light Pavilion

Phase 1 Before the disaster

The daily state of the interior of the public building before the disaster was an indoor shopping street, and the space functioned as a commercial.

Phase 2 In disaster

The function of the indoor commercial street of the public building in the disaster was changed to that of a place to set up disaster relief tents, and the public space such as the activity room on the second floor was changed to a medical room.

Phase 3 After the disaater

As the rains receded after the disaster, people organized themselves to clean up, kill the virus and wait for help.

Assembly Process

Exploded Axonxometric & Materials

Photovoltaic Panel
Polycrystalline silicon photovoltaic conversion efficiency of 14-16% of the current polycrystalline silicon production technology is mature.

Assphalt Shingle
Asphalt shingles have excellent waterproofing properties, thermal insulation properties, can effectively reduce heat absorption in summer and heat dissipation in winter, have good windproof performance.

Timber
Wood has very good mechanical properties, but wood is an organic anisotropic material.

Concrete
Concrete is characterized by abundant raw materials, low price and simple production process.

GLT
Glued laminated timber is an engineered timber system consisting of stacked timbers glued together with the grain running in the same direction, glued laminated timber can be used to create larger and stronger beams

Detail Construction

Illumination Analysis

first floor in winter

first floor in summer

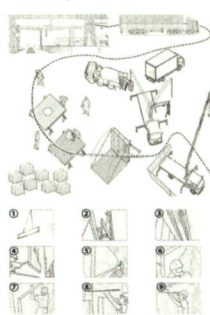

First Plan 1 : 300

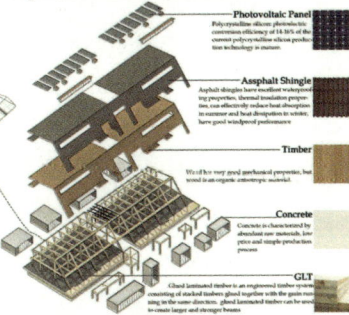

1 kiosk 2 service center 3 clinic 4 rehabilitation room 5 children's playroom
6 accessible toilet 7 toilet 8 Indoor commercial street 9 electricity distribution room

Second Plan 1 : 300

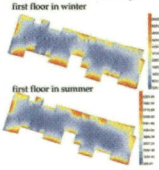

1 sun room 2 multi-function hall 3 tearoom 4 reading room 5 office
6 storeroom 7 lake viewing platform

South Elevation 1 : 300

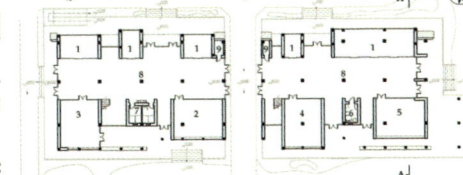

North Elevation 1 : 300

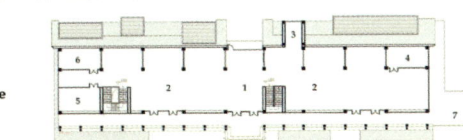

Wind Simulation

First Generation House Type

Second Generation House Type

Third Generation House Type

Public Service Center

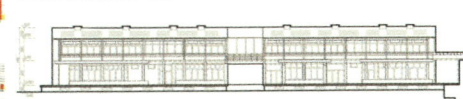

Carbon Emisson Data Table

First Generation House Type

Second Generation House Type

Third Generation House Type

Public Service Center

Comfort Comparative Analysis Chart

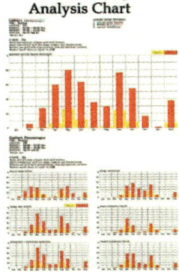

After a comparative analysis, we found that the three houses have more natural ventilation, significantly lower carbon emissions and better comfort after the passive design.

Active Technology Design

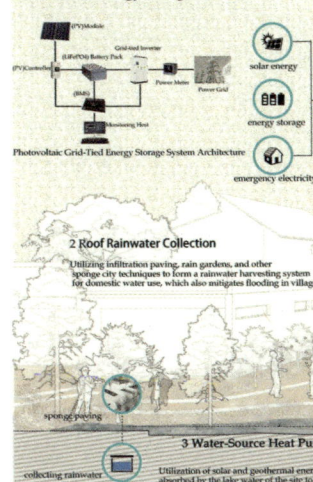

Photovoltaic Grid-Tied Energy Storage System Architecture

1 Solar PV & Energy Storage
- solar energy
- energy storage
- emergency electricity

2 Roof Rainwater Collection
Utilizing infiltration paving, rain gardens, and other sponge city techniques to form a rainwater harvesting system for domestic water use, which also mitigates flooding in villages

3 Water-Source Heat Pump
Utilization of solar and geothermal energy absorbed by the lake water of the site to form thermal energy resources for domestic hot water use

collecting rainwater / domestic hot water

Passive Technology Design

Roof windproofing in winter

1 Wind
The north-facing roof slopes downward to form a large slope, blocking the northeast wind in winter

Natural ventilation in summer
Hot Pressure Ventilation / wind pressure ventilation / ventilated raised floor

By controlling the shape of the building to obtain more sun exposure, and at the same time combining with the exhibition space to set up a sun room to enhance indoor comfort.

2 Light — Sun room

Perspective Section A-A

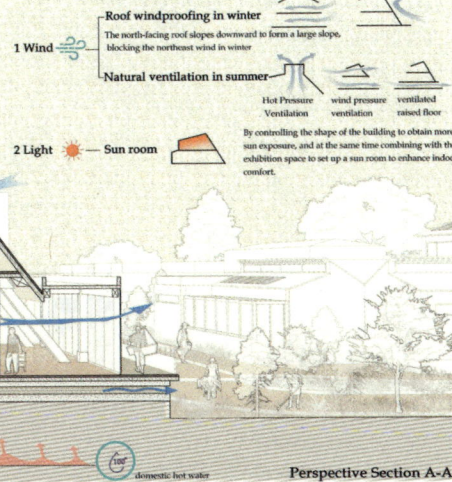

综合奖・三等奖・保定市清苑区李八庄村建设项目
Comprehensive Awards － Third Prize － Libazhuang Village Project，Qingyuan District，Baoding City

注册号：110568

Register Number：110568

项目名称：岸芷汀兰

Entry Title：Fragrant Orchids by the Riverbank

作者：戴含真、黄玥、李可欣、王雅静

Authors：Dai Hanzhen, Huang Yue, Li Kexin, and Wang Yajing

作者单位：厦门大学

Authors from：Xiamen University

指导教师：石峰

Tutor：Shi Feng

指导教师单位：厦门大学

Tutor from：Xiamen University

■ Location Analysis

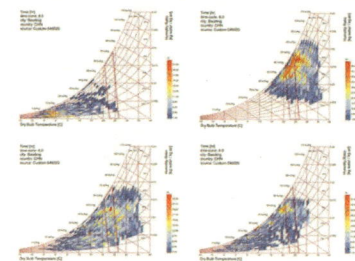

Geographical location: Baoding City is located in the central western part of Hebei Province, at the eastern foot of the Taihang Mountains, and in the western part of the Jizhong Plain.

Terrain and landforms: Baoding City is located at the eastern foot of the northern Taihang Mountains, in the western part of the Jizhong Plain, with a terrain that slopes from northwest to southeast.

■ Historical Evolution

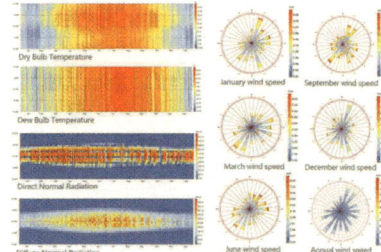

■ Meteorological Data

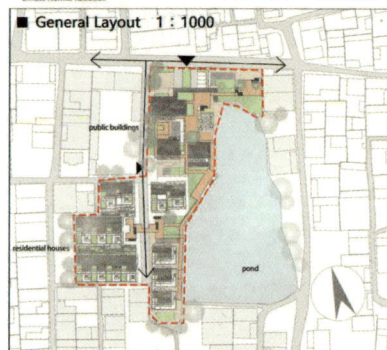

Dry Bulb Temperature

Dew Bulb Temperature

Direct Normal Radiation

Diffuse Normal Radiation

January wind speed　September wind speed

March wind speed　December wind speed

June wind speed　Annual wind speed

■ Logic Generation

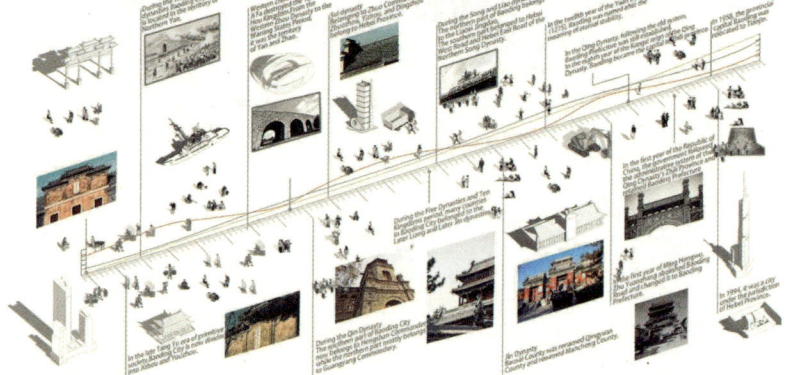

1.The botl land srca oftg sie is 9300m², with a drainage and landscaping pond on the east side.

2.Divide the site into four parts, including residential buildings, public buildings, evacuation squares, and rain gardens.

3.The venue layout gradually transitions from private to public from south to north and from west to east.

4.Residential clusters are arranged in a row on the east side and three groups on the west side, with a public space left in the middle.

5.The main body of public buildings is divided into four parts, adopting a courtyard style layout, with each block connected by corridors.

6.Undertaking the drainage of three areas within the site, ultimately converging into two drainage points.

7.More detailed drainage planning has been established, integrating the drainage of residential areas, transition areas, and public building areas.

8.Finally, based on site planning and architectural layout, set up a landscape route along the pond.

■ General Layout 1：1000

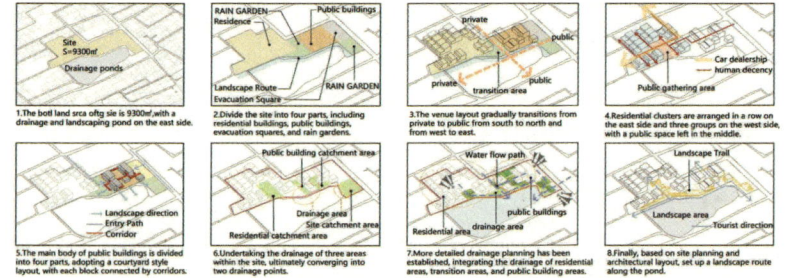

public buildings

residential houses

pond

■ Site Planning Analysis

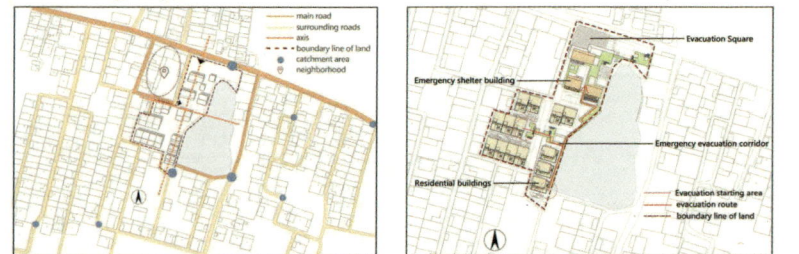

main road
surrounding roads
axis
boundary line of land
catchment area
neighborhood

Evacuation Square

Emergency shelter building

Emergency evacuation corridor

Residential buildings

Evacuation starting area
evacuation route
boundary line of land

专家点评：

该方案总体布局比较合理，对场地和公共廊道空间的处理有趣，在设计中融入了有益健康的亲生物元素，营造出与自然的紧密联系，提升了居住者的幸福感。流线布局的连通性实现了无缝通行和出入，促进了整个空间内的高效通行。对在地性的水灾害影响的考虑也较为充分，尤其是雨水花园的设计不仅能促进日常使用，还为提高日常防灾意识提供了新思路，将功能性与教育性有机结合。该作品建筑形式过度通透、不太适应当地气候环境。

Experts' Commentary：

The overall layout of the scheme is well-conceived, with an intriguing treatment of the site and public corridor spaces. The design integrates healthy biophilic elements to foster a strong connection with nature, thereby enhancing residents' well-being. The circulation layout promotes seamless connectivity, facilitating efficient movement throughout the space. The scheme demonstrates sufficient consideration of the local flood impact. In particular, the rain garden design can support daily functionality while also serving as an innovative approach to raising disaster prevention awareness, effectively merging functionality with educational value. However, the architectural form exhibits excessive transparency, which compromises its adaptability to the local climate.

■ **Site Water Flow Analysis** ■ **Flood Prevention Area Drainage and Measures**

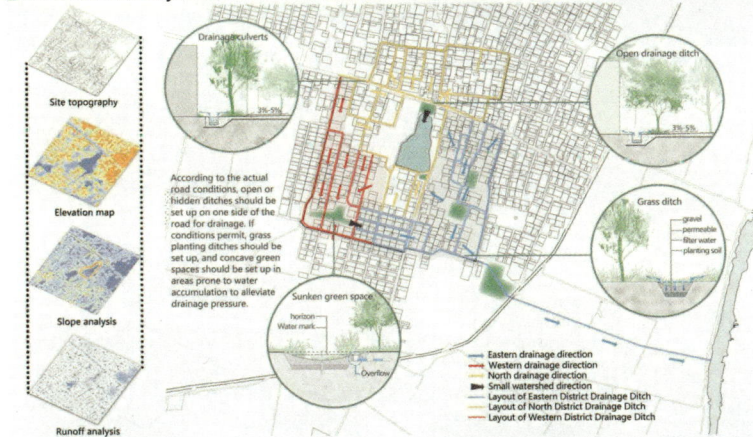

■ **Drainage Within the Venue**

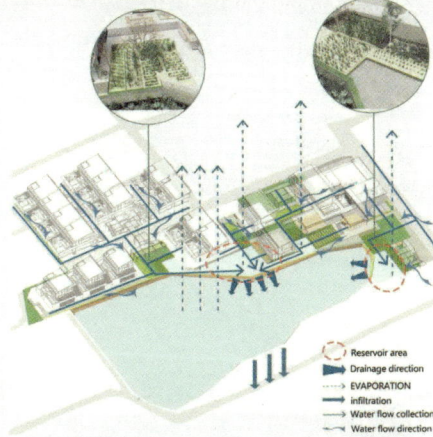

Site topography

Elevation map

Slope analysis

Runoff analysis

Drainage culverts

According to the actual road conditions, open or hidden ditches should be set up on one side of the road for drainage. If conditions permit, grass planting ditches should be set up, and concave green spaces should be set up in areas prone to water accumulation to alleviate drainage pressure.

Sunken green space

Open drainage ditch

Grass ditch
- gravel
- permeable
- filter water
- planting soil

Eastern drainage direction
Western drainage direction
North drainage direction
Small watershed direction
Layout of Eastern District Drainage Ditch
Layout of North District Drainage Ditch
Layout of Western District Drainage Ditch

Reservoir area
Drainage direction
EVAPORATION
infiltration
Water flow collection
Water flow direction

■ **Vegetation Analysis**

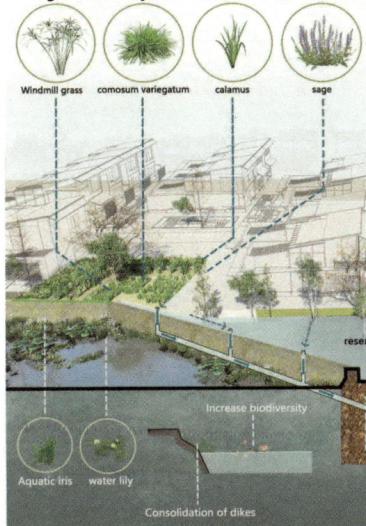

Windmill grass   comosum variegatum   calamus   sage

■ **Periodic Maintenance**

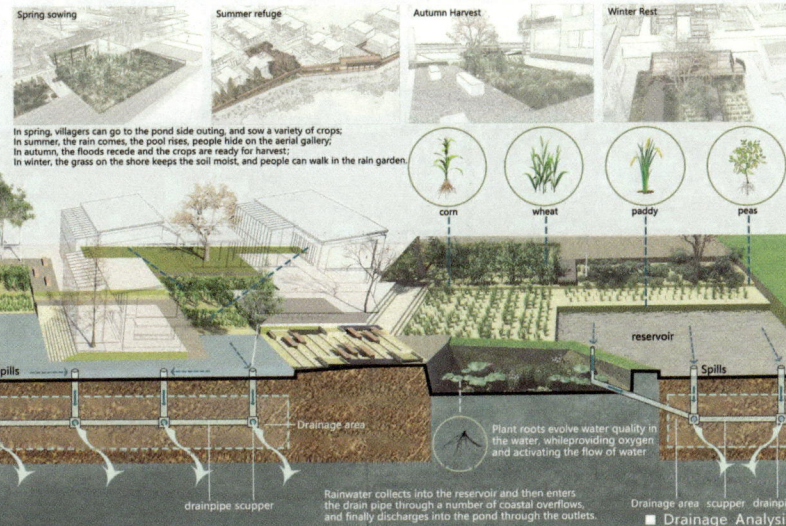

Spring sowing   Summer refuge   Autumn Harvest   Winter Rest

In spring, villagers can go to the pond side outing, and sow a variety of crops;
In summer, the rain comes, the pool rises, people hide on the aerial gallery;
In autumn, the floods recede and the crops are ready for harvest;
In winter, the grass on the shore keeps the soil moist, and people can walk in the rain garden.

corn   wheat   paddy   peas

Aquatic iris   water lily

Consolidation of dikes

Increase biodiversity

reservoir

Spills

Drainage area

drainpipe scupper

Plant roots evolve water quality in the water, while providing oxygen and activating the flow of water

reservoir

Spills

Drainage area   scupper   drainpipe

Rainwater collects into the reservoir and then enters the drain pipe through a number of coastal overflows, and finally discharges into the pond through the outlets.

■ **Drainage Analysis**

■ **First Floor Plan of Public Buildings** 1：300   ■ **Residential Housing Type Analysis Diagram** 1：200 ■ **Construct Nodes**   ■ **Explosion Analysis Diagram**

second floorn plan of A

first floorn plan of House A
HouseA 128.5㎡

second floorn plan of House A

second floorn plan of A

first floorn plan of House B
HouseB 157.0㎡

second floorn plan of House B

■ **Second Floor Plan of Public Buildings** 1：300

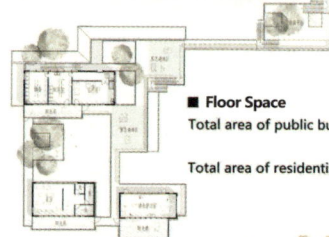

■ **Floor Space**
Total area of public buildings
1184.3㎡
Total area of residential buildings
2055.5㎡

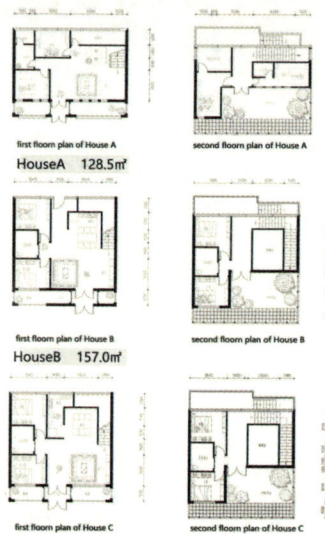

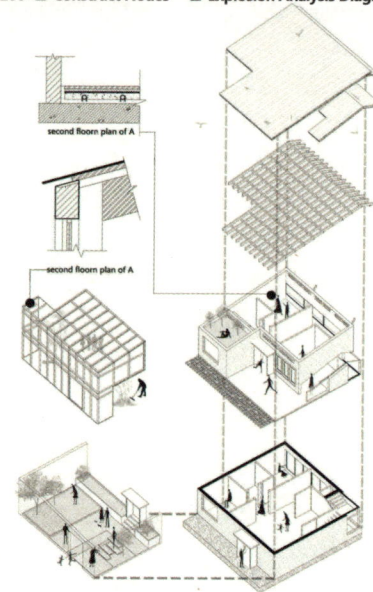

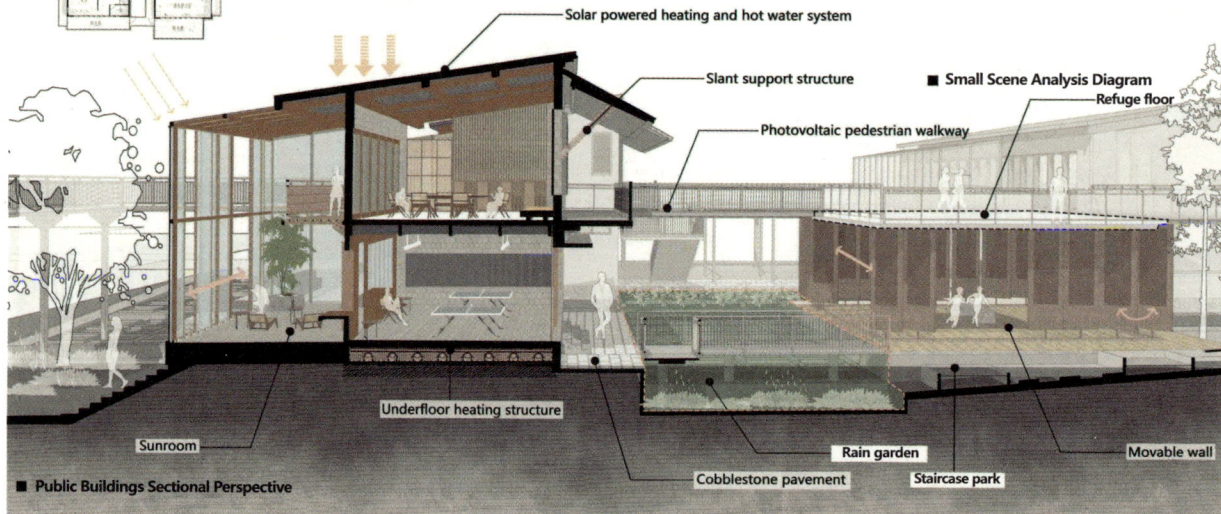

first floorn plan of House C
HouseB 157.0㎡

second floorn plan of House C

Solar powered heating and hot water system

Slant support structure

■ **Small Scene Analysis Diagram**

Refuge floor

Photovoltaic pedestrian walkway

Sunroom

Underfloor heating structure

Rain garden

Movable wall

Cobblestone pavement

Staircase park

■ **Public Buildings Sectional Perspective**

■ Residential Solar Photovoltaic System

Solar photovoltaic panels
Material: Monocrystalline silicon solar cell
Measures to improve power generation efficiency:
install an overhead layer on the roof for ventilation
and coolingSingle slope roof,
reasonable inclination angle of photovoltaic panels

System composition
Solar panels, solar controllers
Battery (group), inverter

Solar power
generation system

Solar water heating system
System composition: collector,
thermal storage tank, circulating water pump
Waterway pipelines and control systems

Material: Thin film photovoltaic glass
Combining with the sunroom to create
a good indoor thermal environment
Increase power generation

Sunshine room and
photovoltaic glass

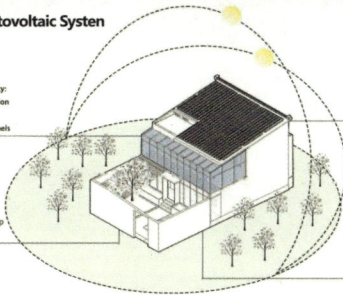

Solar photovoltaic panels

Sunshine room and photovoltaic glass

Solar power generation system

Solar water heating system

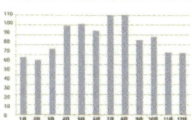
First year photovoltaic power generation map (mwh)

25 year photovoltaic power generation map (mwh)

Explanation of Photovoltaic Power Generation Capacity

The Baoding area has a high level of sunshine radiation and good solar
energy resources. The sloping roof of this building fully utilizes solar
photovoltaic panels.This building includes 720 square meters of photovoltaic
panels and 700 square meters of photovoltaic glass, with peak electricity
generation in July and August. The first year of photovoltaic power generation
is about 1050mwh. Due to the loss of photovoltaic panels, the annual power
generation will decrease by 0.5% to 1% as the number of years increases.
The total power generation in 25 years is approximately 22500mWh.

■ Ventilation and Lighting in Residential Buildings

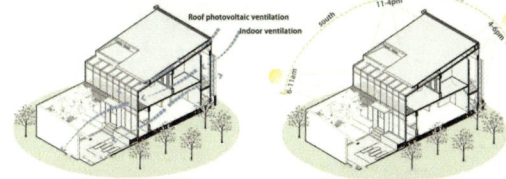
Roof photovoltaic ventilation
Indoor ventilation
Unit 1 | Unit 1

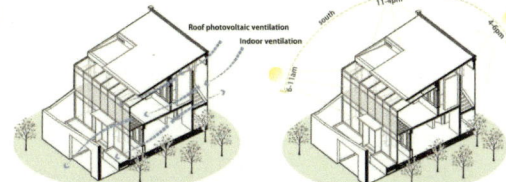
Roof photovoltaic ventilation
Indoor ventilation
Unit 2 | Unit 2

■ Light Environment Simulation

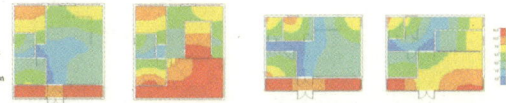

Analysis of lighting on the first floor in Unit 1
Analysis of lighting on the second floor in Unit 1
Analysis of lighting on the first floor in Unit 2
Analysis of lighting on the second floor in Unit 2

■ Carbon Sequestration Measures

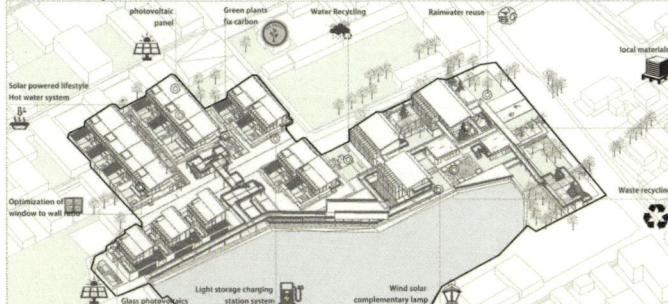

photovoltaic panel
Green plants fix carbon
Water Recycling
Rainwater reuse
local materials
Solar powered lifestyle
Hot water system
Optimization of window to wall ratio
Glass photovoltaics
Light storage charging station system
Wind solar complementary lamp
Waste recycling

■ Carbon Sink

Green plants	Annual COD fixed amount (kg·d·a)-	area (㎡)	month ct of years-	Unit building area-Carbon fixed amount(kg/㎡)-
Mixed planting area with dense flowers and plants-	30-	150-	1080-	
Grass flower garden, natural wild grass-	6-	300-	240-	
total-			50-	1320-

Full lifecycle carbon emissions

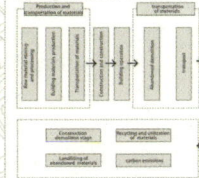

■ Carbon Emissions During Operation Phase

electric power-	subclass-	consume power- (kWh/㎡)-	carbon emission factor- (kgCO2/kWh)-	Carbon emission(t)-	Carbon per unit area-Emissions volume(kg/㎡)-
cooling-(Ex)-	module air conditioning-	3128	0.8843-	4113	2745
	Total cooling supply-	3128			2745
heating-(Eh)-	Unit heat pump-	8143	0.8843-	10747	7165
	Heating total-	8143			7165
Air conditioning (En)-	New exhaust system-	272-	0.8843-	940-	241
	Total number of fans-	272-			241
	lighting-	245-	0.8843-	940-	217
	Socket equipment-	1807-	0.8843-	6220-	1590
other-(Eo)-	Domestic hot water-	0-			0 (Deducting solar energy)
	Other-total-	113-			99
Fossil fuel-	subclass-	consume power- (kWh/㎡)-	carbon emission factor- (kgCO2/T)-	Carbon emission(t)-	Carbon per unit area-Emissions volume(kg/㎡)-
	methane-	0-		0	0
Renew-able	subclass-	consume power- (kWh/㎡)-	carbon emission factor- (kgCO2/kWh)-	Carbon emission(t)-	Carbon per unit area-Emissions volume(kg/㎡)-
renewable energy(Er)-	Solar hot water(Es)-	93	0.8843	23514	81
	photovoltaic(Ep)-	4497			13357
	wind power(Ew)-	36			32
	total-	24624			21775
Total carbon emissions from building operation				459	359

综合奖·优秀奖·张家口市
怀安县二堡子村建设项目
Comprehensive Awards –
Honorable Mention – Erpuzi
Village Project，Huai'an
County，Zhangjiakou City

注册号：110055
Register Number：110055
项目名称：韬光谷仓
Entry Title：Energy Barn in the Field
作者：李思奕、吴韵蕾、辛芃
Authors：Li Siyi，Wu Yunlei，Xin Peng
作者单位：福州大学
Authors from：Fuzhou University
指导教师：郑媛、黄斯
Tutors：Zheng Yuan，Huang Si
指导教师单位：福州大学
Tutors from：Fuzhou University

CODE: 110055

Energy Barn in the Field

韬光谷仓

"风-光-热" 多能互补韧性系统

"Wind-Light-Heat" multi-energy
complementary toughness system

01

Design Instruction

设计旨在构建一座融合安全感与舒适感两个维度的韧性建筑。平时作为村民的社交休闲活动空间，兼顾新兴农村旅游中接待站的角色，灾难发生时，能快速转变为生存保障，有效抵御外界冲击的避难所。

以"谷仓"为设计主题，构建"风-光-热"多能互补韧性系统，将自然能量转化为建筑内部流动的绿色能源，减少建筑对外部能源系统的依赖。在节能设计上，以腔体导控技术为核心，灵活组合阁楼、阳光房、双层幕墙等空间，以"表皮腔""内置腔""共生腔"多种类型腔体植入建筑内部，最大化获取自然气候资源。在应对自然灾害上，利用装配式建造与免震地基技术减少地震波对建筑的影响。

风吹稻浪，颗粒归仓。如同过去人们依靠谷仓抵御严冬的坚韧精神在现代的重生，这一系列对外界力量的有效吸纳、存储、转化，诠释了一座以柔克刚、化险为夷的建筑，一座伫立在田野上的"保障"的象征。

The design aims to create a resilient building that combines the dimensions of security and comfort. Usually as a space for social and leisure activities of villagers, taking into account the role of reception station in emerging rural tourism; When a disaster occurs, it can be quickly transformed into a survival guarantee and an effective refuge against external shocks.

Taking "barn" as the design theme, the "Wind-Light-Heat" multi-energy complementary toughness system is constructed to convert natural energy into green energy flowing inside the building and reduce the building's dependence on external energy systems. In terms of energy saving design, with cavity guidance and control technology as the core, flexible combination of attic, sun room, double curtain wall and other Spaces, with a variety of types of "skin cavity", "built-in cavity", "symbiotic cavity" implanted inside the building, maximize access to natural climate resources. In response to natural disasters, prefabricated construction and seismic-free foundation technology are used to reduce the impact of seismic waves on buildings.

The wind blows the rice waves, the particles return to the warehouse. Like the rebirth in modern times of the tenacious spirit of people who relied on barns to resist the harsh winter in the past, this series of effective absorption, storage and transformation of external forces interprets a building that is soft and tough, and a symbol of "security" standing on the field.

Concept Generating

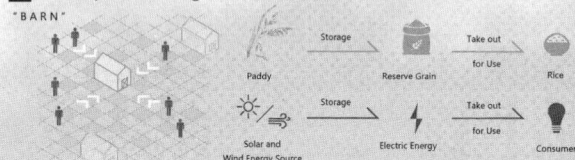

"BARN"

Surrounding enviroment elements

Erbaozi Village is located in Erbaozi Village, Huaian County, Zhangjiakou City, Hebei Province. It is a national renewable energy demonstration zone and capital water resources and ecological conservation zone in Zhangjiakou region. The region is rich in wind and solar energy resources and has a natural advantage for the development of solar buildings. The base is located in the planning area of the village meeting room, the south side is close to the road on the west side, the north side is the lake on the east side, the traffic is convenient and the vision is wide. The terrain is high in the east and low in the west, the site is gentle, and there is no need to preserve buildings and plants.

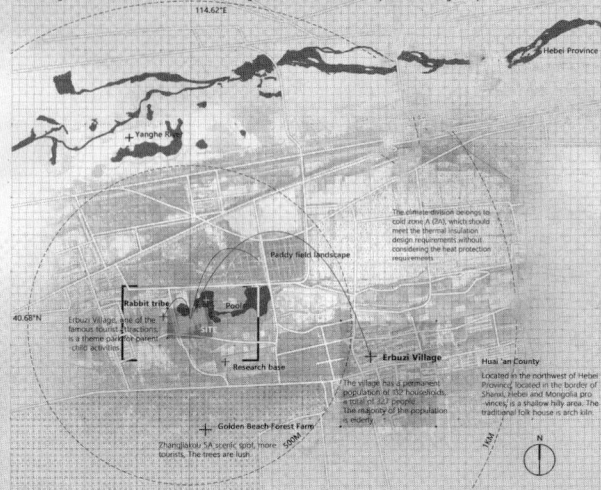

Wind-solar Thermal Energy Conversion Schematic

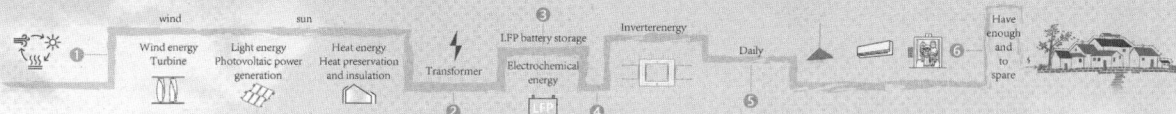

CODE: 110055

Energy Barn in the Field

韬光谷仓

"风-光-热" 多能互补韧性系统

"Wind-Light-Heat" multi-energy
complementary toughness system

02

Technical and economic index

site area	8262.80㎡	
Building area	activity center	1235.50㎡
	Homestay area	587.50㎡
Floor area	1372.00㎡	
Floor area ratio	0.22	
Greening rate	32.40%	
Motor vehicle parking space	13 (Including one accessibility)	
Non-motor vehicle parking space	18	

Site Plan 1：500

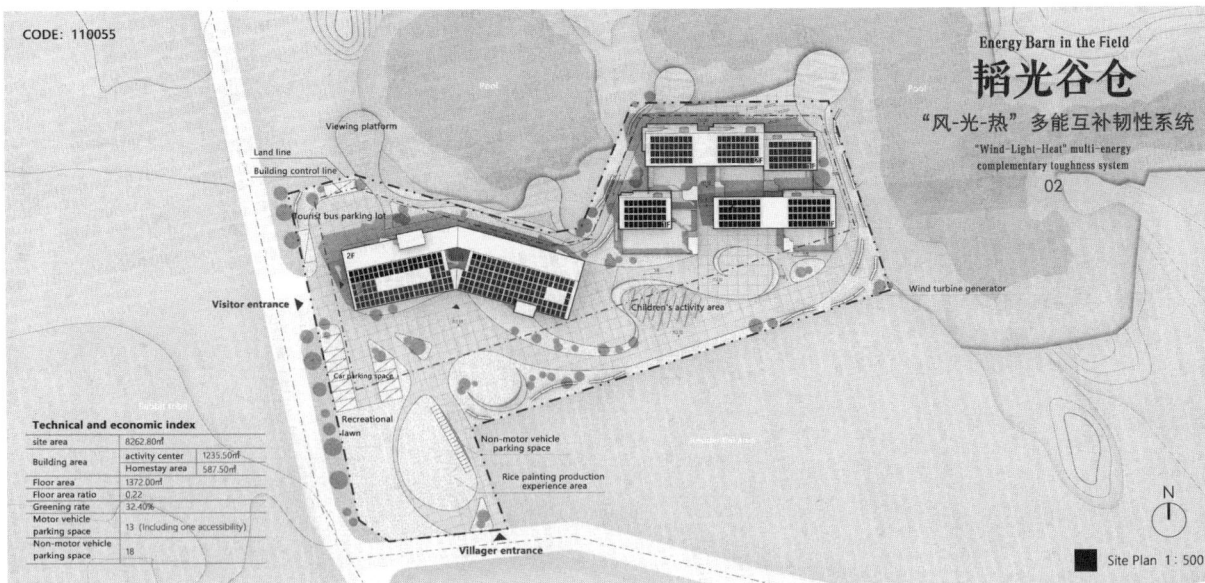

Climate Analysis (Based on Ladybug and paper data)

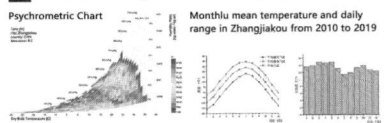

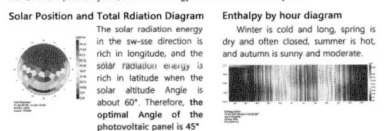

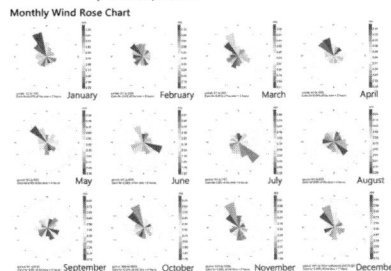

Psychrometric Chart

Monthlu mean temperature and daily range in Zhangjiakou from 2010 to 2019

Zhangjiakou is located at 113°50 '-116°30' east longitude and 39°30 '-42°10' north latitude, with an altitude of about 760m. The average annual sunshine hours in the region are 2500-2700 hours, the annual total radiation reaches 5500-6100MJ/square meter, and **the solar energy resources are rich**. At the same time, the average annual wind speed in the region is about 3.5m/s in the past ten years, and **wind energy resources are also very rich**.

Solar Position and Total Rdiation Diagram

The solar radiation energy in the sw-sse direction is rich in longitude, and the solar radiation energy is rich in latitude when the solar altitude Angle is about 60°. Therefore, **the optimal Angle of the photovoltaic panel is 45°**

Enthalpy by hour diagram

Winter is cold and long, spring is dry and often closed, summer is hot, and autumn is sunny and moderate.

Monthly Wind Rose Chart

January   February   March   April

May   June   July   August

September   October   November   December

Outdoor Wind Environment

The activity center adopts a folding block to block the northwest wind in winter and guide the southeast wind in summer. The partially dispersed layout ensures good daylight and ventilation. According to the analysis results of outdoor wind environment, wind turbine power generation devices are installed in the east, north and south sides of the site with better wind resources.

Winter   Summer   Transition season

Site Design Analysis

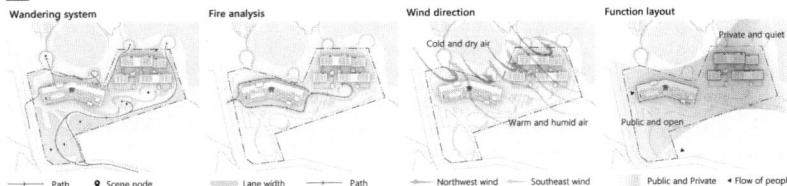

Wandering system   Fire analysis   Wind direction   Function layout

Cold and dry air

Private and quiet

Warm and humid air

Public and open

→ Path   ● Scene node   Lane width   → Path   → Northwest wind   → Southeast wind   Public and Private   → Flow of people

Graphic Technical Drawing (Activity center building) 1：300

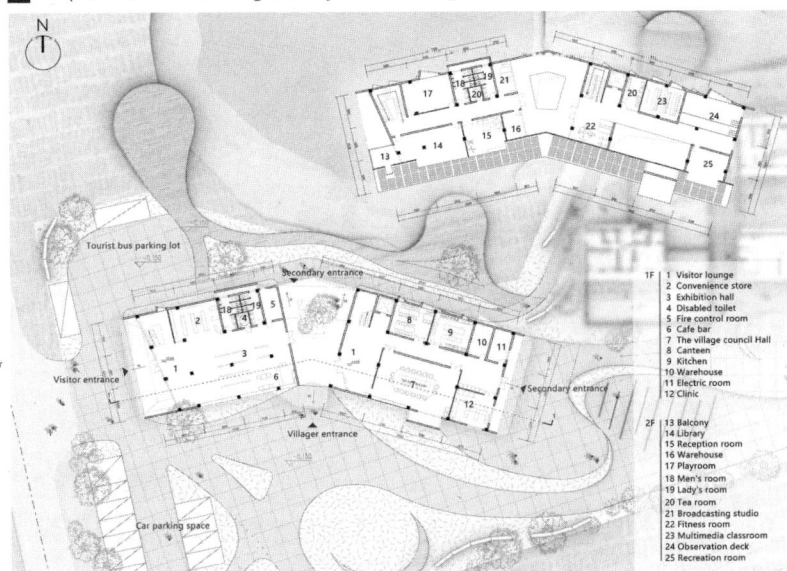

Tourist bus parking lot

Secondary entrance

Visitor entrance

Villager entrance

Car parking space

Secondary entrance

1F
1 Visitor lounge
2 Convenience store
3 Exhibition hall
4 Disabled toilet
5 Fire control room
6 Cafe bar
7 The village council Hall
8 Canteen
9 Kitchen
10 Warehouse
11 Electric room
12 Clinic

2F
13 Balcony
14 Library
15 Reception room
16 Warehouse
17 Playroom
18 Men's room
19 Lady's room
20 Tea room
21 Broadcasting studio
22 Fitness room
23 Multimedia classroom
24 Observation deck
25 Recreation room

Application of Energy-saving Technology

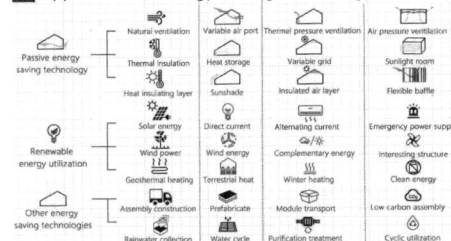

Passive energy saving technology	Natural ventilation	Variable air port	Thermal pressure ventilation	Air pressure ventilation
	Thermal insulation	Heat storage	Variable grid	Sunlight room
	Heat insulating layer	Sunshade	Insulated air layer	Flexible baffle

Renewable energy utilization	Solar energy	Direct current	Alternating current	Emergency power supply
	Wind power	Wind energy	Complementary energy	Interesting structure
	Geothermal heating	Terrestrial heat	Winter heating	Clean energy

Other energy saving technologies	Assembly construction	Prefabricate	Module transport	Low carbon assembly
	Rainwater collection	Water cycle	Purification treatment	Cyclic utilization

Cavity Implantation Schematic

The concept of architectural cavity originates from the study of biological morphology. The building cavity is a micro-structure connected with the outside under the main space of the building, which can be used as the intermediary between the interior space and the outside space, avoiding the dissipative contact between the interior space and the outdoor environment. This scheme realizes the stability of the indoor environment of the building through five cavity forms, avoiding the construction and operation at the cost of high energy consumption.

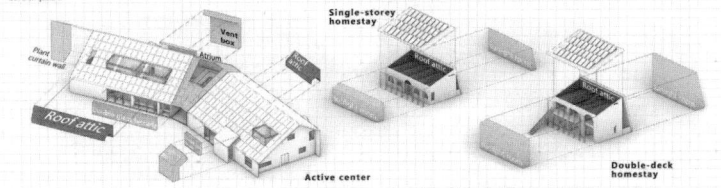

Vent box

Atrium

Single-storey homestay

Plant curtain wall

Roof attic

Active center

Double-deck homestay

CODE: 110055

Energy Barn in the Field

韬光谷仓

"风-光-热" 多能互补韧性系统

"Wind-Light-Heat" multi-energy
complementary toughness system

03

Heat preservation and insulation

Thermal insulation and ventilation

Wind turbine generator

1 The village council Hall
2 Corridor
3 Kitchen
4 Multimedia classroom

5 Guest room
6 living room
7 Canteen
8 Winter room

Earthquake-proof part
Shock isolation device
Solid-soil core packing

Profile perspective

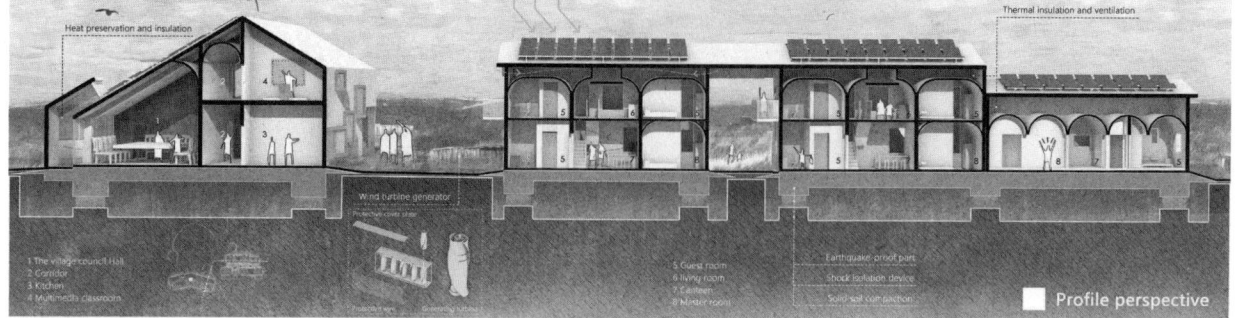

Passive Design

The cavity design of the roof is similar to that of the attic, which can play the role of heat insulation during the day and heat preservation at night.

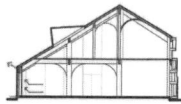

In summer, through the air port set on the box, the outdoor air is imported, and the indoor hot air is taken away by the chimney effect.

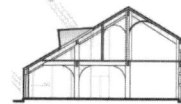

In winter, the top floor is heated by glass tiger Windows; The cavities in the south facade curtain wall are designed to store heat and release it to the interior.

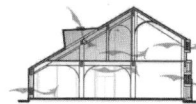

By controlling the height of the air outlet on the envelope structure, good wind conduction effect can be obtained.

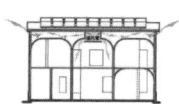

In the summer, the design of the roof attic can be combined with air vents to generate wind pressure on the upper level and expel hot air from the room.

In winter, the cavity design near the roof can play the role of absorbing heat during the day and retaining heat at night.

Graphic Technical Drawing （Residential hostel） 1：250

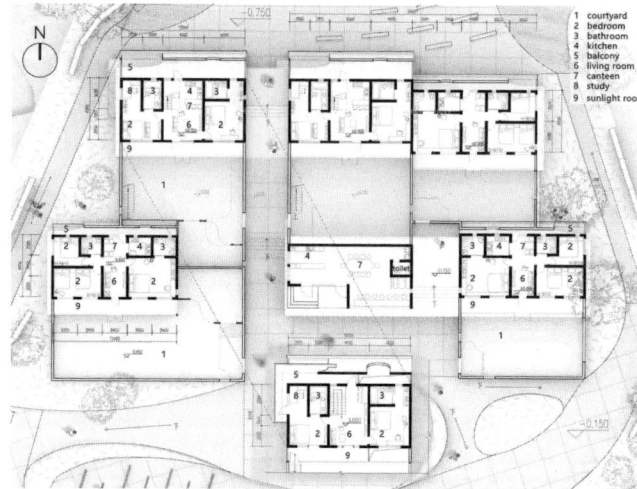

1 courtyard
2 bedroom
3 bathroom
4 kitchen
5 balcony
6 living room
7 canteen
8 study
9 sunlight room

Assembly Decomposition Scheme

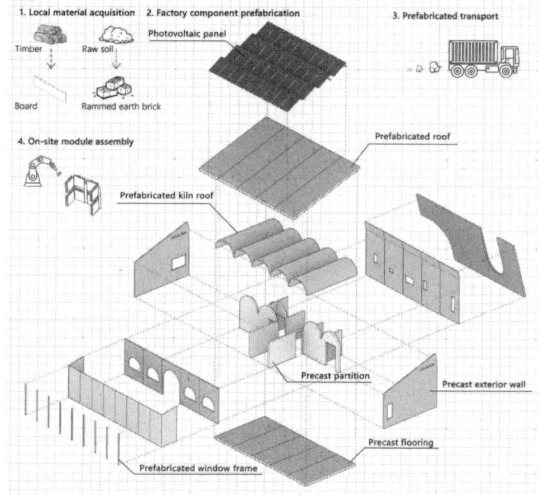

1. Local material acquisition
Timber
Board
Raw soil
Rammed earth brick

2. Factory component prefabrication
Photovoltaic panel

3. Prefabricated transport

4. On-site module assembly

Prefabricated kiln roof

Prefabricated roof

Precast partition

Precast exterior wall

Precast flooring

Prefabricated window frame

Profile Drawing 1：300

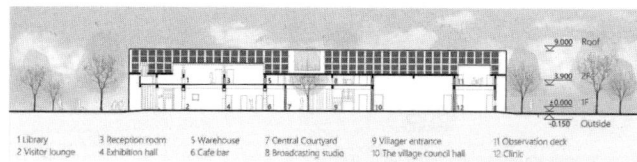

9.000 Roof
3.900 2F
±0.000 1F
-0.150 Outside

1 Library
2 Visitor lounge
3 Reception room
4 Exhibition hall
5 Warehouse
6 Cafe bar
7 Central Courtyard
8 Broadcasting studio
9 Villager entrance
10 The village council hall
11 Observation deck
12 Clinic

Constructed Specification

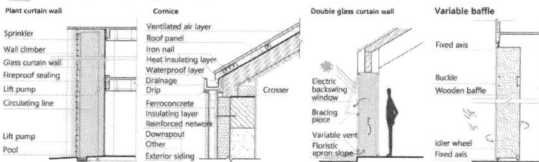

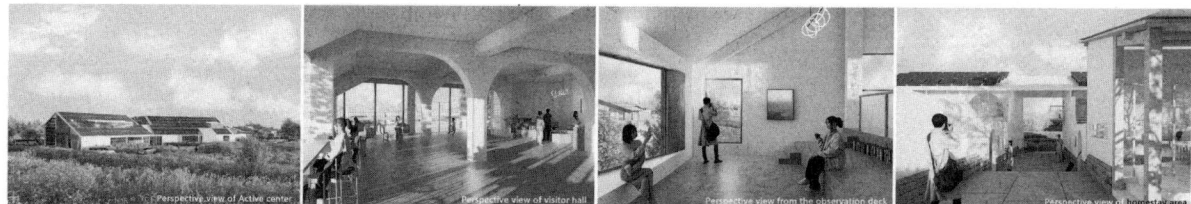

Plant curtain wall
Sprinkler
Wall climber
Glass curtain wall
Fireproof sealing
Lift pump
Circulating line
Lift pump
Pool

Cornice
Ventilated air layer
Roof panel
Iron nail
Heat insulating layer
Waterproof layer
Drainage
Drip
Ferroconcrete
Insulating layer
Reinforced network
Downspout
Other
Exterior siding

Double glass curtain wall
Electric backswing window
Bracing piece
Variable vent
Floristic apron slope

Variable baffle
Fixed axis
Buckle
Wooden baffle
Idler wheel
Fixed axis

Perspective view of Active center

Perspective view of visitor hall

Perspective view from the observation deck

Perspective view of homestay area

Energy Barn in the Field
稻光谷仓
"风-光-热" 多能互补韧性系统
"Wind-Light-Heat" multi-energy complementary toughness system
04

Park
P

External cavity
Buffer loft
Atrium space

Plant curtain wall

Photovoltaic panel

Photovoltaic panel

Cadmium telluride

Viewing platform

External cavity

Sun room

Double glass curtain wall

Straw handmade experience area
Children's activity area
Wind turbine
Ventilated attic

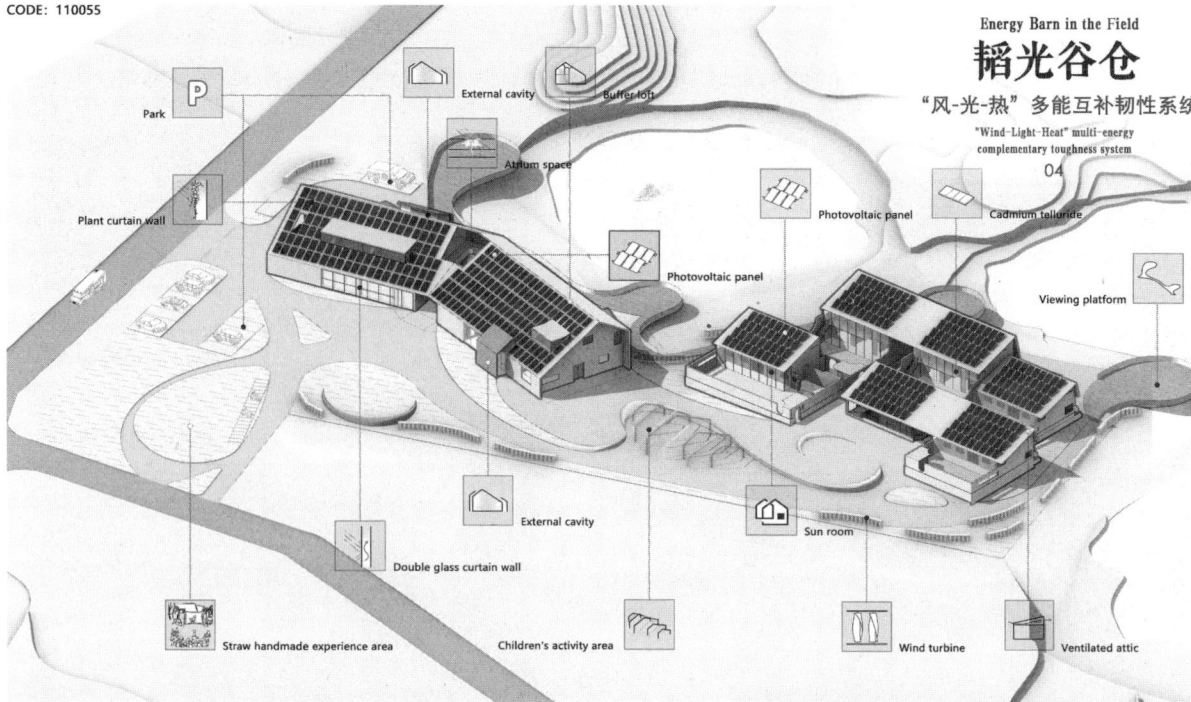

Functional Layout of Disaster Plane & Escape Path

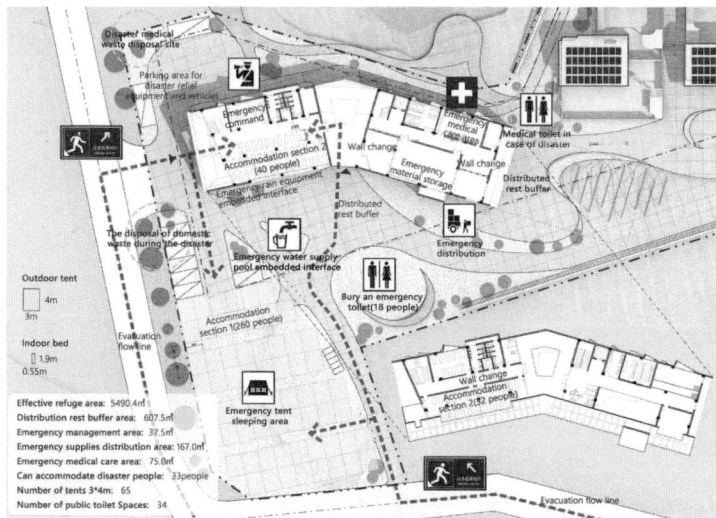

Disaster medical waste disposal site
Parking area for disaster relief equipment and vehicles
Emergency command
Emergency medical care area
Medical toilet in case of disaster
Accommodation section (40 people)
Wall change
Emergency material storage
Distributed rest buffer
Emergency rain equipment embedded interface
Distributed rest buffer
The disposal of domestic waste during the disaster
Emergency water supply pool embedded interface
Emergency distribution
Outdoor tent
4m
3m
Bury an emergency toilet(18 people)
Accommodation section 1(260 people)
Indoor bed
1.9m
0.55m
Evaluation flow line
Wall change
Accommodation section 2(32 people)
Emergency tent sleeping area
Evacuation flow line

Effective refuge area: 5490.4㎡
Distribution rest buffer area: 607.5㎡
Emergency management area: 37.5㎡
Emergency supplies distribution area: 167.0㎡
Emergency medical care area: 75.0㎡
Can accommodate disaster people: 33people
Number of tents 3*4m: 65
Number of public toilet Spaces: 34

Structural Decomposition Diagram & Material Selection

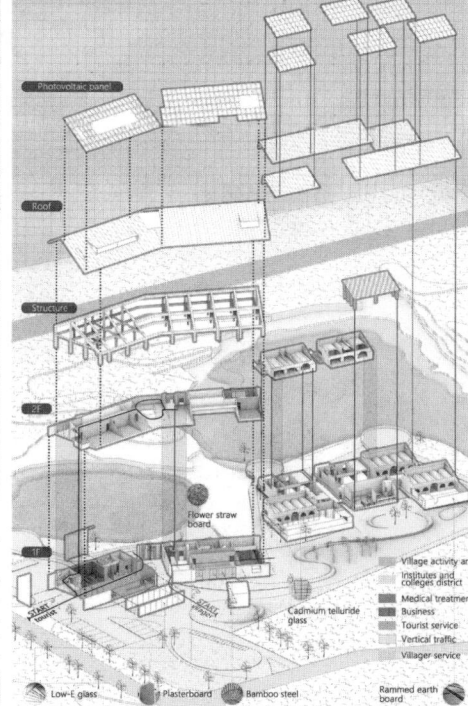

Photovoltaic panel
Roof
Structure
Flower straw board
Village activity area
Institutes and colleges district
Medical treatment
Business
Tourist service
Vertical traffic
Villager service
Cadmium telluride glass
Low-E glass
Plasterboard
Bamboo steel
Rammed earth board

Energy Consumption Calculation

Full life cycle carbon emissions	Carbon emission (tCO$_2$/m²)	Life cycle carbon emissions per unit area (tCO$_2$/m²)	Average annual lifetime carbon emissions per unit area [tCO$_2$/(m²·a)]
Production & transportation	782.7	463.7	9.3
Build	4.1	3.2	0.1
Operation	-25778.5	-39223.0	-784.5
Disassemble	2.8	2.2	0.1
Carbon sink	-1252.5	-809.0	-1252.5
Total	-26241.4	-39562.9	-2027.5

Carbon emissions during operation	Annual operating equivalent power consumption (kWh/a)	Energy consumption	Annual operating carbon emissions (kgCO$_2$/a)	Annual operating carbon emissions per unit area (kgCO$_2$/m²·a)	Carbon emission factor (kgCO$_2$/ unit dosage)
Heating	165602.3	165602.3	94443.0	70.6	0.57
Air conditioner	3346.9	3346.9	1908.8	1.4	0.57
Illumination	19872.8	19872.8	11333.5	8.5	0.57
Renewable	-1092853.5	-1092853.5	-623254.3	-465.6	0.57
Total	-904031.5	-904031.5	-515569.0	-385.1	--

| Photovoltaic module installation area (m²) | 755.3 | Comprehensive efficiency coefficient | 0.85 | Photovoltaic power generation (kWh) | 495635 |

The carbon emission of the whole life cycle of the building is negative, meeting the demand of green/zero carbon buildings; And photovoltaic capacity can meet the needs of use.

South Elevation 1：300

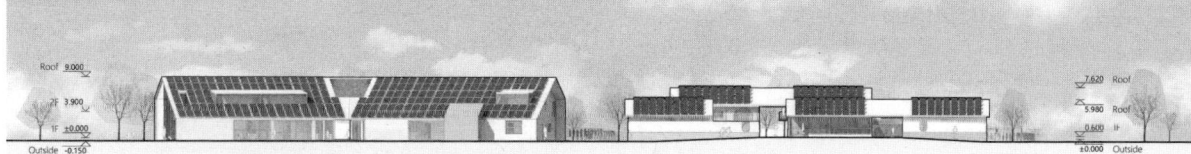

Roof 9.000
2F 3.900
1F ±0.000
Outside -0.150
7.620 Roof
5.980 Roof
0.600 1F
±0.000 Outside

综合奖 · 优秀奖 · 张家口市
怀安县二堡子村建设项目
Comprehensive Awards −
Honorable Mention − Erpuzi
Village Project, Huai'an
County, Zhangjiakou City

注册号：110071
Register Number：110071
项目名称：风廊稻居
Entry Title：Breezy Corridor & Straw
　　　　　　Houses
作者：杨睿涵、白慕瑶、李鑫鑫
Authors：Yang Ruihan, Bai Muyao, and
　　　　　Li Xinxin
作者单位：北京交通大学
Authors from：Beijing Jiaotong University
指导教师：张文
Tutor：Zhang Wen
指导教师单位：北京交通大学
Tutor from：Beijing Jiaotong University

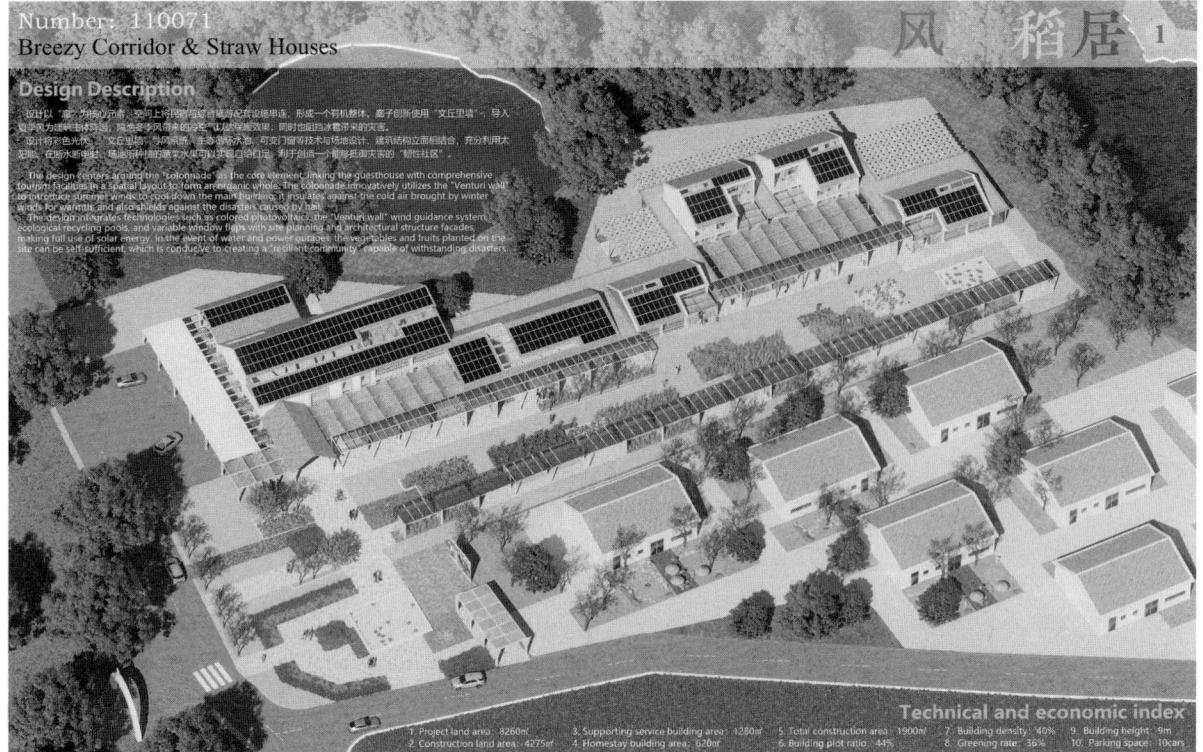

Number：110071
Breezy Corridor & Straw Houses

风 稻 居

Design Description

The design centers around the "colonnade" as the core element, linking the guesthouse with comprehensive tourism facilities in a spatial layout to form an organic whole. The colonnade innovatively utilizes the "Venturi wall" to introduce summer winds to cool down the main building. It insulates against the cold air brought by winter winds for warmth, and also shields against the disasters caused by hail.

The design integrates technologies such as colored photovoltaics, the "Venturi wall" wind guidance system, ecological recycling pools, and variable window flaps with site planning and architectural structure facades, making full use of solar energy. In the event of water and power outages, the vegetables and fruits planted on the site can be self-sufficient, which is conducive to creating a "resilient community" capable of withstanding disasters.

Technical and economic index
1. Project land area: 8260㎡　3. Supporting service building area: 1280㎡　5. Total construction area: 1900㎡　7. Building density: 40%　9. Building height: 9m
2. Construction land area: 4275㎡　4. Homestay building area: 620㎡　6. Building plot ratio: 44%　8. Greening rate: 36%　10. Parking space: 10cars

Location Analysis

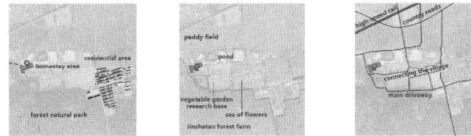

Crowd and Behavior Analysis

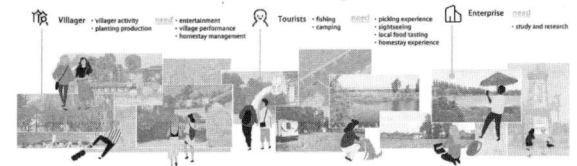

Climate Analysis & Space Generation

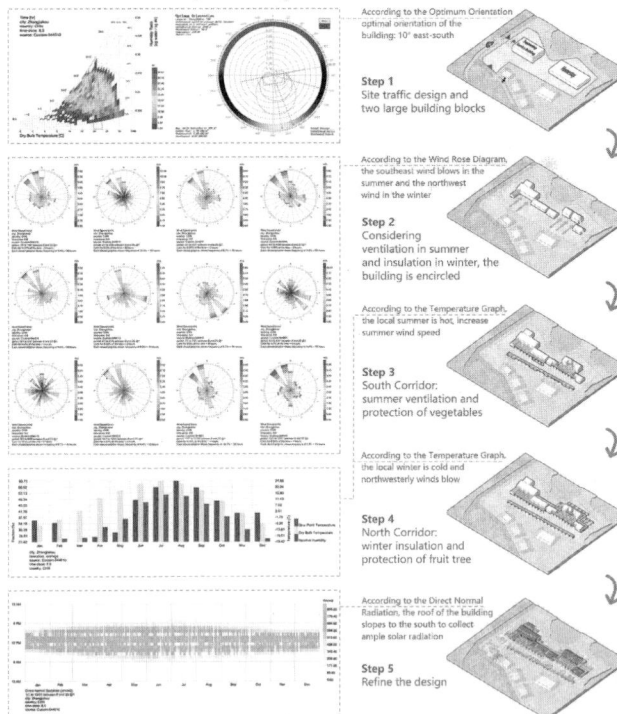

According to the Optimum Orientation, optimal orientation of the building: 10° east-south

Step 1
Site traffic design and two large building blocks

According to the Wind Rose Diagram, the southeast wind blows in the summer and the northwest wind in the winter.

Step 2
Considering ventilation in summer and insulation in winter, the building is encircled

According to the Temperature Graph, the local summer is hot, increase summer wind speed

Step 3
South Corridor: summer ventilation and protection of vegetables

According to the Temperature Graph, the local winter is cold and northwesterly winds blow

Step 4
North Corridor: winter insulation and protection of fruit tree

According to the Direct Normal Radiation, the roof of the building slopes to the south to collect ample solar radiation

Step 5
Refine the design

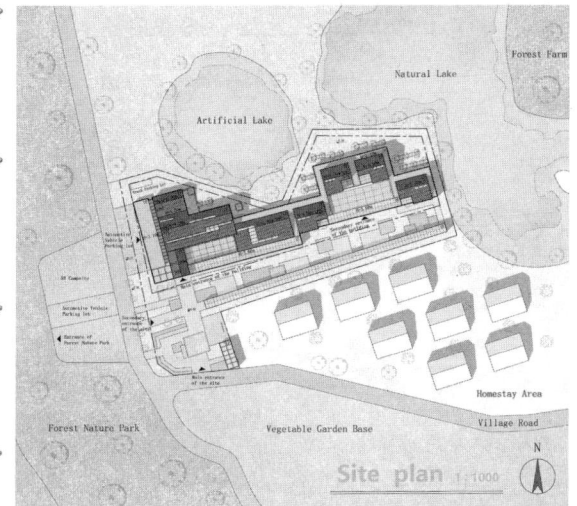

Site plan 1/1000

Winter prevailing winds

Rainwater recycling

Apricot tree picking area

Wind guiding curtain wall

Apple tree picking area

Peach tree picking area

Soybean planting area

Summer prevailing winds

Beet planting area

Wheat planting area

Potato planting area

Lake water collection system

Plant irrigation

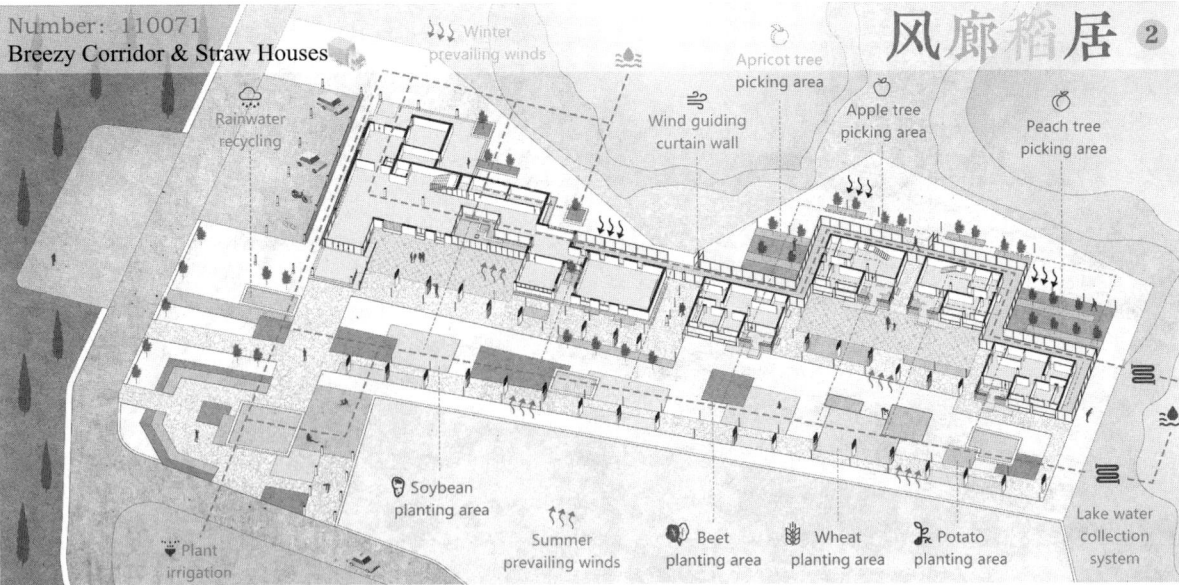

Technical System

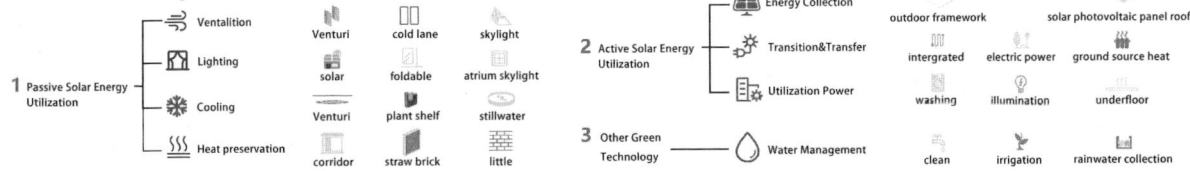

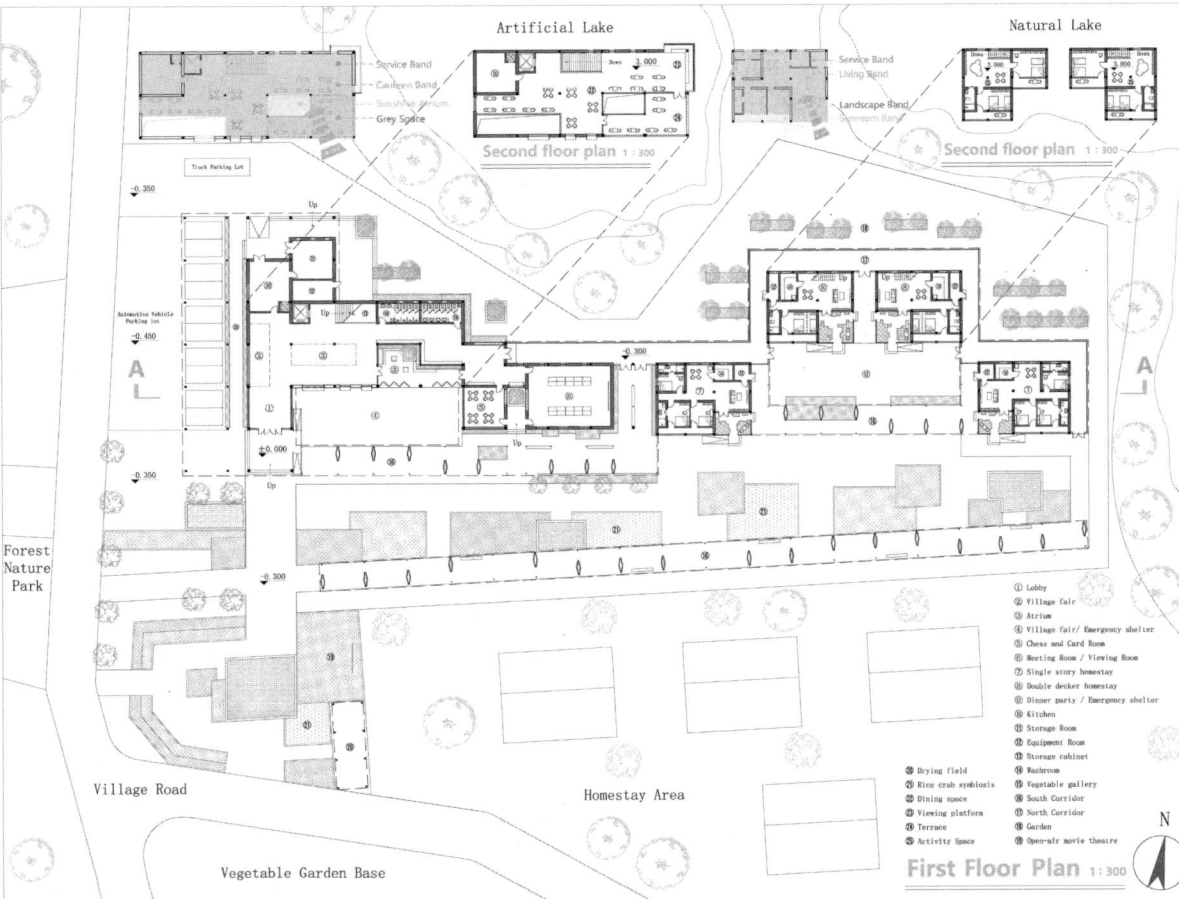

1 Passive Solar Energy Utilization
- Ventalition
- Lighting
- Cooling
- Heat preservation

Venturi
cold lane
skylight

solar
foldable
atrium skylight

Venturi
plant shelf
stillwater

corridor
straw brick
little

2 Active Solar Energy Utilization
- Energy Collection
- Transition&Transfer
- Utilization Power

3 Other Green Technology
- Water Management

outdoor framework
solar photovoltaic panel roof

intergrated
electric power
ground source heat

washing
illumination
underfloor

clean
irrigation
rainwater collection

Artificial Lake

Natural Lake

Service Band
Canteen Band
Sunshine Atrium
Grey Space

Service Band
Living Band
Landscape Band
Sunroom Band

Second floor plan 1 : 300

Second floor plan 1 : 300

Truck Parking Lot

Automotive Vehicle Parking lot

Forest Nature Park

Village Road

Homestay Area

Vegetable Garden Base

① Lobby
② Village fair
③ Atrium
④ Village fair/ Emergency shelter
⑤ Chess and Card Room
⑥ Meeting Room / Viewing Room
⑦ Single story homestay
⑧ Double decker homestay
⑨ Dinner party / Emergency shelter
⑩ Kitchen
⑪ Storage Room
⑫ Equipment Room
⑬ Storage cabinet

⑳ Drying field
㉑ Rice crab symbiosis
㉒ Dining space
㉓ Viewing platform
㉔ Terrace
㉕ Activity Space

⑭ Washroom
⑮ Vegetable gallery
⑯ South Corridor
⑰ North Corridor
⑱ Garden
⑲ Open-air movie theatre

First Floor Plan 1 : 300

N

Number: 110071
Breezy Corridor & Straw Houses

风廊稻居 ③

straw bricks   solar panels   three-dimensional planting   colored solar panels

bathroom

floor heating

pond   roof rainwater   rain garden   infiltration   irrigation

sewage collection   sewage treatment   clean water tank   storage tank

water source heat pump

Village fair & Theme Restaurant

Indoor space of homestays

Main entrance square

The north corridor in summer

Summer courtyard space

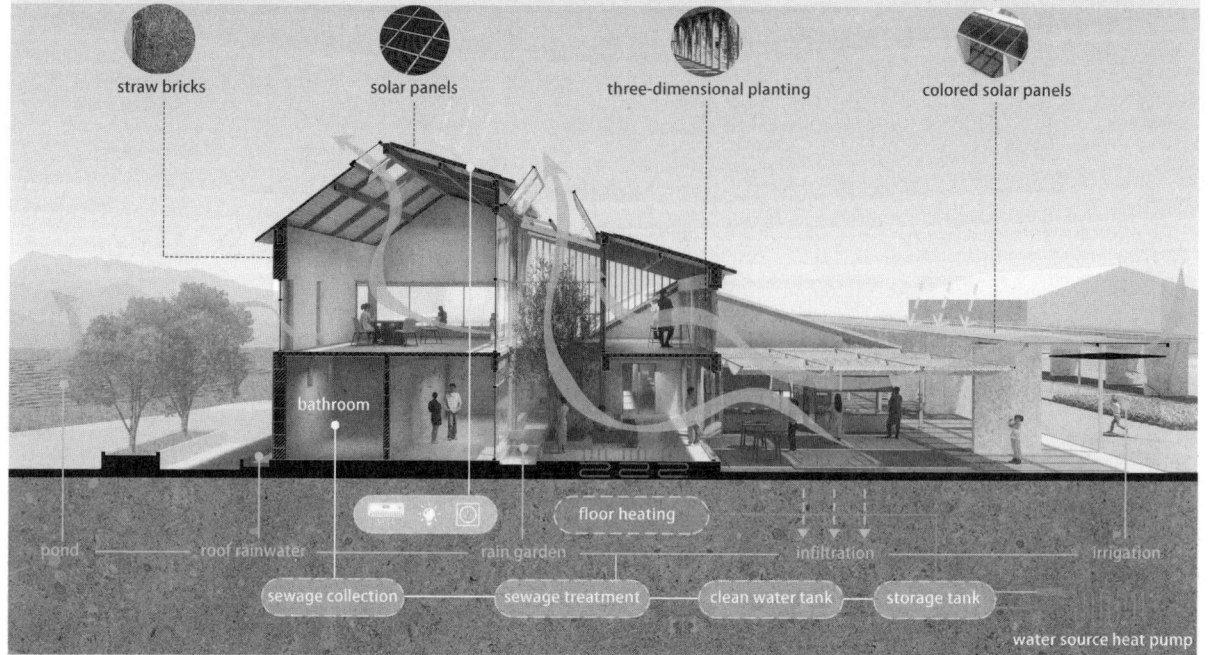

Sunlight Simulation

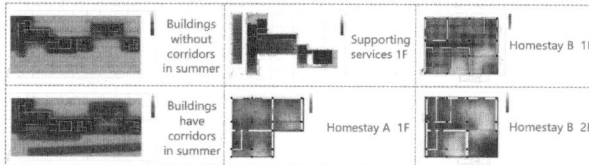

Buildings without corridors in summer

Supporting services 1F

Homestay B 1F

Buildings have corridors in summer

Homestay A 1F

Homestay B 2F

Energy Consumption and Carbon Emissions

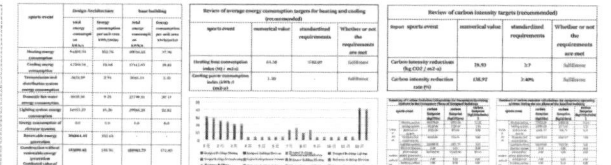

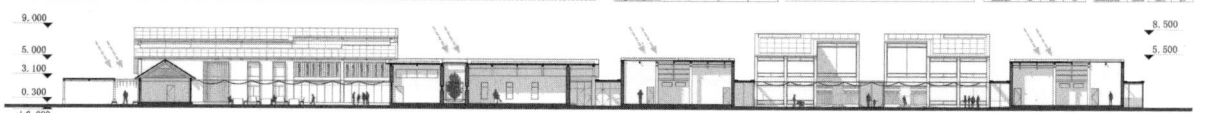

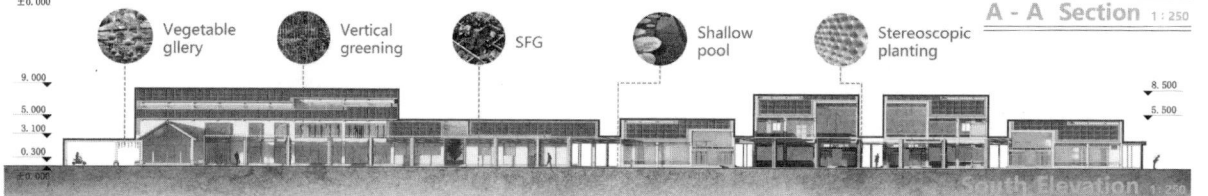

9.000
5.000
3.100
0.300
±0.000

8.500
5.500

A - A Section 1:250

Vegetable gllery   Vertical greening   SFG   Shallow pool   Stereoscopic planting

9.000
5.000
3.100
0.300
±0.000

8.500
5.500

South Elevation 1:250

Comparative Analysis of Corridors

Summer

Glass folding door open | Sun visor opening | "Venturi" board

Winter

Glass folding door closed | Sun visor closed

North Corridor | **South Corridor**

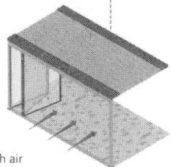 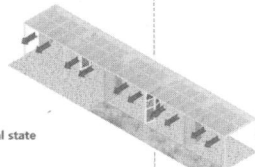

Summer
bring in fresh air

close foldable doors

Normal state

pull the "Venturi" board out

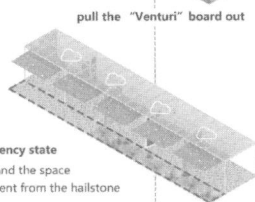

Winter
prevent from
northeast monsoon

Emergency state
1. expand the space
2. prevent from the hailstone

channel guide
top pivot
top pivot
of window
load-bearing
column

southest
monsoon

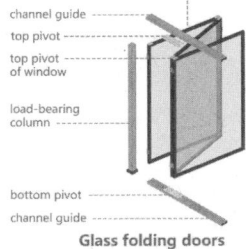 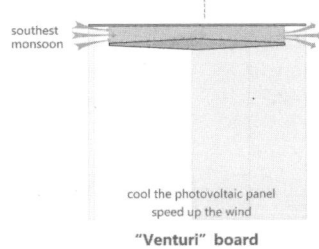

bottom pivot
channel guide

cool the photovoltaic panel
speed up the wind

Glass folding doors | **"Venturi" board**

Homestay Structure

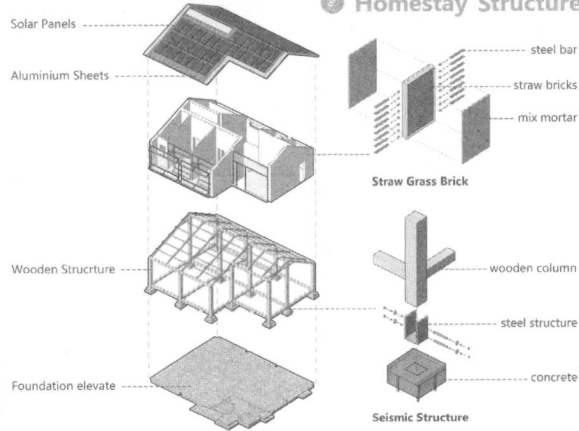

Solar Panels
Aluminium Sheets
steel bar
straw bricks
mix mortar

Straw Grass Brick

Wooden Strucrture
wooden column
steel structure
concrete

Foundation elevate

Seismic Structure

Corridor Structure

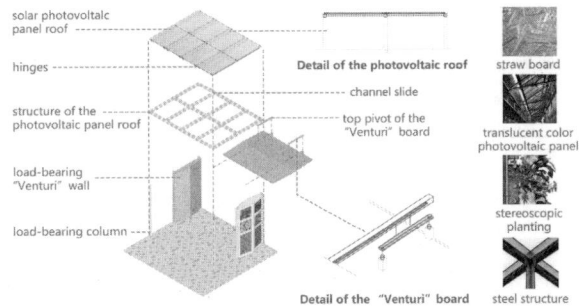

solar photovoltalc panel roof
hinges
structure of the photovoltaic panel roof
load-bearing "Venturi" wall
load-bearing column

Detail of the photovoltaic roof
channel slide
top pivot of the "Venturi" board

straw board
translucent color photovoltaic panel
stereoscopic planting
steel structure

Detail of the "Venturi" board

Courtyard

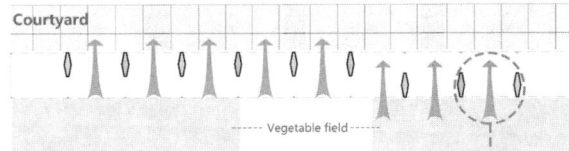

Vegetable field

"Venturi" wall
1. speed up the wind to cool the yard
2. load-bearing

Material of wall
⊡ stereoscopic planting
⊡ straw wall

Function
Formation

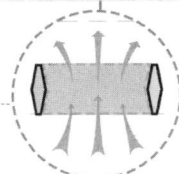

057

110077

日出下的谷仓
Barn at Sunrise

设计汲取当地稻田元素，赋予谷仓全新的光能和应急储能意义。将公共建筑沿狭长地块布置，形成与远山呼应的狭长体量，并在地块西北和东南侧布置民宿，提供水景和山景不同主题的景观，形成丰富的街区空间。

设计注重被动式手法，通过谷仓热压通风、绿化屋顶隔热等实现低碳节能，并通过坡屋顶结合当地充足的光照条件进行光伏发电等主动式技术，满足韧性设计要求。在民宿搭建方面创新性地采用模块化榫卯拼接，实现快速搭建和减震设计。

The design draws on the elements of the local rice paddies to give the barn a new meaning of light energy and emergency storage. Arranging the public buildings along the long and narrow plot to form a long and narrow volume echoing the distant mountains, and arranging the lodgings on the northwest and southeast sides of the plot to provide landscapes with different themes of water and mountain views to form a rich neighborhood space.

The design focuses on passive methods, realizing low-carbon and energy-saving through barn thermal pressure ventilation, green roof insulation, etc., and active technologies such as photovoltaic power generation through sloped roofs combined with sufficient local light conditions to meet the requirements of resilient design. The innovative use of modular mortise and tenon joints in the construction of the lodging house realizes rapid construction and vibration-damping design.

综合奖·优秀奖·张家口市怀安县二堡子村建设项目
Comprehensive Awards – Honorable Mention – Erpuzi Village Project，Huai'an County，Zhangjiakou City

注册号：110077
Register Number：110077
项目名称：日出下的谷仓
Entry Title：Barn at Sunrise
作者：蔡安琪、邢依明、王晨曦、林俊杰
Authors：Cai Anqi, Xing Yiming, Wang Chenxi, and Lin Junjie
作者单位：福州大学
Authors from：Fuzhou University
指导教师：崔育新
Tutor：Cui Yuxin
指导教师单位：福州大学
Tutor from：Fuzhou University

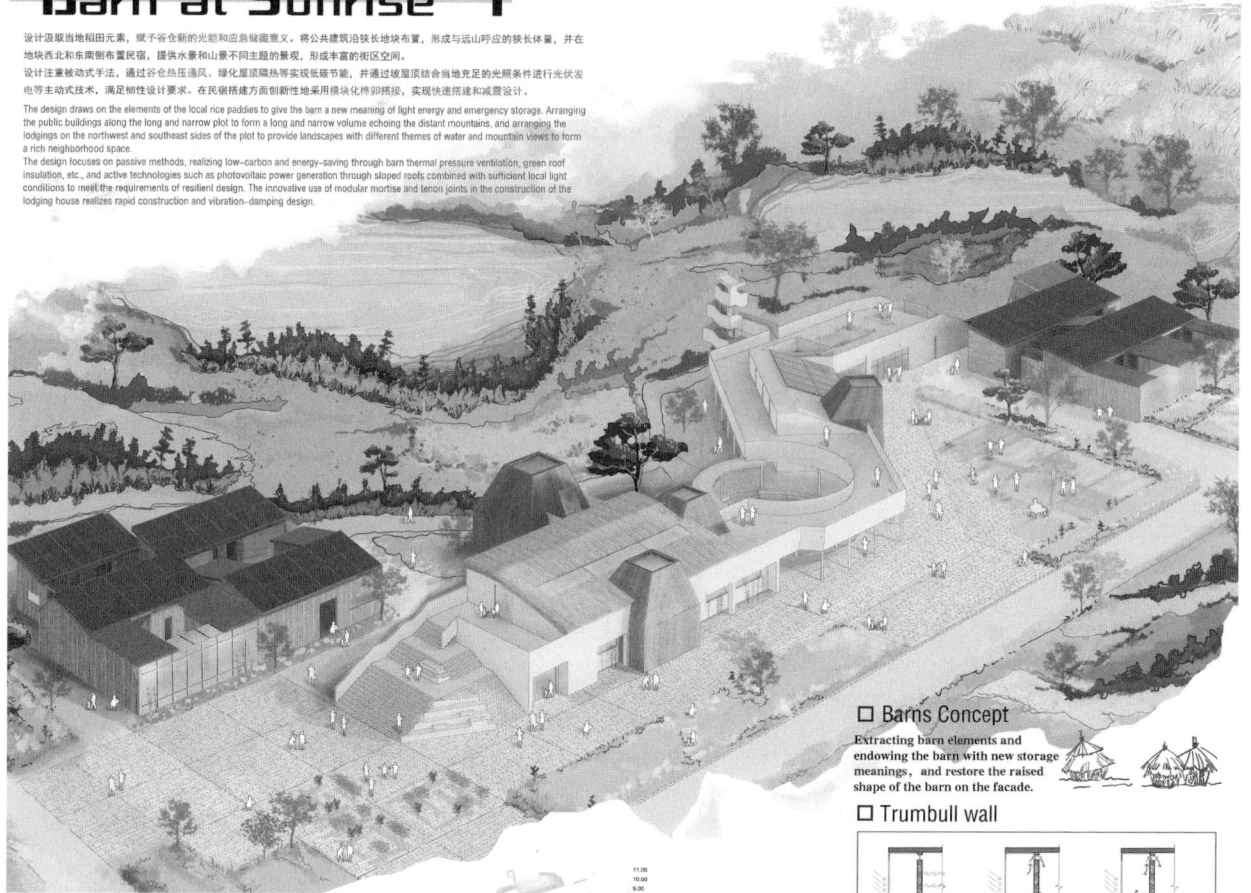

☐ Barns Concept
Extracting barn elements and endowing the barn with new storage meanings, and restore the raised shape of the barn on the facade.

☐ Trumbull wall

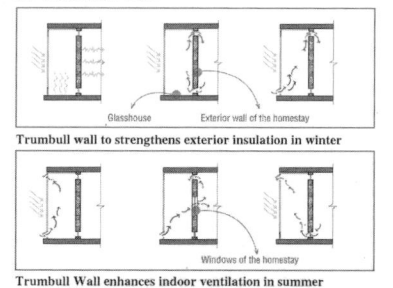

Glasshouse / Exterior wall of the homestay

Trumbull wall to strengthens exterior insulation in winter

Windows of the homestay

Trumbull Wall enhances indoor ventilation in summer

☐ Sunshine Simulation

Analysis of sunshine duration and sunshine intensity in Hebei Province

Energy-saving design using solar photovoltaic panels through insolation analysis.

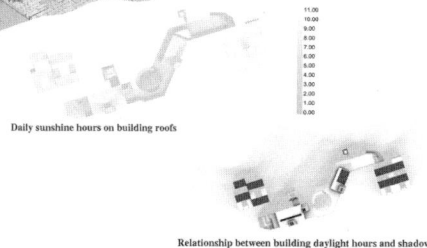

Daily sunshine hours on building roofs

Relationship between building daylight hours and shadows

☐ Wind Simulation

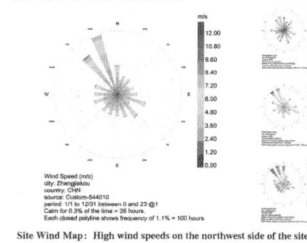

Wind Speed (m/s)
city: Zhangjiakou
country: CHN
source: Custom-544010
period 1/1 to 12/31 between 0 and 23 @1
Calm for 0.3% of the time = 26 hours.
Each closed polyline shows frequency of 1.1% = 100 hours.

Site Wind Map: High wind speeds on the northwest side of the site

☐ Ventilation System

The chimney well is designed in the form of a barn to direct air flow inside the showroom and restaurant through the difference in building heights of the different spaces; and high windows are provided to enhance air building ventilation. Simultaneously utilizing high windows to create rich light and shadow changes for the second floor attic space.

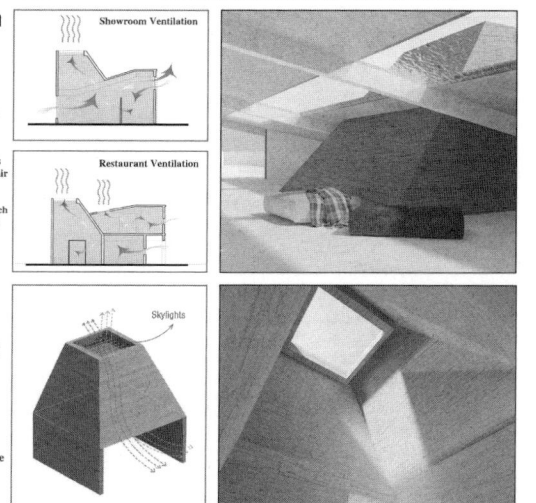

Showroom Ventilation

Restaurant Ventilation

☐ Dumidity Simulation

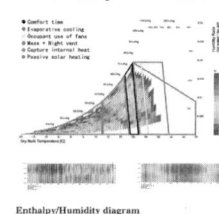

- Comfort time
- Evaporative cooling
- Occupant use of fans
- Mass + Night vent
- Capture internal heat
- Passive solar heating

Enthalpy/Humidity diagram

☐ Energy Consumption Simulation

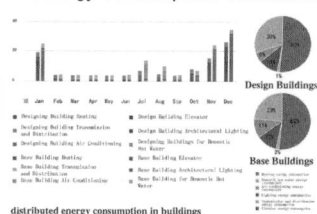

Design Buildings

Base Buildings

distributed energy consumption in buildings

☐ Thermal Pressure Ventilation

Skylights

Through the shape of the barn, the height difference between different spaces is utilized to create natural ventilation under thermal pressure and enhancing ventilation in public buildings.

110077

日出下的谷仓
Barn at Sunrise 2

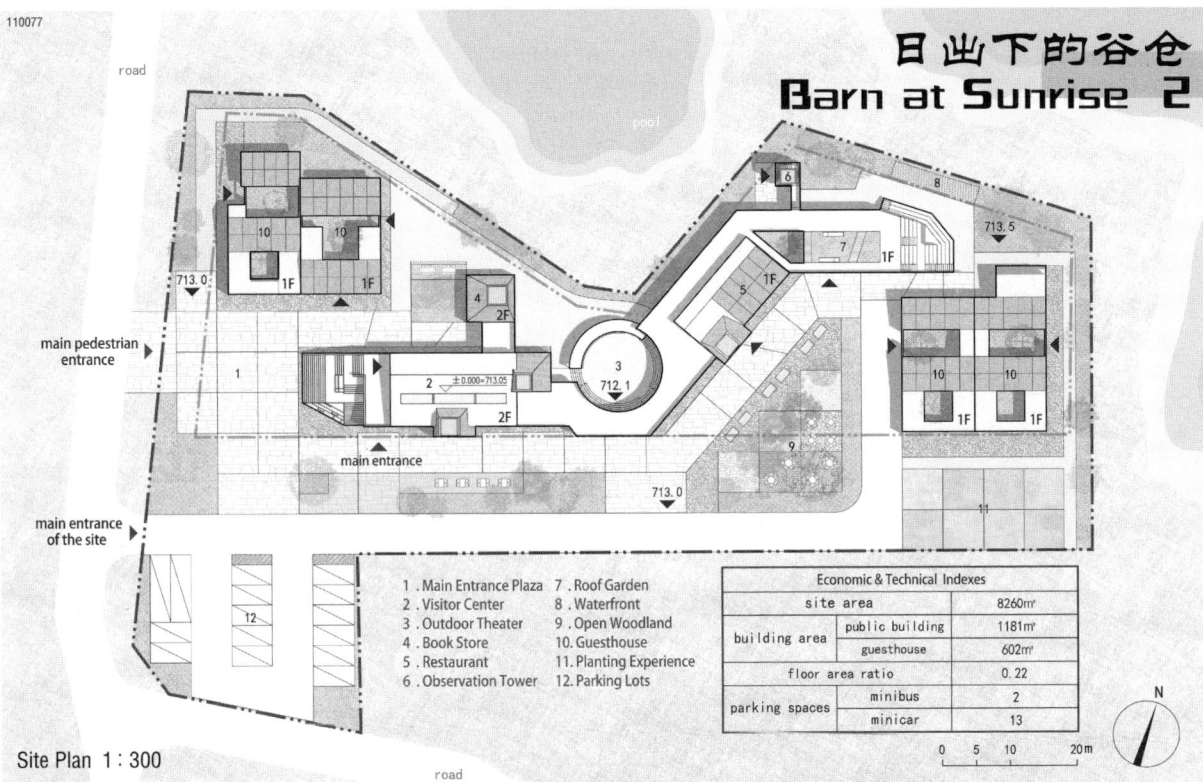

road

pool

10 10

713.0 1F 1F

4 2F

±0.000=713.05 2F

main pedestrian entrance

main entrance

main entrance of the site

3 712.1

6

7 1F

5 1F 1F

713.5

8

10 10

1F 1F

9

11

12

713.0

1. Main Entrance Plaza	7. Roof Garden
2. Visitor Center	8. Waterfront
3. Outdoor Theater	9. Open Woodland
4. Book Store	10. Guesthouse
5. Restaurant	11. Planting Experience
6. Observation Tower	12. Parking Lots

Economic & Technical Indexes		
site area		8260㎡
building area	public building	1181㎡
	guesthouse	602㎡
floor area ratio		0.22
parking spaces	minibus	2
	minicar	13

N

0 5 10 20m

Site Plan 1:300

road

□ Scenario Envisioning

Open-air Amphitheater 露天剧场，舞台形态转换
Design of a circular amphitheater, combining an integral outdoor promenade with theater seating to create a changeable theater space where people can hold different performances, including traditional soft rice-planting songs and modern music festivals.

Bazaars and Boulevards 特色商铺，街巷互动
Design the boulevard plaza on the south side of the theater as the front guiding space of the theater, set up a unified bazaar on the plaza, and form two sides of the street with the restaurants and other public buildings on the north side, so that tourists can freely roam around in the street or take a rest on the boulevard plaza.

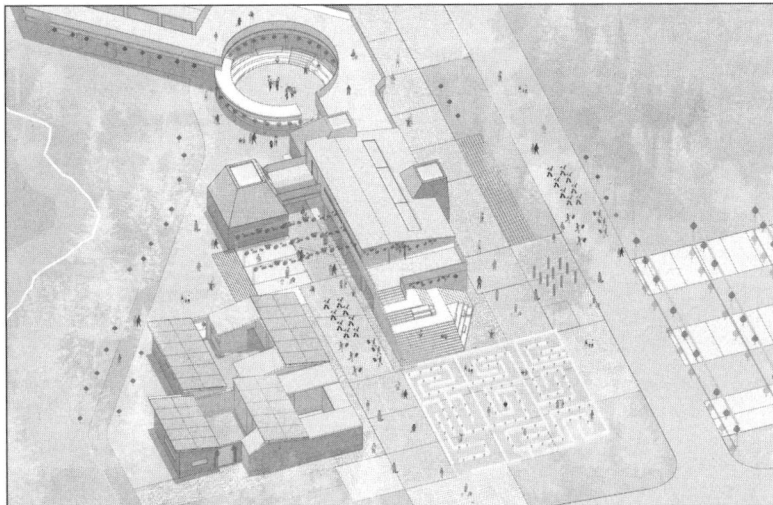

Spring Festival Gala 春节庆典，九曲黄河灯，特色民俗
Using the two squares and theaters on the site, a group Chinese New Year Gala event is held, where tourists and local residents join together in the squares to hold local traditional events such as the Jiuqu Yellow River Lanterns, soft rice-planting songs, and street parades.

Farmland Experience Zone 种植体验
Eight different farmland experience areas are set up on the south side of the B&B, where visitors can experience real vegetable planting and picking activities, and the picked vegetables can be brought back to the homestay to make local traditional food with the host family.

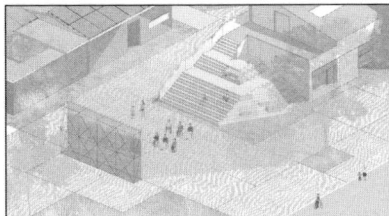

Open Air Cinema 露天影院
Design large steps on the visitor center, which can be used as a common rest area for visitors to have leisure exchanges during weekdays, and temporarily build a movie curtain in the front square at night, so that tourists and local residents can watch open-air movies and carry out exchanges between them.

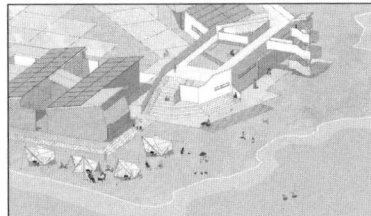

Trails and Watchtowers 瞭望塔，滨水景观
Utilizing the good landscape conditions of the north waterfront and rice paddies, designing recreational trails, recreational spaces and watchtowers, visitors can either play along the north trails throughout the site, or climb the watchtowers to view the distant rice paddies.

059

110077

日出下的谷仓
Barn at Sunrise 3

1. Visitor Center
2. Specialties Showroom
3. Outdoor Theater
4. Book Store
5. Restaurant
6. Observation Tower
7. Medical Clinic
8. Emergency Stores
9. Gueststay1
10. Gueststay2
11. Gueststay3
12. Gueststay4
13. Bazaars
14. Big Steps
15. Resting Area

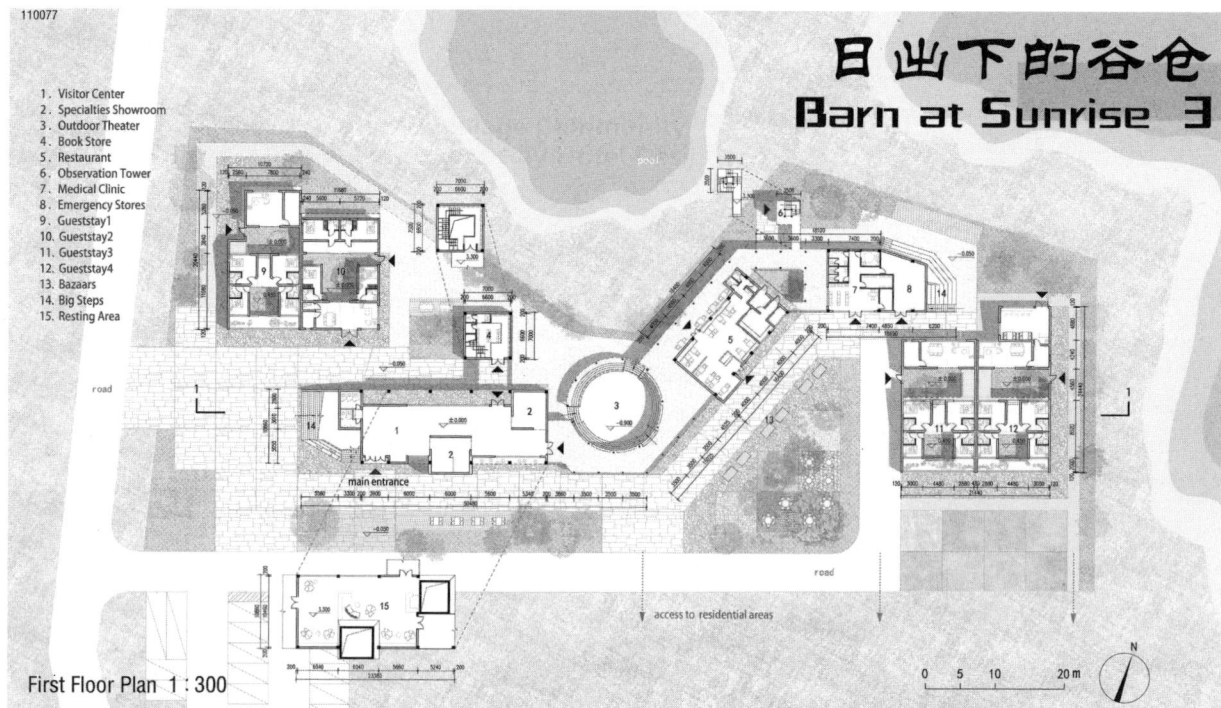

main entrance

access to residential areas

First Floor Plan 1 : 300

0 5 10 20 m

□ Emergency Site

Distribution of sites for emergency use in case of disaster

Use the exhibition hall in the Visitor Center as an indoor temporary disaster shelter, and temporarily set up outdoor shelters such as tents in the outdoor plaza.

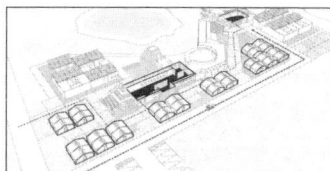

Simulation of emergency use scenarios in case of disaster

Simulation of an outdoor temporary tenting scenario and a simulation of an ambulance route through the site driveway into the medical area.

□ Technology Distribution

Solar panels photovoltaic power generation

Placement of solar photovoltaic panels on the sloped roofs of lodgings, utilizing photovoltaic power generation to provide green electricity for lodgings and public buildings.

Barn Thermal Pressure Ventilation

Chimney wells in the form of barns were designed in the showroom and dining room sections to provide additional light and heat pressure ventilation to the space.

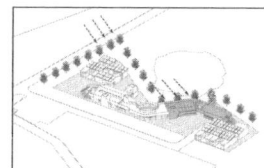

Protected from the cold northwest winds

The south side lodges use the north side public buildings to shield the northwestern side from the cold winds, and the north side lodges use mature trees.

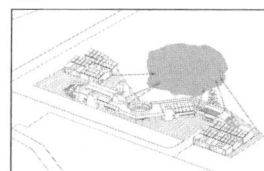

Rainwater harvesting and drainage systems

A roof drainage system is designed on the sloped roof to utilize rainwater harvesting to divert water to the green roof and the pond on the north side.

Trumbull Wall and Promenade

At the B&B utilized the room walls and the south glass room as a special Trumbull wall to enhance insulation and ventilation. The public buildings are connected by a promenade.

Site greening and roof greening

Design patches of green space on the north side of the site and green roofs on the medical roofs to enhance summer heat evaporation from the site.

Section1-1 1 : 300

North Elevation 1 : 300

South Elevation 1 : 300

日出下的谷仓
Barn at Sunrise 4

□ Construction Details

Photovoltaic Panel
Waterproofing Membrane
10cm Wooden Roofs
Cross-beam
Gutter
Modular Structural Walls
Modularized Structural
Damping Structures
Wall-to-foundation
Connectors
Shock-absorbing
Foundations

Photovoltaic Panel

Gutter

Glasshouse

□ Construction Details

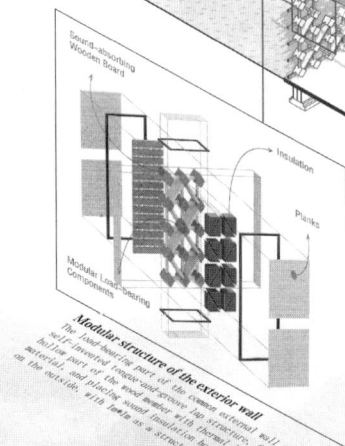

Sound-absorbing
Wooden Board

Insulation

Planks

Modular Load-bearing
Components

Modular structure of the exterior wall

The load-bearing part of the common external wall adopts self-inverted tongue-and-groove lap structure, filling the hollow part of the wood material with thermal insulation material, and placing sound insulation boards and planks on the outside, with lath as a structural module.

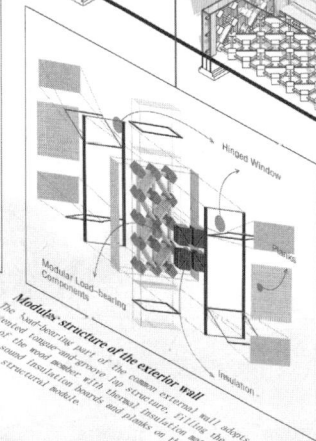

Hinged Window

Planks

Modular Load-bearing
Components

Insulation

Module's structure of the exterior wall

The load-bearing part of the common external wall adopts self-inverted tongue-and-groove lap structure, filling the hollow part of the wood material with thermal insulation material, and placing sound insulation boards and planks on the outside, with lath as a structural module.

□ Modular Lap

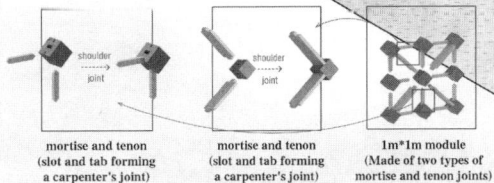

shoulder
joint

mortise and tenon
(slot and tab forming
a carpenter's joint)

shoulder
joint

mortise and tenon
(slot and tab forming
a carpenter's joint)

1m*1m module
(Made of two types of
mortise and tenon joints)

光谷·动脉 （1）
Optical Valley - Venous

综合奖·优秀奖·张家口市
怀安县二堡子村建设项目
Comprehensive Awards −
Honorable Mention − Erpuzi
Village Project，Huai'an
County，Zhangjiakou City

注册号：110182
Register Number：110182

项目名称：光谷·动脉
Entry Title：Optical Valley − Venous

作者：吴凯翔、徐云凡、王德旌、
　　　梅嘉懿、李晗
Authors：Wu Kaixiang, Xu Yunfan,
　　　Wang Dejing, Mei Jiayi,
　　　and Li Han

作者单位：苏州科技大学
Authors from：Suzhou University of
　　　Science and Technology

指导教师：刘长春、金雨蒙
Tutors：Liu Changchun, Jin Yumeng

指导教师单位：苏州科技大学
Tutors from：Suzhou University of Science
　　　and Technology

□ Design statement

Location Analysis

Contextual Analysis

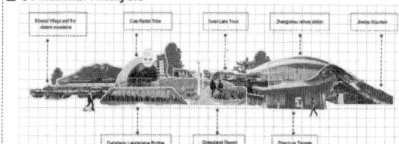

Climate Analysis

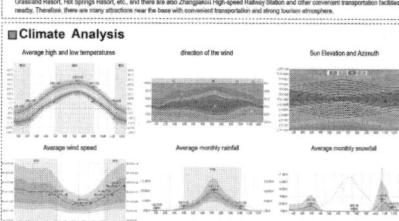

Average high and low temperatures　　direction of the wind　　Sun Elevation and Azimuth

Average wind speed　　Average monthly rainfall　　Average monthly snowfall

Base Analysis

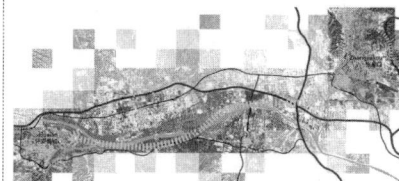

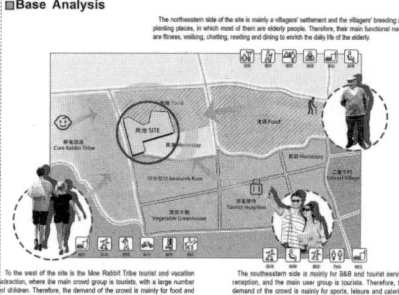

Site Plan 1 : 800

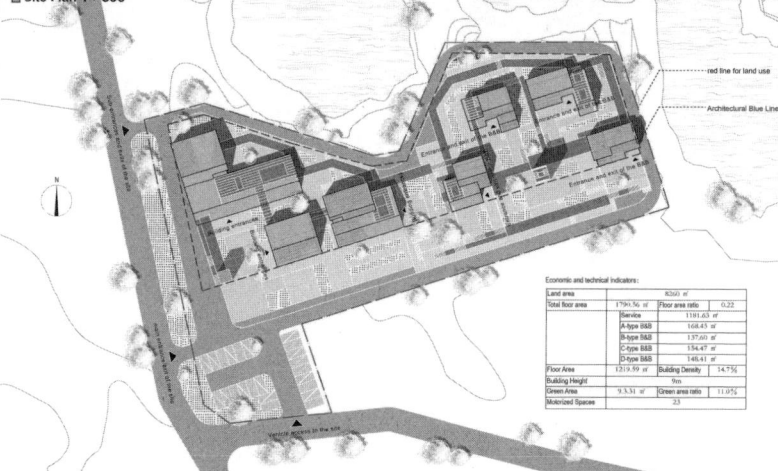

red line for land use

Architectural Blue Line

Concept Generation

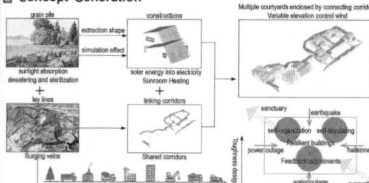

Physical Evolution

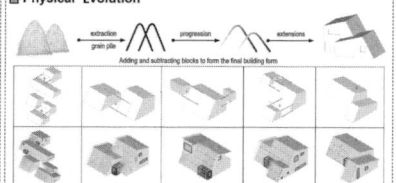

Program Derivation

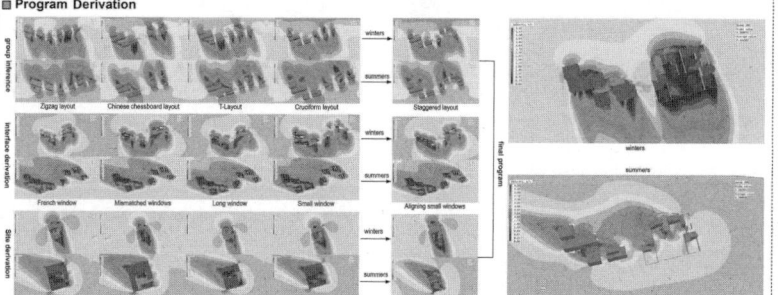

光谷·动脉 (2)
Optical Valley - Venous

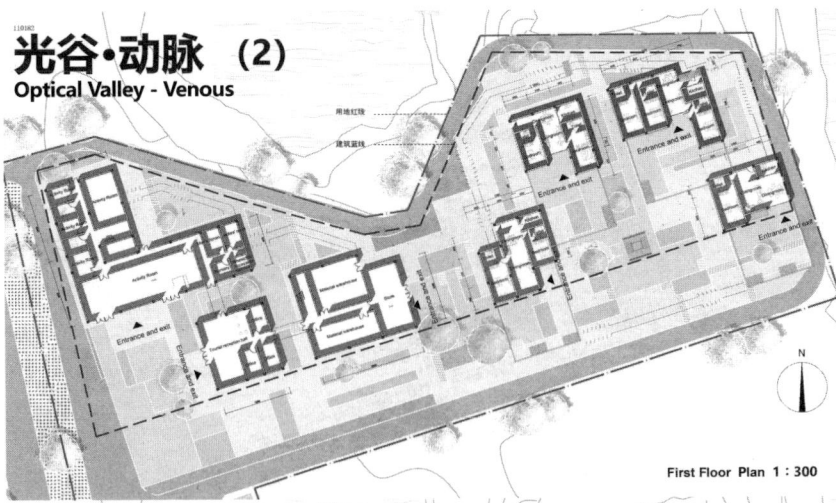

用地红线
建筑蓝线

Entrance and exit

N

First Floor Plan 1:300

Second Floor Plan 1:300

◾ Design Process Diagram

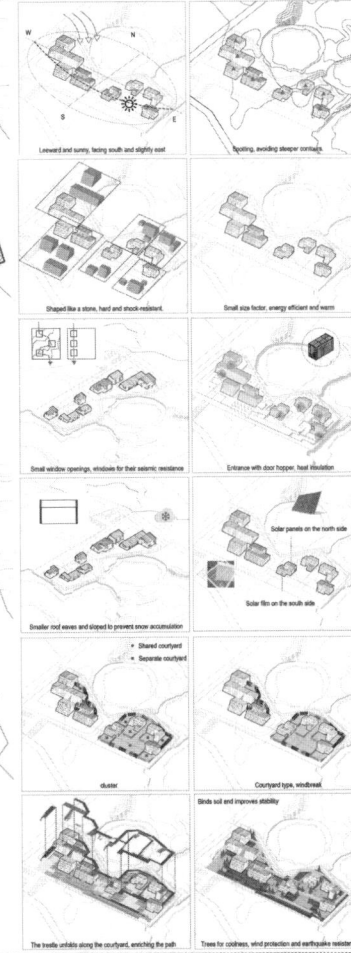

Leeward and sunny, facing south and slightly east

Sloping, avoiding steeper contours

Shaped like a stone, hard and shock-resistant.

Small size factor, energy efficient and warm

Small window openings, windows for their seismic resistance

Entrance with door hopper, heat insulation

Solar panels on the north side

Smaller roof eaves and sloped to prevent snow accumulation

Solar film on the south side

• Shared courtyard
• Separate courtyard

cluster

Courtyard type, windbreak

Binds soil and improves stability

The trestle unfolds along the courtyard, enriching the path

Trees for coolness, wind protection and earthquake resistance

◾ Toughness Design

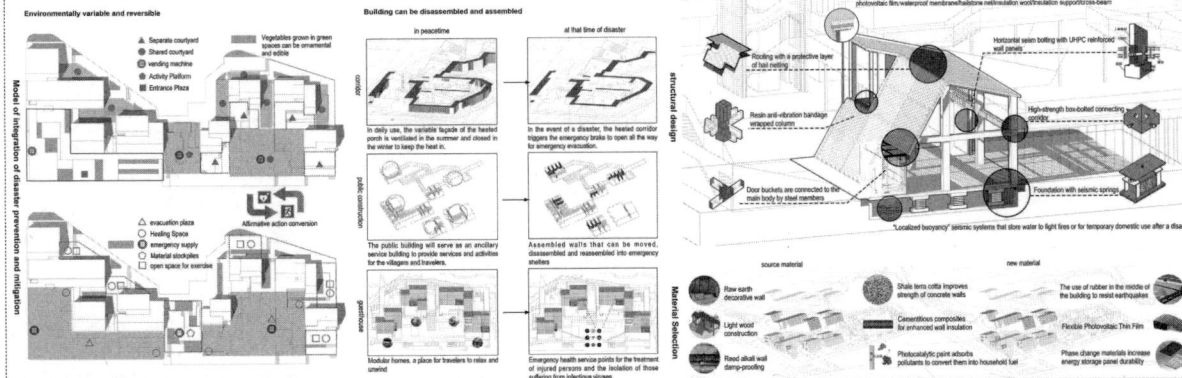

Environmentally variable and reversible

▲ Separate courtyard
● Shared courtyard
◉ vending machine
■ Activity Platform
■ Entrance Plaza

Vegetables grown in green spaces can be ornamental and edible

Model of integration of disaster prevention and mitigation

△ evacuation plaza
○ Healing Space
● emergency supply
□ Material stockpiles
□ open space for exercise

Alternative action conversion

Building can be disassembled and assembled

in peacetime

at that time of disaster

In daily use, the variable facade of the heated porch is ventilated in the summer and closed in the winter to keep the heat in.

In the event of a disaster, the heated corridor triggers the emergency brake to open all the way for emergency evacuation.

The public building will serve as an ancillary service building to provide services and activities for the villagers and travelers.

Assembled walls that can be moved, disassembled and reassembled into emergency shelters

Modular homes, a place for travelers to relax and unwind

Emergency health service points for the treatment of injured persons and the isolation of those suffering from infectious viruses

structural design

Roofing with a protective layer of hail netting

Resin anti-vibration bandage wrapped column

Door buckets are connected to the main body by steel members

Horizontal seam bolting with UHPC reinforced wall panels

High-strength box-bolted connecting corner

Foundation with seismic springs

"Localized buoyancy" seismic systems that store water to fight fires or for temporary domestic use after a disaster

Material Selection

source material

Raw earth decorative wall

Light wood construction

Photocatalytic paint absorbs pollutants to convert them into household fuel

Reed alkali wall damp-proofing

new material

Shale terra cotta improves strength of concrete walls

Cementitious composites for enhanced wall insulation

The use of rubber in the middle of the building to resist earthquakes

Flexible Photovoltaic Thin Film

Phase change materials increase energy storage panel durability

◾ South Elevation 1:250

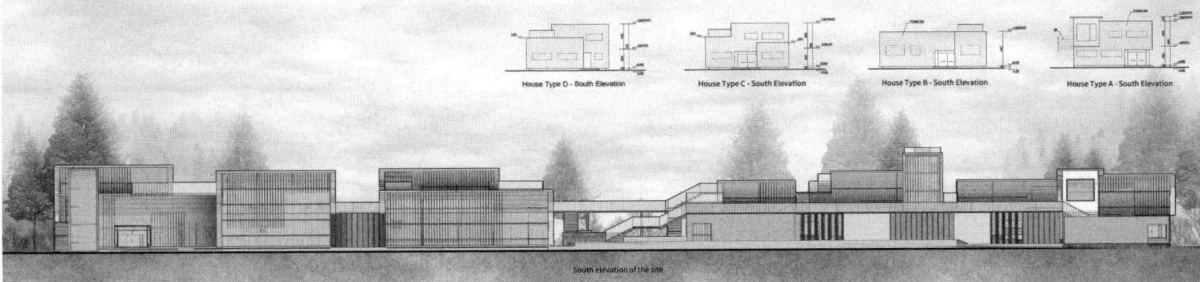

House Type D - South Elevation

House Type C - South Elevation

House Type B - South Elevation

House Type A - South Elevation

South elevation of the site

光谷·动脉 （3）
Optical Valley - Venous

Exploded Diagram

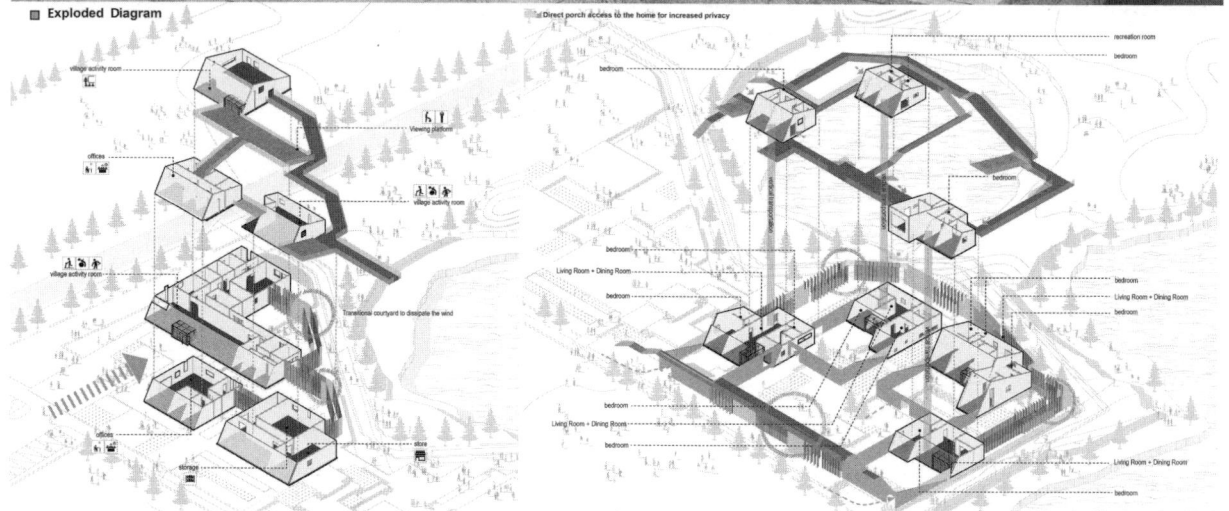

Direct porch access to the home for increased privacy

Passive Green Integration

Transitional Space Design

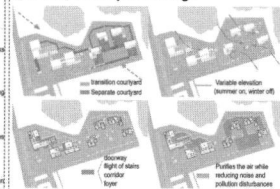

Disaster Prevention Planting

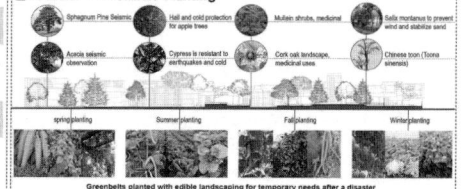

Greenbelts planted with edible landscaping for temporary needs after a disaster

Section View 1 : 60

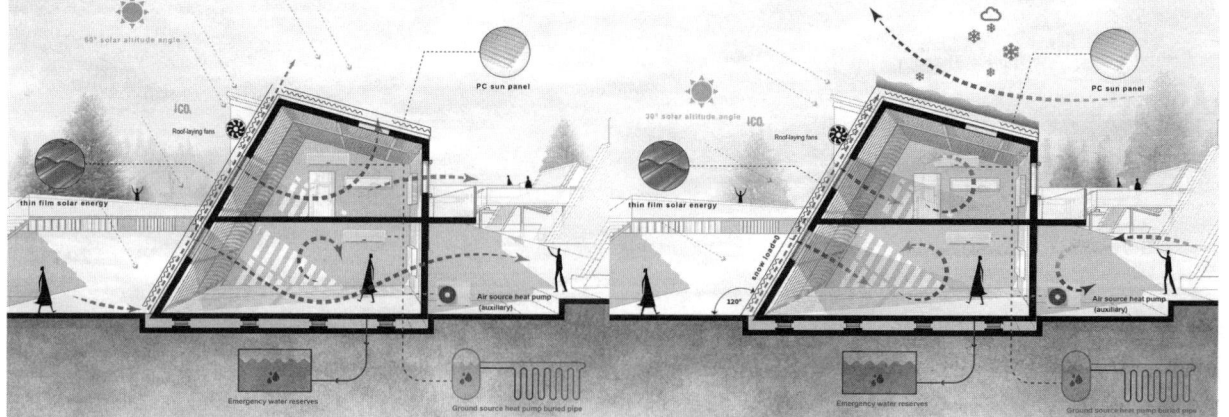

110182

光谷・动脉 （4）
Optical Valley - Venous

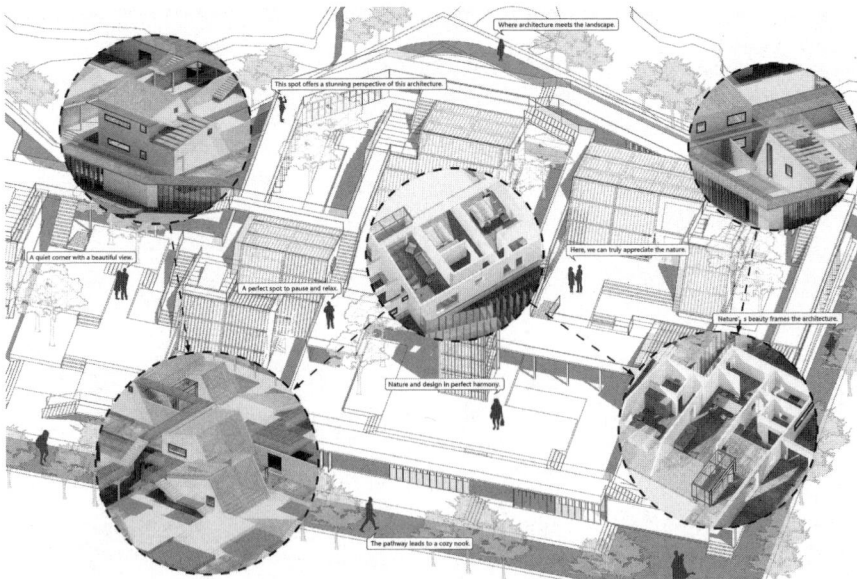

Simulation Analysis

(Residential Building Energy Efficiency Calculation Report (Taking Type A as an Example))

Design Basis:

Main structure and thermal performance parameters of the exterior wall:

Balance comparison:

Body shape coefficients:

Roof structure and thermal performance parameters:

External window structure and thermal performance parameters:

Thermal bridge column structure and thermal performance parameters:

Internal wall construction and thermal performance parameters:

Design/benchmark energy consumption ratio:

Monthly Heating and Cooling Load Diagram for Design/Benchmark Buildings:

Energy consumption calculation result:

Design/benchmark building monthly energy consumption chart:

Monthly renewable energy generation:

Calculation results of primary energy consumption and carbon emission of the building:

Proportion of carbon emissions during the production stage of building materials:

Proportion of carbon emissions during the transportation phase of building materials:

Review of carbon emission intensity indicators during the construction use phase:

Snow Recycling

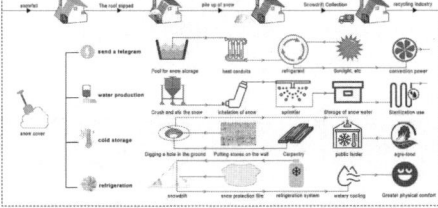

Water Recycling

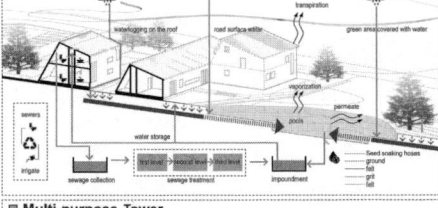

Multi-purpose Tower

production capacities

microclimate regulation

rainwater harvesting

recreation with a view

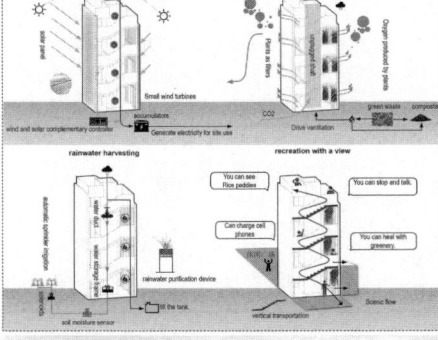

Optical Storage Direct Flexible Technology

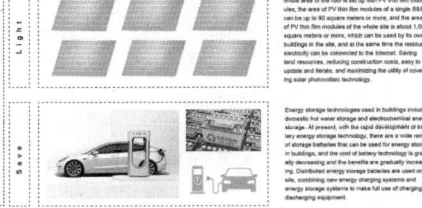

Light

Save

D.C.

Flexible

Solar photovoltaic technology used in the building. The whole area of the roof is set up with PV thin film modules, the area of PV thin film modules of a single B&B can be up to 90 square meters or more, and the area of PV thin film modules of the whole site is about 1,000 square meters or more, which can be used by its own buildings in the site, and at the same time the residual electricity can be connected to the Internet. Saving land resources, reducing construction costs, easy to update and iterate, and maximizing the utility of covering solar photovoltaic technology.

Energy storage technologies used in buildings include domestic hot water storage and electrochemical energy storage. At present, with the rapid development of battery energy storage technology, there are a wide range of storage batteries that can be used for energy storage in buildings, and the cost of battery technology is gradually decreasing and the benefits are gradually increasing. Distributed energy storage batteries are used on the site, combining new energy charging systems and energy storage systems to make full use of charging and discharging equipment.

Low-voltage DC power distribution system adopted in the building, with the increasing proportion of DC power supply and loads used in the building, the LED lamps and lanterns, air conditioners, televisions and other digital equipment currently need to consider frequency conversion and DC loads. Low-voltage DC power distribution system not only reduces the AC/DC conversion between the end DC equipment and the power grid, but also lifts the limitation of frequency and voltage in the power distribution system.

The flexible power technology adopted in buildings is the ultimate goal of "optical storage and direct flexibility". Through the optimization of building equipment control strategy, improve the flexibility of household appliances and other measures, can make the building power system from the past rigid load into a flexible load, more conducive to adapt to different types of building end equipment and user-side demand, fully stimulate the flexibility of flexible regulation of building equipment not only to regulate the peak power load, but also conductive to promote the process of the future of China's power market.

Active Solar Technology

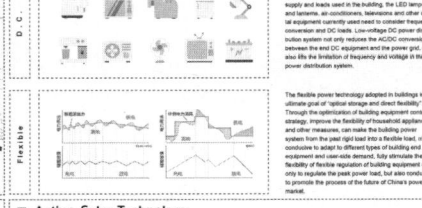

Electrochemical storage battery packs

Solar Thin Film Southbound Main Coverage

Solar thin-film northbound auxiliary coverage

Electrochemical storage battery packs

Two-way energy storage new energy charging equipment

Lighting Calculations

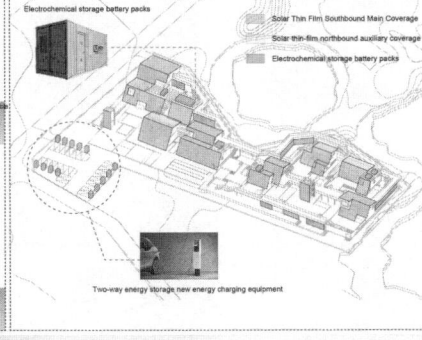

Residential indoor lighting - green building point by point simulation judgment conclusion (take A house type as an example):

1, residential building indoor main functional space at least 60% area proportion area, its lighting illuminance value is not less than 300lx hours on average not less than 8h/d, meet the "green building evaluation standard" GB / T50378 - 2019, the requirements of the 5.2.8 scoring items, scored 9 points.

2, this project Lighting uniformity all meets the requirements of the General Specification for the Built Environment GB 55016 - 2021

3, the bedrooms and living rooms (halls) of this project have glare control measures, and the ratio of the maximum lighting coefficient to the average lighting coefficient is up to the standard, which meets the requirements of the scoring item 5.2.8 of the "Green Building Evaluation Standard" GB / T50378 - 2019, and can score 3 points

North Elevation 1 : 250

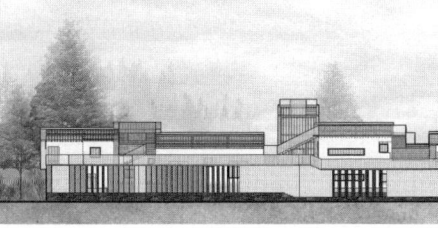

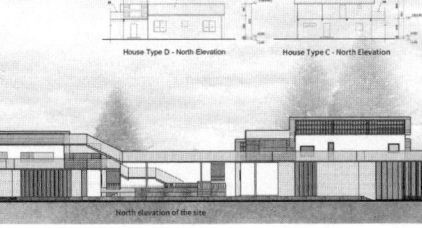

House Type D - North Elevation   House Type C - North Elevation   House Type B - North Elevation   House Type A - North Elevation

North elevation of the site

Where architecture meets the landscape

This spot offers a stunning perspective of this architecture.

A quiet corner with a beautiful view.

A perfect spot to pause and relax.

Here, we can truly appreciate the nature.

Nature and design in perfect harmony.

Nature's beauty frames the architecture.

The pathway leads to a cozy nook.

110229

阳光·绿野营地Ⅰ
Sunshine · Green Field Campsite

综合奖·优秀奖·张家口市
怀安县二堡子村建设项目
Comprehensive Awards –
Honorable Mention – Erpuzi
Village Project, Huai'an
County, Zhangjiakou City

注册号：110229
Register Number：110229
项目名称：阳光·绿野营地
Entry Title：Sunshine—Green Field
Campsite
作者：张淑娴、张云帆、王祎铭
Authors：Zhang Shuxian, Zhang Yunfan,
and Wang Yiming
作者单位：厦门大学
Authors from：Xiamen University
指导教师：石峰、贾令堃
Tutors：Shi Feng, Jia Lingkun
指导教师单位：厦门大学
Tutors from：Xiamen University

Design Description

本设计以"阳光营地"为主题，呼应场地四周的自然环境，同时考虑地震等自然灾害的应急疏散。民居单体与公共服务建筑依据场地和环境特点，灵活分散布局，既减少了对生态环境的干扰，又在灾害发生时便于快速疏散，增强社区的韧性。本方案将太阳能光伏板与建筑屋顶结合，同时，采用种植屋面、雨水收集及地源热泵等被动设计，提升能效，减少碳足迹，实现低碳可持续发展的绿色乡村目标。

This design takes "Sunshine Camp" as the theme, echoing the natural environment around the site, while considering emergency evacuation for natural disasters such as earthquakes. Residential units and public service buildings are flexibly and dispersedly arranged based on the characteristics of the site and environment, which reduces interference with the ecological environment and facilitates rapid evacuation in the event of disasters, enhancing the resilience of the community. This plan combines solar photovoltaic panels with building roofs, and adopts passive designs such as planting roofs, rainwater collection, and ground source heat pumps to improve energy efficiency, reduce carbon footprint, and achieve the goal of low-carbon and sustainable development in green rural areas.

Climate Analysis

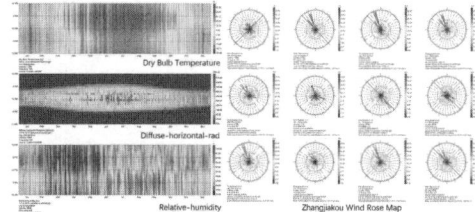

Dry Bulb Temperature
Diffuse-horizontal-rad
Relative-humidity
Zhangjiakou Wind Rose Map

The project is located in Erbaozi Village, Huai'an County, Zhangjiakou City, Hebei Province, with an East Asian continental monsoon climate. The annual average temperature is 8.3 °C, with a maximum temperature of 37.2 °C and a minimum temperature of -22.6 °C. The regional sunshine hours are 2800-3100 hours, and the total solar radiation is 1500-1700 kilowatt hours per square meter, indicating abundant solar energy.

Energy Saving Strategies in Cold Regions

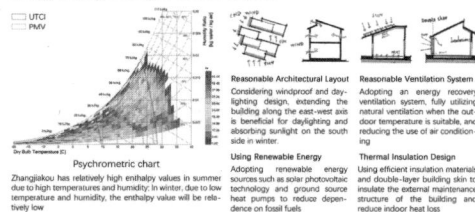

UTCI
PMV

Psychrometric chart

Zhangjiakou has relatively high enthalpy values in summer due to high temperatures and humidity; In winter, due to low temperature and humidity, the enthalpy value will be relatively low

Reasonable Architectural Layout
Considering windproof and daylighting design, extending the building along the east-west axis is beneficial for daylighting and absorbing sunlight on the south side in winter

Reasonable Ventilation System
Adopting an energy recovery ventilation system, fully utilizing natural ventilation when the outdoor temperature is suitable, and reducing the use of air conditioning

Using Renewable Energy
Adopting renewable energy sources such as solar photovoltaic technology and ground source heat pumps to reduce dependence on fossil fuels

Thermal Insulation Design
Using efficient insulation materials and double-layer building skin to insulate the external maintenance structure of the building and reduce indoor heat loss

Site in Normal Times and Disaster Times

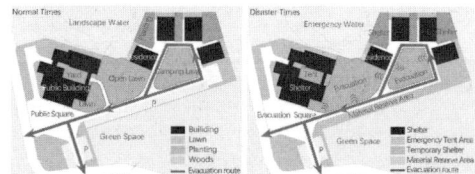

Normal Times | Disaster Times

Design Concept

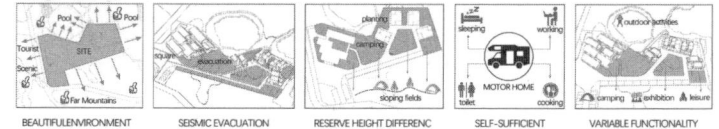

BEAUTIFULENVIRONMENT | SEISMIC EVACUATION | RESERVE HEIGHT DIFFERENCE | SELF-SUFFICIENT | VARIABLE FUNCTIONALITY

SWOT Analysis

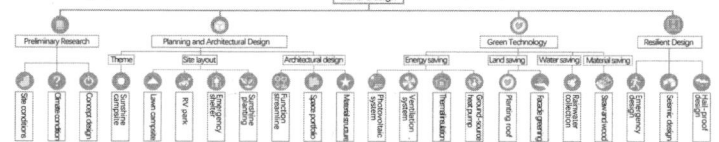

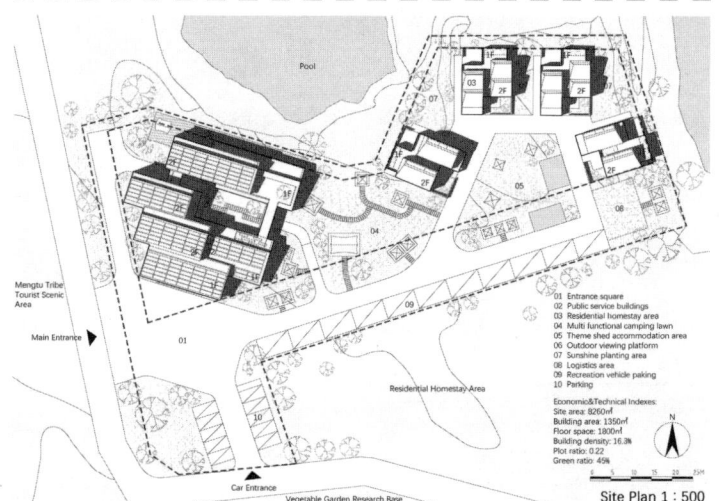

nature + environment + ecology

SWOT Analysis

Rural Design

Site Plan

01 Entrance square
02 Public service buildings
03 Residential homestay area
04 Multi functional camping lawn
05 Theme shed accommodation area
06 Outdoor viewing platform
07 Sunshine planting area
08 Logistics area
09 Recreation vehicle paking
10 Parking

Economic&Technical Indexs:
Site area: 8260㎡
Building area: 1350㎡
Floor space: 1800㎡
Building density: 16.3%
Plot ratio: 0.22
Green ratio: 45%

Site Plan 1：500

110229

阳光·绿野营地 II
Sunshine - Green Field Campsite

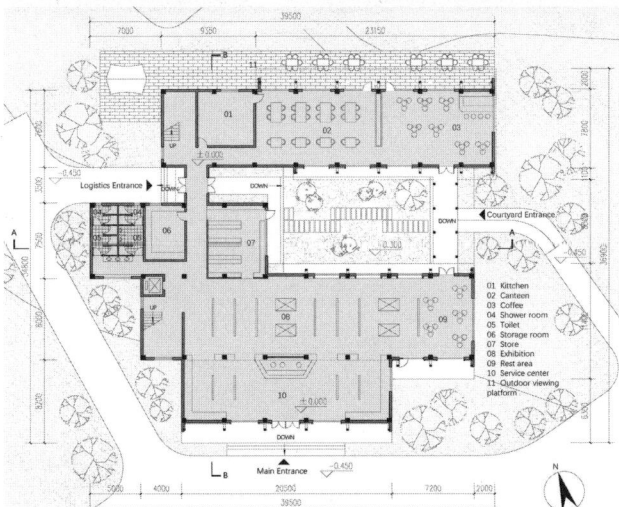

Public Service Building First Floor Plan 1:200

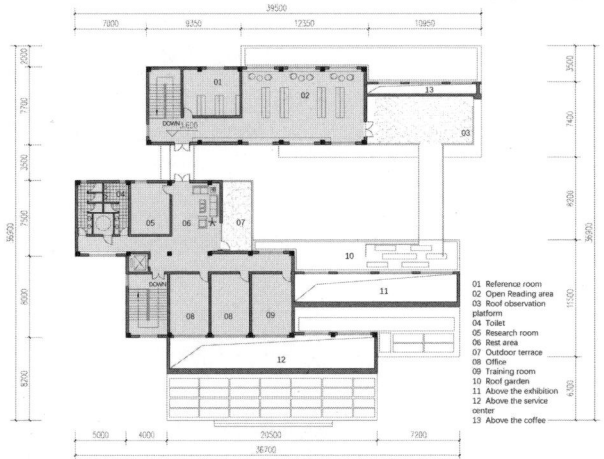

Logistics Entrance
Courtyard Entrance
Main Entrance
±0.000
−0.450

01 Kittchen
02 Canteen
03 Coffee
04 Shower room
05 Toilet
06 Storage room
07 Store
08 Exhibition
09 Rest area
10 Service center
11 Outdoor viewing platform

Public Service Building Second Floor Plan 1:200

01 Reference room
02 Open Reading area
03 Roof observation platform
04 Toilet
05 Research room
06 Rest area
07 Outdoor terrace
08 Office
09 Training room
10 Roof garden
11 Above the exhibition
12 Above the service center
13 Above the coffee

Public Service Building Design Development

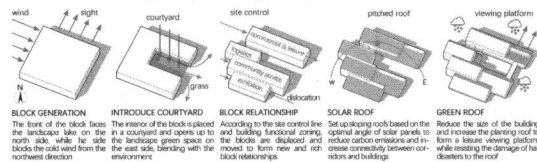

wind | sight | courtyard | site control | pitched roof | viewing platform

BLOCK GENERATION
The front of the block faces the landscape on the north side while he side blocks the cold wind from the northwest direction

INTRODUCE COURTYARD
The interior of the block is placed in a courtyard and opens up to the landscape green space on the east side, blending with the environment

BLOCK RELATIONSHIP
According to the site control line and building functional zoning, the blocks are displaced and moved to form new and rich block relationships.

SOLAR ROOF
Set up sloping roofs based on the optimal angle of solar panels to reduce carbon emissions and increase connectivity between corridors and buildings

GREEN ROOF
Reduce the size of the building and increase the planting roof to form a leisure viewing platform while resisting the damage of hail disasters to the roof

---

Explosion Diagram Of Public Service Building

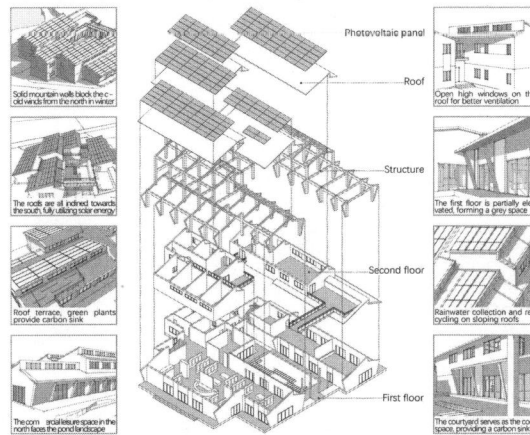

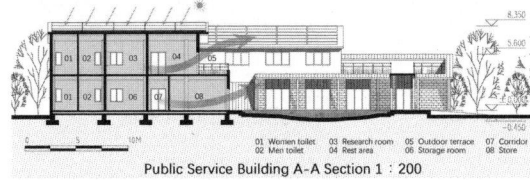

Photovoltaic panel
Roof
Structure
Second floor
First floor

Solid mountain walls block the cold winds from the north in winter

The roofs are all inclined towards the south, fully utilizing solar energy

Roof terrace, green plants provide carbon sink

The com arcial leisure space in the north faces the pond landscape

Open high windows on the roof for better ventilation

The first floor is partially elevated, forming a grey space

Rainwater collection and recycling on sloping roofs

The courtyard serves as the core space, providing a carbon sink

---

Public Service Building A-A Section 1:200

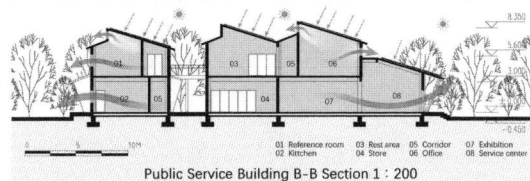

01 Women toilet
02 Men toilet
03 Research room
04 Rest area
05 Outdoor terrace
06 Storage room
07 Corridor
08 Store

Public Service Building B-B Section 1:200

01 Reference room
02 Kitchen
03 Rest area
04 Store
05 Corridor
06 Office
07 Exhibition
08 Service center

阳光·绿野营地 III
Sunshine - Green Field Campsite

Generation of Residential Building Blocks

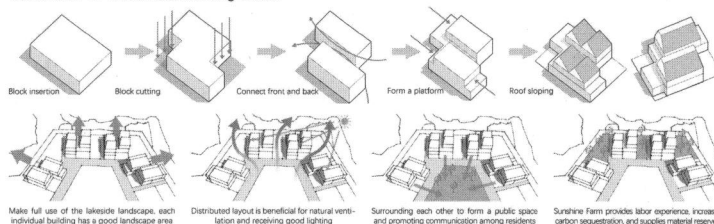

Block insertion | Block cutting | Connect front and back | Form a platform | Roof sloping

Make full use of the lakeside landscape, each individual building has a good landscape area | Distributed layout is beneficial for natural ventilation and receiving good lighting | Surrounding each other to form a public space and promoting communication among residents | Sunshine Farm provides labor experience, increases carbon sequestration, and supplies material reserves

Explosion Diagram of Residential Buildings

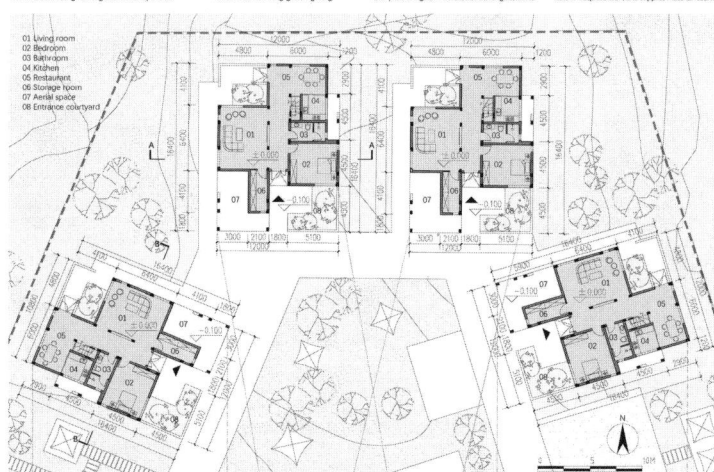

Skylight — Opening skylights on the roof is beneficial for summer hot pressure ventilation and increases natural lighting

Roof — Sloping roof facing south, fully utilizing solar photovoltaic power generation

Second floor — Planting roofs, thermal insulation, and providing carbon sinks

Structure — The second floor block protrudes to form a gray space under the eaves

— Increase slant support to improve the stability of the building, which is conducive to earthquake resistance

First floor — The courtyard wall should avoid complete transparency in the north-south direction

01 Living room
02 Bedroom
03 Bathroom
04 Kitchen
05 Restaurant
06 Storage Room
07 Aerial space
08 Entrance courtyard

01 Terrace
02 Rooms
03 Bathroom
04 Corridor

Residential Buildings First Floor Plan 1 : 200

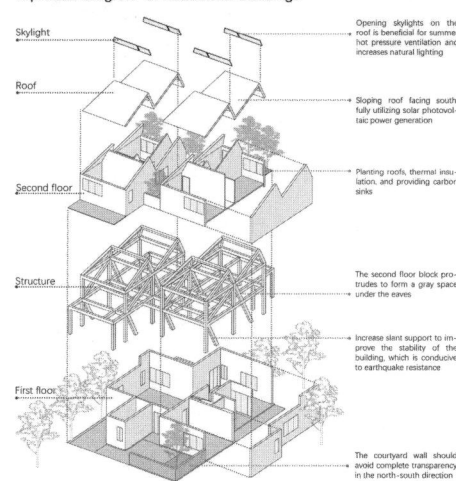

Residential Buildings Second Floor Plan 1 : 200

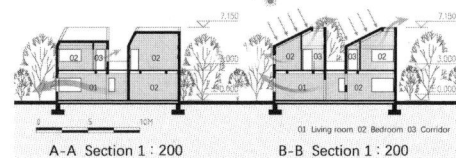

01 Living room 02 Bedroom 03 Corridor

A-A Section 1 : 200   B-B Section 1 : 200

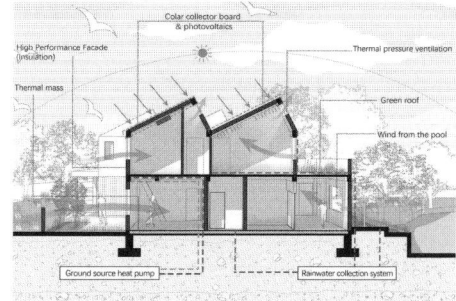

Colar collector board & photovoltaics
High Performance Facade (insulation)
Thermal mass
Thermal pressure ventilation
Green roof
Wind from the pool
Ground source heat pump
Rainwater collection system

Perspective Analysis of Residential Buildings

阳光·绿野营地 IV Sunshine - Green Field Campsite

Colar collector board & photovoltaics

Thermal pressure ventilation

Thermal mass

Green roof

Double-skin

Courtyard thermal pressure ventilation effect

High performance facade (Insulation)

Heat pump unit

Ground-coupled system

Rain garden recharge

Grey water reuse

Rainwater cistern

Irrigation — Domestic water — Fire water

Perspective View of Public Service Building

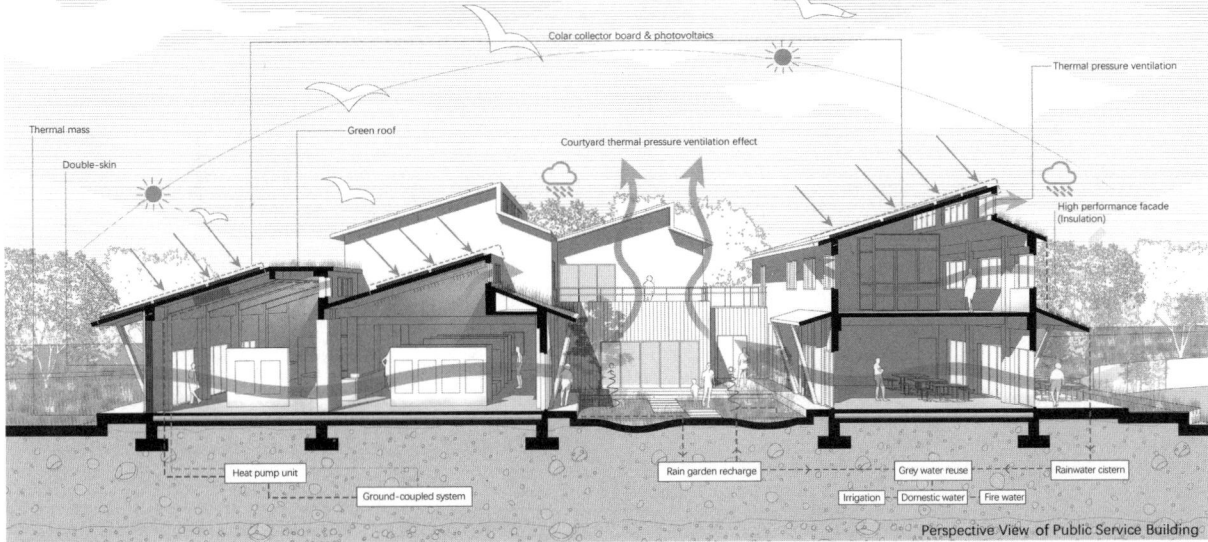

Ground-source System

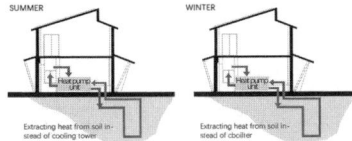

SUMMER

WINTER

Heatpump unit

Heatpump unit

Extracting heat from soil instead of cooling tower

Extracting heat from soil instead of cboiler

Double-layer Epidermal System

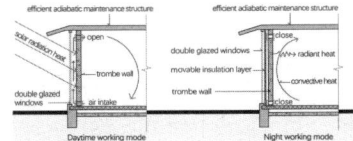

efficient adiabatic maintenance structure

efficient adiabatic maintenance structure

solar radiation heat

open

double glazed windows

movable insulation layer

trombe wall

close

PV + radiant heat

convective heat

double glazed windows

air intake

trombe wall

close

Daytime working mode

Night working mode

Indoor Lighting Analysis

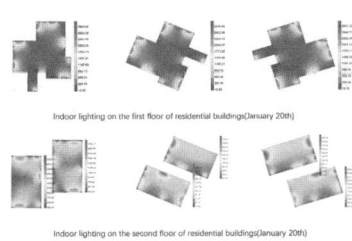

Indoor lighting on the first floor of residential buildings(January 20th)

Indoor lighting on the second floor of residential buildings(January 20th)

Sunshine Planting

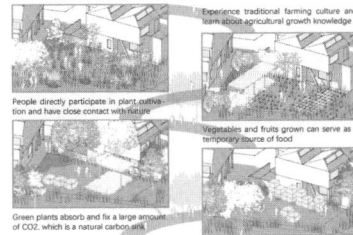

Experience traditional farming culture and learn about agricultural growth knowledge

People directly participate in plant cultivation and have close contact with nature

Vegetables and fruits grown can serve as a temporary source of food

Green plants absorb and fix a large amount of CO_2, which is a natural carbon sink

Hail Resistant Analysis

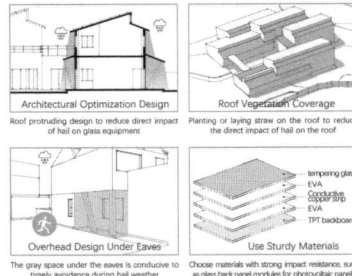

Architectural Optimization Design
Roof protruding design to reduce direct impact of hail on glass equipment

Roof Vegetation Coverage
Planting or laying straw on the roof to reduce the direct impact of hail on the roof

Overhead Design Under Eaves
The gray space under the eaves is conducive to timely avoidance during hail weather

Use Sturdy Materials
Choose materials with strong impact resistance, such as glass back panel modules for photovoltaic panels

tempering glass
EVA
Conductive copper strip
EVA
TPT backboard

Seismic Resistance Analysis

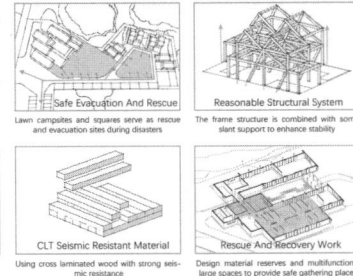

Safe Evacuation And Rescue
Lawn campsites and squares serve as rescue and evacuation sites during disasters

Reasonable Structural System
The frame structure is combined with some slant support to enhance stability

CLT Seismic Resistant Material
Using cross laminated wood with strong seismic resistance

Rescue And Recovery Work
Design material reserves and multifunctional large spaces to provide safe gathering places

Sunlight Analysis

January 20th June 22nd December 22nd June 22nd December 22nd

January 20th June 22nd December 22nd June 22nd December 22nd

Calculation results of carbon emissions during the use chase of the building

Project	Public service building			Residential buildings		
	power consumption emissions (kWh/㎡)	Carbon emissions (t)	Carbon emissions per unit area(kg/㎡)	power consumption emissions (kWh/㎡)	Carbon emissions (t)	Carbon emissions per unit area(kg/㎡)
HVAC	1572	1665	1390	2282	1652	2018
Heating	3155	3340	2790	4007	2910	3552
Illumination	730	773	645	242	176	214
Else	1951	2065	1725	1033	752	926
Renewable energy	8856	9376	7832	7862	5696	6952
Total	/	0	0	/	0	0

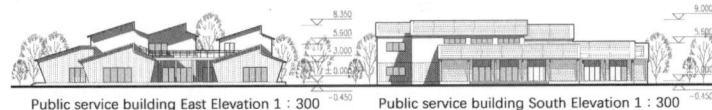

Public service building East Elevation 1 : 300

Public service building South Elevation 1 : 300

综合奖 · 优秀奖 · 张家口市
怀安县二堡子村建设项目
Comprehensive Awards —
Honorable Mention — Erpuzi
Village Project, Huai'an
County, Zhangjiakou City

注册号：110270
Register Number：110270

项目名称：湖栖 · 夯居
Entry Title：Lakeside Earthen Retreat

作者：吴琳歆、徐凯、黄新航、黄可杰
Authors：Wu Linxin，Xu Kai，
Huang Xinhang，and Huang Kejie

作者单位：福州大学
Authors from：Fuzhou University

指导教师：邱文明
Tutor：Qiu Wenming

指导教师单位：福州大学
Tutor from：Fuzhou University

CODE:110270

Lakeside Earthen Retreat · I

湖栖 · 夯居

Erbaozi Village Construction Project in Huai'an County, Zhangjiakou City

Regional Characteristics Analysis

Swan Kiln Loess Materials / Brick and Wood Construction / High Roll-top Roof

Arch Space / Climate Regulation by Plants / Central Courtyard

Design Strategy

Architectural Design: Service Center, Zigzag Layout, Visitors and Villagers, Guesthouse, Townhouse, Different Sets, Site, Linking Corridors, Waterside Walkway

Seismic Technology: Symmetry Seismic, Independent Foundation, Straw Lightweight Partition Wall, Reinforced Concrete + Rammed Earth

Green Technology: proactive technologies, Solar Panel, Solar Bed, Photovoltaic Sunroom, Passive Technologies, Rammed Earth Wall Insulation, Straw Insulation

Design Description

项目位于张家口市怀安县二堡子村。该项目拾取当地的民居风格进行设计，分别加入水边栈道、连廊，以加强建筑与水体和建筑之间的联系。在太阳能技术方面，项目设置了太阳能板、光伏连廊、光伏阳光房、太阳能炕，以利用太阳能。在低碳技术方面，项目使用了较多的当地秸秆与夯土作为建筑材料，并且运用多种蓄热保温技术以促进节能。在韧性设计方面，项目考虑了平时和受灾时的建筑功能转变，并且采用联排布局、独立基础构造等方式，以利于抗震。

The project is located in Erbaozi Village, Huai'an County, Zhangjiakou City. The project picks up the style of the local houses in the area, and incorporates waterfront walkways and corridors to strengthen the connection between the buildings and the water body and the buildings. In terms of solar technology, solar panels, photovoltaic corridors, photovoltaic sunrooms, and solar kangs are installed to utilize solar energy. In terms of low carbon technology, the project uses more local straw and rammed earth as building materials, and utilizes a variety of thermal storage and insulation technologies to promote energy conservation. In terms of resilient design, the project takes into account the change of building functions in times of peace and disaster, and adopts a townhouse layout and independent foundation construction to facilitate earthquake resistance.

Population Structure Analysis

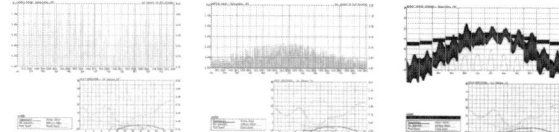

Age Composition of the Household Population in Erbaozi Village

Proportion of Different Populations in Erbaozi Village

The percentage of the household population in Erbaozi Village that is over 60 years old is 26.9%, which is higher than the aging rate in rural China. The total household population is 402.

Climate Analysis

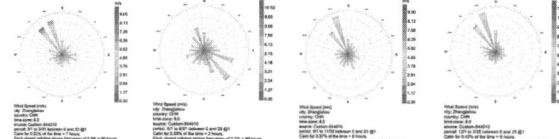

From the table, it is clear that there are relatively abundant solar energy resources available in the area; the area has cold winters and high summer temperatures. Thermal insulation in winter must be taken into account, with due regard for summer insulation.

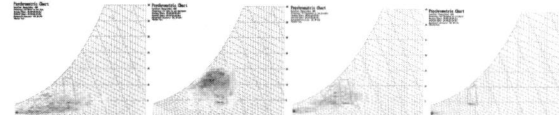

From the wind rose map, it can be seen that the local spring, fall and winter prevails northwest wind, while the summer is dominated by southeast wind. Therefore, attention should be paid to the windproof performance of the building in the northwest direction.

From the enthalpy-humidity diagram data, it can be seen that there are more sampling points in the local thermal comfort range in summer, fewer in spring and fall, and the least in winter, which leads to the conclusion that the local thermal comfort is better in summer, worse in spring and fall, and the worst in winter.

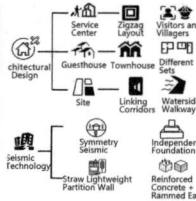

As can be seen from the above figure, the more appropriate strategies for enhancing local thermal comfort, especially in winter, are passive solar as well as wall thermal storage.

Site Analysis

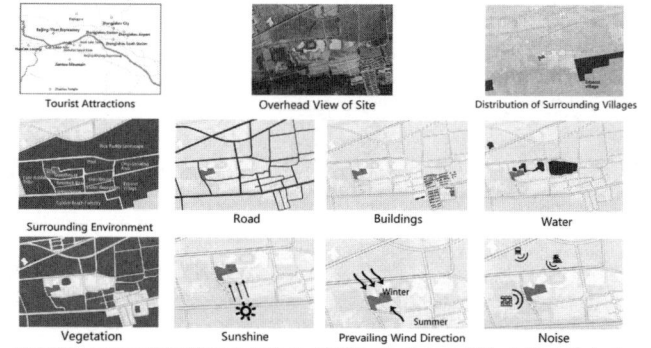

Tourist Attractions / Overhead View of Site / Distribution of Surrounding Villages

Surrounding Environment / Road / Buildings / Water

Vegetation / Sunshine / Prevailing Wind Direction / Noise

Located in the northwest of Erbaozi Village, Huai'an County, at 114.62° east longitude and 40.68° north latitude, it is close to the highway. There are paddy fields to the north of the base, and the Dayang River to the north of the paddy fields. The south side is supported by the Jinshatan Forest Farm, and there are large tracts of woodland and farmland on the east and west sides. The west side is adjacent to the Cute Rabbit Tribe Tourist Scenic Area.

Logic Generation

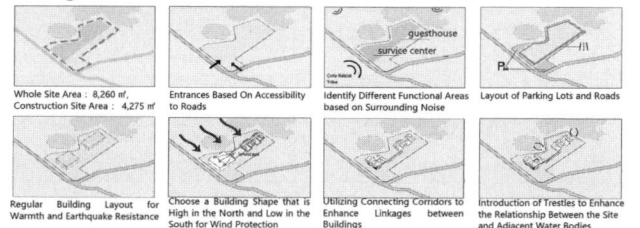

Whole Site Area : 8,260 ㎡,
Construction Site Area : 4,275 ㎡

Regular Building Layout for Warmth and Earthquake Resistance

Entrances Based On Accessibility to Roads

Choose a Building Shape that is High in the North and Low in the South for Wind Protection

Identify Different Functional Areas based on Surrounding Noise

Utilizing Connecting Corridors to Enhance Linkages between Buildings

Layout of Parking Lots and Roads

Introduction of Trestles to Enhance the Relationship Between the Site and Adjacent Water Bodies

CODE:110270

Lakeside Earthen Retreat · II

湖栖·夯居

Erbaozi Village Construction Project in Huai'an County, Zhangjiakou City

Site Plan 1 : 750

■ Major Technical-economic Indices
Land area: 8260m²
Occupied area: 1311m²
Total building area: 1898m²
Service center building area: 632m²
Homestay building area: 1266m²
Building density: 15.87%
Greening rate: 26%
Floor area ratio: 0.23

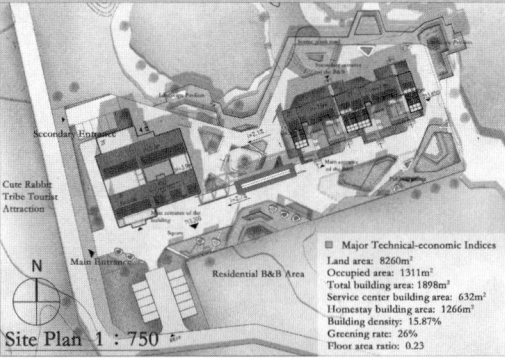

■ Carbon Emission Index Table

Calculation Results of Carbon Emissions during the Use Phase of the Building

Tourist Service Center	Design building		Reference building	
	Carbon footprint[kgCO₂]	Carbon emissions per unit area[kgCO₂/m²]	Carbon footprint[kgCO₂]	Carbon emissions per unit area[kgCO₂/m²]
Carbon Footprint	50979.09	40.26	56964.7	44.99
Renewable Energy	-44677.14	-35.29	0	0
Building Carbon Sinks	-17913.9	-14.15	0	0
Sum	-11611.95	9.17	56964.7	44.99

Calculation Results of Carbon Emissions during the Use Phase of the Building

Guesthouse	Design Architecture		Reference building	
	Carbon footprint[kgCO₂]	Carbon emissions per unit area[kgCO₂/m²]	Carbon footprint[kgCO₂]	Carbon emissions per unit area[kgCO₂/m²]
Carbon Footprint	59655.44	94.39	71348.96	112.89
Renewable Energy	-41149.52	-65.11	0	0
Building Carbon Sinks	-19756.32	-31.26	0	0
Sum	-1250.4	-1.98	71348.96	112.89

	Total Building Energy Consumption[kgCO₂]	Renewable Energy[kgCO₂]	Building Carbon Sinks[kgCO₂]
Carbon Footprint	110634.53	-85826.66	-37670.22
Net Carbon Emissions from the Site		-12862.35	

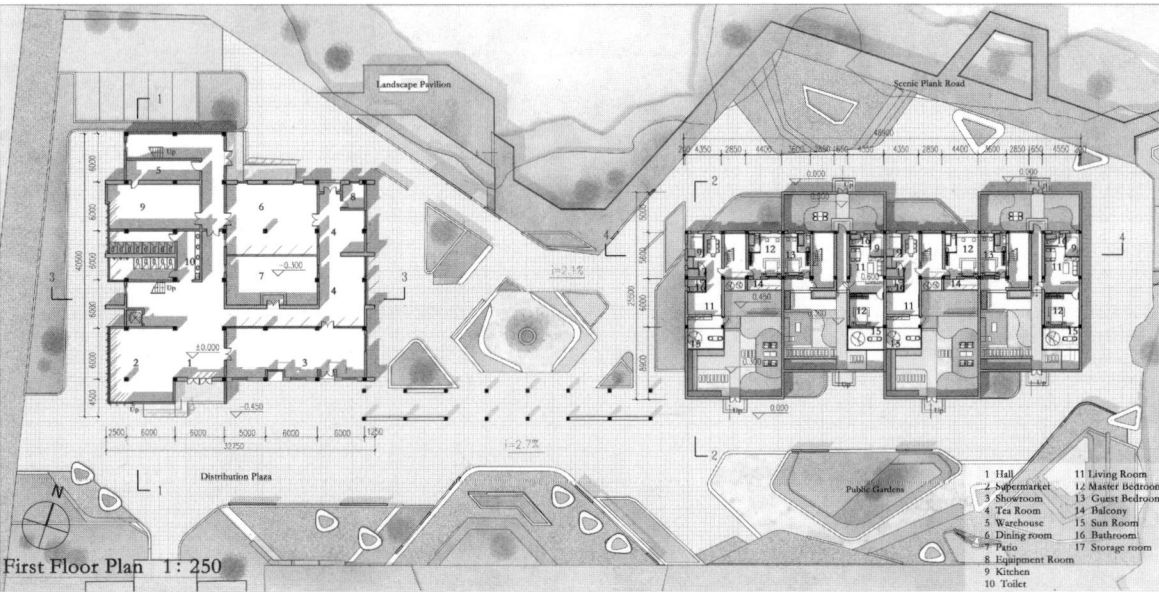

First Floor Plan 1 : 250

1 Hall
2 Supermarket
3 Showroom
4 Tea Room
5 Warehouse
6 Dining room
7 Patio
8 Equipment Room
9 Kitchen
10 Toilet
11 Living Room
12 Master Bedroom
13 Guest Bedroom
14 Balcony
15 Sun Room
16 Bathroom
17 Storage room

■ Streamline Analysis

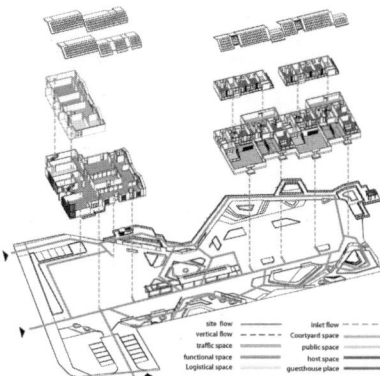

site flow
vertical flow
traffic space
functional space
logistical space
inlet flow
Courtyard space
public space
host space
guesthouse place

■ Homestay Features Analysis

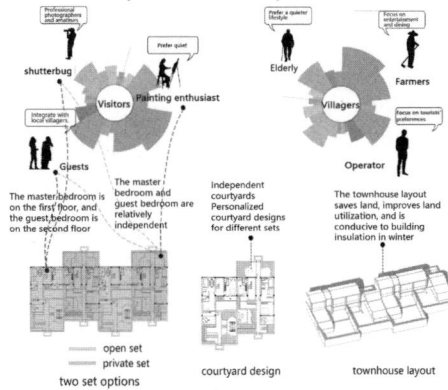

Professional photographers and amateurs — Prefer quiet

shutterbug

Painting enthusiast

Visitors

Prefer a quieter lifestyle — Elderly

Villagers — Farmers

Focus on entertainment and dining

Focus on tourists' preferences

Operator

Integrate with local villagers

Guests

The master bedroom is on the first floor, and the guest bedroom is on the second floor

The master bedroom and guest bedroom are relatively independent

Independent courtyards Personalized courtyard designs for different sets

The townhouse layout saves land, improves land utilization, and is conducive to building insulation in winter

open set
private set
two set options

courtyard design

townhouse layout

■ Sunlight Simulation

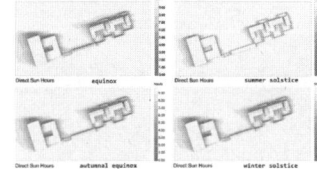

Direct Sun Hours — equinox
Direct Sun Hours — summer solstice
Direct Sun Hours — autumnal equinox
Direct Sun Hours — winter solstice

■ Wind Simulation

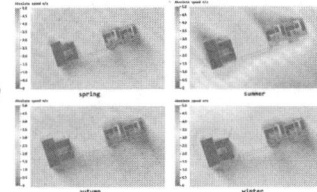

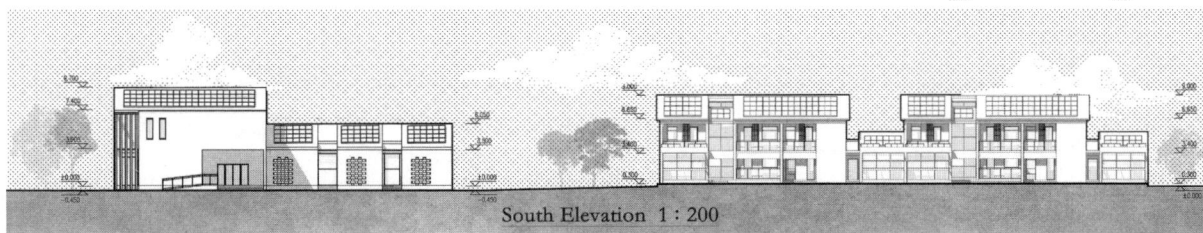

spring
summer
autumn
winter

South Elevation 1 : 200

CODE:110270

Lakeside Earthen Retreat · III

湖栖·夯居

Erbaozi Village Construction Project in Huai'an County, Zhangjiakou City

Seismic Analysis

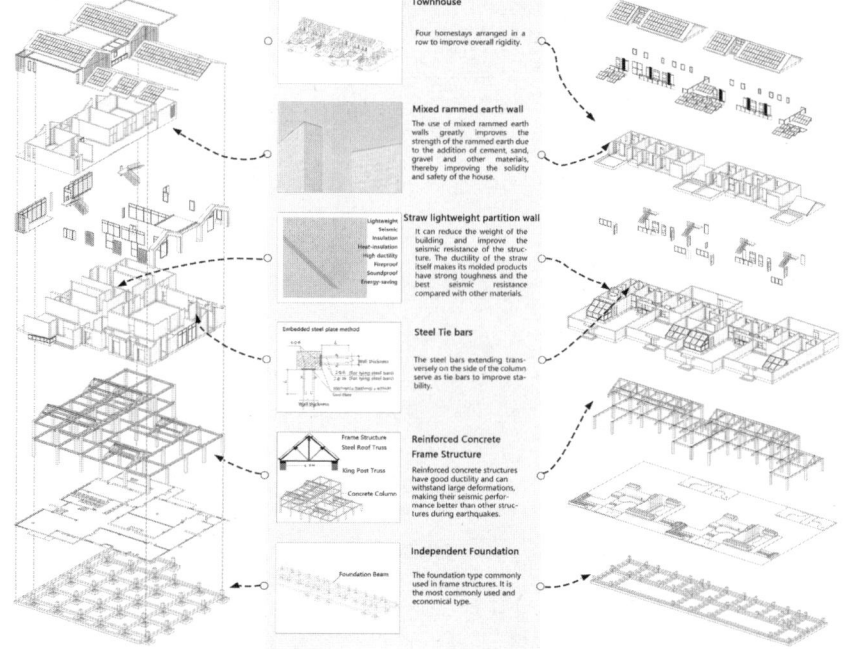

Townhouse
Four homestays arranged in a row to improve overall rigidity.

Mixed rammed earth wall
The use of mixed rammed earth walls greatly improves the strength of the rammed earth due to the addition of cement, sand, gravel and other materials, thereby improving the solidity and safety of the house.

Straw lightweight partition wall
It can reduce the weight of the building and improve the seismic resistance of the structure. The ductility of the straw itself makes its molded products have strong toughness and the best seismic resistance compared with other materials.

Lightweight
Seismic
Insulation
Heat-insulation
High ductility
Fireproof
Soundproof
Energy-saving

Embedded steel plate method

Steel Tie bars
The steel bars extending transversely on the side of the column serve as tie bars to improve stability.

Frame Structure
Steel Roof Truss
King Post Truss
Concrete Column

Reinforced Concrete Frame Structure
Reinforced concrete structures have good ductility and can withstand large deformations, making their seismic performance better than other structures during earthquakes.

Foundation Beam

Independent Foundation
The foundation type commonly used in frame structures. It is the most commonly used and economical type.

Site Emergency Design

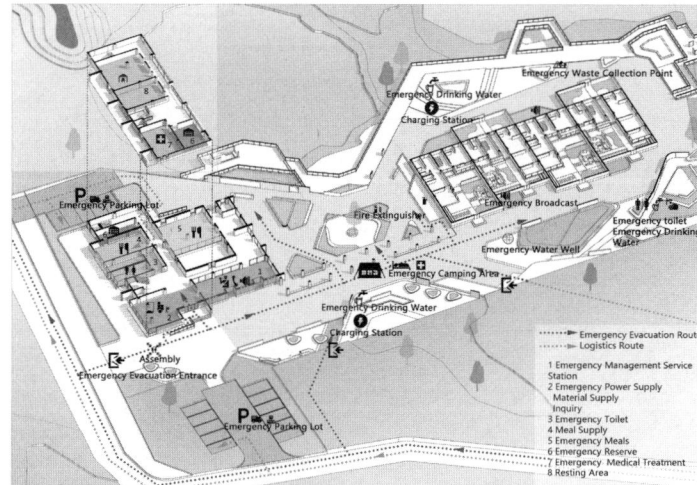

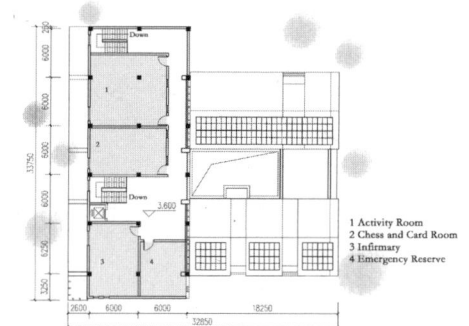

Design Strategy: Create a multifunctional, integrated emergency shelter.

Resilient Space
Functional Variability
Connectivity
Combining Disaster Relief with Disaster
Proximity Principle

1) Villager and Tourism Supporting Service Buildings - Emergency Command, Inquiry, Material Supply, etc.
2) Venue Distribution Square, Pocket park - Shed Accommodation Area, Emergency Reception Area, Garbage Storage

Providing a comfortable and saft shelter environment

Site partitioning to provide privacy

Open, convenient for quick evacuation of refugees

Equipped with basic leisure and emergency facilities

The tent area can accommodate up to 67 tents of about 12 square meters. Taking into account the public facilities, it can accommodate about 720 people.

Before the Disaster:
Entertain Entertainment Shopping Tourism Gather Communicate

Disaster:
Emergency Shelter Emergency Medical Rescue Assembly Area for Makeshift Tents Emergency Goods Supply Emergency Command

Emergency Waste Collection Point
Emergency Drinking Water
Charging Station
Emergency Parking lot
Fire Extinguisher
Emergency Broadcast
Emergency Water Well
Emergency toilet
Emergency Drinking Water
Emergency Camping Area
Emergency Drinking Water
Charging Station
Assembly
Emergency Evacuation Entrance
Emergency Parking Lot

- - - - Emergency Evacuation Route
▷ ▷ Logistics Route
1 Emergency Management Service Station
2 Emergency Power Supply Material Supply Inquiry
3 Emergency Toilet
4 Meal Supply
5 Emergency Meals
6 Emergency Reserve
7 Emergency Medical Treatment
8 Resting Area

Second Floor Plan(Service Center) 1 : 250

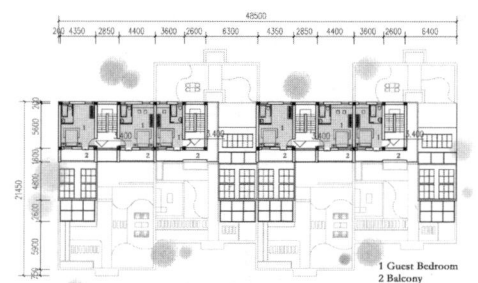

1 Activity Room
2 Chess and Card Room
3 Infirmary
4 Emergency Reserve

Second Floor Plan(Homestay) 1 : 250

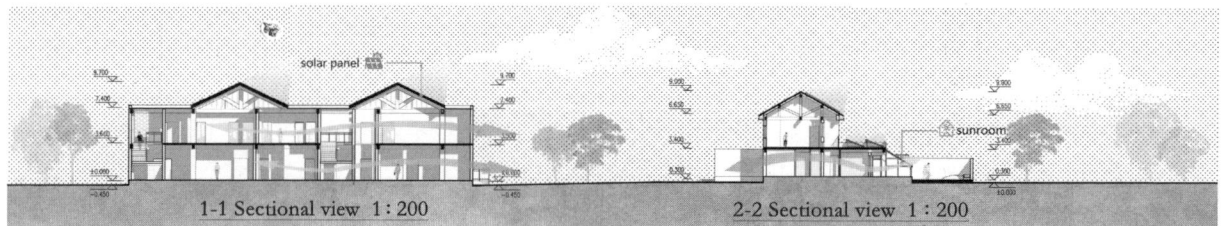

1 Guest Bedroom
2 Balcony

solar panel

1-1 Sectional view 1 : 200

sunroom

2-2 Sectional view 1 : 200

Lakeside Earthen Retreat · IV

湖栖 · 夯居
Erbaozi Village Construction Project in Huai'an County, Zhangjiakou City

Solar Energy Technology Applications

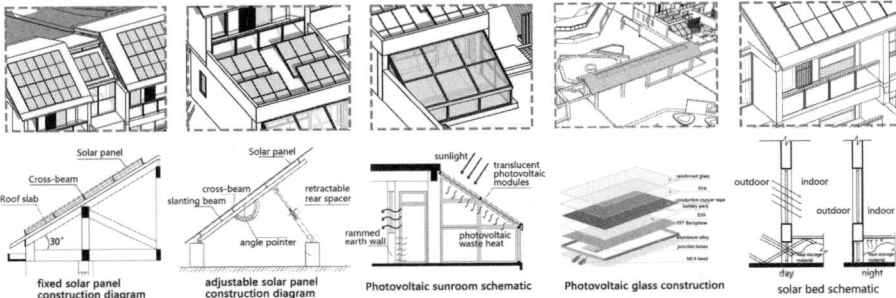

fixed solar panel construction diagram

30° is close to the optimal PV power generation angle in Zhangjiakou: 36.75°

adjustable solar panel construction diagram

This can be adjusted to the angle of optimal reception of light according to changes in sunlight.

Photovoltaic sunroom schematic

A photovoltaic sun room is set up in the south of the homestay, using translucent photovoltaic modules.

Photovoltaic glass construction

A photovoltaic colonnade is installed between the B&B and the service center. The top of the colonnade is a photovoltaic glass module.

solar bed schematic

Place materials with good heat storage capacity inside the kang. During the winter daytime, the heat storage materials convert solar energy into heat energy, and emit heat at night for heating.

Insulation Structure Analysis

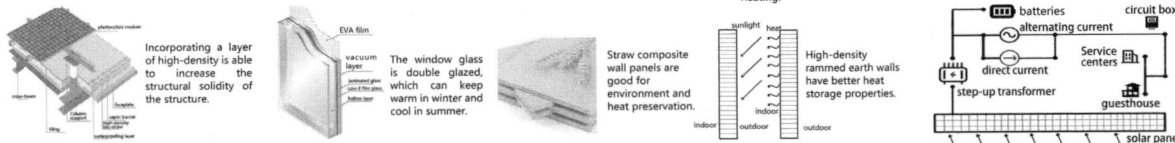

Incorporating a layer of high-density is able to increase the structural solidity of the structure.

The window glass is double glazed, which can keep warm in winter and cool in summer.

Straw composite wall panels are good for environment and heat preservation.

High-density rammed earth walls have better heat storage properties.

Energy Caculation

Electricity consumption index is taken as 50W/m².

The total floor area is 1898 ㎡, and the average power consumption is 8 hours.

1898 ㎡ × 50W/m² × 8h = 759200w·h = 759 kW·h

Total solar panel area of 369 ㎡, each ㎡ of solar panels produce 6 kW·h of electricity per day

369 ㎡ × 6 = 2214 kW·h

After deducting 30% loss, there are 1550 kW·h left > 759 kW·h.

Therefore, only relying on solar panels can meet the daily electricity demand of the service center and the guesthouse.

Green Building Strategy Analysis

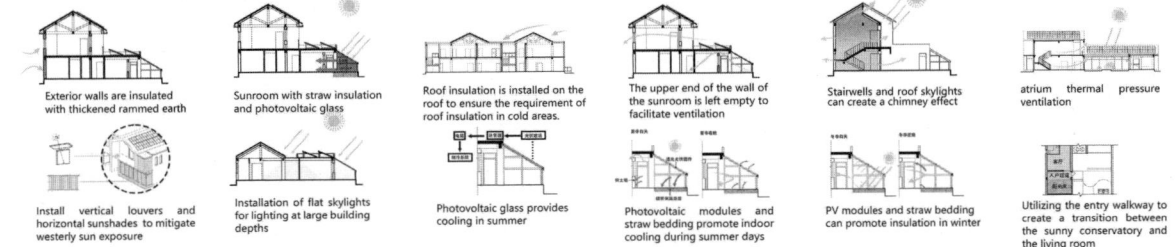

Exterior walls are insulated with thickened rammed earth

Sunroom with straw insulation and photovoltaic glass

Roof insulation is installed on the roof to ensure the requirement of roof insulation in cold areas.

The upper end of the wall of the sunroom is left empty to facilitate ventilation

Stairwells and roof skylights can create a chimney effect

atrium thermal pressure ventilation

Install vertical louvers and horizontal sunshades to mitigate westerly sun exposure

Installation of flat skylights for lighting at large building depths

Photovoltaic glass provides cooling in summer

Photovoltaic modules and straw bedding promote indoor cooling during summer days

PV modules and straw bedding can promote insulation in winter

Utilizing the entry walkway to create a transition between the sunny conservatory and the living room

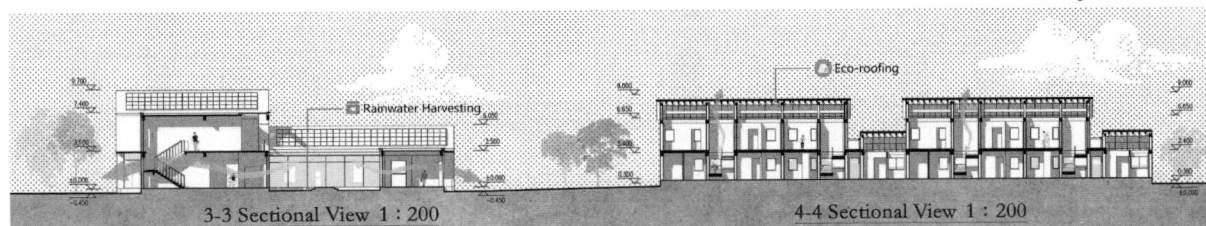

3-3 Sectional View 1 : 200

4-4 Sectional View 1 : 200

编号：110435

01

综合奖·优秀奖·张家口市怀安县二堡子村建设项目
Comprehensive Awards − Honorable Mention − Erpuzi Village Project，Huai'an County，Zhangjiakou City

注册号：110435
Register Number：110435
项目名称：烟囱·院落
Entry Title：Stack − Courtyard
作者：邓戴子阳、王磷萱、林仪铭、
     倪诗睿、莫旭阳
Authors：Deng Daiziyang, Wang Linxuan,
     Lin Yiming, Ni Shirui, and
     Mo Xuyang
作者单位：南京工业大学
Authors from：Nanjing Tech University
指导教师：倪震宇
Tutor：Ni Zhenyu
指导教师单位：南京工业大学
Tutor from：Nanjing Tech University

烟囱·院落
基于空间调节的村落重构

项目以烟囱和院落为意象，重构了烟囱和院落，构烟囱重构成了供暖和通风的构件，以当地传统民居为母体构成院落的意象，顺应季节季风风向，建筑整体建筑旋转27°，进行多面采光，充分利用太阳能，考虑到当地冬采暖和夏季制冷时间都较长，项目利用厚重的夯土墙增强其保温能力，减少其能量损耗，得益于当地丰富的太阳能资源，屋顶设置光伏电板，用当地原有条件，设置太阳能光伏、导风烟囱、供热地暖、夯土、木构、光伏导风板、扶壁、导风板片墙、集水系统，从主动和被动两个方面进行设计，塑造一个基于空间调节的村落重构。

The project uses chimneys and courtyards as symbols, reconstructing them. The chimney is reconstructed as a heating and ventilation component, taking the traditional rural dwellings of the local area as the mother body to form the image of a courtyard. The building is rotated 27° to align with the seasonal monsoon wind direction, allowing for multi-faceted lighting and maximum utilization of solar energy. Considering that the winter heating and summer cooling periods in the area are long, the project uses a thick rammed earth wall to enhance its insulation ability and reduce its energy loss. Thanks to the abundant solar energy resources in the area, solar panels are installed on the roof. Using the existing conditions in the area, solar photovoltaic, chimney ventilation, heating floor heating, rammed earth, wooden structure, photovoltaic ventilation board, buttress, and wind-blocking wall panels, and rainwater collection systems are used to design both actively and passively, shaping a village reconstruction based on space regulation.

Economic and technical indicators
Project site area：8260㎡
Construction land area：4275㎡
Gross floor area：1834㎡
Building height：9m

General Plan　1 : 500

Climate Analysis

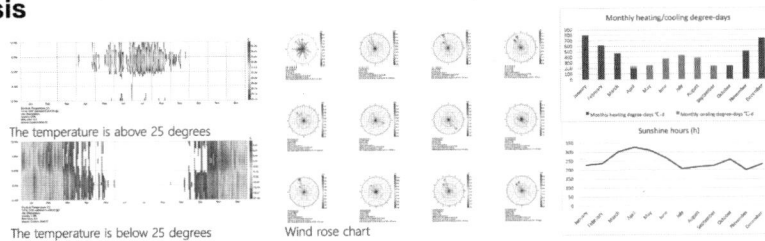

Monthly heating/cooling degree-days

The temperature is above 25 degrees

The temperature is below 25 degrees

Wind rose chart

Sunshine hours (h)

Block Generation

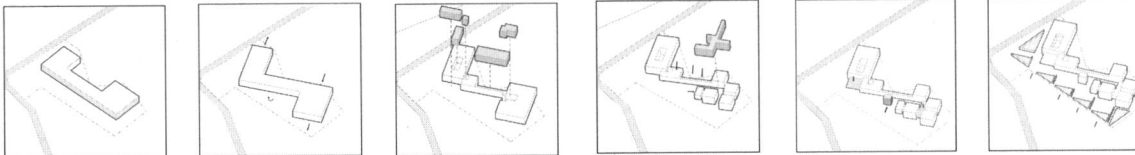

Design Strategy

Site-related issues

During winter, the northwest monsoon brings temperatures below freezing

The summer heat reaches temperatures exceeding 30 degrees Celsius

Hail may occasionally fall

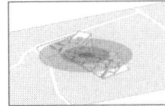

The region is prone to seismic disasters.

Electricity	Energy Consumption (kWh/m²)	Carbon Emission Factor (kgCO2/kWh)	Carbon Emissions (tCO2)
Cooling Supply (Qc)	200	0.581	227.332
Heating Supply (Qh)	630	0.581	838.007
Air Conditioning Fan (EY)	0	0.581	0
Illumination	746	0.581	1340.941
Socket Device	0	0.581	0
Others (Qo)	1430	0.581	2448.2

Carbon emissions from building operations

Proposed solutions

The northern facade features rammed earth walls, which serve to shield against the wind and provide insulation

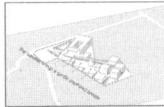

The architecture and partition walls are oriented towards the southeast, allowing the wind to flow through

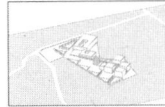

Establish meteorological monitoring stations to mitigate the threat of hailstorms

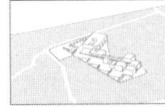

Utilize a steel-wood composite structure, incorporating traditional local techniques of buttressing for seismic resistance.

Area of Photovoltaic Panels (m²)	1900
Unit area Power generation parameters	0.4
Efficiency of the Photovoltaic System	0.8
Degradation Correction Factor for Photovoltaic Cell Performance	0.9
Annual Power Supply (kWh/a)	309024
Carbon Emission Factor (kg CO2/kWh)	0.581
Potential Reduction in Carbon Emissions (t CO2/year)	179.543

Photovoltaic power generation

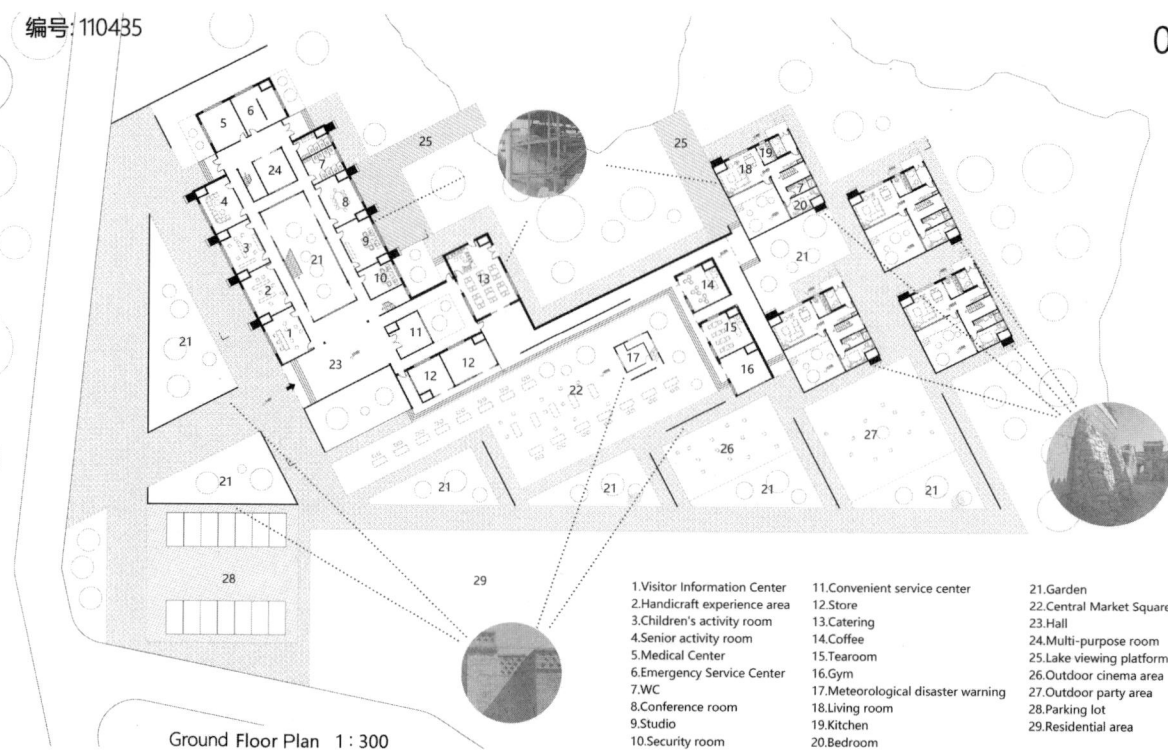

Ground Floor Plan 1:300

1.Visitor Information Center
2.Handicraft experience area
3.Children's activity room
4.Senior activity room
5.Medical Center
6.Emergency Service Center
7.WC
8.Conference room
9.Studio
10.Security room
11.Convenient service center
12.Store
13.Catering
14.Coffee
15.Tearoom
16.Gym
17.Meteorological disaster warning
18.Living room
19.Kitchen
20.Bedroom
21.Garden
22.Central Market Square
23.Hall
24.Multi-purpose room
25.Lake viewing platform
26.Outdoor cinema area
27.Outdoor party area
28.Parking lot
29.Residential area

Profiling Strategies

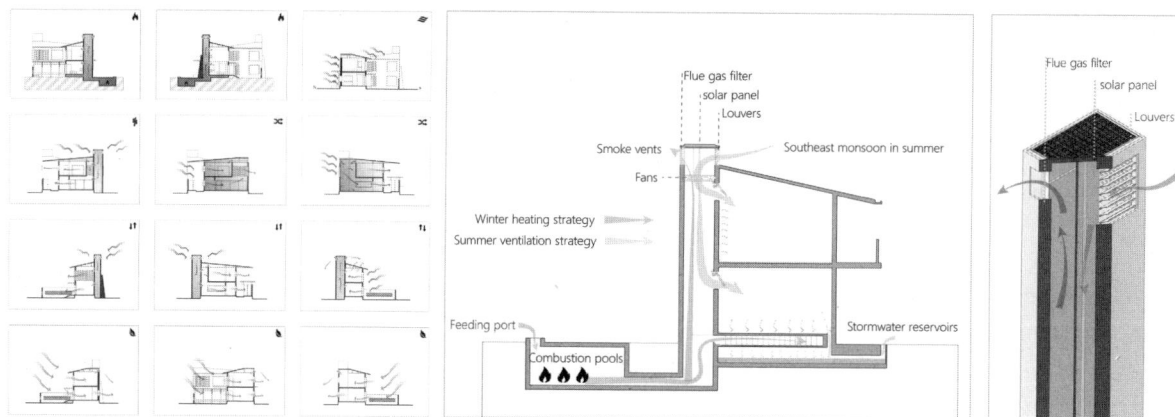

Sectional View

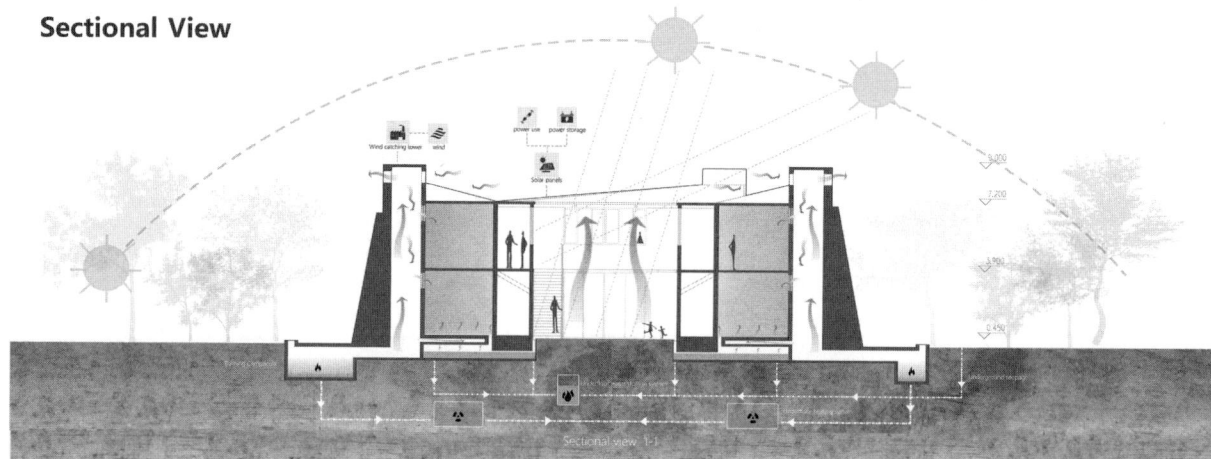

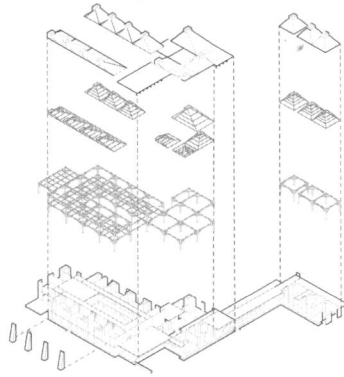
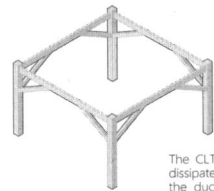

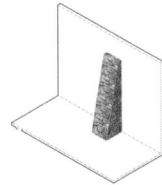

Second Floor Plan 1:300   Third Floor Plan 1:300

1.Reading room
2.Library
3.Village History Exhibition Hall
4.Performing arts room
5.Multi-purpose room
6.Cultural creativity
7.WC
8.Art Gallery
9.Bedroom
10.Observation deck

Detailed Explanation of the Seismic

CLT

KES

The CLT shear wall system can dissipate seismic energy through the ductile deformation of the connecting joints, and has good seismic performance.

The KES system can be used as a connector between CLT beams and columns, and has good structural performance.

The buttresses are inspired by a structure of traditional houses in Hebei Province that resist earthquakes.

Buttresses are used to resist the thrust of the coupon or arch and enhance the stability of the wall.

Elevation

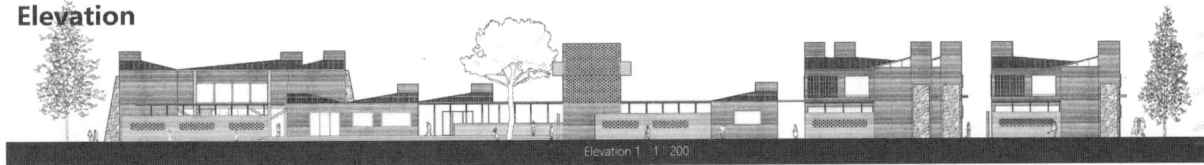

Elevation 1   1:200

Elevation 2   1:200

CODE:110465

固土・登临
Solid Soil-Boarding

综合奖・优秀奖・张家口市怀安县二堡子村建设项目
Comprehensive Awards － Honorable Mention － Erpuzi Village Project, Huai'an County, Zhangjiakou City

注册号：110465
Register Number：110465

项目名称：固土・登临
Entry Title：Solid Soil － Boarding

作者：刘学、厉相栋、李吉星
Authors：Liu Xue, Li Xiangdong, and Li Jixing

作者单位：山东建筑大学
Authors from：Shandong Jianzhu University

指导教师：魏瑞涵、侯世荣
Tutors：Wei Ruihan, Hou Shirong

指导教师单位：山东建筑大学
Tutors from：Shandong Jianzhu University

■ Design Description

本方案从韧性设计角度出发，对当地居民生活习性及民居的特征进行调研，通过对当地城墙、烽火台的现代化转译，在实现保温隔热的同时，结合天窗、双层墙及太阳能光伏板对建筑进行热压通风，以促进建筑内外空气交换。另外，我们为居民提供了一条可自由攀登建筑的路线，使人们能够体验到更丰富的空间环境。

This design takes a resilience approach, conducting research on the living habits and characteristics of local residents and their dwellings. It modernizes the local city walls and beacon towers while achieving thermal insulation and heat ventilation through the use of skylights, double walls, and solar photovoltaic panels. Additionally, it provides residents with a route to freely ascend the building, enabling them to experience a richer spatial environment.

■ Design Intent

■ Meteorological Data

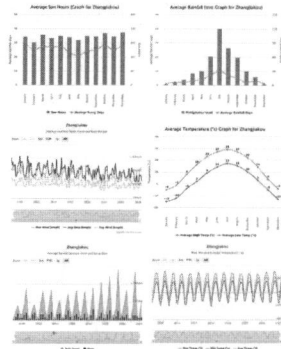

■ Design Ideas

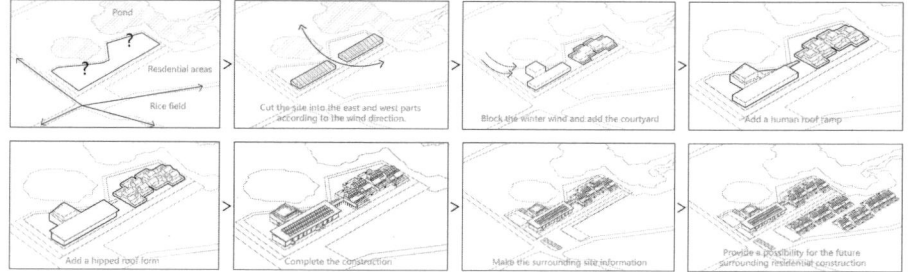

■ Master plan

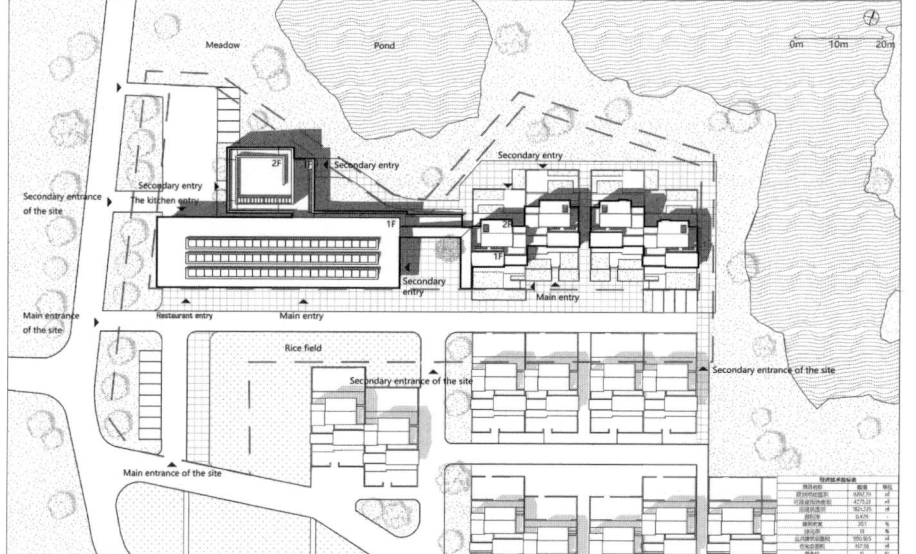

固土・登临
Solid Soil-Boarding

■ First Floor Plan 1：250

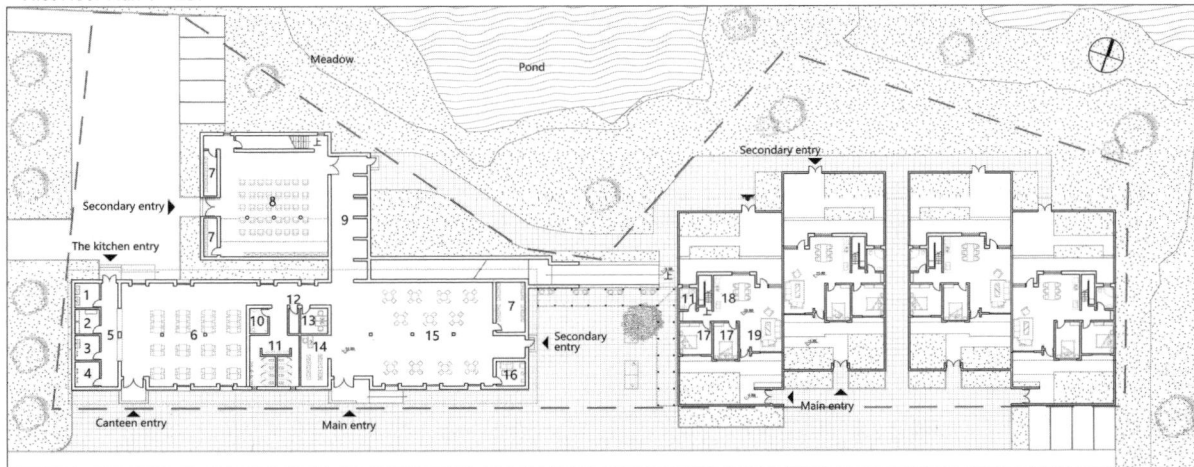

Meadow
Pond

Secondary entry

Secondary entry

The kitchen entry

7

8

7

9

1
2
3
4
5
6
10
11
12
13
14
15
16
7
Secondary
entry

11
18
17
17
19

Main entry

Canteen entry
Main entry

■ Second Floor Plan 1：250

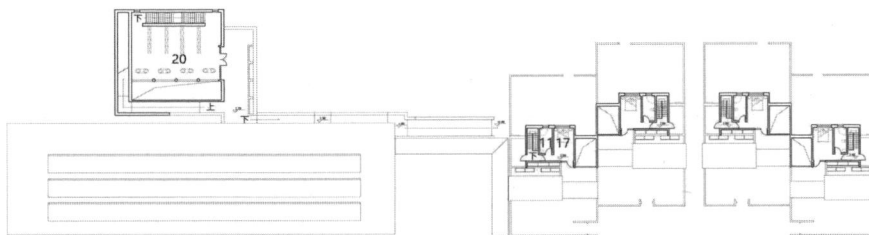

20

17

1 Wash and disinfectant house	11 toilets
2 Food bank	12 Equipment room
3 Rough processing of food	13 Management office
4 Thermal processing of food	14 Convenience stores
5 Sales window	15 Public recreation
6 Canteen	16 Pantry
7 Material storage	17 Bedchamber
8 Screening room	18 Restaurant
9 Exhibition area	19 Living room
10 Accessible toilets	20 Reading room

■ South Elevation 1：300   ■ Eest Elevation 1：300

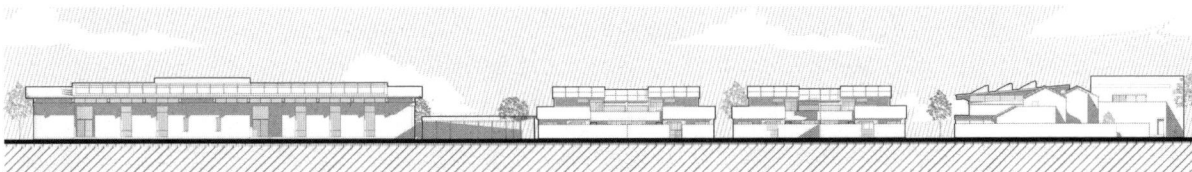

CODE:110465

■ Toughness Design Description

In terms of structure ,the focus is on making the rammed earth wall resistant to earthquakes,thus forming a rammed earth frame tube with certain earthquake resistance capabilities. In the post-disaster aspect, the original use space is changed to meet the post-disater code requirements, and an out door simple device is introduced to provide an outdoor use area. The indoor isolation area also considers the partiton device.

建筑使用阶段能耗计算结果		
	单个住宅能耗	
项目	总能耗 kW·h/a	单位面积能耗 kW·h/(m²·a)
供暖能耗	1.82	0.01
供冷能耗	1532.16	11.43
输配系统能耗	423.76	3.16
生活热水能耗	0.0	0.0
照明系统能耗	2295.02	17.11
电梯系统能耗	0.0	0.0
可再生能源发生量	3923814.62	29059.90
不含可再生能源发电的供	4252.76	31.71
热及耗电综合值		
建筑能耗综合值	-1916138	-98.14
	公共部分建筑	
项目	总能耗 kW·h/a	单位面积能耗 kW·h/(m²·a)
供冷能耗	1938.55	1.01
输配系统能耗	436.42	0.23
生活热水能耗	0.00	0.00
照明系统能耗	12784.1	6.6
电梯系统能耗	0.00	0.00
可再生能源发生量	1538334218.7	798259.7
不含可再生能源发电的综	15759.40	7.87
合热及耗电综合值		
建筑能耗综合值	-107475.41	-55.77

■ Technical Overview

固土·登临
Solid Soil - Boarding

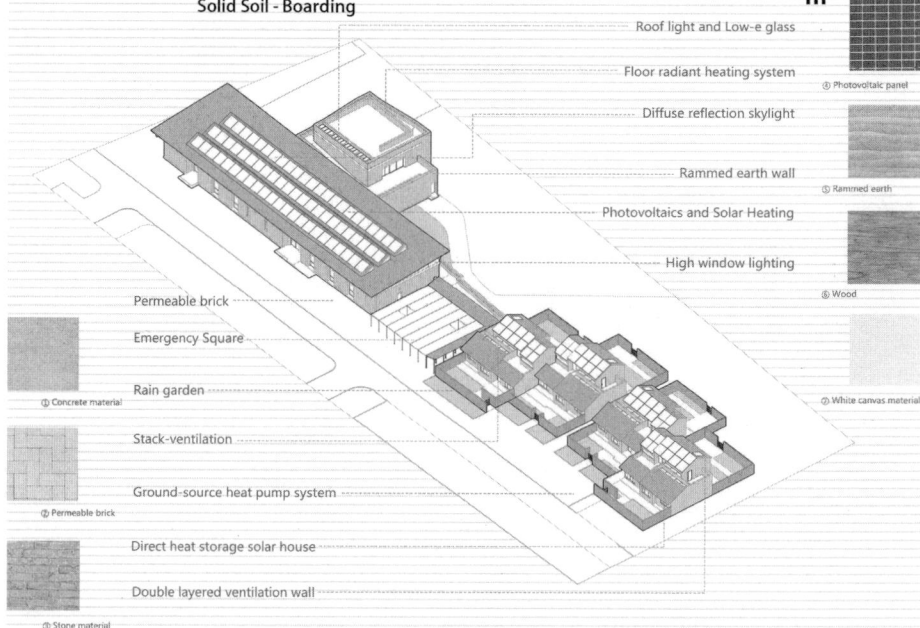

- Roof light and Low-e glass
- Floor radiant heating system
- Diffuse reflection skylight
- Rammed earth wall
- Photovoltaics and Solar Heating
- High window lighting
- Permeable brick
- Emergency Square
- Rain garden
- Stack-ventilation
- Ground-source heat pump system
- Direct heat storage solar house
- Double layered ventilation wall

① Concrete material
② Permeable brick
③ Stone material

④ Photovoltaic panel
⑤ Rammed earth
⑥ Wood
⑦ White canvas material

■ Device Description

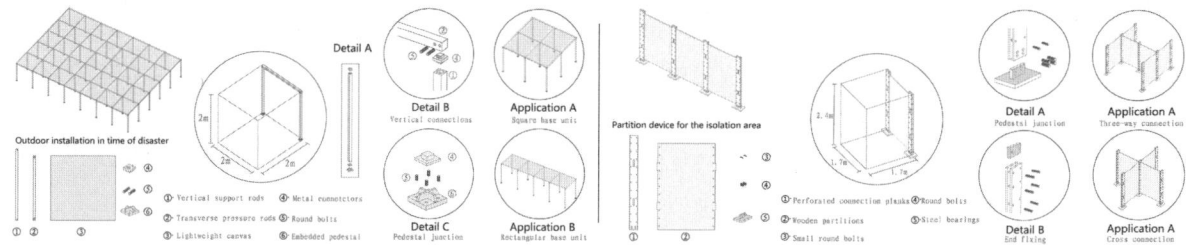

Outdoor installation in time of disaster

① Vertical support rods
② Transverse pressure rods
③ Lightweight canvas
④ Metal connectors
⑤ Round bolts
⑥ Embedded pedestal

Detail A
Detail B
Vertical connections
Application A
Square base unit
Detail C
Pedestal junction
Application B
Rectangular base unit

Partition device for the isolation area

① Perforated connection planks
② Wooden partitions
③ Small round bolts
④ Round bolts
⑤ Steel bearings

Detail A
Pedestal junction
Application A
Three-way connection
Detail B
End fixing
Application A
Cross connection

■ Structural Seismic

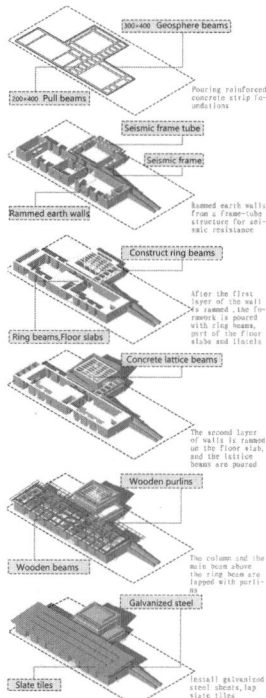

- 300*400 Geosphere beams
- 200*400 Pull beams — Pouring reinforced concrete strip foundations
- Seismic frame tube
- Seismic frame
- Rammed earth walls — Rammed earth walls from a frame-tube structure for seismic resistance
- Construct ring beams — After the first layer of the wall is rammed ,the formwork is poured with ring beams, part of the floor slabs and lintels
- Ring beams,Floor slabs
- Concrete lattice beams — The second layer of walls is rammed on the floor slab, and the lattice beams are poured
- Wooden purlins
- Wooden beams — The column and the main beam above the ring beam are lapped with purlins
- Galvanized steel
- Slate tiles — Install galvanized steel sheets, lap slate tiles

固土·登临
Solid Soil-Boarding

IV

■ Climate Analysis

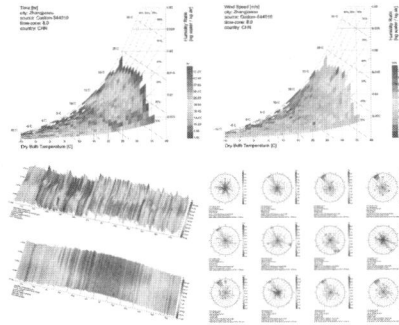

The winter in Zhangjiakou is cold and long, while the summer is hot; Affected by the climate of the Mongolian Plateau, the Zhangjiakou region is dominated by southeast winds in summer and northwest winds with high wind speeds in winter. The thermal zone belongs to the cold (A) zone, which requires both winter insulation and summer insulation.

■ Analysis Of Sunroom

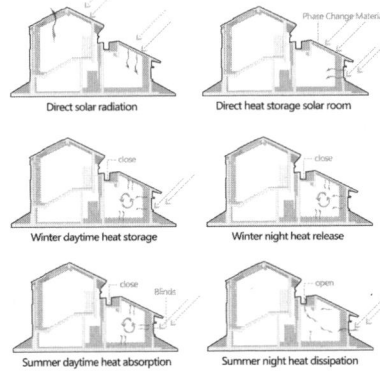

Direct solar radiation | Direct heat storage solar room
Winter daytime heat storage | Winter night heat release
Summer daytime heat absorption | Summer night heat dissipation

The region has abundant solar energy resources, large temperature differences between day and night, and good ventilation conditions at night. Adequate sunshine provides sufficient energy source for phase change materials to store heat in winter, large temperature difference between day and night provides conditions for solid-liquid circulation of phase change materials, and good nighttime ventilation provides climate guarantee for phase change materials to store cold in summer. This is conducive to the utilization of phase change materials for heat storage and release, and improves the efficiency of use.

■ Post-disaster Plane

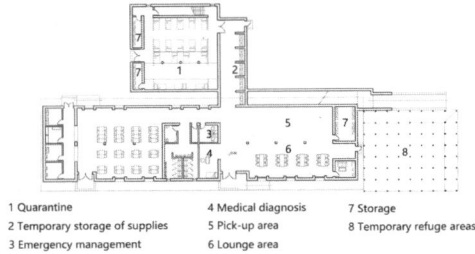

1 Quarantine
2 Temporary storage of supplies
3 Emergency management
4 Medical diagnosis
5 Pick-up area
6 Lounge area
7 Storage
8 Temporary refuge areas

■ Typical Day Sunshine Simulation

Vernal equinox day | Summer Solstice Day | Autumn Equinox Day | Winter Solstice Day

■ Outdoor Wind Environment Simulation

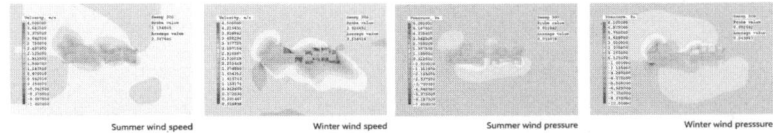

Summer wind speed | Winter wind speed | Summer wind pressure | Winter wind presssure

Due to the abundant summer sunshine in the region, we fully considered the natural lighting in the south direction during the design, and designed the roof slope of this building and placed photovoltaic panels according to the local latitude. Simultaneously design eaves for shading and insulation. In summer, the site fully utilizes the lake land breeze for ventilation, and at the same time, to cope with the strong winter breeze, we use thick rammed earth walls for insulation and heat preservation.

■ Comparison of Building Energy Consumption

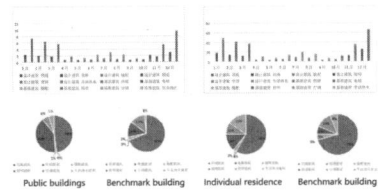

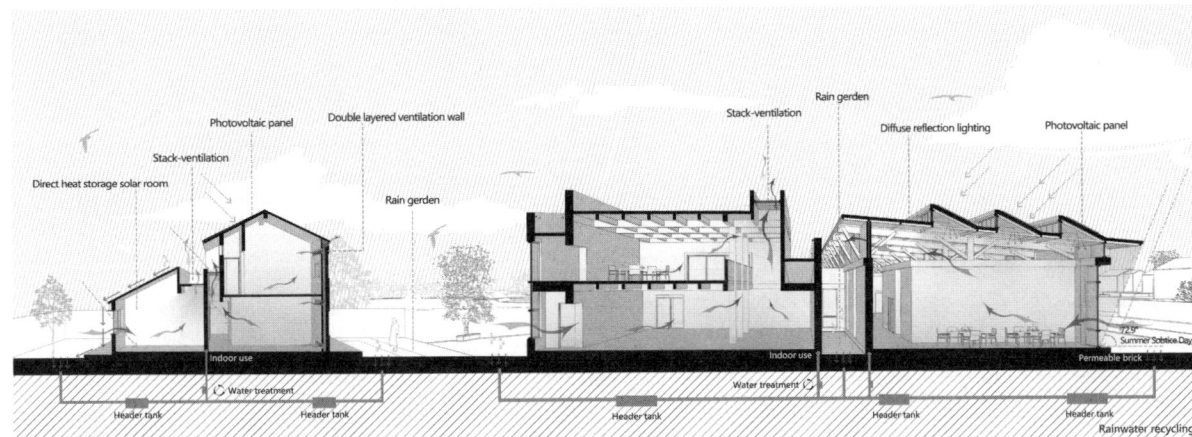

Public buildings | Benchmark building | Individual residence | Benchmark building

Direct heat storage solar room | Stack-ventilation | Photovoltaic panel | Double layered ventilation wall | Rain gerden | Stack-ventilation | Rain gerden | Diffuse reflection lighting | Photovoltaic panel

Indoor use | Water treatment | Header tank | Header tank | Indoor use | Water treatment | Header tank | Header tank | Header tank | Summer Solstice Day | Permeable brick | Rainwater recycling

综合奖·优秀奖·张家口市怀安县二堡子村建设项目
Comprehensive Awards − Honorable Mention − Erpuzi Village Project，Huai'an County，Zhangjiakou City

注册号：110500
Register Number：1105000

项目名称：坡起拾光·绿韵融居
Entry Title：Rise and Shine：Green Melody Melds

作者：罗逸文、李佳玉、许青芽、黄楠、郭纪壮
Authors：Luo Yiwen, Li Jiayu, Xu Qingya, Huang Nan, and Guo Jizhuang

作者单位：长安大学
Authors from：Chang'an University

指导教师：夏博
Tutor：Xia Bo

指导教师单位：长安大学
Tutor from：Chang'an University

作品编号：110500

Rise and Shine: Green Melody Melds ·

01

坡起拾光·绿韵融居

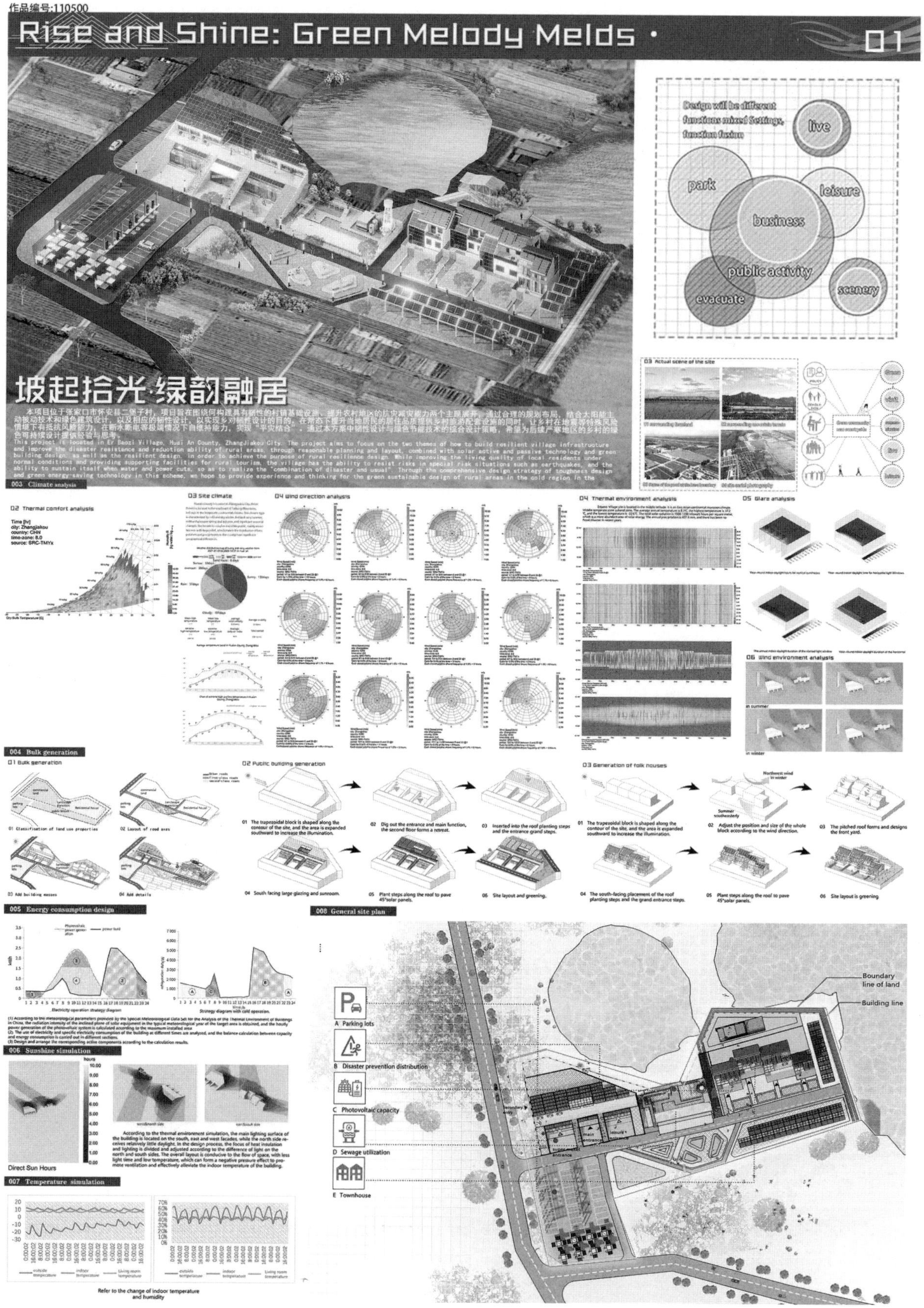

Damage analysis

The impact of the earthquake on Erbaozi Village, Hebei Province

According to the available information on the damage caused by the earthquake, Erbaozi Village in Hebei Province experienced severe damage to houses after the 6.2 magnitude earthquake. While the main information is focused on the affected areas near Jishishan County, similar affected villages are facing the same challenges.

Injuries and Losses: Casualties were reported in some villages. In Sibaozi Village, for example, 13 people were injured and one person was killed.

Infrastructure: The houses and infrastructure of the village were largely damaged in the earthquake, resulting in many residents temporarily living in tents or prefabricated houses. The government has allocated supplies to help villagers survive the winter, including warm materials such as stoves and cotton clothes.

Relief measures: The government and all sectors of society acted quickly to provide emergency tents, daily necessities, and began to build prefabricated houses. By the end of December, more than 10,000 prefabricated houses are expected to be built in the affected areas to ensure that the affected people can stay

Sponge City

roof planting

Water-tolerant plants

Green ladders

Rainwater catchment

pool

Filtration & Purification

Ecological Aquatic Plant Area

Natural climate Control system

Green roof filtration

Rainwater catchment

water storage

Service water

Irrigation water

Rainwater catchment

Disaster Prevention and Emergency Design

Disaster prevention and evacuation sites

Disaster prevention and relief roads

Emergency facilities

Disaster prevention barrier

Site Hazard Analysis

peacetime	versatile modular unit	peacetime	plantation	peacetime	viewing platform	peacetime	furniture
Disaster		Disaster	emergency tents	Disaster	water tower	Disaster	bed

East Elevation

South Elevation

作品编号:110500

Rise and Shine: Green Melody Melds ·

residential accommodation

plantation

Business

Water treatment station

Square

Main entrance

Parking lot

Small market

Secondary entrance

A A

Paking lot

Acommodation

Square&Water

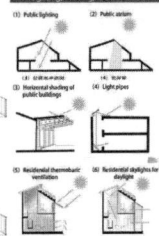

Plan of the First Floor of a Residence

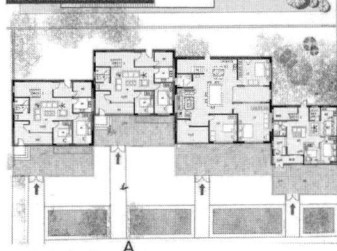

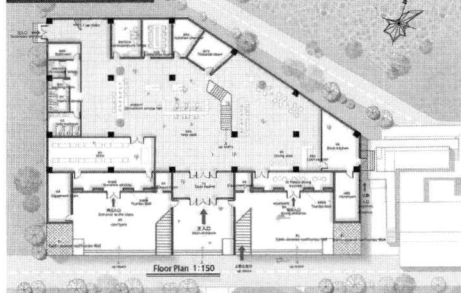

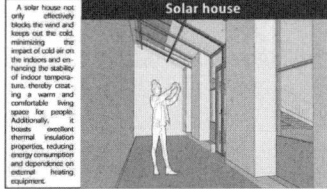

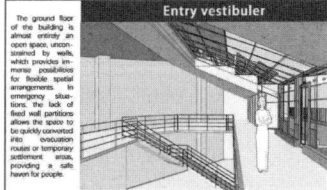

01 Trumbert Wall

(1) Winter daytime
(2) Winter nights
(3) Summer daytime
(4) Summer nights

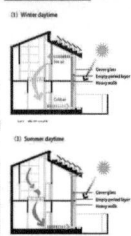

02 Solar house

(1) Winter daytimeA
(2) Winter daytimeB
(3) Winter nights
(4) Summer daytimeA
(5) Summer daytimeB
(6) Summer nights

(1) Public lighting
(2) Public atrium
(3) Horizontal shading of public buildings
(4) Light pipes
(5) Residential thermbuler ventilation
(6) Residential skylights for daylight

Plan of the First Floor of a Residence

Floor Plan 1:150

Second floor plan 1:150

Solar house

In severely cold regions, the establishment of an anteroom not only provides a buffer space between indoors and outdoors, but also significantly blocks the wind and keeps out the cold, effectively reducing the intrusion of cold air and thus enhancing the stability of indoor temperature.

Atrium

The atrium can effectively introduce natural light, not only bringing ample illumination to indoor spaces but also significantly reducing the dependence on artificial lighting, thus achieving remarkable energy-saving effects. In winter, the daylighting design of the atrium maximizes the utilization of solar energy, raising indoor temperatures and reducing heating requirements.

Solar house

A solar house not only effectively blocks the wind and keeps out the cold, minimizing the impact of cold air on the indoors and enhancing the stability of indoor temperature, thereby creating a warm and comfortable living space for people. Additionally, it boasts excellent thermal insulation properties, reducing energy consumption and dependence on external heating equipment.

Overall Strategy

Roof Planting

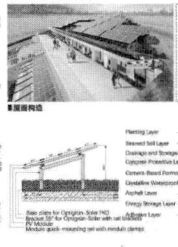

Indoor Atrium

Site Greening

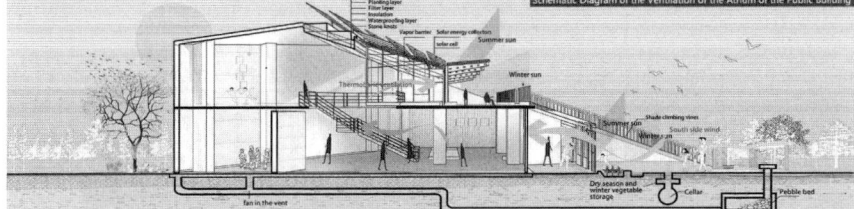

Entry vestibuler

The ground floor of the building is almost entirely an open space, unconstrained by walls, which provides immense possibilities for flexible spatial arrangements. In emergency situations, the lack of fixed wall partitions allows the space to be quickly converted into evacuation routes or temporary settlement areas, providing a safe haven for people.

Schematic Diagram of the Ventilation of the Atrium of the Public Building

CODE : 99-110547

本方案主题为"暖廊韧影"。以被动式太阳能技术为设计出发点，建筑局部升高形成通风光塔，利用集热暖廊将各建筑连接贯通。该暖廊不仅可以综合蓄热墙体解决日间得热蓄能与夜晚放热供能，房间隔温等功能，也可以在应急疏散过程中形成安全的逃生通道，并且墙体可拆卸，通过简易搭建可形成避难所，充分实现建筑节能的同时保证了该场地的韧性需求。在主动技术上，利用太阳能光伏系统与建筑屋顶进行一体化设计，引入雨水收集、中水系统，与水源热泵共同发挥水循环的作用，微阅建筑能源自维护。此外，在场地的东南侧设有应急疏散避难安全区，该区域预埋救灾棚基坑，方便快速构成避难单元。暖廊在实时状态可为避难单元供热供暖供能。

The theme of this proposal is "Disaster-Resistant Heat-Collecting Corridor". Starting from passive solar energy technology, the design features a locally elevated ventilation and light tower, using a heat-collecting warm corridor to connect the buildings. The warm corridor not only integrates heat-storing walls for daytime heat storage and nighttime heat release, room insulation, and emergency evacuation routes, but can also be dismantled to form temporary shelters, fully meeting the energy-saving and resilience requirements of the site.
On the active technology side, integrated design of the solar photovoltaic system with the building roof, rainwater collection, greywater system, and water source heat pump are utilized to achieve building energy self-sufficiency. Additionally, an emergency evacuation shelter is located on the southeast side of the site, with pre-buried foundation pits for disaster relief shelters, enabling rapid formation of refuge units. In times of disaster, the warm corridor can provide heating and energy supply to the refuge units.

暖廊韧影1
Disaster-Resistant Heat-Collecting Corridor

综合奖・优秀奖・张家口市怀安县二堡子村建设项目
Comprehensive Awards − Honorable Mention − Erpuzi Village Project, Huai'an County, Zhangjiakou City

注册号：110547
Register Number：110547
项目名称：暖廊韧影
Entry Title：Disaster − Resistant Heat − Collecting Corridor
作者：崔珂、吴静、张梦龙、蒋贤阳、田泽轩
Authors：Cui Ke, Wu Jing, Zhang Menglong, Jiang Xianyang, and Tian Zexuan
作者单位：西南民族大学、重庆大学
Authors from：Southwest Minzu University, Chongqing University
指导教师：熊健吾、张埕
Tutors：Xiong Jianwu, Zhang Yin
指导教师单位：西南民族大学
Tutors from：Southwest Minzu University

◻ Disaster Analysis

Frequency map of different kinds of disasters in Zhangjiakou city, Hebei province

◻ Site Resources

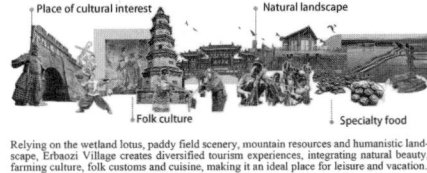

Place of cultural interest
Natural landscape
Folk culture
Specialty food

Relying on the wetland lotus, paddy field scenery, mountain resources and humanistic landscape, Erbaozi Village creates diversified tourism experiences, integrating natural beauty, farming culture, folk customs and cuisine, making it an ideal place for leisure and vacation.

◻ Site Population

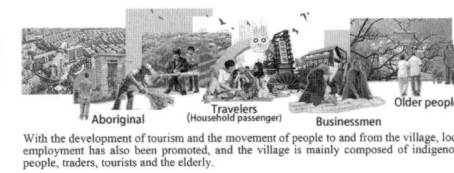

Aboriginal
Travelers (Household passenger)
Businessmen
Older people

With the development of tourism and the movement of people to and from the village, local employment has also been promoted, and the village is mainly composed of indigenous people, traders, tourists and the elderly.

◻ Site Environment

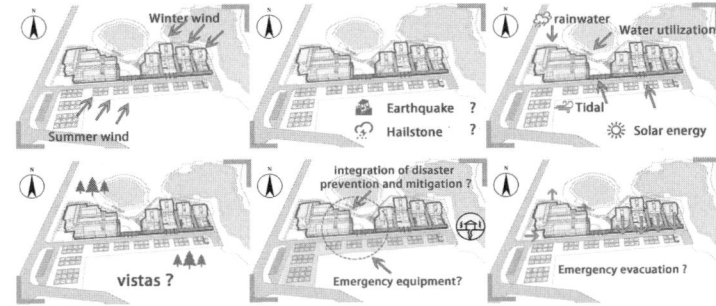

Winter wind
Summer wind
Earthquake ?
Hailstone ?
vistas ?
integration of disaster prevention and mitigation ?
Emergency equipment?
rainwater
Water utilization
Tidal
Solar energy
Emergency evacuation ?

◻ Climate Simulation:

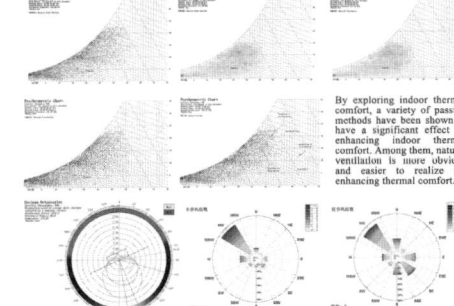

By exploring indoor thermal comfort, a variety of passive methods have been shown to have a significant effect on enhancing indoor thermal comfort. Among them, natural ventilation is more obvious and easier to realize for enhancing thermal comfort.

Through the preliminary analysis of the climate of Zhangjiakou, it can be obtained that the south-east is the best orientation for the building in the area, which can get enough light. Meanwhile, through the exploration of indoor thermal comfort, a variety of passive methods have a significant effect on improving indoor thermal comfort.

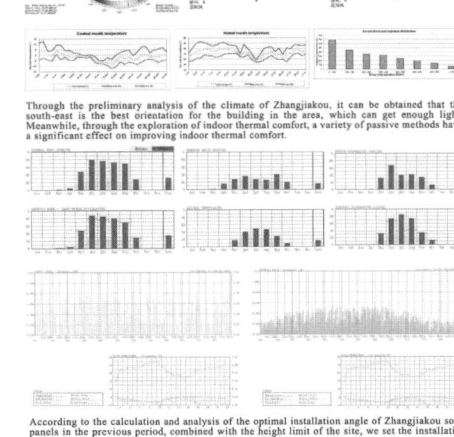

According to the calculation and analysis of the optimal installation angle of Zhangjiakou solar panels in the previous period, combined with the height limit of the site, we set the installation angle of solar panels on the roof of the building at 38 degrees.

◻ Integrated Technology Strategies

Active Solar Energy Utilization
Solar system
Solar collector:
Heating system
Hot water system
Water source heat pump
Photovoltaic system:
Photovoltaic panel
Photovoltaic glass
Thermal insulation
Trumbull Wall
Phase change material
Passive Solar Energy Utilization
Architectural form
Trumbull Wall
Window design
Sunshade
Natural air, heat
Ventilation design
Breezeway
Sunroom
Other Green Technologies
Rainwater harvesting systems
Central water system
Disaster prevention technology
Rainwater harvesting systems
Rainwater harvesting systems
evacuation corridor
Solar Energy

CODE : 99-110547

暖廊韧影2

Disaster-Resistant Heat-Collecting Corridor

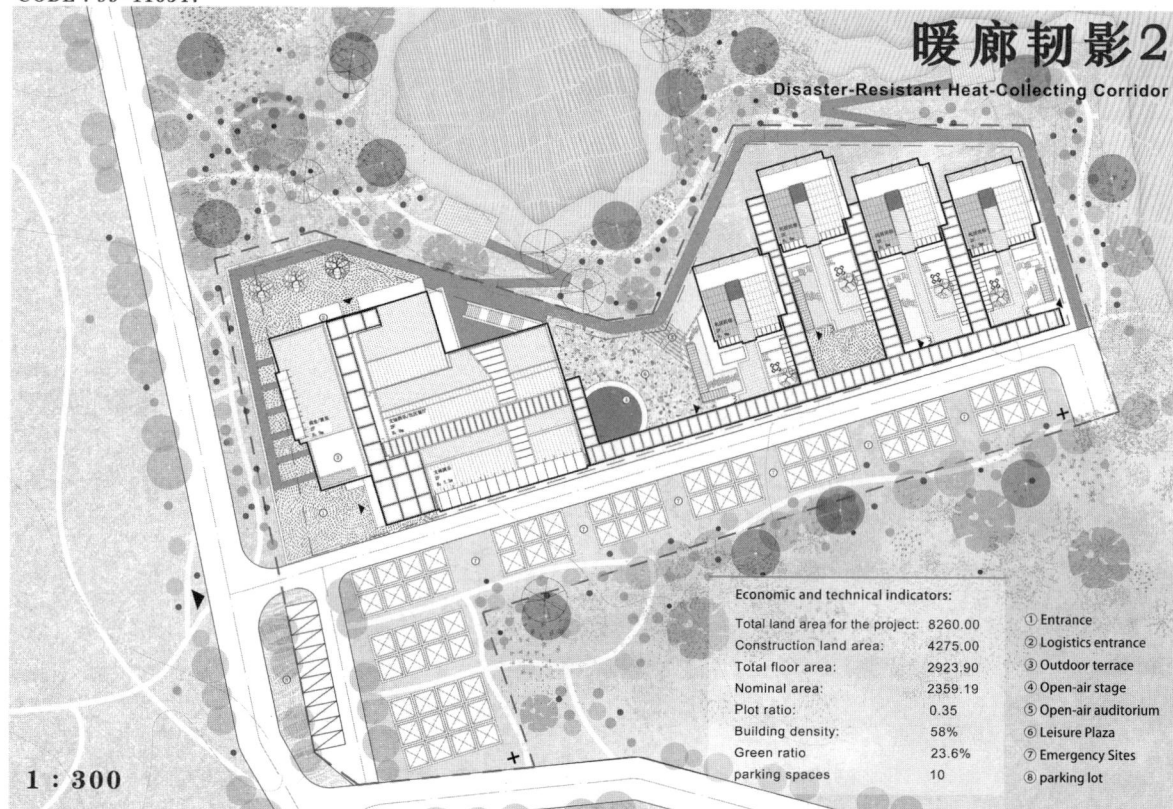

1 : 300

Economic and technical indicators:

Total land area for the project: 8260.00
Construction land area: 4275.00
Total floor area: 2923.90
Nominal area: 2359.19
Plot ratio: 0.35
Building density: 58%
Green ratio: 23.6%
parking spaces 10

① Entrance
② Logistics entrance
③ Outdoor terrace
④ Open-air stage
⑤ Open-air auditorium
⑥ Leisure Plaza
⑦ Emergency Sites
⑧ parking lot

■ Public Building 1F

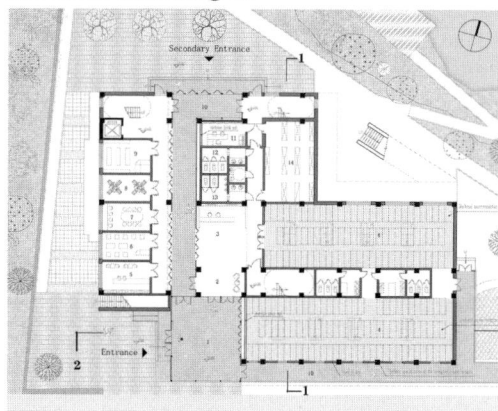

■ Residence 1F

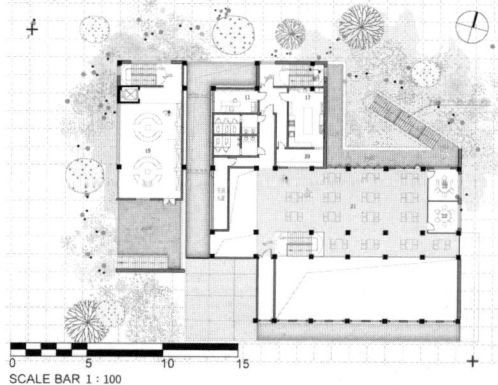

■ Functional Flow Lines

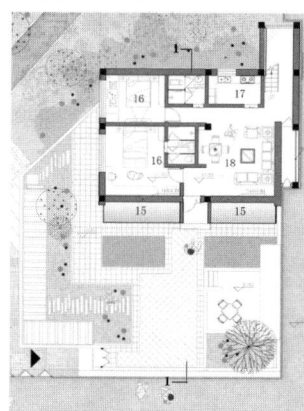

■ Public Building 2F

SCALE BAR 1 : 100

■ Residence 2F

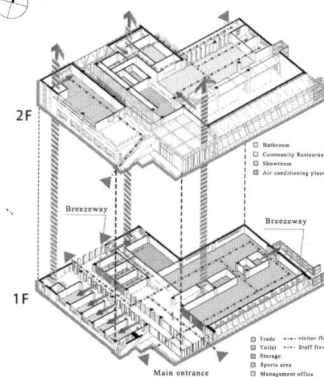

1 Foyer/Collector Sunroom
2 Resting area
3 Visitors' Desk
4 Exercise area/emergency evacuation area
5 Leisurely Water Bar
6 Cultural and Creative Store
7 Cultural room
8 chess room
9 Convenience store
10 Collective heating/ventilation corridors
11 Management office
12 Women's Restroom
13 Men's Restroom
14 Storage/emergency storage
15 Sunroom
16 Bedroom
17 Kitchen
18 Living room
19 Experience/Demonstration
20 Serve a meal
21 Community Restaurant
22 Private room
23 B&B rooms
24 Corridors
25 Overhead

■ Sample Node

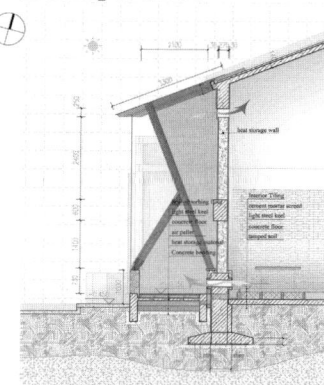

CODE : 99-110547

☐ **Section 1-1 1：50**

暖廊韧影3

Disaster-Resistant Heat-Collecting Corridor

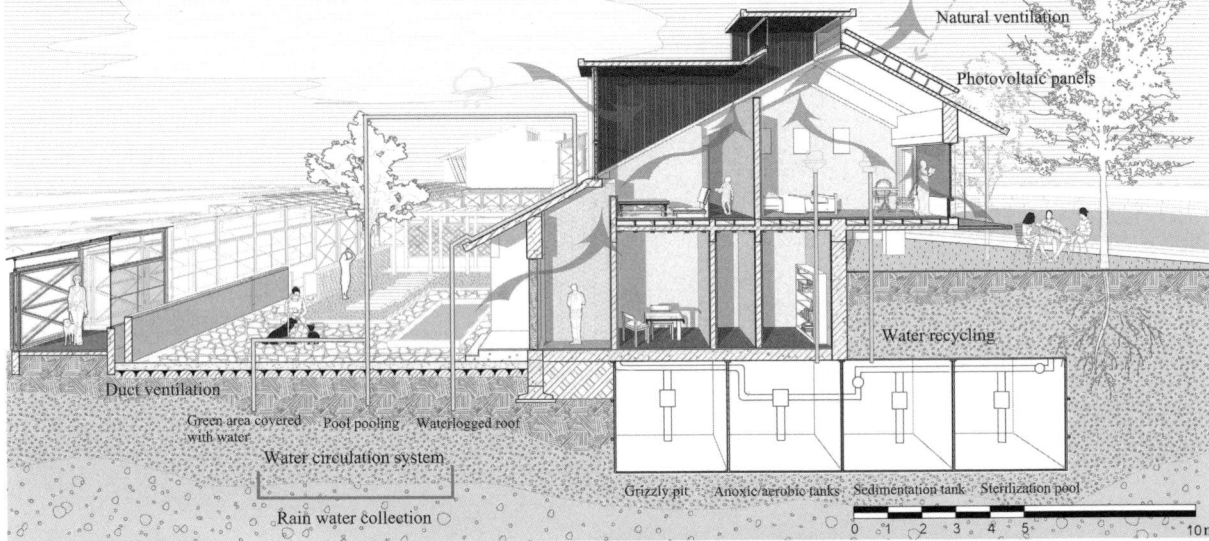

Natural ventilation

Photovoltaic panels

Water recycling

Duct ventilation

Green area covered with water   Pool pooling   Waterlogged roof

Water circulation system

Grizzly pit   Anoxic/aerobic tanks   Sedimentation tank   Sterilization pool

Rain water colleetion

0 1 2 3 4 5   10m

☐ **Framework**

opening for ventilation

Photovoltaic glass

Column as Heat Source

Corridor

Trombe Wall

Ventilation brick

Foundation system

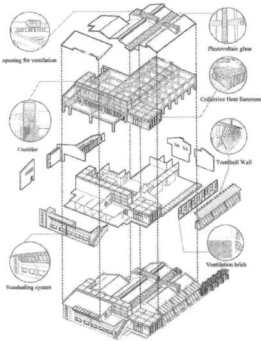

☐ **Emergency Evacuation**

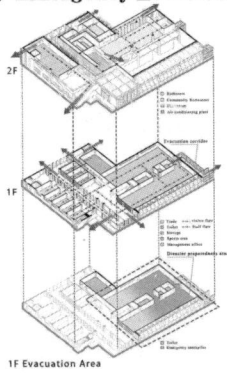

2F

1F

Evacuation corridor

Disaster preparedness area

1F Evacuation Area

☐ **Heated Corridor Units**

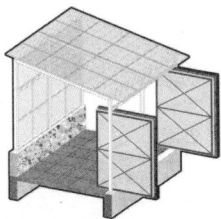

Structure System:
Steel Frame Fabricated Structure

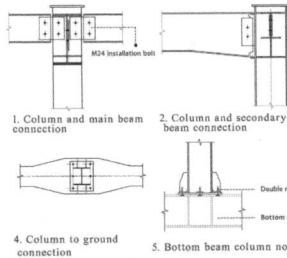

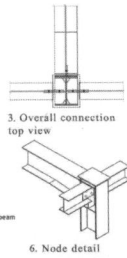

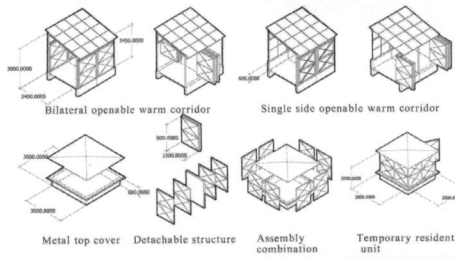

M24 installation bolt

1. Column and main beam connection
2. Column and secondary beam connection
3. Overall connection top view

Double nut

Bottom steel beam

4. Column to ground connection
5. Bottom beam column node
6. Node detail

Bilateral openable warm corridor   Single side openable warm corridor

Metal top cover   Detachable structure   Assembly combination   Temporary residential unit

☐ **Peacetime Disaster Conversion**

CODE：99-110547

暖廊韧影 4
Disaster-Resistant Heat-Collecting Corridor

By fully integrating the ventilation corridors, photovoltaic panels and the building, the project realizes a better wind environment and self-sustainability of the building's energy consumption. In terms of building energy consumption, the community activity center consumes 330,000kWh annually, which can be self-sustained by laying photovoltaic panels on the roof of the building and the warm corridor. In the B&B building, the annual power consumption is 3.6%kWh, which can be self-sustained by laying photovoltaic panels on the roof of the B&B and the warm corridor. In terms of ventilation, the building layout is set up in the direction of the main incoming wind. By relying on the building itself for wind shielding and wind blocking, a better winter wind blocking effect is realized. In the internal space of the building, thanks to the ventilation corridor design of the community activity center, the natural ventilation inside the building is good. At the same time, for the B&B building, thanks to the setting of through-height space, the indoor ventilation effect is improved while creating the architectural atmosphere, which better meets the requirements of B&B indoor ventilation and air exchange.

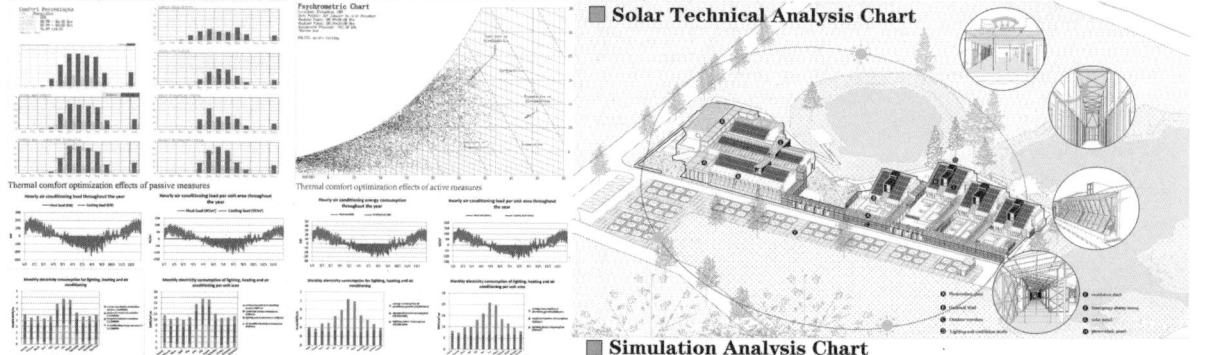

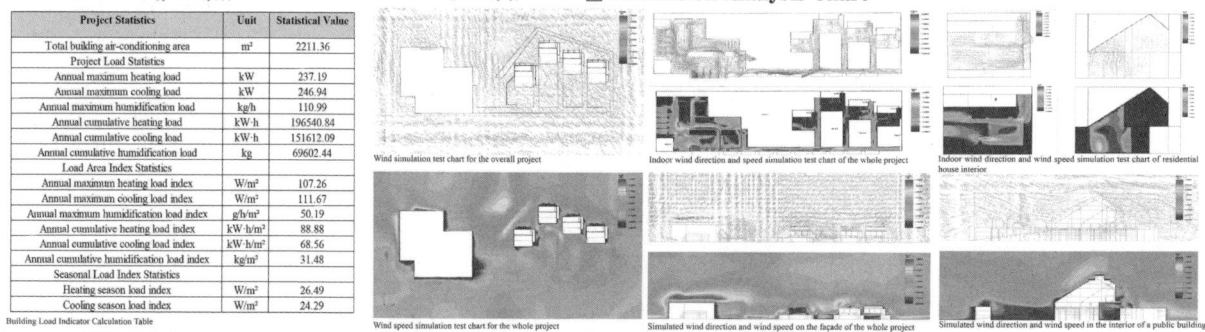

Thermal comfort optimization effects of passive measures

Thermal comfort optimization effects of active measures

■ Solar Technical Analysis Chart

■ Simulation Analysis Chart

Project Statistics	Unit	Statistical Value
Total building air-conditioning area	m²	2211.36
Project Load Statistics		
Annual maximum heating load	kW	237.19
Annual maximum cooling load	kW	246.94
Annual maximum humidification load	kg/h	110.99
Annual cumulative heating load	kW·h	196540.84
Annual cumulative cooling load	kW·h	151612.09
Annual cumulative humidification load	kg	69602.44
Load Area Index Statistics		
Annual maximum heating load index	W/m²	107.26
Annual maximum cooling load index	W/m²	111.67
Annual maximum humidification load index	g/h/m²	50.19
Annual cumulative heating load index	kW·h/m²	88.88
Annual cumulative cooling load index	kW·h/m²	68.56
Annual cumulative humidification load index	kg/m²	31.48
Seasonal Load Index Statistics		
Heating season load index	W/m²	26.49
Cooling season load index	W/m²	24.29

Building Load Indicator Calculation Table

Wind simulation test chart for the overall project

Indoor wind direction and speed simulation test chart of the whole project

Indoor wind direction and wind speed simulation test chart of residential house interior

Wind speed simulation test chart for the whole project

Simulated wind direction and wind speed on the façade of the whole project

Simulated wind direction and wind speed in the interior of a public building

■ 1-1 Section 1：100

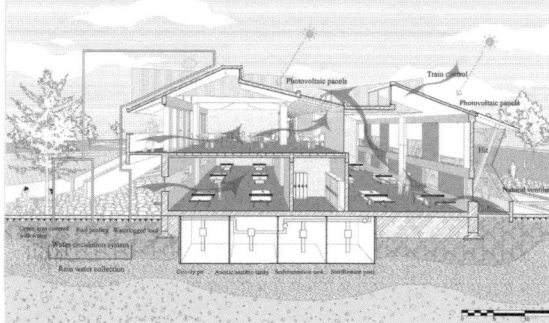

■ 2-2 Section 1：100

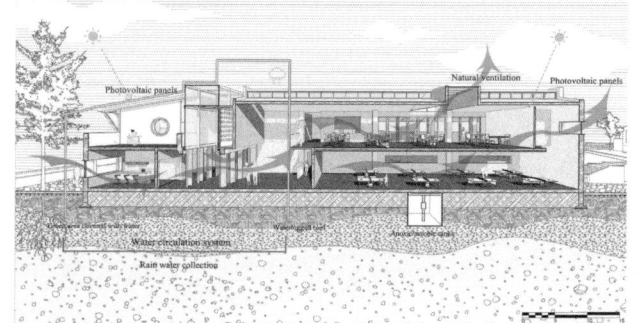

Towards a Solar Organic Architecture
走向太阳能有机建筑 1

AERIAL VIEW

DESIGN GENERATION

HISTORICAL EVOLUTION

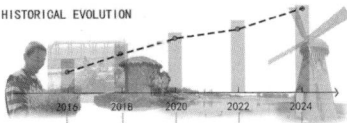

| 2016 | 2018 | 2020 | 2022 | 2024 |

Erbuzi Village and Prun Pure Ecology Cooperation with Agricultural Limited Company.Developing green and organic industries Leisure and tourism industry creating The Straw Culture of Northern and Southern China Industrial park.

Erbuzi Village Supporting the development of the tourism industry increasing income for impoverished households, in practice Stable poverty alleviation for the brand of "Rice Fragrance Countryside".

Huai'an County takes Erbuzi Village as its starting point, inter grating the local antique water resources with the long-standing agricultural culture, and constantly setting the company for planting organic rice.

Erbuzi Village has introduced Dezhong Ecological Industry Development, and all 1622 acres of rice fields have been transferred to the company for planting organic rice.

Huai'an County focuses on the "four focuses" and the achievements of poverty alleviation have been continuously consolidated and expanded, laying a solid foundation for rural revitalization.

HOMESTAY DESIGN

Generally, homestay residences lack private space, with several rooms sharing this same public open space, which results in a lack of personal and private activity space for tourists during their stay.

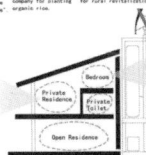

DESGIN PURPOSE

有机建筑观点的核心是自然、人和建筑的和谐共处。有机建筑的观点强调"形成建筑复杂整体的各个子系统，应该是相互依赖的、有内在联系的。"和"建筑应该有生、死、生长等自然过程的生命性特征。"在当代环境保护和"碳中和"目标下，有机建筑理论为当今的生态建筑实践提供了新的启发。

At the heart of the organic architectural perspective is the harmonious coexistence of nature, people and architecture. The perspective of organic architecture emphasizes that "the subsystems that form a complex whole of architecture should be interdependent and intrinsically related." And "Architecture should have the life-like characteristics of natural processes such as birth, death, and growth." . Under the contemporary goal of environmental protection and carbon neutrality, the theory of organic architecture provides new inspiration for today's ecological architecture practice.

在我的设计方案，以"让可再生能量系统与建筑生成和运行深度融合，走向新的太阳能有机建筑"为核心设计概念。围绕这个核心理论，提出了以下的设计原则：

In my design plan, the core design concept is "to deeply integrate the renewable energy system with the generation and operation of the building, and move towards a new solar-powered organic building". Focusing on this core theory, the following design principles are proposed:

1. 建筑形式的生成以可再生能源利用为导向。
1. The generation of building form is oyriented towards the use of renewable energy.

2. 能量系统的设备布局与空间系统和建筑构造有机结合。
2. The equipment layout of the energy system is organically combined with the space system and the building structure.

3. 让设备系统的日常维护得更简单，保证系统的寿命得以延长。
3. Make the daily maintenance of the equipment system simpler and ensure that the life of the system can be extended.

4. 能量系统、空间系统和结构系统均有可生长性，可以方便的扩展更改。
4. The energy system, space system and structural system are all growable and can be easily expanded or changed.

DESGIN CONCEPT

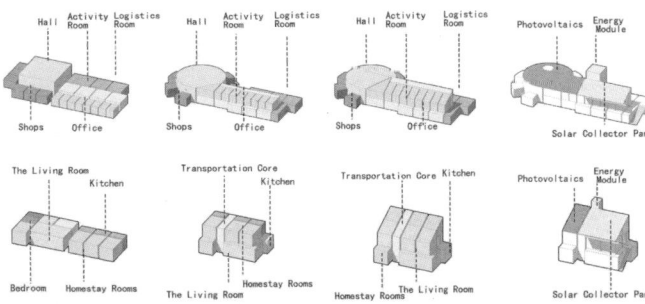

The village activity center of this plan aims to create a comprehensive complex that integrates emergency shelter, village activities, and tourism economy to better serve tourists and local residents, as well as respond to sudden

This design aims to enhance the privacy of residents and provide a better accommodation experience while ensuring the full utilization of various functions of the homestay.
Both the bedroom and private living room have better views, allowing for better interaction with the surrounding scenery.

MASTER PLAN 1:500

Floorage:Activity Center 1100 ㎡
Homestay 237 ㎡x4
Total Area:2048 ㎡
Land Area:8262 ㎡
Plot Ratio:0.25
Parking Space:26

综合奖·优秀奖·张家口市怀安县二堡子村建设项目
Comprehensive Awards － Honorable Mention － Erpuzi Village Project, Huai'an County, Zhangjiakou City

注册号：110632
Register Number：110632
项目名称：走向太阳能有机建筑
Entry Title：Towards a Solar Organic Architecture
作者：白苏日吐、李子涵
Authors：Bai Suritu, Li Zihan
作者单位：天津大学、内蒙古工业大学
Authors from：Tianjin University, Inner Mongolia University of Technology
指导教师：朱丽
Tutor：Zhu Li
指导教师单位：天津大学
Tutor from：Tianjin University

110632

Towards a Solar Organic Architecture
走向太阳能有机建筑 2

HUMAN VISUAL RENDERING

Sectional Drawing

D-D　　　C-C　　　B-B　　　A-A

First Floor Plan 1 : 200

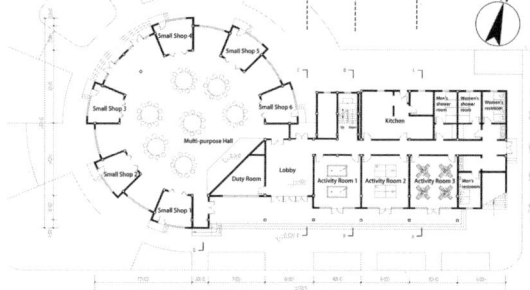

Construction

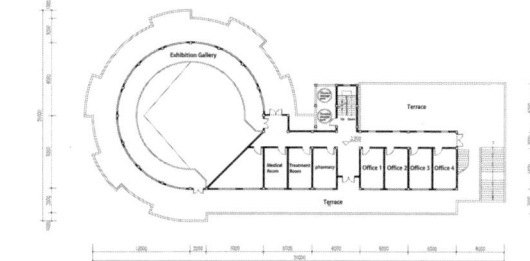

Site:
Huaian, Hebei (2A)
SHAPE FACTOR:
0.4
Window Ratio:
S 0.35
N 0.20
E 0.08
W 0.29
AREA: 1100㎡

Energy Consuption

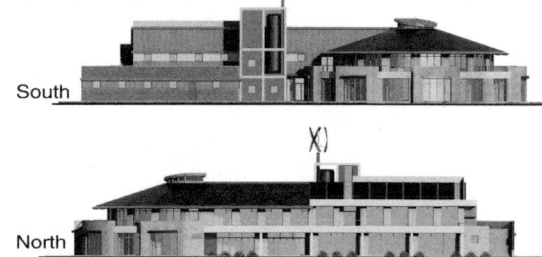

Heating period:
11.01—3.31 (NO Cooling)
Heating Energy consumption intensity:
51.96kWh/㎡
Comprehensive energy consumption intensity:
80.61kWh/㎡
non-heating energy consumption intensity:
54.51kWh/㎡

Second Floor Plan 1 : 200

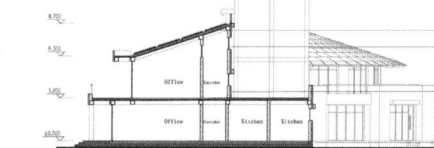

Solar Photovoltaic (PV) System

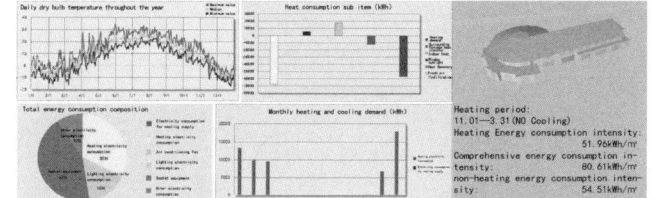

Monthly sunshine hours (h)

Total solar irradiance (kWh/㎡)　　First year power generation (MWh)

PV panel:Single crystal silicon
Peak power per panel:260W
Number of photovoltaic panels:152
Total power:39.52kw
Total irradiance in the horizontal plane:1494.1kWh/㎡.a
Total annual electricity generation: 50.7MWh
Proportion of non-heating energy consumption provided by renewable energy:84%

Elevation View 1 : 200

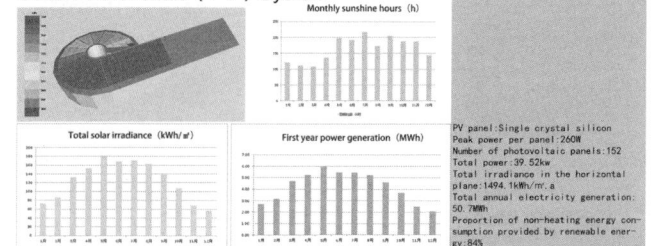

South

Solar Thermal System

A:Domestic hot water collector plates
plates area:9.43㎡
Running time:365d
Hot water demand:50941.83MJ
Solar panels collect heat:54832MJ
Proportion of domestic hot water provided by renewable energy:100%

B:Domestic hot water collector plates
plates area:114.63㎡
Running time:150d (11.01-3.31)
heating demand:51.96kWh/㎡X1100㎡
Solar panels collect heat:198296MJ
Proportion of heating provided by renewable energy:96%

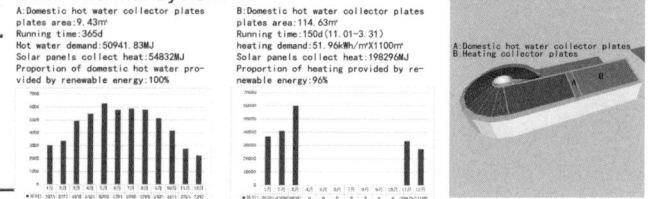

A:Domestic hot water collector plates
B:Heating collector plates

North

Towards a Solar Organic Architecture
走向太阳能有机建筑 3

HUMAN VISUAL RENDERING

First Floor Plan 1：200

Sectional Drawing

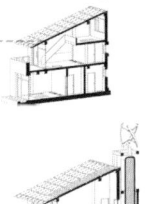

Construction

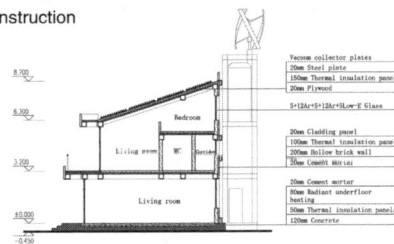

Vacuum collector plates
20mm Steel plate
150mm Thermal insulation panels
20mm Plywood

5+12Ar+5+12Ar+5Low-E Glass

20mm Cladding panel
100mm Thermal insulation panels
200mm Hollow brick wall
30mm Cement mortar

20mm Cement mortar
80mm Radiant underfloor heating
50mm Thermal insulation panels
120mm Concrete

D-D

Bedroom

Living room   WC   Corridor

Living room

Site:
Huaian,Hebei (2A)
SHAPE FACTOR:
0.59
Window Ratio:
S 0.42
N 0.18
E 0.00
W 0.00
AREA: 237㎡

Second Floor Plan 1：200

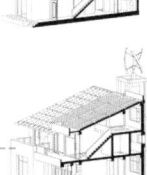

C-C

Energy Consupion

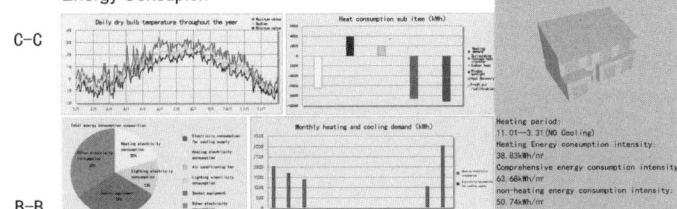

Daily dry bulb temperature throughout the year

Heat consumption sub item (kWh)

Monthly heating and cooling demand (kWh)

Heating period:
11.01-3.31(NO Cooling)
Heating Energy consumption intensity:
38.83kWh/㎡
Comprehensive energy consumption intensity:
63.68kWh/㎡
non-heating energy consumption intensity:
50.74kWh/㎡

Third Floor Plan 1：200

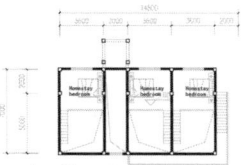

B-B

Solar Photovoltaic (PV) System

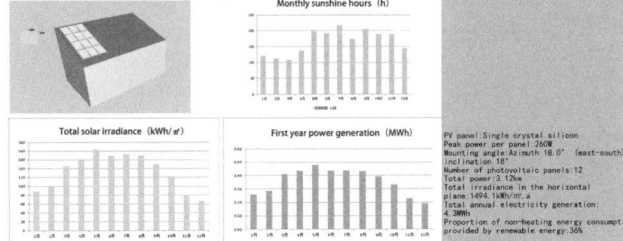

Monthly sunshine hours (h)

Total solar irradiance (kWh/㎡)

First year power generation (MWh)

PV panel:Single crystal silicon
Peak power per panel:260W
Mounting angle:Azimuth 16.0° (east-south)
inclination 18°
Number of photovoltaic panels:12
Total power 3.12kw
Total irradiance in the horizontal plane:1494.1kWh/㎡.a
Total annual electricity generation:
4.30Wh
Proportion of non-heating energy consumption provided by renewable energy:36%

Elevation View 1：200

A-A

South   West

Solar Thermal System

A:Domestic hot water collector plates
plates area:4.21㎡
Running time:365d
Hot water demand:30565.10MJ
Solar panels collect heat:24438MJ
Proportion of domestic hot water provided by renewable energy:80%

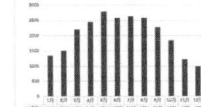

B:Domestic hot water collector plates
plates area:56.58㎡
Running time:150d(11.01-3.31)
Heating demand:38.83kWh/㎡X237㎡
Solar panels collect heat:97869MJ
Proportion of heating provided by renewable energy:100%

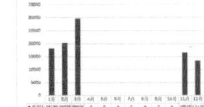

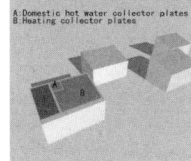

A:Domestic hot water collector plates
B:Heating collector plates

North   East

110632

Towards a Solar Organic Architecture
走向太阳能有机建筑 4

Ventilation & Air Tempeaeture（without AC）

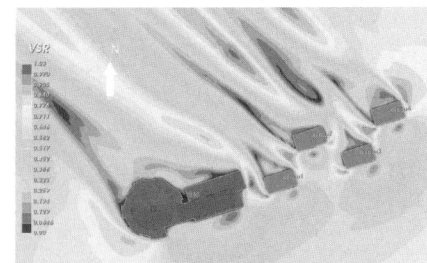

Summer design day（7.21）

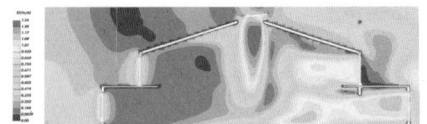

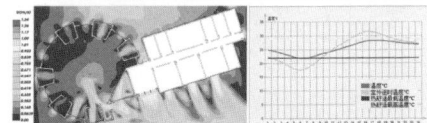

Summer design day（7.21）

By implementing a hot press ventilation strategy, the indoor air environment and temperature (without air conditioning) of the multifunctional auditorium can meet the requirements of thermal comfort

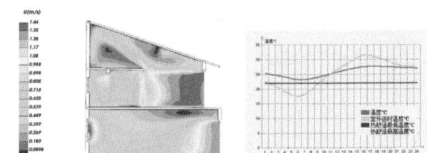

Summer design day（7.21）

By using a cross-sectional strategy, the indoor air environment and temperature (without air conditioning) of the living room and bedroom in the guest room can meet the requirements of thermal comfort

Space Flexibility

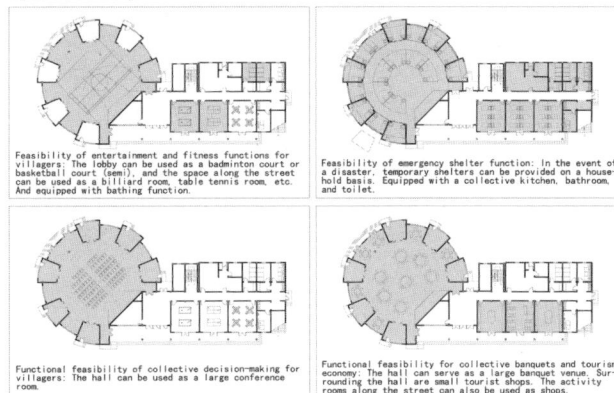

Feasibility of entertainment and fitness functions for villagers: The lobby can be used as a badminton court or basketball court (semi), and the space along the street can be used as a billiard room, table tennis room, etc. And equipped with bathing function.

Feasibility of emergency shelter function: In the event of a disaster, temporary shelters can be provided on a household basis. Equipped with a collective kitchen, bathroom, and toilet.

Functional feasibility of collective decision-making for villagers: The hall can be used as a large conference room.

Functional feasibility for collective banquets and tourism economy: The hall can serve as a large banquet venue. Surrounding the hall are small tourist shops. The activity rooms along the street can also be used as shops.

Energy Flexibility

The village activity center adopts three renewable energy systems, including wind power generation system, solar energy collection system, and solar photovoltaic power generation system, to cope with the energy shortage caused by disasters.

Vertical axis wind turbine:
Total power:1kw
Annual power generation:6MWh

Solar thermal system
Area: 124㎡
Collect heat:253128MJ

Photovoltaic system
Total power:39.52kw
Total annual electricity generation:50.7MWh

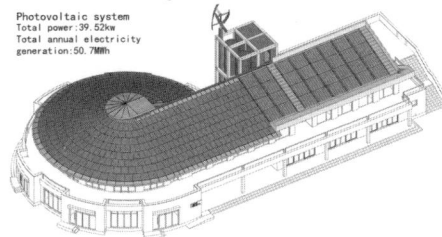

Structure Flexibility

The energy module and building structure are independent of each other and arranged in parallel. Its advantage is that it provides great convenience for energy system modification and spatial adjustment.

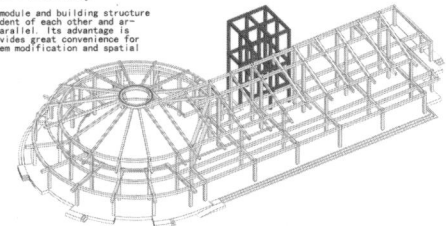

Energy Systiem

The clever integration of renewable energy systems into building structures and spaces is the focus of our design. Its challenges are:
1. Rational layout of renewable energy types and values according to site conditions and functional requirements.
2. Find a reasonable spatial location and volume for the components of the energy system.

Vertical axis wind turbine:
Total power:0.6kw
Annual power generation: 3.6MWh

Solar thermal system
Photovoltaic area:60.79㎡
Solar panels collect heat:122307MJ
Proportion of domestic hot water provided by renewable energy:80%
Proportion of heating provided by renewable energy:100%

Photovoltaic system:
Total power:3.12kw
Total irradiance in the horizontal plane:1494.1kWh/㎡.a
Total annual electricity generation:4.3MWh

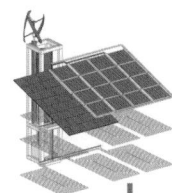

Construct Logic

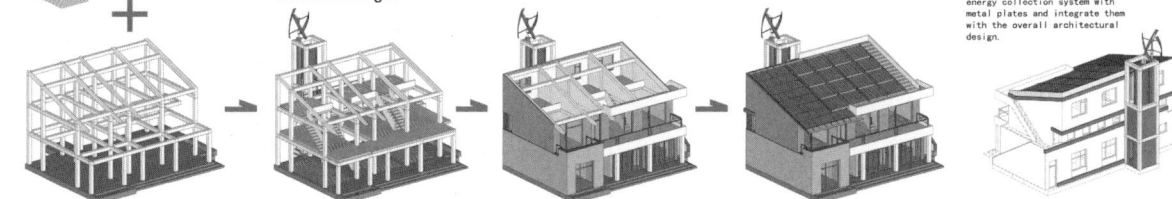

Wrap the pipes of the solar energy collection system with metal plates and integrate them with the overall architectural design.

廊・稻 1
——民居及公共服务建筑设计
Corridor - Paddy: Residential and Public Service Building Design

综合奖・优秀奖・张家口市怀安县二堡子村建设项目
Comprehensive Awards － Honorable Mention － Erpuzi Village Project，Huai'an County，Zhangjiakou City

注册号：110685
Register Number：110685

项目名称：廊・稻——民居及公共服务建筑设计
Entry Title：Corridor—Paddy：Residential and Public Service Building Design

作者：李露昕、文亚、朱娜
Authors：Li Luxin，Wen Ya，and Zhu Na

作者单位：西安建筑科技大学
Authors from：Xi'an University of Architecture and Technology

指导教师：陈敬、王芳
Tutors：Chen Jing，Wang Fang

指导教师单位：西安建筑科技大学
Tutors from：Xi'an University of Architecture and Technology

Design Description

作品"廊・稻"，在方案设计中结合场地现状条件、外部形体的呈现充分挖掘二堡子村的自身特色，如稻田、山脉、剪纸文化。内部空间的营造汲取周边聚落、合院的肌理。为充分融入环境，方案以院落为起点，采用文化塑形、空间赋能的设计策略，将当地人文因素、文化背景赋于设计方案。

我们的概念是用一个公共廊道将散落的房屋串联，人们可以在屋顶平台休憩赏欣，也可以在廊道上欣赏远处的稻田，时有停顿、时而游走，廊道与房屋之间形成大大小小的院子，这些功能形态各异的院子在现有场地内交相呼应，和层次丰富的坡屋顶房屋互相穿插组合，创造出趣味十足的空间体验。

"Corridor - Rice" combines the current conditions of the site in the scheme design, and fully explores the characteristics of Erbaozi Village, such as rice fields, mountains and paper-cutting culture. The construction of the interior space draws from the surrounding settlements and the texture of the courtyard. In order to fully integrate into the environment, the scheme takes the courtyard as the starting point, adopts the design strategy of cultural shape and space empowerment, and encodes the local humanistic factors and cultural background into the design scheme.

Our concept is to connect the scattered houses with a public corridor. People can relax and enjoy on the roof platform, or enjoy the rice fields in the distance on the corridor, sometimes stopping and sometimes walking. Large and small courtyards are formed between the corridors and the houses. These courtyards with different functions and forms echo each other in the existing site, and are interspersed and combined with the multi-level sloped roof houses to create an interesting space experience.

Location and Site Analysis

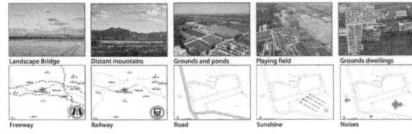

Conceptual Analysis

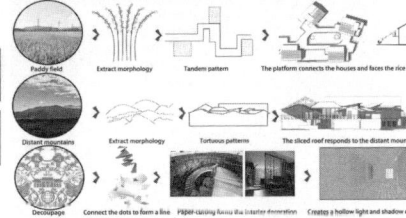

Block Generation Analysis

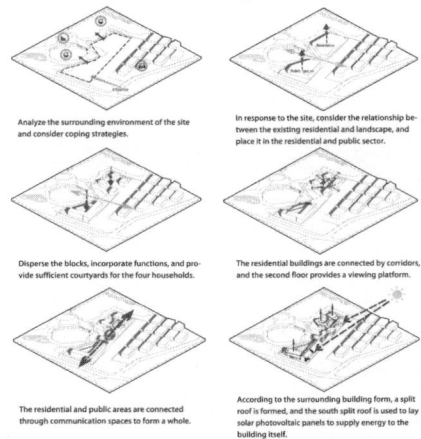

Analyze the surrounding environment of the site and consider coping strategies.

In response to the site, consider the relationship between the existing residential and landscape, and place it in the residential and public sector.

Disperse the blocks, incorporate functions, and provide sufficient courtyards for the four households.

The residential buildings are connected by corridors, and the second floor provides a viewing platform.

The residential and public areas are connected through communication spaces to form a whole.

According to the surrounding building form, a split roof is formed, and the south split roof is used to lay solar photovoltaic panels to supply energy to the building itself.

Site Plan

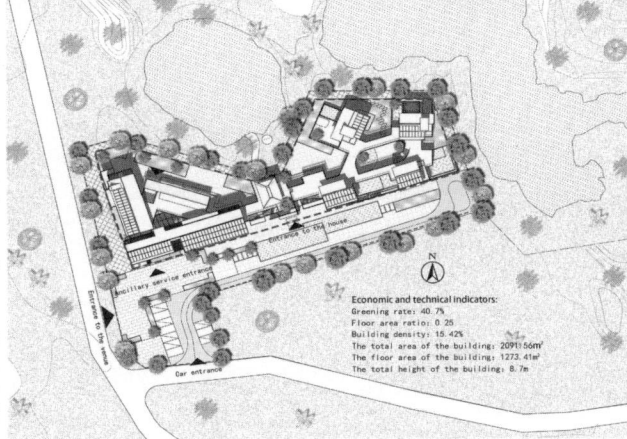

Economic and technical indicators:
Greening rate: 40.7%
Floor area ratio: 0.25
Building density: 15.4%
The total area of the building: 2091.56m²
The floor area of the building: 1273.41m²
The total height of the building: 8.7m

Site Plan Analysis

Landscape analysis

Functional analysis

Fire lanes

Section1-1

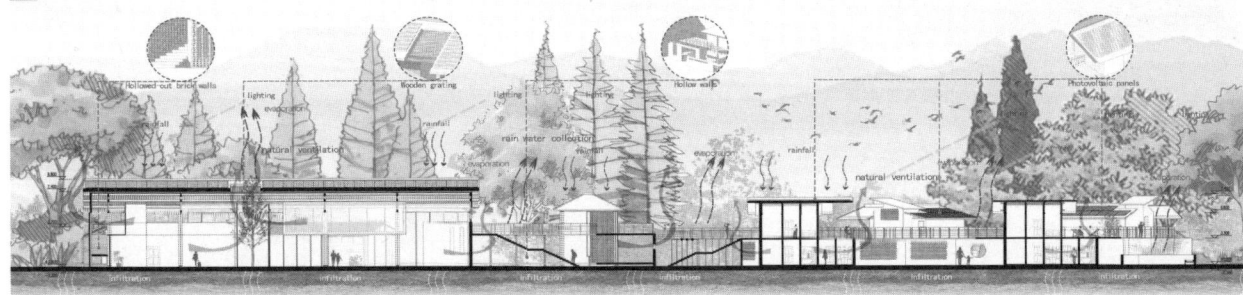

编号:110685

廊·稻2
—民居及公共服务建筑设计
Corridor-Paddy: Residential and Public Service Building Design

Climate Analysis

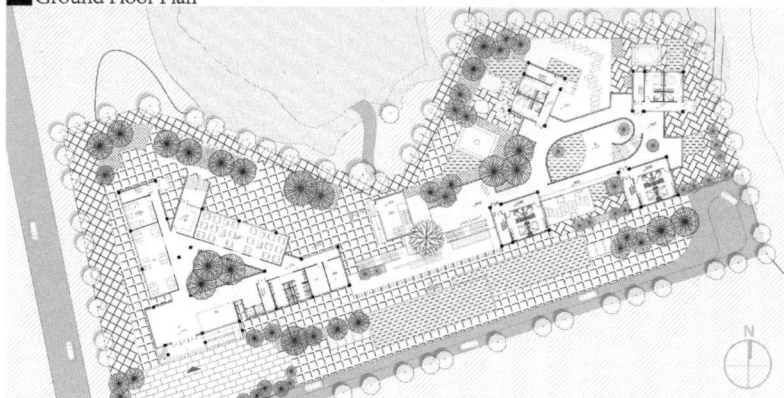

Temperature and Orientation Analysis

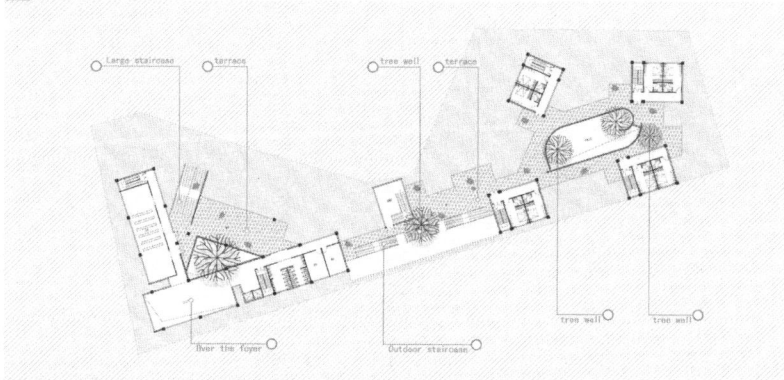

Ground Floor Plan

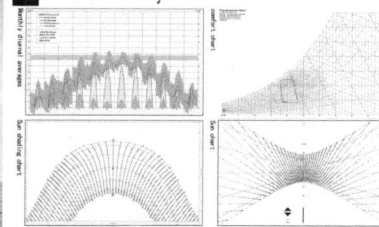

Monthly Winds Rose

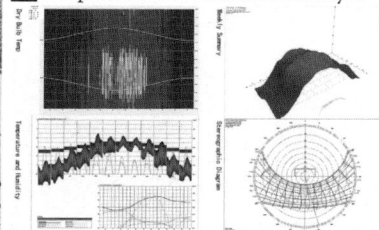

Floor Plan of the Second Floor

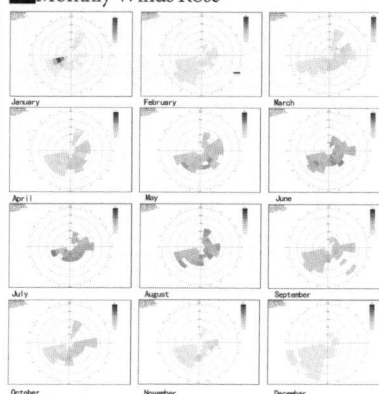

Exploded View of the Structure

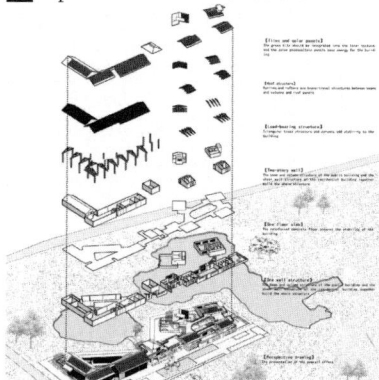

Elevation

廊 · 稻 3
——民居及公共服务建筑设计
Corridor-Paddy: Residential and Public Service Building Design

Node Space

Node Space

Node Space

Node Space

Hail-resistant Tectonic Analysis

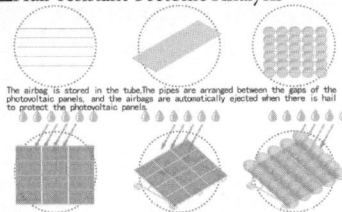

The airbag is stored in the tube. The pipes are arranged between the gaps of the photovoltaic panels, and the airbags are automatically ejected when there is hail to protect the photovoltaic panels.

Seismic Tectonic Analysis

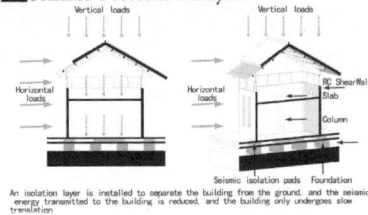

An isolation layer is installed to separate the building from the ground, and the seismic energy transmitted to the building is reduced, and the building only undergoes slow translation.

Solar Energy Saving Analysis

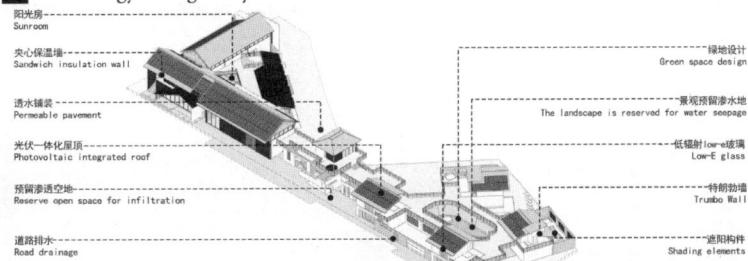

阳光房
Sunroom

夹心保温墙
Sandwich insulation wall

透水铺装
Permeable pavement

光伏一体化屋顶
Photovoltaic integrated roof

预留渗透空地
Reserve open space for infiltration

道路排水
Road drainage

绿地设计
Green space design

景观预留渗水地
The landscape is reserved for water seepage

低辐射low-e玻璃
Low-E glass

特朗勃墙
Trumbo Wall

遮阳构件
Shading elements

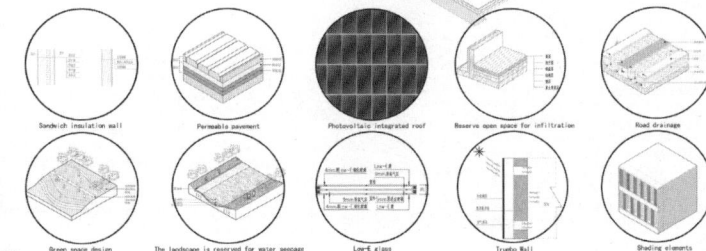

Section 2-2

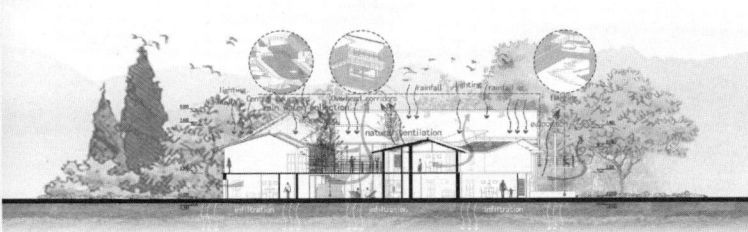

编号：110685

廊·稻4
——民居及公共服务建筑设计
Corridor-Paddy: Residential and Public Service Building Design

Green Building Design Guidance

1. 太阳能应用：张家口光资源丰富，设计安装光伏板为主要电力，屋顶及开放空间尽量大化光照利用，逆变器转换供电，含照明、供暖等；太阳能集热器供热水，智能控制按需供应，提升能效。

2. 低碳技术：采用竹材、秸秆基秸等低碳建材，优化建筑设计促自然通风采光，减少能耗；绿色种植墙美化环境，净化空气；雨水收集与节水器具减少水资源消耗。

3. 韧性设计：结构抗震抗风，加固地基；能源备用系统确保电网中断时运营；活动中心兼应急避难所，提供紧急住宿、医疗等服务，增强防灾能力。

1. Solar energy application: Zhangjiakou is rich in light resources, the design and installation of photovoltaic panels are the main power, the roof and open space to maximize the utilization of light, inverter conversion of power supply, including lighting, heating and so on; Solar collector heating water, intelligent control of on-demand supply, improve energy efficiency.

2. Low-carbon technology: Adopt low-carbon building materials such as bamboo and straw bricks to optimize building design to promote natural ventilation and lighting and reduce energy consumption; Green planting wall beautify the environment and purify the air; Rainwater harvesting and water saving devices reduce water consumption.

3. Toughness design: structural seismic and wind resistance, strengthen the foundation; An energy backup system ensures operation in the event of a grid outage, The activity center doubles as an emergency shelter, providing emergency accommodation, medical care and other services to enhance disaster prevention capabilities.

Architectural Techniques Analysis

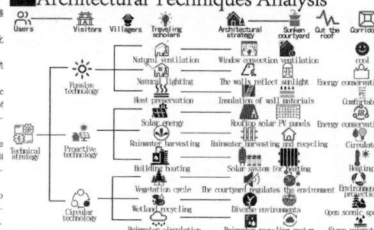

Carbon Emissions

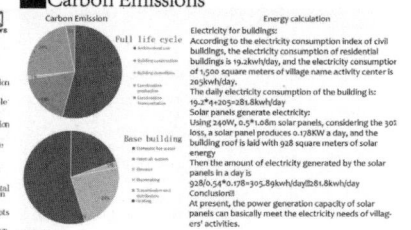

Electricity for buildings:
According to the electricity consumption index of civil buildings, the electricity consumption of residential buildings is 19.2kwh/day, and the electricity consumption of 1,500 square meters of village name activity center is 20.5kwh/day.
The daily electricity consumption of the building is:
19.2*4+205*28.1.8kwh/day
Solar panels generate electricity:
Using 240W, 0.5*1.08m solar panels, considering the 30% loss, a solar panel produces 0.178KW a day, and the building roof is laid with 928 square meters of solar energy
Then the amount of electricity generated by the solar panels in a day is
928/0.54*0.178=305.89kwh/day$\boxed{}$281.8kwh/day
Conclusion:
At present, the power generation capacity of solar panels can basically meet the electricity needs of villagers' activities.

Daylighting and Radiation Analysis

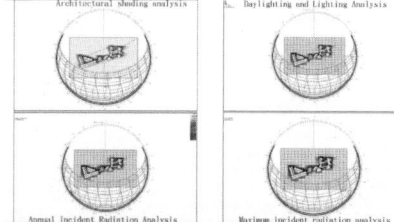

Optimised Lighting and Ventilation

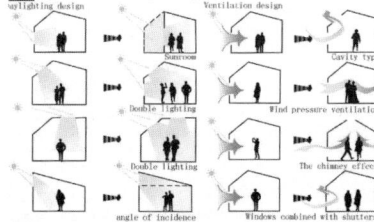

Winter Insulation Strategies

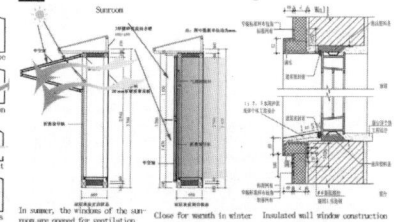

Rainwater Ecosystems Analysis

Emergency Evacuation Analysis

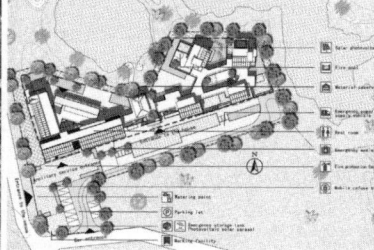

Emergency Evacuation Scenario

Rainwater Recycling System

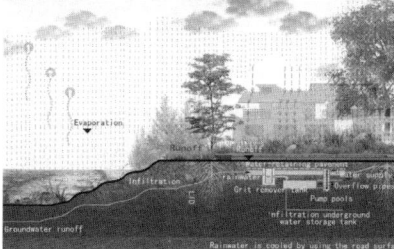

Facade

编号：110698
张家口市怀安县二堡子村的韧性乡村设计
Resilient rural design in Erbaozi Village, Huai'an County, Zhangjiakou City

Lodging—Resistant Village

浸谷光盒 1

阳光・乡村韧性
2024 台达杯国际太阳能建筑设计竞赛获奖作品集

■ 设计说明

乡村建筑的生长应当汲取自然的能量。稻田为建筑创造自然的景观背景，也为设计带来"抗倒伏"的场所精神。设计高度重视建筑本身防灾减灾能力，希望建筑不是只有美价值的"花架子"，而是能真真切切为居民带来抵御自然风险的助力。建筑体块生成采用最稳定的三角形，倾斜的屋面似低垂的稻穗一般，沉稳有力，可极大地提高建筑抵御地震、风雪的能力，也适应了高效的太阳能光伏板角度。此外，公建部分的布局回环互掩，四栋民居相互依靠也进一步增强了风险抵抗能力。

技术层面上，设计既应用阳光房、深色太阳能晶硅板、变相蓄热墙等常规技术，还充分结合本土元素，设计出保暖、环保的装配式稻草墙板。稻草墙板性能优良且具有可快速再生的特点，村民可以零碳生产，自产自足，快速实现灾后重建。浸谷光盒（Lodging- Resistant Village）选取水稻元素，充分考虑脆弱的乡村生态，因地制宜地选用、创新绿色建筑技术，是对美好稻乡意向的回应，是给乡村振兴建设的答卷，也是对村庄"抗倒伏"能力的保障。

■ Design Concept

The growth of rural architecture should draw upon the energy of nature. Rice fields provide a natural landscape backdrop for the buildings and bring forth the site spirit of "lodging resistance" to the design.The design attaches great significance to the disaster prevention and mitigation capabilities of the buildings themselves, hoping that the buildings are not merely "decorative" with only aesthetic value but can genuinely offer assistance to residents in resisting natural risks.The generation of the building blocks adopts the most stable triangular shape. The inclined roof is like a drooping rice ear, stable and powerful, significantly enhancing the building's ability to withstand earthquakes, wind, and snow, and also adapting to the efficient angle of solar photovoltaic panels. Additionally, the layout of public buildings is interwoven and concealed, and the four residential buildings' mutual reliance further strengthens the risk resistance capacity.From a technical perspective, the design not only employs conventional techniques such as sunrooms, dark solar crystalline silicon panels, and phase-change heat storage walls but also fully integrates local elements to design warm-keeping and environmentally friendly prefabricated straw wall panels. The straw wall panels have excellent performance and the characteristic of rapid regeneration. Villagers can produce them in a zero-carbon manner, being self-sufficient and achieving post-disaster reconstruction rapidly.Lodging-resistant Village selects rice elements, fully considers the fragile rural ecology, and selectively and innovatively employs green building technologies based on local conditions. It is a response to the aspiration of a beautiful rice village, an answer to rural revitalization construction, and a guarantee of the village's "lodging resistance" capacity.

■ Pre-site Analysis

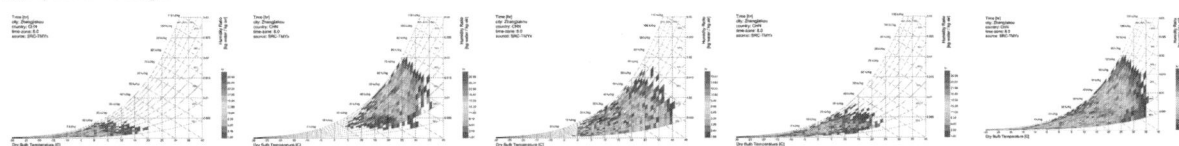

■ Conceptual Analysis

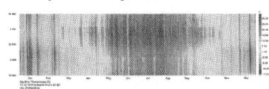

■ Psychrometric Diagram

■ Dry bulb Temperature

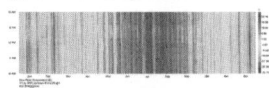

■ Sun Altitude Angle

■ Wind Environment

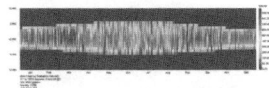

■ Dew Point Temperature

■ Direct Normal Radiation

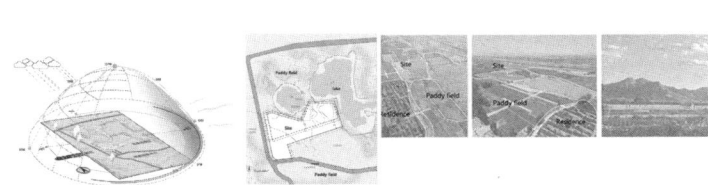

■ Carbon Emissions Table

建筑运行阶段碳排放计算结果

项目	年均耗电量（kWh/a）	建筑寿命（年）	电力碳排放因子（kgCO₂e/kW·h）	碳排放量（kgCO₂e）
暖通空调照明碳排放量	65702.35	50	0.8843	2905052.74
电梯碳排放量	0.00			0.00
可再生能源减碳量	-66615.76			-2945415.99
合计				-40363.25

■ Summary

Based on the enthalpy-humidity chart and the hourly dry-bulb temperature chart for different seasons, it is clear that the site experiences cold winters and hot summers, with significant temperature differences between the two. Similarly, the charts indicate that the site is relatively humid in summer and dry in winter, which necessitates effective insulation for buildings during winter and proper ventilation measures in summer. From the wind rose diagram, it can be seen that the site experiences prevailing northwesterly winds throughout the year. While it is important to ensure that the wind does not negatively impact daily life, it should also be utilized as part of the building's energy strategy, making natural wind an exploitable resource. In summary, appropriate passive design strategies for the region include evaporation cooling, natural ventilation, thermal storage walls, and the insulation of sunrooms. Additionally, based on the hourly solar radiation chart, the site enjoys abundant solar radiation throughout the year. Therefore, it is advisable to incorporate large, sloped photovoltaic roof panels, oriented based on the solar altitude angle, to optimize solar energy capture.

综合奖・优秀奖・张家口市怀安县二堡子村建设项目
Comprehensive Awards — Honorable Mention — Erpuzi Village Project, Huai'an County, Zhangjiakou City

注册号：110698
Register Number：110698

项目名称：浸谷光盒
Entry Title：Lodging — Resistant Village

作者：田如俊、谢佳利、黄恒瑞、龙我在、侯竹筠
Authors：Tian Rujun, Xie Jiali, Huang Hengrui, Long Wozai, and Hou Zhuyun

作者单位：重庆大学
Authors from：Chongqing University

指导教师：张海滨
Tutor：Zhang Haibin

指导教师单位：重庆大学
Tutor from：Chongqing University

■ Rendering of a Rice Field Scenery from a Human Viewpoint

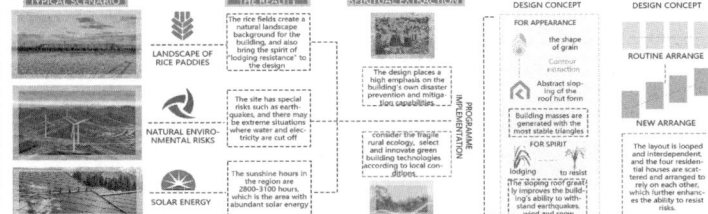

编号：110698

张家口市怀安县二棒子村的韧性乡村设计
Resilient rural design in Erbanzi Village, Huai'an County, Zhangjiakou City

Lodging–Resistant Village

浸谷光盒 2

■ Logic Generation

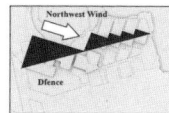

STEP 1 :
In order to block the cold northwest wind, the building faces northwest to form a fortress.

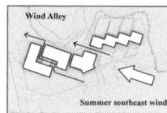

STEP 2 :
In order to create a cross-hall breeze in summer, the building opens to the southeast.

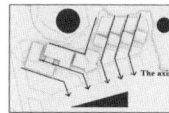

STEP 3 :
The axis of the building is folded towards various important landscape directions.

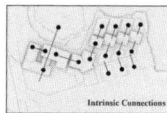

STEP 4 :
The landscape, buildings, yards and green areas are connected at all levels.

■ Block Generation

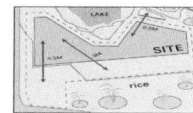

Original site :
The original site is flat inside, surrounded by two lakes and rice fields, with good landscape resources.

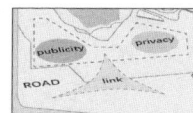

Public-private division :
The eastern transportation lines are difficult to reach, so from west to east, there is a transition from public to private.

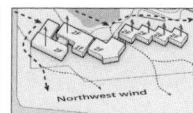

Raise the block:
The volume is designed to block the cold northwest wind while allowing for ventilation in summer.

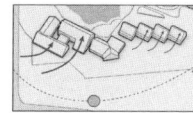

Pull up the slope :
In order to achieve better building lighting and solar energy absorption, the slope roof is pulled out.

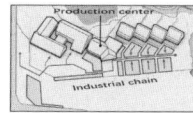

Special planting :
Special crop plantations are designed on site to form a key prefabricated straw wall industry chain.

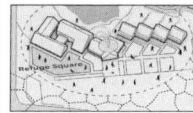

Overall toughness :
The overall design approach gives the entire site a "toughness" like straw.

■ Total Flat Analysis

■ Evacuation Route Analysis

■ Sunshine Simulation

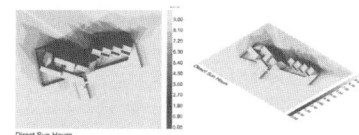

Direct Sun Hours

Adapt to light condition:
The layout of the building is open to the southwest, with solar panels placed on the light side to get long hours of sunlight. In addition, the angled roof Angle will further improve the efficiency of solar energy capture.

■ Wind Simulation

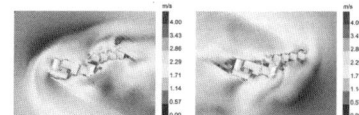

Adapt to wind condition:
The layout of the building is open to the southwest, with solar panels placed on the light side to get long hours of sunlight. In addition, the angled roof Angle will further improve the efficiency of solar energy capture.

■ Plan 1 : 500

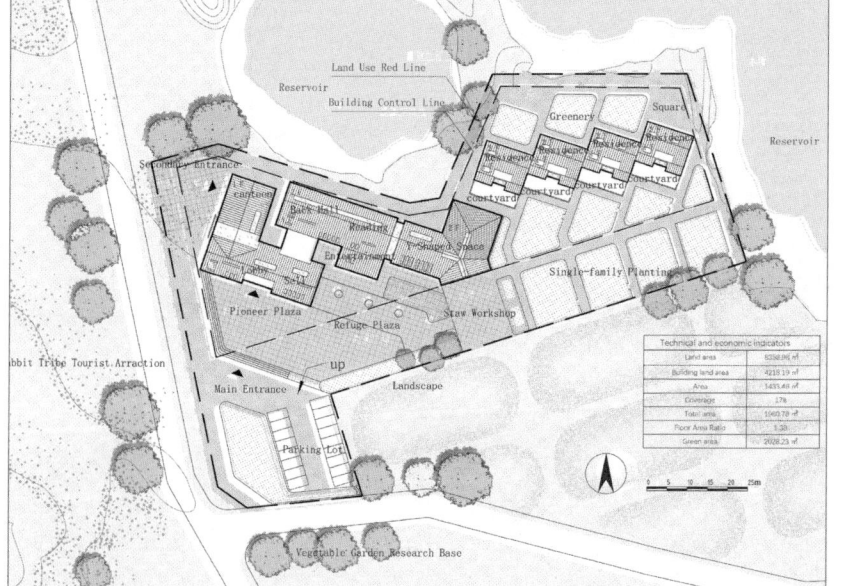

Lodging–Resistant Village

Ground floor plan / 1:250

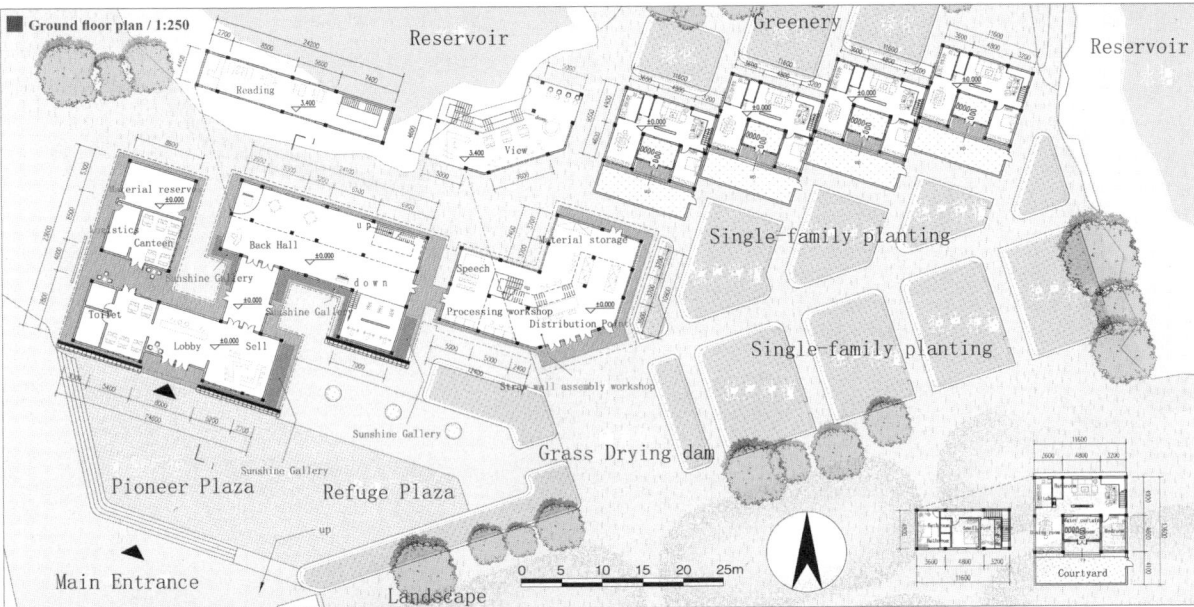

Reservoir

Greenery

Reservoir

Reading

View

Material reserve

Logistics

Canteen

Back Hall

Sunshine Gallery

Toilet

Lobby   Sell

Material storage

Speech

Processing workshop

Distribution Point

Straw wall assembly workshop

Single-family planting

Single-family planting

Grass Drying dam

Sunshine Gallery

Pioneer Plaza   Refuge Plaza

Sunshine Gallery

Main Entrance

Landscape

Courtyard

0 5 10 15 20 25m

Aerial View

Exploded Axonometric & Materials

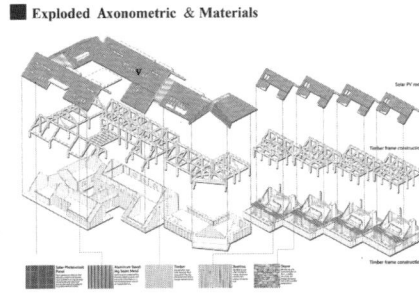

Rendering of a Person's Viewpoint

Perspective Section

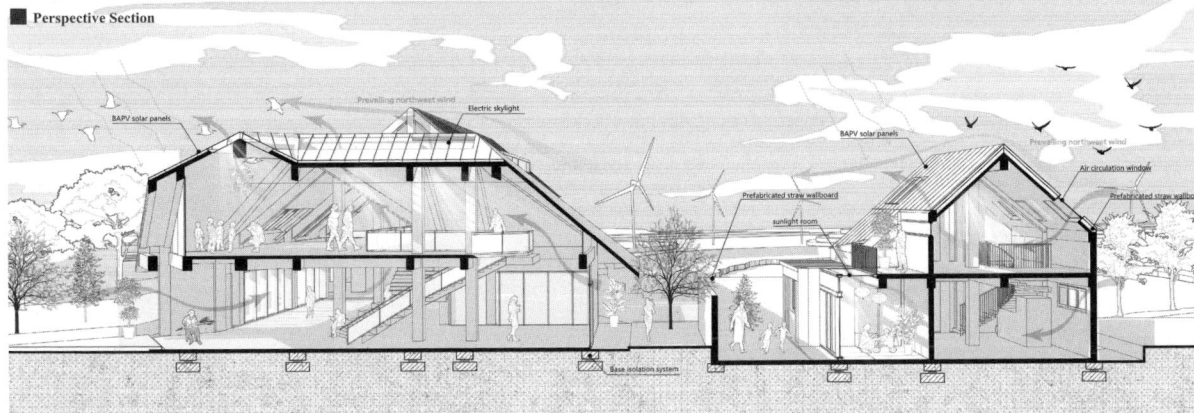

BAPV solar panels

Prevailing northwest wind

Electric skylight

BAPV solar panels

Prevailing northwest wind

Air circulation window

Prefabricated straw wallboard

sunlight room

Prefabricated straw wallboard

Base isolation system

编号：110698
张家口市怀安县二堡子村的韧性乡村设计
Resilient rural design in Erbaozi Village, Huai'an County, Zhangjiakou City

Lodging–Resistant Village

浸谷光盒 4

■ Rendering of a Person's Viewpoint

■ Toughness Design of Straw

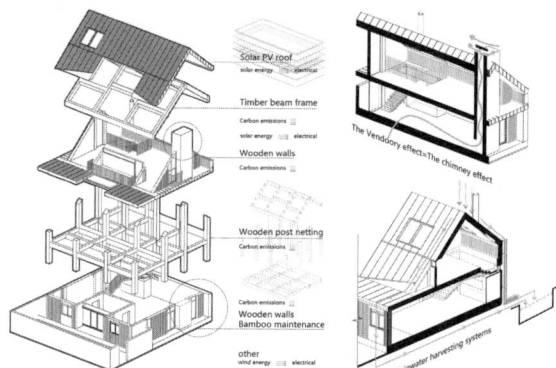

① Carbon fixation by photosynthesis ② Harvest
③ Straw collection and preprocessing (making straw insulation board)
④ Assembly of straw wall components in the workshop
⑤ Use of the straw wall for routine maintenance
⑥ Logistics for straw wall products export (selling)

■ Sun Room Explosion Diagram

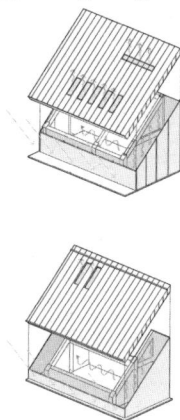

Solar PV roof
solar energy   electrical

Timber beam frame
Carbon emissions

solar energy   electrical

Wooden walls
Carbon emissions

Wooden post netting
Carbon emissions

Wooden walls
Bamboo maintenance

other
wind energy   electrical

The Venddory effect=The chimney effect

Rainwater harvesting systems

Sun room detail design:
It is a reasonable way to collect rainwater on the ground and rooftop, and then drain the rainwater into the rainwater collection system through the rainwater pipe, and then go through a series of filtration and purification to achieve the purpose of reuse.

The chimney effect. The vertical exhaust air duct has an unobstructed circulation space from the bottom to the top, and there is a density difference between the inlet air and the top air, so the air spreads or discharges quickly along the channel. This is a

■ Prefabricated Composite Straw Exterior Wall Panels

Carbon reduction calculation: (Energy saved per square meter of target wall compared to traditional wall in one year)

E (Emission reduction)=Q1-Q2 Q=(T2-T1)*k*t

T2-T1 is the absolute value of the indoor and outdoor temperature difference during the heating period in Zhangjiakou, K, which is 10 in the reference standard calculation; t is the annual heating time, s. Assuming that the annual heating period is 100 days, t=8.64x106s. The heat transfer coefficient k value of ordinary walls is 1.62 W/(m2·k), while the k value of self-made walls is calculated to be 0.46 W/(m2·k) (The following calculation shows)

E (Emission reduction)=(T2-T1)*(k1-k2)*t=10x(1.62-0.46)x8.64x106=6.048x10^7(J/m²)

■ Technology Strategy

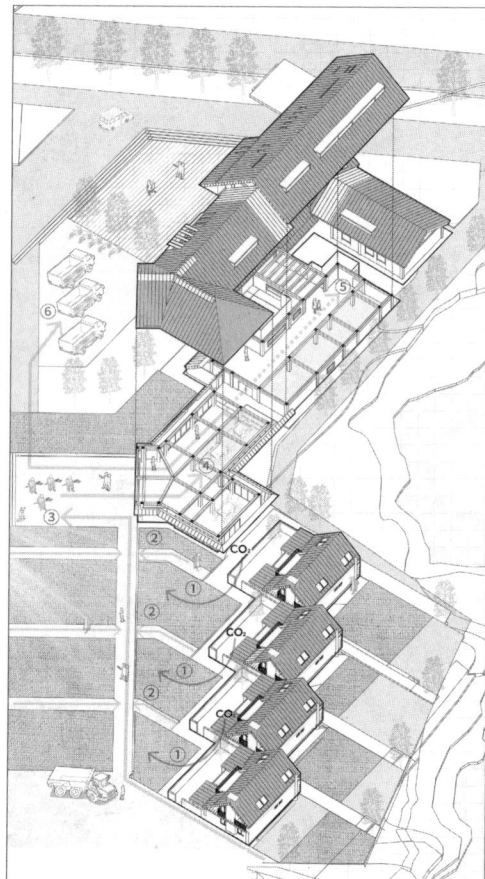

External glass structure:
The external attached glass structure is built based on the principles of the greenhouse effect. After the temperature inside the greenhouse rises, the long-wave radiation emitted helps trap heat, allowing little to no heat to escape through the glass or plastic film to the outside. The main heat loss in a greenhouse occurs through convection and conduction. If measures such as sealing and insulation are taken, this heat loss can be reduced. Additionally, if heat storage devices are installed indoors, the excess heat can be stored for later use.

In the summer, the heat can be quickly dissipated by opening the ventilating windows at the top of the glass enclosure.

Sun Room:
Sunrooms also utilize the principle of the greenhouse effect. Sunrooms, through transparent glass or other materials, allow sunlight to enter. The shortwave radiation from the sun passes through the glass and heats the interior surfaces, such as the floor and objects inside. These heated surfaces then emit longwave radiation, releasing heat.

Solar skylight:
A solar skylight helps create a stack effect. In summer, sunlight heats the air at the top of the room through the skylight, causing the warm air to escape through the vent windows at the top, while cooler air enters the interior. In winter, the top vent is closed to reduce the loss of warm air, thereby helping to retain heat inside and improve indoor insulation.

■ 1—1 Sectional Drawing / 1 : 200

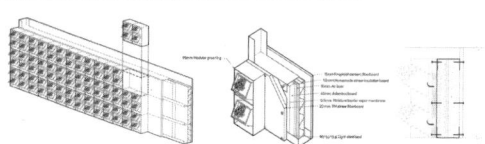

■ South Elevation / 1 : 200

Box^n Dual-Phase Green Dwelling

小盒 n 方
二元绿居

综合奖·优秀奖·张家口市
怀安县二堡子村建设项目
Comprehensive Awards —
Honorable Mention — Erpuzi
Village Project, Huai'an
County, Zhangjiakou City

注册号：110734
Register Number：110734
项目名称：小盒 n 方，二元绿居
Entry Title：Box^n Dual — Phase Green
Dwelling
作者：庞志祥、张春杰、武雨晨、
黄向文、刘纪源
Authors：Pang Zhixiang, Zhang Chunjie,
Wu Yuchen, Huang Xiangwen,
and Liu Jiyuan
作者单位：大连理工大学
Authors from：Dalian University of Technology
指导教师：李国鹏
Tutor：Li Guopeng
指导教师单位：大连理工大学
Tutor from：Daliang University of Technology

In the design of building resilience, we have enhanced the capacity to withstand natural disasters such as earthquakes by adopting modular and modularized designs, which endow buildings with high flexibility to meet diverse functional requirements before and after disasters, including emergency health services and emergency shelters. At the technical level, we have integrated active technologies such as solar photovoltaic and ground-source heat pumps, in conjunction with site resources, to achieve green energy efficiency; simultaneously, passive technologies such as rainwater harvesting and ventilation fans are employed to ensure good ventilation and insulation of the buildings. During the construction process, modular assembly technology enables rapid erection and dismantling of buildings, improving construction efficiency, reducing construction waste, and achieving environmental sustainability.

在建筑韧性设计中，我们强化了应对地震等自然灾害的能力，采用模块化和模数化设计，赋予建筑高度灵活性，满足灾前灾后多样化功能需求，包括紧急卫生服务与应急避难所等。技术层面，结合场地资源，我们融入太阳能光电、地源热泵等主动技术，实现绿色节能；同时，采用雨水回收、通风风扇等被动技术，确保建筑通风良好且保温。建造过程中，模块化装配式技术使得建筑快速搭建与拆除，提升施工效率，减少建筑垃圾，实现环保可持续。

Technology Uses

Active technology — Solar photovoltaic / Air Source Heat Pump Systems / Ground source heat pump system / Wind power

Passive technology — Rainwater recycling system / Additional sunroom / Train control / Additional window design

Other technologies — Combination of fan and ventilation

Program Generation

STEP 1
Determine site redlines and traffic relationships.

STEP 2
Clear functional zoning based on logical analysis.

STEP 3
Design monolithic buildings and creating courtyard spaces.

STEP 4
Improve the architectural details and decide on the final plan.

General Plan 1:600

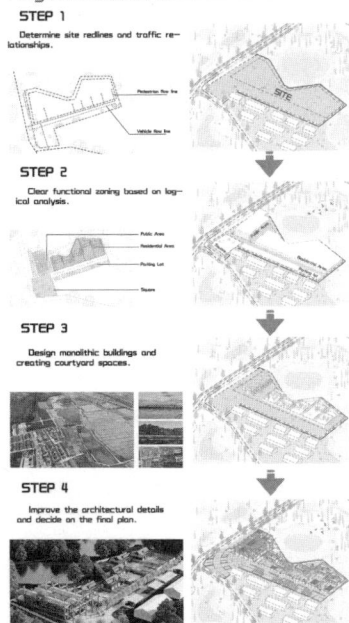

Homestay area	127.47 ㎡
Public floor area	1162.61 ㎡
Gross floor area	1389.03 ㎡
Building density	9.71%
Plot ratio	0.82
Ratio of green space	32.60%
Land area	8650 ㎡
Construction land area	4275 ㎡
Number of parking Spaces	17 个

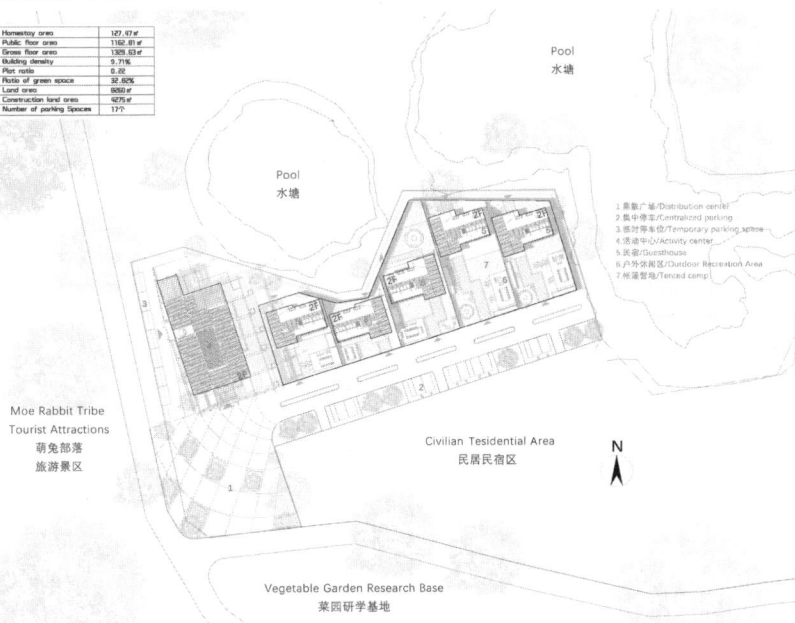

Pool 水塘

Pool 水塘

Moe Rabbit Tribe Tourist Attractions
萌兔部落 旅游景区

Civilian Tesidential Area
民居民宿区

1. 集散广场/Distribution center
2. 集中停车/Centralized parking
3. 临时等车位/Temporary parking space
4. 活动中心/Activity center
5. 民宿/Guesthouse
6. 户外休闲区/Outdoor Recreation Area
7. 帐篷营地/Tenced camp

N

Vegetable Garden Research Base
菜园研学基地

作品编号: 110734

Plan of the Activity Center

Floor Plan of the Activity Center 1 : 200

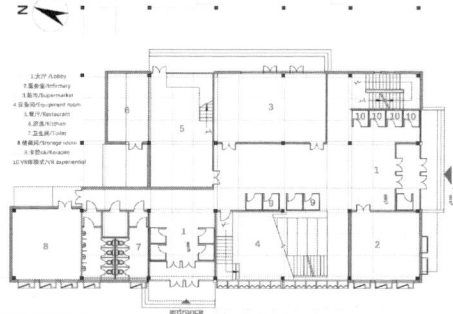

N

1.大厅/Lobby
2.器物室/Infirmary
3.超市/Supermarket
4.设备间/Equipment room
5.餐厅/Restaurant
6.厨房/Kitchen
7.卫生间/Toilet
8.储藏间/Storage room
9.水族馆/Aquarium
10.VR体验馆/VR experience

entrance

entrance

Plan of the second floor of the activity center 1 : 200

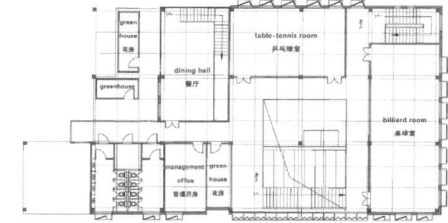

green house 花房

table-tennis room 兵乓球室

dining hall 餐厅

greenhouse 花房

management office 管理办公

green house 花房

billiard room 桌球室

Basic Modulor Space Analysis

The basic spatial module size for architecture-
1.5×1.5×3 micro-unit.
Based on a 02 grid size, a 1.51.5 architectural unit can be arranged and freely moved around. Its function can be easily replaced, making it suitable for use as a single-person study room, as well as for office work and micro-experiences such as VR and AR experiences.

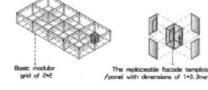

Basic modulor grid of 2×2

The replaceable facade template /panel with dimensions of 1×3.3meters

Basic Spatial Module Size for Architecture-
6×4 Basic Module Space
Using six 2×2 units for the layout, the space can basically fulfill both office and double occupancy residential needs.

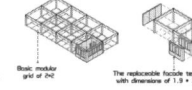

Basic modulor grid of 2×2

The replaceable facade template/panel with dimensions of 1.9 × 3.15 meters.

The basic spatial module size for orchitecture-
10×6 basic module space.
Using 15 2×2 units for the layout, the space can generally accommodate reading areas, office spaces, and also serve as a venue for small gatherings.

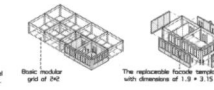

Basic modulor grid of 2×2

The replaceable facade template/panel with dimensions of 1.9 × 3.15 meters.

The basic spatial module size for orchitecture-
6×6 basic module space.
Using 5 2×2 units for the layout, the space can generally accommodate typical office spaces while also serving as a utility room and storage area.

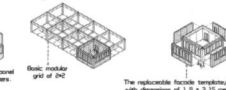

Basic modulor grid of 2×2

The replaceable facade template/panel with dimensions of 1.9 × 3.15 meters.

The basic spatial module size for orchitecture –
6×3 basic module space
Using 3 2×2 units for the layout, the space can basically accommodate double occupancy for lodging, as well as the residential and isolation needs of post-disaster field hospitals.

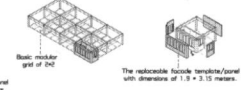

Basic modulor grid of 2×2

The replaceable facade template/panel with dimensions of 1.9 × 3.15 meters.

Site Emergency Design

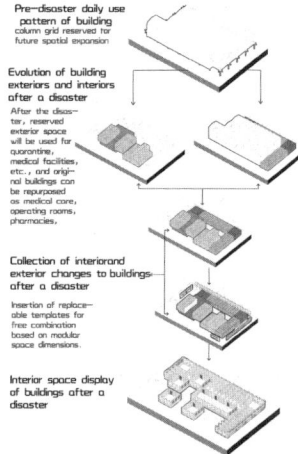

Pre-disaster daily use pattern of building
column grid reserved for future spatial expansion

Evolution of building exteriors and interiors after a disaster
After the disaster, reserved exterior space will be used for quarantine, medical facilities, etc., and original buildings can be repurposed as medical care, operating rooms, pharmacies.

Collection of interior and exterior changes to buildings after a disaster
Insertion of replaceable templates for free combination based on modular space dimensions.

Interior space display of buildings after a disaster

2-4 Activity Center Design

Analysis of Multi-Space and Multi-Function Conversion

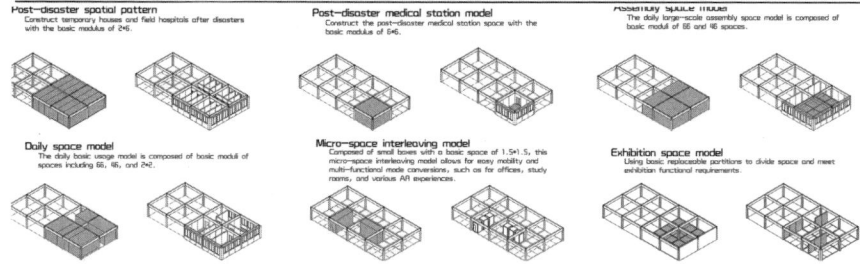

Post-disaster spatial pattern
Construct temporary houses and field hospitals after disasters with the basic modulus of 2×6.

Post-disaster medical station model
Construct the post-disaster medical station space with the basic modulus of 6×6.

Assembly space model
The daily large-scale assembly space model is composed of basic modulus of 8×8 and 4×6 spaces.

Daily space model
The daily basic usage model is composed of basic modulus of spaces including 6×6, 4×6, and 2×2.

Micro-space interleaving model
Composed of small boxes with a basic space of 1.5×1.5, this micro-space interleaving model allows for easy mobility and multi-functional mode conversions, such as for offices, study rooms, and various AR experiences.

Exhibition space model
Using basic replaceable partitions to divide space and meet exhibition functional requirements.

Replaceable Facade Module

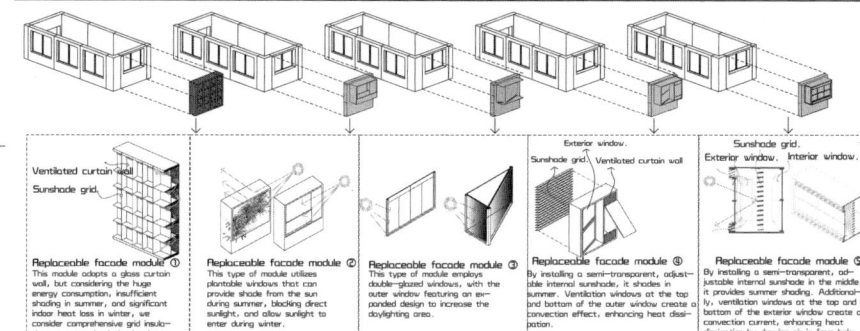

Ventilated curtain wall
Sunshade grid.

Exterior window.
Sunshade grid. Ventilated curtain wall

Sunshade grid.
Exterior window. Interior window.

Replaceable facade module ①
This module adopts a glass curtain wall, but considering the huge energy consumption, insufficient shading in summer, and significant indoor heat loss in winter, we consider comprehensive grid insula-

Replaceable facade module ②
This type of module employs plantable windows that can provide shade from the sun during summer, blocking direct sunlight, and allow sunlight to enter during winter.

Replaceable facade module ③
This type of module employs double-glazed windows, with the outer window featuring an expanded design to increase the daylighting area.

Replaceable facade module ④
By installing a semi-transparent, adjustable internal sunshade, it shades in summer. Ventilation windows at the top and bottom of the outer window create a convection effect, enhancing heat dissipation.

Replaceable facade module ⑤
By installing a semi-transparent, adjustable internal sunshade in the middle, it provides summer shading. Additionally, ventilation windows at the top and bottom of the exterior window create a convection current, enhancing heat

Seismic Analysis

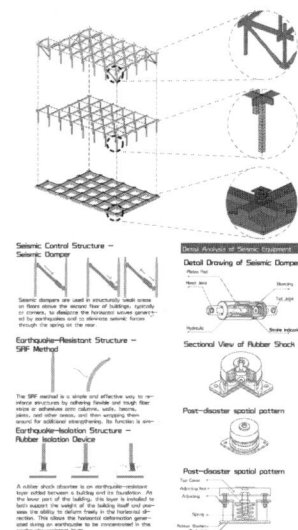

Seismic Control Structure – Seismic Damper
Seismic dampers are used in structurally weak areas on floors above the second floor of buildings, typically at corners, to dissipate the horizontal seismic energy transmitted by earthquakes and to eliminate seismic forces through the spring in the damper.

Earthquake-Resistant Structure – SRF Method
The SRF method is a simple and effective way to reinforce structures by adhering flexible and tough fiber straps or adhesives onto columns, walls, beams, plates, and other areas, and then wrapping them around for additional strengthening. Its function is similar to the ability to deform freely in the horizontal direction. This allows the horizontal deformation generated during an earthquake to be concentrated in the earthquake-resistant layer.

Earthquake-Isolation Structure – Rubber Isolation Device
A rubber shock absorber is an earthquake-resistant layer added between a building and its foundation. At the lower part of the building, this layer is installed to both support the weight of the building itself and possess the ability to deform freely in the horizontal direction. This allows the horizontal deformation generated during an earthquake to be concentrated in the earthquake-resistant layer.

Detail Analysis of Seismic Equipment

Detail Drawing of Seismic Damper

Sectional View of Rubber Shock Absor

Post-disaster spatial pattern

Post-disaster spatial pattern

Plan of the Homestay

■ First Floor Plan 1 : 100

Magazine store
杂物院

15 14 12 13 13 12 11

18 17 16 16 16

19 19 19

entrance
Tent camp
帐篷营地

11 储藏间/Storage room
12 设备间/Equipment room
13 卫生间/Toilet
14 玄关/Porch
15 餐厅/Dining room
16 卧室/Bedroom
17 厨房/Kitchen
18 起居室/Living room
19 附加阳光间/Additional sun room

Planting pot
种植盆

Outdoor party
室外聚会

■ Loft Floor Plan 1:150

The concept of core space and transitional space is proposed in this architectural scheme. By analyzing the natural environment and climate of Zhangjiakou and the demand for use, the bedrooms, kitchens and bathrooms, equipment, and storage functions are placed in the core space, while the dining room, living room, and balcony functions are placed in the transitional space. Zhangjiakou is a cold region with warm summers and cold winters, and the building can be used in different modes in summer and winter due to the climate characteristics of the region.

summertime
The outdoor climate is warm and the main use of the building is core + transitional spaces.

winters
In winter, the transition space serves as a sunroom and airroom in the south direction of the bedrooms, providing heat and thermal insulation to the bedrooms in the core space.

■ 1-1 Section

Living room   Bedroom Bedroom Bedroom

■ 2-2 Section

■ Elevation plan

■ Elevation plan

3-4   Homestay Residence Design

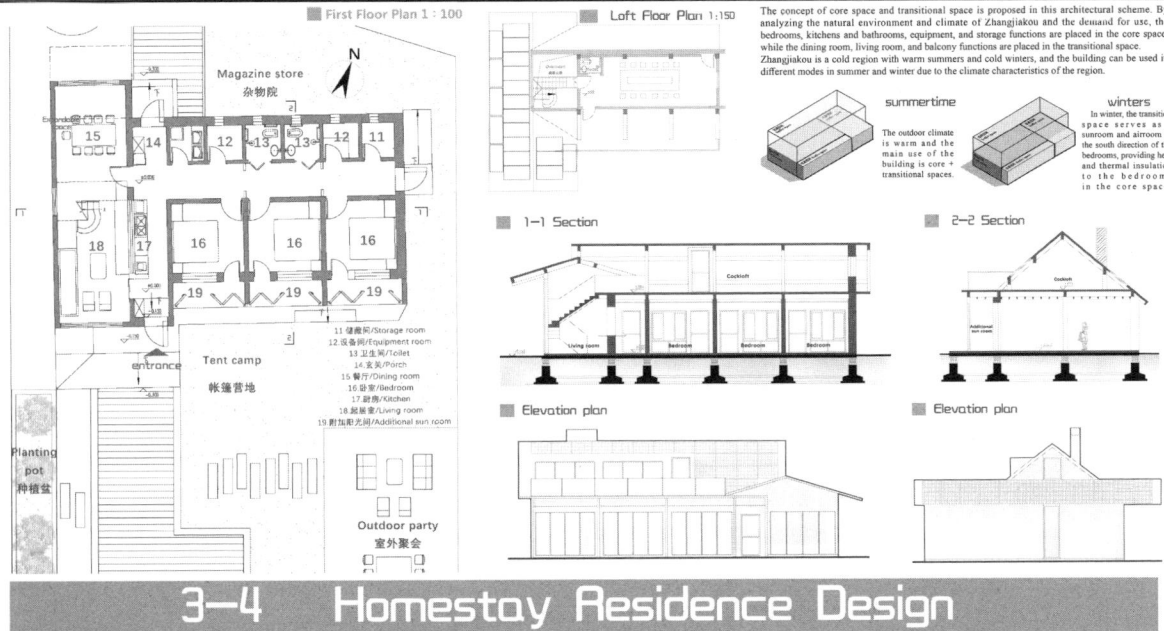

Functional Space Diversity for Activities

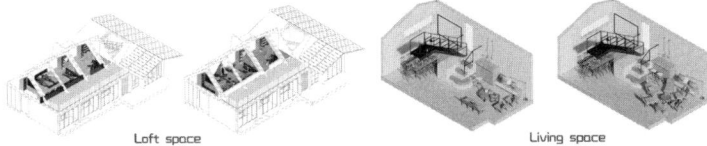

Loft space

Living space

Functional Space Diversity in Homestay

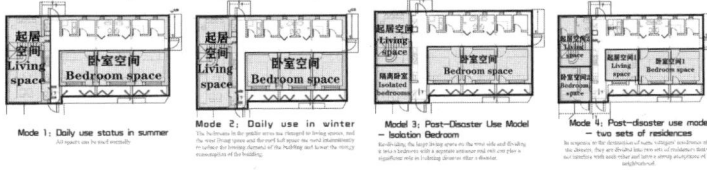

起居空间
Living space

卧室空间
Bedroom space

Mode 1: Daily use status in summer
All spaces can be used normally.

起居空间
Living space

卧室空间
Bedroom space

Mode 2: Daily use in winter

起居空间
Living space

卧室空间
Bedroom space

Mode 3: Post-Disaster Use Model – Isolation Bedroom

起居空间
Living space

卧室空间
Bedroom space

Mode 4: Post-disaster use mode – two sets of residences

■ Main Structure

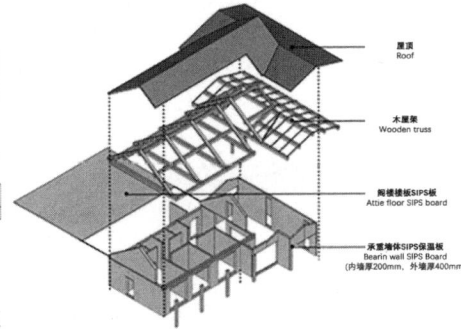

屋顶
Roof

木屋架
Wooden truss

阁楼楼板SIPS板
Attic floor SIPS board

承重墙体SIPS保温板
Bearing wall SIPS Board
(内墙厚200mm, 外墙厚400mm)

Transportation and On-Site Construction

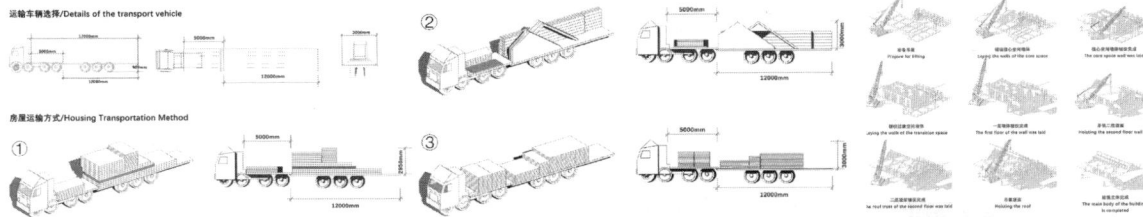

运输车辆选择/Details of the transport vehicle

房屋运输方式/Housing Transportation Method

Homestay Energy Efficient Design

Homestay Water System Diagram

The hot water is supplied by air source heat pump near the water equipment room, and the hot water is supplied by KitchenAid on the kitchen side which is farther away, which greatly shortens the distance of hot water pipeline. An emergency water tank is installed for short-term emergency use in case of disastrous weather.

- Water supply pipes 给水管
- Hot water supply pipes 热水给水管
- Hot water return pipes 热水回水管
- Drainpipe 排水管

Homestay HVAC System

By placing the air-source heat pump outside unit in the additional sunroom, the passive heating and heat pump are reasonably combined to complement each other's advantages. The air source heat pump system utilizes passive heating to obtain a suitable working environment, and at the same time stores the excess heat from passive heating during the day and uses it for heating at night, achieving the purpose of stable heating and energy saving.

- Water supply pipes 给水管
- Hot water supply pipes 热水给水管
- Hot water return pipes 热水回水管
- Drainpipe 排水管

Kitchen Prefabricated Modules

Modular prefabrication of kitchen, bathroom, and other sections that contain a lot of plumbing greatly reduces the speed of construction.

Solar Active and Passive Technologies

Passive Solar Design – Summer

Passive Solar Design – Winter

Equipment incorporating passive design – Summer

Equipment incorporating passive design – Winter

4-4 Green, Low Carbon, Energy Efficient Design

Sunshine room with different

The skylight opens in summer and can be used as a sunroom.

The skylight opens in summer and can be used as an open greenhouse.

The skylight is closed in winter and can be used as a meditation room.

The skylight is closed in winter and can be used as a greenhouse.

Fan energy saving system

The fan blows downward.

The fan blows upward.

Bernoulli's principle

When the skylight is opened in summer, the fan blows downward, the indoor downwind speed becomes larger, the pressure becomes smaller, which is more conducive to the indoor dirty air discharge and strengthen ventilation.

When the skylight is opened in summer, the fan blows upward, the indoor upward wind speed becomes larger, the pressure becomes smaller, which is more conducive to the outdoor natural wind entering the room and strengthening ventilation.

Rain garden construction

Wind Power Maximization Analysis

The wind turbine remains stationary and is ready to turn towards the windward.

The prevailing wind direction in the local summer is southeast wind. The wind turbine will now

The prevailing wind direction in the local winter is northwest wind. The wind turbine will now

Node Structure

Building Simulation Analysis

Baseline Building Electricity Demand

Area and Building Power Generation

The building generates significantly more electricity than it needs, and both the activity center and the residence are energy self-sufficient.

Carbon Emission Simulation Results for Homestay

Carbon Emission Simulation Results for Activity Centers

Serial number	Project name	Conclusion
1	Shape factor	Fulfillment
2	Airtightness rating of external building windows	Fulfillment
3	Average heat transfer coefficient of external walls	Fulfillment
4	Window-to-wall area ratio	Fulfillment
5	Heat transfer coefficient of external windows	Fulfillment

光合聚落 01

110051

李八庄防洪韧性设计

·设计说明： 河北省保定市李八庄韧性家园重建计划旨在为受洪水内涝影响的村庄打造一个可持续、生态友好的居住环境。项目包含 14 套民居和村民公共服务建筑。总体规划强调与自然和谐共生，通过延伸周边道路，形成规整的居住单元，提升交通便捷性与社区凝聚力。特别设计的雨水蓄积区有效管理洪水，减少内涝风险。住宅设计以暖廊为核心，不仅作为室内的交通空间，更是绿色建筑理念的体现，通过自然通风和采光，降低能耗，提升居住舒适度。公共服务建筑采用光伏屋顶，覆盖大面积太阳能板，确保在洪水期间也能保障电力供应，增强社区的能源自给能力。本设计旨在构建一个洪水适应性强、环境友好、能源自足的新型乡村社区。

·Design Description: The Libazhuang Resilience Home Reconstruction Plan in Baoding City, Hebei Province aims to create a sustainable and eco-friendly living environment for villages affected by floods and waterlogging. The project includes 14 sets of residential buildings and public service buildings for villagers. The overall plan emphasizes harmonious coexistence with nature, by extending surrounding roads to form orderly residential units, enhancing transportation convenience and community cohesion. Specially designed rainwater storage areas effectively manage floods and reduce the risk of waterlogging. Residential design revolves around warm corridors, which not only serve as indoor transportation spaces but also embody the concept of green building. Through natural ventilation and lighting, energy consumption is reduced and living comfort is improved. Public service buildings adopt photovoltaic roofs, covering large areas of solar panels to ensure power supply during floods and enhance the community's energy self-sufficiency. This design aims to build a new type of rural community with strong flood adaptability, environmental friendliness, and energy self-sufficiency.

综合奖·优秀奖·保定市清苑区李八庄村建设项目
Comprehensive Awards − Honorable Mention − Libazhuang Village Project, Qingyuan District, Baoding City

注册号：110051
Register Number：110051
项目名称：光合聚落
Entry Title：Photosynthetic Settlement
作者：刘菊影、胡瑞鹏、黄新鹏、李乐乐
Authors：Liu Juying, Hu Ruipeng, Li Xinpeng, and Li Lele
作者单位：合肥工业大学
Authors from：Hefei University of Technology
指导教师：王旭
Tutor：Wang Xu
指导教师单位：合肥工业大学
Tutor from：Hefei University of Technology

■ Location Analysis

Rural waterlogging
Rural street
Traditional village streetscape
Traditional Residential Layout

·The project is located in the center of Libazhuang Village, Beidian Township, Qingyuan County, Baoding City, Hebei Province, with a longitude of 115.49°E and a latitude of 38.71°N. It is 20 kilometers away from the center of Baoding City.

·Due to the large size of the village, concentrated construction land, flat site, and proximity to rivers, there have been multiple occurrences of farmland flooding and village waterlogging during flood seasons in recent decades.

·Li Bazhuang Village belongs to the warm temperate monsoon continental climate, with a significant monsoon climate and distinct four seasons.

·The transportation around the project site is relatively convenient, and the entire village has undergone gas to coal transformation. There are basic water and electricity facilities, but there is no sewer network to alleviate the problem of waterlogging during the rainy season. The riverbank between the village embankment and the Qingshui River can buffer the disasters caused by the rising water level of the Qingshui River during the flood season. It is usually used as farmland by villagers.

■ Design Strategy

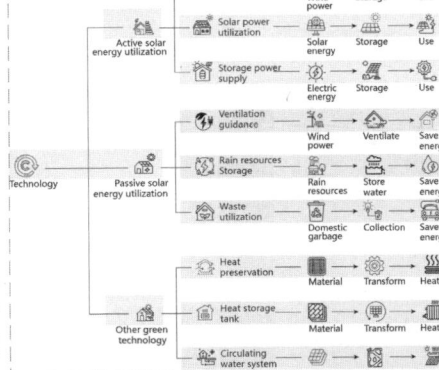

Technology

Active solar energy utilization
- Wind energy utilization: Wind power → Storage → Use
- Solar power utilization: Solar energy → Storage → Use
- Storage power supply: Electric energy → Storage → Use

Passive solar energy utilization
- Ventilation guidance: Wind power → Ventilate → Save energy
- Rain resources Storage: Rain resources → Store water → Save energy
- Waste utilization: Domestic garbage → Collection → Save energy

Other green technology
- Heat preservation: Material → Transform → Heat
- Heat storage tank: Material → Transform → Heat
- Circulating water system

■ Base General Layout

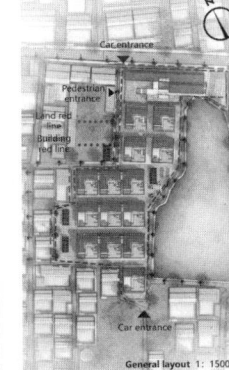

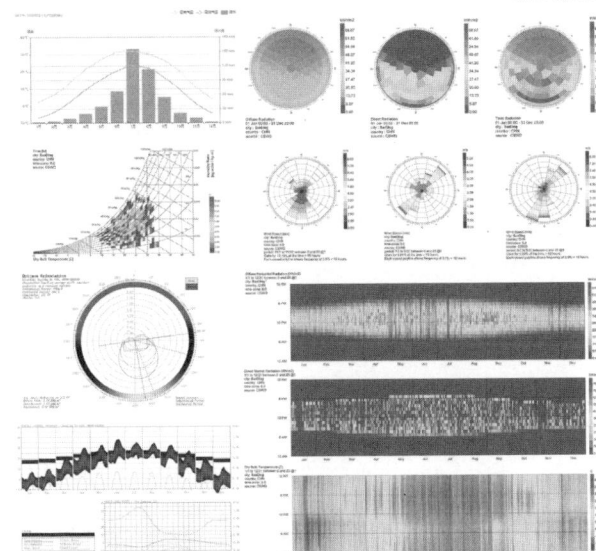

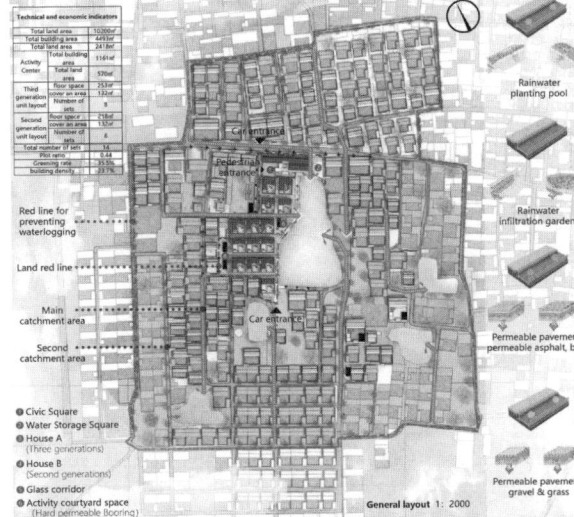

Car entrance
Pedestrian entrance
Land red line
Building red line
Car entrance

General layout 1：1500

■ General Layout

Technical and economic indicators		
Total land area		10,200m²
Total building area		4493m²
Total land area		241m²
Activity Center	Total building area	
	Total land area	590m²
Third generation unit layout	Total floor space	253m²
	cover on area	132m²
	Number of sets	
Second generation unit layout	Floor space	218m²
	cover on area	132m²
	Number of sets	
Total number of sets left		14
Plot ratio		0.44
Greening rate		35.5%
building density		34.5%

Red line for preventing waterlogging
Land red line
Main catchment area
Second catchment area

- Civic Square
- Water Storage Square
- House A (Three generations)
- House B (Second generations)
- Glass corridor
- Activity courtyard space (Hard permeable flooring)

Car entrance
Pedestrian entrance
Car entrance

General layout 1：2000

■ Road Design

Rainwater planting pool
Rainwater infiltration garden
Permeable pavement: permeable asphalt, brick
Permeable pavement: gravel & grass

光合聚落 02
110051
韧性 生态 应急 可持续 低能耗

Planar Analysis

1-Entrance Hall Space
2-Waiting Space
3-Medical Space
4-Toilet
5-Guard Room
(Emergency Command Center)
6-Citizen Service Center
7-Community Restaurant
8-Cold Food Storage
9-Non-staple Food Warehouse
10-Employee Bathing
(Emergency Bathing Water)
11-Staff Lounge
12-Operation Room
(Emergency Cooking)
13-Disaster Prevention Warehouse
14-Space Under The Tree
15-Public Open Space
16-Entrance Square

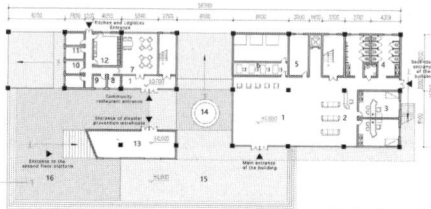

1-Aisle Space
2-Teaching Space
3-Self-Study Room
4-Toilet
5-Courtyard Space
6-Above The Entrance Hall
7-Reading Space
8-Activity Room
9-Music Room
10-Gymnasium
11-Public Platform
12-Entrance Steps
13-Roof Staircase Platform
14-Rest Platform

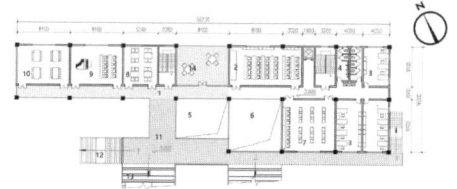

First floor plan 1 : 300

Second Floor Plan 1 : 300

Spatial Resilience

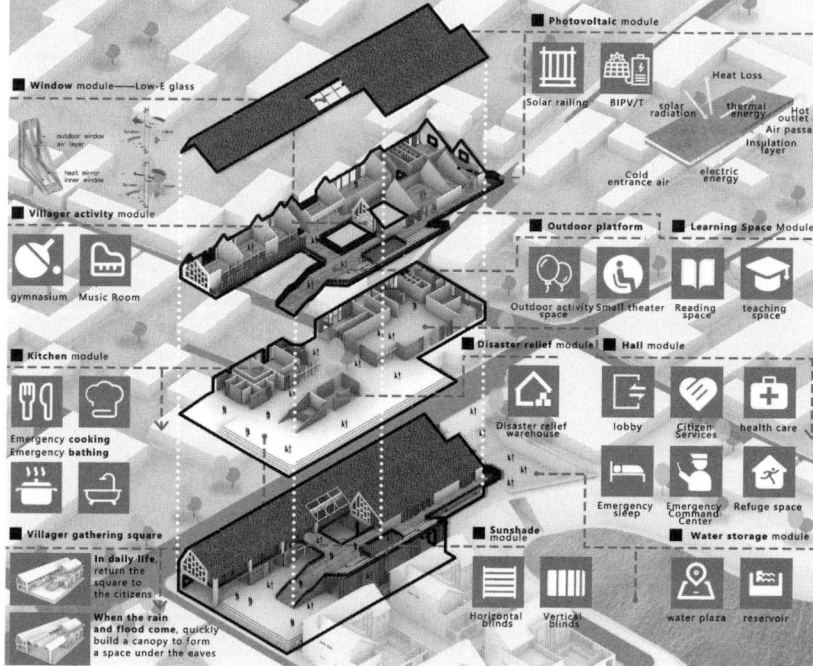

■ Window module——Low-E glass

outdoor window air module
heat mirror inner window

■ Villager activity module
gymnasium   Music Room

■ Kitchen module
Emergency cooking
Emergency bathing

■ Villager gathering square
In daily life, return the square to the citizens
When the rain and flood come, quickly build a canopy to form a space under the eaves

■ Photovoltaic module
Solar railing   BIPV/T   solar radiation
Heat Loss   thermal energy
Hot outlet air
Air passage
Insulation layer
Cold entrance air   electric energy

■ Outdoor platform   ■ Learning Space Module
Outdoor activity space   Small theater   Reading space   teaching space

■ Disaster relief module   ■ Hall module
Disaster relief warehouse   lobby   Citizen Services   health care
Emergency sleep   Emergency Command Center   Refuge space

■ Sunshade module   ■ Water storage module
Horizontal blinds   Vertical blinds   water plaza   reservoir

■ Before & After   ■ During

Elevation

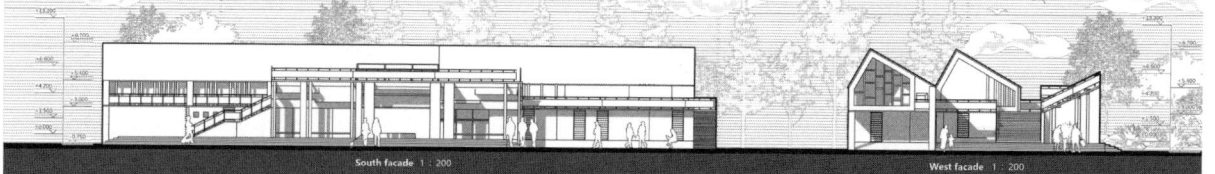

South facade 1 : 200

West facade 1 : 200

光合聚落 03

韧性 绿色 暖廊 可持续 低能耗

110051

Glass corridor

BIPV/T

Green House

Garage entrance

Main entrance

wind

summer door open

winter door close

sun

BIPV/T

Green House

Garage entrance

Main entrance

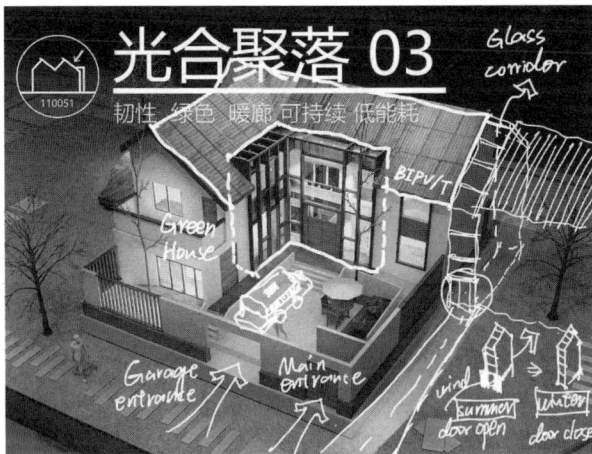

House A (Three generations)

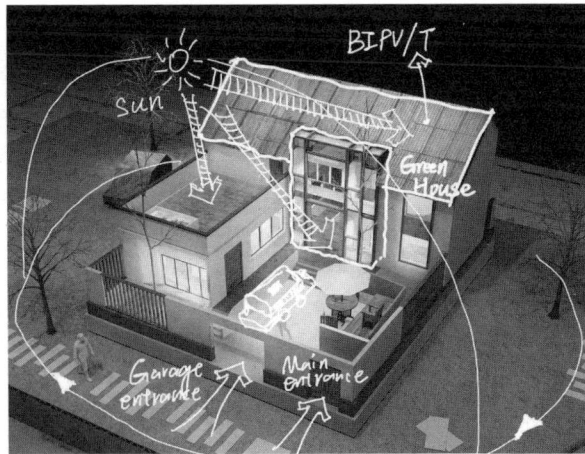

House B (Second generations)

1-courtyard space
2-Wooden Platform
3-screen wall
4-Warm Corridor
5-restaurant
6-kitchen
7-boiler room
8-toilet
9-Guest Room
10-living room
11-Storage room
12-Elderly room
13-study
14-bedroom
15-master room
16-Outdoor platform
17-lounge

Car entrance Pedestrian entrance

First floor plan 1 : 150

Second floor plan 1 : 150

1-courtyard space
2-Wooden Platform
3-screen wall
4-Warm Corridor
5-restaurant
6-kitchen
7-boiler room
8-toilet
9-Guest Room
10-living room
11-Storage room
12-Elderly room
13-study
14-bedroom
15-master room
16-Outdoor platform
17-lounge

Car entrance Pedestrian entrance

First floor plan 1 : 150

Second floor plan 1 : 150

Exploded View

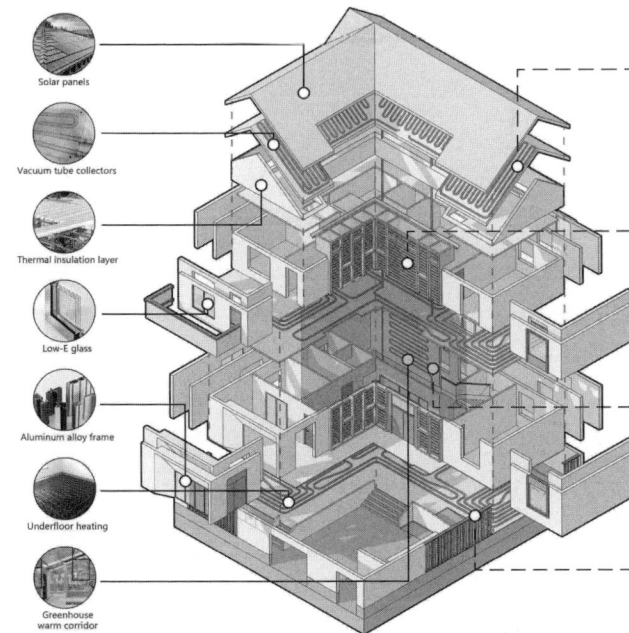

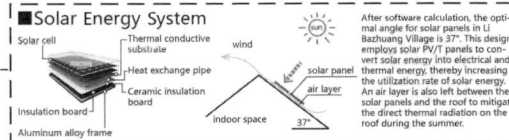

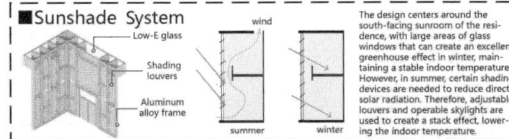

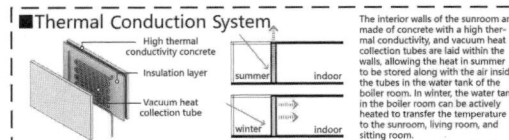

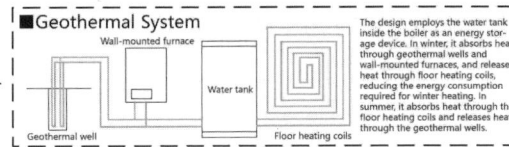

Solar panels

Vacuum tube collectors

Thermal insulation layer

Low-E glass

Aluminum alloy frame

Underfloor heating

Greenhouse warm corridor

Solar Energy System

Solar cell
Thermal conductive substrate
Heat exchange pipe
Ceramic insulation board
Insulation board
Aluminum alloy frame

wind
solar panel
air layer
indoor space
37°

After software calculation, the optimal angle for solar panels in Li Bazhuang Village is 37°. This design employs solar PV/T panels to convert solar energy into electrical and thermal energy, thereby increasing the utilization rate of solar energy. An air layer is also left between the solar panels and the roof to mitigate the direct thermal radiation on the roof during the summer.

Sunshade System

Low-E glass
Shading louvers
Aluminum alloy frame

wind
summer
winter

The design centers around the south-facing sunroom of the residence, with large areas of glass windows that can create an excellent greenhouse effect in winter, maintaining a stable indoor temperature. However, in summer, certain shading devices are needed to reduce direct solar radiation. Therefore, adjustable louvers and operable skylights are used to create a stack effect, lowering the indoor temperature.

Thermal Conduction System

High thermal conductivity concrete
Insulation layer
Vacuum heat collection tube

summer indoor
winter indoor

The interior walls of the sunroom are made of concrete with a high thermal conductivity, and vacuum heat collection tubes are laid within the walls, allowing the heat in summer to be stored along with the air inside the tubes in the water tank of the boiler room. In winter, the water tank in the boiler room can be actively heated to transfer the temperature to the sunroom, living room, and sitting room.

Geothermal System

Wall-mounted furnace
Water tank
Geothermal well
Floor heating coils

The design employs the water tank inside the boiler as an energy storage device. In winter, it absorbs heat through geothermal wells and wall-mounted furnaces, and releases heat through floor heating coils, reducing the energy consumption required for winter heating. In summer, it absorbs heat through the floor heating coils and releases heat through the geothermal wells.

Elevation and Section

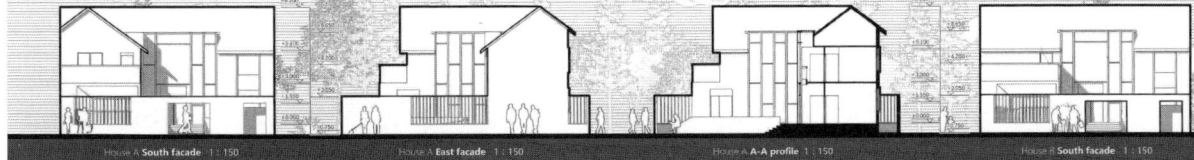

House A South facade 1 : 150

House A East facade 1 : 150

House A A-A profile 1 : 150

House B South facade 1 : 150

110051

光合聚落 04

风环境 日照 辐射热 系统 碳排放

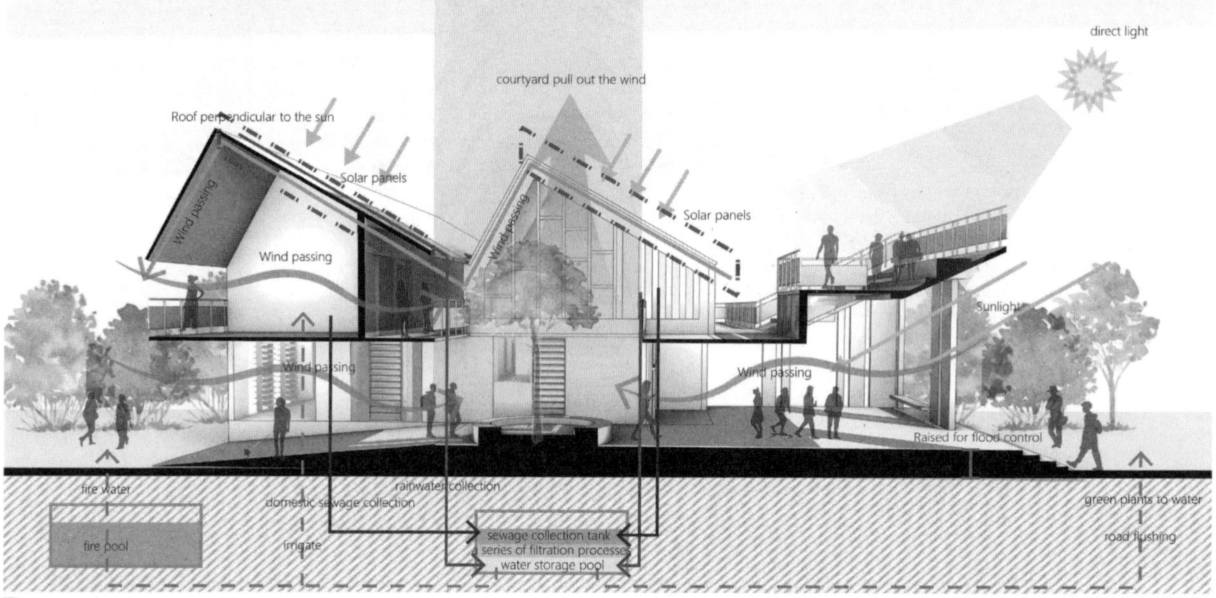
direct light
courtyard pull out the wind
Roof perpendicular to the sun
Solar panels
Solar panels
Wind passing
Wind passing
Wind passing
Wind passing
Wind passing
Sunlight
Raised for flood control
fire water
fire pool
domestic sewage collection
rainwater collection
irrigate
sewage collection tank
series of filtration processes
water storage pool
green plants to water
road flushing

Integrated System Diagram

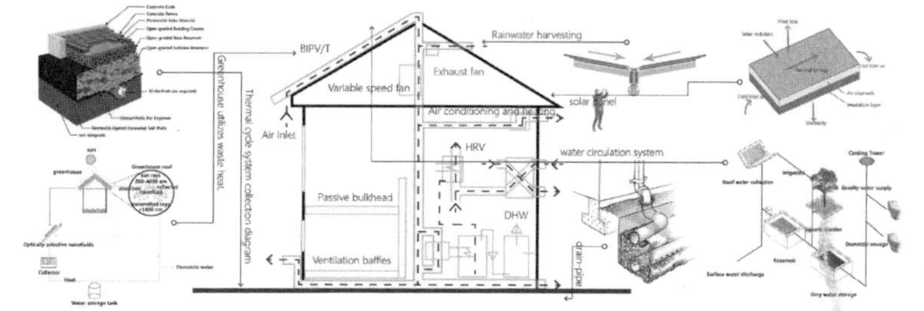
BIPV/T
Rainwater harvesting
Exhaust fan
Variable speed fan
Air conditioning and heating
Air inlet
HRV
Passive bulkhead
DHW
Ventilation baffles
Greenhouse utilizes waste heat
Thermal cycle system collection diagram
solar panel
water circulation system
drain pipe

Carbon Emissions Calculation

Building operation phase carbon emission table							
Project		Community Center		House A		House B	
		Carbon emission quantity ($kgCO_2$)	Carbon emission per unit area	Carbon emission quantity ($kgCO_2$)	Carbon emission per unit area	Carbon emission quantity ($kgCO_2$)	Carbon emission per unit area
Carbon emission	Heating	610.56	2.41	128.73	0.51	110.92	0.51
	Non-heating	2053.08	8.11	1442.86	5.70	1243.25	5.70
carbon reduction items	Renewable energy	4368.95	17.27	1277.47	5.05	1100.74	5.05
	Carbon sink	139.20	0.55	1.15	0.00	0.99	0.00
Total		-1844.51	-7.29	292.97	1.16	252.44	1.16

Wind Environment Analysis

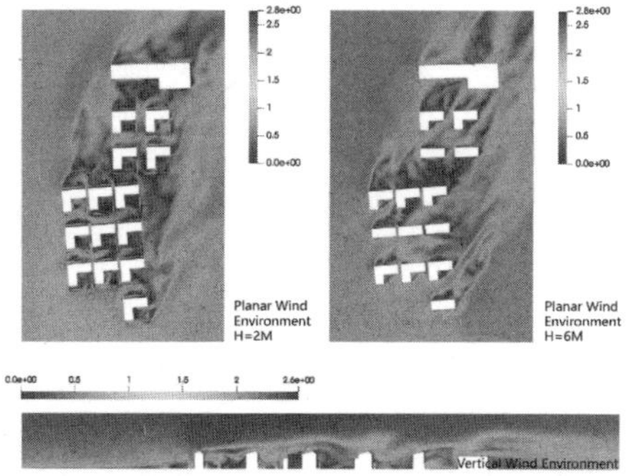
Planar Wind Environment H=2M
Planar Wind Environment H=6M
Vertical Wind Environment

Outdoor Sunlight

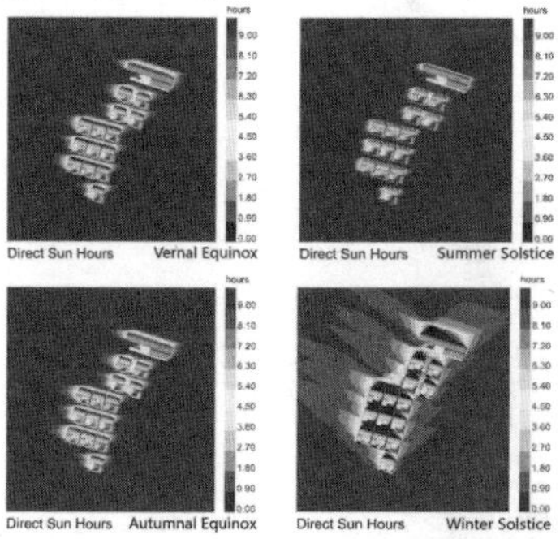
Direct Sun Hours Vernal Equinox
Direct Sun Hours Summer Solstice
Direct Sun Hours Autumnal Equinox
Direct Sun Hours Winter Solstice

Outdoor Radiant Heat

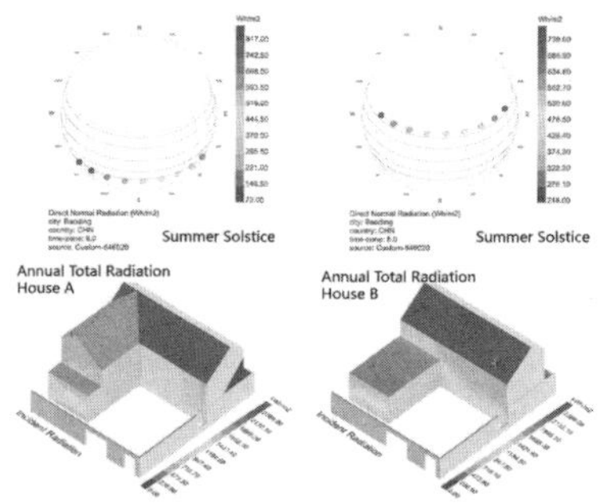
Summer Solstice
Summer Solstice
Annual Total Radiation House A
Annual Total Radiation House B

Summer Indoor Daylighting

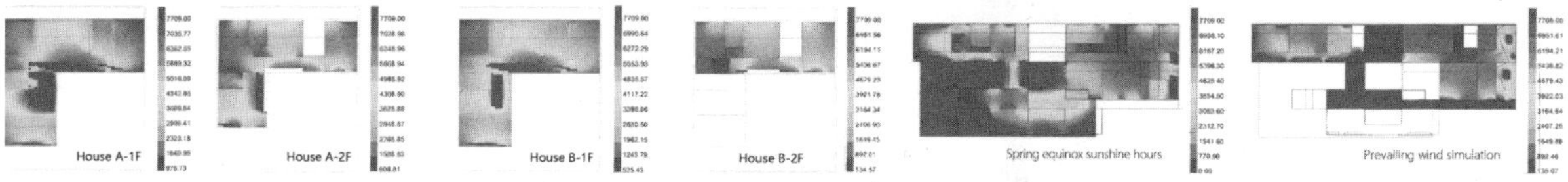
House A-1F
House A-2F
House B-1F
House B-2F
Spring equinox sunshine hours
Prevailing wind simulation

Winter Indoor Daylighting

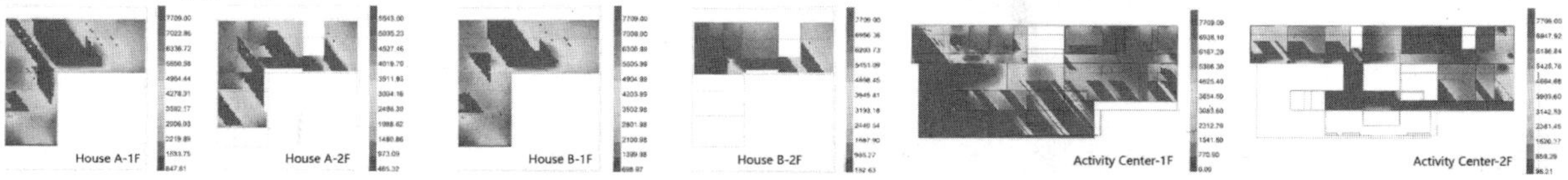
House A-1F
House A-2F
House B-1F
House B-2F
Activity Center-1F
Activity Center-2F

河北保定市清苑区李八村阳光乡村韧性设计
NO. 100-110092

雨·虹细胞 1
Rain—Rainbow Cells

阳光·乡村韧性
2024 台达杯国际太阳能建筑设计竞赛获奖作品集

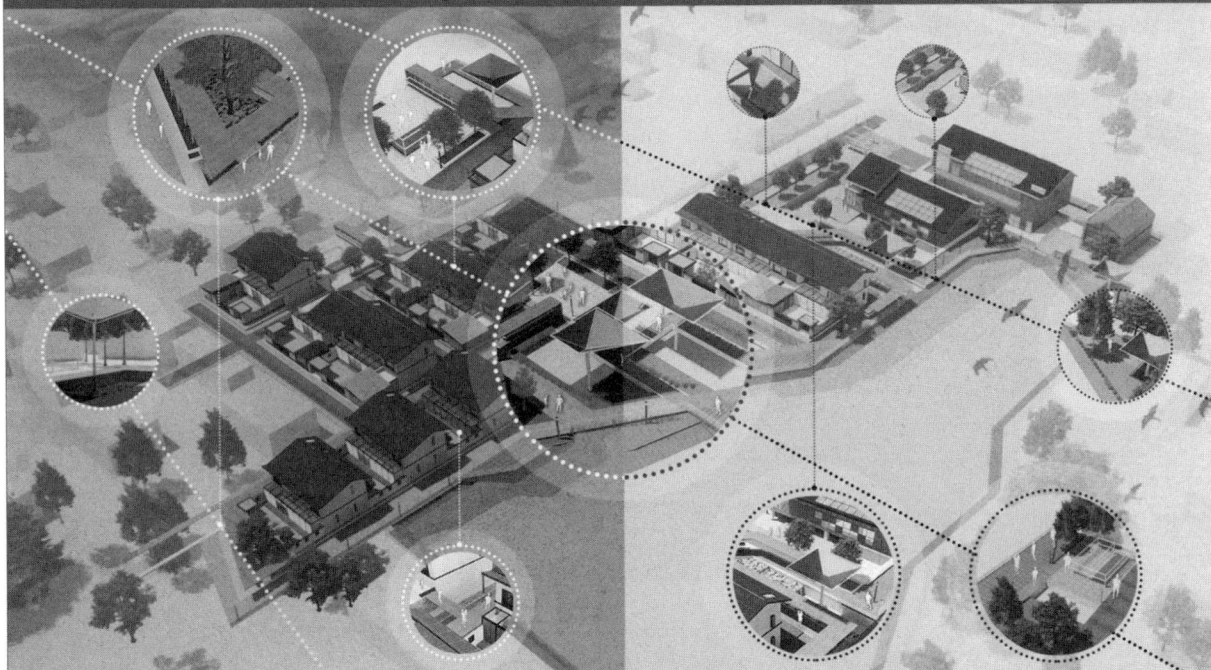

综合奖·优秀奖·保定市清苑区李八庄村建设项目
Comprehensive Awards — Honorable Mention — Libazhuang Village Project, Qingyuan District, Baoding City

注册号：110092
Register Number：110092

项目名称：雨·虹细胞
Entry Title：Rain — Rainbow Cells

作者：张蕾、高原、韩旭、刘梦飞
Authors：Zhang Lei, Gao Yuan, Han Xu, and Liu Mengfei

作者单位：苏州科技大学
Authors from：Suzhou University of Science and Technology

指导教师：金雨蒙、刘长春
Tutors：Jin Yumeng, Liu Changchun

指导教师单位：苏州科技大学
Tutors from：Suzhou University of Science and Technology

Design Specification

设计以细胞为概念，置入雨虹细胞，构建整体的防灾抗灾系统，利用分散的绿地系统和雨量监测系统分担村落泄洪压力，同时利用风、光、水等清洁能源满足应急状态下能源自维持系统。技术层面上同时建筑利用腔体空间将蓄热保温和雨水系统结合，同时将太阳能晶硅板、蓄热天窗、可调节百叶、相变蓄热墙等技术与传统建筑形象结合。

The design is based on the concept of cells, incorporating the "Rainbow Cell" to build an overall disaster prevention and mitigation system. It utilizes a decentralized green space system and rainfall monitoring system to share the flood discharge pressure of the village, while also harnessing clean energies such as wind, solar, and water to sustain energy self-sufficiency during emergency situations. Technically, the architecture integrates thermal storage and heat insulation with the rainwater system using cavity spaces. Additionally, the design blends advanced technologies such as solar silicon panels, thermal storage skylights.

Historical development

Historical development

Severe rainstorm weather
The persistent intense heavy rain led to severe waterlogging within the village.

Villages have poor adaptability to rainstorm
Water accumulation in the low-lying areas of the village, and the houses of many villagers have been flooded.

Flood disaster
In recent decades, frequent flooding is farmland and waterlogging in villages have occurred during the flood season, every waterlogging within the village.

Lack of emergency measures
In the event of extreme situations such as water and power outages, there is no ability to sustain oneself.

The Concept of the Rain·Rainbow Cell

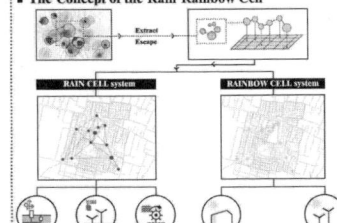

Extract Escape

RAIN CELL system

RAINBOW CELL system

Rainbow cells are distributed throughout the green space system centered on rural public spaces and individual residences. Each cell coordinates with one another to form a complete rainbow system. Through early warning, drainage, storage, and reuse of rainwater, as well as the utilization of solar energy, the system meets the requirements of rural resilience design.

General Plan

主入口一 Main Entrance One
入口广场 Entrance Plaza
车行入口 Vehicle Entrance
建筑红线 Building Line
用地红线 Site Boundary Line
次入口 Secondary Entrance
水塘 Pond
绿地 Green Area
次入口 Secondary Entrance
主入口二 Main Entrance Two

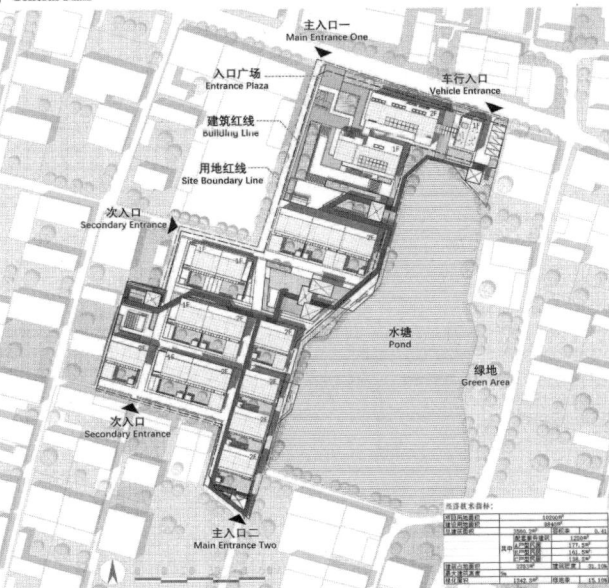

Integrated Flood Management Plan

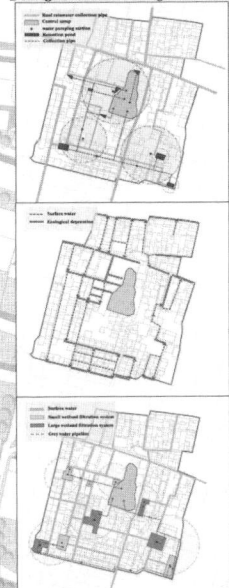

location Analysis

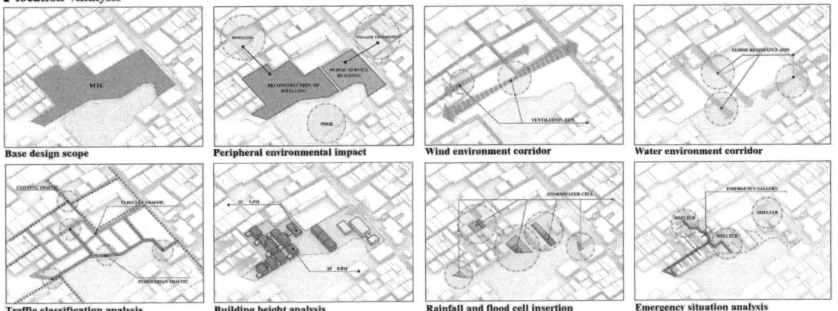

Base design scope
Peripheral environmental impact
Wind environment corridor
Water environment corridor
Traffic classification analysis
Building height analysis
Rainfall and flood cell insertion
Emergency situation analysis

河北保定市清苑区李八村阳光乡村韧性设计
NO. 100-110092

雨 · 虹细胞 2
Rain-Rainbow Cells

Stormwater Cell Filter System 雨细胞过滤系统

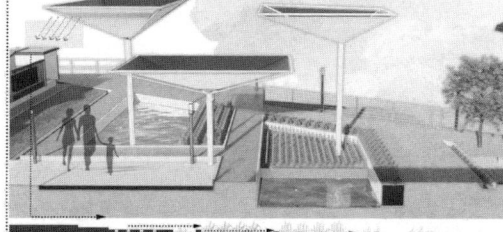

Permeable pavement　Sod ditch　Primary purification　Secondary purification　Tertiary purification

The mode of Combining Disaster with Peace 平灾结合模式

The real-time regulation of rainwater storage and discharge by the rainwater cells, while monitoring the village's risk coefficient to serve as an early warning.

Corridors serve the function of temporarily evacuating people, green spaces are transformed into temporary shelters, and the rural service center can be used as a temporary resettlement site.

Post-disaster, the ecological filtration system purifies rainwater in stages to maintain the normal functioning of rural areas.

Per-catastrophe　　In-catastrophe　　Post-catastrophe

Layered Drainage System
分层排水系统

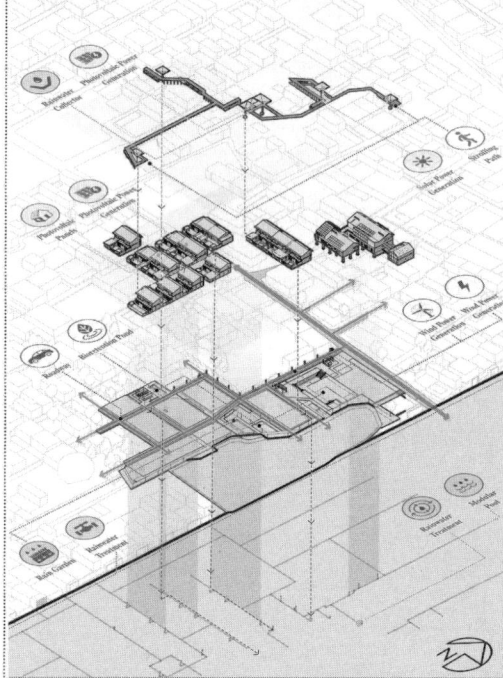

Village Rain Cell Resilience System
乡村雨 · 细胞韧性系统 —— The stormwater network of point-line-surface organic series is constructed

Technology system —— spatial structure analysis

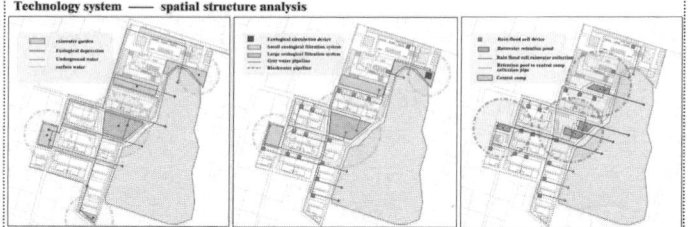

Rainwater management system　　Sewage management system　　Building stormwater control system setup

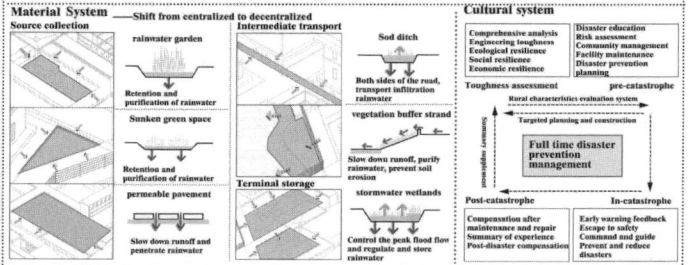

Material System —— Shift from centralized to decentralized

Source collection　　Intermediate transport

rainwater garden

Retention and purification of rainwater

Sunken green space

Retention and purification of rainwater

permeable pavement

Slow down runoff and penetrate rainwater

Sod ditch

Both sides of the road, transport infiltration rainwater

vegetation buffer strand

Slow down runoff, purify rainwater, prevent soil erosion

Terminal storage

stormwater wetlands

Control the peak flood flow and regulate and store rainwater

Cultural system

Comprehensive analysis
Engineering toughness
Ecological resilience
Social resilience
Economic resilience

Disaster education
Risk assessment
Community management
Facility maintenance
Disaster prevention planning

Toughness assessment　　pre-catastrophe

Rural characteristics evaluation system

Targeted planning and construction

Full time disaster prevention management

Post-catastrophe　　In-catastrophe

Compensation after maintenance and repair
Summary of experience
Post-disaster compensation

Early warning feedback
Escape to safety
Command and guide
Prevent and reduce disasters

Village Rainbow Cell resilience system
乡村虹 · 细胞韧性系统 Hydroelectric Wind and solar complementary power generation

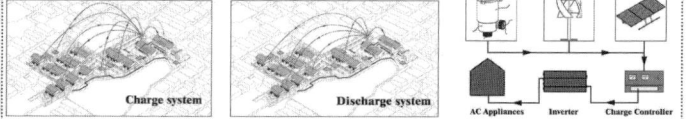

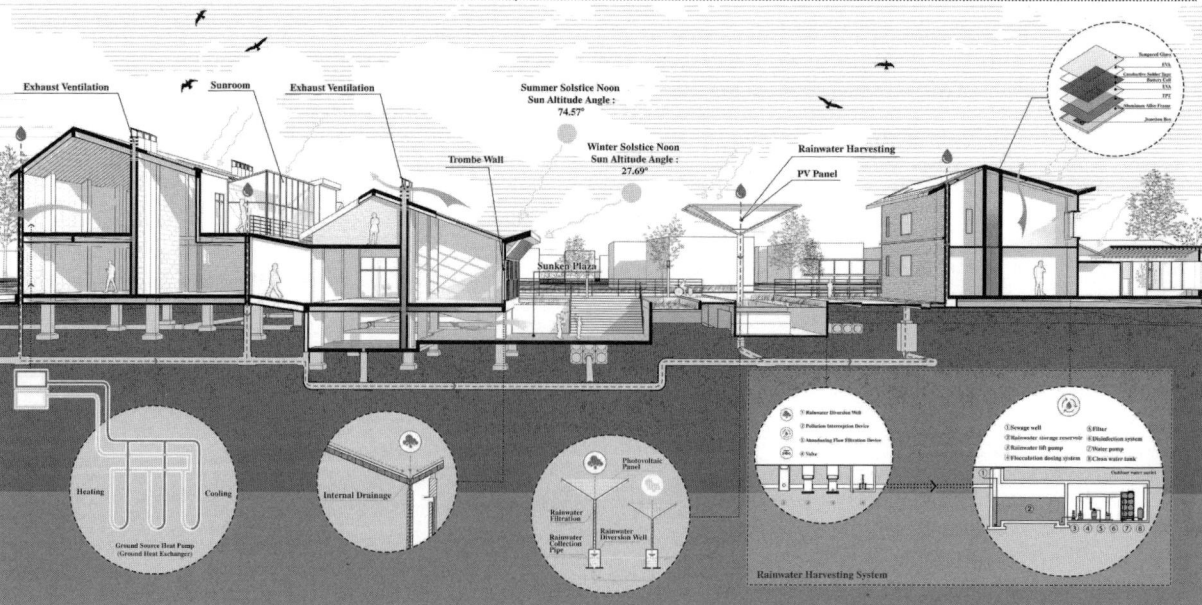

hydroelectric generation wind power generation photovoltaic power

Charge system　　Discharge system

AC Appliances　Inverter　Charge Controller

Exhaust Ventilation　Sunroom　Exhaust Ventilation

Summer Solstice Noon
Sun Altitude Angle : 74.57°

Winter Solstice Noon
Sun Altitude Angle : 27.69°

Trombe Wall

Rainwater Harvesting

PV Panel

Sunken Plaza

Heating　Cooling

Ground Source Heat Pump
(Ground Heat Exchanger)

Internal Drainage

Photovoltaic Panel

Rainwater Filtration

Rainwater Collection Pipe　Rainwater Diversion Well

Rainwater Harvesting System

河北保定市清苑区李八村阳光乡村韧性设计
NO. 100-110092

雨·虹细胞 3
Rain-Rainbow Cells

阳光·乡村韧性
2024 台达杯国际太阳能建筑设计竞赛获奖作品集

Public Building Ichnography

- First Floor Plan （880㎡）

- Second Floor Plan （400㎡）

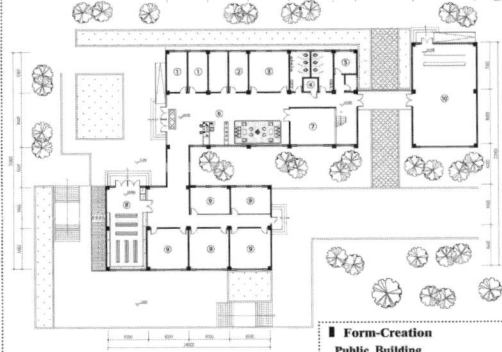

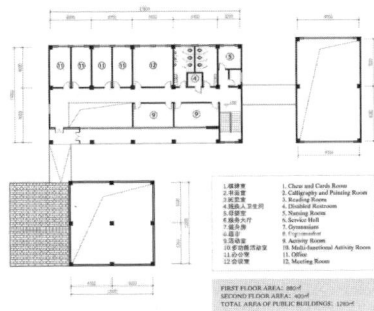

- Lower Ground Floor Plan

1. 棋牌室　　1. Chess and Cards Room
2. 书画室　　2. Calligraphy and Painting Room
3. 阅览室　　3. Reading Room
4. 残疾人卫生间　4. Disabled Restroom
5. 护理室　　5. Nursing Room
6. 服务大厅　6. Service Hall
7. 健身房　　7. Gymnasium
8. 活动室　　8. Party-member
9. 活动室　　9. Activity Room
10. 多功能活动室　10. Multi-functional Activity Room
11. 办公室　　11. Office
12. 会议室　　12. Meeting Room

FIRST FLOOR AREA: 880㎡
SECOND FLOOR AREA: 400㎡
TOTAL AREA OF PUBLIC BUILDINGS: 1280㎡

Public Building Ichnography

| Solar Energy | Photovoltaic Panel |
| Drainage System | Ventilation |

Form-Creation

Public Building

Step 1 | Step 2
Step 3 | Step 4

Residential house

Step 1 | Step 2
Step 3 | Step 4

Residential House Type Map

- Unit Type One

TYPE A
TYPE B
TYPE C
CHILDLESS FAMILY
NUCLEAR FAMILY
BLENDED FAMILY
7 HOUSEHOLDS
3 HOUSEHOLDS
4 HOUSEHOLDS

- Unit Type Two

- Unit Type THREE

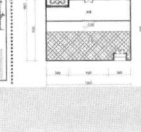

Exploded View Diagram

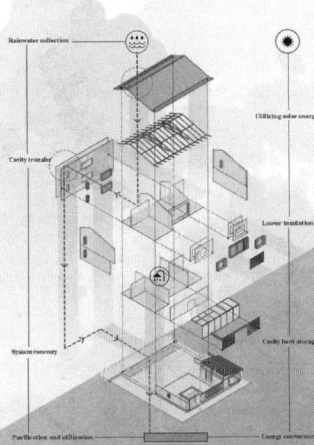

Rainwater collection
Utilizing solar energy
Cavity transfer
Lower insulation
System runway
Cavity heat storage
Purification and utilization
Energy conversion

Passive Solar Energy Utilization

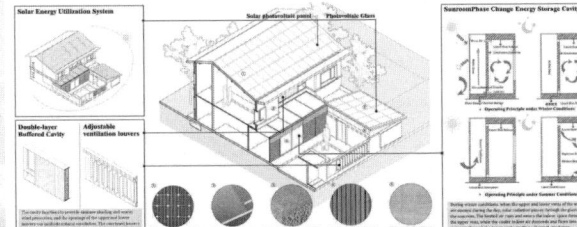

Solar Energy Utilization System
Solar photovoltaic panel
Photovoltaic Glass
Double-layer Buffered Cavity
Adjustable ventilation louvers

Solar photovoltaic panel | Photovoltaic Glass | Functionally Ceramic BAR | Wood Composite Panel | Polycarbonate Sheet

Sunroom/Phase Change Energy Storage Cavity

Elevation Drawing

河北保定市清苑区李八村阳光乡村韧性设计
NO. 100-110092

雨 · 虹细胞 4
Rain-Rainbow Cells

Question
SOS
Emergency
Gathering
Substation
Wind-heating device

Emergency   Solar energy   Communication

Solar energy   Sun room

Rainwater harvesting

Hydropower

Emergency

Solar energy   Sun room   PV-roof

Solar energy   Gathering   Windbreak wall   Louver   Courtyard   Communication

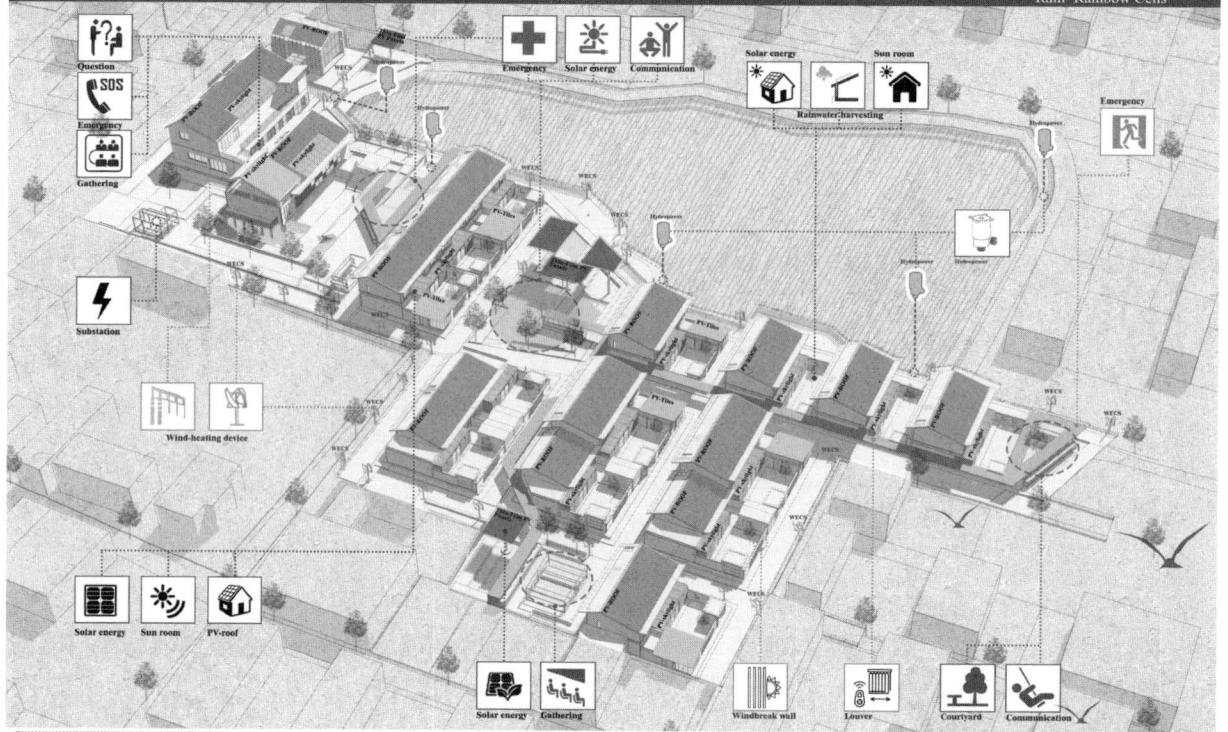

Environmental Illuminance Analysis

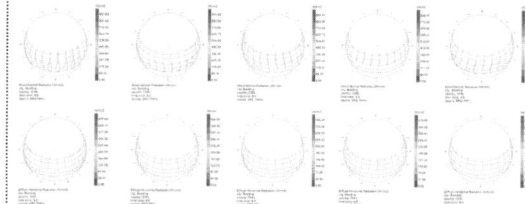

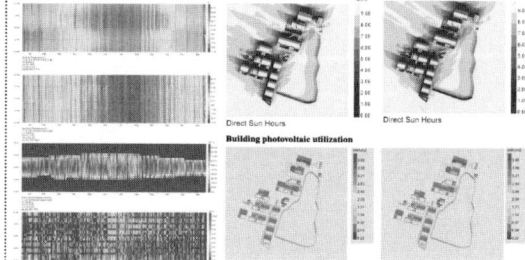

Environmental Illuminance Analysis

Direct Sun Hours   Direct Sun Hours

Building photovoltaic utilization

Wind Environment Analysis

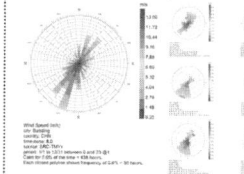

Environmental Comfort Analysis

Wind Environment Simulation

夏季1.5M风模拟   夏季3M风模拟   冬季1.5M风模拟   冬季3M风模拟

Node Detail Drawing

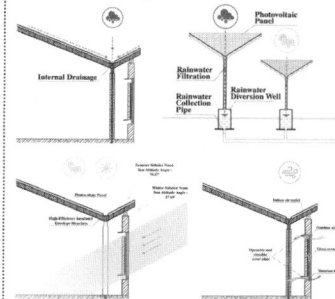

Internal Drainage
Photovoltaic Panel
Rainwater Filtration
Rainwater Diversion Well
Rainwater Collection Pipe

Energy Consumption Simulation Analysis

Residential Building Energy Saving Calculation Report (Taking Type II Apartment as an Example)

Exterior Wall Construction and Thermal Performance Parameters

Roof Construction and Thermal Performance Parameters

External Window Construction and Thermal Performance Parameters

Energy consumption calculation results:

Building Use Phase Carbon Emission Intensity Indicator Review:

Summary of Carbon Emission Calculation Results

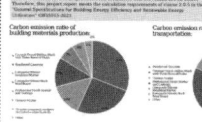

Building primary energy consumption and carbon emission calculation results.

The design building and benchmark building monthly cooling and heating load diagrams.

Designed Building and Benchmark Building Monthly Energy Consumption Charts

Renewable energy monthly generation capacity

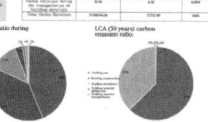

Chart of Proportions of Various Energy Consumption!

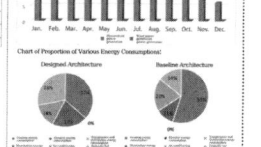

Designed Architecture   Baseline Architecture

Carbon emission ratio of building materials production.   Carbon emission ratio during transportation.   LCA (50 years) carbon emission ratio.

作品编号:110328

■ Preliminary Analysis

■ 设计说明:

项目位于河北省保定市清苑县北店乡李八庄村中心，因多种原因，近几十年汛期多次出现农田被淹和村子内涝问题。本方案秉持尊重自然的原则，顺应水的变化，营造"水进人退，水退人进"的韧性公共活动空间。设计通过梳理李八村庄土地资源与本底肌理，规划七大公共活动区域并用一条乡村环道将其串联，同时采用"公共设施+雨水花园"的手法对活动区域进行韧性设计，再搭配乡村洪涝预警系统，形成一条集防洪、生态、景观、游憩与智慧于一体的环道。在解决洪涝灾害的同时，同时也为村民提供更多相遇与交流的可能。

Design Statement:

The project is located in Li Bazhuang Village, Beidian Township, Qingyuan County, Baoding City, Hebei Province. Over the past few decades, the village has faced frequent farmland flooding and waterlogging during the rainy season. This plan respects natural processes and adapts to water changes, aiming to solve flooding issues while creating public spaces for villagers. The design organizes village land resources, establishing seven public activity areas linked by a rural loop road. Using a "public facilities + rain garden" approach and a flood early warning system, the loop road integrates flood control, ecology, landscape, recreation, and smart technology, addressing flood problems while enhancing villagers' quality of life and social interaction.

综合奖·优秀奖·保定市清苑区李八庄村建设项目
Comprehensive Awards — Honorable Mention — Libazhuang Village Project, Qingyuan District, Baoding City

注册号:110328
Register Number:110328
项目名称:顺水·回见
Entry Title:Along the Stream — Retrospect

作者:王玺凌、黄皓晖、唐雪晴、侯思琪
Authors:Wang Xiling, Huang Haohui, Tang Xueqing, and Hou Siqi
作者单位:重庆大学
Authors from:Chongqing University
指导教师:曾旭东、黄海静
Tutors:Zeng Xudong, Huang Haijing
指导教师单位:重庆大学
Tutors from:Chongqing University

■ Problem Analysis

Flood Disaster
The village has suffered from farmland flooding and waterlogging during the flood season. At its worst, the roads near the potholes were completely submerged, with 500 mm of water on the roads. Direct economic losses from various disasters were about 5.9 million yuan.

Deficient Planning
As a natural village, the village has a complex situation of concentrated building land, a high proportion of idle green space, and a lack of public service facilities. It needs to be carried out as a whole in the village's overall planning.

Poor Transportation
Due to the river on the east side, the overall road in the village falls to form a smooth traffic pattern on all sides. Moreover, due to the concentration of buildings and the numerous internal roads and alleys, the village is fragmented, resulting in a complex and inconvenient traffic environment on the village roads.

Poor Rural Aesthetics
The building structures in the village are mostly made of bricks and concrete. Many buildings are too old and need to be repaired. In addition, some villagers have renovated their houses by themselves and installed solar roofs. The overall appearance of the village is poor.

■ Solution Strategy

- Drainage network
- Rain garden
- Flood early warning system
- Land use optimization
- Ecological landscape
- Addition of public service facilities
- Pavement repair
- Permeable paving
- Enhancement of the road network
- Building restoration
- Local materials
- Traditional functional space

■ Master Plan Layout

① Village activity center
② New village houses
③ Village market
④ Village rest station
⑤ Village mini-theater
⑥ Village canteen
⑦ Village library
⑧ Children's playground
⑨ Ring road

RESILIENCE VIBRANCY CONNECTIVITY BEAUTY

顺水·回见

0 10 20 50 100 m

1/4

作品编号:110328

顺水·回见

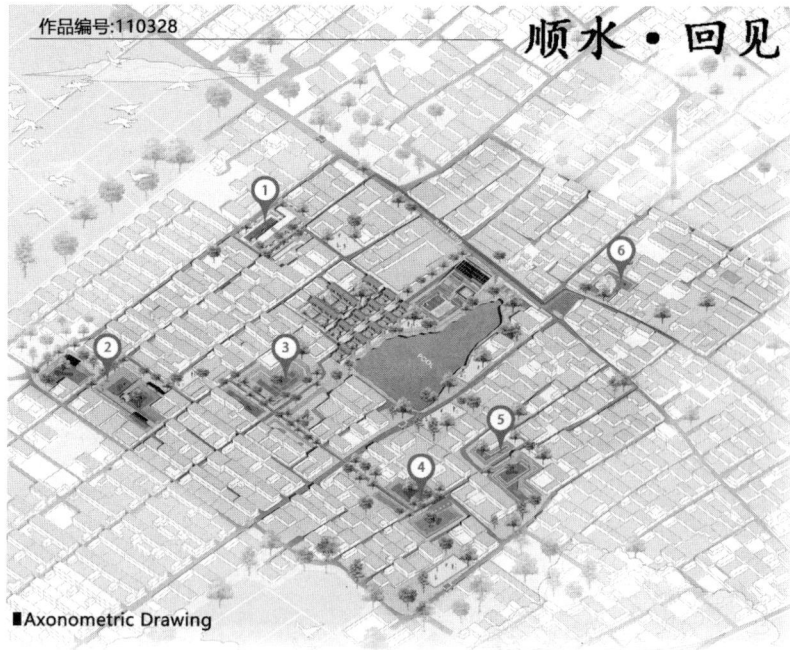

■Axonometric Drawing

① Village market
② Village rest station
③ Village mini-theater
④ Village canteen
⑤ Village library
⑤ Children's playground

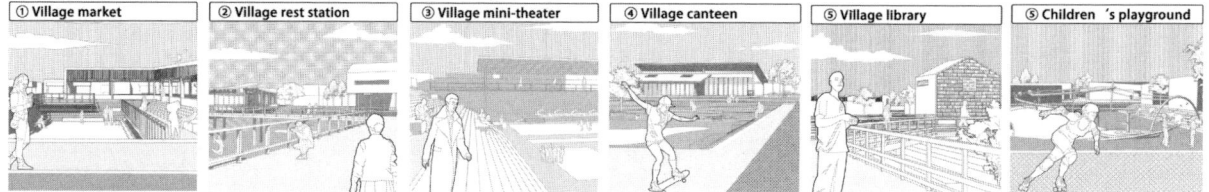

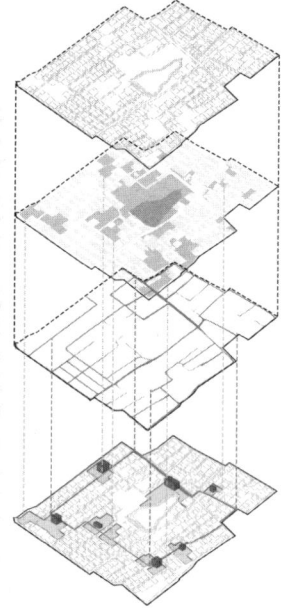

■Generate logic diagram

land conditions
The building layout is very concentrated and the available public space is not effectively used.

idle green space
Based on the location and size of the green space, the value of idle green space is judged and suitable green space is selected for renovation.

road classification
Classify roads according to their location, utilization rate, and width, and extract the most commonly used and easy-to-use roads.

generate loop
Ring roads are generated based on the selected green space and road conditions, rain gardens are designed for the green space, and the most suitable public service facilities are set up according to the location of each area.

■Special Report — Rural Flood Control and Electronic Early Warning System

Introduction

Water level sensors and rainfall monitors will be installed around the village ring road and in the seven rain gardens to monitor real-time rainfall and soil moisture. This data will be uploaded to the command center at the village community center, where it will be analyzed alongside historical data and weather models to assess flood risks. Based on rainfall levels, the system will categorize warnings into three stages: stable, alert, and emergency evacuation. Warnings will be promptly communicated through SMS, broadcasts, and mobile apps to notify the government and residents to take evacuation or property protection measures.

■ System Architecture and Design

IoT device side	Rain gauge	Liquid level meter	Electronic water gauge	Flow Meter
Data acquisition module	Video data collection and processing	Precipitation data collection and processing	Sensor data collection and processing	Pipeline network data collection and processing
	Digital geographic information	Refined precipitation forecast	Ground water detection	Pipeline water level monitoring
Data processing module	Cloud processing platform based on IoT			
	Cloud Case Library	Cloud real-time database	Cloud plan library	
Warning release module	Flood warning release platform			
	Early warning display	Information Query	Disaster Management	Data maintenance
Decision support module	Dispatch and command	Department Coordination	Graded Response	Social mobilization

■ Three Stages of Implementation and Response

○ **Stable stage** rainfall : 0-200mm
The rain garden appears in the state of a sunken square.

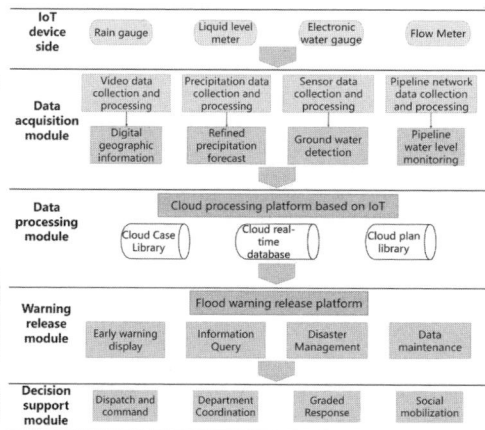

○ **Persistence stage** rainfall : 200-500mm
The rain garden appears in the state of a pool landscape.

○ **Emergency evacuation stage** rainfall : ≥500mm
The rain garden becomes an emergency evacuation site.

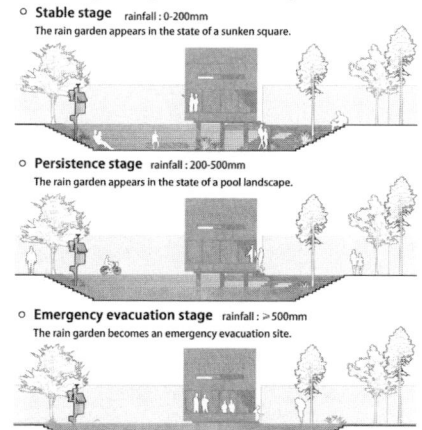

■Road Profile

Elevated pedestal · Seepage pavement · Roadside culverts · Rain gardens · Solar panel · Seepage pavement

Surface runoff · infiltrate · Surface runoff · infiltrate · Underground runoff · Infiltration

2/4

作品编号:110328

顺水·回见

■ Volume Generation

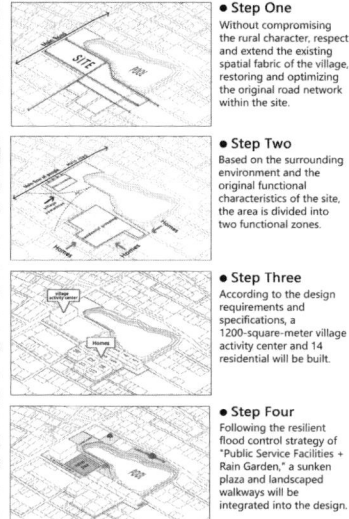

● Step One
Without compromising the rural character, respect and extend the existing spatial fabric of the village, restoring and optimizing the original road network within the site.

● Step Two
Based on the surrounding environment and the original functional characteristics of the site, the area is divided into two functional zones.

● Step Three
According to the design requirements and specifications, a 1200-square-meter village activity center and 14 residential will be built.

● Step Four
Following the resilient flood control strategy of "Public Service Facilities + Rain Garden," a sunken plaza and landscaped walkways will be integrated into the design.

■ Economic and Technical Index

Construction land area Total building area	8840㎡	Total floor area of residential buildings	2030㎡
Total floor area of public buildings	3217㎡	Total floor area of residential buildings for three generations	172㎡
Land area of public buildings	1187㎡	Land area of residential buildings for three generations	275㎡
The occupied area of public buildings	1169㎡	Number of households in residential buildings for three generations	5households
Number of floors in public buildings	2floors	Total floor area of residential buildings for two generations	130㎡
Height of public buildings	9m	Land area of residential buildings for two generations	204㎡
Building density	50%	Number of households in residential buildings for two generations	9households
Greening rate	25%	Number of floors in residential buildings	2floors
Number of parking spaces	9	Height of residential buildings	8m

■ 设计说明

本设计旨在尊重并延续乡村原有的空间肌理,结合场地周边环境和本底功能特征,与蓄水池形成"两轴两区"的总平面布局。村民活动中心遵循"公共服务设施+雨水花园"的韧性规划策略,打造集功能与景观于一体的活力村中心。住宅区与场地西侧和南侧的现有民居紧密衔接,形成组团关系,坚持乡村风格布局一致性与整体性的原则。通过提升传统民居的功能与品质,确保新建住宅与周边现有住宅和谐统一。

Design Description:

The design aims to respect and continue the village's original spatial fabric, integrating the site's surrounding environment and functional characteristics to form a master plan of "two axes and two zones" around the reservoir. The village activity center follows the resilient planning strategy of "Public Service Facilities + Rain Garden," creating a vibrant village center that combines functionality and landscape. The residential area seamlessly connects with the existing homes on the west and south sides of the site, forming a cluster relationship. By enhancing the functionality and quality of traditional homes, the design ensures the new residences are in harmony with the existing ones, maintaining consistency in style without differentiation.

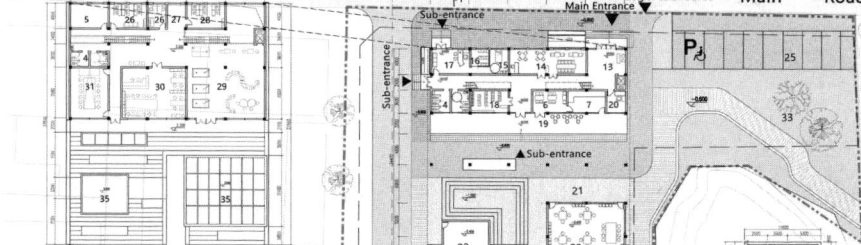

■ Village Activity Center 2F Plan 1:300

■ Functional Replacement of Village Activity Center 2F Plan 1:300

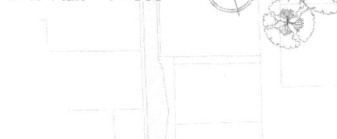

■ Second Gemertation Home 2F Plan 1:300

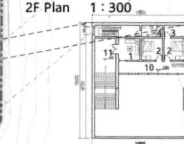

■ 1F Plan 1:300

Pool

■ Three-Genertation Home 2F Plan 1:300

1 Fireplace	13 Foyer	25 Parking Space
2 Bedroom	14 Senior Center	26 Office
3 Checkroom	15 Sickroom	27 Information Wall
4 Restroom	16 Drugstore	28 Information Centre
5 Cupboard	17 Health Office	29 Recreation
6 Cafeteria	18 Store	30 Library
7 Bathroom	19 Outdoor Dining	31 Council Chamber
8 Screen Wall	20 Duty Office	32 Emergency Living Modules
9 Courtyard	21 Pathway	33 Park
10 Sun Porch	22 Plant Room	34 Square
11 Entrance Platform	23 Chess Room	35 Roof
12 Vestibule	24 Sunken Garden	

■ Functional Circulation Analysis

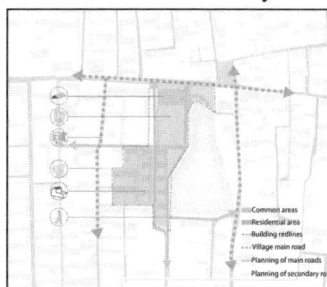

- Common areas
- Residential area
- Building redlines
- Village main road
- Planning of main roads
- Planning of secondary roads

■ Spatial form Analysis

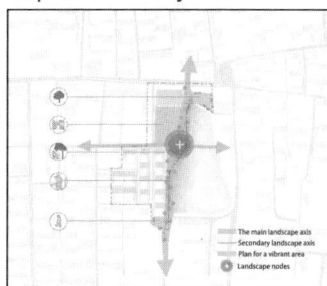

- The main landscape axis
- Secondary landscape axis
- Plan for a vibrant area
- Landscape nodes

3/4

Building Conponents

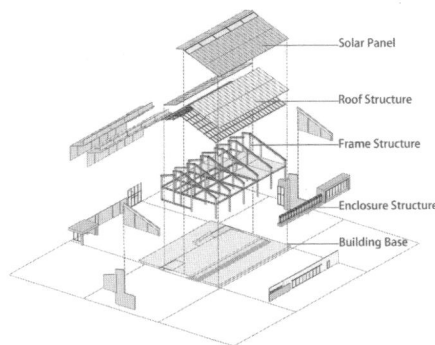

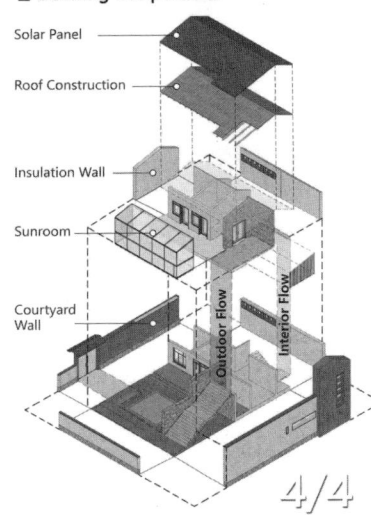

- Solar Panel
- Roof Structure
- Frame Structure
- Enclosure Structure
- Building Base

Bulk Generation

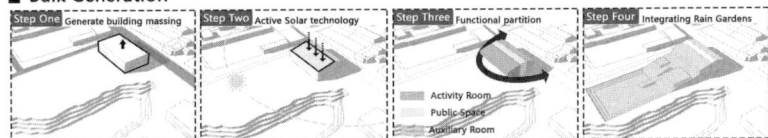

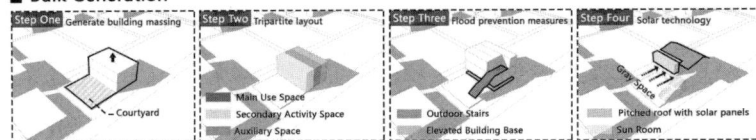

| Step One Generate building massing | Step Two Active Solar technology | Step Three Functional partition | Step Four Integrating Rain Gardens |

- Activity Room
- Public Space
- Auxiliary Room

Technical Strategy

Hot pressure ventilation
Cross-draft

Building ventilation
The combination of different spaces creates a chimney effect and thermobaric ventilation.

Activity · Traffic · Auxiliary

Covertible
Activity
Base

Three-stage layout
Direct and indirect combination, phase change exothermy.

Radiation benefit
Direct and indirect combination, phase change exothermy.

Control room

Active technology
Photovoltaic integration combined with ground source heat pump.

Daylighting Analysis

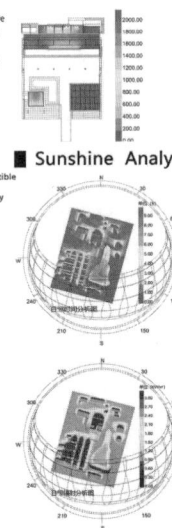

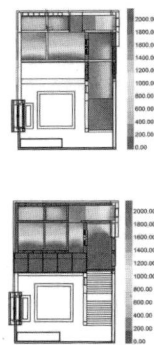

Carbon Emission

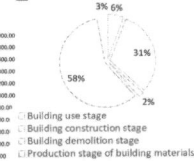

- Building use stage
- Building construction stage
- Building demolition stage
- Production stage of building materials

Main Modular Types

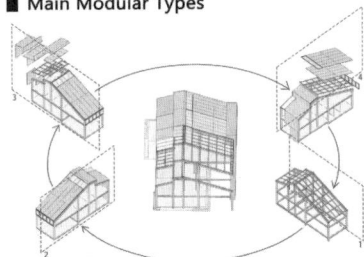

1-Modular Structure 2-Basic Module 3-Technical Module1 4-Technical Module2

Sunshine Analysis

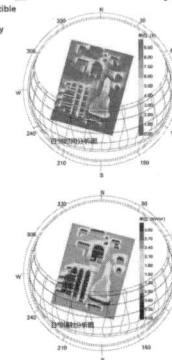

Wind Environment

Building Self-Sustainment

Category	
Total Area of Solar Panels (m²)	800.41
Annual Energy Conversion (kW·h/year)	221475.63
Annual Electricity Consumption (kW·h/year)	93415.20
Annual Electricity Storage (kW·h/year)	128060.43
Post-disaster Consumption (kW·h/day)	245.28
Availability of Offset	Yes

Village Activity Centre	
Total Area of Solar Panels (m²)	301.10
Annual Energy Conversion (kW·h/year)	83482.49
Annual Electricity Consumption (kW·h/year)	53000.00
Annual Electricity Storage (kW·h/year)	30482.49
Post-disaster Consumption (kW·h/day)	100.73
Availability of Offset	Yes

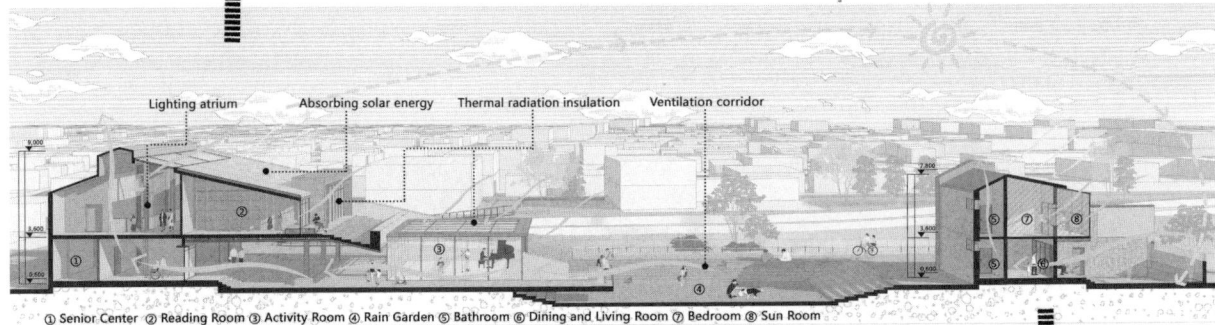

Lighting atrium   Absorbing solar energy   Thermal radiation insulation   Ventilation corridor

① Senior Center ② Reading Room ③ Activity Room ④ Rain Garden ⑤ Bathroom ⑥ Dining and Living Room ⑦ Bedroom ⑧ Sun Room

Bulk Generation

| Step One Generate building massing | Step Two Tripartite layout | Step Three Flood prevention measures | Step Four Solar technology |

Courtyard

- Main Use Space
- Secondary Activity Space
- Auxiliary Space

- Outdoor Stairs
- Elevated Building Base

- Pitched roof with solar panels
- Sun Room

Grey space

Residential Technical Strategy

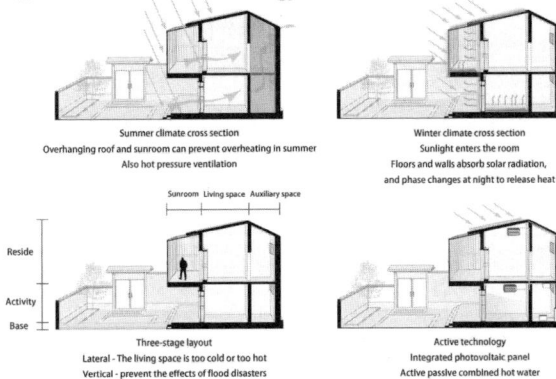

Summer climate cross section
Overhanging roof and sunroom can prevent overheating in summer
Also hot pressure ventilation

Winter climate cross section
Sunlight enters the room
Floors and walls absorb solar radiation,
and phase changes at night to release heat

Sunroom Living space Auxiliary space

Reside
Activity
Base

Three-stage layout
Lateral - The living space is too cold or too hot
Vertical - prevent the effects of flood disasters

Active technology
Integrated photovoltaic panel
Active passive combined hot water

Daylighting Analysis

Building Conponents

- Solar Panel
- Roof Construction
- Insulation Wall
- Sunroom
- Courtyard Wall
- Outdoor Flow
- Interior Flow

4/4

100-110524

洪水逃跑地图01
Map for Leaking Flood

Design Concept

设计旨在延续并创新当地乡村肌理，提出"网"（Web）作为核心概念，确定建筑整体布局，利用"网"所具有的紧密连结性、分岔性，创建不同层面的立体网进行诠释，创造独特的、具备灾时自持能力的空间序列。

节能技术方面，本设计以技术带动产业发展为理念，采用被动优先，主动优化的方式。利用当地主要种植的玉米作物，进行废料回收再利用。使用秸秆编织、轻钢骨架搭建的技术制作预制模块化墙体，并且将该墙体组合，形成灾时供人们逃生。配送物资的应急连通网，该预制模块化墙体表面结合绿色种植技术，形成种植复合墙体，吸收热辐射，提升室内舒适性。由于该材料来自于当地且造价低廉，编织技术作为推广，可带动当地制造业发展，实现技术与产业的息息相关。

The design aims to continue and innovate the local rural fabric, and proposes the "Web" as the core concept to determine the overall layout of the building. Using the "network" has: close connection, bifurcation, create different levels of three-dimensional network for interpretation, create a unique spatial sequence with disaster self-sustaining ability.

In terms of energy saving technology, this design takes technology to drive industrial development as the concept, and adopts passive priority and active optimization. The corn crop, which is mainly grown locally, is used for waste recycling. The technology of straw weaving and light steel skeleton construction is used to make prefabricated modular walls, and the walls are combined to form an emergency connection network for people to escape and distribute materials in times of disaster. The prefabricated modular wall surface is combined with green planting technology to form a planting composite wall that absorbs heat radiation and improves indoor comfort. Because the material comes from the local and the cost is low, weaving technology as a promotion, can promote the development of local manufacturing industry, and realize the close relationship between technology and industry.

Site Status

Regional Waterlogging Design Network

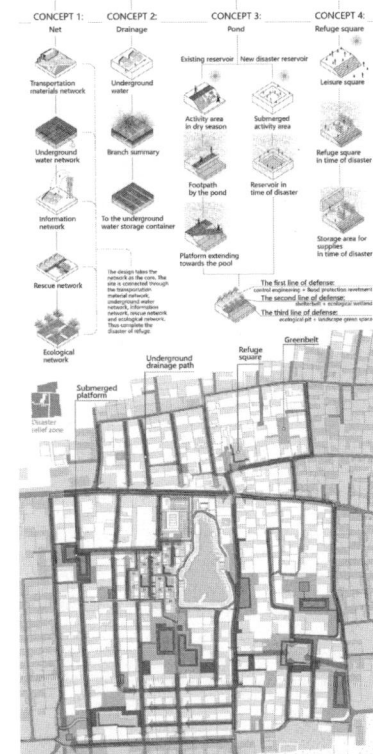

CONCEPT 1: Net
CONCEPT 2: Drainage
CONCEPT 3: Pond
CONCEPT 4: Refuge square

General Plan 1 : 500

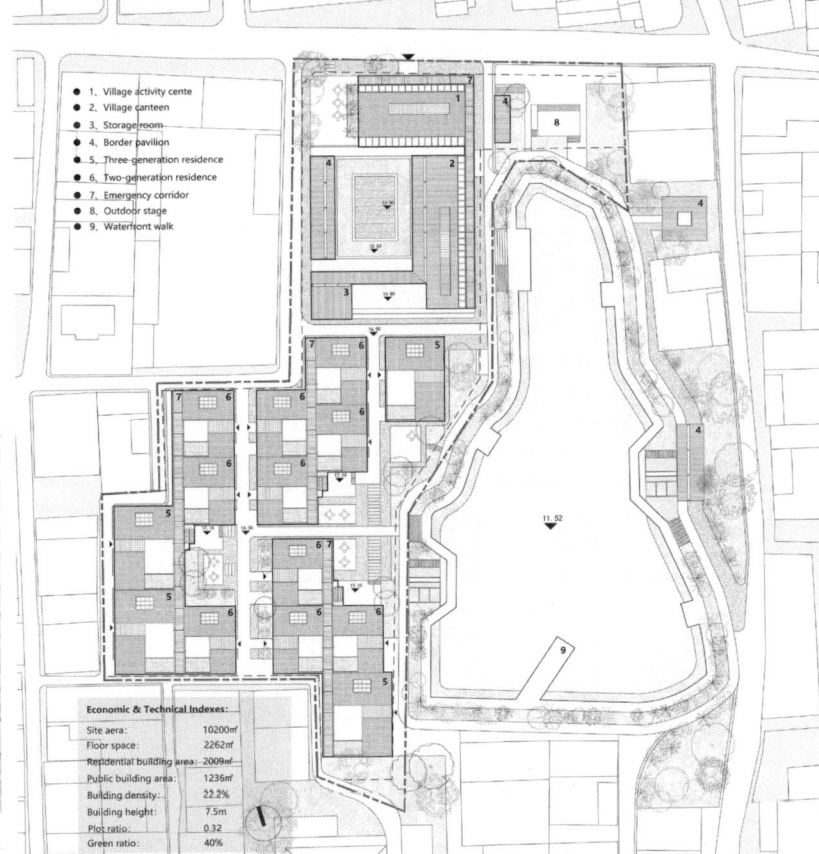

1. Village activity centre
2. Village canteen
3. Storage room
4. Border pavilion
5. Three-generation residence
6. Two-generation residence
7. Emergency corridor
8. Outdoor stage
9. Waterfront walk

Economic & Technical Indexes:	
Site aera:	10200m²
Floor space:	2262m²
Residential building area:	2009m²
Public building area:	1236m²
Building density:	22.2%
Building height:	7.5m
Plot ratio:	0.32
Green ratio:	40%

综合奖·优秀奖·保定市清苑区李八庄村建设项目
Comprehensive Awards － Honorable Mention － Libazhuang Village Project, Qingyuan District, Baoding City

注册号：110524
Register Number：110524

项目名称：洪水逃跑地图
Entry Title：Map for Leaking Flood

作者：郭裕旸、王业飞、刘雨森、
　　　周一婷、徐杨阳
Authors：Guo Yuyang, Wang Yefei,
　　　Liu Yusen, Zhou Yiting, and
　　　Xu Yangyang

作者单位：重庆大学
Authors from：Chongqing University

指导教师：黄海静
Tutor：Huang Haijing

指导教师单位：重庆大学
Tutor from：Chongqing University

100-110524
洪水逃跑地图02
Map for Leaking Flood

Public Area First Floor Plan 1：300

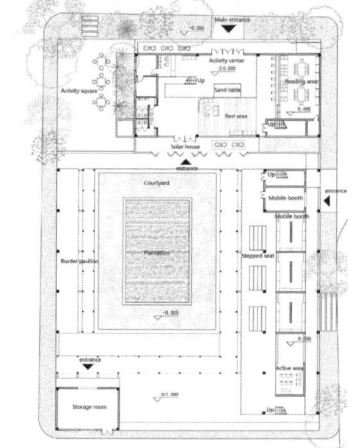

Public Area Second Floor Plan 1：300

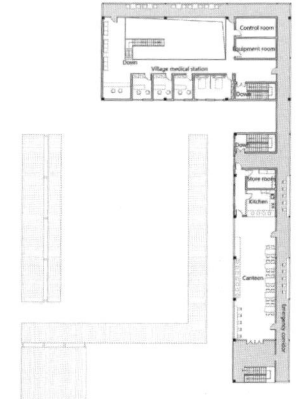

Site Structure Analysis

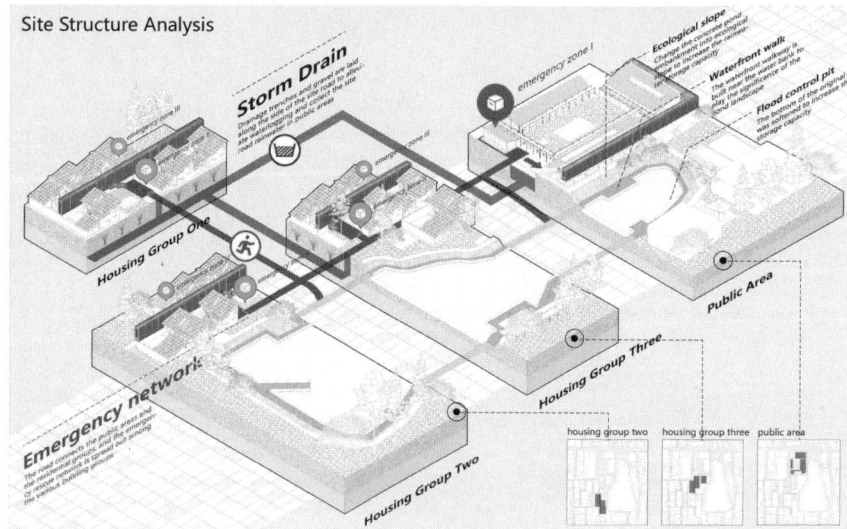

Storm Drain

Drainage trenches and gravel are laid along the side of the road to allow site to rainwater collected in the area road rainwater in public areas

Ecological slope
Change the concrete pond embankment into ecological slope to increase the ecological and landscape capacity.

Waterfront walk
The waterfront walk is built near the water can to play the significance of the land landscape of the

Flood control pit
The bottom of the original pit was deepened to increase the storage capacity

emergency zone I

Housing Group One

Emergency network
The road connects the public areas and the residential return, and the emergency to rescue channel is served between the various building various and

Housing Group Two

Housing Group Three

Public Area

housing group two | housing group three | public area

Emergency Corridor：
During a disaster: The corridor connects several households in the cluster and plays the role of material allocation and exchange in emergency situations.

Resource Distribution and Rescue Platform：
The platform is at a higher elevation to support the temporary storage and accumulation of relief supplies. Moreover, the platform is connected with the emergency corridor.

Dwelling Entrance：
Set up a front yard buffer at the entrance. Open windows to the east and west for lighting.

Water-friendly facilities：
During the rainy season, villagers can get close to the water on the steps. And during the dry season, it can be used as a staircase leading to the bottom of the pool, and the bottom of the pool can be used as a limited-time activity venue to prevent it from being abandoned.

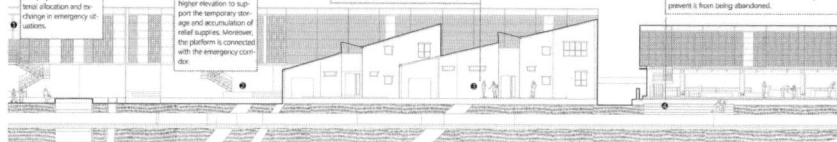

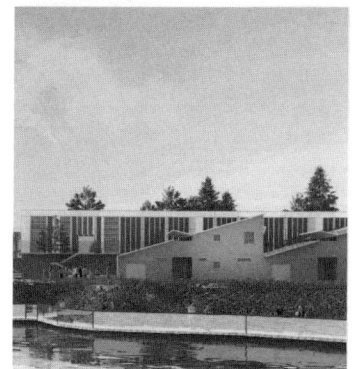

100-110524
洪水逃跑地图03
Map for Leaking Flood

House Plan 1 : 150
Two-generation residence

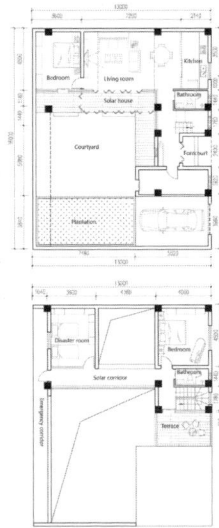

Bedroom · Living room · Kitchen
Solar house · Bathroom
Courtyard · Front hall
Plantation

Disaster room · Bedroom
Solar corridor · Bathroom
Terrace

Axonometric Explosion Pattern

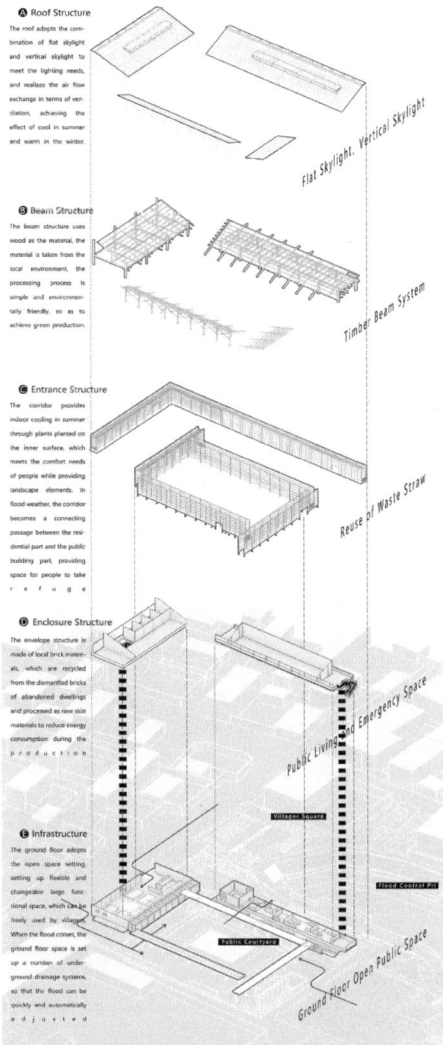

A Roof Structure
The roof adopts the combination of flat skylight and vertical skylight to meet the lighting needs, and realizes the air flow exchange in terms of ventilation, achieving the effect of cool in summer and warm in the winter.

Flat Skylight, Vertical Skylight

B Beam Structure
The beam structure uses wood as the material, the material is taken from the local environment, the processing process is simple and environmentally friendly, so as to achieve green production.

Timber Beam System

C Entrance Structure
The corridor provides indoor cooling in summer through plants planted on the inner surface, which meets the comfort needs of people while providing landscape elements. In flood weather, the corridor becomes a connecting passage between the residential part and the public building part, providing space for people to take refuge.

Reuse of Waste Straw

D Enclosure Structure
The envelope structure is made of local brick materials, which are recycled from the dismantled bricks of abandoned dwellings and processed as new skin materials to reduce energy consumption during production.

Public Living and Emergency Space

E Infrastructure
The ground floor adopts the open space setting, setting up flexible and changeable large functional space, which can be freely used by villagers. When the flood comes, the ground floor space is set up a number of underground drainage systems, so that the flood can be quickly and automatically adjusted.

Villager Square
Flood Control Pit
Public Courtyard
Ground Floor Open Public Space

Folk House Group

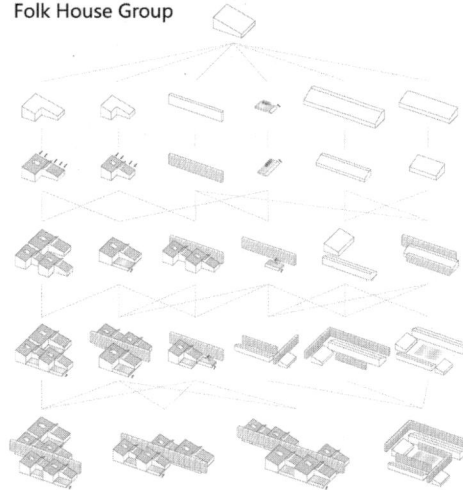

Structure Analysis

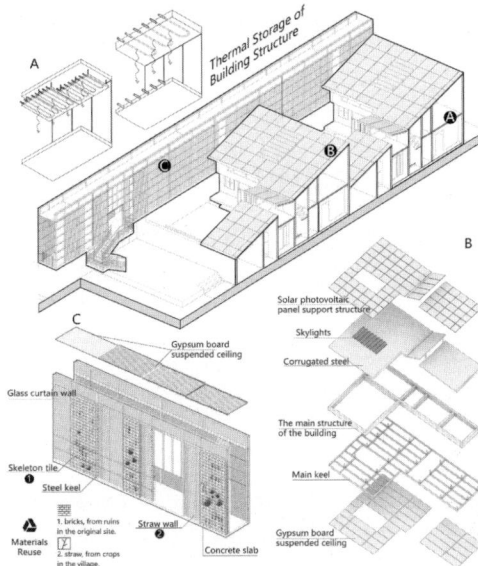

Thermal Storage of Building Structure

A
B
C

Gypsum board suspended ceiling
Glass curtain wall
Skeleton tile
1 Steel keel
Materials Reuse
1. bricks, from ruins in the original site.
2. straw, from crops in the village.
Straw wall 2
Concrete slab

Solar photovoltaic panel support structure
Skylights
Corrugated steel
The main structure of the building
Main keel
Gypsum board suspended ceiling

Three-generation residence

Bedroom · Living room · Bathroom
Solar house · Front hall · Kitchen
Courtyard · Storeroom
Plantation · Parking space

Disaster room · Bedroom
Reading corner · Bathroom
Solar corridor · Terrace

100-110524

洪水逃跑地图 04
Map for Leaking Flood

Climate Visualization
Enthalpy-humidity diagram

Wind Rose

The relative humidity is relatively average throughout the year, with mainly dry air. Since the higher the temperature, the higher the relative humidity, so the relative humidity of the site is higher in summer and lower in spring and winter. At the same time, it can also be analyzed in conjunction with the seasonality of rainfall. In summer, rainfall is more concentrated at the site, resulting in higher air humidity and even urban waterlogging.

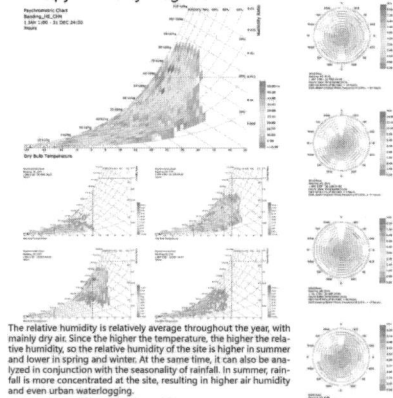

Radiation Simulation Diagram

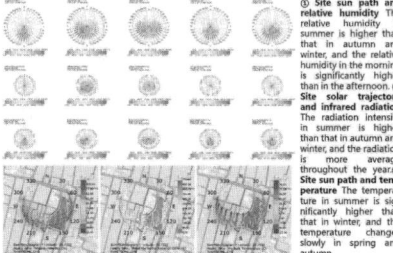

① **Site sun path and relative humidity** The relative humidity in summer is higher than that in autumn and winter, and the relative humidity in the morning is significantly higher than in the afternoon. ② **Site solar trajectory and infrared radiation** The radiation intensity in summer is higher than that in autumn and winter, and the radiation is more average throughout the year. ③ **Site sun path and temperature** The temperature in summer is significantly higher than that in winter, and the temperature changes slowly in spring and autumn.

Carbon Emission Calculation Results

Calculation results of carbon emissions during building operation phase

Project	Average annual electricity consumption (kWh/a)	Construction age (year)	Electricity Carbon Emission Factor (kgCO2z/ kWh)	Carbon emissions (kgCO2e)
HVAC lighting carbon emissions	441445.98			195186665
Elevator carbon emissions	0	50	0.8843	0
Renewable energy carbon reduction	-307162.8			-135612012
total				5937463.45

Renewable energy energy production calculation results

Photovoltaic power generation (kWh /a)	wind power (kWh /a)	Solar heating (kWh/a)	Solar cooling(kWh/a)	Solar domestic hot water (kWh /a)	total (kWh /a)
79662327	438.75	750.23	901.12	310.16	80102593

Passive Solar Energy Utilization

① **Passive energy saving using sunlight** First, facing south can make full use of the sun's rays, keep the house warm in winter and cool in summer, and allow full air circulation in the house. Second, in winter, by setting up a sunroom, the heat energy of the sun is stored in the sunroom, thereby reducing heating energy consumption. Third, use sunshades to prevent solar heat radiation from entering the room in summer, thereby reducing cooling energy consumption. ② **Passive energy saving using natural wind** On the one hand, skylights are opened in the vertical direction to create a chimney effect through hot-pressed air supply. On the other hand, the pressure is formed horizontally through the inner courtyard to enhance ventilation. ③ **Passive energy saving through resource recycling** On the one hand, the flexible surface is used to purify rainwater for irrigation. On the other hand, the temperature difference between the green space and the building is used to form a green island wind effect and enhance ventilation.

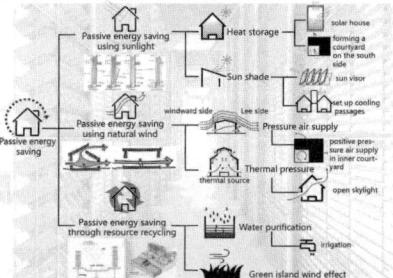

Water Footprint

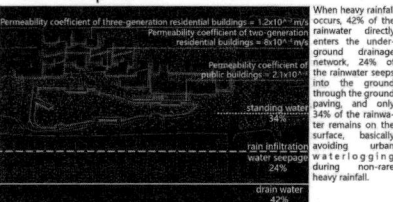

Permeability coefficient of three-generation residential buildings = 1.2×10^-7 m/s
Permeability coefficient of two-generation residential buildings = 8×10^-6 m/s
Permeability coefficient of public buildings = 2.1×10^-8 m/s

When heavy rainfall occurs, 42% of the rainwater directly enters the underground drainage network, 24% of the rainwater seeps into the ground through the ground paving, and only 34% of the rainwater remains on the surface, basically avoiding urban waterlogging during non-rare heavy rainfall.

standing water 34%
rain infiltration water seepage 24%
drain water 42%

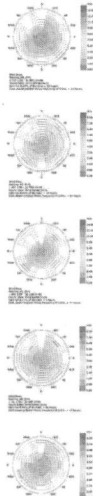

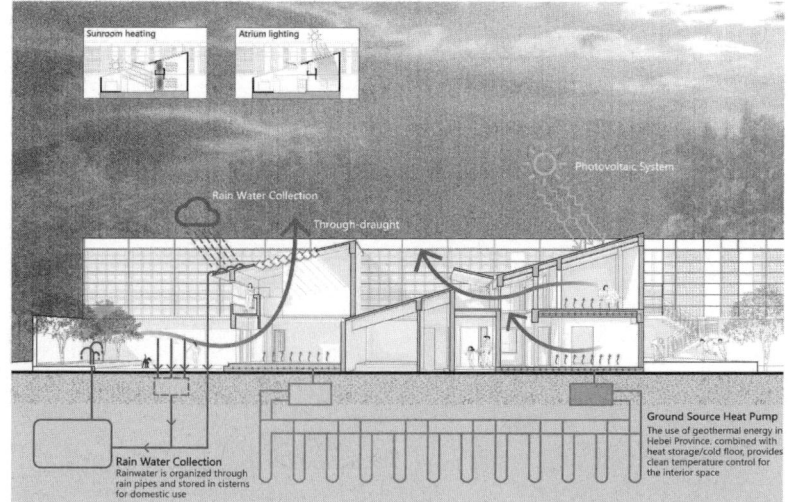

Light Environment and Wind Environment Analysis
Overall

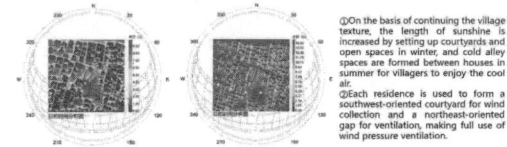

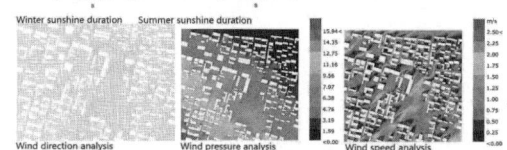

①On the basis of continuing the village texture, the length of sunshine is increased by setting up courtyards and open spaces in winter, and cold alley spaces are formed between houses in summer for villagers to enjoy the cool air. ②Each residence is used to form a southwest-oriented courtyard for wind collection and a northeast-oriented gap for ventilation, making full use of wind pressure ventilation.

Winter sunshine duration   Summer sunshine duration

Wind direction analysis   Wind pressure analysis   Wind speed analysis

Residential Area

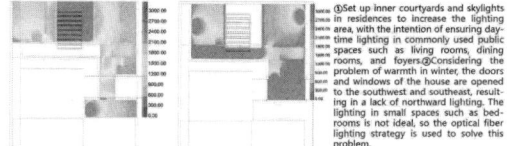

①Set up inner courtyards and skylights in residences to increase the lighting area, with the intention of ensuring daytime lighting in commonly used public spaces such as living rooms, dining rooms, and foyers. ②Considering the problem of warmth in winter, the doors and windows of the house are opened to the southwest and southeast, resulting in a lack of northward lighting. The lighting in small spaces such as bedrooms is not ideal, so the optical fiber lighting strategy is used to solve this problem.

Active Solar Energy Utilization

① **Photovoltaic panels** Photovoltaic panels convert solar energy into electrical energy and store it, which can be used in various aspects such as cooling, heating, lighting, and mechanical operation to reduce energy consumption. ② **Optical fiber** Optical cables can directly introduce external sunlight into the room, increase indoor natural lighting and illumination, and reduce the energy consumption required for lighting. And the optical cable can effectively solve the problem of insufficient illumination caused by buildings not opening windows facing north. ③ **Converting wind energy into electricity** Baoding has abundant wind energy. Windmills are used to convert the kinetic energy of the wind into electrical energy and store it to solve part of the villagers' daily electricity needs and emergency electricity needs during disasters.

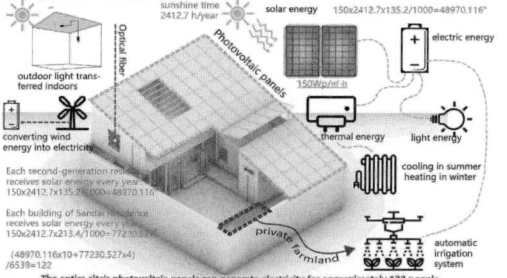

Summarize： The entire site's photovoltaic panels can generate electricity for approximately 122 people.Photovoltaic panels can basically meet the daily electricity needs of the village, and wind energy will be stored and supplied to public areas and emergency power in disasters.

sunshine time 2412.7 h/year   solar energy 150×2412.7×135.2/1000=48970.116*

150Wp/m² h

Each second-generation residential building receives solar energy every year：
150×2412.7×135.2/1000=48870.116

Each building of Sandai Residence receives solar energy every year：
150×2412.7×213.4/1000=77230.527

(48970.116x10+77230.527×4)/6539=122

The entire site's photovoltaic panels can generate electricity for approximately 122 people.

converting wind energy into electricity

electric energy   thermal energy   light energy

cooling in summer heating in winter

automatic irrigation system

Water Management

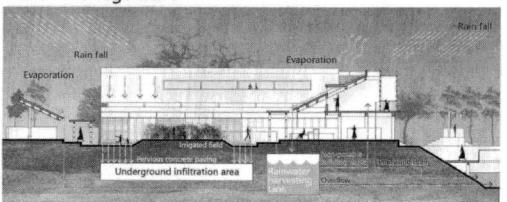

Rain fall   Evaporation   Evaporation

Underground infiltration area

Characteristic Space Technology Analysis - Aerial Tunnel
Daylighting simulation

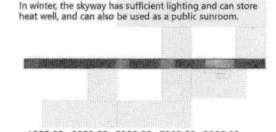

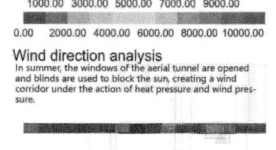

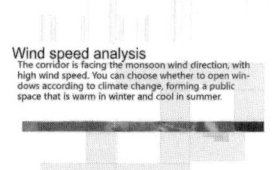

In winter, the skyway has sufficient lighting and can store heat well, and can also be used as a public sunroom.

1000.00 3000.00 5000.00 7000.00 9000.00
0.00 2000.00 4000.00 6000.00 8000.00 10000.00

Wind direction analysis

In summer, the windows of the aerial tunnel are opened and blinds are used to block the sun, creating a wind corridor under the action of heat pressure and wind pressure.

Wind speed analysis

The corridor is facing the monsoon wind direction, with high wind speed. You can choose whether to open windows according to climate change, forming a public space that is warm in winter and cool in summer.

0.50 1.50 2.50 3.50 4.50 M/S
<0.00 1.00 2.00 3.00 4.00 5.00<

Wind pressure analysis

The wind pressure is larger at the southern end of the corridor and then decreases rapidly, providing conditions for the wind pressure to supply air.

1.59 4.76 7.94 11.11 14.29
<0.00 3.17 6.35 9.52 12.70 15.87<

Universal Thermal Climate Index

The corridor is made of transparent glass and can be used as a sunroom in winter or as a greenhouse for indoor planting. The average temperature throughout the year is on the high side.

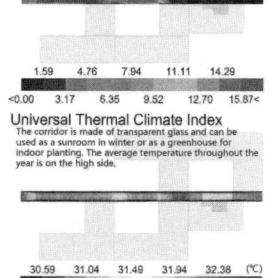

30.59 31.04 31.49 31.94 32.38 (℃)
30.37 30.82 31.26 31.71 32.16 32.61

Summer Comfort Analysis

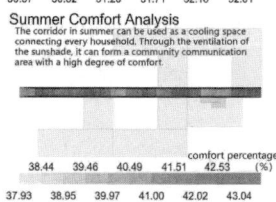

The corridor in summer can be used as a cooling space connecting every household. Through the ventilation of the sunshade, it can form a community communication area with a high degree of comfort.

comfort percentage
38.44 39.46 40.49 41.51 42.53 (%)
37.93 38.95 39.97 41.00 42.02 43.04

Sunroom heating   Atrium lighting

Photovoltaic System

Rain Water Collection

Through-draught

Ground Source Heat Pump
The use of geothermal energy in Hebei Province. combined with heat storage/cold floor, provides clean temperature control for the interior space

Rain Water Collection
Rainwater is organized through rain pipes and stored in cisterns for domestic use

■ Site environment

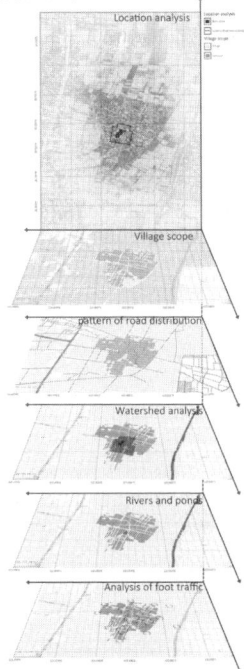

Location analysis

Village scope

pattern of road distribution

Watershed analysis

Rivers and ponds

Analysis of foot traffic

FloodCarry Continent 01
水亦载洲

■ Climate analysis

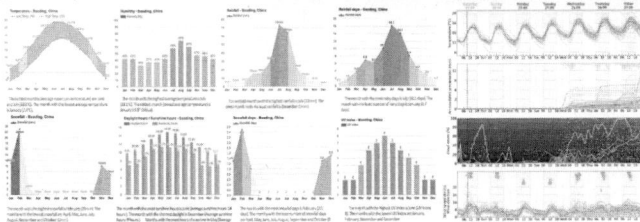

■ Volume analysis

step1:the site and environment

step2:Class 1 flood control

step3:Secondary Flood Storage and Public Places

step4:Level 3 Prediction of Floods and Traffic Nuclear

step5:Residential and public function layout

step6:Improve the courtyard form and transportation

■ Design Description

The project is located in Libazhuang Village, Qingyuan District, Baoding City, Hebei Province, and is designed with a three-level flood control strategy due to the severe seasonal flood disaster in the local area. The first level is flood control, which solves the flood problem at each point of the village, the second level is flood storage, which improves the water storage carrying capacity of the region, and the third level is flood prevention, which visualizes the flood as an alarm bell. In terms of functional zoning, the public building is placed in the middle of the site, which is to improve the convenience of villagers in the north-south direction, and to provide a safe place to evacuate. The roads on the site are made of flexible paving of plants and hollow bricks, and the root system is transferred to form a recharge relationship with the floating farm to realize the utilization of water resources. The combination of light and water is embodied in a light well in linear space. It serves as a channel for wind and light, as well as a venue for flood forecasting and cultural events. The most distinctive feature of this project is the combination of water, light, heat and wind to solve practical local problems and universal construction paradigms.

Master Plan 1 : 1500

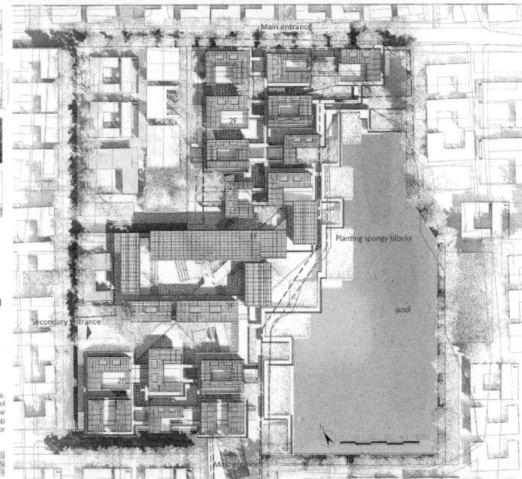

Main entrance

Secondary entrance

Planting spongy blocks

pool

综合奖·优秀奖·保定市清苑区李八庄村建设项目
Comprehensive Awards – Honorable Mention – Libazhuang Village Project, Qingyuan District, Baoding City

注册号：110537
Register Number：110537
项目名称：水亦载洲
Entry Title：Flood Carry Continent
作者：汪郑政、崔佳琳、焦佳喜
Authors：Wang Zhengzheng, Cui Jialin, and Jiao Jiaxi
作者单位：新疆大学
Authors from：Xinjiang University
指导教师：樊辉、袁萍
Tutors：Fan Hui, Yuan Ping
指导教师单位：新疆大学
Tutors from：Xinjiang University

Flood Carry Continent 02
水亦载洲

Floor plan of the second floor 1:700

■ Sunlight and Enthalpy Zologenography Analysis

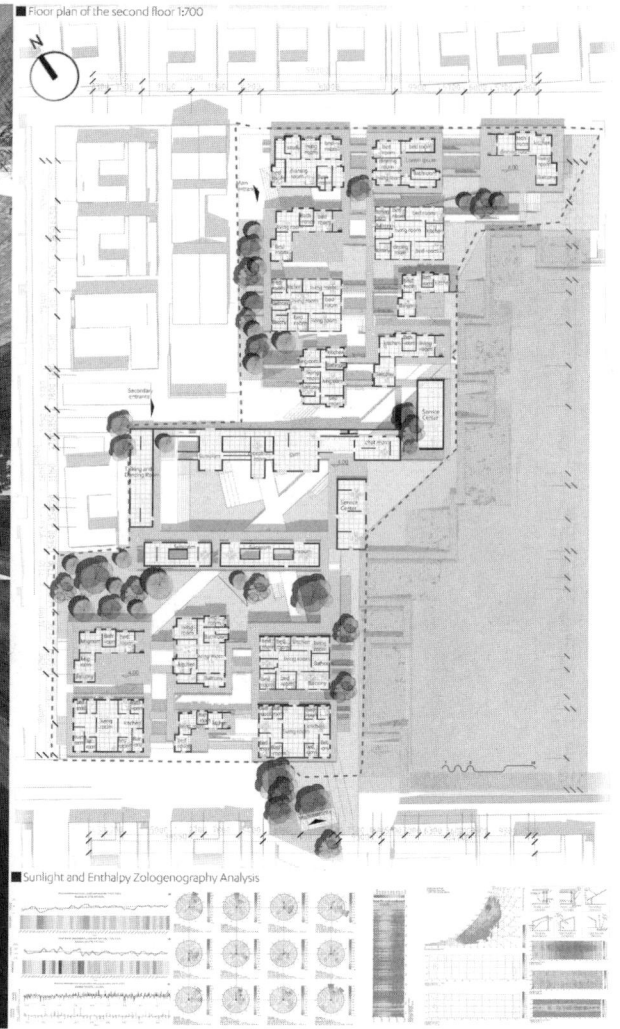

■ Analysis of single dwelling types and energy savings

Unit 1 Plan

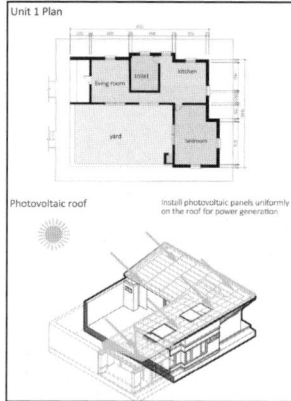

Photovoltaic roof

Install photovoltaic panels uniformly on the roof for power generation.

Unit 2 Plan

Cellular reservoir

The roof is organized for drainage, and finally converges to a reservoir.

Unit 3 Plan

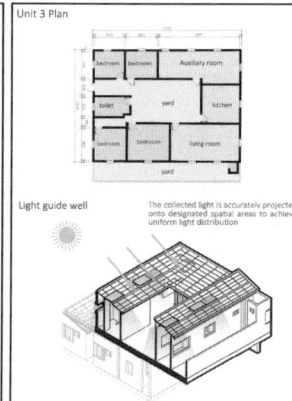

Light guide well

The collected light is accurately projected onto designated spatial areas to achieve uniform light distribution.

Unit 4 Plan

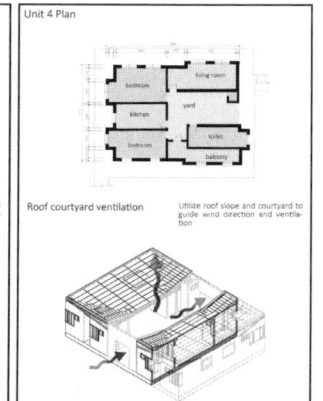

Roof courtyard ventilation

Utilize roof slope and courtyard to guide wind direction and ventilation.

■ Reservoir structure and west section 1 : 300

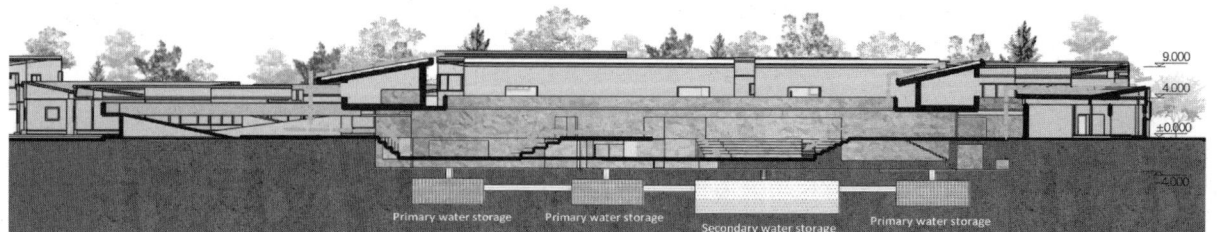

Primary water storage　　Primary water storage　　Secondary water storage　　Primary water storage

FloodCarry Continent 03
水亦载洲

■ Floor plan of the first floor 1 : 1000

■ Floor plan of the negative floor 1 : 1000

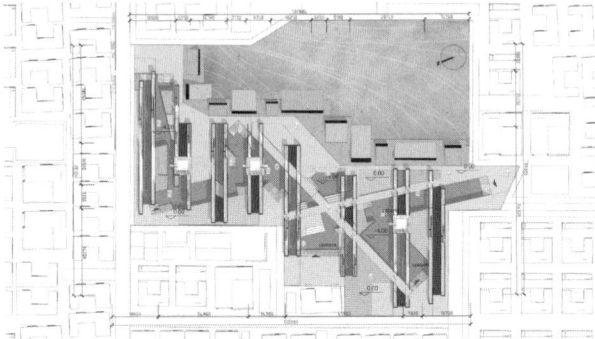

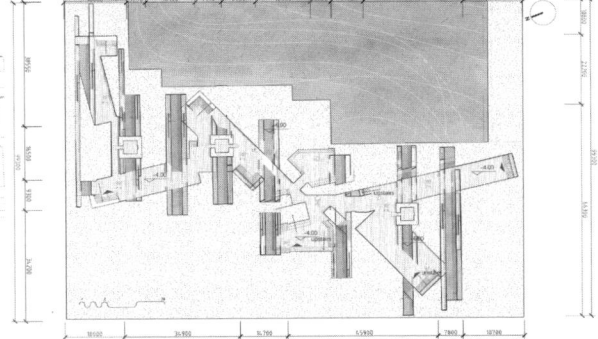

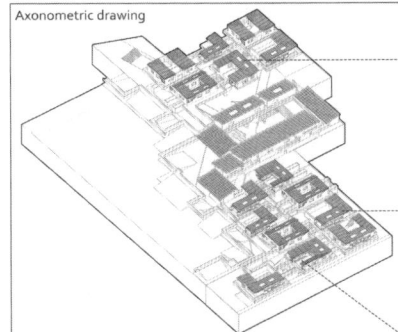

Axonometric drawing

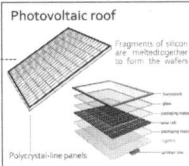

Photovoltaic roof

Fragments of silicon are melted together to form the wafers.

Polycrystal-line panels

Light guide well

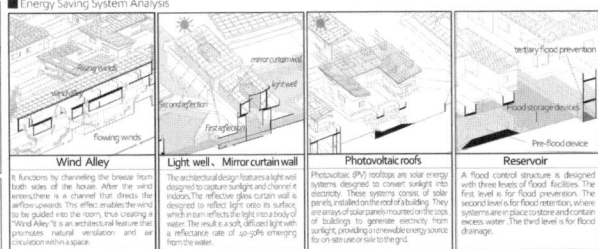

Accurately project the collected light onto the designated spatial area through a light collector, light guide, and diffuser.

(1)Main keel
(2) Sub keel
(3) Sub frame
(4) Glass panel
(5) Elastic strip
(5a)Gusset plate

Ventilation system

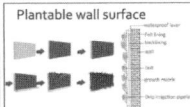

Plantable wall surface

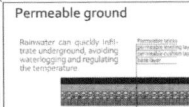

Permeable ground

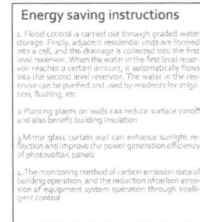

Rainwater can quickly infiltrate underground, avoiding waterlogging and regulating the temperature.

Energy saving instructions

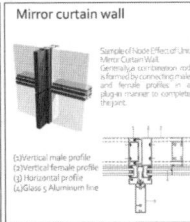

1. Flood control is carried out through graded water storage. Firstly, adjacent residential units are formed into a cell, and the drainage is collected into the first level reservoir. When the water in the first level reservoir reaches a certain amount, it automatically flows into the second level reservoir. The water in the reservoir can be purified and used by residents for irrigation, flushing, etc.
2. Planting plants on walls can reduce surface runoff and also benefit building insulation.
3. Mirror glass curtain wall can enhance sunlight reflection and improve the power generation efficiency of photovoltaic panels.
4. The monitoring method of carbon emission data of building operation, and the reduction of carbon emission of equipment system operation through intelligent control.

Mirror curtain wall

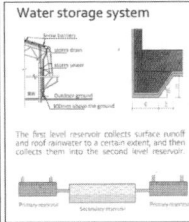

Sample of Node Effect of Unit Mirror Curtain Wall
Cermaltype combination rod is formed by connecting male and female profiles in a plug-in manner to complete the joint.

(1)Vertical male profile
(2)Vertical female profile
(3) Horizontal profile
(4)Glass 5 Aluminum line

Water storage system

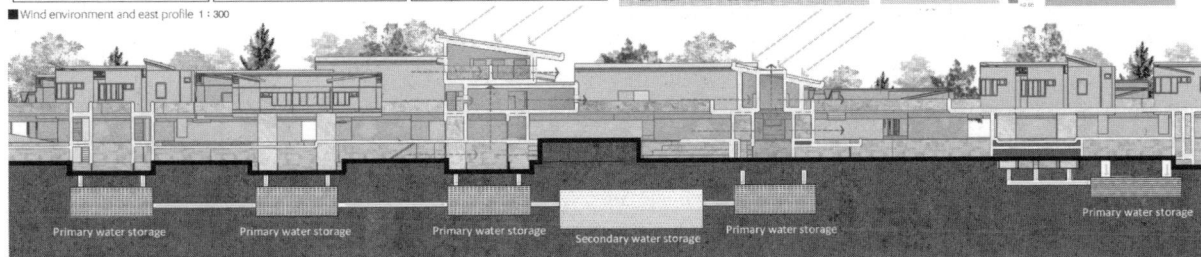

The first level reservoir collects surface runoff and roof rainwater to a certain extent, and then collects them into the second level reservoir.

Primary reservoir   Secondary reservoir

■ Energy Saving System Analysis

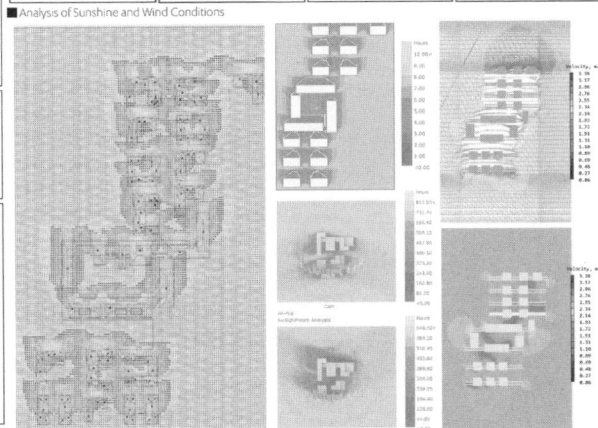

Wind Alley
It functions by channeling the breeze from both sides of the house. After the wind enters, there is a channel that directs the airflow upwards. This effect enables the wind to be guided into the room, thus creating a "Wind Alley". It is an architectural feature that promotes natural ventilation and air circulation within a space.

Light well、Mirror curtain wall
The architectural design features a light well designed to capture sunlight and channel it indoors. The reflective glass curtain wall is designed to reflect light onto its surface, which in turn reflects the light into a body of water. The result is a soft, diffused light with a reflectance rate of 40-50% emerging from the water.

Photovoltaic roofs
Photovoltaic (PV) rooftops are solar energy systems designed to convert sunlight into electricity. These systems consist of solar panels, installed on the roof of a building. They are arrays of solar panels mounted on the tops of buildings to generate electricity from sunlight, providing a renewable energy source for on-site use or sale to the grid.

Reservoir
A flood control structure is designed with three levels of flood facilities. The first level is for flood prevention. The second level is for flood retention, where systems are in place to store and contain excess water. The third level is for flood drainage.

■ Analysis of Sunshine and Wind Conditions

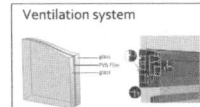

■ Wind environment and east profile 1 : 300

Primary water storage   Primary water storage   Primary water storage   Secondary water storage   Primary water storage   Primary water storage

Flood Carry Continent 04
水亦载洲

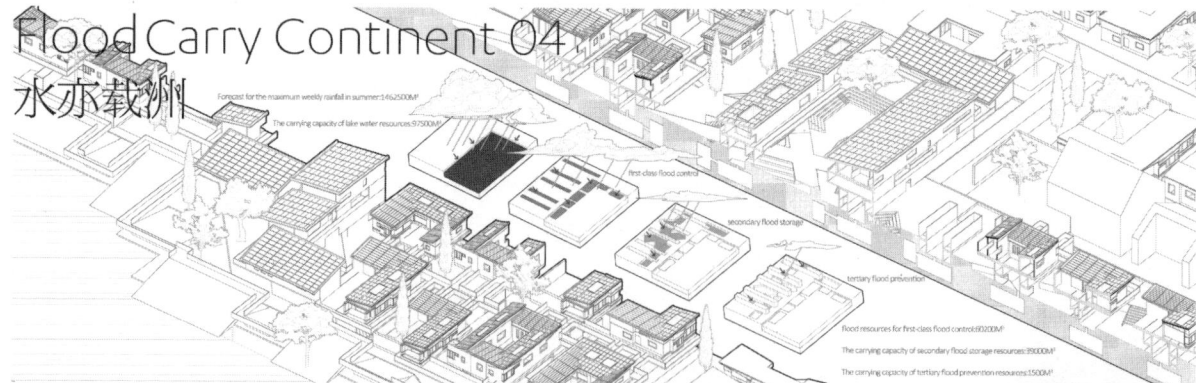

Forecast for the maximum weekly rainfall in summer:1462500M³

The carrying capacity of lake water resources:97500M³

first-class flood control

secondary flood storage

tertiary flood prevention

Flood resources for first-class flood control:60200M³

The carrying capacity of secondary flood storage resources:39000M³

The carrying capacity of tertiary flood prevention resources:1500M³

■ Building Resilience Design Analysis Diagram

Above-ground landscape pool

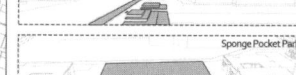

Underground water storage

Above-ground bypass water features

Sponge Pocket Park

Planting a water-storing roof

Above-ground landscape pool

Underground water storage

Above-ground bypass water features

Sponge Pocket Park

Planting a water-storing roof

■ Energy-saving system

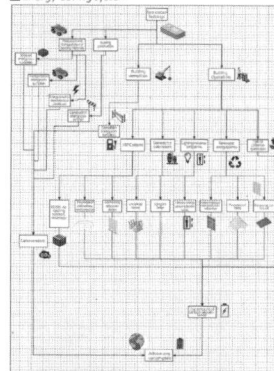

■ Carbon emission analysis

■ Material Analysis

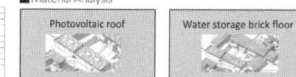

Photovoltaic roof

Part of the roof installs photovoltaic power generation.

By using solar panels to convert sunlight into electricity, photovoltaic modules can absorb most of the sunlight and convert it into electricity, reducing direct sunlight on the roof.

Water storage brick floor

Most of the ground uses water storage bricks.

The back of the upper brick has a storage channel, and the surface of the lower brick has a water storage tank.

Planting wall

The south wall is a plantable wall surface.

Plant wall is a wall made of green plants, which utilizes the water-retaining ability of plant roots to the growth environment.

Mirror glass curtain

Mirror glass curtains are used on the walls near the water storage area to increase light reflection.

The glass curtain wall is composed of mirror glass and ordinary glass, with insulated glass filled with dry air or inert gas in the building envelope or decorative structure.

古院绿廊，水韵新乡 01
Ancient Courtyard and Green Corridor, Water Charm of New Village

■ Design Specification

设计提取河北传统建筑中的廊院与水元素，构建与水共生的传统生态聚落。

大范围红线内以Gis水文分析为数据支撑，设计情景防洪，针对暴雨、大雨和中小雨三种降雨情况分别规划防涝排泄、生态沟渠与源头控制系统三级径流，构建弹性排水景观系统。

小范围红线内植入点、线、面三类海绵体构建生态海绵景观系统，设计漂浮廊道串联起民居与村民活动中心，作为平时的观景平台与健身环道，灾时的人群疏散与物资运输线。

设计针对性运用了光伏瓦发电、热缓冲空间、地源热泵等低碳策略，达到建筑运行阶段零碳、全生命周期低碳的绿建设计目标。

The design extracts corridor, courtyard and water elements from traditional buildings in Hebei to build a traditional ecological settlement symbiosis with water.

In a large range of red lines, Gis hydrological analysis was used as the data support to design scenario flood control, and the three-level runoff of flood prevention and drainage, ecological ditch and source control system were planned respectively for the three types of rainfall, heavy rain and moderate and light rain, and the elastic drainage landscape system was constructed.

In a small range, three types of spongy bodies, including points, lines and surfaces, were implanted in the red line to construct an ecological sponge landscape system, and floating corridors were designed to connect residential houses and villagers' activity centers, which were used as viewing platforms and fitness rings in ordinary times, as well as the streamline of crowd evacuation and material transportation in times of disaster.

The design targeted the use of photovoltaic tile power generation, heat buffer space, modular design, ground source heat pump and other low carbon strategy, to achieve the building operation phase of zero carbon, the whole life cycle of low carbon green construction plan goals.

■ Climate Strategy

■ Wind-rose diagram

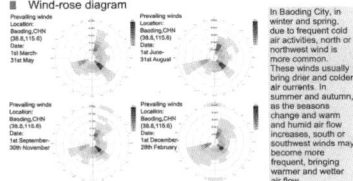

In Baoding City, in winter and spring, due to frequent cold air activities, north or northwest wind is more common. These winds may bring drier and colder air currents. In summer and autumn, as the seasons change and warm and humid air flow increases, south or southwest winds may become more frequent, bringing warmer and wetter air flow.

■ Enthalpy autonogram

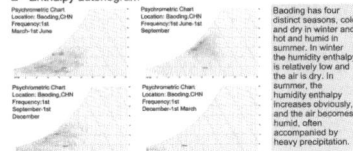

Baoding has four distinct seasons, cold and dry in winter and hot and humid in summer. In winter the humidity enthalpy is relatively low and the air is dry. In summer, the humidity enthalpy increases obviously, and the air becomes humid, often accompanied by heavy precipitation.

■ Rainfall analysis

The precipitation in Baoding City is unevenly distributed in time and space, with concentrated precipitation in summer and relatively less in winter.

■ Green building strategy analysis

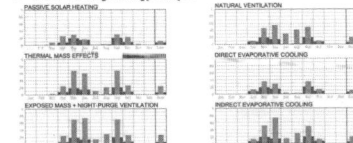

According to the analysis, in terms of green building strategy, we should pay attention to the thermal mass effect and exposed mass+ night - purge ventilation. In addition, we should pay attention to the natural ventilation and indirect evaporative cooling of the scheme.

■ Site Analysis

Strengths
·Abundant land resources
·Solid agricultural foundation
·Abundant labor resources

Weaknesses
·The drainage system is not perfect
·Single industrial structure
·Serious economic loss

Opportunities
·Policy support
·Infrastructure construction
·Diversified development

Threats
·Frequent natural disasters
·Resources and environmental pressure

■ Crowd Analysis

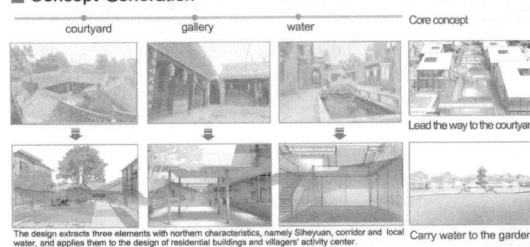

The elderly in Li Bazhuang account for a relatively large proportion, so the design should focus on the needs of the elderly.

17.75%   57.56%   24.69%
——The data comes from the seventh population census of Baoding City

Activities for the elderly for a day

Demand   Design strategy
The central corridor can be fully utilized as a health trail. 50%
Public courtyards are set up in residential areas to create social spaces. 30%
A small fitness area is added to the public area to strengthen the body. 20%

■ Concept Generation

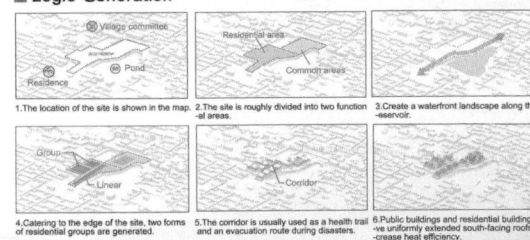

courtyard   gallery   water   Core concept

Lead the way to the courtyard

Carry water to the garden

The design extracts three elements with northern characteristics, namely Siheyuan, corridor and local water, and applies them to the design of residential buildings and villagers' activity center.

■ Logic Generation

1.The location of the site is shown in the map.
2.The site is roughly divided into two function -al areas.
3.Create a waterfront landscape along the reservoir.
4.Catering to the edge of the site, two forms of residential groups are generated.
5.The corridor is usually used as a health trail and an evacuation route during disasters.
6.Public buildings and residential buildings have uniformly extended south-facing roofs to increase heat efficiency.

综合奖·优秀奖·保定市清苑区李八庄村建设项目
Comprehensive Awards – Honorable Mention – Libazhuang Village Project, Qingyuan District, Baoding City

注册号：110690
Register Number：110690

项目名称：古院绿廊，水韵新乡
Entry Title：Ancient Courtyard and Green Corridor, Water Charm of New Village

作者：舒情、李子晴、何柳莹、谭扬深
Authors：Shu Qing, Li Ziqing, He Liuying, and Tan Yangshen

作者单位：厦门大学
Authors from：Xiamen University

指导教师：石峰
Tutor：Shi Feng

指导教师单位：厦门大学
Tutor from：Xiamen University

■ A Comprehensive Guide to Green Building Technology

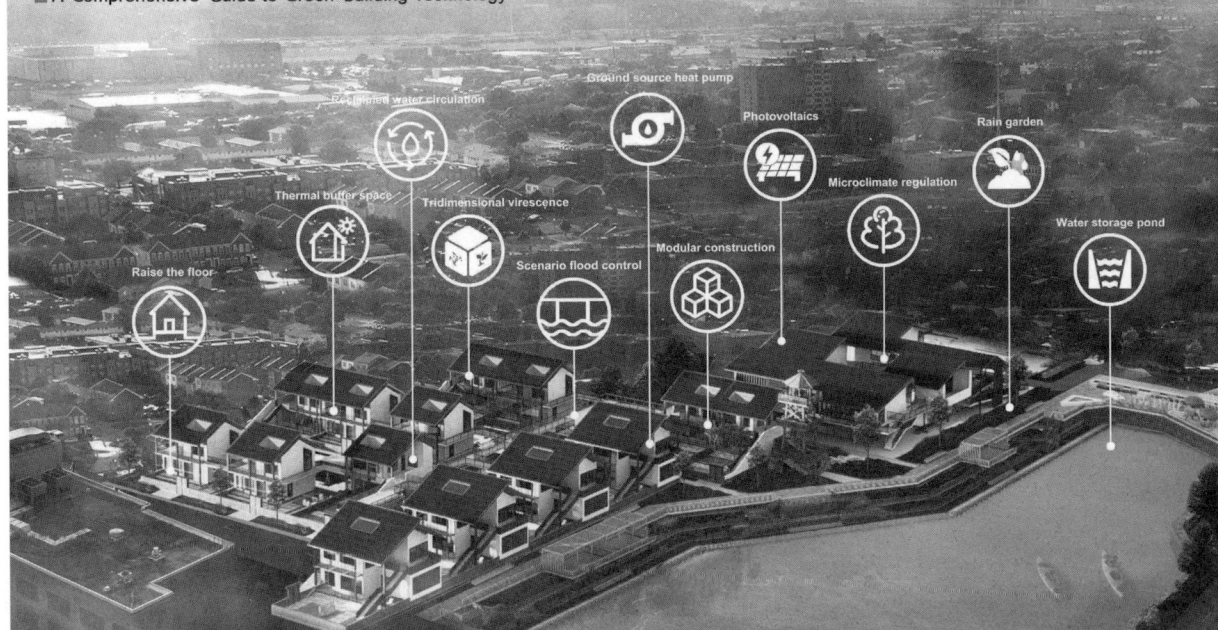

Regimened water circulation
Ground source heat pump
Photovoltaics
Rain garden
Thermal buffer space
Tridimensional virescence
Microclimate regulation
Water storage pond
Modular construction
Scenario flood control
Raise the floor

110690

古院绿廊，水韵新乡 02
Ancient Courtyard and Green Corridor, Water Charm of New Villag

residential houses   trail

extreme water level once in 20 years

Summer high water level 2m

Summer normal water level 0.70m

According to the height of the water level, the three-layer platform type revetment is designed. Trees, shrubs, herbs and floating plants are planted from high to low respectively. The rainwater is filtered and purified layer by layer, and facilities such as hydrophilic walking platform, viewing platform and riverside seat are added to create an ecological, diversified and humane elastic revetment landscape.

■ Hydrologic Analysis ■ Large Scale Drainage Planning

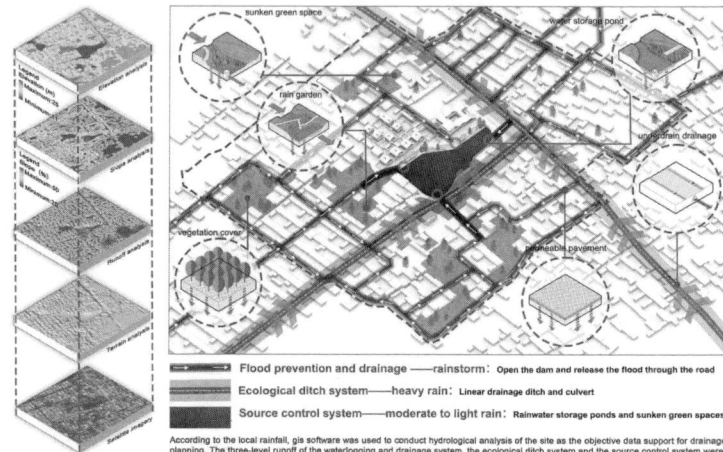

sunken green space
water storage pond
rain garden
underdrain drainage
vegetation cover
permeable pavement

Flood prevention and drainage——rainstorm: Open the dam and release the flood through the road
Ecological ditch system——heavy rain: Linear drainage ditch and culvert
Source control system——moderate to light rain: Rainwater storage ponds and sunken green spaces

According to the local rainfall, gis software was used to conduct hydrological analysis of the site as the objective data support for drainage planning. The three-level runoff of the waterlogging and drainage system, the ecological ditch system and the source control system were respectively planned for the three rainfall conditions of exceeding the standard rainstorm, heavy rain and moderate and light rain, and the elastic drainage landscape system was constructed.

■ Situational Flood Control

Light rain condition
Drainage through stormwater cisterns and lower green spaces.

Heavy rain condition
The anqupaishui road is flanked by culverts through which water is drained, while other roads are drained through grass ditches where the road and the greenery border.

Excessive rainstorm condition
Open the dam and release water through the road into the reservoir.

■ Rain Garden Node Design for Public Courtyards

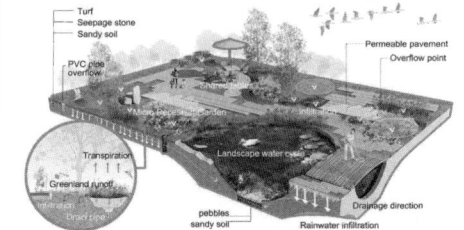

■ Roaming Streamline and Landscape Nodes

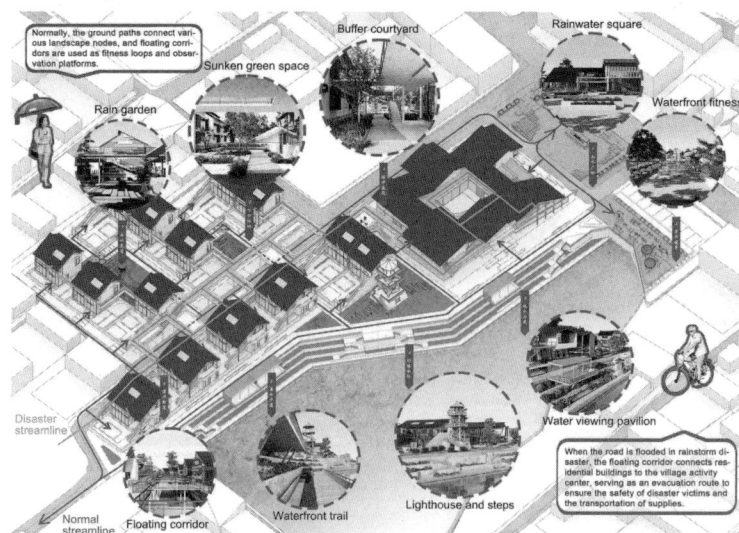

■ General Layout 1:500

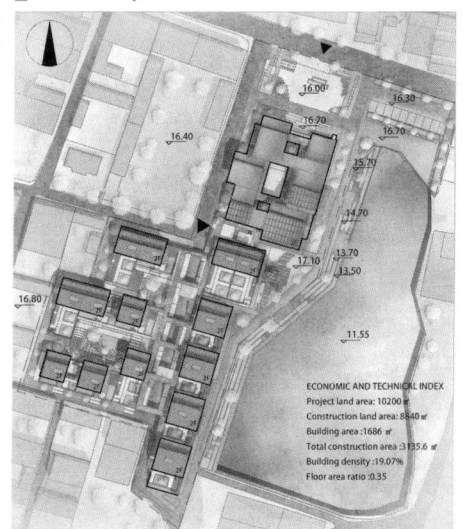

ECONOMIC AND TECHNICAL INDEX
Project land area: 10200 ㎡
Construction land area: 8840 ㎡
Building area:1686 ㎡
Total construction area :3135.6 ㎡
Building density :19.07%
Floor area ratio :0.35

古院绿廊，水韵新乡 03

Ancient Courtyard and Green Corridor, Water Charm of New Village

Waterfront view of residential buildings

Planting area in private courtyard

Public courtyard

Leisure area in private courtyard

Modular Construction （residence）

Stair
Storage room
Master bedroom
Corridor
Room for the elderly

Living room | Dining room | Kitchen | Studying room | Second bedroom | Toilet

Floor Plan 1：200 （residence）

The design of the residential units adopts the elements of traditional houses in North China, and each one is equipped with cout -yards, warm corridors, su -n rooms, shadow walls. The waterfront apartment considers three generatio -ns in one house and add -s a room for the old.In co -nsideration of evacuation in the event of disasters, s -tairs leading to the terrac -e are installed in the apar -tments.

Room Type A （130.5㎡）

Room Type C （124.4㎡）

Finishing layer
Reinforcement layer
Fireproof layer
Waterproofing layer
Straw core
Box beam
Waterproofing layer
Reinforcement layer
Finishing layer

Wall detail

Room Type B （175.9㎡）

Thermal Buffer Space （residence）

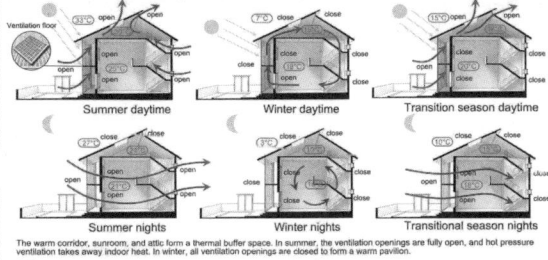

Summer daytime | Winter daytime | Transition season daytime

Summer nights | Winter nights | Transitional season nights

The warm corridor, sunroom, and attic form a thermal buffer space. In summer, the ventilation openings are fully open, and hot pressure ventilation takes away indoor heat. In winter, all ventilation openings are closed to form a warm pavilion.

Rooftop Photovoltaic power Generation of the Project

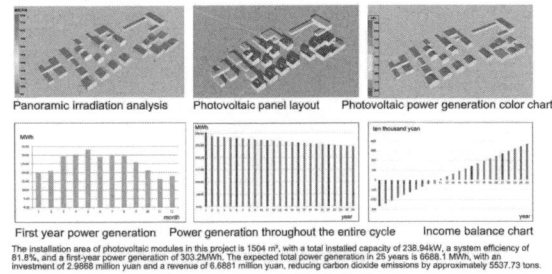

Panoramic irradiation analysis | Photovoltaic panel layout | Photovoltaic power generation color chart

First year power generation | Power generation throughout the entire cycle | Income balance chart

The installation area of photovoltaic modules in this project is 1504 m², with a total installed capacity of 238.94kW, a system efficiency of 81.8%, and a first-year power generation of 303.2MWh. The expected total power generation in 25 years is 6688.1 MWh, with an investment of 2.9888 million yuan and a revenue of 6.6881 million yuan, reducing carbon dioxide emissions by approximately 5537.73 tons.

Green Building Strategy （residence）

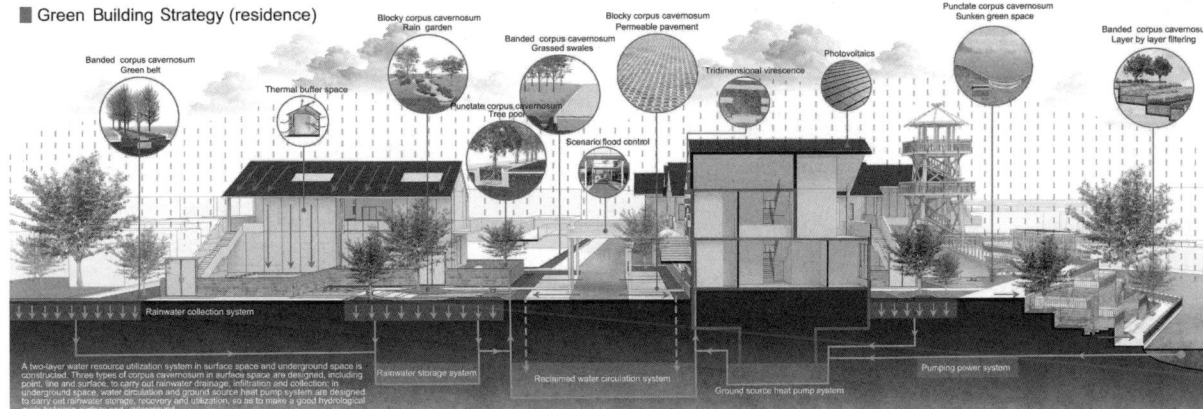

Banded corpus cavernosum Green belt
Thermal buffer space
Blocky corpus cavernosum Rain garden
Punctate corpus cavernosum Tree pool
Banded corpus cavernosum Grassed swales
Blocky corpus cavernosum Permeable pavement
Scenario flood control
Tridimensional virescence
Photovoltaics
Punctate corpus cavernosum Sunken green space
Banded corpus cavernosum Layer by layer filtering

Rainwater collection system
Rainwater storage system
Reclaimed water circulation system
Ground source heat pump system
Pumping power system

A two-layer water resource utilization system in surface space and underground space is constructed. Three types of corpus cavernosum in surface space are designed, including point, line and surface, to carry out rainwater drainage, infiltration and collection; in underground space, water circulation and ground source heat pump system are designed to carry out rainwater storage, recovery and utilization, so as to make a good hydrological cycle between surface and underground.

127

110690

古院绿廊，水韵新乡 04

Ancient Courtyard and Green Corridor, Water Charm of New Village

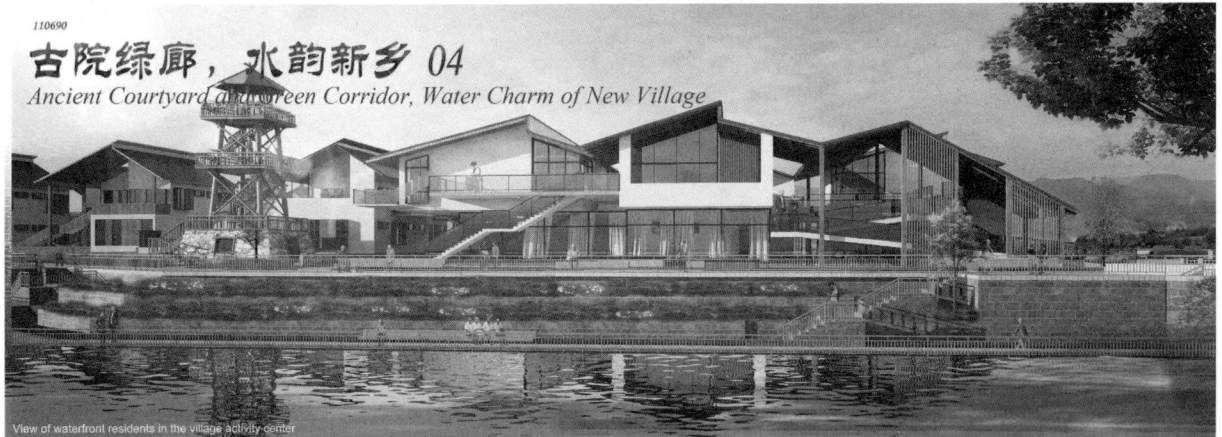
View of waterfront residents in the village activity center

Second floor of the water courtyard

First floor of the forecourt

First floor of the water courtyard

■ Ground Floor Plan 1：300 (village activity center)

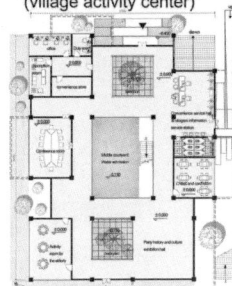

■ Second Floor Plan 1：300 (village activity center)

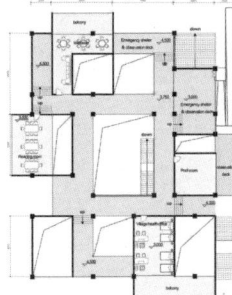

■ Bulk Generation (village activity center)

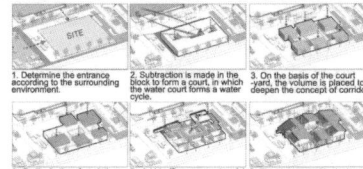

1. Determine the entrance according to the surrounding environment.
2. Subtraction is made in the block to form a court, in which the water court forms a water cycle.
3. On the basis of the court-yard, the volume is placed to deepen the concept of corridor.
4. On the bottom foundation, the volume is pushed and pulled according to the corridor and courtyard.
5. Add traffic space and enrich the concept of corridors.
6. The sloping roof is designed according to the sun orientation, and photovoltaic panels are placed on the south side.

■ Explosive Resolution Diagram (village activity center)

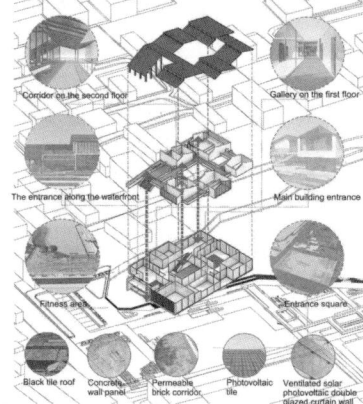

Corridor on the second floor
Gallery on the first floor
The entrance along the waterfront
Main building entrance
Fitness area
Entrance square
Black tile roof
Concrete wall panel
Permeable brick corridor
Photovoltaic tile
Ventilated solar photovoltaic double glazed curtain wall

■ Building Skin Technology (village activity center)

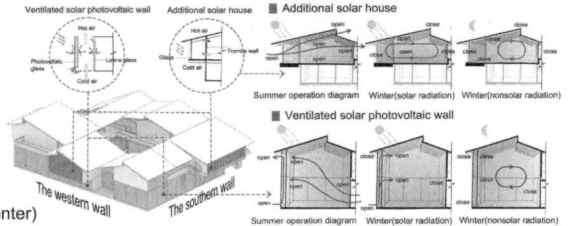

Ventilated solar photovoltaic wall
Additional solar house

■ Additional solar house
Summer operation diagram | Winter(solar radiation) | Winter(nonsolar radiation)

■ Ventilated solar photovoltaic wall
The western wall | The southern wall
Summer operation diagram | Winter(solar radiation) | Winter(nonsolar radiation)

The first floor of the south, west and east walls of the village activity center is designed as a ventilated solar photovoltaic double-layer glass curtain wall, and the second floor is designed as an additional sunlight room.

■ Energy Saving Analysis

Energy saving project name	Calculation	Standard	Conclusion
Body shape coefficient	0.37	≤0.40	Satisfying criteria
Air tightness level of building exterior windowst	6	≥6	Satisfying criteria
Outer door air tightness level	4	≥4	Satisfying criteria
Airtightness level of transparent curtain wall	3	≥3	Satisfying criteria
Roof heat transfer coefficient	0.29	≤0.40	Satisfying criteria
Average heat transfer coefficient of the roof	0.29	≤0.40	Satisfying criteria
Average heat transfer coefficient of exterior wall	0.41	≤0.45	Satisfying criteria
Effective ventilation area ratio	0.15	≥0.1	Satisfying criteria
External window heat transfer coefficient	1.70	≤2.50	Satisfying criteria
External window solar heat gain coefficient	0.36	≤0.52	Satisfying criteria
Visible light transmittance of external windows	0.60	≥0.60	Satisfying criteria
Thermal resistance of ground insulation layer	1.21	≥0.60	Satisfying criteria

■ Carbon Emission Analysis of the Project

Summary of carbon emission calculation results of equipment operation system in the design building use stage

	Item	Carbon emission (kgCO₂e)	Carbon emission (kgCO₂e/a)	Carbon emission (kgCO₂e/(m²a))
Carbon emission term	Heating system	436648.15	8732.92	5.88
	Cooling system	211275.94	4225.58	2.85
	Transmission and distribution system	13790.39	275.81	0.19
	Domestic hot water system	0.00	0.00	0.00
	Lighting system	981101.85	19622.04	13.21
	Elevator system	0.00	0.00	0.00
Carbon reduction term	Photovoltaic power generation	2279055.98	45581.12	30.69
	Wind power generation	0.00	0.00	0.00
	Building carbon sink	93550.00	1871.00	1.26
	Total	0.00	0.00	0.00

Note:The above calculation results are based on the floor area.

Summary of carbon emission calculation results

Item	Carbon emission kgCO₂	Carbon emissions per unit area kgCO₂/m²	Specific gravity
Building unit-phase carbon emissions	0	0.00	0
Building site-phase carbon emission	38154.16	12.11	6.78%
Carbon emissions during building demolition	76308.84	24.23	13.56%
Carbon emissions in building materials production stage	427246.87	135.63	75.89%
Carbon emissions from building materials transportation	21222.06	6.74	3.77%
Total carbon emissions	563331.93	178.71	100%

According to calculations, the village activity center meets the corresponding requirements of the "Design Standard for Energy Efficiency of Public Buildings" (GB50189-2015).

According to calculations, The project has achieved the green building goal of zero carbon during the operation phase and low-carbon throughout the entire lifecycle.

■ Green Building Strategy (village activity center)

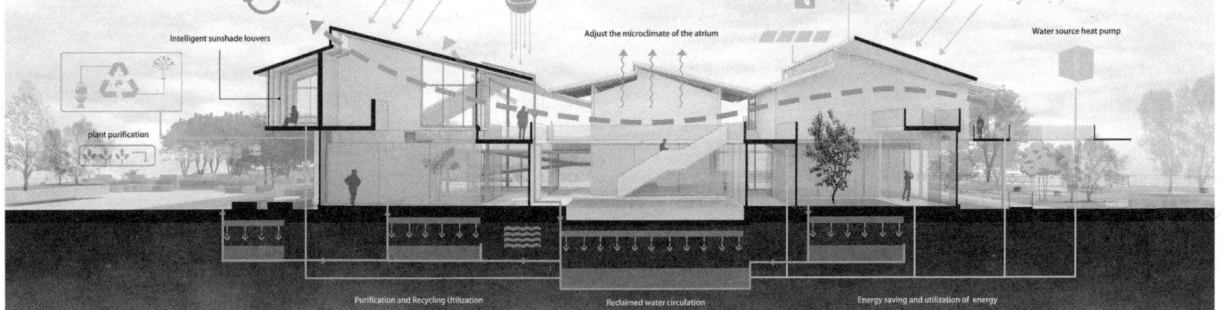

Air convection cycle
rainwater collection
Solar photovoltaic power generation
Intelligent sunshade louvers
Adjust the microclimate of the atrium
Water source heat pump
plant purification
Purification and Recycling Utilization
Reclaimed water circulation
Energy saving and utilization of energy

归园—The Returning Garden

可持续的地域建筑设计及技术体系
SUSTAINABLE REGIONAL ARCHITECTURE DESIGN AND TECHNOLOGY SYSTEM

怀安县二堡子村民宿民居设计
Design of B&B and residence in Erpuzi Village, Huai'an County

110005

综合奖・入围奖・张家口市
怀安县二堡子村建设项目
Comprehensive Awards −
Nomination − Erpuzi Village
Project, Huai'an County,
Zhangjiakou City

注册号：110005
Register Number：110005
项目名称：归园
Entry Title：The Returning Garden
作者：李嘉杭、车佩玉、韦彦滢
Authors：Li Jiahang, Che Peiyu, and
　　　Wei Yanying
作者单位：西安建筑科技大学
Authors from：Xi'an University of Archi-
　　　tecture and Technology
指导教师：陈敬、王芳
Tutors：Chen Jing, Wang Fang
指导教师单位：西安建筑科技大学
Tutors from：Xi'an University of Architecture
　　　and Technology

1 Store 2 Office 3 Activity Room 4 Rescue Material Storage
5 Open Activity Space 6 Courtyard 7 Viewing Platform

Structure & Spatial Units

Entity Empty

Pavilion Room

Colonnade

Colonnade

Door

Form a centripetal overall layout and grey space under the eaves in the form of a sloping roof

The structure of the framework system, with a basic framework that is transparent on all sides, creates a natural and open environment, while the enclosed space creates a private and enclosed space. The flexible application of this structure enables ancient architecture to adapt to the environment and meet people's living and activity needs.

The bottom floor of the building is elevated, incorporating "nature" into the elements of the building itself, achieving a poetic dwelling for users, and the mutual infiltration of orderly architecture and natural environment

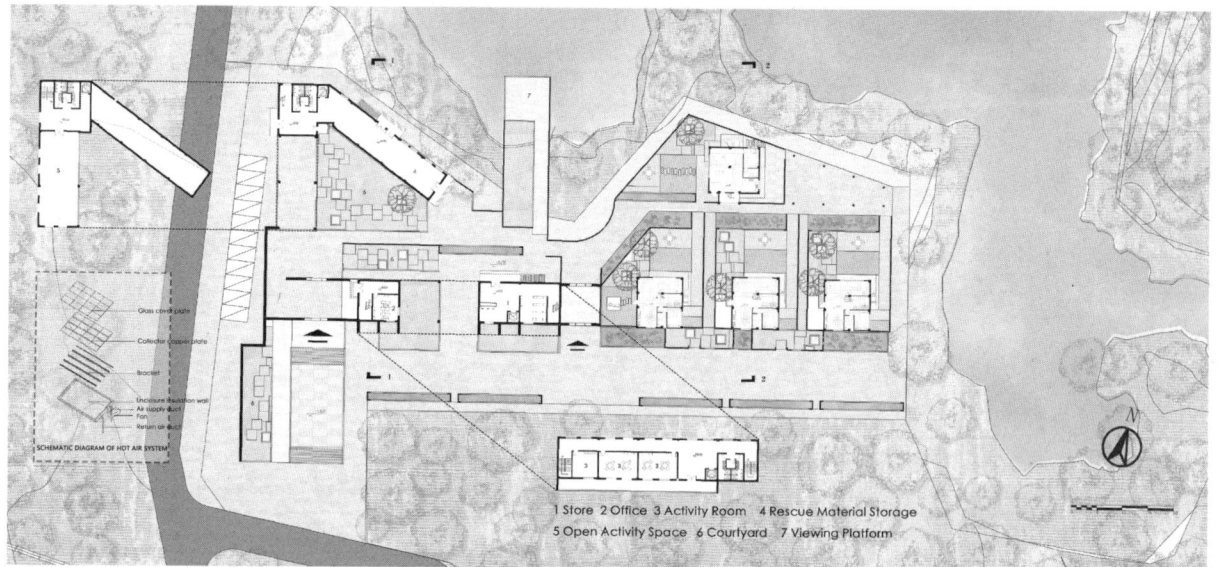

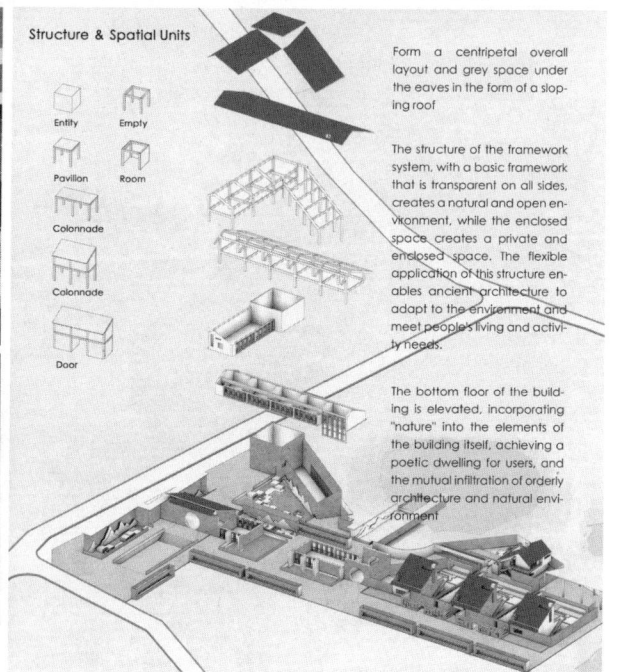

In summer,the blinds are open, stuck top covering the roof provides insulation.

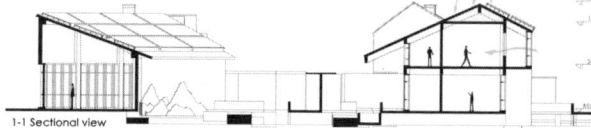

1-1 Sectional view

In winter, the blinds are closed, and the corridor benefits as an additional sunroom.

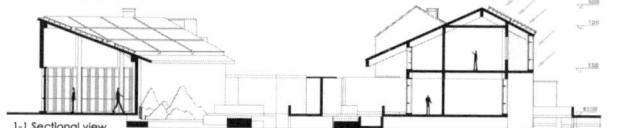

1-1 Sectional view

【Unit A】

【Self occupancy+homestay】
Building area: 1F 90 ㎡ 2F 74 ㎡

Architectural Lighting Analysis

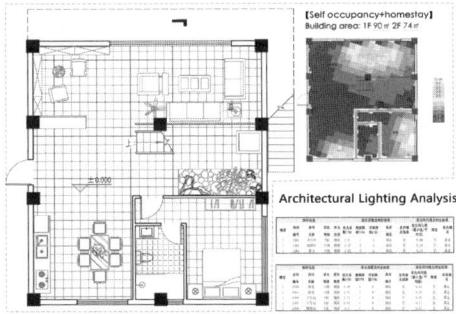

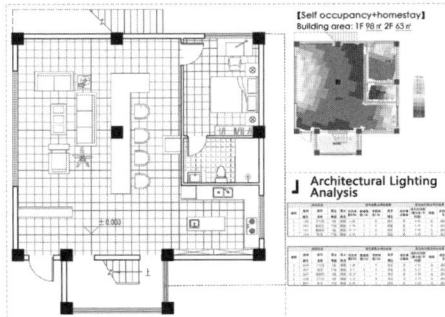

First floor Plan 1 : 100

Building Energy Efficiency Analysis

Second floor Plan 1 : 100

【Lighting】

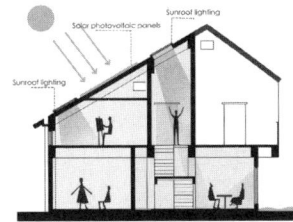

The living room is equipped with a large and spacious window facing north to ensure indoor lighting, while the second floor of the staircase is equipped with a south facing skylight in the bedroom to increase indoor lighting.

【Insulation】

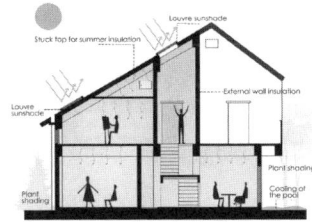

The cypress trees on the west and south sides can provide shade in summer, combined with cooling by the four sided water pools. The skylight louvers prevent direct sunlight from entering the room in summer. The stuck top can provide insulation during summer.

【Heat preservation】

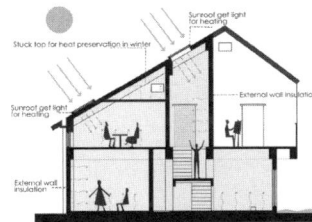

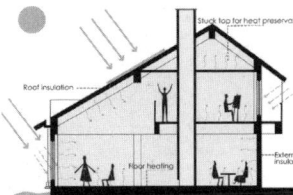

The sunshade of the skylight and the louvers will not create shading in winter, and the opening and closing direction of the louvers is suitable for light to enter, striving for the maximum amount of direct sunlight. The stuck top can provide insulation during winter.

【Air circulation】

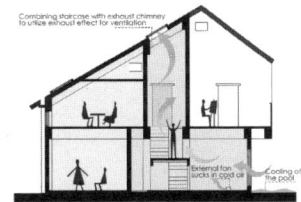

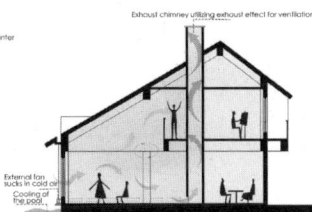

In summer, open the west ventilation door, and the external fan can suck in cold air with lower water surface temperature. The cold air enters the room for cold and hot gas exchange; At the same time, utilizing the thermal pressure effect of the chimney to extract air, driving indoor air flow and increasing indoor ventilation.

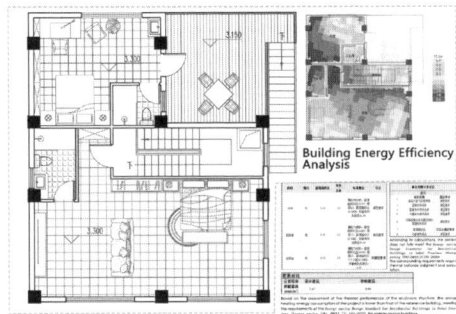

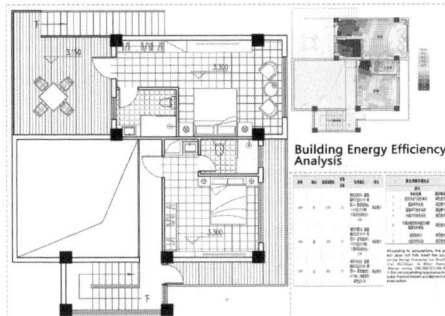

Unit A-Living room and courtyard

Unit A-Second floor bedroom

Unit B-Living room and courtyard

Unit B-Second floor bedroom

【Unit B】

【Self occupancy+homestay】
Building area: 1F 98 ㎡ 2F 63 ㎡

Architectural Lighting Analysis

First floor Plan 1 : 100

Building Energy Efficiency Analysis

Second floor Plan 1 : 100

【Lighting】

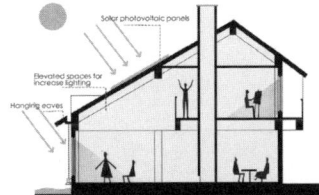

Large windows are installed on the west and east sides of the living room to ensure indoor lighting, and combined with the eaves to prevent direct sunlight and exposure to the west. Set up a sunlit staircase and elevated space facing south to increase indoor lighting.

【Insulation】

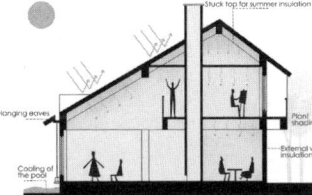

Plants on the west and east sides can provide shade in summer, combined with cooling in the west pool. Extend the eaves to avoid direct summer sunlight entering the room. The stuck top can provide insulation during summer.

【Heat preservation】

The overhanging scale of the eaves will not create shading in winter, and large windows and south facing sunlit stairwells can maximize the use of solar radiation for heating. The stuck top can provide insulation during winter.

【Air circulation】

In summer, open the west ventilation door, and the external fan can suck in cold air with lower water surface temperature. The cold air enters the room for cold and hot gas exchange; At the same time, utilizing the thermal pressure effect of the chimney to extract air, driving indoor air flow and increasing indoor ventilation.

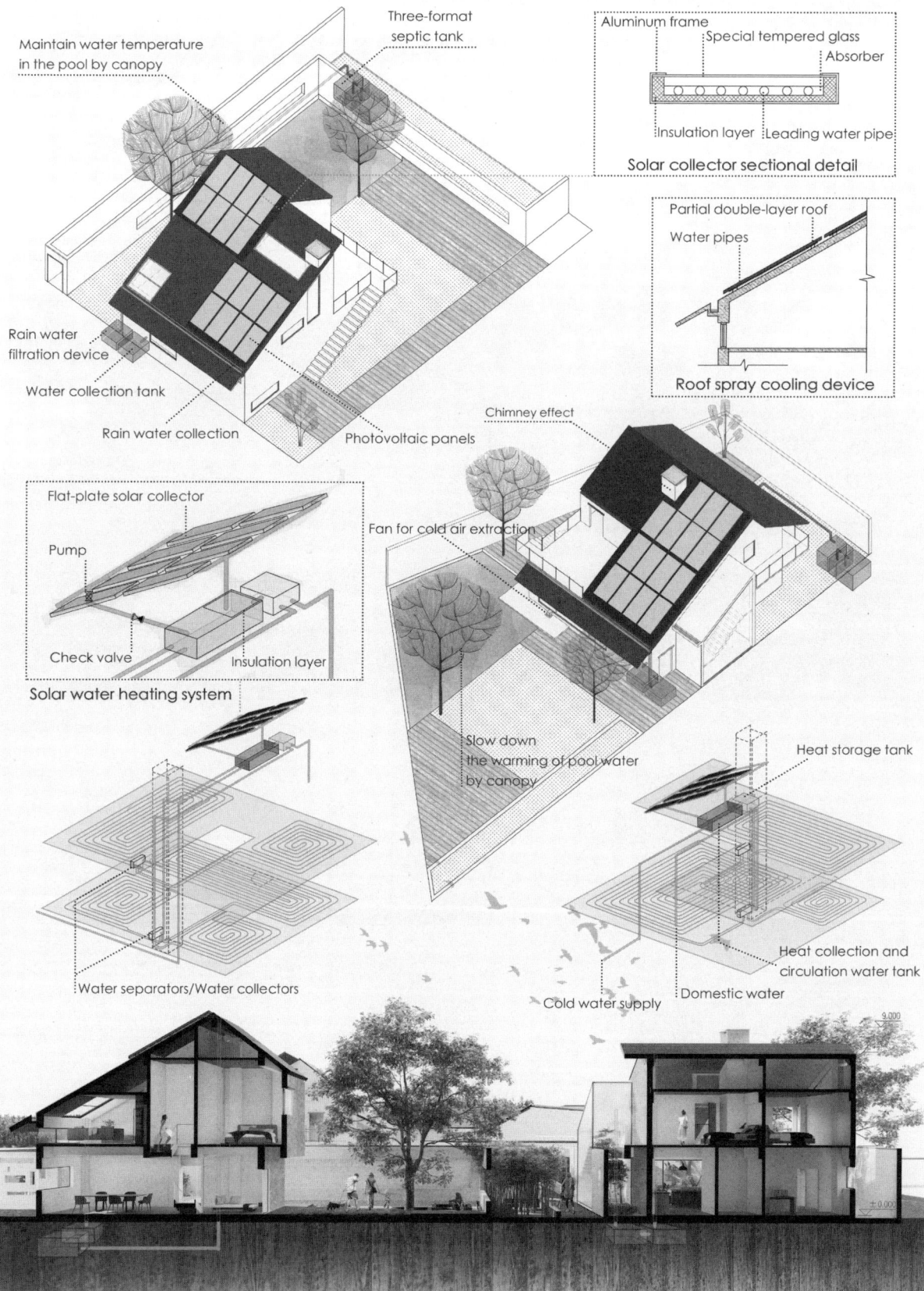

Maintain water temperature in the pool by canopy

Three-format septic tank

Aluminum frame

Special tempered glass

Absorber

Insulation layer

Leading water pipe

Solar collector sectional detail

Rain water filtration device

Water collection tank

Rain water collection

Photovoltaic panels

Partial double-layer roof

Water pipes

Roof spray cooling device

Chimney effect

Flat-plate solar collector

Pump

Fan for cold air extraction

Check valve

Insulation layer

Solar water heating system

Slow down the warming of pool water by canopy

Heat storage tank

Heat collection and circulation water tank

Water separators/Water collectors

Cold water supply

Domestic water

风栖光院

Breeze & Light Haven 01

设计说明

(本设计借鉴河北民居庭院围合院落为中心的布局形式，使建筑大部分空间基本处于阳光...保证南向的采光。一方面，建筑采用空间可变式设计，在村居活动中心部分，设计结合空气流动及对减灾等灾害的应对，屋顶屋面采用玻璃顶的形式，屋面可开启玻璃顶的形式。打开时，可满足夏季通风、关闭时，结合可移动的轻质墙体，形成封闭式玻璃阳光房以采暖、关闭减少散热，为暖可以有效减少通过建筑外表面的热流失。同时，院内的庭院可作为灾区使用。满足夏季避雨的夏季避雨需求。在民宿部分，采用轻质可移动墙体，在灾害来临可予快速分割为独立户型立面使用，确保私密性。另一方面，建筑采用的可变式设计，可在灾害来临时，根据所需的功能迅速能够改变并为应灾功能，比如观景塔能够转变为观景塔。观景餐饮转化为灾变大厅。最后，建筑设置屋顶太阳能板、太阳能路灯、雨水收集、透水地面等，结合地边资源等波源，满足应灾时能源、食品需要。)

■ Design Specification

The design draws inspiration from the layout of traditional Hebei residential courtyards, with the courtyard as the central focus. Most of the building spaces face south to ensure adequate southern sunlight. On one hand, the building adopts a flexible spatial design. In the village activity center, the design addresses climate and disaster responses, such as earthquakes, by using a glass roof over the courtyard. The roof is operable, allowing ventilation in the summer when open, and forming a closed sunroom when shut, in combination with movable lightweight walls. This setup provides heating while minimizing heat loss through the building's exterior surface. Additionally, the enclosed courtyard can serve as accommodation in disaster-stricken areas, meeting the needs for rain protection in the summer and heating in the winter. In the guesthouse section, lightweight movable walls are used, enabling the space to be quickly divided into two independent units during emergencies, ensuring privacy. On the other hand, the building is designed with functional flexibility, allowing it to be rapidly adapted for disaster response. For example, the viewing tower can be converted into an observation tower, and the scenic restaurant can be turned into a communal dining hall. Lastly, the building incorporates rooftop solar panels, solar streetlights, rainwater collection, and permeable pavements, integrating nearby garden resources to ensure energy, food, and accommodation needs are met during disaster situations.

■ Concept Extraction

1	2	3	4
Winter garden	Lane	Courtyard	Village

■ Design Highlight

Winter windows closing | Summer windows opening
Atrium in Summer | Atrium in Winter
Garden | Solar umbrela

■ General Plan

Economic and Technical Indicators

Gross floor area: 1300㎡
Building footprint: 1080㎡
Ratio of green space: 38%
Number of floors: 2floor
Building height: 9m

Entrance to Medical area
Secondary Entrance
Main Entrance
Exhibition Entrance
Garage Entrance
Square Entrance
Entrance to Residential area

0 5 10 15 20 25M

Landscape sight analysis

- - Pedestrian route
— vehicle route

■ Elevation and Section View

courtyard | lobby

West elevation 1 : 250

A-A section 1 : 250

综合奖・入围奖・张家口市
怀安县二堡子村建设项目
Comprehensive Awards –
Nomination – Erpuzi Village
Project, Huai'an County,
Zhangjiakou City

注册号：110069
Register Number：110069
项目名称：风栖光院
Entry Title：Breeze & Light Haven
作者：浮英媛、胡安达、袁林、陈钺、
方皓宇
Authors：Fu Yingyuan, Hu Anda,
Yuan Lin, Chen Yue, and
Fang Haoyu
作者单位：昆明理工大学
Authors from：Kunming University of
Science and Technology
指导教师：谭良斌
Tutor：Tan Liangbin
指导教师单位：昆明理工大学
Tutor from：Kunming University of Science
and Technology

■ Volume Generation Analysis

Rice field view/Water view
Pond
Pond
Quiet
2.00
Activity Transition
±0.00
Building Line
Boundary line of land
Road
Pre-site element

courtyard
courtyard
Factor extraction

Partially lifting the second floor

Reduce the volume according to the physical environment and form

Add walls, short walls, variable walls

Inserted function

Layout of square, garden and parking lot

Pitched roof
(with open skylights and solar panels)

风栖光院

Breeze & Light Haven 02

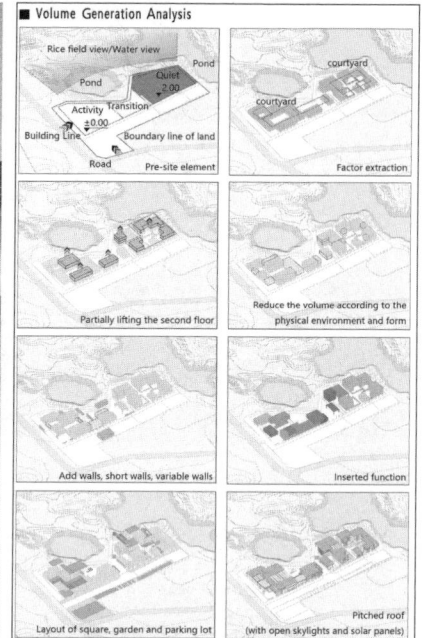

■ Ground Floor Plan 1:300

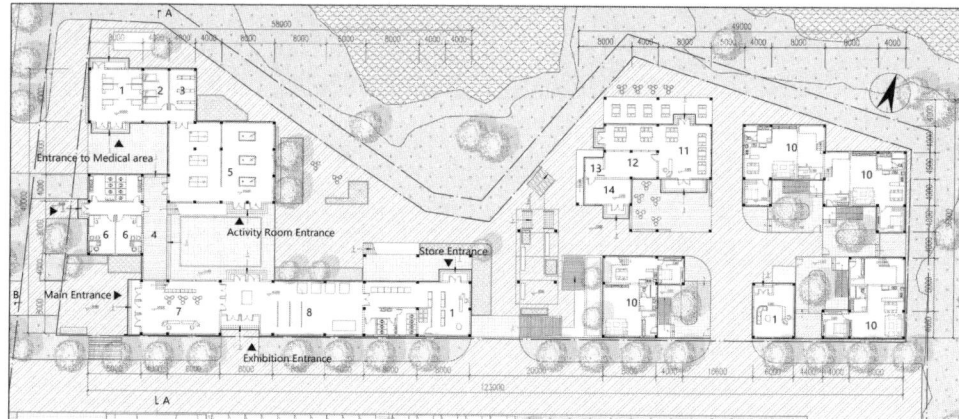

Entrance to Medical area

Activity Room Entrance

Store Entrance

Main Entrance

Exhibition Entrance

■ Function Setting/traffic Layout

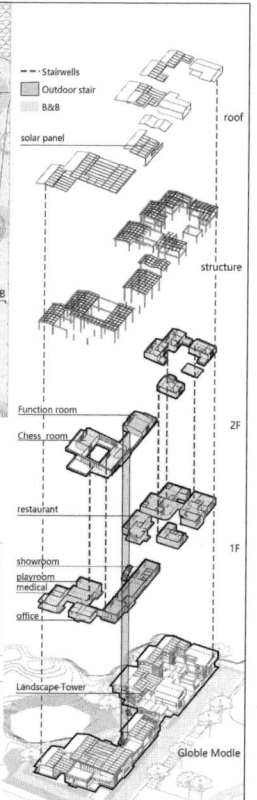

Stairwells
Outdoor stair
B&B

solar panel

roof

structure

Function room

Chess room

restaurant

showroom
playroom
medical
office

Landscape Tower

Globle Modle

2F

1F

■ Second Floor Plan 1:300

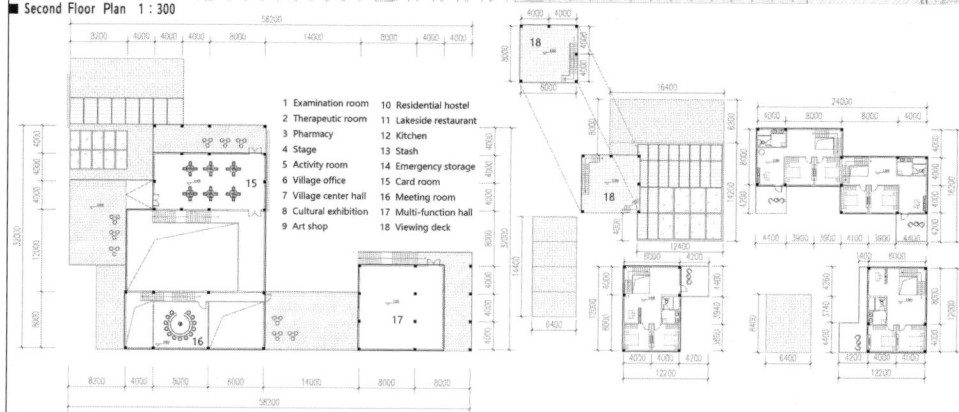

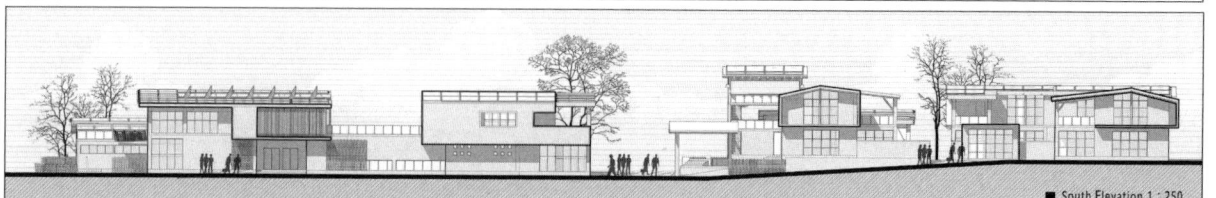

1 Examination room
2 Therapeutic room
3 Pharmacy
4 Stage
5 Activity room
6 Village office
7 Village center hall
8 Cultural exhibition
9 Art shop
10 Residential hostel
11 Lakeside restaurant
12 Kitchen
13 Stash
14 Emergency storage
15 Card room
16 Meeting room
17 Multi-function hall
18 Viewing deck

■ South Elevation 1:250

风栖光院

Breeze & Light Haven_03

■ Low Carbon Strategy + Emergency Variable Design Analysis

Normal State

summertime (heat dissipation)
- Courtyard + Ventilation gallery
- Skylight can be opened above the courtyard
- Terrace + grey space
- Pool + green plants on site
- Electrically controlled Angle blinds

wintertime (Heating/insulation)
- The courtyard ensures the building's need for natural lighting
- The courtyard can be closed to form a large sunroom
- External wall 240mm with internal insulation
- Large window openings on the south side of the building

Avoid adverse factors
- The west opening window area is small (Avoid the western sun light)
- Parapet (Avoid instantaneous strong winds)

Emergency State

Earthquake

Aseismatic
- Small body size factor/regular geometry
- Dispersed layout
- 4/8m standard column span, easy to assemble construction and maintenance
- Solid storage silo + viewing platform, emergency change warning tower

Spatial variation
- Movable sliding wall(The outdoor space is transformed into an indoor space for a large number of villagers to stay in emergency)
- Simple push-pull wall(Meet the requirements of a multi-purpose household in an emergency)

Self-feeding system

Water resources
- rainwater collection system(Sloping roof + reservoir)
- Artificial pools purify fresh water

Electric energy
- The roof is covered with solar panels

Food resources
- Existing paddy field transportation
- Emergency grain storage

Flood

Flood channel + artificial pool
- Large area of green planting in the site, permeable pavement
- Second floor platform for villagers to stay/rescue evacuation

The open skylight

The large sunroom

The Parapet

HELP!!

The second floor to avoid flood disaster

The Reserve bin

Large south facing window opening

■ Standard Column Span

■ Function Replacement

1 Medical emergency area
2 Refuse collection area
3 Emergency accommodation area
4 Disaster response deployment
5 Sale of living goods
6 Outdoor loading bay
7 Public canteen
8 Emergency storage area
9 Supplies distribution area
10 Rescue vehicles Parking

■ Structural Analysis of Variable Push-pull Wall

■ Technical Analysis of Sponge Village

Ecological wetland
Roof runoff
Surface runoff
Roof runoff
Rainwater collection tank
Rainwater collection tank
Permeable pavement
Roof runoff
Artificial lake
Surface runoff
Roof runoff
Recessed green space
Recessed green space
Artificial lake

■ B-B Section 1 : 250

Meeting room
Multi-function hall
Residential hostel
Residential hostel
Residential hostel
Village center hall
Cultural exhibition

风栖光院
Breeze & Light Haven 04

■ Construction Joint of Skylight Opening in Courtyard Sunroom

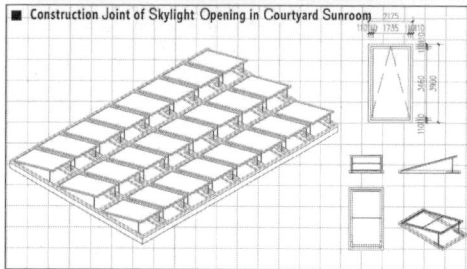

■ Roof Solar Panel Construction Joints

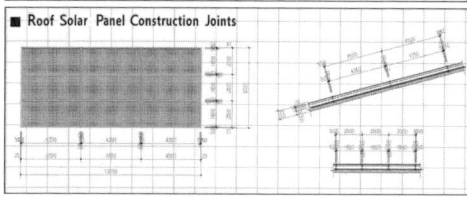

■ Climatic Analysis

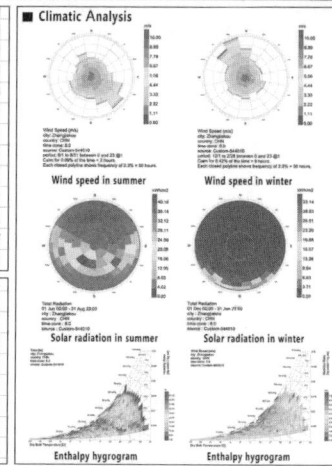

Wind speed in summer / Wind speed in winter

Solar radiation in summer / Solar radiation in winter

Enthalpy hygrogram / Enthalpy hygrogram

■ Wind Velocity Analysis

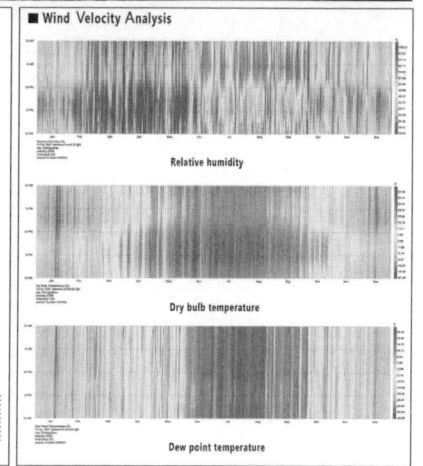

Relative humidity

Dry bulb temperature

Dew point temperature

■ UTCI Analysis

Wintertime / Summertime

■ Wind Environment Simulation

Summertime

Wintertime

The wind field environment in the site is good, the outdoor activity area is all avoided without wind area, the shape of the middle part of the building is open, so that the north and south winds of the building are continuous, there is no obvious vortex area, and the wind speed of the courtyard space of the building B&B is moderate, which is convenient for outdoor activities.

■ Solar Radiation Simulation

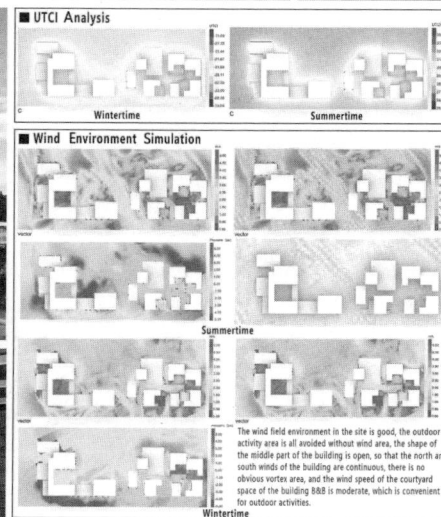

Wintertime

Summertime

■ Sunshine Simulation

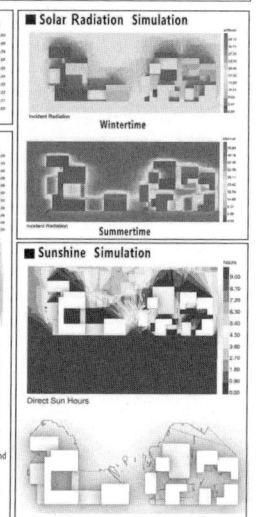

Direct Sun Hours

■ Carbon Emission Analysis

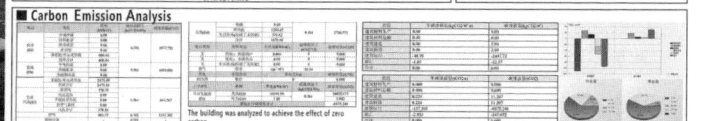

The building was analyzed to achieve the effect of zero carbon.

编号：110084

SOCIAL IMPACT

4 X B&B HOTEL + SERVICE CENTER < SOCIAL CONDENSER

EXTERIOR SURFACE

4 X B&B HOTEL + SERVICE CENTER > SHARED PERIMETER

FLEXIBILITY

4 X B&B HOTEL + SERVICE CENTER = MEGA PACK

设计要点1：为什么要集成民宿与村民服务中心？
KEY POINT1: WHY INTEGRATING THE B&B HOETEL AND VILLAGE SERVICE CENTER INTO A COMPLEX?

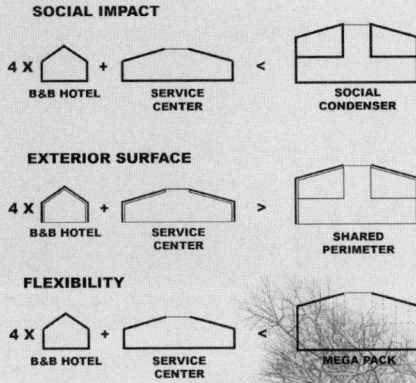

联系城市与乡村的窗口

随着我国社会主要矛盾转化为人民日益增长的美好生活需要和不平衡不充分的发展之间的矛盾，乡村民宿的设计应着眼于加强城市与乡村的联系，把民宿当作城市与乡村相互发现、相互了解的窗口。我们应极力避免对于看乡村振兴的旗号把乡村民宿变成富裕人群的新殖界，加剧城市与乡村、富裕与贫穷的割裂。将民宿与村民服务中心在空间上的集成有助于创造更多交流的机会，同时也搭建了一个乡村向外界展现自我的舞台。

A WINDOW BETWEEN URBAN AND RURAL AREA

As the main contradiction in our society transforms into the contradiction between the people's growing needs for a better life and the unbalanced and inadequate development, the design of rural Bed&Breakfast hotel should focus on strengthening the connection between cities and villages, and treat it as a window for cities and villages to discover and understand each other. We should try our best to avoid turning rural B&B hotel into new colony for wealthy people under the banner of rural revitalization, exacerbating the division between cities and villages, and between rich and poor. The spatial integration of B&B hotel and village service center helps create more opportunities for communication, and also builds a stage for the village to show itself to the outside world.

寒冷地区的建筑保温

本项目位于张家口市怀安县二堡子村，年平均气温8.3℃，最低气温-22.6℃。将民宿与村民服务中心集成一体有助于减少暴露于严寒环境的外围护结构，从而减少建筑形体系数。

THERMAL INSULATION IN COLD CLIMATE

This project is located in Erbaozi Village, Zhangjiakou City, with an average annual temperature of 8.3℃ and a minimum temperature of -22.6℃. Integrating the B&B hotel with the village service center helps to reduce the surface exposed to the severe cold environment.

设施的共享与建筑的灵活性（韧性）

集成的民宿和村民服务中心可与民宿客人共享村民服务中心内的书吧、餐厅等服务设施，给村民带来经济收入。同时，可根据后期运营需要对空间功能进行灵活转换，避免村民服务中心的低频使用而造成投资的浪费。面临灾害，集成了一体的建筑空间有利于资源的统一调配，通过建筑的灵活性增强建筑在灾害中的服务韧性。

FACILITY SHARE AND BUILDING FLEXIBILITY(RESILIENCE)

The integrated B&B hotel and village service center can share the service facilities such as the book bar and restaurant in the village service center with the B&B hotel guests, bringing economic income to the villagers. At the same time, the space function can be flexibly converted according to the needs of later operations to avoid the waste of investment caused by the low-frequency use of the village service center. In the face of disasters, the integrated building space is also conducive to the unified allocation of resources, and the flexibility of the building can enhance the service resilience of the building in disasters.

综合奖·入围奖·张家口市怀安县二堡子村建设项目
Comprehensive Awards – Nomination – Erpuzi Village Project, Huai'an County, Zhangjiakou City

注册号：110084
Register Number：110084
项目名称：云影
Entry Title：The Cloud Shadow
作者：杨文祺、邝微
Authors：Yang Wenqi, Kuang Wei
作者单位：加州大学伯克利分校
Authors from：University of California, Berkeley

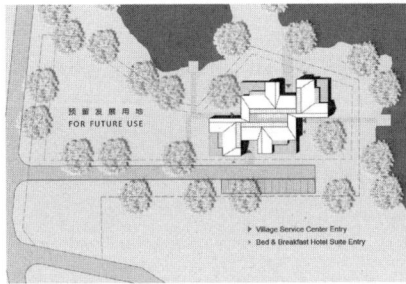

FOR FUTURE USE 预留发展用地

▶ Village Service Center Entry
▶ Bed & Breakfast Hotel Suite Entry

张家口 二堡子村
Erbaozi Village, Zhangjiakou

设计前提

根据中华人民共和国《建筑气候划分标准》GB 50178—93，我国共划分为7个建筑气候区。张家口二堡子村处于第Ⅰ建筑气候区（严寒地区）与第Ⅱ建筑气候区（寒冷地区）的交界。据第三章建筑气候特征和建筑基本要求：

一、建筑物应满足冬季防寒、保温、防冻等要求。

二、总体规划、单体设计和构造处理应使建筑满足冬季日照和防御寒风的要求；建筑物应采取减少外露面积，加强冬季密闭性并注意夏季通风利用和太阳能等节能措施；结构上应考虑气温年较差大及大风的不利影响。

BASIS OF DESIGN

According to GB 50178-93 "Standards for Building Climate Zoning" of the People's Republic of China, China is classified into 7 building climate zones. Erbaozi Village in Zhangjiakou is located at the junction of Building Climate Zone I (severe cold area) and Building Climate Zone II (cold area). According to Chapter 3 <Building Climate Characteristics and Basic Building Requirements>:

1. Buildings should meet the requirements of winter cold protection, heat preservation, and frost protection.

2. The overall planning, building design and construction detail should enable the building to meet the requirements of winter sunshine and wind protection, the building should take energy-saving measures such as reducing the exposed area, strengthening the airtightness in winter, and taking into account summer ventilation and solar energy utilization; the structure should consider the adverse effects of large annual temperature differences and strong winds.

光伏板 PV Panel
光伏玻璃 Solar Glass
首层村民服务中心 1F Village Service Center
二层民宿 2F Bed&Breakfast hotel
村民服务中心入口 Main Entry
民宿露台 Hotel Exterior Deck
民宿入口 Hotel Suite Entry
停车场（13车位） Parking (13 stalls)

Site Area	8262 ㎡
Total Building Area	1634 ㎡
Bed&Breakfast Hotel (4Suites)	600 ㎡ +240 ㎡
Interior (each suite)	150 ㎡
Deck (each suite)	60 ㎡
Village Service Center	1034 ㎡
Core Function Space	488 ㎡
Multi-Purpose Atrium	382 ㎡
Restroom/Storage	164 ㎡
Building Footprint	1034 ㎡
Coverage	12.5%
FAR	0.2
Parking Stalls	13

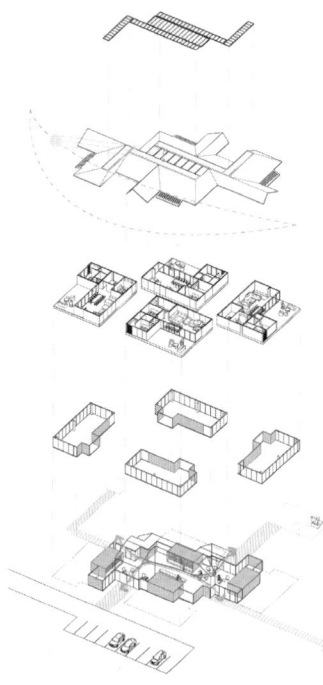

5. 光伏板和光伏幕墙 ——
在民宿屋面安装光伏板并养采光天窗设计为光伏幕墙，集团直接的太阳能辐射获得间歇的光伏发电。
PV panel & Solar glass
PV panels are installed on the Bed & Breakfast hotel roof. Meanwhile, solar glass are used as the skylight window glass to balance the need of direct solar heat gain and indirect power generation.

4. 采光天窗 ——
通过对采光天窗的设置，多功能大厅得以实现在冬季通过玻璃以辐为主的节能策略。
Skylight
By incorporating skylights, the multi-purpose atrium can depend on passive solar energy to improve energy efficiency throughout the winter.

3. 民宿空间 ——
位于建筑二层；包括 4 套民宿。
每套民宿包含一个客餐厅、三个独立卧室，两个卫生间及一个室外露台。每个民宿对内可朝瞰多功能大厅，对外可观赏远眺美好风景。
Bed & Breakfast hotel
Located on the second floor, it includes 4 Bed & Breakfast hotel units. Each unit includes a living and dinning space, three bedrooms, two restrooms and an exterior deck. All the units would have a view towards the inside into the atrium, and a view to the outside nature.

2. 核心功能空间 ——
位于建筑首层，包括民宿接待、儿童课后自习、小型演讲报告、民宿餐饮及零售、村民服务和快递收发等功能空间。核心功能空间两侧直接连接着多功能大厅，也从内连接多功能大厅。
Core Function Space
Located on the ground floor, it serves as a space for Bed & Breakfast hotel reception, after-school homework, small auditorium, dining for Bed & Breakfast hotel, retail, villager service and express delivery.

1. 多功能大厅空间 ——
位于建筑首层；由核心功能空间围合而成，也是连接各功能空间的交通空间。多功能大厅在东西方向形成贯穿的通道，形成夏季穿堂风，并在南北方向分别连接建筑主入口和北处的滨水景观。
Multi-Purpose Atrium
Located on the ground floor, the Multi-Purpose Atrium is surrounded and serves as a celebrated place that connects each zone. It is a linear space that connects east and west to nature as a way to remove excess heat from the building during the summer. It also leads south and north to the main entry and waterfront, respectively.

设计要点2：空间的分级与能源的利用
KEY POINT2: SPATIAL CLASSIFICATION AND ENERGY USE

我们是否有必要按相同的标准对整个民宿和村民服务中心进行温度调节？从建筑节能的角度，是否可以根据使用属性对不同的空间进行针对性的设计以减少能源的不必要浪费？
本方案根据不同空间对能源的使用需求将建筑空间分为三个层级：民宿空间、社区核心功能空间、多功能大厅空间。
民宿空间 —— 24 小时采暖，满足客居使用需求。
社区核心功能空间 —— 根据使用需求，合理控制建筑面积，以避免铺张而低效的空调和能源使用。通过控制进深和增大开窗面积的方式实现 100% 自然采光，避免白天室内照明对能源的浪费，并使建筑在冬季吸取更多太阳能。
多功能大厅空间 —— 此空间作为非正式交流空间，主动空调为辅的设计策略。通过屋顶采光天窗的设计，将此空间打造成阳光房，

Is it necessary to heat the entire B&B hotel and village service center to the same standard? Is it possible to apply distinct design strategies to various spaces according to their unique characteristics in order to minimize energy waste from the standpoint of energy conservation?
The building is categorized into three types of spaces in this design: Bed & Breakfast hotel suites, core function space and Multi-Purpose Atrium.
BED & BREAKFAST HOTEL SUITES - 24h heating available to satisfy the hotel room needs.
CORE FUNCTION SPACE - The square footage is reasonably controlled based on the spatial needs to avoid extravagant and inefficient air conditioning and energy use. By controlling the depth and increasing the window area, 100% natural lighting is achieved to avoid energy waste from indoor lighting during the day and enable the building to absorb more solar energy in winter.
MULTI-PURPOSE ATRIUM - This space, serving informal use, heavily relies on passive heating as the main heating source while using air conditioning as supplemental. The space is designed as a green house with skylight, so that it can fully absorb solar energy in winter and use the through-draught to take away the heat in summer, so as to avoid inefficient use of energy.

SECTION A-A

EAST ELEVATION

设计要点3：多功能大厅 - 面向世界的一扇窗
KEY POINT3: MULTI-FUNCTIONAL ATRIUM - A WINDOW TO THE WORLD

通过对民宿和村民服务中心的叠加与咬合，本方案创造了一个让乡村往向世界展现自我的机会，同时也赋予了民宿客人参与开融入乡村文化的机会。多功能大厅作为核心功能空间围合而成的非正式的交流空间，如同村落里的街道，为村民提供休息、阅读和媒体和集会的场地，同时也是乡村农产品、手工艺和民俗文化遗产的展示的平台；对内，游客可在民宿中俯瞰大厅，感受乡村文化生活并参与其中；对外，游客可在民宿楼台欣赏乡村美美丽的风景，实现对乡村全方位、多角度的体验。

By superimposing and integrating the B&B hotel and the village service center, the building allows the village to showcase itself to the world while enabling B&B guests to participate in and integrate into rural culture. The multi-purpose atrium serves as an informal communication space surrounded by the core function space, similar to a village street, providing villagers with a place to relax, read, engage, and gather. It also serves as a showcase for rural agriculture products, handicrafts, and other forms of cultural heritage. Internally, tourists may view the atrium from the hotel suites, feel the rural culture, and participate in it; externally, tourists can enjoy the lovely landscape of the countryside from the hotel suites' terrace, gaining an immersive experience of the countryside from both cultural and natural perspectives.

民宿接待
Reception

景观餐厅
Cafe

二层民宿
Bed&Breakfast hotel

LEVEL 1
1. Villager Service
2. B&B Hotel Reception
3. Auditorium
4. Cafe
5. Kitchen
6. Multi-Purpose Atrium
7. Restrooms
8. Storage

LEVEL 2
1. Living Room
2. Open Kitchen
3. Dinning
4. Bedroom
5. Master Bedroom
6. Deck

0   5   10   15 m

编号：110084

设计要点4：云影 - 阳光的采集和直射光的柔化
KEY POINT4: CLOUDS-SUNLIGHT COLLECTION AND SOFTENING

张家口 11 月至 3 月平均气温小于 5℃，月平均度日数为 3600 ℃·d。本方案通过屋顶天窗和使用热工性能好的玻璃（高得热系数和低导热系数）等方式，使大量太阳能进入并停留在室内以提高室内舒适度。与此同时，进入多功能大厅的低角度光线则需要适当的遮阳处理。为了增加使用的灵活性，本方案通过可活动的悬挂遮阳板来柔化由东西方向进入室内的低角度光线。优如云影，使悬挂遮阳板的空间就成为以空间的一道景观。

The average temperature in Zhangjiakou from November to March is less than 3℃, and the monthly heating degree days are 3600℃·d. The design uses roof skylights and glass with good thermal performance(high SHGC and Low U Value) to allow a large amount of solar energy to enter and stay in the room to improve indoor comfort. At the same time, the low-angle sunlight entering the multi-functional atrium requires appropriate shading. In order to increase the flexibility of use, automated light baffles are used to diffuse the low-angle sunlight entering the atrium from the east and west, and the space is energized by the rhythm of these hanging baffles as light filtering through the clouds.

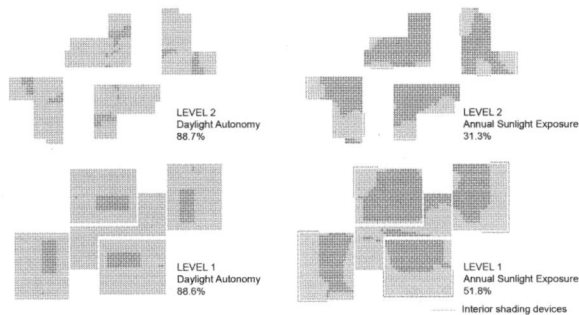

LEVEL 2
Daylight Autonomy
88.7%

LEVEL 2
Annual Sunlight Exposure
31.3%

LEVEL 1
Daylight Autonomy
88.6%

LEVEL 1
Annual Sunlight Exposure
51.8%

Interior shading devices

本方案采用 300Lux/8am-6pm 的 50% 时间作为计算标准，经测算，首层和二层分别可达到 88.6% 和 88.7% 的日光照明，代表绝大部分空间在不借助人工照明的情况下，一半以上时间可达到 300Lux 的照度，大大减少了室内照明对能源的消耗。同时，建筑通过可活动的室内遮阳的方式，灵活解决太阳眩光。

300Lux / 50% from 8am to 6pm over the year is used as the standard for simulation. According to the simulation, the first floor and the second floor can reach 88.6% and 88.7% of daylight autonomy respectively, which means that most spaces can reach 300Lux illuminance for more than half of the time throughout the year without artificial lighting, greatly reducing the energy consumption of indoor lighting. At the same time, the building uses movable indoor sunshades to flexibly solve the problem of solar glare.

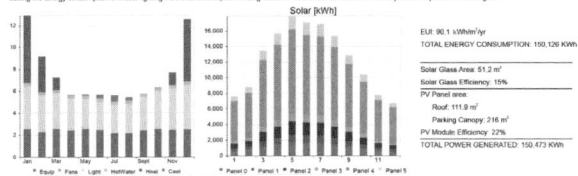

Solar [kWh]

EUI: 90.1 kWh/m²/yr
TOTAL ENERGY CONSUMPTION: 150,126 KWh

Solar Glass Area: 51.2 m²
Solar Glass Efficiency: 15%
PV Panel area:
Roof: 111.9 m²
Parking Canopy: 216 m²
PV Module Efficiency: 22%
TOTAL POWER GENERATED: 150,473 KWh

建筑的能耗约为 15 万 kW·h，通过光伏玻璃和光伏板产生的总发电量为 15 万 kW·h，建筑可达到能源生产和利用的平衡。

注：为应对当地的冰雹灾害风险，本方案在停车区域铺设光伏板以抵御冰雹对车辆的伤害，同时增加发电总量。

The building's energy consumption is about 150,000 kWh, and the total power generation through PV glass and PV panels is 150,000 kWh, so the building can achieve a balance between energy production and utilization.

Note: To respond to the hail risk, PV panels are installed above the parking stalls to prevent the damage to the vehicles and increase the total power generation.

光伏板
PV panel

光伏玻璃
Solar glass

夏季最大太阳高度角
Maximum solar altitude in summer

钢筋混凝土墙
保温砂浆
陶板
Concrete wall
Insulation mortar
Terracotta plate

活动遮阳
Light baffle

LOW-E 中空玻璃
Double panel LOW-E glass
SHGC: 0.65
U Value: 1.5
TVis: 0.77

仿木金属吊顶
Wood print metal ceiling

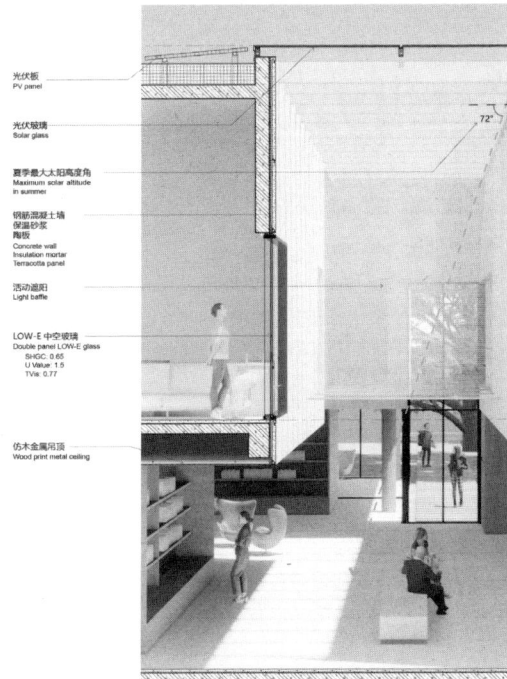

向阳而生
Born to the Sun

110088

Design of Public Service Center in Erpuzi Village, Hebei
河北二堡子村公共服务中心设计 01

Annotate:
1 Public Service Center
2 Guesthouse
3 B&B Courtyard
4 Landscape Pool
5 Landscape Tree Pool
6 Main Entrance
7 Parking lots
8 New energy charging station
9 Flower bed
10 Landscape node
11 Pools
12 Moe Rabbit Tribe Tourist Attractions
13 Residences and lodgings (not built)
14 Vegetable Garden Research Base
15 Village road

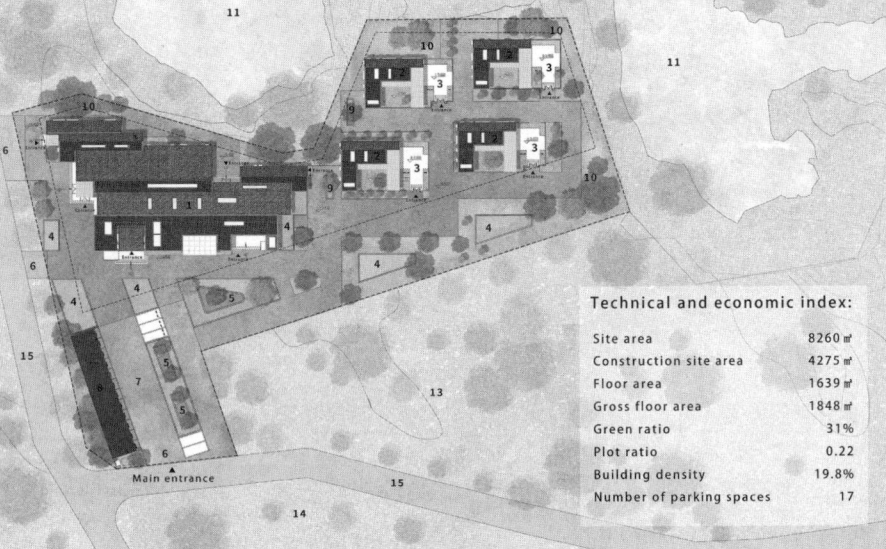

General layout 1：500

Main entrance

Technical and economic index:

Site area	8260 ㎡
Construction site area	4275 ㎡
Floor area	1639 ㎡
Gross floor area	1848 ㎡
Green ratio	31%
Plot ratio	0.22
Building density	19.8%
Number of parking spaces	17

综合奖·入围奖·张家口市怀安县二堡子村建设项目
Comprehensive Awards — Nomination — Erpuzi Village Project, Huai'an County, Zhangjiakou City

注册号：110088
Register Number：110088

项目名称：向阳而生
Entry Title：Born to the Sun

作者：莫宇豪、秦斯玲
Authors：Mo Yuhao, Qin Siling

作者单位：桂林理工大学
Authors from：Guilin University of Technology

指导教师：朱文霜
Tutor：Zhu Wenshuang

指导教师单位：桂林理工大学
Tutor from：Guilin University of Technology

Site Analysis

The project site is located within Erbaozi Village, Huai'an County, Zhangjiakou City, Hebei Province, with a longitude of 114.62° E and a latitude of 40.68° N. It is close to the Beijing-Tibet Expressway and the Beijing-New Zealand Expressway, and the high-speed railway line passes through from the north side of the site in an east-west direction. It is 35 kilometres from Zhangjiakou City and 200 kilometres from Beijing city centre. It is in close proximity to the villagers' residential areas of Erbaozi Village and Beiguojiayao Village, within a straight line distance of 1km. There are patches of paddy fields in the area of Erbaozi Village, the Dayang River in the north, the Jinshitan Forest in the south, and large areas of woodland and farmland in the east and west. There are already tourist attractions such as Moe Rabbit Tribe, Swan Lake Town, Dajinggou, Zhaohua Temple, Grassland Resort and Hot Spring Resort within 50km around the site. The village has a household population of 164 households and 402 people. The existing resident population is 132 households with 327 people.

Climate Analysis

Rose of the Moon Winds Chart

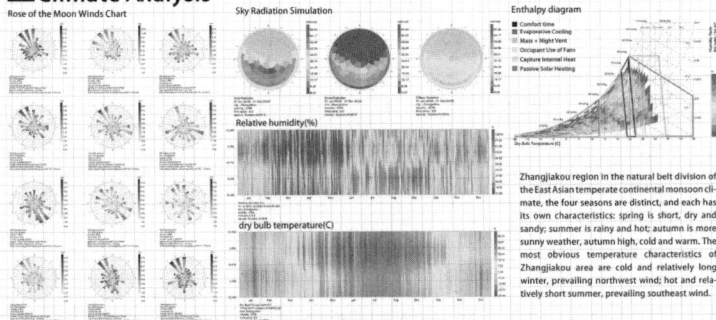

Sky Radiation Simulation

Enthalpy diagram

Relative humidity(%)

dry bulb temperature(C)

Zhangjiakou region in the natural belt division of the East Asian temperate continental monsoon climate, the four seasons are distinct, and each has its own characteristics: spring is short, dry and sandy; summer is rainy and hot; autumn is more sunny weather, autumn high, cold and warm. The most obvious temperature characteristics of Zhangjiakou area are cold and relatively long winter, prevailing northwest wind; hot and relatively short summer, prevailing southeast wind.

Design Concept

本方案位于河北张家口市二堡子村，为了最大限度获得当地丰富的太阳能资源，方案初期便优先考虑了被动式节能技术的运用，围绕"向阳而生"的主题，置入了"斜屋"的概念。建筑同时结合了自然通风、压力通风、微气候循环、冬夏季遮阳转换及雨水收集等被动技术，以及太阳能光伏系统、地源热泵等主动技术，配备智能环境监测控制系统从而实现绿色、生态的公共服务中心。

This project is located in Erbaozi Village, Zhangjiakou City, Hebei Province. In order to maximize access to the abundant solar energy resources in the area, the project prioritizes the use of passive energy-saving technologies at the beginning of the project, and centers around the theme of "Born to the sun", and incorporates the concept of the "slanted house". Solar energy. The building also combines passive technologies such as natural ventilation, pressure ventilation, microclimate circulation, winter and summer shading conversion and rainwater collection, as well as active technologies such as solar photovoltaic systems and ground-source heat pumps, and is equipped with an intelligent environmental monitoring and control system to realize a green and ecological public service center.

contextual Analysis

Erbanzi Village, HuaFan County, Zhangjiakou City, has a longitude of 114.62° E, a latitude of 40.68° N, and a distance of 35km from Zhangjiakou City, which is in an advantageous location with convenient traffic.

Erbaozi village through the development of green organic industry and leisure and tourism industry, successfully changed the village's industrial structure of a single status quo, to achieve the diversification of economic development.

Erbaozi Village has unique natural endowments and a favourable geographical location. Because of its rich history, culture and geographical advantages, it has become the "Little Jiangnan on the Seas".

The forest theme park of Mengrabbit Tribe, modelled on high-quality boutique parks, is an outdoor family park that combines 'nature', 'fun' and 'learning' into one, where all ages can learn and play together. Outdoor parent-child paradise.

In the Folk Culture Park, crafts with different characteristics, such as Winter Olympic culture, farming culture, Go culture and folk culture, which are designed and woven with straw, bring a colourful visual impact.

Crowd Analysis

WHO?

$E = mc^2$

villagers　travelers　scholar　office worker

What do they need?

villagers	travelers	scholar	office worker
We all talk and relax and meet at the entrance of the village after our daily labor. Therefore, we hope that the new service center will have a recreation area, chess and cards room, gymnasium and other entertainment places, and also a reading room and other cultural places.	Erbaozi village of rice field town set of sightseeing tours, ecological leisure, sports experience and other functions in one, 100 miles of Yanghe wetland green ecological corridor of the important node, we have come to the name. We hope that the service center can provide a total of catering shopping and accommodation services.	Huai'an Erbaozi Village has a long history, rich natural landscape and deep cultural heritage. Research and study activities are held here on a regular basis. We hope that the service center can provide bright and spacious briefing room for lectures, etc., as well as catering and lodging services.	We are the staff responsible for the operation of the service center and the revitalization of the countryside, and we need spacious and bright offices and meeting rooms. We hope that the service center can have a software system that is easy to manage the building hardware and facilities.

向阳而生
Born to the Sun

110088

河北二堡子村公共服务中心设计 02

Design of Public Service Center in Erpuzi Village, Hebei

First floor plan 1:500

1 Lobby
2 Reception
3 Office
4 Cupboard
5 Multi-purpose hall
6 Rest area
7 Bar area
8 Management office
9 Reading room
10 Library
11 Medical room
12 Restrooms
13 Gymnasium
14 Ping-pong room
15 Chess room
16 Cafeteria
17 Dining Room Kitchen
18 Specialty store
19 Convenience store
20 Souvenir store
21 Repository
22 Foyer
23 Corridors
24 B&B Guest Restaurant
25 B&B Room
26 B&B Bathroom

Smart Building

photovoltaic system · Wind speed monitoring · Temperature monitoring · lighting control · louvre control · ground source heat pump · Sunshade control · Light monitoring · Intelligent Control System · Sunroof control · Air monitoring · Heating and cooling control · New energy charging station

Building Resilience Strategy

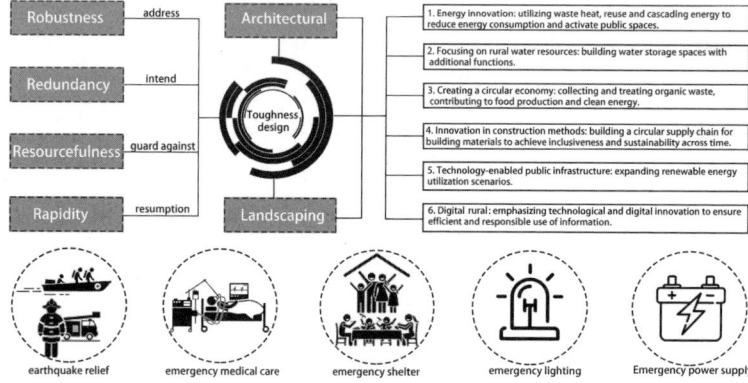

Robustness — address
Redundancy — intend
Resourcefulness — guard against
Rapidity — resumption

Architectural — Toughness design — Landscaping

1. Energy innovation: utilizing waste heat, reuse and cascading energy to reduce energy consumption and activate public spaces.

2. Focusing on rural water resources: building water storage spaces with additional functions.

3. Creating a circular economy: collecting and treating organic waste, contributing to food production and clean energy.

4. Innovation in construction methods: building a circular supply chain for building materials to achieve inclusiveness and sustainability across time.

5. Technology-enabled public infrastructure: expanding renewable energy utilization scenarios.

6. Digital rural: emphasizing technological and digital innovation to ensure efficient and responsible use of information.

earthquake relief · emergency medical care · emergency shelter · emergency lighting · Emergency power supply

A study of the Relationship between Local Building Forms and Climate

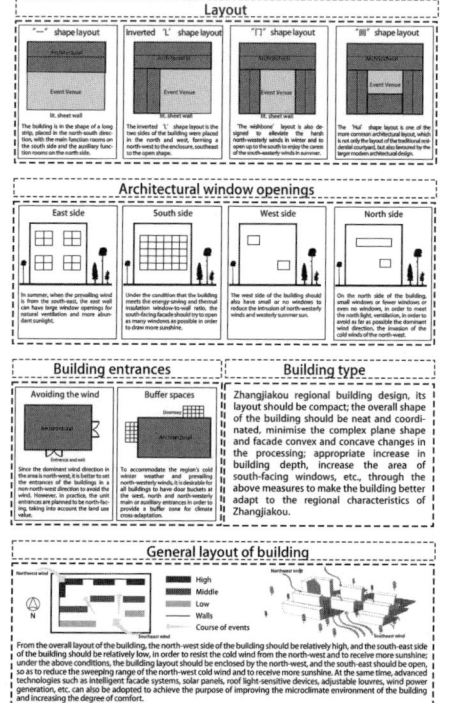

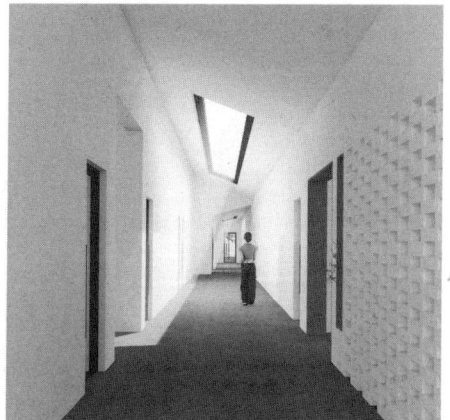

Layout

"—" shape layout	Inverted "L" shape layout	"T" shape layout	"回" shape layout

Architectural window openings

East side	South side	West side	North side

Building entrances

Avoiding the wind · Buffer spaces

Building type

Zhangjiakou regional building design, its layout should be compact; the overall shape of the building should be neat and coordinated, minimise the complex plane shape and facade convex and concave changes in the processing; appropriate increase in building depth, increase the area of south-facing windows, etc., through the above measures to make the building better adapt to the regional characteristics of Zhangjiakou.

General layout of building

High · Middle · Low · Walls · Course of events

From the overall layout of the building, the north-west side of the building should be relatively high, and the south-east side of the building should be relatively low, in order to resist the cold wind from the north-west and to receive more sunshine; under the above conditions, the building layout should be enclosed by the north-west, and the south-east should be open, so as to reduce the sweeping range of the north-west cold wind and to receive more sunshine. At the same time, advanced technologies such as intelligent facade systems, solar panels, roof light-sensitive devices, adjustable louvres, wind power generation, etc. can also be adopted to achieve the purpose of improving the microdimate environment of the building and increasing the degree of comfort.

Roof Angle Simulation

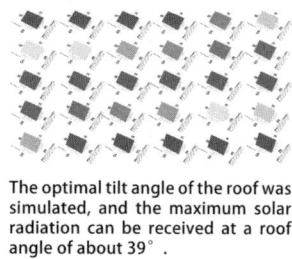

The optimal tilt angle of the roof was simulated, and the maximum solar radiation can be received at a roof angle of about 39°.

Diagram of the Design Process

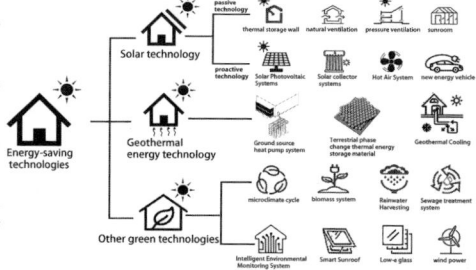

Solar technology — passive technology: thermal storage wall, natural ventilation, pressure ventilation, sunroom; proactive technology: Solar Photovoltaic Systems, Solar collector systems, Hot Air System, new energy vehicle

Geothermal energy technology — Ground source heat pump system, Terrestrial phase change thermal energy storage material, Geothermal Cooling

Energy-saving technologies

Other green technologies — microclimate cycle, biomass system, Rainwater Harvesting, Sewage treatment system, Intelligent Environmental Monitoring System, Smart Sunroof, Low-e glass, wind power

向阳而生

Born to the Sun

Design of Public Service Center in Erpuzi Village, Hebei
河北二堡子村公共服务中心设计 03

■ Enwegy Calculation

Power Consumption Calculation

Based on indicators of electrical loads in residential buildings,
Hourly electricity consumption in small public buildings:40-80w/㎡
Hourly electricity consumption in B&Bs: 40-70w/㎡.
Based on a public service center with an area of 1227 ㎡, the size of the
B&B is 621 ㎡,the average length of electricity use is 8 hours

Annual electricity consumption of public service centers:
    1127 ㎡×60w/㎡×8h×365=214970.4 kW·h

Annual electricity consumption of B&Bs:
    621 ㎡×55w/㎡×8h×365=99732.6 kW·h

Total annual electricity consumption: 314703 kW·h

Calculation of Power Generation

Based on the efficiency of solar power generation,Power generation efficiency
of solar photovoltaic panels:200w/㎡,Power Generation Efficiency of Thin Film
Photovoltaics:62w/㎡.
Based on a solar panel area is 856.92 ㎡,photovoltaic thin film area is 240.31 ㎡:

Annual electricity generation from solar panels:
    1739.12kW·h/㎡×200w/㎡×856.92m²=298057.342 kW·h

Photovoltaic thin-film annual power generation:
    1739.12kW·h/㎡×62w/㎡×240.31m²=25866.834kW·h

Annual power generation: 323924 kW·h>314703kW·h

conclusion:
The power generation capacity of solar photovoltaic panels and thin-film
photovoltaics basically meets the electricity demand of public service centers
and lodgings, realizing building self-sustainability.

■ Design Ideas

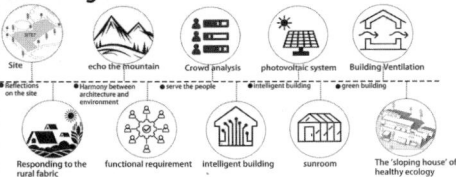

■ Design Process

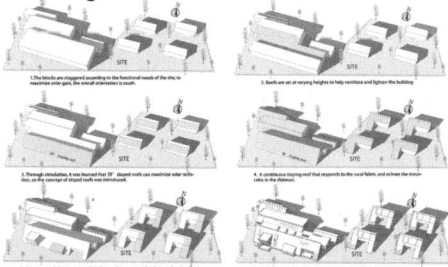

■ West Elevation

■ South Elevation

110088

向阳而生
Born to the Sun

Design of Public Service Center in Erpuzi Village, Hebei 河北二堡子村公共服务中心设计 04

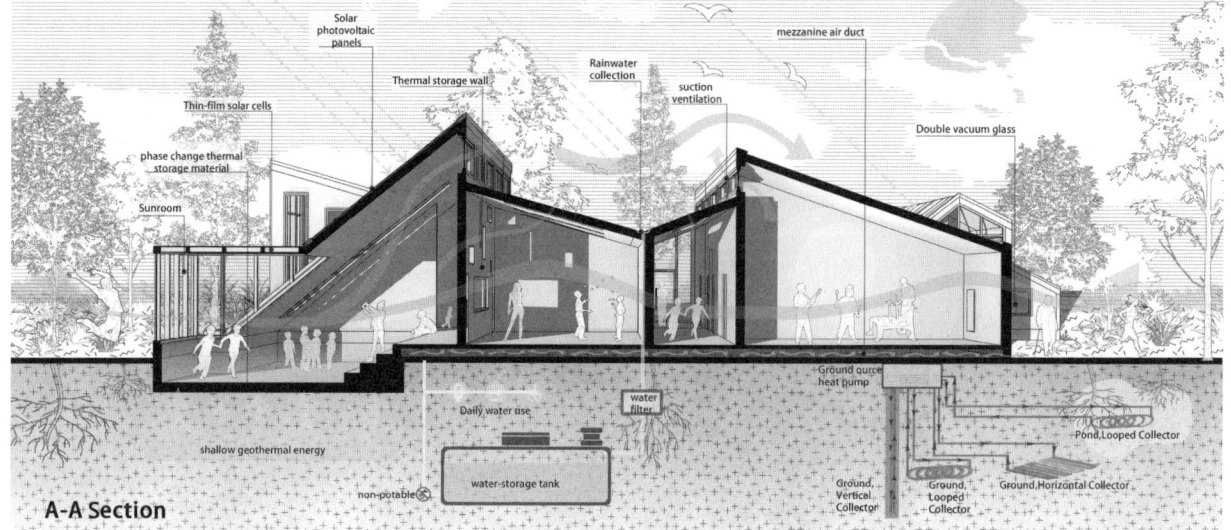

A-A Section

Labels in section: Solar photovoltaic panels; Thin-film solar cells; phase change thermal storage material; Sunroom; Thermal storage wall; Rainwater collection; suction ventilation; mezzanine air duct; Double vacuum glass; Ground qurce heat pump; Daily water use; water filter; shallow geothermal energy; water-storage tank; non-potable; Pond,Looped Collector; Ground,Vertical Collector; Ground,Looped Collector; Ground,Horizontal Collector

Technological Strategy

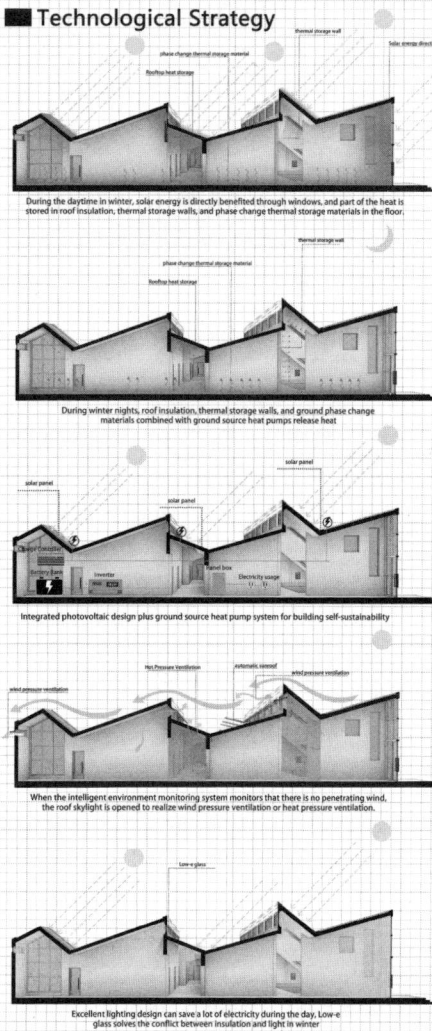

During the daytime in winter, solar energy is directly benefited through windows, and part of the heat is stored in roof insulation, thermal storage walls, and phase change thermal storage materials in the floor.

During winter nights, roof insulation, thermal storage walls, and ground phase change materials combined with ground source heat pumps release heat

Integrated photovoltaic design plus ground source heat pump system for building self-sustainability

When the intelligent environment monitoring system monitors that there is no penetrating wind, the roof skylight is opened to realize wind pressure ventilation or heat pressure ventilation.

Excellent lighting design can save a lot of electricity during the day. Low-e glass solves the conflict between insulation and light in winter

Exploded Diagram

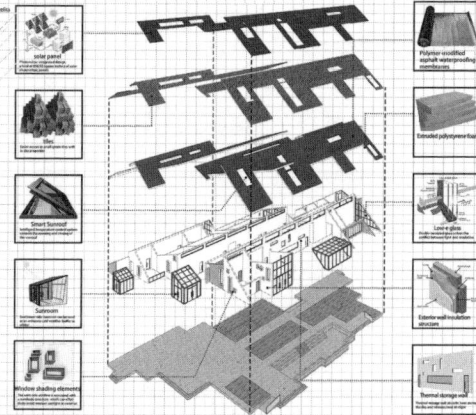

Lighting Analysis

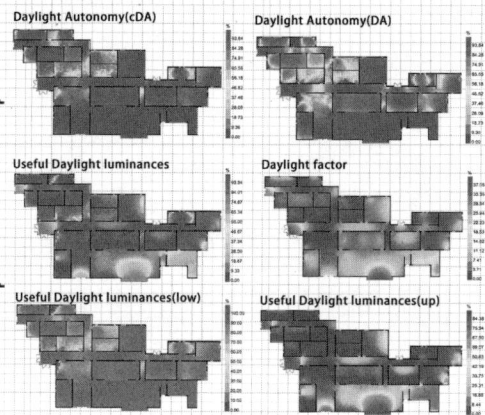

Daylight Autonomy(cDA)

Daylight Autonomy(DA)

Useful Daylight luminances

Daylight factor

Useful Daylight luminances(low)

Useful Daylight luminances(up)

The simulation shows that without artificial lighting, the spaces in the building are able to meet the lighting needs for a certain period of time, and the interior spaces are naturally well lit. Reduces reliance on artificial lighting and improves energy efficiency performance.

Wind Environment Simulation

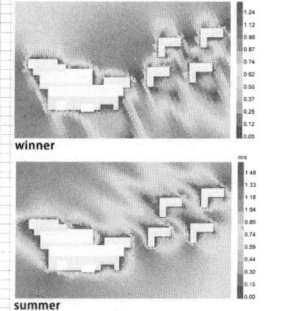

winner

summer

Sunshine Simulation

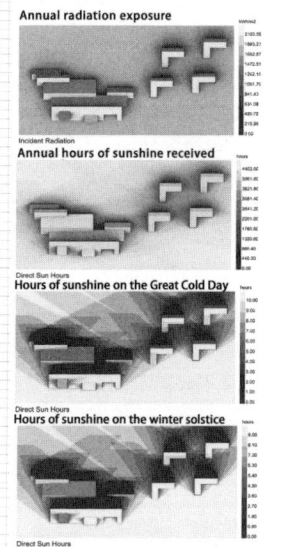

Annual radiation exposure

Incident Radiation

Annual hours of sunshine received

Hours of sunshine on the Great Cold Day

Direct Sun Hours

Hours of sunshine on the winter solstice

Direct Sun Hours

The simulation shows that the building can receive abundant solar energy resources to meet the demand of photovoltaic power generation. And it meets the basic sunshine requirements of the big cold day and winter solstice day

编号：110170

稻香绿苑·1
Environmentally Friendly Yard with Rice Fragrance

Design Description
设计从社会、生态、经济和技术层面充分探索二堡子村韧性的可能性，形成四个层面相互耦合的以特色稻田种植的设计方案，并在此基础上形成一套完备的储水储电储粮系统，场地公共建筑还可以变化为村民应急避难所。

设计使二堡子村在岁月静好时可以利用稻田种植业吸引游客前来，形成旅游产业，改善当地生活，在灾害来临时可以到场地进行避难，维持正常的供水供电和供粮，等待救援和home重建。

The design fully explores the possibility of resilience in Erbaozi Village from the perspectives of society, ecology, economy, and technology, forming a design scheme for characteristic rice field planting that is coupled at four levels. Based on this, a complete water, electricity, and grain storage system is formed, and the public buildings on the site can also be transformed into emergency shelters for villagers.

The design enables Erbaozi Village to attract tourists through rice paddy planting during peaceful times, forming a tourism industry and improving local living standards. In the event of disasters, the village can seek refuge on the site, maintain normal water and electricity supply, and food supply, waiting for rescue and home reconstruction.

Resilience Strategy

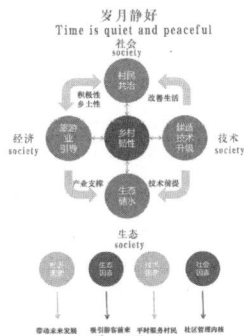

岁月静好 Time is quiet and peaceful
社会 society / 经济 society / 技术 society / 生态 society

灾害来临 Disaster is coming
一套完备应急供电供水供粮系统 Energy storage technology
储水技术 Water storage technology
储粮系统 Grain storage system
场地功能可变化为村民应急避难所 The site function can be changed to become an emergency shelter for villagers
空间 Space / 交通 Traffic / 建筑 Architecture / 能源 Energy

Climate Analysis
全年温度/Annual temperature / 相对湿度/relative humidity / 焓湿图/Psychrometric Chart
逐月风玫瑰图/Monthly Wind Rose Chart
日照与建筑朝向/White light and building orientation
风向与建筑高度 Wind direction and building height
风向季节变化 Seasonal variation of wind direction
风速变化分析 Analysis of Wind Speed Changes
建筑布局 architectural composition

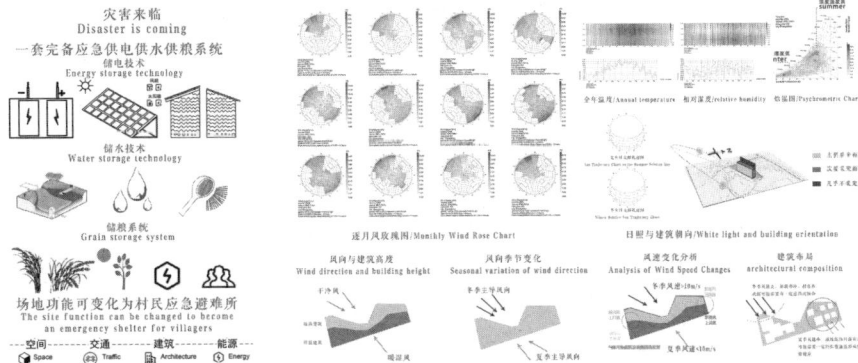

General Layout
POOL / POOL
Second Entrance / Paddy / Logistic Entrance / Main Entrance
经济技术指标

Analysis of General Layout

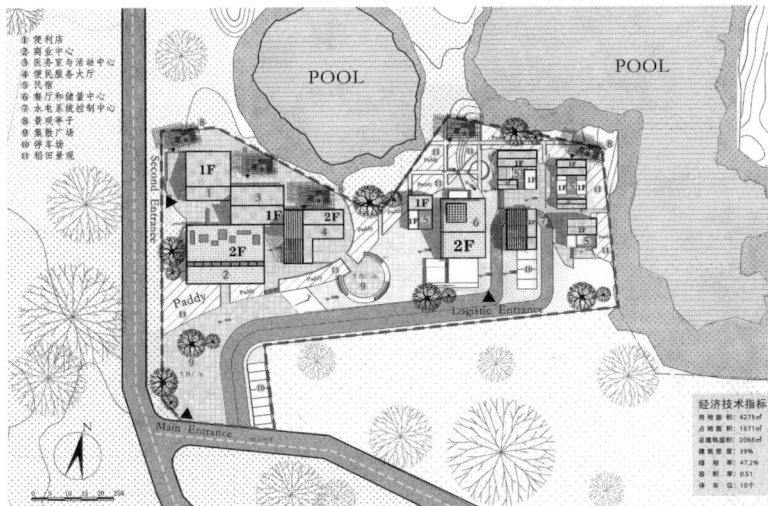

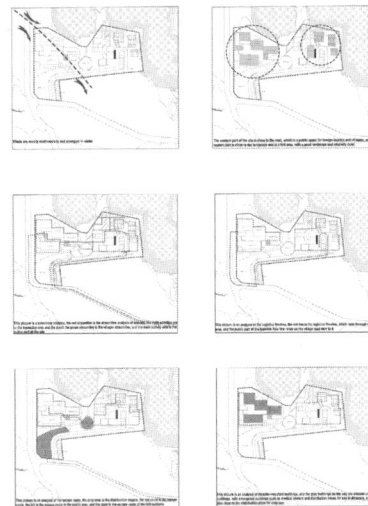

综合奖·入围奖·张家口市怀安县二堡子村建设项目
Comprehensive Awards – Nomination – Erpuzi Village Project, Huai'an County, Zhangjiakou City

注册号：110170
Register Number：110170
项目名称：稻香绿苑
Entry Title：Environmentally Friendly Yard with Rice Fragrance
作者：段有斌、刘志敏、唐诗意、李成才、陈浩、王思凡
Authors：Duan Youbin, Liu Zhimin, Tang Shiyi, Li Chengcai, Chen Hao, and Wang Sifan
作者单位：昆明理工大学
Authors from：Kunming University of Science and Technology
指导教师：陆莹、毛志睿
Tutors：Lu Ying, Mao Zhirui
指导教师单位：昆明理工大学
Tutors from：Kunming University of Science and Technology

编号：110170

稻香绿苑·II
Environmentally Friendly Yard with Rice Fragrance

建筑功能 Building Function		老年活动中心 Senior activity Centre	
a商业中心 a.Business Center		e1.会议室 e1.Meeting Room	
a1.餐饮 a1.Food and Beverage		e2.棋牌室 e2.Chess room	
a2.休息厅 a2.Lounge		e3.活动室 e3.Activity room	
a3.手工制作 a3.Handmade		f.室外凉亭 f.Outdoor pavilion	
a4.手工展示 a4.Manual display		村民防灾减灾宣传栏 Earthquake prevention and disaster prevention publicity solven for villagers	
a5.特色售卖 a5.Handmade		F1.入口 F1.Entrance	
a6.村史馆 a6.Village History Hall		F2.便民服务大厅 F2.Convenience Service Hall	
a7.村民活动中心 a7.Village Product Exhibition		F3.村民会议室 F3.Village Meeting Room	
a8.休息厅 a8.The Lounge		F4.村级信息服务站 F4.Village Information service station	
b.便利店 b.Convenience stores		F5.办公 F5.Office	
b1.服务台 b1.Information Desk		g.水务系统控制中心 g.Hydropower system control center	
b2.货站 b2.Cargo station		g1.安防 g1.Security booth	
b3.书吧 b3.Book bar		g2.水务系统控制平台 g2.Hydropower system control platform	
c.卫生所 c.Toilet		g3.屋顶花园 g3.Roof garden	
c1.母婴室 c1.Mother and baby room		h.粮食储藏室 h.Grain and storage room	
c2.储藏室 c2.Storage room		h1.厨房 h1.Kitchen	
c3.配药 c3.Med-ical slicing		h2.外卖区 h2.Takeout counter	
d.服务台 d.Service desk		h3.就餐区 h3.Dining area	
d1.候诊区 d1.Waiting area		h4.私人观景 h4.Private rooms	
d2.问诊室 d2.Ask the consulting room		h5.私人观景室 h5.Private viewing room	
d3.药品储藏室 d3.Medicine storage room		i.民宿 i.Homestay	
d4.配药区 d4.Medicine pick-up area		j.院落 j.Courtyard	
d5.点滴室 d5.Drip Room			

0 5 10 15 20 25m

■ Analysis of Residential Courtyards

冀北怀安院落民居
Courtyard Residential Buildings in Northern Hebei

平面类型 Plane Type	"口"字院 character courtyard	"日"字院 character courtyard	"目"字院 character courtyard
原始民居 Primitive dwellings			
民居形制 Residential form			
设计演变 Design Evolution			

场地气候适应性民宿设计
Climate adaptive homestay design for the venue

院落类型 courtyard Type	稻香小院 fragrance of rice	宜居小院 liveable	舒逸小院 Comfortable	畅谈小院 chatting
民宿空间 Homestay Space				

民宿生成 Homestay generation

模数生成 modulus　功能划分 Functional division　节能瓦屋面 Energy saving tile roof　庭院通风换气 Courtyard ventilation

集水装置 Water collection device　积雨雨径 Rainy path　屋顶积光 Roof light accumulation　屋顶防冰雹 Roof hail prevention

■ Analysis of Site homestay

i1:稻香小院
观景客卧　北门稻香风光
闲暇小憩　围合庭院

i2:宜居小院
室外棋室　冬正房、夏南房
北卧室稻田景观　围合庭院

i3:舒逸小院
观景卧室　北面稻田
室外棋室　东正房、夏南房

i4:畅谈小院
屋顶花园　二层楼梯
稻田引入
公共餐厅

稻香绿苑·III
Environmentally Friendly Yard with Rice Fragrance

■ Energy Supply System and Structural Analysis

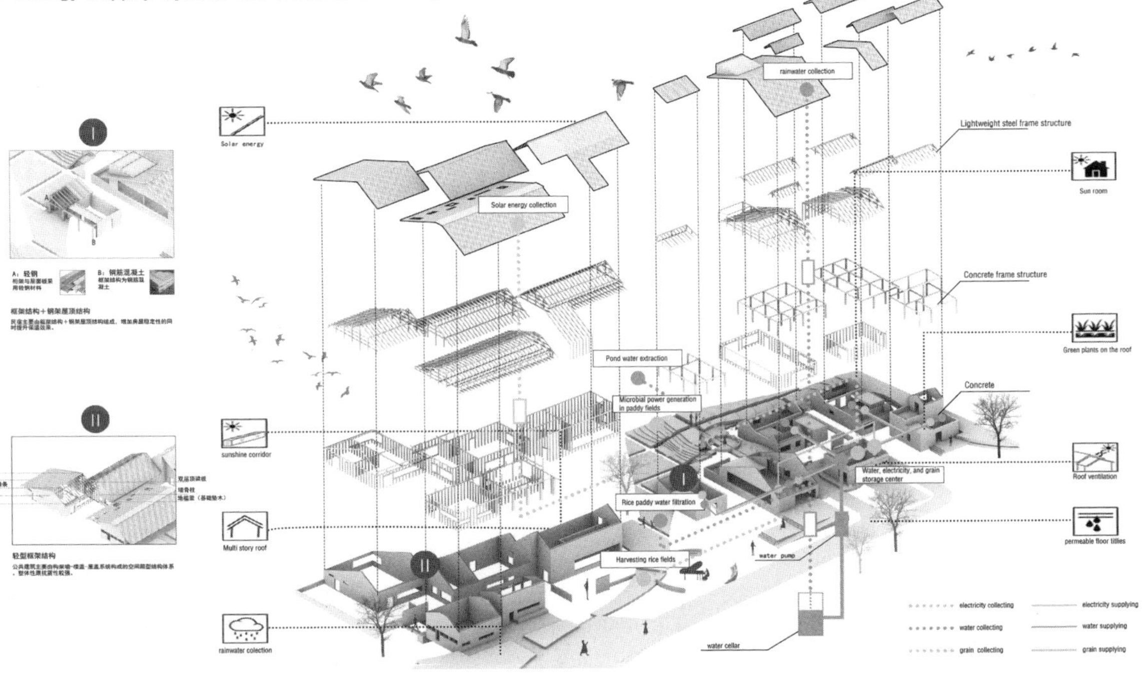

■ Section 2-2 and Passive Energy Supply System Analysis

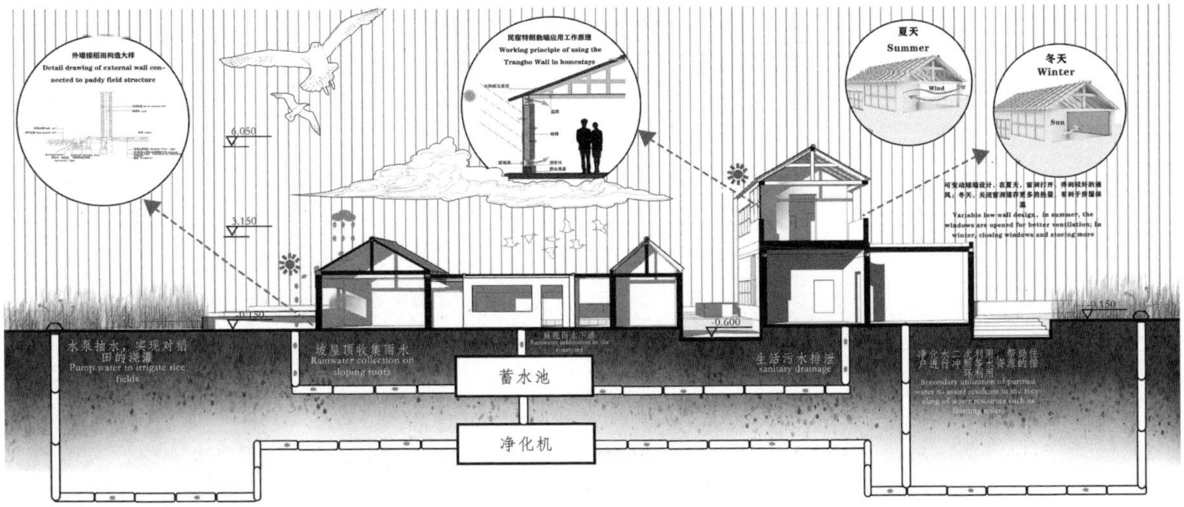

蓄水池

净化机

■ Festival Rendering

播种插秧节
Planting and Transplanting Festival

稻田丰收节
Rice Harvest Festival

冬季雪景节
Winter Snow Festival

稻香绿苑·IV
Environmentally Friendly Yard with Rice Fragrance

■ Section 1-1 and Analysis of Active Energy Supply System

屋檐出挑，可以避免太阳光直射到屋内，同时使太阳光经折射进入下沿较大的窗口，增加采光，保证室内舒适程度
The eaves are protruding, which can prevent direct sunlight from entering the room and refract it into the larger window at the bottom, increasing sunlight and ensuring indoor comfort

建筑采用蔗架结构，在屋顶上采用新型太阳能瓦片，经过特殊构造搭接，形成完整的太阳能瓦屋顶，在白天太阳充足的时候进行太阳能吸收，并最终实现对太阳能的利用
The building adopts a steel frame structure and new solar tiles are used on the roof. Through special construction and overlapping, a complete solar tile roof is formed, which absorbs solar energy during the day when the sun is abundant, and ultimately realizes the utilization of solar energy

连续的大坡屋顶实现对太阳能技术的大面积铺设，避免太阳光的直射，同时在屋顶进行断续的开天窗，进行采光
Continuous sloping roofs enable large-scale installation of solar energy technology to avoid direct sunlight, while intermittent skylights are installed on the roof for natural lighting

①.屋顶坡度分析 Roof Slope Analysis

②.屋顶挂瓦大样 Detailed Drawing of Roof Tile Hanging

③.屋顶太阳能瓦大样 Roof Solar Tile Sample

■ Seismic Analysis and Energy-saving Calculation

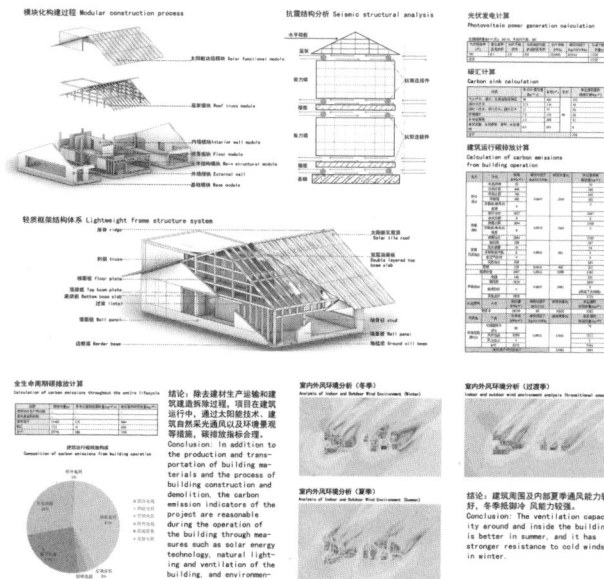

模块化构造过程 Modular construction process

抗震结构分析 Seismic structural analysis

光伏发电计算 Photovoltaic power generation calculation

碳汇计算 Carbon sink calculation

建筑运行碳排放计算 Calculation of carbon emissions from building operation

轻质框架结构体系 Lightweight frame structure system

全生命周期碳排放计算 Calculation of carbon emissions throughout the entire lifecycle

结论：除去建材生产运输和建筑建造拆除过程，项目去建筑运行中，通过太阳能技术、建筑自然采光通风以及环境景观等措施，碳排放指标综合理。
Conclusion: In addition to the production and transportation of building materials and the process of building construction and demolition, the carbon emission indicators of the project are reasonable during the operation of the building through measures such as solar energy technology, natural lighting and ventilation of the building, and environmen-

室内外风环境分析（冬季）
Analysis of Indoor and Outdoor Wind Environment (Winter)

室内外风环境分析（过渡季）
Indoor and outdoor wind environment analysis (transitional season)

室内外风环境分析（夏季）
Analysis of Indoor and Outdoor Wind Environment (Summer)

结论：建筑周围及内部夏季通风能力较好，冬季抵御冷风能力较强。
Conclusion: The ventilation capacity around and inside the building is better in summer, and it has stronger resistance to cold winds in winter.

■ South Elevation View

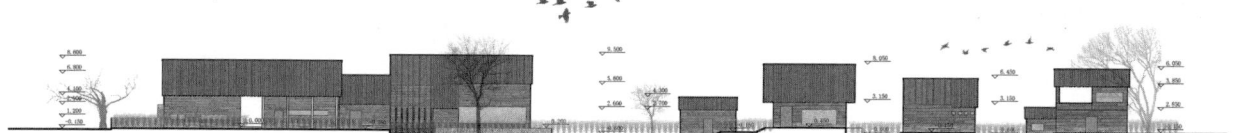

设计说明 DESIGN SPECIFICATION

基于严寒地区气候特征，并对防灾减灾及乡村韧性等问题思考，依据建筑节能的基本要求，以"联结"为设计出发点，自一而多，建立韧性与节能联系。

设计基于"构件-模块-箱体-建筑"系统组合设计策略，将围护构件集成模块化以方便快速搭建，以应对紧急情况；利用丰富的太阳能资源等构建能源核，以方便快速插拔、及时供电；同时利用策略可快速组装优势，植入复合腔体，通过对腔体位置、功能变化，提高建筑的舒适性，最后结合主动式技术的整体系统达到建筑的低碳节能。

Based on the climate characteristics of cold regions, and on the issues of disaster-prevention and reduction and rural resilience, according to the basic requirements of building energy conservation, with "connection" as the design starting point, from one to many, to establish the connection between toughness and energy conservation. The design is based on the system combination design strategy of "component-module-box-building", and the enclosure components are integrated and modular to facilitate quick construction and cope with emergency situations. Make use of abundant solar energy resources to build energy cores, so as to facilitate quick insertion and timely power supply; At the same time, the strategy can be used to quickly assemble the advantages, implant the composite chamber, improve the comfort of the building by changing the position and function of the chamber, and finally combine the overall system of active technology to achieve low carbon and energy saving of the building.

综合奖·入围奖·张家口市怀安县二堡子村建设项目
Comprehensive Awards – Nomination – Erpuzi Village Project，Huai'an County，Zhangjiakou City

注册号：110233
Register Number：110233
项目名称：稻野织链
Entry Title：Weaving Wildness & Paddy Fields
作者：罗丽峰、徐嘉琳、余悦、武玉洁、毛穿有、齐修远
Authors：Luo Lifeng，Xu Jialin，Yu Yue，Wu Yujie，Mao Chuanyou，and Qi Xiuyuan
作者单位：苏州大学
Authors from：Suzhou University
指导教师：韩冬辰、孙磊磊、吴国栋
Tutors：Han Dongchen，Sun Leilei，and Wu Guodong
指导教师单位：苏州大学
Tutors from：Suzhou University

• CONCEPT

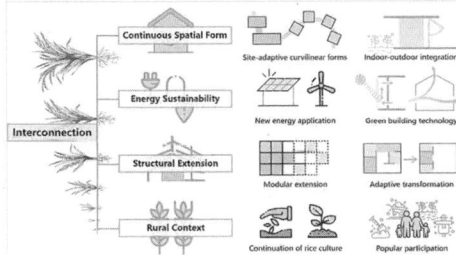

• SITE ANALYSIS

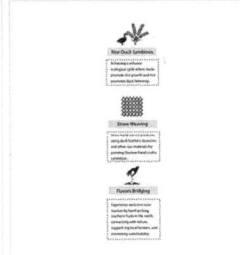

• CLIMATIC ANALYIS

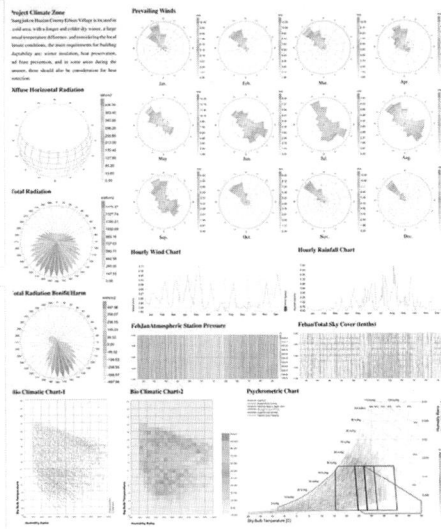

• TECHNOLOGY

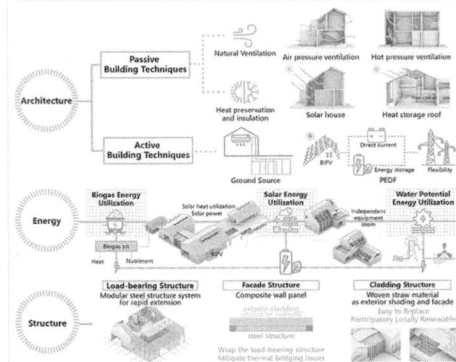

• RESILIENCE

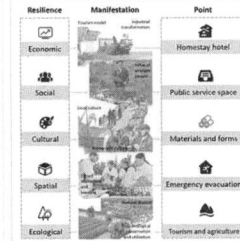

• PROJRCT DESIGN

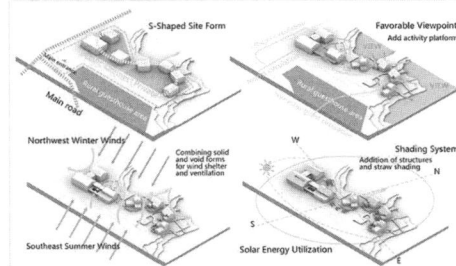

SITE PLAN

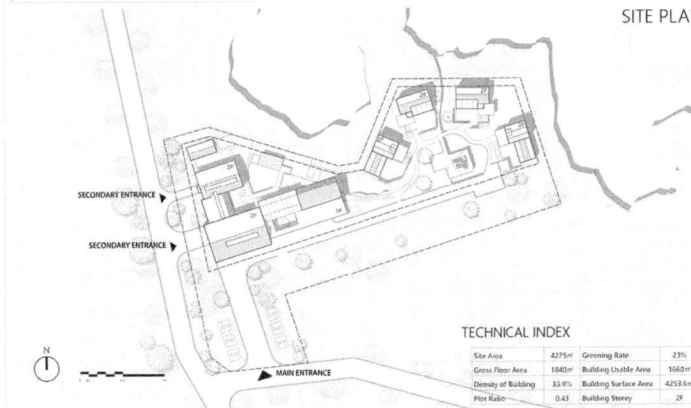

TECHNICAL INDEX

Site Area	4275m²	Greening Rate	23%
Gross Floor Area	1840m²	Building Usable Area	1660m²
Density of Building	33.9%	Building Surface Area	4253.6㎡
Plot Ratio	0.43	Building Storey	2F

PLAN

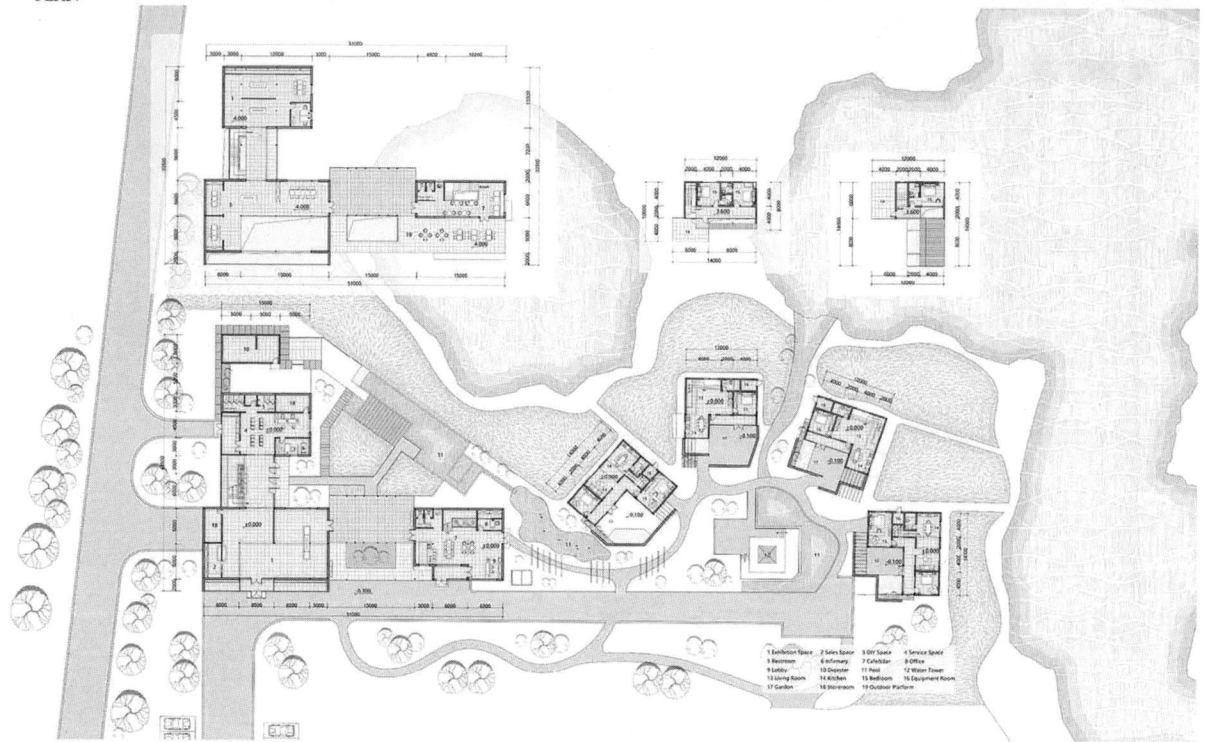

1 Exhibition Space 2 Sales Space 3 DIY Space 4 Service Space
5 Restroom 6 Infirmary 7 Cafe&Bar 8 Office
9 Lobby 10 Depoter 11 Pool 12 Water Tower
13 Living Room 14 Kitchen 15 Bedroom 16 Equipment Room
17 Carbin 18 Storeroom 19 Outdoor Platform

Regular	Emergencies	Society	Ecology	Economy

Regular

Service Space

Cafe&Bar

Exhibition Space

Expansion Process

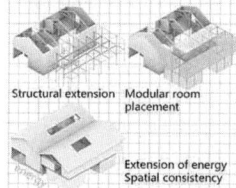

Structural extension   Modular room placement

Extension of energy
Spatial consistency

Emergencies

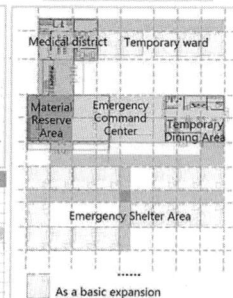

Medical district   Temporary ward

Material Reserve Area   Emergency Command Center   Temporary Dining Area

Emergency Shelter Area

As a basic expansion

Through structural extensions and continuous energy connections, quickly expand emergency spaces that can accommodate approximately 400 people from the entire village for temporary shelter in emergencies.

Society

Sociality: Villagers participate
Villagers participate in construction and maintenance work.

Interactivity: Activation of crowd vitality
Villagers interact with tourists to enhance crowd interaction and establish developmental relationships.

Provinciality: Continuation of human relations
Courtyard-style residential layout

Ecology

Localized Building Materials
Handwoven straw as part of the building enclosure and facade materials.

Low-Carbon   Accessible

Ecological Planting
Wetland conservation and sustainable agriculture go hand in hand.

Rice-Duck Symbiosis
Achieving a virtuous ecological cycle where ducks promote rice growth and rice promotes duck fattening.

Economy

Cultural Monetization
Establish a development model integrating rice field art, cultural display, memory homestay, leisure and sight-seeing

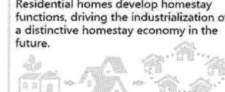

Experience   Product   Living

Enterprise Cooperation
Enterprises enter to drive industrial transformation

B&b economic development
Residential homes develop homestay functions, driving the industrialization of a distinctive homestay economy in the future.

Prefabricated Design Strategy

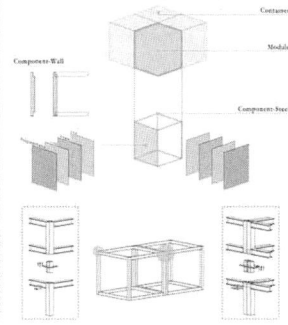

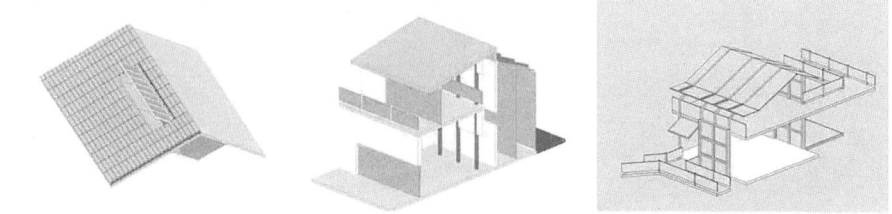

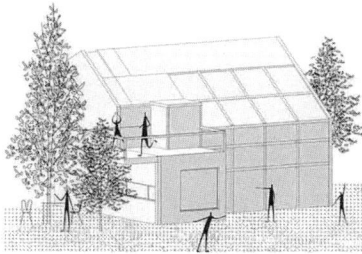

Container

Module

Components-Wall

Components-Steel

Buffer Space

Regulate indoor microclimate through buffer spaces to reduce outdoor climate impacts and achieve thermal insulation

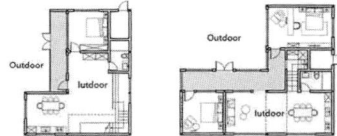

Outdoor

lutdoor

Outdoor

lutdoor

Solar House

Open the skylight

in summer

in winter

Open the skylight

Hot Pressure Ventilation

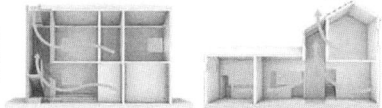

Air Pressure Ventilation

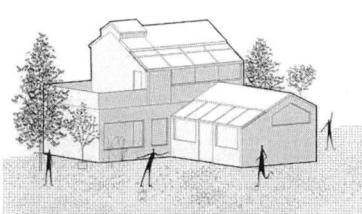

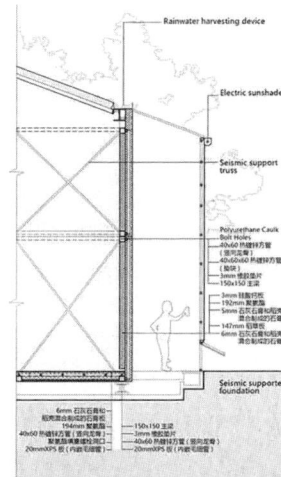

Rainwater harvesting device

Electric sunshade

Seismic support truss

Polyurethane Caulk
Bolt Holes
40x60 热镀锌方管（室内龙骨）
40x50x50 热镀锌方管（龙骨）
3mm 镀锌垫片
150x150 主梁

挂面找找找
192mm 聚苯板
5mm 石灰石膏和稻壳混合制成的石膏板
142mm 岩棉板
6mm 石灰石膏和稻壳混合制成的石膏卷材

Seismic supported foundation

6mm 石灰石膏和
稻壳混合制成的石膏板
194mm 聚苯板
40x60 热镀锌方管（室内龙骨）
聚氨酯填缝螺栓孔
20mmXPS 板（内侧龙骨端）

150x150 主梁
3mm 镀锌垫片
40x60 热镀锌方管（室内龙骨）
20mmXPS 板（内侧龙骨端）

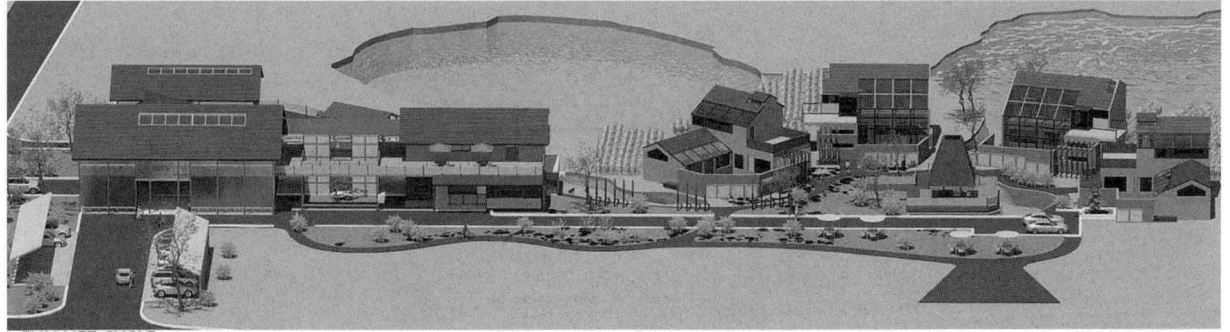

• FULLLIFE CYCLE

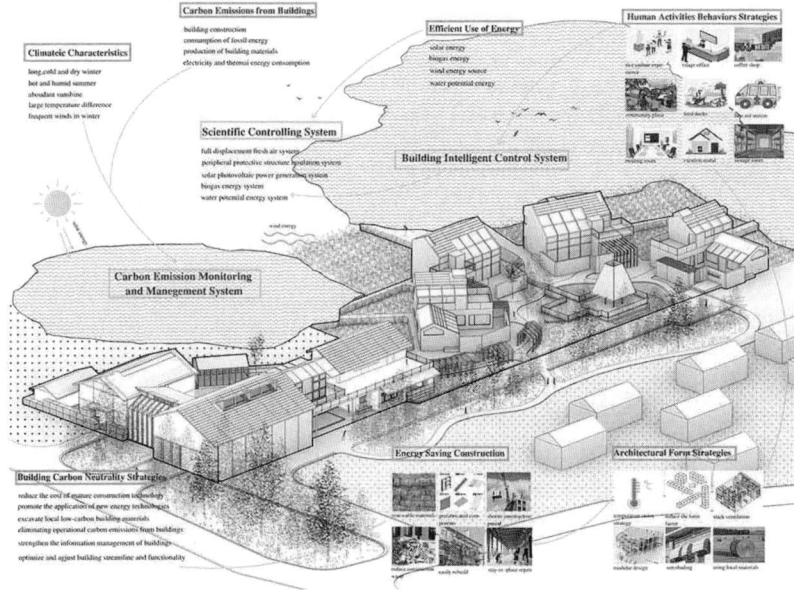

Climateic Characteristics
long,cold and dry winter
hot and humid summer
abundant sunshine
large temperature difference
frequent winds in winter

Carbon Emissions from Buildings
building construction
consumption of fossil energy
production of building materials
electricity and thermal energy consumption

Efficient Use of Energy
solar energy
biogas energy
wind energy source
water potential energy

Human Activities Behaviors Strategies

Scientific Controlling System
full-displacement fresh air system
peripheral protective structure insulation system
solar photovoltaic power generation system
biogas energy system
water potential energy system

Building Intelligent Control System

Carbon Emission Monitoring and Management System

Building Carbon Neutrality Strategies
reduce the cost of mature construction technology
promote the application of new energy technologies
excavate local low-carbon building materials
eliminating operational carbon emissions from buildings
strengthen the information management of buildings
optimize and adjust building streamline and functionality

Energy Saving Construction

Architectural Form Strategies

太阳能光伏发电计算结果
Solar Photovoltaic Power Generation Calculation Results

component count	component area (m²)	Total Horizontal Radiation(kwh/(m²·a))	System Power Generation Effiesency Rate(%)	Theoretical First-Year Power Output(mwh)	Theoretical Carbon Reduction(t)
359	588	1494.1	86	277.6	202.76

建筑使用阶段碳排放计算结果
Calculation results of carbon emissions during service stage

Project		Amount of carbon emissions (kgCO₂/a)	Amount of carbon emissions per unit area(kgCO₂/m²·a)
Carbon emission	HVAC	144272	4.28
	hot water	5950	0.178
	lighting	12275	1.56
	other system	54160	1.61
Carbon reduction	renewable energy	-158371	-4.70
	carbon sink	-15051	-0.446
Summation		43235	1.28

使用阶段可再生能产量设计计算结果
Calculation results of renewable energy production during service stage

Photovoltaics kWh/(m²a)	36.25
Solar power supply Warm/Cold kWh/(m²a)	21.06
Solar Thermal Collector kWh/(m²a)	18.7
Biomethane energy kWh/(m²a)	31.85

建筑能耗技术指标计算结果
Calculation results of building energy consumption technical indicators

Project	Figures
Comprehensive energy saving rate (%)	75.42
Utilization of renewable energy (%)	73.10
Energy efficiency rate of building body (%)	19.95
Heating cooling and lighting Average energy consumption index kWh/(m²a)	18.75
Reduction in carbon emission intensity (kgCO₂/m²a)	9.5

After software simulation and calculation, the project technology we designed can meet the requirements of Technical standard for nearly zero energy buildings. Therefore,we can achieve ultra-low energy consumption buildings.

• ENERGY

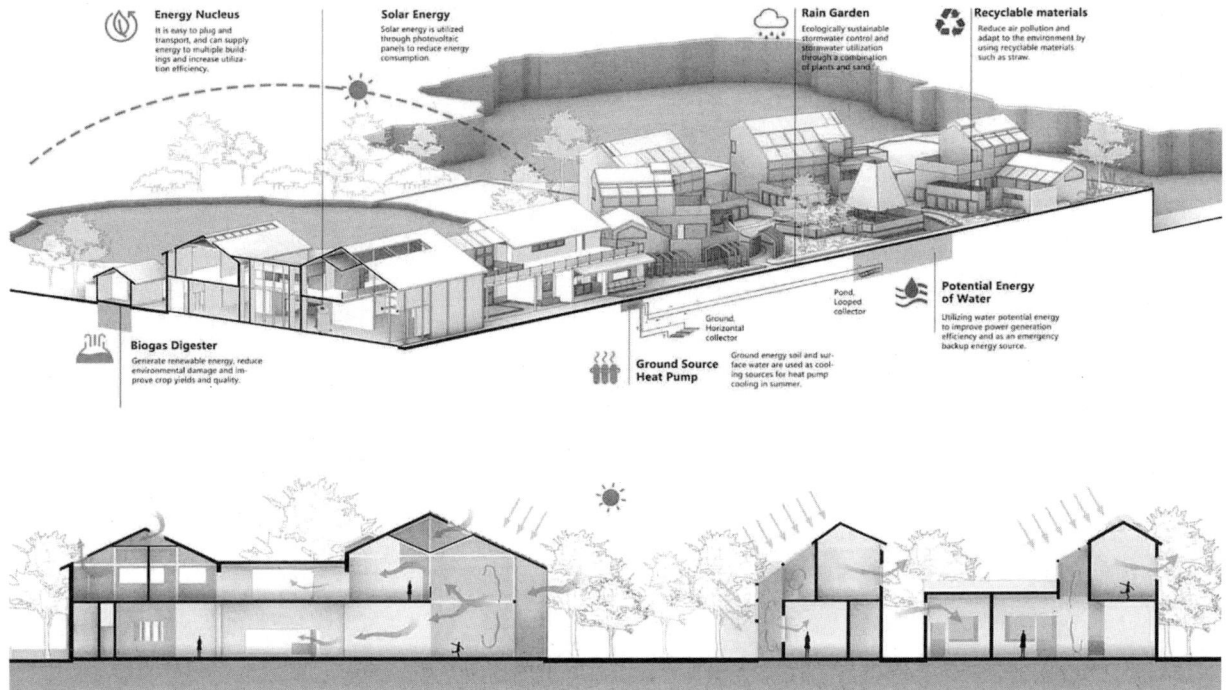

Energy Nucleus
It is easy to plug and transport, and can supply energy to multiple buildings and increase utilization efficiency.

Solar Energy
Solar energy is utilized through photovoltaic panels to reduce energy consumption.

Rain Garden
Ecologically sustainable stormwater control and stormwater utilization through a combination of plants and sand.

Recyclable materials
Reduce air pollution and adapt to the environment by using recyclable materials such as straw.

Biogas Digester
Generate renewable energy, reduce environmental damage and improve crop yields and quality.

Ground Source Heat Pump
Ground energy soil and surface water are used as cooling sources for heat pump cooling in summer.

Potential Energy of Water
Utilizing water potential energy to improve power generation efficiency and as an emergency backup energy source.

重构——传统体系下的二堡子村适应性设计 01
Reconstruct Adaptive Design of Erpuzi Village under Traditional System

Design statement

本设计位于河北省张家口市怀安县二堡子村内，为提升当地居民的居住品质，拟建设相关配套设施与商业民宿。设计以"韧性"提角出发，选择以传统的场地布局形式，迎合场地风向、日照条件，塑造可持续的生态建筑，通过传统建筑木构的变形来呼应场地原有的抗震、防冰雹的需求，提升整体建筑的抗风险性。传统木构以及坡屋顶的形式具有地域特色，且更够有效融入周边场地肌理，令建筑融于景，立于景，是对物理空间的低碳重构，为居民社交的场所赋能。

This design is located within Erpuzi Village, Huai'an County, Zhangjiakou City, Hebei Province, aiming to enhance the living quality of local residents through the construction of related supporting facilities and commercial homestays. The design adopts a "resilience" perspective, choosing a traditional site layout that aligns with the prevailing wind directions and sunlight conditions to create a sustainable ecological building. By transforming traditional wooden structures, the design responds to the existing requirements for earthquake resistance and hail protection, thereby increasing the overall resilience of the building.

The use of traditional wooden structures and pitched roofs not only reflects regional characteristics but also integrates effectively with the surrounding context, allowing the architecture to blend into and stand out as part of...

综合奖·入围奖·张家口市怀安县二堡子村建设项目
Comprehensive Awards － Nomination － Erpuzi Village Project，Huai'an County，Zhangjiakou City

注册号：110253
Register Number：110253
项目名称：重构——传统体系下的二堡子村适应性设计
Entry Title：Reconstruct Adaptive Design of Erpuzi Village under Traditional System
作者：赵楠、郭佳音、李珑玲
Authors：Zhao Nan, Guo Jiayin, and Li Longling
作者单位：西安建筑科技大学
Authors from：Xi'an University of Architecture and Technology
指导教师：李帆、周志菲
Tutors：Li Fan, Zhou Zhifei
指导教师单位：西安建筑科技大学
Tutors from：Xi'an University of Architecture and Technology

Status Analysis

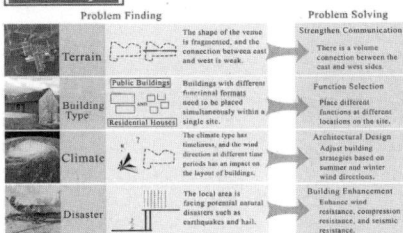

Strategic Analysis

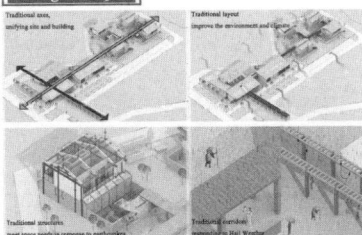

Site plan 1：500

Land Area	8260 ㎡	Building Density	19%
Construction Land Area	4275 ㎡	Number of Parking Spaces	14
Total Building Area	1891 ㎡	Homestay I Building Area	138 ㎡
Floor Area Ratio	0.23	Homestay II Building Area	205 ㎡
Green Space Ratio	45%	Country parlor Building Area	1200 ㎡

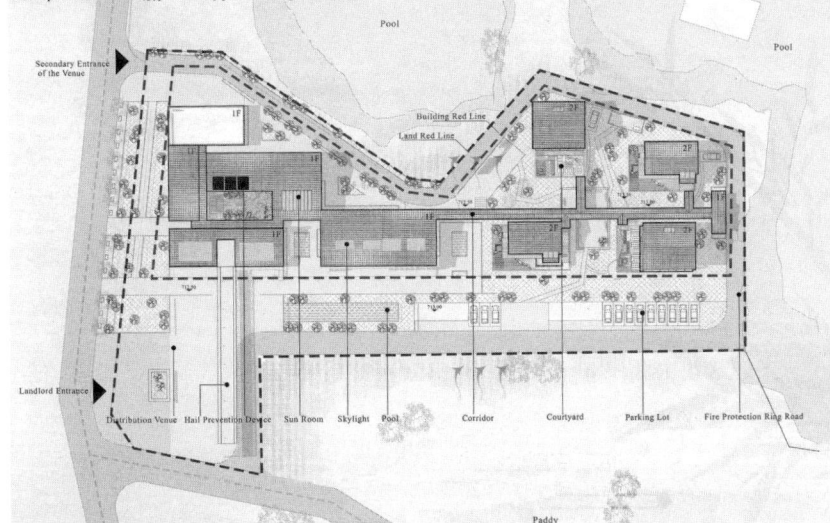

Meteorological Analyses

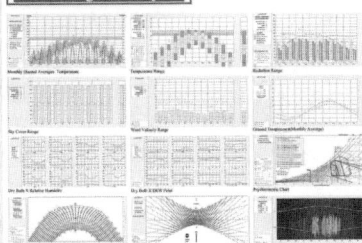

Rose of the Moon Wind

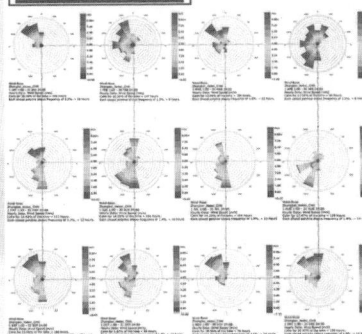

Ground Floor Plan 1:300

Lakes

Building Red Line

Land Red Line

2nd Floor Plan 1:300

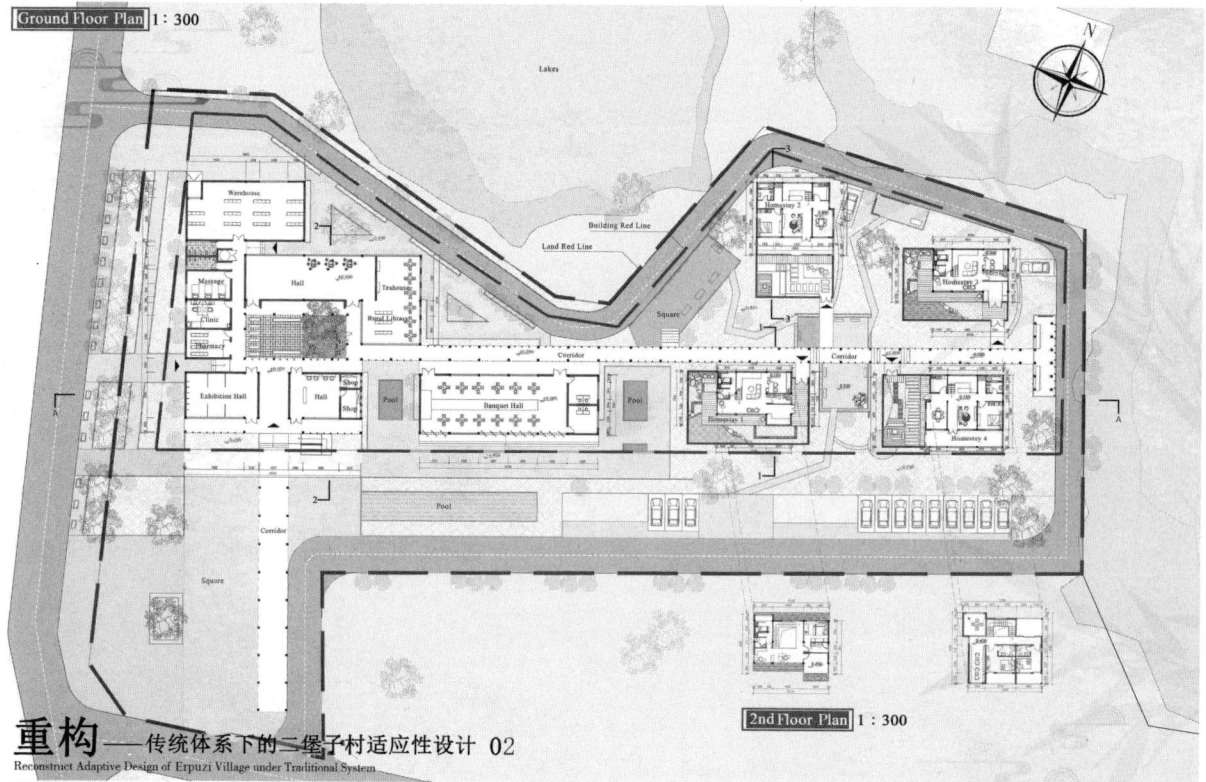

重构——传统体系下的二堡子村适应性设计 02
Reconstruct Adaptive Design of Erpuzi Village under Traditional System

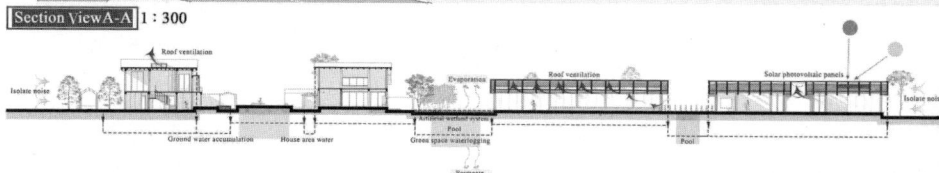

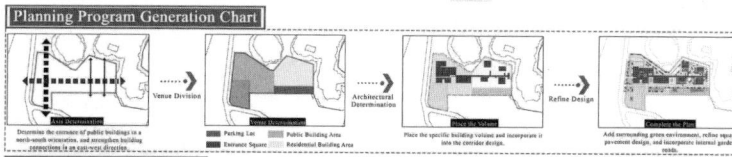

Section View A-A 1:300

Roof ventilation

Isolate noise

Roof ventilation

Evaporation

Roof ventilation

Solar photovoltaic panels

Isolate noise

Ground water accumulation

House area water

Artificial wetland overpass

Pool

Green space waterlogging

Pool

Fountain

Scene Effect Diagram

Planning Program Generation Chart

Venue Division

Architectural Determinations

Refine Design

Parking Lot   Public Building Area
Entrance Square   Residential Building Area

Site Suitability Analysis

A number of evacuation plazas have been set up on the site and are connected to the fire fighting pipeline to ensure that people can quickly reach a safe location in case of an earthquake.

The wind environment of the site is harmonized through measures such as ecological ponds, ecological landscaping and adjustable doors to make the environment suitable.

Solve the problem of light environment and thermal comfort of the site through solar photovoltaic panels, sunrooms, ecological hedges, solar power generators, etc.

Site Environmental Analysis

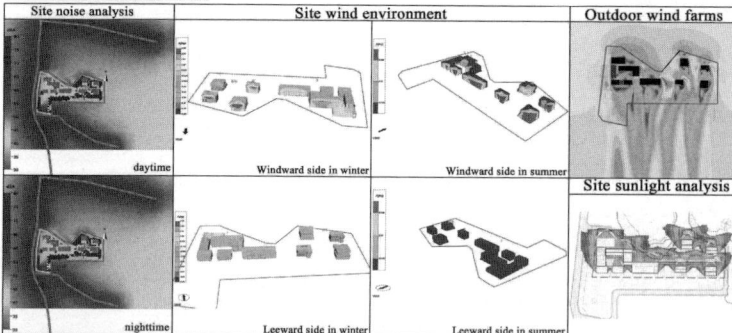

Site noise analysis	Site wind environment		Outdoor wind farms

daytime

nighttime

Windward side in winter

Leeward side in winter

Windward side in summer

Leeward side in summer

Site sunlight analysis

Elevation Plan 1:200

Country Parlor People View

重构——传统体系下的二堡子村适应性设计 03
Reconstruct Adaptive Design of Erpuzi Village under Traditional System

Country Parlor section view 2-2 | 1 : 200

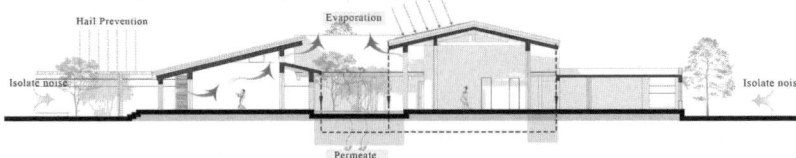

Hail Prevention
Evaporation
Isolate noise
Isolate noise
Permeate

Carbon Emission Indicator Table

A summary of the carbon emission calculation results of the equipment operation system in the design and building use phase

Project		Carbon emissions (kgCO₂e)	Carbon emissions (kgCO₂e/a)	Carbon emissions [kgCO₂e/(m²a)]
Carbon emission items	Heating system	1407316.24	28146.32	33.54
	Cooling system	82740	1654.8	1.97
	Transmission and distribution systems	97699.59	1954.8	2.33
	Lighting system	2696786.27	53953.73	64.27
Carbon reduction items	Photovoltaic power generation	4226144.42	84522.89	100.72
	Wind power	131.43	2.63	0
	Building carbon sinks	177095.00	3541.9	4.22
total		0	0	0

Notes:1. The above calculations are based on the floor area (public building)

Review of carbon intensity indicators during the building use phase

Project	numerical value	Standard requirements	Whether the requirements are met
Reduction in carbon intensity	41.06	≥7	Yes
Carbon Intensity Reduction Rate (%)	107.41%	≥40%	Yes
Conclusions of the review	The technology of this project meets the requirements of Article 2.0.3 of the General Code for Building Energy Conservation and Renewable Energy Utilization requirements		

A summary of the carbon emission calculation results of the equipment operation system in the design and building use phase

Project		Carbon emissions (kgCO₂e)	Carbon emissions (kgCO₂e/a)	Carbon emissions [kgCO₂e/(m²a)]
Carbon emission items	Cooling system	89630.33	1792.61	2.17
	Transmission and distribution systems	30199.74	603.99	0.73
	Domestic hot water system	95886.2	1901.72	2.3
	Lighting system	1007264.75	20145.29	24.36
Carbon reduction items	Photovoltaic power generation	816346.41	16326.93	19.74
	Wind power	109.54	2.19	0
	Building carbon sinks	57819	1156.38	1.4
total		347906.03	6958.12	8.41

Notes:1. The above calculations are based on the floor area (public building)

Review of carbon intensity indicators during the building use phase

Project	numerical value	Standard requirements	Whether the requirements are met
Reduction in carbon intensity (kgCO₂/m²a)	103.21	≥7	Yes
Carbon Intensity Reduction Rate (%)	95.83%	≥40%	Yes
Conclusions of the review	The technology of this project meets the requirements of Article 2.0.3 of the General Code for Building Energy Conservation and Renewable Energy Utilization requirements		

Indoor Environmental Analysis

Country Parlor

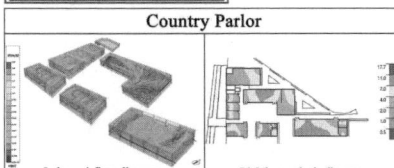

Indoor airflow diagram | Lighting analysis diagram

Homestay

Homestay I | Ground floor

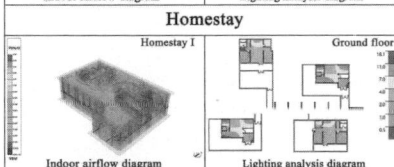

Indoor airflow diagram | Lighting analysis diagram

Homestay II | Second floor

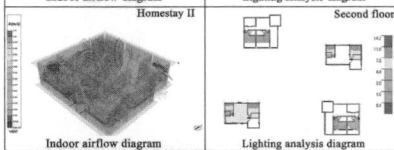

Indoor airflow diagram | Lighting analysis diagram

Earthquake Preparedness

Normal State	Disaster Prevention Status

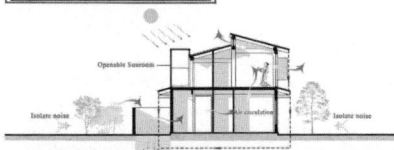

The building ensure daily leisure activities and even use of both large and small spaces.

Disaster strikes, the large space is used as a centralized shelter, and the original medical facilities are converted into first aid.

Homestay Technological Strategy

Solar radiation benefits	Rainwater Harvesting

Homestay I | Homestay I

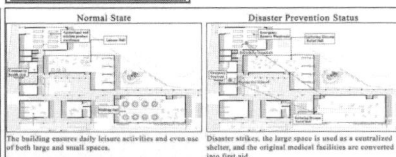

Homestay II | Homestay II

Construction Pattern Drawing | 1 : 5

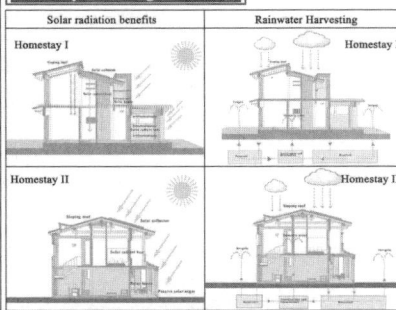

10-thick 1:2.5 cement mortar plastering
15-thick 1:3 cement mortar leveling
30 thick mineral wool insulation
15-thick cement white stone slurry scouring stone plastering
15-thick 1:3 cement mortar leveling

paper mache plaster
20 thick 1:3 lime mortar primed and applied in layerspaper mache plaster
20-thick 1:2.5 cement mortar plastering
100 thick concrete
100-thick triplex soil bedding-concrete rammer
25-thick 1:2 cement mortar skirting boards
±0.000

10-thick cement mortar plaster
60 thick concrete
60 thick crushed brick base concrete rammer

20 thick 1:3 cement mortar leveling, lay a layer of linoleum and then apply two layers of asphalt oil vapor barrier on top of the linoleum

Homestay I Section View 1-1 | 1 : 200

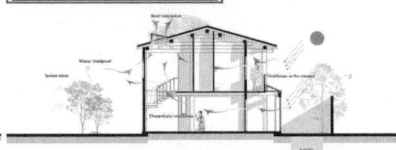

Operable louvers
Isolate noise
Isolate noise
Aquifer

Homestay II Section View 3-3 | 1 : 200

Homestay I People View

Homestay II People View

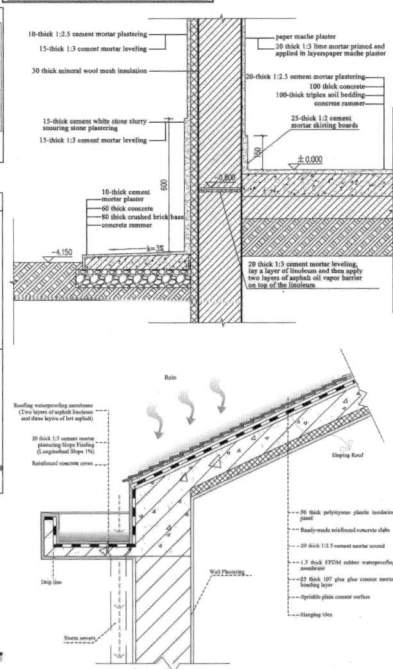

综合奖·入围奖·张家口市
怀安县二堡子村建设项目
Comprehensive Awards −
Nomination − Erpuzi Village
Project，Huai'an County，
Zhangjiakou City

注册号：110324
Register Number：110324
项目名称：激活·弹性·乐园
Entry Title：Activate！Elastic Paradise
作者：胡梓钺、徐静如
Authors：Hu Ziyue，Xu Jingru
作者单位：中国矿业大学
Authors from：China University of Mining
　　　　　　and Technology
指导教师：朱冬冬
Tutor：Zhu Dongdong
指导教师单位：中国矿业大学
Tutor from：China University of Mining
　　　　　　and Technology

编号：110324
张家口怀远县二堡子韧性乡村设计
Zhangjiakou City Huai'an County
Erpuzi resilience village design

激活·弹性·乐园 I
Activite！Elastic Paradise

经济技术指标：
总建筑面积：1618 平方米　绿地率：35%　容积率：0.07　建筑密度：0.043
设计说明：
　设计主题为激活·弹性·乐园，意在将其打造成具有应急快速可变性的村民活动中心和代表民宿的集合场地，在充分考虑当地气候、地形的基础上进行场地的规划设计。让成员具有台堡落地性的创新。在建筑技术方面，运用了模块化装配式的玻璃构件、可拆卸可组装可变形的应急集装箱等，结合冬季两季阳光房设计、太阳能光伏板的计算使用等太阳能利用方法，达到应急可变、绿色可持续、节约成本的效果。

1 Location

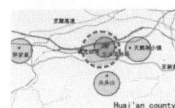

Huai'an county

2 Design Mind Map

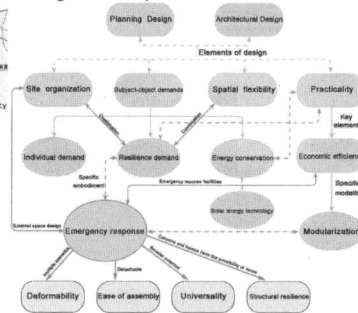

3 Site Zoning

We first analyzed the building and road features around the site to determine the main flow of people. The location of the building layout is determined according to the sunshine law, and the site road is planned according to the behavioral characteristics of the two types of building users.

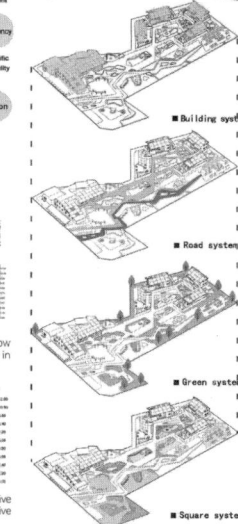

■ Building system
■ Road system
■ Green system
■ Square system

4 Schematic Design Description

The theme of the the design is activation, flexibility and paradise, aiming to build it into a villagers' activity center with emergency and rapid variability and a gathering site representing B&B. The planning and design of the site are carried out on the basis of fully considering the local climate and terrain to achieve innovation with reasonable landing. In terms of building technology, the use of modular assembly building components, detachable, assembler and deformable emergency containers, combined with winter and summer sunshine room design, solar photovoltaic panel calculation and use of solar energy utilization methods, to achieve emergency variable, green sustainable, cost-saving effects.

6 Scheme Testing

■ Wind environment analysis and determination

For the layout of the four homestays, we first made these four guesses, and placed them in the wind field of the prevailing wind direction in winter for simulation, to ensure that the outdoor activity venue has stable wind pressure in winter.

① Traditional determinant
② Along the northern boundary

For the traditional determinant, it is too uniform, and can not reflect the characteristics of the site layout.

This arrangement forms an open square in the residential area, making it difficult to correspond with public building.

③ Site-adjusted determinant
④ Final scheme

The lack of interconnectedness of homestays destroys the integrity. Meanwhile, the wind speed and wind pressure of the site are large, which is not conducive to residents' activities.

This scheme, not only encloses the central square serving the homestay, but also connects with the square of the public building. Meanwhile, it has the uniqueness of the site.

■ Sunshine analysis and inspection

After selecting the site layout, we simulated the winter solstice, summer solstice, and winter and summer sunshine to ensure that the southern homestay would not interfere with the northern homestays.

Summer　Summer solstice　Winter　Winter solstice

5 Climate Analysis

■ H-D chart

Psychrometric chart　Solar radiation H-D chart　Wind H-D chart

In Zhangjiakou, under conditions of relatively high humidity, temperatures are often concentrated below 10° C, especially in winter. This reflects the condensation of moisture in the air during the cold season, resulting in higher relative humidity.

■ Wind rose

Spring　Summer　Autumn　Winter

When solar radiation intensity is high (between 15 ° C and 25 ° C), it is usually accompanied by lower relative humidity. Reduced wind speed typically occurs during the warmer summer months (above 20 ° C), when relative humidity is lower, and with lower wind speeds, evaporation is not significant under stable air conditions.

7 Solar Landscape Perspectives

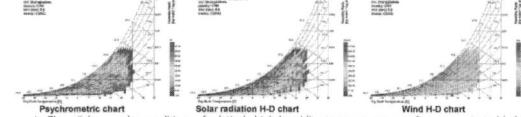

8 Elevations 1：200

Public building's west elevation　Public building's south elevation

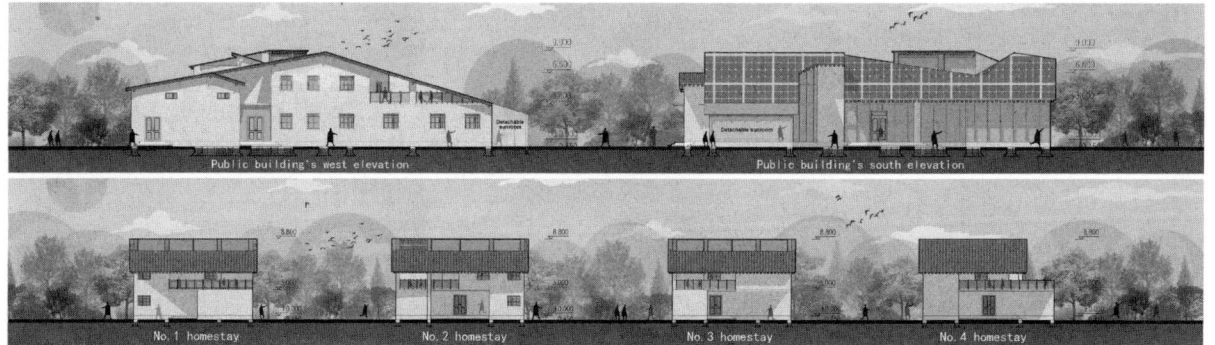

No. 1 homestay　No. 2 homestay　No. 3 homestay　No. 4 homestay

9 1st floor Plan 1 : 250

Sec-entrance

Main entrance

No. 2 Homestay

No. 4 Homestay

1 Motor vehicle parking
2 Non-motor vehicle parking
3 Emergency container

Site boundary

Building line

Sec-entrance
of site

Main entrance
of site

Main entrance
of site

Secondary entrance of site

N

10 General plan 1:800

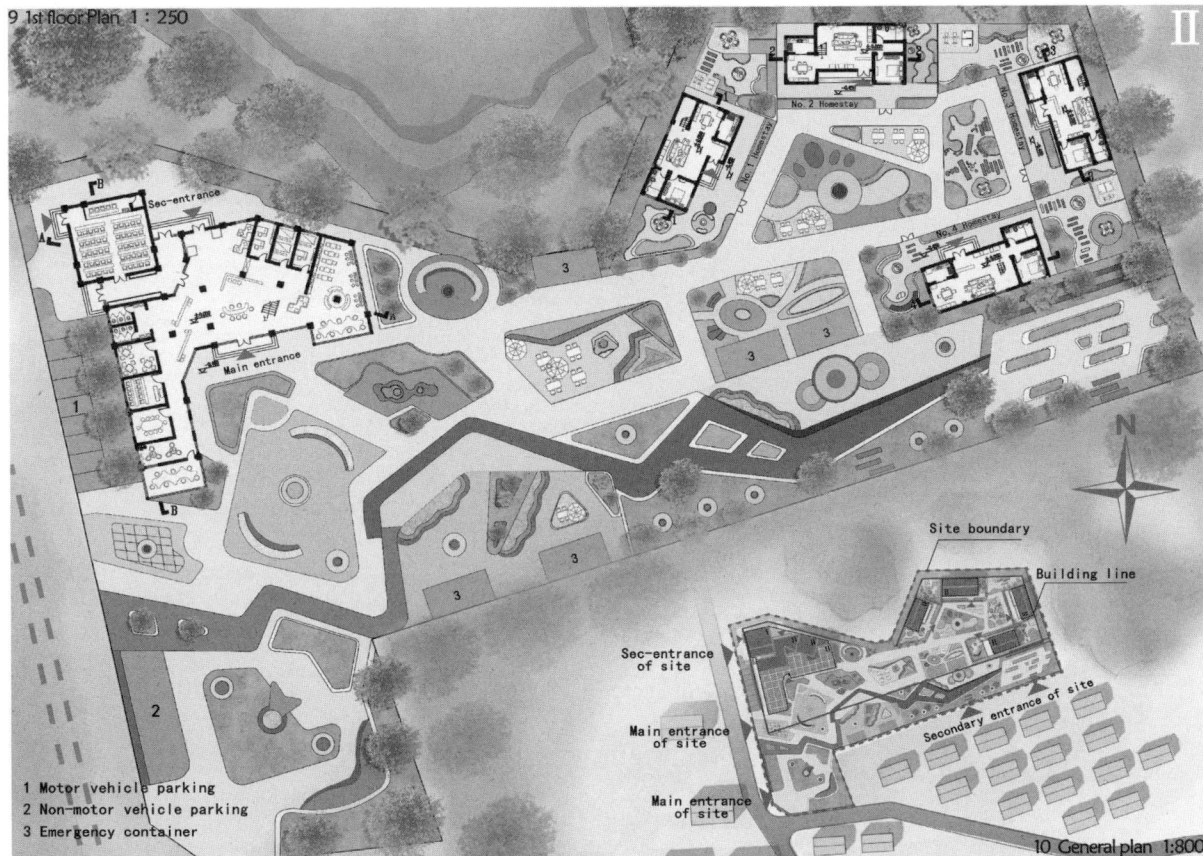

11 Public Building's 2nd Floor Plan 1 : 200

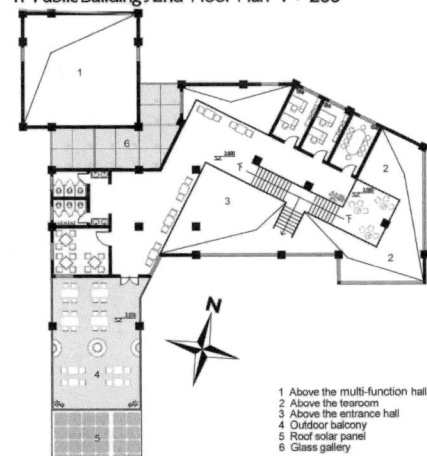

1 Above the multi-function hall
2 Above the tearoom
3 Above the entrance hall
4 Outdoor balcony
5 Roof solar panel
6 Glass gallery

12 Homestays' 2nd Floor Plan 1 : 100

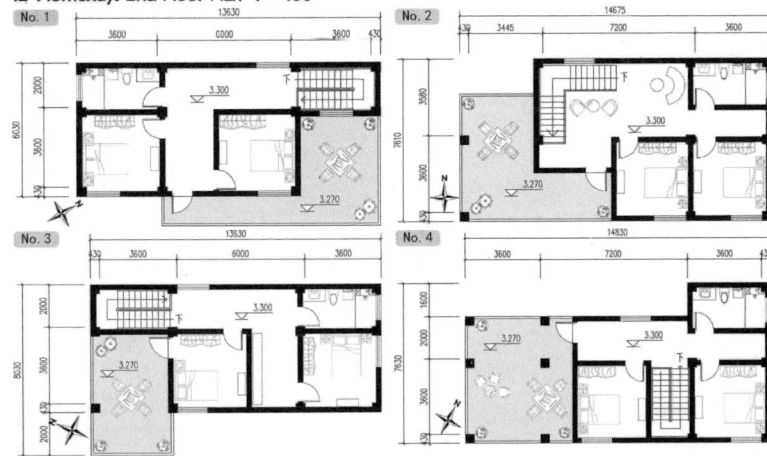

No. 1

No. 2

No. 3

No. 4

13 Sections 1 : 200

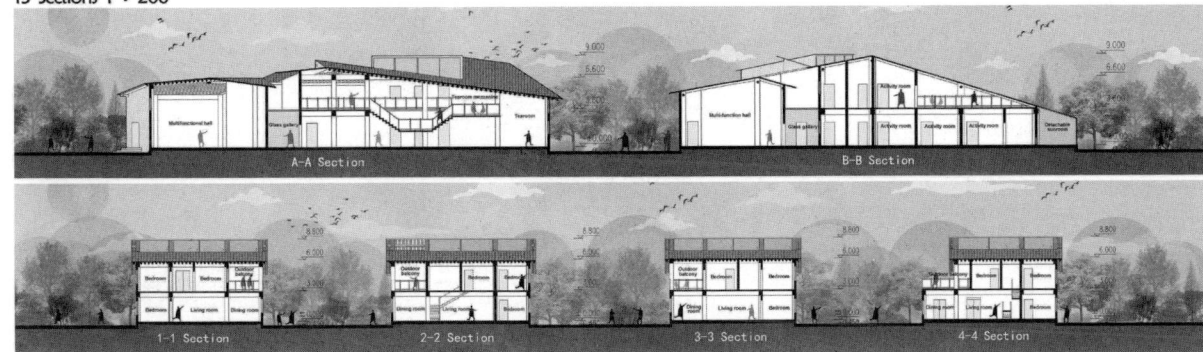

A-A Section

B-B Section

1-1 Section

2-2 Section

3-3 Section

4-4 Section

III

14 Axonometrical Drawing

Solar photovoltaic panel

Roof

Wood

Roof structure

Reinforced concrete

Beam-column consruction

WC

office

activity room

hall

veranda

infirmary

WC

observation

activity room

tearoom (half-floor)

sunlight room (detachable)

tearoom

15 Resilience Design Mind Map

Site planning — Emergency square / Sleeping on power generation / Gathering square / Solar power generation

Infrastructure — Emergency containers / Normal time / Emergency response / Emergency recovery

Resilience design

Load-bearing structure — Seismic structure / Split column / Independent foundation / Elastic and strong

Reuse of building materials / Local resources / Unified module / Fast and convenient

Variable functions

Before the disaster — Communication + Sale of goods + Coffee&brunch — Material storage

During the disaster — Mobility — Contact station

After the disaster — Temporary ward + Medical support — Material distribution

16 Site Resilience Analysis

■ Emergency container

■ Solar street lamp

■ Evacuation square

17 Emergency Container Analysis

Normal: Closed—Open

Open: Normal—Emergency

Variable components

Movable wall

Change of emergency function

Change of emergency function

Normal: closure—storage containers
Internal storage :emergency supplies
External storage:additional sports facilities

Normal: opening—entertainment container
Interior :exquisite store layout
Exterior: featured landscape

Emergency state— mobile first-aid container
Emergency contact station, temporary ward,
material distribution, and so on

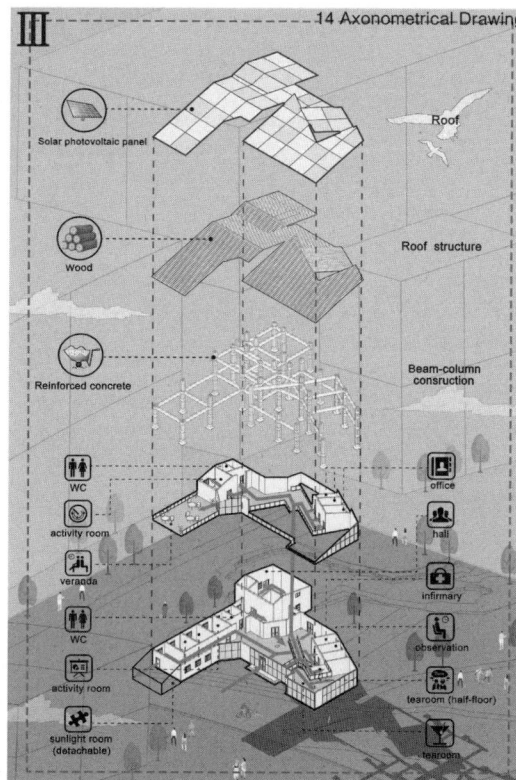

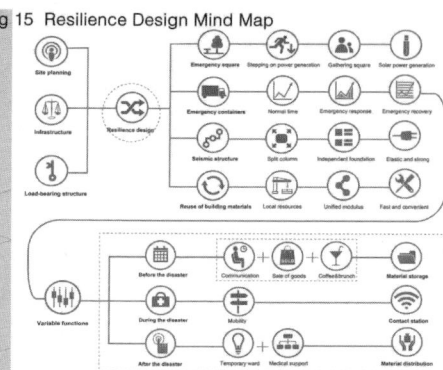

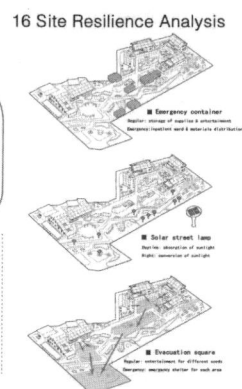

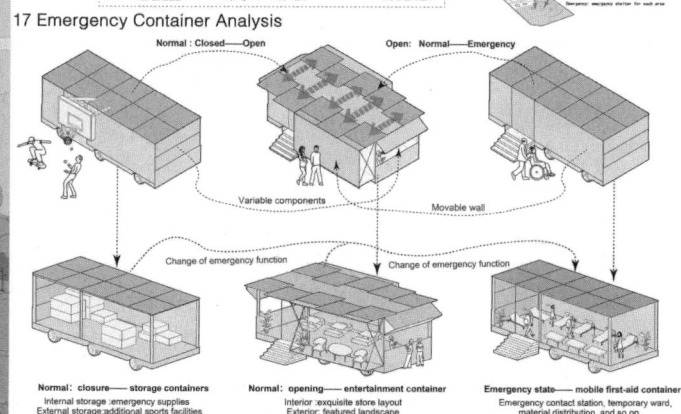

18 Seismic Column Structure Analysis

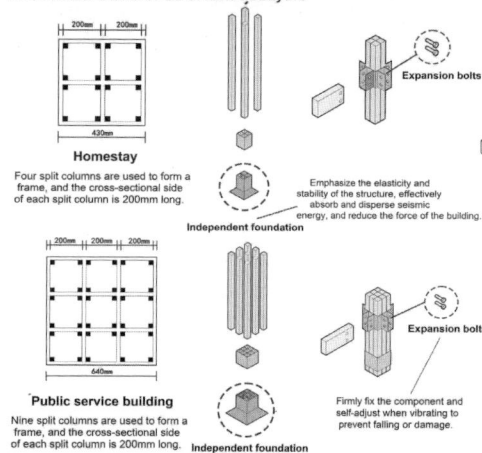

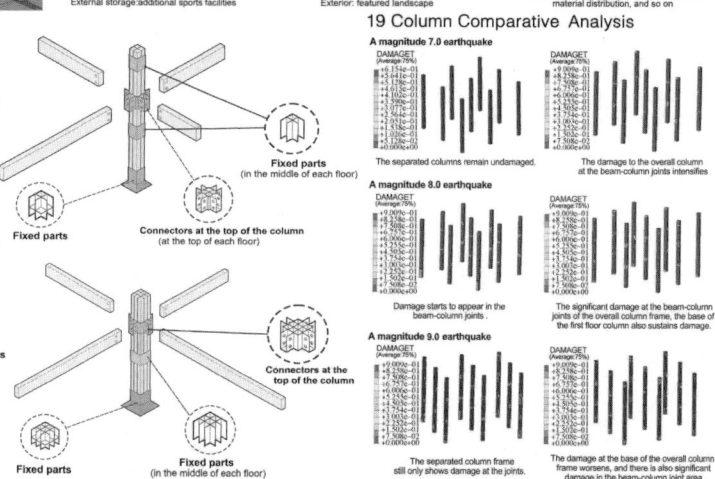

200mm 200mm
430mm

Homestay
Four split columns are used to form a frame, and the cross-sectional side of each split column is 200mm long.

Independent foundation

Emphasize the elasticity and stability of the structure, effectively absorb and disperse seismic energy, and reduce the force of the building.

Expansion bolts

Fixed parts

Fixed parts (in the middle of each floor)

Connectors at the top of the column (at the top of each floor)

200mm 200mm 200mm
640mm

Public service building
Nine split columns are used to form a frame, and the cross-sectional side of each split column is 200mm long.

Independent foundation

Expansion bolts

Firmly fix the component and self-adjust when vibrating to prevent falling or damage.

Connectors at the top of the column

Fixed parts

Fixed parts (in the middle of each floor)

19 Column Comparative Analysis

A magnitude 7.0 earthquake

The separated columns remain undamaged.

The damage to the overall column at the beam-column joints intensifies

A magnitude 8.0 earthquake

Damage starts to appear in the beam-column joints .

The significant damage at the beam-column joints of the overall column frame, the base of the first floor column also sustains damage.

A magnitude 9.0 earthquake

The separated column frame still only shows damage at the joints.

The damage at the base of the overall column frame worsens, and there is also significant damage in the beam-column joint area.

20 Sectional Perspective Analysis

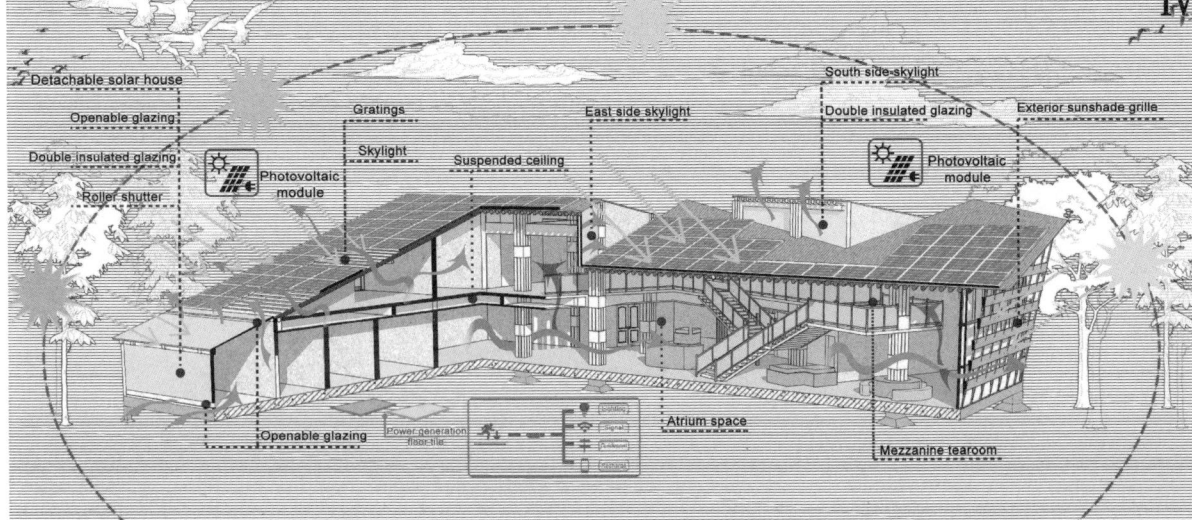

Detachable solar house
Openable glazing
Double insulated glazing
Roller shutter
Photovoltaic module
Gratings
Skylight
Suspended ceiling
East side skylight
South side-skylight
Double insulated glazing
Photovoltaic module
Exterior sunshade grille
Openable glazing
Power generation floor tile
Atrium space
Mezzanine tearoom

21 Energy Generation

One square meter of solar panels generates 6kwh per day, the public building solar panel area is about 327 square meters, the four sets of residential solar panels area is about 267 square meters, the public building solar panels can produce 1,962kW · h of energy a day,the four sets of residential solarpanels can produce a total of 1,602kW · h of energy, minus 30% loss, in a day, The solar panel capacity of public buildings is1373.4kW · h , and the total capacity of four sets of residential accommodation is 1121.4 kW · h.

22 Energy Consumption

For public building, the energy consumption per square meter per hour is 50 watts, the floor area is 1003 square meters, and the energy consumption of public buildings is 1203.6kwh per day.

For homestays, the energy consumption per square meter per hour is 35 watts, the total floor area of four homestays is 615 square meters, and the energy consumption of four homestays is 516.6kwh in one day.

23 Homestays' Wind Field Test

Due to the different layout of the four homestays, we have simulated the indoor air velocity, air direction and surface wind pressure of the four homestays respectively under the prevailing wind direction in summer, so as to ensure that each homestay has indoor drafts.

① Streamline test

② Wind speed test

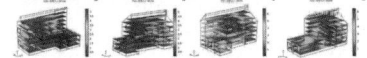

③ Wind pressure test

24 Public Building's Beam Module Analysis

For economic and construction considerations, the size types of beams in public buildings are minimized, and most of the beams in public buildings are controlled in 3600, 7200 and 8400 sizes according to the module.
1M=1200mm

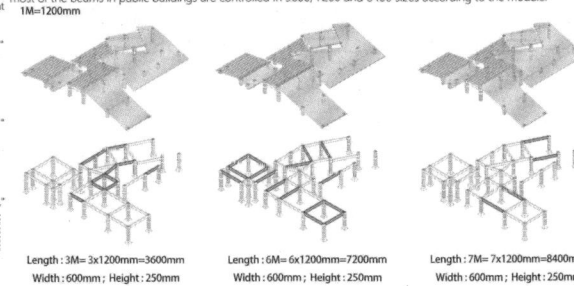

Length : 3M= 3x1200mm=3600mm
Width : 600mm; Height : 250mm

Length : 6M= 6x1200mm=7200mm
Width : 600mm; Height : 250mm

Length : 7M= 7x1200mm=8400mm
Width : 600mm; Height : 250mm

28 Partial Perspectives

25 Homestays' Room Module Analysis

Taking into account that the four guesthouses are located in different site conditions, we did not use a uniform floor plan, but integrated the orientation and location of the indoor rooms to modularize the treatment of the interior rooms. The volume of the indoor rooms in each guesthouse is the same, but the combination is different. This treatment method not only retains the uniqueness of each guesthouse, but also saves economic costs in construction.

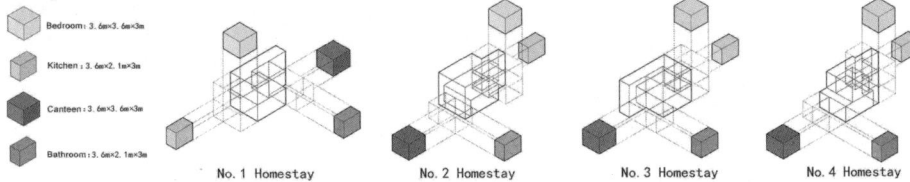

Bedroom : 3. 6m×3. 6m×3m
Kitchen : 3. 6m×2. 1m×3m
Canteen : 3. 6m×3. 6m×3m
Bathroom : 3. 6m×2. 1m×3m

No. 1 Homestay
No. 2 Homestay
No. 3 Homestay
No. 4 Homestay

26 Detachable Sunroom Installation Analysis

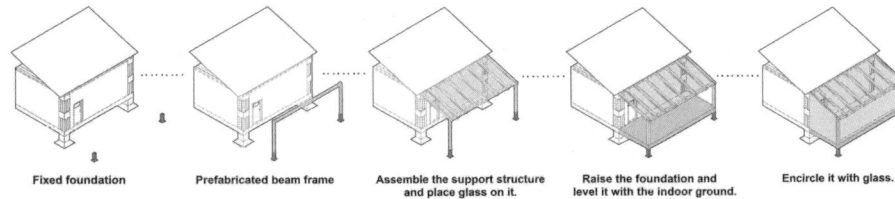

Fixed foundation
Prefabricated beam frame
Assemble the support structure and place glass on it.
Raise the foundation and level it with the indoor ground.
Encircle it with glass.

27 Detachable Sunroom Seasonal Analysis

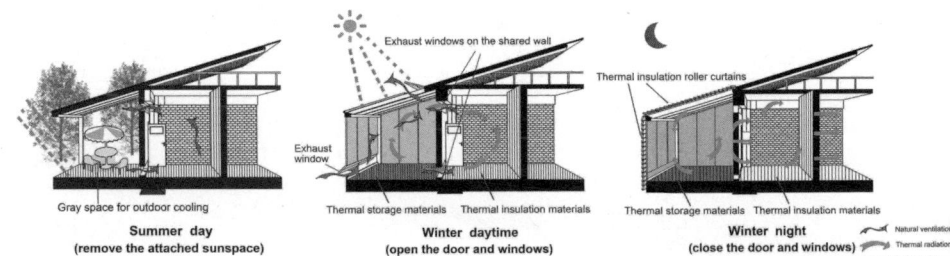

Exhaust windows on the shared wall
Thermal insulation roller curtains
Exhaust window
Thermal storage materials
Thermal insulation materials
Thermal storage materials
Thermal insulation materials

Gray space for outdoor cooling

Summer day
(remove the attached sunspace)

Winter daytime
(open the door and windows)

Winter night
(close the door and windows)

Natural ventilation
Thermal radiation
Solar radiation

综合奖·入围奖·张家口市怀安县二堡子村建设项目
Comprehensive Awards － Nomination － Erpuzi Village Project，Huai'an County，Zhangjiakou City

注册号：110335
Register Number：110335

项目名称：夜雪初霁·稻麦弥望
Entry Title：Night's Snow Lifts to Reveal Endless Yellow Fields，Where Beauty Makes a Mournful Echo of a Vanished Past

作者：张菱桐、张奕瑾、汤语婷、骆博奥
Authors：Zhang Lingtong, Zhang Yijin, Tang Yuting, and Luo Boao

作者单位：福州大学
Authors from：Fuzhou University

指导教师：王炜
Tutor：Wang Wei

指导教师单位：福州大学
Tutor from：Fuzhou University

夜雪初霁·稻麦弥望1

Night's Snow Lifts to Reveal Endless Yellow Fields,
Where Beauty Masks a Mournful Echo of a Vanished Past.

Position Analysis

Erpuzi village as seen from the farm

Internal streets

Landscape Inabashi

Location Analysis

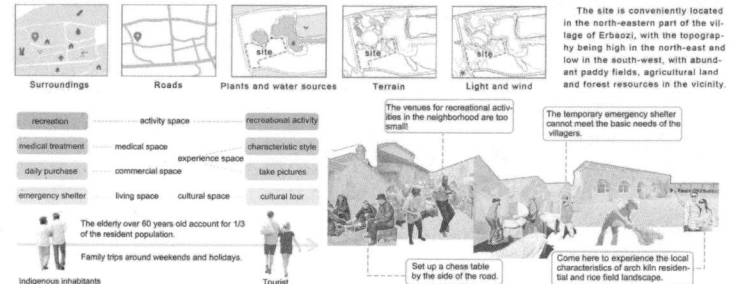

Surroundings | Roads | Plants and water sources | Terrain | Light and wind

The site is conveniently located in the north-eastern part of the village of Erbaozi, with the topography being high in the north-east and low in the south-west, with abundant paddy fields, agricultural land and forest resources in the vicinity.

recreation — activity space — recreational activity
medical treatment — medical space — characteristic style
daily purchase — commercial space — experience space — take pictures
emergency shelter — living space — cultural space — cultural tour

The elderly over 60 years old account for 1/3 of the resident population.

Family trips around weekends and holidays.

indigenous inhabitants | Tourist

The venues for recreational activities in the neighborhood are too small!

The temporary emergency shelter cannot meet the basic needs of the villagers.

Set up a chess table by the side of the road.

Come here to experience the local characteristics of arch kiln residential and rice field landscape.

Climate Analysis

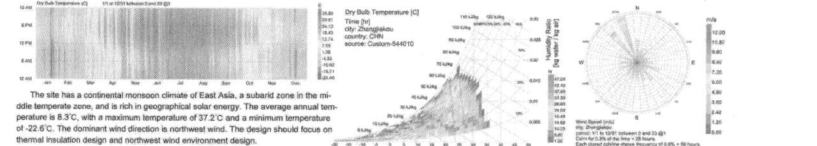

The site has a continental monsoon climate of East Asia, a subarid zone in the middle temperate zone, and is rich in geographical solar energy. The average annual temperature is 8.3℃, with a maximum temperature of 37.2℃ and a minimum temperature of -22.6℃. The dominant wind direction is northwest wind. The design should focus on thermal insulation design and northwest wind environment design.

Wind Rose Chart by Month
Wind Speed(m/s)

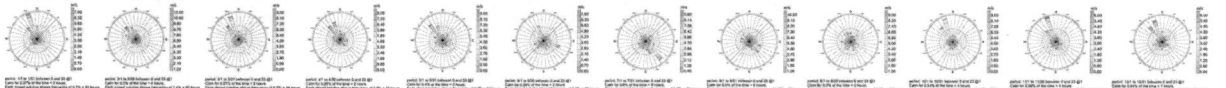

Design Specification

　　本方案利用主动式和被动式相结合的节能方式，同时贯彻防灾减灾规划，保证建筑具备抗风险及兼应急避难场所的功能，将绿色、生态、有韧性、可持续的理念融入乡村建设中。设计提取了怀安县传统民居碹窑的拱形门洞形式，承袭了当地特色的建筑文化，并对乡村常见的秸秆、蔚柴等材料进行改性处理，形成建筑墙板、屋面、窗框的保温材料。我们希望通过民居民宿和村民服务中心的设计，能够提升居民的生活品质，为建设安全、可持续、富有活力的农村社区提供新思路。

　　This proposal utilizes a combination of active and passive energy-saving methods, while implementing disaster prevention and mitigation planning to ensure that the building has the functions of a risk-resistant and emergency shelter, and integrating the concepts of green, ecological, resilient, and sustainable into the construction of the countryside. The design extracts the arched doorway form of the Swan Kiln, a traditional residential house in Huai'an County, which inherits the local characteristics of the architectural culture, and modifies materials such as straw and firewood, which kare common in the countryside, to form the thermal insulation materials for the wall panels, roofs and window frames of the building. We hope that the design of the residential lodging and the villagers' service center will enhance the quality of life of the residents and provide new ideas for building a safe, sustainable and vibrant rural community.

Technological Application

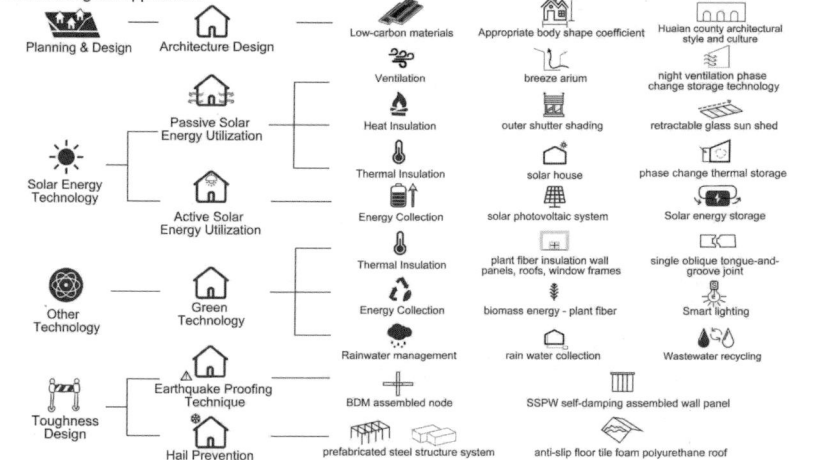

Planning & Design — Architecture Design — Low-carbon materials | Appropriate body shape coefficient | Hualan county architectural style and culture

Ventilation — breeze arium — night ventilation phase change storage technology

Passive Solar Energy Utilization — Heat Insulation — outer shutter shading — retractable glass sun shed

Solar Energy Technology — Thermal Insulation — solar house — phase change thermal storage

Active Solar Energy Utilization — Energy Collection — solar photovoltaic system — Solar energy storage

Green Technology — Thermal Insulation — plant fiber insulation wall panels, roofs, window frames — single oblique tongue-and-groove joint

Other Technology — Energy Collection — biomass energy - plant fiber — Smart lighting

Rainwater management — rain water collection — Wastewater recycling

Earthquake Proofing Technique — BDM assembled node — SSPW self-damping assembled wall panel

Toughness Design — Hail Prevention — prefabricated steel structure system — anti-slip floor tile foam polyurethane roof

夜雪初霁·稻麦弥望 2

Night's Snow Lifts to Reveal Endless Yellow Fields,
Where Beauty Masks a Mournful Echo of a Vanished Past.

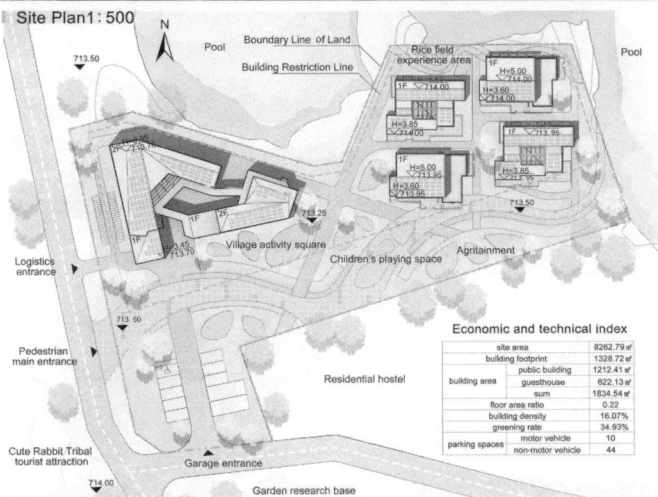

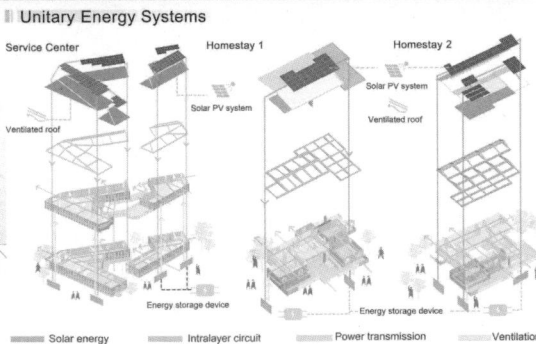

Site Plan 1:500

- Boundary Line of Land
- Building Restriction Line
- Rice field experience area
- Pool
- Village activity square
- Children's playing space
- Agritainment
- Residential hostel
- Logistics entrance
- Pedestrian main entrance
- Cute Rabbit Tribal tourist attraction
- Garage entrance
- Garden research base

Economic and technical index

site area		8262.79 ㎡
building footprint		1328.72 ㎡
building area	public building	1212.41 ㎡
	guesthouse	622.13 ㎡
	sum	1834.54 ㎡
floor area ratio		0.22
building density		16.07%
greening rate		34.93%
parking spaces	motor vehicle	10
	non-motor vehicle	44

Site Plan During Disaster 1:800

- mobile toilet (4 stalls)
- emergency tents (56 people)
- direct water intake from pools
- pool
- SPF roofs resist hail
- road for fire trucks
- medical and health rescue area
- emergency tents (56 people)
- emergency material storage area
- fire relief set-up
- emergency relief set-up (earthquake,hailstone)
- crowd buffer zone
- emergency material storage area
- mobile toilet (4 stalls)
- emergency tents (32 people)
- household waste disposal during disasters
- medical waste disposal during disasters
- emergency tents (72 people)
- crowd buffer zone

Emergency Relief Process

- disaster warning
- safe evacuation
- temporary accommodation
- deliver materials
- post-disaster reconstruction
- provide food
- transport the wounded
- set up medical teams

Emergency shelter related indicators

effective refuge area	3646.96 ㎡
crowd evacuation buffer zone	690.43 ㎡
medical and health rescue area	107.32 ㎡
capable of accommodating disaster victims	252 people
number of tents	54
public restroom stalls	42

Block Generation

- Function Module Placement
- Shape Cutting Combination
- Roofing Contact Group
- Morphology Refinement
- Functional Partitions
- Road Planning
- Site Refinement
- Continued Refinement

Unitary Energy Systems

Service Center / Homestay 1 / Homestay 2

- Solar PV system
- Ventilated roof
- Energy storage device

Legend: Solar energy | Intralayer circuit | Power transmission | Ventilation

Calculation of Energy Consumption and Carbon Emissions

	Service Center			Homestay		
Phases	CE(tCO₂)	LCCE per unit area(kgCO₂/㎡)	Average annual LCCE per unit area(kgCO₂/(㎡·a))	Phases	CE(tCO₂)	LCCE per unit area(kgCO₂/㎡)	Average annual LCCE per unit area (kgCO₂/㎡·a)
Building materials production	305.67	770.23	11.43	Building materials production	136.03	357.45	7.14
Transportation of building materials	86.92	247.32	3.28	Transportation of building materials	36.67	96.36	1.93
Construction phase	2.81	6.01	1.29	Construction phase	1.21	3.17	0.06
Operational phase	-10658.17	-13406.65	-222.12	Operational phase	-14624.71	-38427.43	-768.55
Dismantling phase	1.95	4.81	-3.61	Dismantling phase	0.83	2.20	0.04
Greening carbon sinks	-926.10	-964.37	-19.34	Greening carbon sinks	-301.35	-791.82	-15.84
Add up the total	-11166.92	-13342.85	-229.05	Add up the total	-14751.31	-38760.08	-775.20

The whole life cycle carbon emission of the building is negative, which meets the demand of green building and zero-carbon building.

Number of photovoltaic module installations	156	Number of photovoltaic module installations	158
Photovoltaic module installation area	249.6 ㎡	Photovoltaic module installation area	252.8 ㎡
Photovoltaic power generation	351836.16kW·h	Photovoltaic power generation	356346.88kW·h

The photovoltaic capacity of the service center is 351836.16kW·h, and the photovoltaic capacity of the homestay is356346.88kW·h.

Energy-saving operation	Energy consumption (kW·h/a)	Annual operational CE (kgCO₂/a)	Annual operational CE per unit area(kgCO₂/㎡·a)	Energy-saving operation	Energy consumption (kW·h/a)	Annual operational CE (kgCO₂/a)	Annual operational CE per unit area(kgCO₂/㎡·a)
heat	70253.00	40065.29	105.27	heat	104954.34	59855.46	62.47
refrigeration	1531.59	873.47	2.30	refrigeration	2009.47	1146.00	1.20
illumination	6373.94	3635.06	9.55	illumination	14651.50	8355.75	8.72
renewable energy	-372242.66	-212289.99	-557.80	renewable energy	-377015.00	-215011.65	-224.42
Add up the total	-294084.12	-167716.18	-440.68	Add up the total	-255399.69	-145654.50	-152.03

Negative carbon emissions during the operational phase of the building.Meet the needs of green and zero-carbon buildings.
The solar photovoltaic panel capacity of the service center and the homestay can meet the electricity demand.

Annotation: Life-Cycle Carbon Emissions (LCCE)　Carbon Emissions (CE)
Optimal sloped total irradiation for stationary PV in Hebei, 2023 : △T=1762.0kWh/㎡　　Combined efficiency factor: K=80%

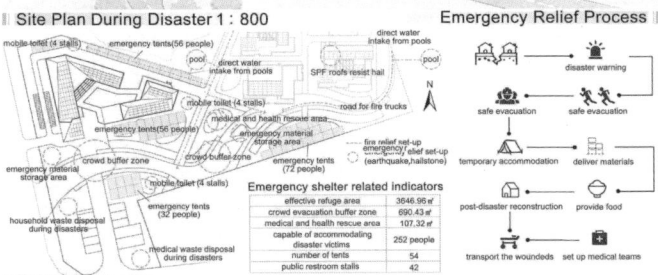

Module Analysis

- bedroom
- kitchen
- restaurant
- Homestay 1
- Homestay 2
- solar house
- toilet
- livingroom
- Modular layout

East Elevation 1:200

East elevation of the Service Center　　East Elevation of the Homestay 1　　East Elevation of the Homestay 2

夜雪初霁 · 稻麦弥望 3

Night's Snow Lifts to Reveal Endless Yellow Fields,
Where Beauty Masks a Mournful Echo of a Vanished Past.

The First Floor Plan of the Villagers' Service Center 1:200

Room name annotation explanation

1	lobby	6	convenience store	11	office
2	reception	7	strong and weak current	12	locker room
3	clinic	8	equipment room	13	food elevator
4	scullery	9	storage room	14	public toilet
5	kitchen	10	cultural and creative shop		

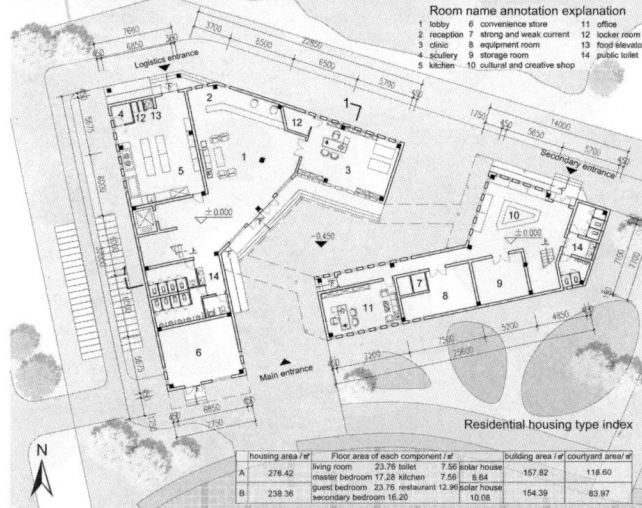

Logistics entrance

Secondary entrance

Main entrance

The First Floor Plan of the Homestay 1:200

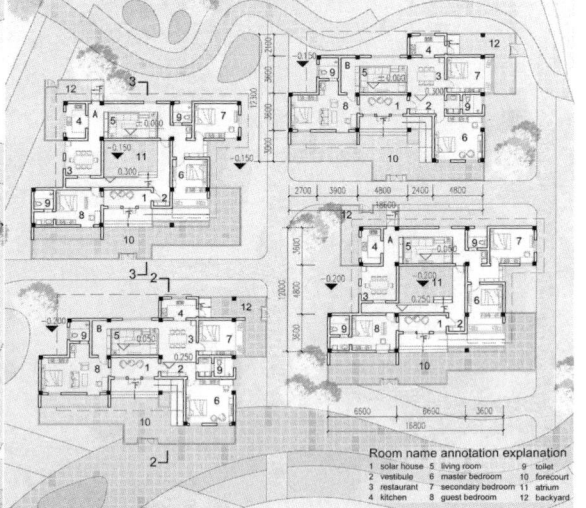

Room name annotation explanation

1	solar house	5	living room	9	toilet
2	vestibule	6	master bedroom	10	forecourt
3	restaurant	7	secondary bedroom	11	atrium
4	kitchen	8	guest bedroom	12	backyard

Residential housing type index

housing area /㎡		Floor area of each component /㎡				building area /㎡	courtyard area/㎡
A	276.42	living room	23.76	toilet	7.56	solar house		157.82	118.60
		master bedroom	17.28	kitchen	7.56	8.64		
		guest bedroom	23.76	restaurant	12.96	solar house		
B	238.36	secondary bedroom	16.20			10.08	154.39	83.97

The Second Floor Plan of the Villagers' Service Center 1:200

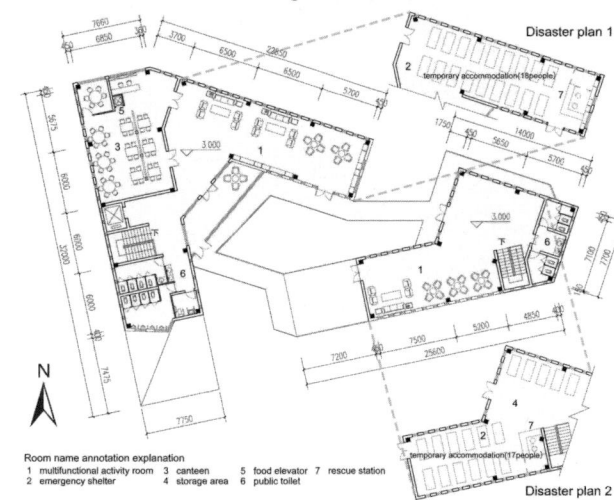

Disaster plan 1

temporary accommodation(18people)

temporary accommodation(17people)

Disaster plan 2

Room name annotation explanation

1	multifunctional activity room	3	canteen	5	food elevator	7	rescue station
2	emergency shelter	4	storage area	6	public toilet		

Climate Simulation

Annual Solar Radiant Heat Analysis

kWh/m2

Wind Environment Simulation (Maximum wind speed)

Analysis of Solar Trajectory, Sunshine Duration (Summer Solstice)

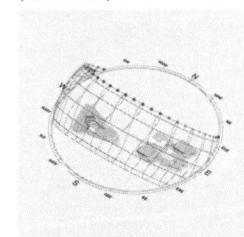

Analysis of Solar Trajectory, Sunshine Duration (Winter Solstice)

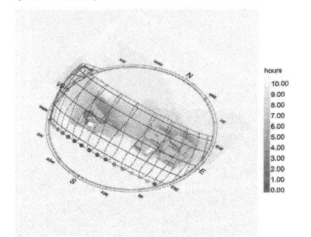

Analysis of Assembled Seismic Structures

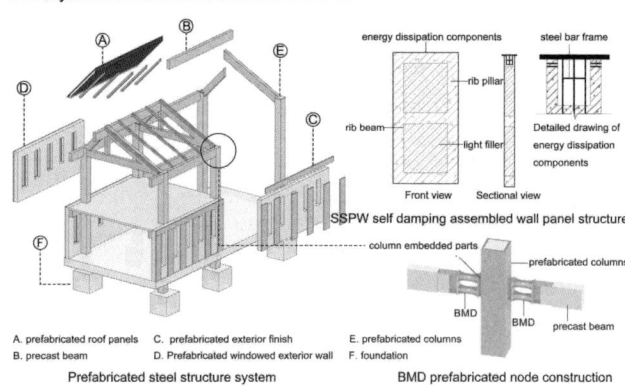

energy dissipation components

steel bar frame

rib pillar

rib beam

light filler

Detailed drawing of energy dissipation components

Front view   Sectional view

SSPW self damping assembled wall panel structure

column embedded parts

prefabricated columns

BMD   BMD   precast beam

A. prefabricated roof panels   C. prefabricated exterior finish   E. prefabricated columns
B. precast beam   D. Prefabricated windowed exterior wall   F. foundation

Prefabricated steel structure system

BMD prefabricated node construction

Analysis of Green Strategies for the Homestay

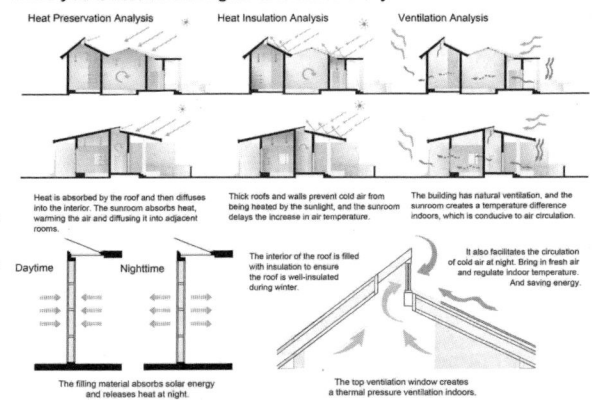

Heat Preservation Analysis   Heat Insulation Analysis   Ventilation Analysis

Heat is absorbed by the roof and then diffuses into the interior. The sunroom absorbs heat, warming the air and diffusing it into adjacent rooms.

Thick roofs and walls prevent cold air from being heated by the sunlight, and the sunroom delays the increase in air temperature.

The building has natural ventilation; and the sunroom creates a temperature difference indoors, which is conducive to air circulation.

Daytime   Nighttime

The interior of the roof is filled with insulation to ensure the roof is well-insulated during winter.

It also facilitates the circulation of cold air at night. Bring in fresh air and regulate indoor temperature. And saving energy.

The filling material absorbs solar energy and releases heat at night.

The top ventilation window creates a thermal pressure ventilation indoors.

South Elevation 1:200

South Elevation of the Service Center   South Elevation of the Homestay 1   South Elevation of the Homestay 2

夜雪初霁·稻麦弥望 4

Night's Snow Lifts to Reveal Endless Yellow Fields,
Where Beauty Masks a Mournful Echo of a Vanished Past.

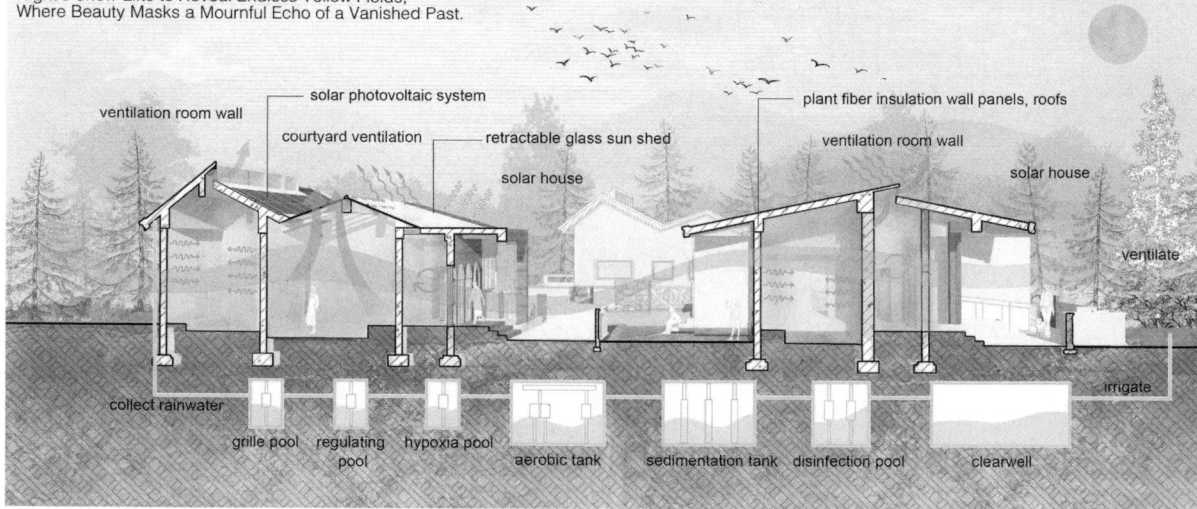

ventilation room wall
solar photovoltaic system
courtyard ventilation
retractable glass sun shed
solar house
plant fiber insulation wall panels, roofs
ventilation room wall
solar house
ventilate
irrigate

collect rainwater

grille pool / regulating pool / hypoxia pool / aerobic tank / sedimentation tank / disinfection pool / clearwell

▌ Analysis of Green Strategies for the Service Center

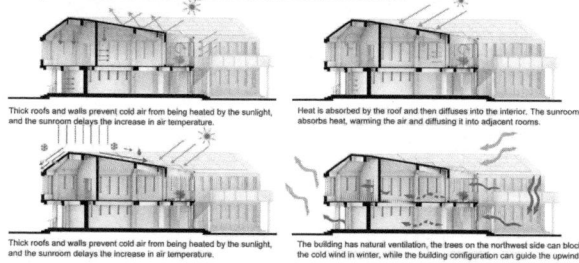

Thick roofs and walls prevent cold air from being heated by the sunlight, and the sunroom delays the increase in air temperature.

Heat is absorbed by the roof and then diffuses into the interior. The sunroom absorbs heat, warming the air and diffusing it into adjacent rooms.

Thick roofs and walls prevent cold air from being heated by the sunlight, and the sunroom delays the increase in air temperature.

The building has natural ventilation, the trees on the northwest side can block the cold wind in winter, while the building configuration can guide the upwind

▌ Analysis of Green Strategies for the Service Center

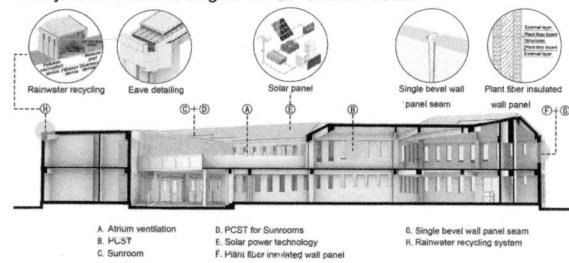

Rainwater recycling / Eave detailing / Solar panel / Single bevel wall panel seam / Plant fiber insulated wall panel

A. Atrium ventilation
B. PCST
C. Sunroom
D. PCST for Sunrooms
E. Solar power technology
F. Plant fiber insulated wall panel
G. Single bevel wall panel seam
H. Rainwater recycling system

▌ Local Specialty Building Materials

1. Zhangjiakou Huaian County has traditional folk dwellings called qiantaiya.
2. Extract and use arch kiln shapeElements into homestay design
3. Use Huaian county characteristic woodTimber as building material
4. House style adopts Huai'an traditional residential style - courtyard style

▌ Hail-resistant Construction

SPF roof is adopted to prevent hail

Anti-slip floor tile foam polyurethane roof

Parapet node

▌ Phase-change Materials(PCMs) Application Analysis

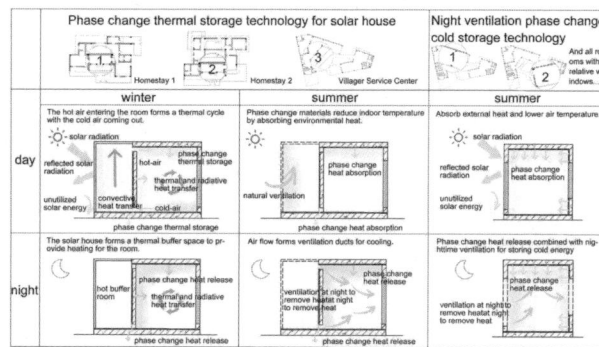

Phase change thermal storage technology for solar house

Night ventilation phase change cold storage technology

Homestay 1 / Homestay 2 / Villager Service Center

And all rooms with relative windows.

	winter	summer	summer
day	The hot air entering the room forms a thermal cycle with the cold air coming out.	Phase change materials reduce indoor temperature by absorbing environmental heat.	Absorb external heat and lower air temperature.
night	The solar house forms a thermal buffer space to provide heating for the room.	Air flow forms ventilation ducts for cooling.	Phase change heat release combined with nighttime ventilation for storing cold energy.

▌ Interior and ExteriorRendering

Center Courtyard of Homestay 2

Emergency Disaster Relief Effectiveness

Outdoor Activity Plaza

North Side Water View

▌ Section View 1:200

Service Center 1-1 Section

Homestay 2-2 Section

Homestay 3-3 Section

110390

结庐窑境

综合奖·入围奖·张家口市
怀安县二堡子村建设项目
Comprehensive Awards —
Nomination — Erpuzi Village
Project，Huai'an County，
Zhangjiakou City

注册号：110390

Register Number：110390

项目名称：结庐窑境

Entry Title：Homes in Cave—like Dwellings

作者：王喆、许馨、关项心、林甄欣、
　　　尹思源

Authors：Wang Zhe，Xu Xin，Guan
　　　　　Xiangxin，Lin Zhenxin，and
　　　　　Yin Siyuan

作者单位：厦门大学

Authors from：Xiamen University

指导教师：贾令堃、石峰

Tutors：Jia Lingkun，Shi Feng

指导教师单位：厦门大学

Tutors from：Xiamen University

● DESIGN DESCRIPTION

本次设计场地位于张家口二堡子村，是个温带湿润半干旱大陆性季风气候，四季分明，夏热冬冷，太阳能丰富。场地主要需的问题是容易遭到冰雹灾害的风险，其次主要兼顾抗震设防和应急避难的需5求。所以本设计认当地传统的密闭民居的拱形态和台院式的繁急为灵感出发，在民窑部分和公建盒子体放的部分采用了半圆拱形屋顶，既能增加了屋顶的抗冰雹能力和对地震的力学性能，也能充分体现对在地文化的思考与传承。同时合院的民俗形式不但能最大限度的保障了房间通风采光和四季的热舒适性，也回应了当地民居的传统气候形态。并且我们在公共建筑的庭院设计了可变的屋结构，来进一步降低冰雹灾害带给给人们的风险。在功能方面，公共建筑及周围广场通过可变的设计，平时可以作为在地文化展览馆和体验场所，并且整合传统产业与公共活动的空间，解决了传统技艺活化展示的问题，又满足了村里共享公共活动的需要。通过整合村中的小作坊提升传统工艺，促进乡村产业复兴与村落共享品牌的建立，强化了村落共同体的意识，而当遇到灾害断水断电时也可以作为应急避难的场所，深刻回应了此次竞赛的主题：阳光·乡村韧性。

The design site is located in Erbaozi Village, Zhangjiakou, which has a temperate humid semi-arid continental monsoon climate with four distinct seasons, hot summers and cold winters, and abundant solar energy. The main problem of the site is the risk of hailstorms, and the second problem is the need for seismic defense and emergency evacuation. Therefore, this design is inspired by the arch form of local traditional kiln dwellings and the courtyard settlement, and adopts a semi-circular arch roof in the lodging part and the public building box block, which not only increases the hail resistance of the roof and the mechanical properties to cope with earthquakes, but also fully reflects the thinking and inheritance of the local culture. At the same time, the form of the residential compound not only maximizes the ventilation and lighting of the rooms and the thermal comfort in all seasons, but also responds to the traditional settlement pattern of the local houses. In the courtyard of the public building, we designed a variable membrane structure to further reduce the risk of hailstorms. In terms of function, the public building and the surrounding plaza can be used as a local cultural exhibition hall and experience place through variable design, and integrating the space for traditional industries and public activities, which solves the problem of revitalizing and displaying the traditional skills, and meets the needs of shared public activities in the village. By integrating small workshops in the village to enhance traditional crafts, it promotes the revitalization of village industries and the establishment of village shared brands, and strengthens the sense of village community. And it can also be used as an emergency shelter when water and electricity are cut off in case of disaster, which profoundly responds to the theme of this competition: Sunshine · Village Resilience.

● LOCATION ANALYSIS

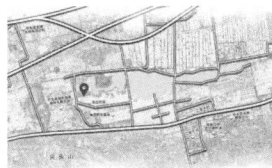

THE PROJECT SITE IS LOCATED IN ERBAOZI VILLAGE, HUAAN COUNTY, ZHANGJIAKOU CITY, HEBEI PROVINCE, AT 114.63°E LONGITUDE AND 40.68°N LATITUDE. IT IS CLOSE TO THE JINGZANG EXPRESSWAY AND JINGXIN EXPRESSWAY, WITH A HIGH-SPEED RAILWAY LINE RUNNING EAST-WEST ACROSS THE NORTHERN SIDE OF THE SITE.

● CULTURAL STUDIES

ZHANGJIAKOU HAS A RICH LOCAL CULTURE, INCLUDING NUMEROUS ANCIENT BUILDINGS AND A VARIETY OF HOMESTAY ACTIVITIES. ITS GEOGRAPHICAL POSITION ADDS UNIQUE CULTURAL SIGNIFICANCE. THE LOCAL ARCHITECTURAL FORMS ARE HIGHLY DISTINCTIVE.

● REGIONAL ELEMENTS TRANSLATION

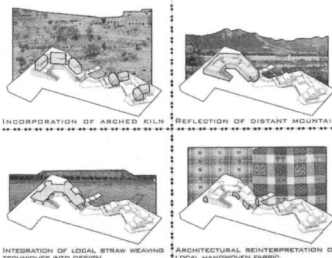

INCORPORATION OF ARCHED KILN

REFLECTION OF DISTANT MOUNTAIN

INTEGRATION OF LOCAL STRAW WEAVING TECHNIQUES INTO DESIGN.

ARCHITECTURAL REINTERPRETATION OF LOCAL HANDWOVEN FABRIC.

● NATURAL DISASTERS ASSESSMENT

● TARGET GROUP ANALYSIS

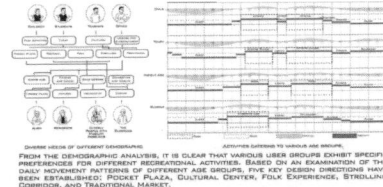

FROM THE DEMOGRAPHIC ANALYSIS, IT IS CLEAR THAT VARIOUS USER GROUPS EXHIBIT SPECIFIC PREFERENCES FOR DIFFERENT RECREATIONAL ACTIVITIES. BASED ON AN EXAMINATION OF THE DAILY MOVEMENT PATTERNS OF DIFFERENT AGE GROUPS, FIVE KEY DESIGN DIRECTIONS HAVE BEEN ESTABLISHED: POCKET PLAZA, CULTURAL CENTER, FOLK EXPERIENCE, STROLLING CORRIDOR, AND TRADITIONAL MARKET.

110390

结庐密境

● SITE PLAN

● MASS DIAGRAM

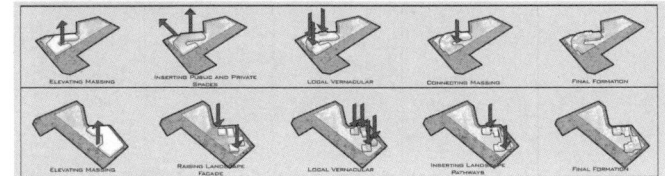

● ZHANGJIAKOU CLIMATE ANALYSIS

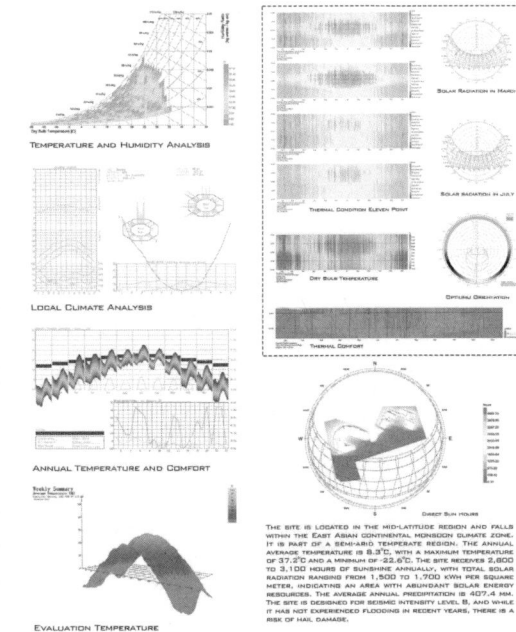

● ENVIRONMENT STRATEGY

● ELEVATION PLAN

THE ELEVATION SHOWCASES THE INTEGRATION OF TRADITIONAL VAULTED CAVE DWELLING FORMS WITH MODERN DESIGN ELEMENTS. THE PUBLIC BUILDING INCORPORATES AN EARTH-COVERED ROOF AND INTRODUCES A SERIES OF CORRIDORS, OFFERING USERS A MORE DYNAMIC ACTIVITY SPACE. AT THE ENTRANCE OF THE PUBLIC BUILDING, THERE IS AN OPEN PLAZA WITH INTEGRATED TREE WELLS, WHICH NOT ONLY BRING NATURE INTO THE SITE BUT ALSO PROVIDE RESTING AREAS FOR VISITORS. IN THE RESIDENTIAL SECTION, THE ELEVATION REVEALS THE USE OF CORRIDORS THAT CORRESPOND TO THE PUBLIC BUILDING, CREATING A VISUAL AND FUNCTIONAL DIALOGUE BETWEEN THE TWO. THE CAVE DWELLING DESIGN IS REINTERPRETED WITH LARGE FLOOR-TO-CEILING GLASS WINDOWS, BREAKING AWAY FROM CONVENTIONAL FORMS. THE CORRIDORS NOT ONLY SERVE TO CONNECT THE VARIOUS BUILDINGS BUT ALSO PROVIDE COVERED WALKWAYS THAT PROTECT THE COURTYARDS FROM THE ELEMENTS. A WATER FEATURE IS INTRODUCED BETWEEN THE PUBLIC BUILDING AND THE RESIDENTIAL AREA, ENHANCING THE LANDSCAPE WHILE CREATING A DIVISION BETWEEN ACTIVE AND QUIET ZONES, AND CONTRIBUTING TO ECOLOGICAL REGULATION.

110390
结庐寄境

● EXPLODSIVE STRUCTURE & MATERIAL

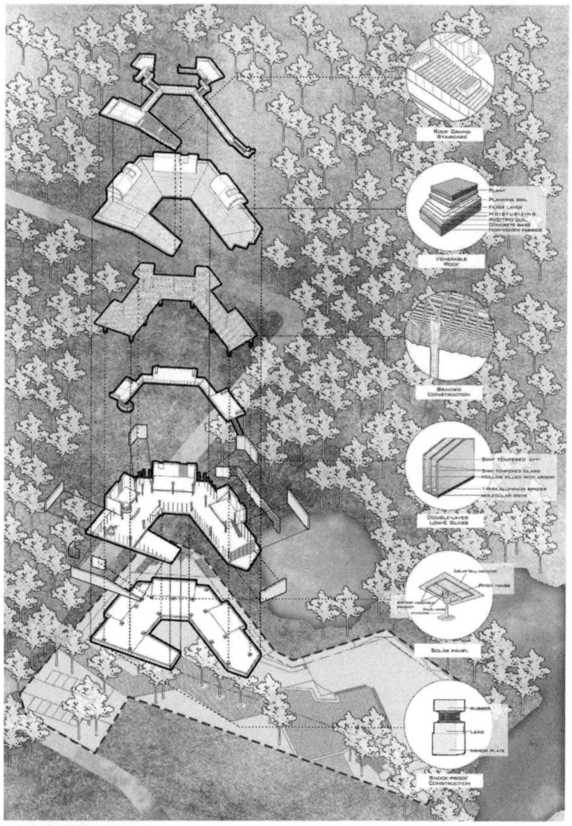
Roof Grand Staircase
Plant
Planting soil
Filter layer
Moisturizing
PVC/TPO coil
Concrete base
Non-woven fabrics
Venerable Roof
Braided Construction
5mm tempered off-
5mm tempered glass
Hollow filled with argon
1.4mm aluminum spacer
Molecular sieve
Double-layer Low-E Glass
Pitch hinge
Solar panel
Rubber
Lead
Armor plate
Shock-proof Construction

● FIRST FLOOR PLAN

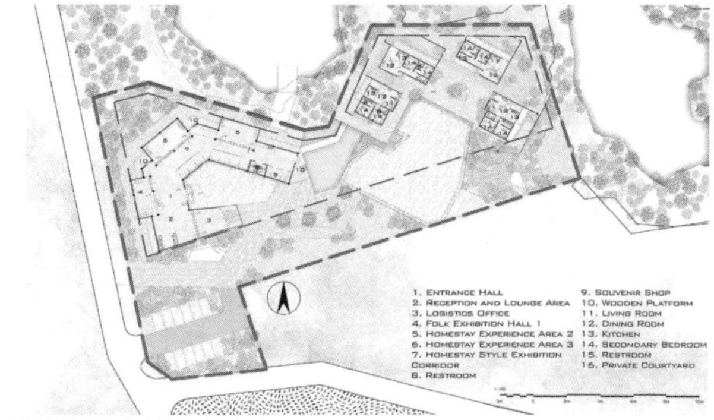
1. Entrance Hall
2. Reception and Lounge Area
3. Logistics Office
4. Folk Exhibition Hall 1
5. Homestay Experience Area 2
6. Homestay Experience Area 3
7. Homestay Style Exhibition Corridor
8. Restroom
9. Souvenir Shop
10. Wooden Platform
11. Living Room
12. Dining Room
13. Kitchen
14. Secondary Bedroom
15. Restroom
16. Private Courtyard

● SECOND FLOOR PLAN

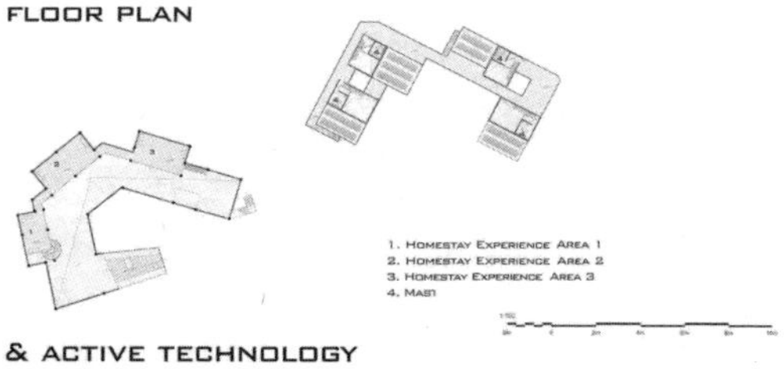
1. Homestay Experience Area 1
2. Homestay Experience Area 2
3. Homestay Experience Area 3
4. Mast

● PASSIVE & ACTIVE TECHNOLOGY

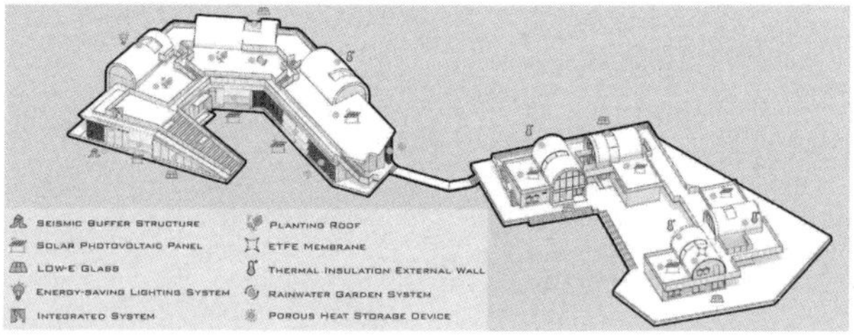
Seismic Buffer Structure
Solar Photovoltaic Panel
Low-E Glass
Energy-saving Lighting System
Integrated System
Planting Roof
ETFE Membrane
Thermal Insulation External Wall
Rainwater Garden System
Porous Heat Storage Device

● SECTION PLAN

THE SECTION CLEARLY DEMONSTRATES THE SPATIAL RELATIONSHIPS WITHIN THE PUBLIC BUILDING AND THE UNDERGROUND SEISMIC-RESISTANT STRUCTURE. THE VAULTED FORM OF THE ARCHITECTURE, THE STRAW-WEAVING STRUCTURES WITHIN THE INTERIOR SPACES, AND THE TEXTURED SURFACE TREATMENTS ALL INCORPORATE LOCAL CULTURAL ELEMENTS. THE PUBLIC BUILDING UTILIZES A MEZZANINE LEVEL, CONNECTED BY A SPIRAL STAIRCASE, WHICH ENHANCES THE SPATIAL LAYERING. UNDER NORMAL CONDITIONS, THE PUBLIC BUILDING PRIMARILY SERVES AS A VENUE FOR EXHIBITIONS AND VARIOUS TEMPORARY CULTURAL EVENTS, WHILE DURING EMERGENCIES, IT PROVIDES A LARGE AND FLEXIBLE REFUGE SPACE.

THE SECTION OF THE RESIDENTIAL BUILDINGS REVEALS THE RELATIONSHIPS BETWEEN INTERNAL SPACES, BETWEEN THE BUILDING UNITS, AND BETWEEN THE ARCHITECTURE, COURTYARDS, AND CORRIDORS. THE DESIGN OF BOTH THE INTERIOR AND EXTERIOR SPACES OF THE RESIDENCES OFFERS A RICH VARIETY OF EXPERIENCES, ENSURING PRIVACY FOR THE INHABITANTS WHILE ALSO PROVIDING DIVERSE SPATIAL QUALITIES.

110390

结庐密境

●FUNCTIONAL REPLACEMENT ANALYSIS

1.ENTRANCE HALL
2.EMERGENCY DINING AREA
3.EMERGENCY FOOD SUPPLY AREA
4.EMERGENCY ACCOMMODATION AREA
5.CULTURAL EXHIBITION AREA
6.EMERGENCY MEDICAL AREA
7.EMERGENCY TREATMENT ROOM
8.TOILET
9.TEMPORARY WARD
10.WOODEN PLATFORM
11.LEISURE ACTIVITY AREA
12.PSYCHOLOGICAL TREATMENT AREA
13.DOUBLE-HEIGHT SPACE

●CONSTRUCTED SPECIFICATION

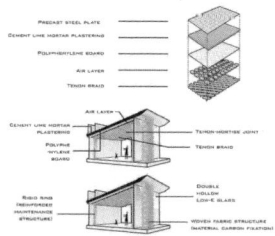

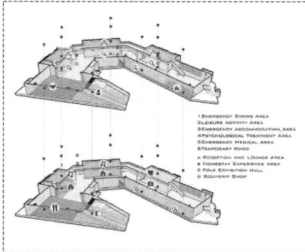

●SECTION STRATEGY

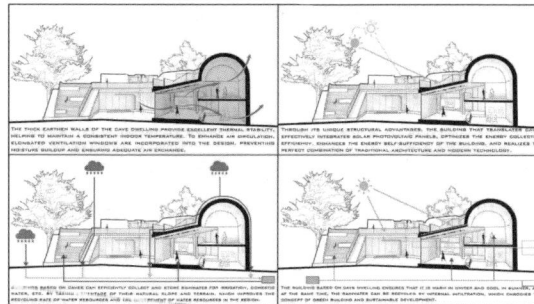

●WATER FOOTPRINT

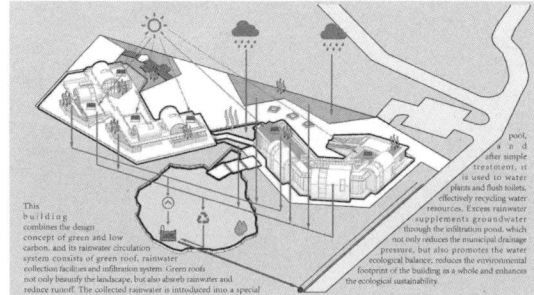

This building combines the design concept of green and low carbon, and its rainwater circulation system consists of green roof, rainwater collection facilities and infiltration system. Green roofs not only beautify the landscape, but also absorb rainwater and reduce runoff. The collected rainwater is introduced into a special pool, and after simple treatment, it is used to water plants and flush toilets, effectively recycling water resources. Excess rainwater supplements groundwater through the infiltration pond, which not only reduces the municipal drainage pressure, but also promotes the water ecological balance, reduces the environmental footprint of the building as a whole and enhances the ecological sustainability.

●Carbon emissions & energy consumption calculations

BUILDING CARBON EMISSION		
	Projects	Carbon Emission
CARBON REDUCTION TERMS	HVAC	2399686.22
	Domestic Hot Water	176240.33
	Illumination	503985.36
	Total	3079911.91
RENEWABLE ENERGY	Renewable Energy	699640.65
	Building Carbon	2380271.26
	Total	3079911.91
TOTAL		0

●PLANT FOOTPRINT

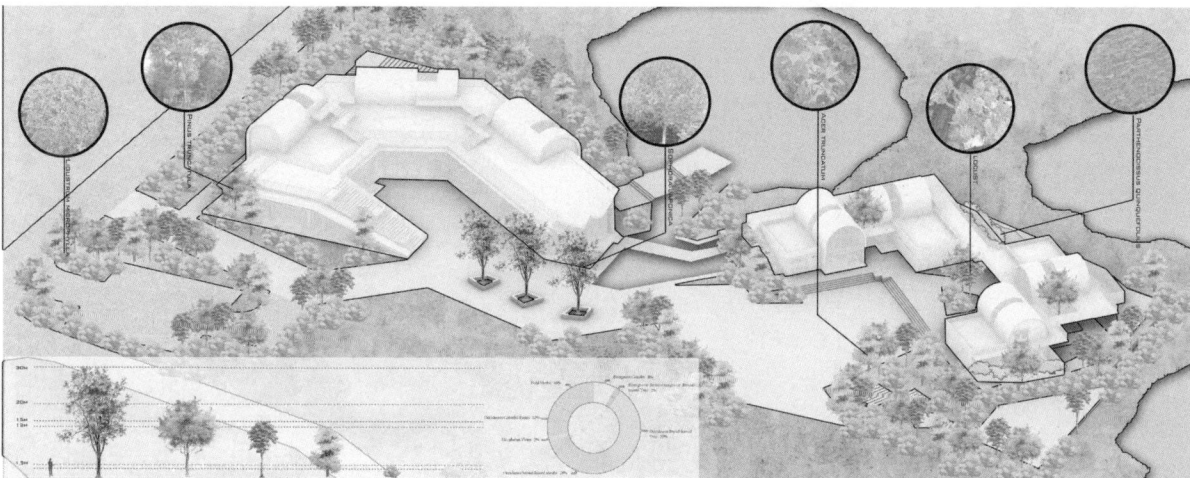

综合奖·入围奖·张家口市怀安县二堡子村建设项目
Comprehensive Awards — Nomination — Erpuzi Village Project，Huai'an County，Zhangjiakou City

注册号：110484
Register Number：110484

项目名称：韧性之光
Entry Title：The Light of Resilience

作者：于子正、周珈屹
Authors：Yu Zizheng, Zhou Jiayi

作者单位：沈阳建筑大学
Authors from：Shenyang Jianzhu University

指导教师：赵钧
Tutor：Zhao Jun

指导教师单位：沈阳建筑大学
Tutor from：Shenyang Jianzhu University

The Light of Resilience | 韧性之光 I

Reasearch on Green Building Based on Wind System

设计受华北地区张家口传统建筑启发,建筑形态采用单坡屋顶 小间距,形成自遮阳,构建冷巷,增强通风的同时将传统建筑中的天井、廊道与腔体相结合,形成新的通风系统,使建筑内部自然通风良好。光伏板雨水收集、浅色墙体深色屋顶与华北传统建筑形象紧密结合合。在达到零碳的同时与张家口传统文化进行对话。

The design is inspired by the traditional architecture of Zhangjiakou in North China, the building form adopts a single slope roof with small pitch, forming a self-shading, constructing cold alleys, enhancing ventilation and at the same time, combining the patio, corridor and cavity of the traditional architecture to form a new ventilation system, so as to make the building's interior well ventilated naturally. Photovoltaic panels for rainwater collection, light-colored walls and dark-colored roofs are closely integrated with the traditional architectural image of North China, achieving zero carbon while engaging in a dialogue with the traditional culture of Zhangjiakou.

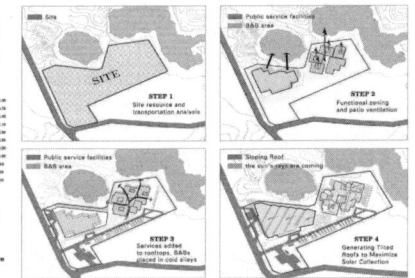

■ Site Analysis

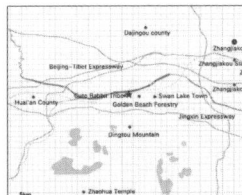

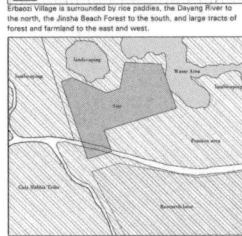

■ Climate Diagnostic

Located in Zhangjiakou, the project aims to create a low-carbon, pleasant rural center and residential accommodation through an analysis of the local climate.

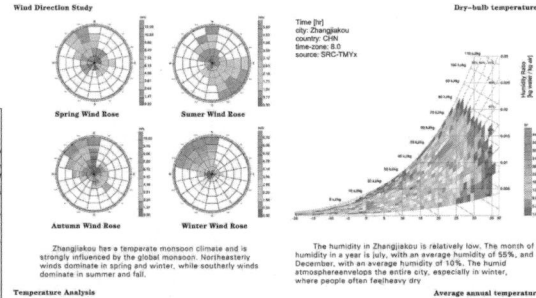

Erpuzi Village is surrounded by rice paddies, the Dayang River to the north, the Jinsha Beach Forest to the south, and large tracts of forest and farmland to the east and west.

There are already tourist attractions such as Moe Rabbit Tribe Grassland Resort and Hot Spring Resort within 50km around the site. Natural landscape on the north side of the site, tourism resources on the west side, research base on the southeast side, and major roads on the west and south sides of the site.

Wind Direction Study

Spring Wind Rose Sumer Wind Rose

Autumn Wind Rose Winter Wind Rose

Zhangjiakou has a temperate monsoon climate and is strongly influenced by the global monsoon. Northeasterly winds dominate in spring and winter, while southerly winds dominate in summer and fall.

Temperature Analysis

Zhangjiakou is located in North China and has a temperate monsoon climate with four distinct seasons throughout the year, with a warm and comfortable spring, a hot and humid summer, a cool and dry fall, and a relatively cold winter with more sub-zero temperatures.

Time [hr]
city: Zhangjiakou
country: CHN
time-zone: 8.0
source: SRC-TMYx

Dry-bulb temperature

The humidity in Zhangjiakou is relatively low. The month of humidity in a year is july, with an average humidity of 55%, and December, with an average humidity of 10%. The humid atmosphereanvelope the entire city, especially in winter, where people often feel heavy dry

Average annual temperature

With little change in temperature throughout the year, abundant rainfall in summer, and few clouds in North China, solar radiation energy resources are abundant, providing important conditions for building energy efficiency and low-carbon design.

■ Form Extraction

We extracted architectural elements from Zhangjiakou residential buildings as design elements for the B&B. For example, single-slope roofs, courtyards, cold alleys, and flower walls, while also drawing on traditional energy-saving measures.

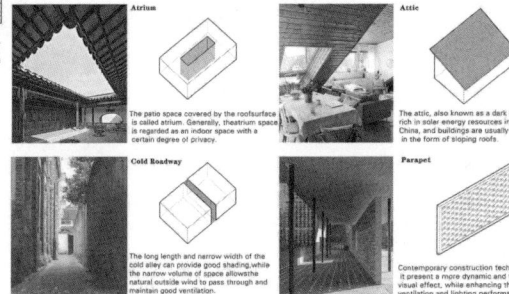

Atrium

The patio space covered by the roofsurface is called atrium. Generally, theatrium space is regarded as an indoor space with a certain degree of privacy.

Attic

The attic, also known as a dark building,is rich in solar energy resources in North China, and buildings are usually constructed in the form of sloping roofs.

Cold Roadway

The long length and narrow width of the cold alley can provide good shading,while the narrow volume of space allowsthe natural outside wind to pass through and maintain good ventilation.

Parapet

Contemporary construction techniques make it present a more dynamic and transparent visual effect, while enhancing the building's ventilation and lighting performance.

STEP 1
Site resource and transportation analysis

STEP 2
Functional zoning and patio ventilation

STEP 3
Services added to rooftops, b&bs placed in cold alleys

STEP 4
Generating Tilted Roofs to Maximize Solar Collection

electricity	category	power consumption (kW·h/m²)	carbon emission factor (kgCO₂/kWh)	carbon emission (tCO₂)
cooling (Ec)	central cooling source	163.28	0.486	1125.312
	cooling water pump	170.19		
	chilled water pumps	133.45		
	cooling Tower	0.00		
	multi-split/unit air conditioner	0.00		
	total cooling supply	466.92		
heating (Eh)	central heat source	0.00	0.486	58.524
	heating water pump	20.16		
	heat source side water pump	0.00		
	multi-split/unit heat pump	0.00		
	total heating	20.16		
air conditioning fan (Ef)	new exhaust	21.35	0.486	98.616
	fan coil unit	8.74		
	multi-connection indoor unit	0.00		
	full air system	0.00		
	total fans	30.09		
	illumination	72.34	0.486	169.28
	socket equipment	–	0.486	–
other (Eo)	elevator	–	0.486	526.532
	ventilator	145.72		
	domestic hot water (excluding solar energy)	0.00		
	total	145.72		
fossil fuels	category	calorie consumption (kWh/m²)	carbon emission factor (tCO₂/TJ)	carbon emission (tCO₂)
bituminous coal Ⅱ	heating:Heat source boiler	18.326	72	35.425
none	heating:Municipal heat	0.00	0.00	0.000
none	domestic hot water (excluding solar energy)	0.00	0.00	0.000
gas	cooking	3.25	32.17	–
other	category		consumption (kg)	carbon emission (tCO₂)
refrigerant	cooling		0	0.000
renewable	category	powered by (kWh/m²)	carbon emission factor(kgCO₂/kWh)	carbon emission reduction (tCO₂)
renewable Energy (Er)	photovoltaic (Ep)	873.61	163.28	1536.681
	wind force (Ew)	0.00		0.000
	total carbon emissions from building operations			-13.635

The Light of Resilience | 韧性之光 Ⅱ

Reasearch on Green Building Based on Wind System

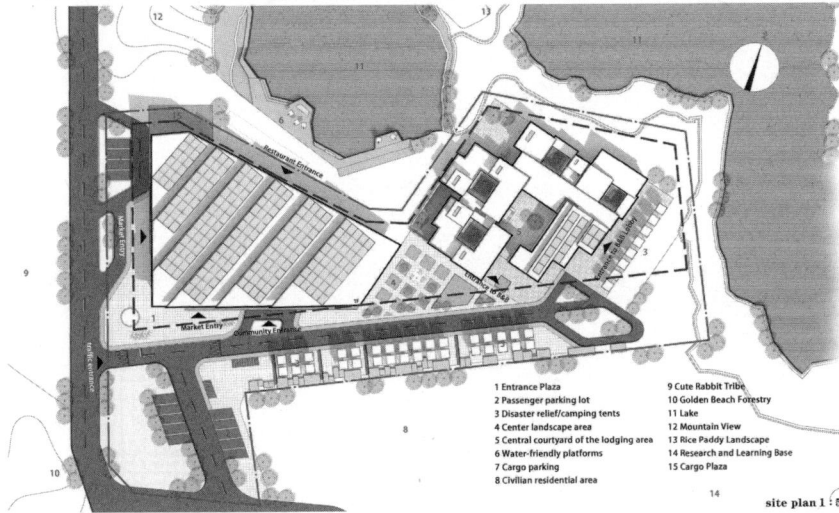

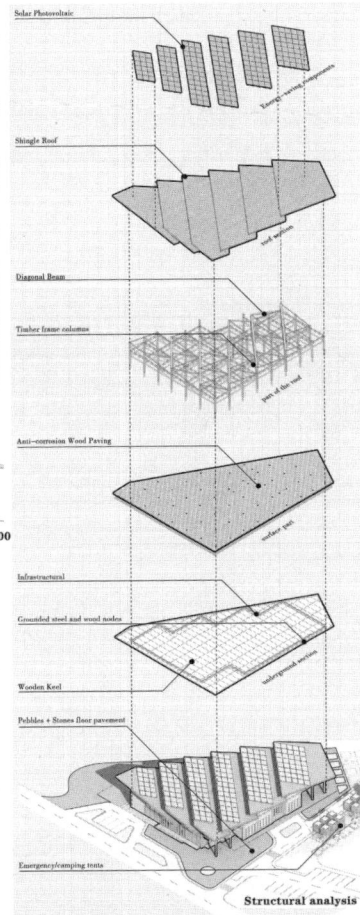

1 Entrance Plaza
2 Passenger parking lot
3 Disaster relief/camping tents
4 Center landscape area
5 Central courtyard of the lodging area
6 Water-friendly platforms
7 Cargo parking
8 Civilian residential area
9 Cute Rabbit Tribe
10 Golden Beach Forestry
11 Lake
12 Mountain View
13 Rice Paddy Landscape
14 Research and Learning Base
15 Cargo Plaza

site plan 1 : 500

Analysis of sunshine hours

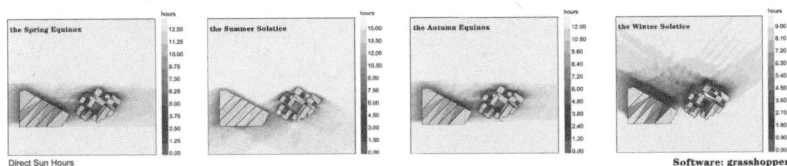

the Spring Equinox
the Summer Solstice
the Autumn Equinox
the Winter Solstice

Direct Sun Hours

Software: grasshopper

Site Design Strategy

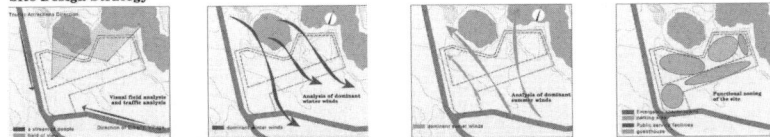

Solar Photovoltaic
Shingle Roof
Diagonal Beam
Timber frame columns
Anti-corrosion Wood Paving
Infrastructural
Grounded steel and wood nodes
Wooden Keel
Pebbles + Stones floor pavement
Emergency/camping trails

Structural analysis

The Light of Resilience | 韧性之光 Ⅲ

Reasearch on Green Building Based on Wind System

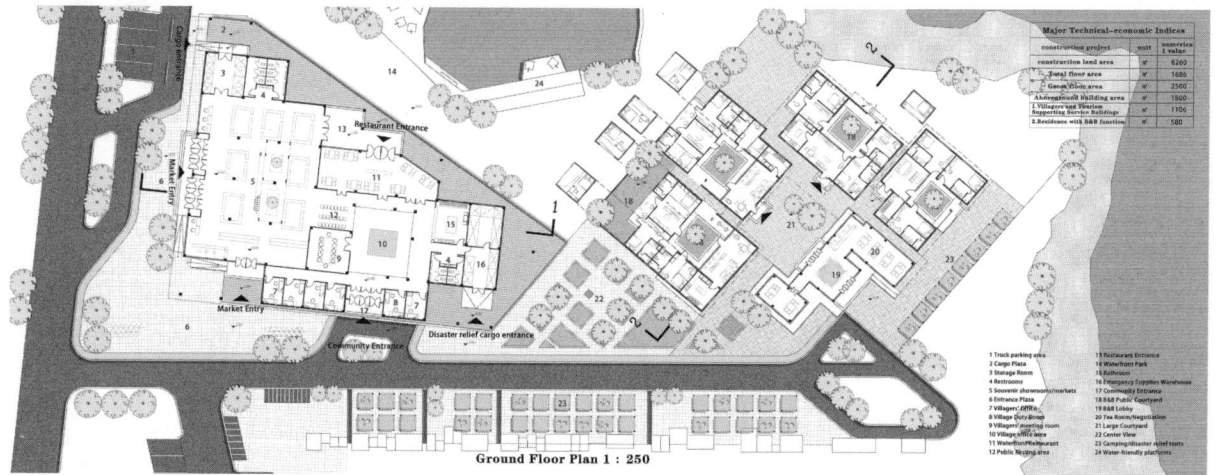

Major Technical-economic Indices		
construction project	unit	numerical value
construction land area	m²	6200
Total floor area	m²	1686
Green floor area	m²	2500
Aboveground Building area	m²	1600
1.Villagers and Tourism Supporting Service Buildings	m²	1106
2.Residence with B&B function	m²	580

1 Truck parking area
2 Cargo Plaza
3 Storage Room
4 Restrooms
5 Souvenir showroom/markets
6 Entrance Plaza
7 Villagers' office
8 Village Duty Room
9 Villagers' meeting room
10 Village office area
11 Waterfront Restaurant
12 Public Nursing area
13 Restaurant Entrance
14 Waterfront Park
15 Bathroom
16 Emergency Supplies Warehouse
17 Community Entrance
18 B&B Lobby
19 B&B Public Courtyard
20 Tea Room/Negotiation
21 Large Courtyard
22 Center View
23 Camping/disaster relief tents
24 Water-friendly platforms

Ground Floor Plan 1 : 250

earthquake resistant construction

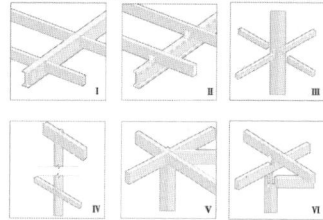

Post–disaster circulatory system analysis

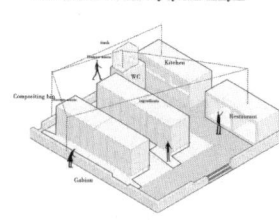

Assembly of disaster relief units

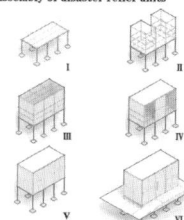

Structural analysis of the disaster relief module

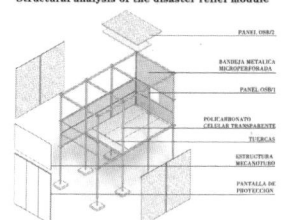

1-1 Section Plan 1 : 250

2-2 Section Plan 1 : 250

The Light of Resilience | 韧性之光 IV

Reasearch on Green Building Based on Wind System

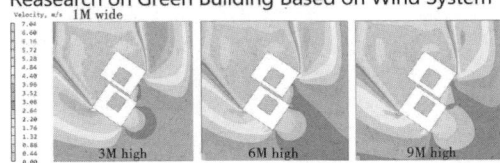

Velocity, m/s 1M wide

3M high | 6M high | 9M high

Velocity, m/s 2.5M wide

3M high | 6M high | 9M high

Velocity, m/s 3M wide

3M high | 6M high | 9M high

Velocity, m/s 5M wide

3M high | 6M high | 9M high

We utilized the "cold alley" principle in conjunction with cavities to obtain the optimal gap width for maximum wind speed through data simulation. We measured three pitch widths of 3M, 6M, and 9M, and three heights of 1M, 2.5M, and 3M in each pitch.

The simulation analysis used the Hebei summer wind, with a wind speed of 2.3 m/s and a wind direction of S–S–E. The simulation analysis used the Hebei summer wind, with a wind speed of 2.3 m/s and a wind direction of S–S–E.

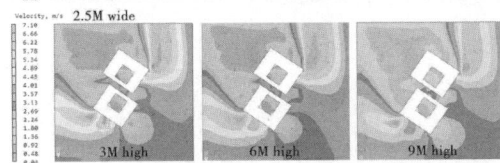

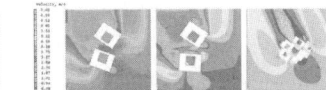

We conducted another angular simulation, taking a gap of 3M width for measurement, and tested the size of three kinds of openings, and finally found that the windspeed in the gap is the largest when the two body blocks are parallel, and the opening will slow down the windspeed in the second half.

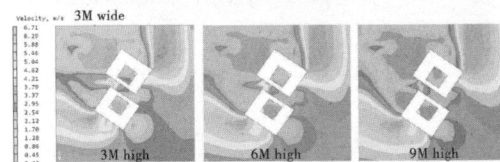

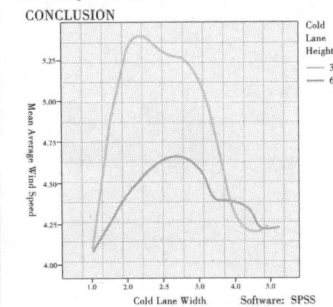

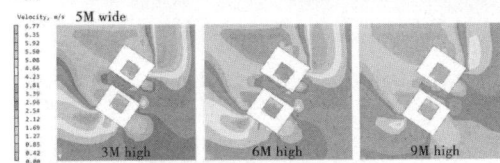

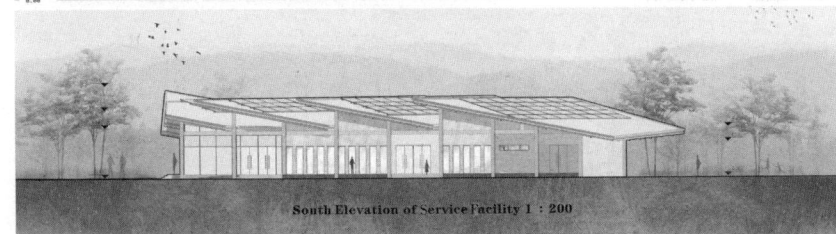

CONCLUSION

Cold Lane Height
— 3
— 6

Mean Average Wind Speed

Cold Lane Width   Software: SPSS

South Elevation of Service Facility 1 : 200

综合奖·入围奖（单项奖·设计创意奖）·张家口市怀安县二堡子村建设项目

Comprehensive Awards − Nomination（Individual Award − Design Creativity Prize）− Erpuzi Village Project，Huai'an County，Zhangjiakou City

注册号：110508

Register Number：110508

项目名称：反向戴森球——高技引导的太阳能韧性乡村设计

Entry Title：Reversed Dyson Sphere: High−Tech Guided Solar Resilient Rural Design

作者：刘桐昊、刘昱辰、李怀宇、刘云娜、张博宣

Authors：Liu Tonghao, Liu Yuchen, Li Huaiyu, Liu Yunna, and Zhang Boxuan

作者单位：重庆大学

Authors from：Chongqing University

指导教师：黄海静、聂诗东

Tutors：Huang Haijing, Nie Shidong

指导教师单位：重庆大学

Tutors from：Chongqing University

°CODE：99-110508

反向戴森球
Reversed Dyson Sphere

——高技引导的太阳能韧性乡村设计
High-Tech Guided Solar Resilient Rural Design

Concept Translation

Centrifugal
Centripetal

Freeman Dyson proposed in 1960 a sphere wrapped around the sun to make full use of sunlight, and the solar thermal power plant is an attempt to this great idea. At present, the solar thermal power station is showing a trend of high efficiency, miniaturization and civil use, and it is much cleaner than the photovoltaic.

Scheme Generation

I. Solar thermal power station   II. Comply with red line   III. Volume variation

IV. Add atriums   V. Set up escape capsule   VI. Set up the heliostat

设计说明

本设计受到戴森球——一种用于捕获恒星几乎全部能量的球形结构的启发，旨在最大化捕获有限地面积内的太阳能。定日镜系统将分散的太阳能反射集中于集热塔顶端，加热熔盐进行发电，能量流动方向恰与戴森球相反。建筑空间布局方面，公共建筑与民宿以集热塔为圆心呈放射状分居场地两侧，二者通过环形廊道相连。设计整合了多项先进技术，包括熔融盐塔式光热发电、抗震分级设计、可变避难逃逸舱及地源热泵系统等。这些高科技手段不仅引导了空间的划分与组合，更提升了建筑在极端环境条件下的自维持能力，确保了村庄的韧性与可持续发展。

Design Description

The design is inspired by the Dyson Sphere—a spherical structure intended to capture nearly all the energy emitted by a star—and aims to maximize solar energy capture within a limited site area. The heliostat system concentrates dispersed solar energy onto the apex of a solar tower, where molten salt is heated for power generation, with the energy flow direction being opposite to that of the Dyson Sphere. In terms of spatial layout, the public building and guesthouse are arranged radially around the solar tower, situated on opposite sides of the site, and connected by a circular corridor. The design integrates multiple advanced technologies, including molten salt tower solar thermal power generation, seismic-resistant hierarchical design, variable emergency escape pods, and the geothermal heat pump system. These high-tech strategies not only guide the spatial organization and configuration, but also enhance the building's self-sustaining capabilities under extreme environmental conditions, ensuring the resilience and sustainable development of the village.

CODE : 99-110508

Reversed Dyson Sphere II
High-Tech Guided Solar Resilient Rural Design

Site Analysis

Erbuzi · Surroundings · Roads · Buildings · Terrain · Sunshine · Noise · Vegetation · Water

Industrial Culture

Woodworking Culture
Erbuzi village has a long history of woodworking.

Almond Cultivation Culture
The Site has a rich planting Culture, with the main variety being the flat almonds.

Climate Simulation

Temprature

Wind Rose

Jan-Mar · Apr-Jun · Jul-Sept · Oct-Dec

Solar Radiation

Jan-Mar · Apr-Jun · Jul-Sept · Oct-Dec

Heliostat Elevation Angle-Time

Winter Solstice - 8:00 · Winter Solstice - 12:00 · Winter Solstice - 16:00
Summer Solstice - 8:00 · Summer Solstice - 12:00 · Summer Solstice - 16:00

Green Building Standard Inspection
After inspection, the carbon emission reduction amount of this design is 17kg CO2 / m2a, and the carbon emission intensity reduction rate is 43.22%. Therefore, the project technology meets the requirements of article 2.0.3 of the General Code for Building Energy Conservation and Renewable Energy Utilization.

Site Plan 1:750

Sun Hours

Spring Solstice · Summer Solstice · Autumn Solstice · Winter Solstice

Incident Radiation

Spring Solstice · Summer Solstice · Autumn Solstice · Winter Solstice

Advantages of Solar Thermal Power

Efficiency · I. Annual Operation Days · II. Maintenance Frequency · III. Life Cycle Production

Disadvantages of Solar Thermal Power

Pollution · IV. Life Cycle Cost · V. Life Cycle Pollution · VI. Life Cycle Damage Rate

CODE : 99-110508

Reversed Dyson Sphere III
High-Tech Guided Solar Resilient Rural Design

Functional Organization

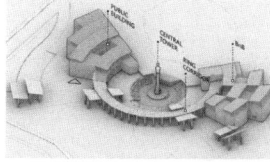

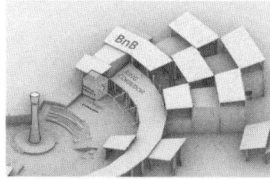

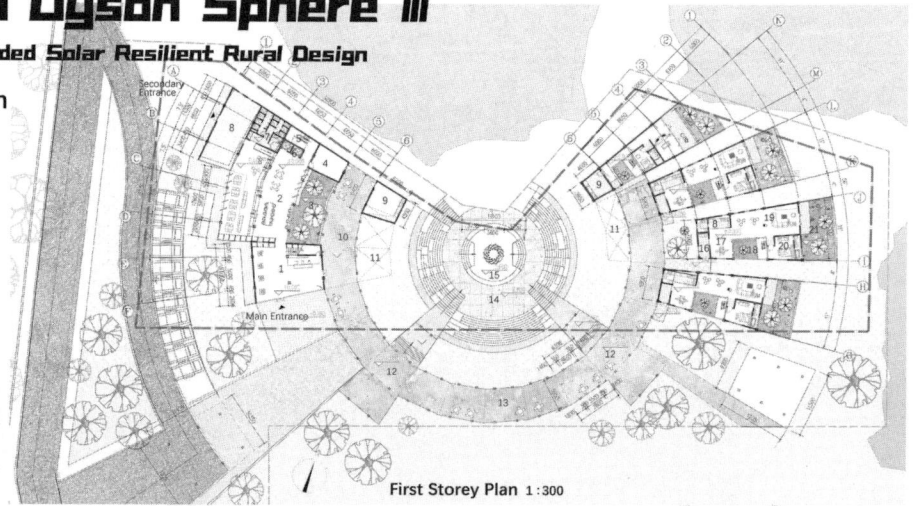

First Storey Plan 1:300

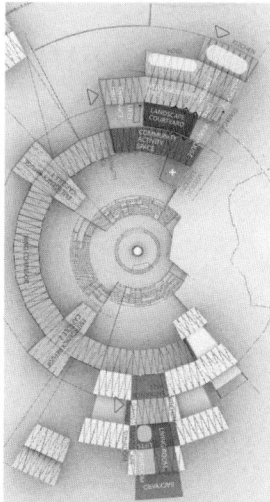

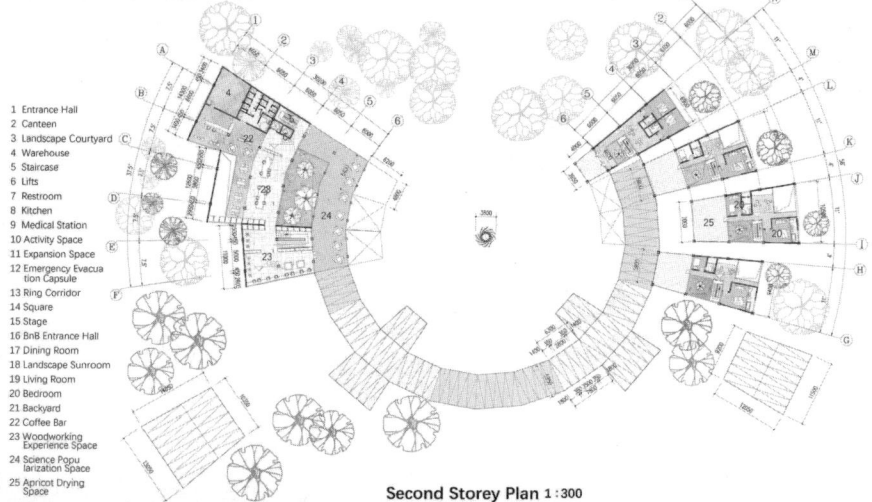

1 Entrance Hall
2 Canteen
3 Landscape Courtyard
4 Warehouse
5 Staircase
6 Lifts
7 Restroom
8 Kitchen
9 Medical Station
10 Activity Space
11 Expansion Space
12 Emergency Evacuation Capsule
13 Ring Corridor
14 Square
15 Stage
16 BnB Entrance Hall
17 Dining Room
18 Landscape Sunroom
19 Living Room
20 Bedroom
21 Backyard
22 Coffee Bar
23 Woodworking Experience Space
24 Science Popularization Space
25 Apricot Drying Space

Second Storey Plan 1:300

High-Tech Overview

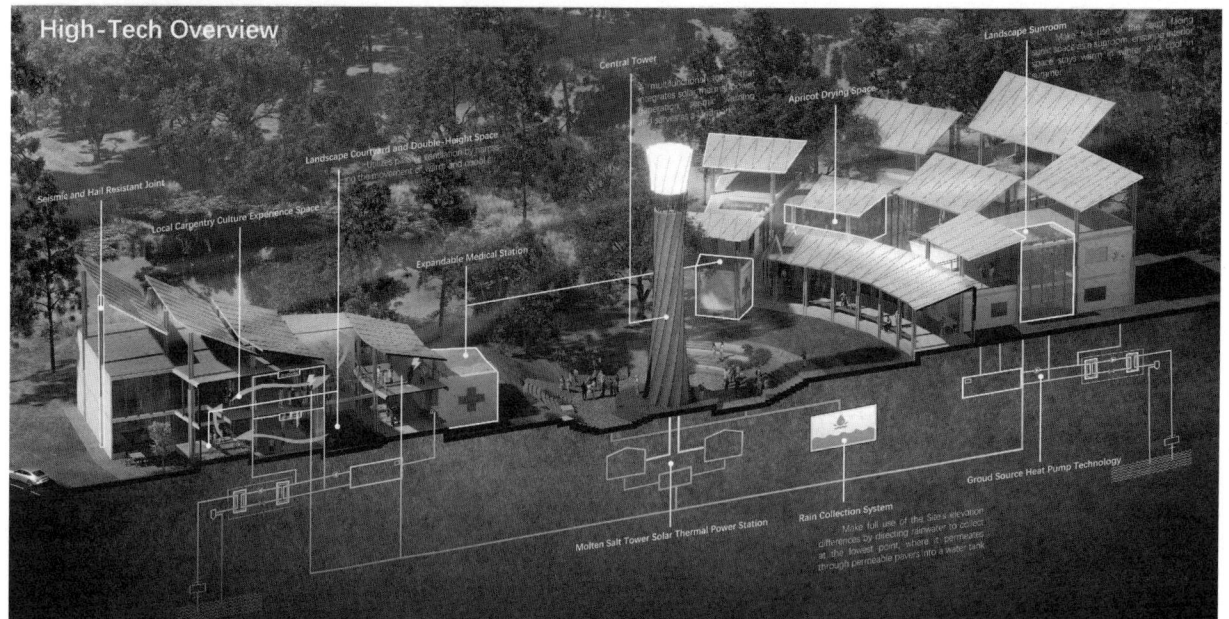

Reversed Dyson Sphere IV
High-Tech Guided Solar Resilient Rural Design

Hierarchical Design of Seismic-Resistant Structures
Seismic-Resistant Structures Layout

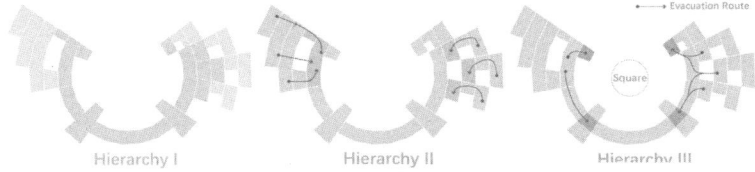

→ Evacuation Route

Square

Hierarchy I   Hierarchy II   Hierarchy III

Seismic-Resistant Structural Analysis

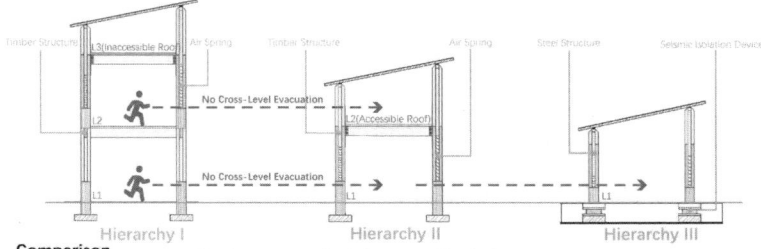

Timber Structure   L3(Inaccesible Roof)   Air Spring   Timber Structure   Air Spring   Steel Structure   Seismic Isolation Device

L2   No Cross-Level Evacuation   L2(Accessible Roof)

L1   No Cross-Level Evacuation   L1   L1

Hierarchy I   Hierarchy II   Hierarchy III

Comparison

	Hierarchy I	Hierarchy II	Hierarchy III
Number of Structural Levels	3	2	1
Structural Self-Weight	Heavy	Medium	Light
Structural Material	Timber	Timber	Steel
Distance from Square	Long	Medium	Short
Special Seismic Design	Air Spring, Timber Joint	Air Spring, Timber Joint	Seismic Isolation Device

Hierarchical Design of Seismic-Resistant Structures utilizes structural weight, material selectio etc to achieve cost-effective and efficient evacuation flow organization, which enhances building safety by minimizing earthquake damage and ensures quick and safe evacuation through optimized pathways. Additionally, It maximizes resource efficiency and adaptability, allowing adjustments for varying needs, ultimately reducing long-term losses and maintenance costs.

Seismic and Hail Resistant Joint

A Seismic-Resistant Structure Combining Heliostat Attitude Control with Energy-Absorbing Rods (**National Utility Model Patent Applied**)

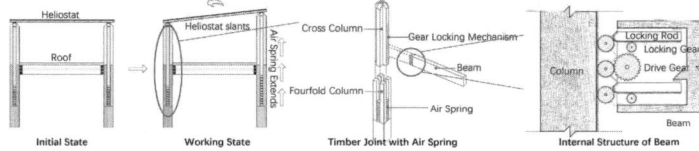

Heliostat   Heliostat slants   Cross Column   Gear Locking Mechanism   Locking Rod

Roof   Air Spring Extends   Locking Gear   Drive Gear

Column   Beam

Fourfold Column   Air Spring

Beam

Initial State   Working State   Timber Joint with Air Spring   Internal Structure of Beam

Variable Space

Emergency Evacuation Capsule

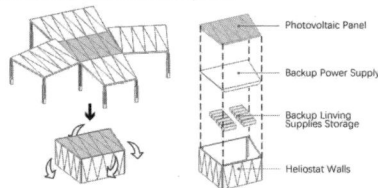

Photovoltaic Panel

Backup Power Supply

Backup Linving Supplies Storage

Heliostat Walls

Expandable Medical Station

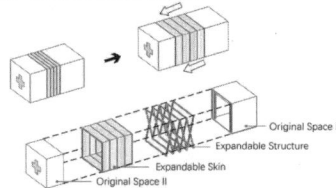

Original Space I

Expandable Structure

Expandable Skin

Expandable Skin

Original Space II

Solar Thermal Power Generation

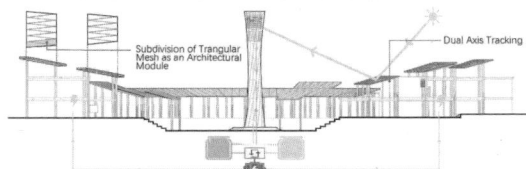

Subdivision of Trangular Mesh as an Architectural Module

Dual Axis Tracking

A molten salt tower solar thermal power station uses heliostats to focus sunlight to heat molten salt at the top of a tower. The stored thermal energy is used for power generation. Its advantages include efficient energy storage, enabling continuous power generation even without sunlight; and minimal environmental impact, making it a green alternative to traditional one.
The overall architectural layout is determined by the fundamental form of the solar thermal power plant, utilizing a regular icositatragon (24-sided polygon) grid to establish the building's scale. To facilitate the mass production of heliostats, a modular subdivision approach using a triangular mesh is employed.

Groud Source Heat Pump Technology

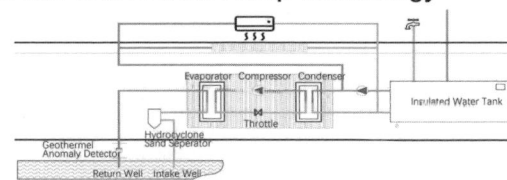

Evaporator Compressor Condenser

Throttle

Insulated Water Tank

Geothermel Anomaly Detector   Hydrocyclone Sand Seperator

Return Well   Intake Well

Below the ground of public buildings and BnBs, a ground source heat pump system is installed, which utilizes the constant temperature characteristics of the underground environment, exchanging heat through buried heat exchange pipes to achieve efficient heating and cooling. Its main advantages include high energy efficiency, operational stability, multifunctional applications, and long system lifespan.
Additionally, a geothermal anomaly detector is installed at the return well. When a geothermal anomaly is detected, which may indicate a precursor to an earthquake, the system will issue an alert through the acoustic and visual alarm devices on the central tower.

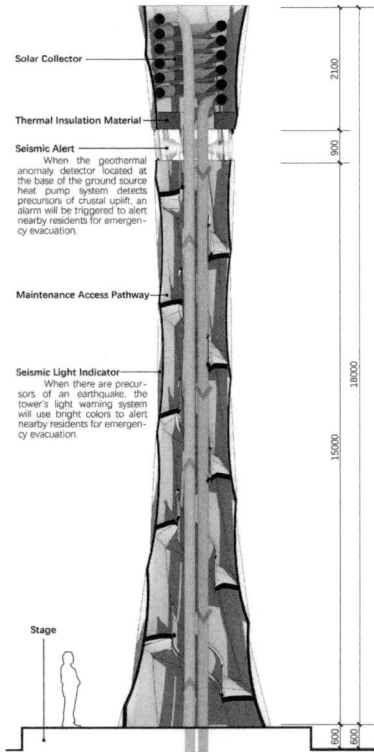

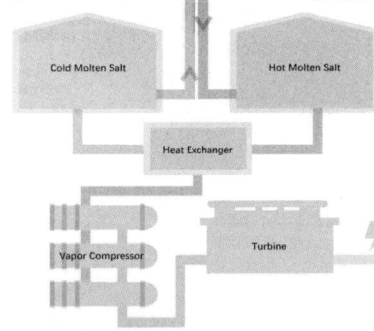

Central Tower

Solar Collector

Thermal Insulation Material

2100

900

Seismic Alert
When the geothermal anomaly detector located at the base of the ground source heat pump system detects precursors of crustal uplift, an alarm will be triggered to alert nearby residents for emergency evacuation.

18000

Maintenance Access Pathway

15000

Seismic Light Indicator
When there are precursors of an earthquake, the tower's light warning system will use bright colors to alert nearby residents for emergency evacuation.

Stage

600 600

Cold Molten Salt	Hot Molten Salt

Heat Exchanger

Vapor Compressor   Turbine

110509

综合奖 · 入围奖 · 张家口市怀安县二堡子村建设项目
Comprehensive Awards − Nomination − Erpuzi Village Project，Huai'an County, Zhangjiakou City

注册号：110509
Register Number：110509
项目名称：光垣 · 暖堡
Entry Title：Light Walls & Warm Fortresses
作者：赵彤山、崔雨洋、田宇琪
Authors：Zhao Tongshan，Cui Yuyang, and Tian Yuqi
作者单位：东南大学
Authors from：Southeast University
指导教师：张彧
Tutor：Zhang Yu
指导教师单位：东南大学
Tutor from：Southeast University

二堡子村公共服务和民宿建筑设计

光 垣 · 暖 堡
Light Walls & Warm Fortresses

Strategy Index

1. Photovoltaic Roof
2. Energy Storage System
3. Water Treatment System
4. Passive Solar Room
5. Light and Air Shaft
6. Windbreak Wall
7. Water-Source Heat Pump
8. Vertical Skylight
9. Trombe Wall
10. Integrated Shading

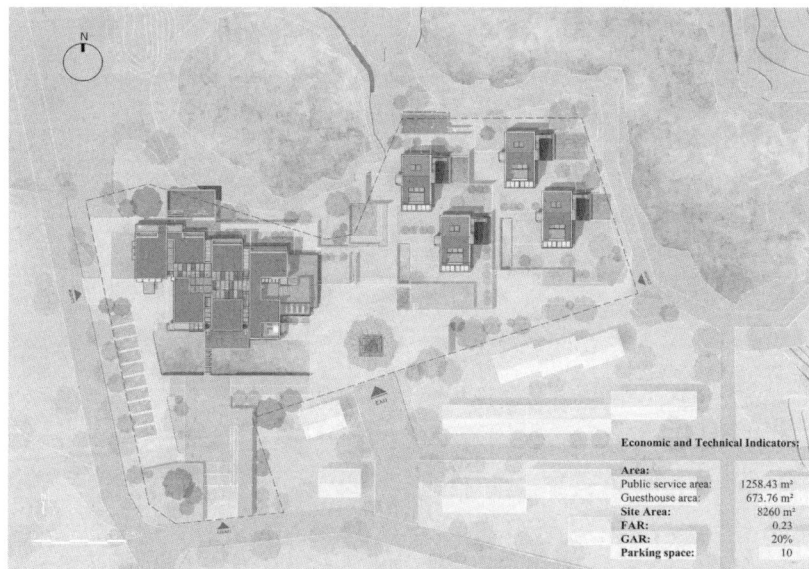

Economic and Technical Indicators

Area:	
Public service area:	1258.43 m²
Guesthouse area:	673.76 m²
Site Area:	8260 m²
FAR:	0.23
GAR:	20%
Parking space:	10

Climate

The site, in rural Zhangjiakou, Hebei, lies in China's "Severely Cold" climate zone. The design focuses on winter performance, using solar energy for insulation and maximizing daylight during low solar altitude, supported by detailed climate data analysis.

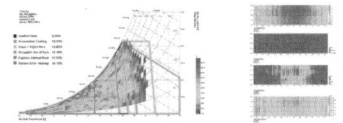

Disaster Resilience

The building will serve as an emergency shelter, requiring structural stability, safety redundancy, and rapid functional conversion. A decentralized structure and spatial layout ensure efficiency and quick evacuation, while the external space must remain open and adaptable for sheltering needs.

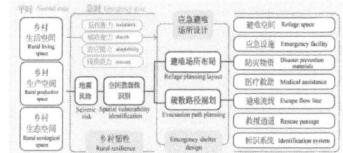

Traditional Experience

Zhangjiakou has a tradition of fortress-like architecture, historically built for military defense. These fortresses featured a solid platform with a "raised courtyard" and lookout spaces, thick walls, and minimal windows for protection. The upper buildings were decentralized for rapid troop mobilization.

Design Inspiration

We draw inspiration from how Zhangjiakou's ancestors defended against enemies to tackle the challenges of cold and earthquakes. Essential equipment for resolving water and electricity issues during emergencies is placed beneath the raised platform, and ascending along this platform leads to a "warm fortress space" on the second level. This space is enclosed by solid building masses to shield against cold winds, while solar energy is stored through a combination of glass roofing, masonry walls, and earth-covered surfaces, providing continuous heating for the building. The building masses are separated by gaps to facilitate rapid evacuation during aftershocks, and seismic joints are incorporated to enhance structural safety and redundancy. The gaps between the masses also allow light to penetrate deep into the building, creating what we call a "light-wall space."

方案设计说明

我们受到当地古堡建筑的启发，学习张家口祖先抵御敌人的方式来抵御寒冷和地震。在公共服务的部分，我们将应急情况下解决水电问题的设备放置于台基之下，沿台基而上，可以进入位于二层的"暖堡空间"。这一空间被建筑体量围合以抵御寒风，并通过玻璃顶、砌体墙和覆土面来储存太阳的能量，延长应急状态下建筑的温暖时间。体量被拉开缝隙，利于余震时的快速疏散，并提高建筑的抗震冗余能力。体量之间的空间也使得光线可以沿进深方向均匀进入建筑，形成"光垣空间"。
在民宿部分，方案延续这一思路，保持克制体型的同时关注竖向空间设计，引入被动式保温和采光策略。并且通过特殊的功能流线设计，实现一栋建筑在灾后重建阶段能安置两个房屋受损的家庭。

Form-giving Process

1. Zoning
public service guesthouse
emergency assembly

2. Arrangement
N
S

3. Adjustment
view view

4. Carving
carving
carving

5. Roof vs Sun
10° 25°

6. Site Planning
courtyard
parking
plaza

110509

Second Floor Plan 1:250

N

First Floor Plan 1:250

Second Floor Plan 1:250

主入口
Maed entrance

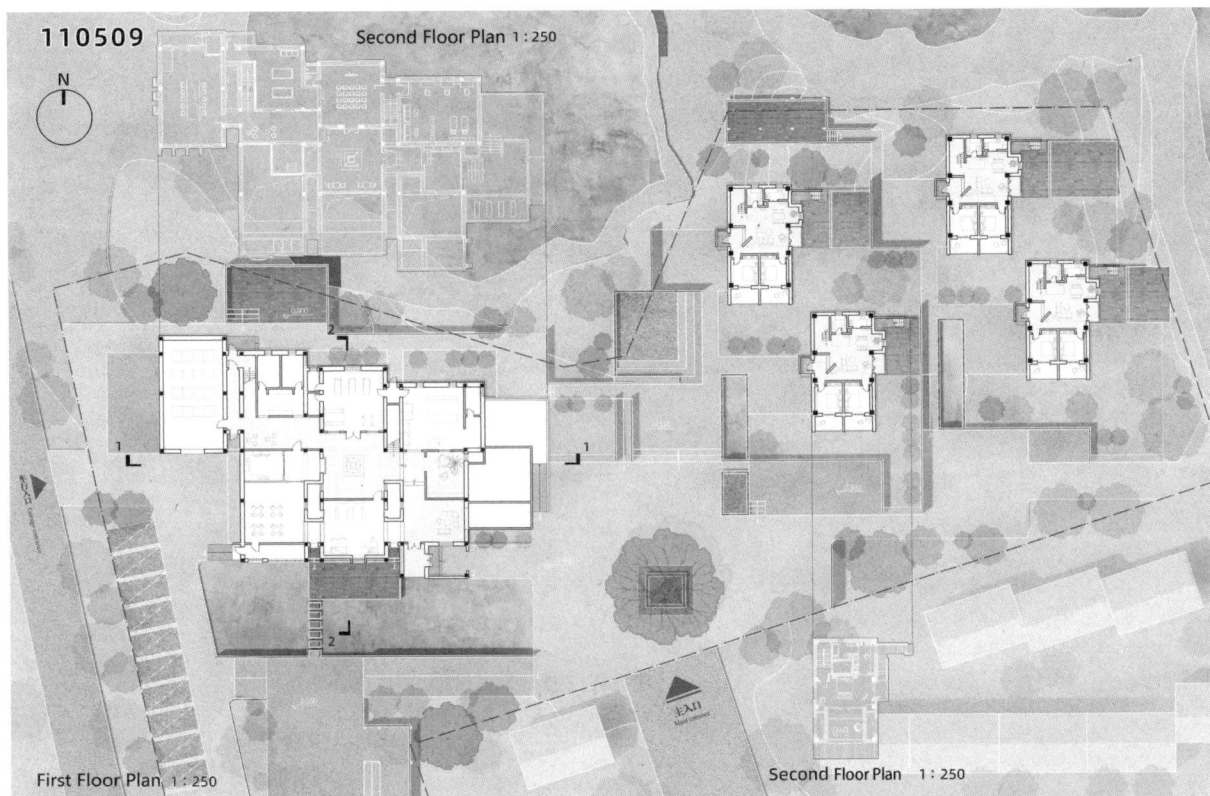

Emergency Shelter Facilities Layout
应急避难设施布局

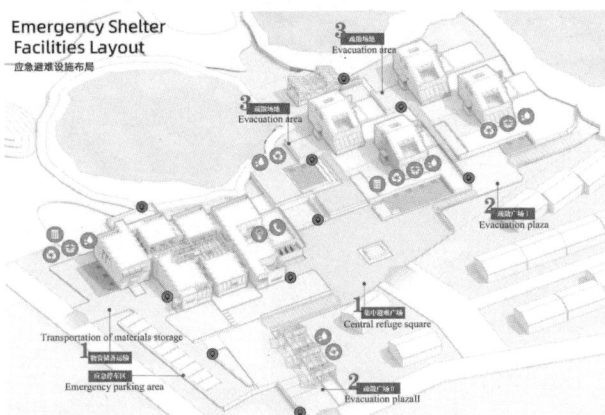

3 疏散场地
Evacuation area

3 疏散场地
Evacuation area

2 疏散广场I
Evacuation plaza

Transportation of materials storage
物资储存运转

1 应急停车区
Emergency parking area

1 集中避难广场
Central refuge square

2 疏散广场II
Evacuation plaza II

Refuse disposal facility	垃圾处理设施	
Health and epidemic prevention	卫生防疫设施	
Emergency supplies	应急物资区域	
Water supply	紧急供水设施	
Emergency communication	应急通信设施	
Guarantee power supply	保障供电设施	
Shelter sign map Evacuation route map	避难场所标志地图 疏散路线地图	

Evacuation place opening sequence 避难场所开放顺序

1 First open 第一顺位开放避难场所

2 Second open 第二顺位开放避难场所

3 Third open 第三顺位开放避难场所

"平时"功能布局
Function layout in normal state

服务中心
Service center

"灾时"功能布局
Functional layout in an emergency

临时避难中心
Temporary shelter center

二堡子村公共服务和民宿建筑设计

光 垣 · 暖 堡
Light Walls & Warm Fortresses

110509

娱乐体闲室
Entertainment

影音室
Audio and video

聚会
Get together

阳光平台
Enjoy the sun

服务流线
Service flow line

游览流线
Tourist flow line

服务流线
Service flow line

共享厨房
Shared kitchen

餐厅
Party restaurant

火塘
Hearth

医务室
Dispensary

服务会客厅
Service meeting room

Construction Detail

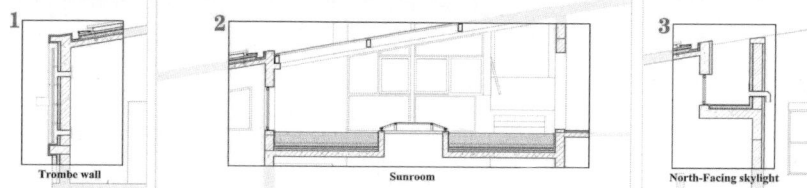

Trombe wall

Sunroom

North-Facing skylight

Spatial Rendering

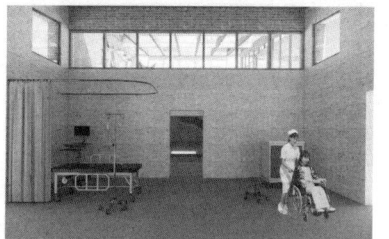

Normal to Emergency Transition Mode

一层功能分区
Layer 1 functional partition

一层功能分区
Layer 1 functional partition

二层功能分区
Layer II functional partition

二层功能分区
Layer II functional partition

"平时"服务中心功能布局
Function layout in Normal State

"灾时"服务中心功能布局
Function layout in Disaster State

功能分区图例
Function partition

公共活动
Public activity

商业服务
Commercial service

娱乐 Entertainment

医疗救援 Medical

餐饮厨房 Catering

卫生盥洗 Rest room

仓储储备 Warehouse

应急宿住
Accommodation

应急管理 Emergency
management

应急卫生 Emergency
hygiene

常用出入口
Regular exit

应急出口
Emergency exit

Section and Elevation

South elevation 1：250

East elevation 1：250

1-1 Section 1：250

2-2 Section 1：250

Emergency Facilities Distribution

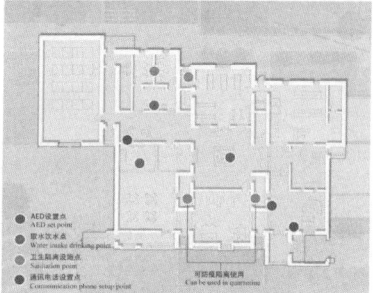

AED设置点
AED set point

取水饮水点
Water intake drinking point

卫生隔离放置点
Sanitation point

通讯布法设置点
Communication phone setup point

可隔离周围使用
Can be used in quarantine

110509

某学院公共服务和民宿建筑设计

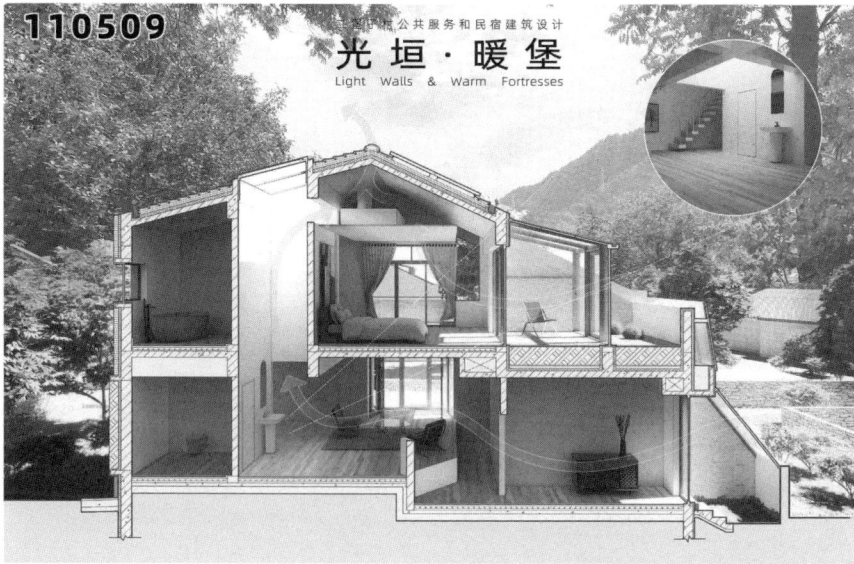

光垣·暖堡
Light Walls & Warm Fortresses

Comparison Simulation

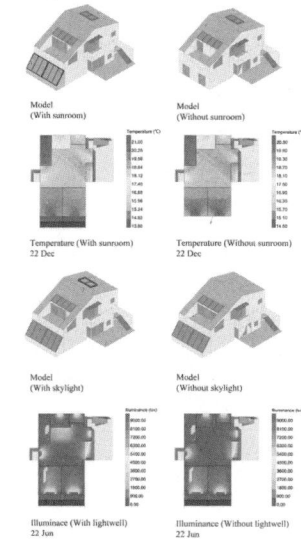

Model
(With sunroom)

Model
(Without sunroom)

Temperature (With sunroom)
22 Dec

Temperature (Without sunroom)
22 Dec

Model
(With skylight)

Model
(Without skylight)

Illuminace (With lightwell)
22 Jun

Illuminace (Without lightwell)
22 Jun

Active Building Technologies

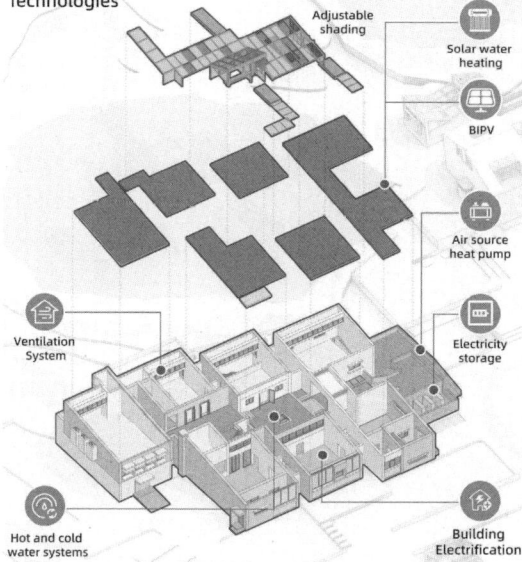

Adjustable shading

Solar water heating

BIPV

Air source heat pump

Electricity storage

Building Electrification

Hot and cold water systems

Ventilation System

Indoor Temperature in Winter

The design solution achieved an average indoor temperature of 20 degrees. In winter, the sunroom functions as a warming room, raising the temperature inside and improve the indoor illumination. On average over the year, the control group was 1.9 degrees Celsius warmer than the average temperature of the con-

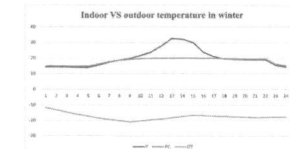

Indoor VS outdoor temperature in winter

Energy Load

Energy balancing reduces the building's energy consumption. Compared to the control group, the sunroom group experienced a 9.7% reduction in load throught the year. In winter, heat loss can be reduced through the warming room. Fire pits can also be used for heating.

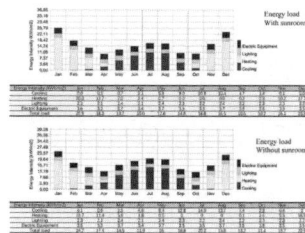

Energy load
With sunroom

Energy load
Without sunroom

Passive Design

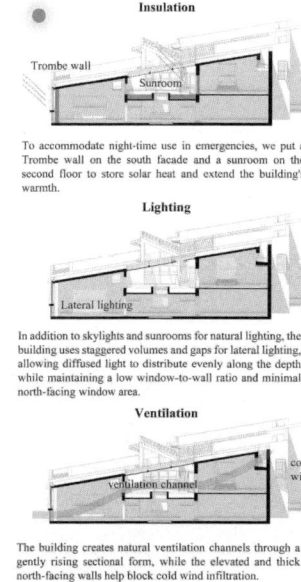

Insulation

Trombe wall
Sunroom

To accommodate night-time use in emergencies, we put a Trombe wall on the south facade and a sunroom on the second floor to store solar heat and extend the building's warmth.

Lighting

Lateral lighting

In addition to skylights and sunrooms for natural lighting, the building uses staggered volumes and gaps for lateral lighting, allowing diffused light to distribute evenly along the depth while maintaining a low window-to-wall ratio and minimal north-facing window area.

Ventilation

ventilation channel
cold wind

The building creates natural ventilation channels through a gently rising sectional form, while the elevated and thick north-facing walls help block cold wind infiltration.

Comparison Simulation （Public Service Building）

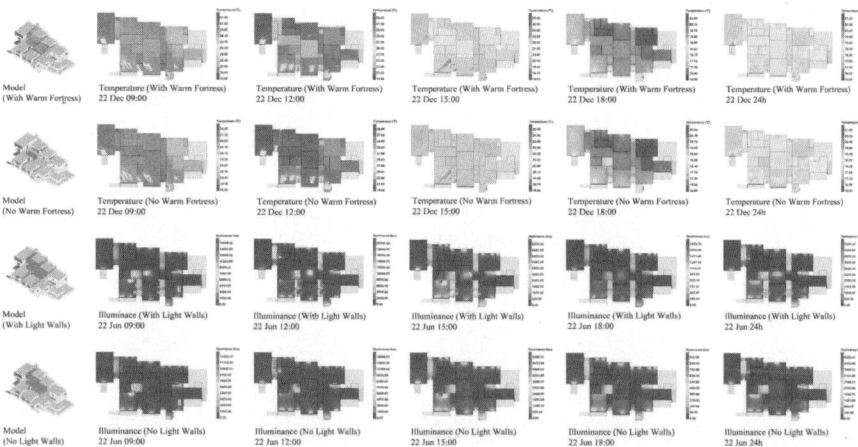

Model (With Warm Fortress)

Temperature (With Warm Fortress) 22 Dec 09:00

Temperature (With Warm Fortress) 22 Dec 12:00

Temperature (With Warm Fortress) 22 Dec 15:00

Temperature (With Warm Fortress) 22 Dec 18:00

Temperature (With Warm Fortress) 22 Dec 24h

Model (No Warm Fortress)

Temperature (No Warm Fortress) 22 Dec 09:00

Temperature (No Warm Fortress) 22 Dec 12:00

Temperature (No Warm Fortress) 22 Dec 15:00

Temperature (No Warm Fortress) 22 Dec 18:00

Temperature (No Warm Fortress) 22 Dec 24h

Model (With Light Walls)

Illuminace (With Light Walls) 22 Jun 09:00

Illuminace (With Light Walls) 22 Jun 12:00

Illuminace (With Light Walls) 22 Jun 15:00

Illuminace (With Light Walls) 22 Jun 18:00

Illuminace (With Light Walls) 22 Jun 24h

Model (No Light Walls)

Illuminace (No Light Walls) 22 Jun 09:00

Illuminace (No Light Walls) 22 Jun 12:00

Illuminace (No Light Walls) 22 Jun 15:00

Illuminace (No Light Walls) 22 Jun 18:00

Illuminace (No Light Walls) 22 Jun 24h

Microclimate Control

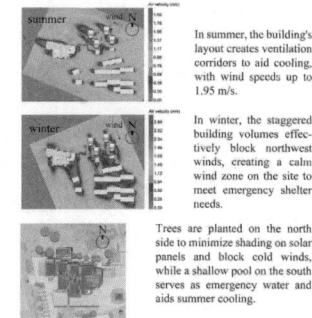

summer

In summer, the building's layout creates ventilation corridors to aid cooling, with wind speeds up to 1.95 m/s.

winter

In winter, the staggered building volumes effectively block northwest winds, creating a calm wind zone on the site to meet emergency shelter needs.

Trees are planted on the north side to minimize shading on solar panels and block cold winds, while a shallow pool on the south serves as emergency water and aids summer cooling.

Carbon Emission Calculation

Trough simulation, the service area's carbon emissions for spring, summer, fall, and winter are 0.84, 0.13, 1.03, and 3.00 kg/m², with an annual average of 1.25 kg/m² and total emissions of 1573.04 kg for 1258.43 m². The guestroom area's emissions are 0.30, 0.12, 0.32, and 0.72 kg/m² per season, averaging 0.37 kg/m² annually, with total emissions of 249.29 kg for 673.76 m². Combined, both areas emit 1822.33 kg of carbon.

综合奖 · 入围奖 · 张家口市
怀安县二堡子村建设项目
Comprehensive Awards –
Nomination – Erpuzi Village
Project，Huai'an County，
Zhangjiakou City

注册号：110514
Register Number：110514
项目名称：原居
Entry Title：Original Habitat
作者：严章鑫、邢茹、李秋予
Authors：Yan Zhangxin, Xing Ru, and
　　　　　Li Qiuyu
作者单位：北京建筑大学
Authors from：Beijing University of Civil
　　　　　　　Engineering and Architecture
指导教师：刘博、李勤
Tutors：Liu Bo，Li Qin
指导教师单位：北京建筑大学
Tutors from：Beijing University of Civil
　　　　　　　Engineering and Architecture

原居 Original Habitat
——Design of Erbaozi Plan in Zhangjiakou, Hebei Province

设计说明 Design Description

历史沿革 Historical Development

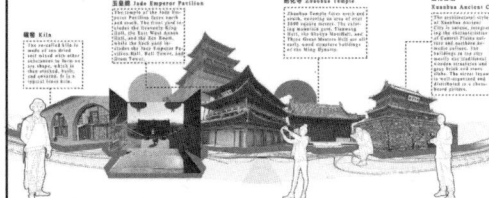

气候分析 Climate Analysis

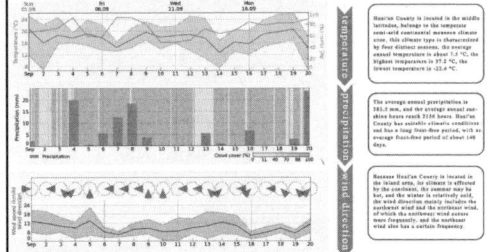

人群分析 Crowd Analysis

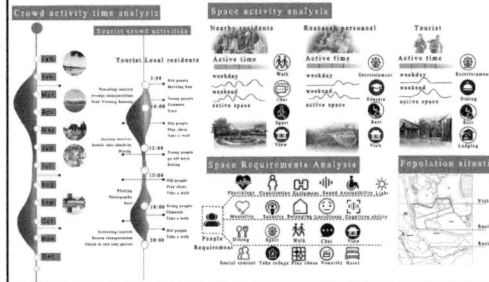

基地分析 Base Analysis

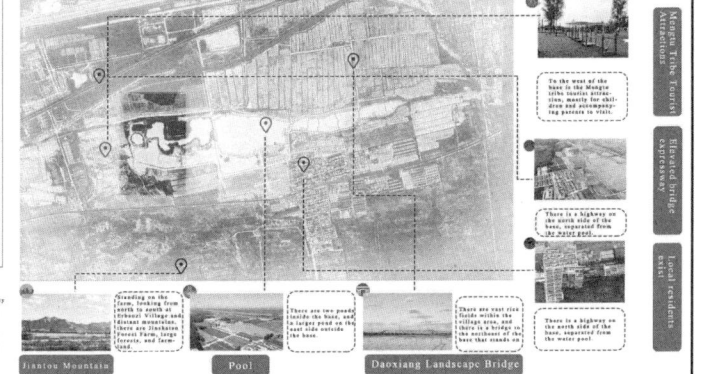

总平面图 General Layout

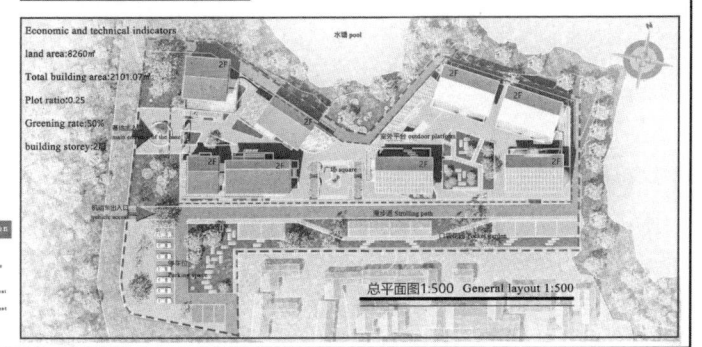

总平面图1:500　General layout 1:500

原居 Original Habitat —— Design of Erbaozi Plan in Zhangjiakou, Hebei Province II

SWOT分析 SWOT analysis

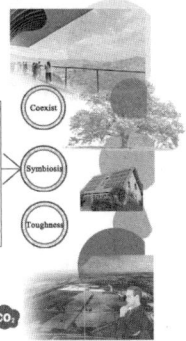

Low-E玻璃分析 Low-E glass analysis

夏季 summertime 冬季 wintertime

平面图 Planc graph

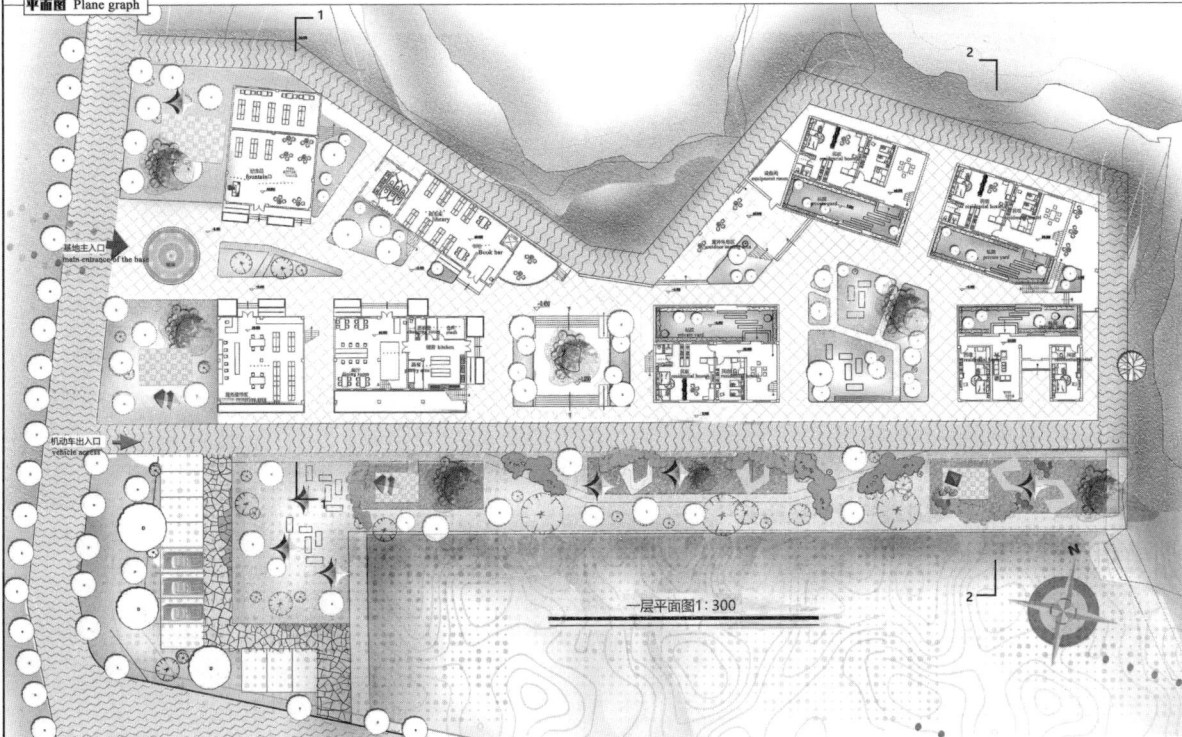

基地主入口 main-entrance of the base

机动车出入口 vehicle access

一层平面图1:300

N

立面图 Elevation

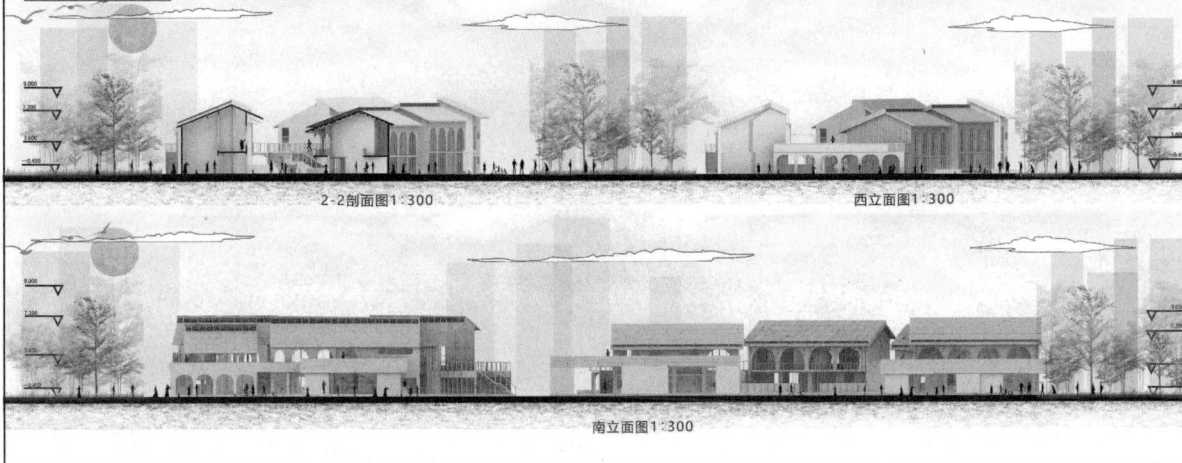

2-2剖面图1:300   西立面图1:300

南立面图1:300

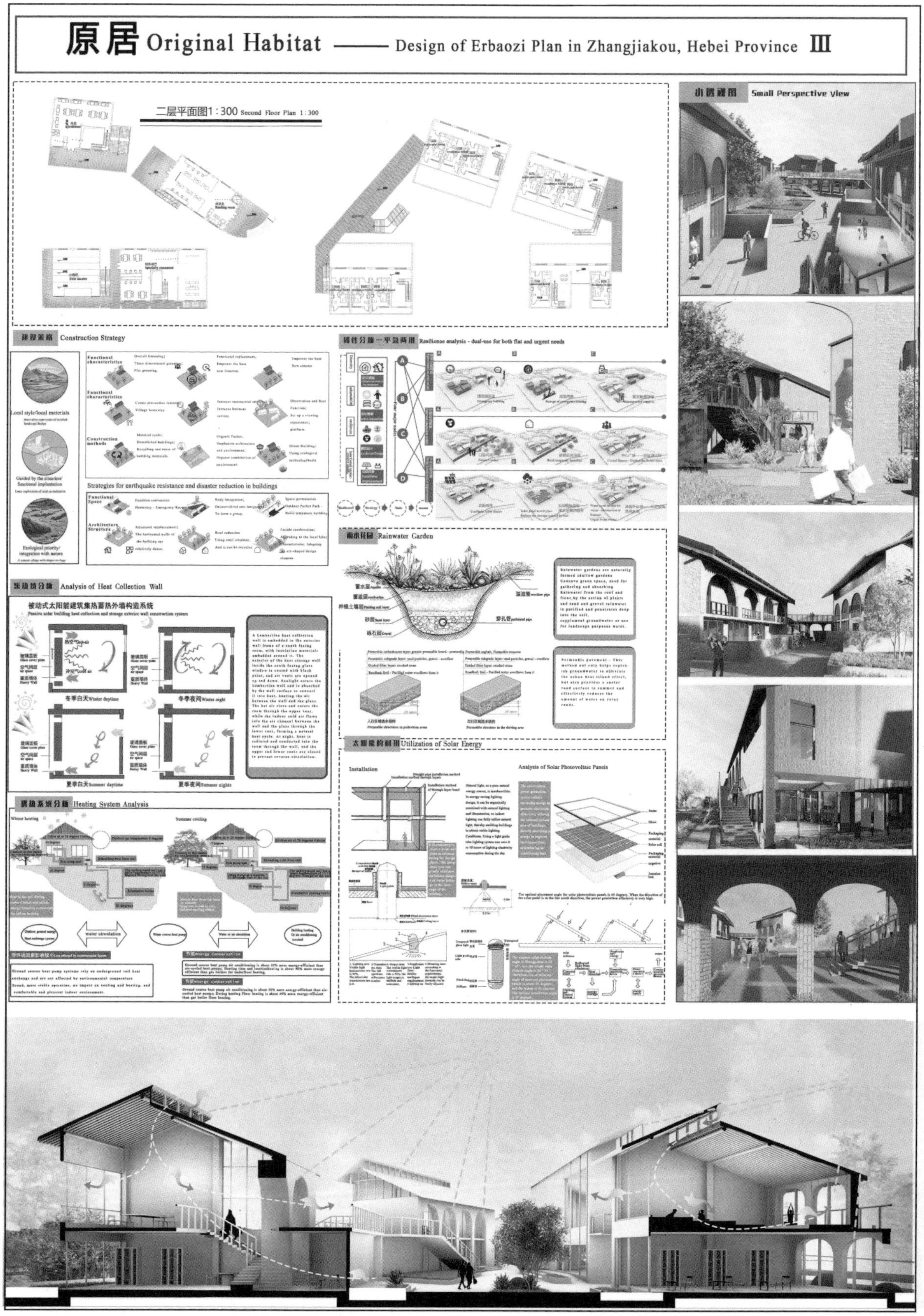

原居 Original Habitat —— Design of Erbaozi Plan in Zhangjiakou, Hebei Province III

二层平面图1:300 Second Floor Plan 1:300

小透视图 Small Perspective View

建设策略 Construction Strategy

Local style/local materials

Guided by the situation/functional implantation

Ecological priority/integration with nature

Strategies for earthquake resistance and disaster reduction in buildings

韧性分析—平急两用 Resilience analysis - dual-use for both flat and urgent needs

集热墙分析 Analysis of Heat Collection Wall

被动式太阳能建筑集热蓄热外墙构造系统
Passive solar building heat collection and storage exterior wall construction system

冬季白天 Winter daytime
冬季夜间 Winter night
夏季白天 Summer daytime
夏季夜间 Summer nights

雨水花园 Rainwater Garden

太阳能的利用 Utilization of Solar Energy

Installation

Analysis of Solar Photovoltaic Panels

供热系统分析 Heating System Analysis

Winter heating
Summer cooling

原居 Original Habitat —— Design of Erbaozi Plan in Zhangjiakou, Hebei Province   IV

功能流线分析 Function and Streamline Analysis

Functional analysis

Floor plan function room
Second floor plan function room

Public service activity area
Residential area

Streamline analysis

Sourcer
Library
Service reception area
Dining room

Residential hostel
Residential hostel
Residential hostel
Residential hostel

Logistics flow line
Accommodation visitor flow line
Tourist activity flow line

技术指标表 Technical Index Sheet

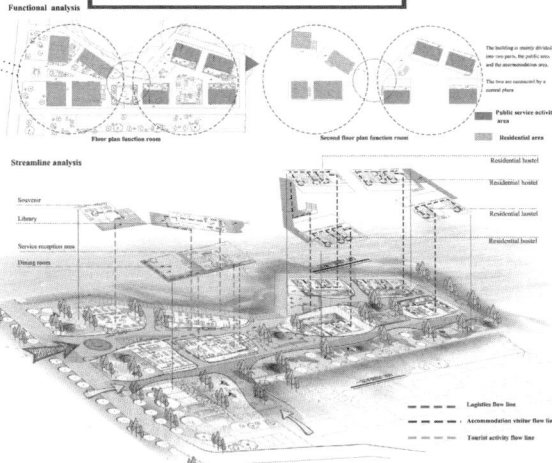

Technical index review

Energy consumption calculation report

Review of energy efficiency index of near zero energy/ultra-low energy buildings

Item	Numerical value	Standard requirements	Whether the requirements are met
Comprehensive value of building energy	0.0	<30.00kWh/(m²·a)	Be satisfied
Renewable Energy utilization rate (%)	369.43	≥10%	Be satisfied
Performance index of building body	Annual heat consumption for heating(kWh/(m²·a))	14.72	<15.00	Be satisfied
	Annual heat consumption for cooling(kWh/(m²·a))	0.0	<5.03	Be satisfied
	Building air tightness rate(change frequency N50)	0.50	<0.60	Be satisfied
Conclusions of the review	The technology of the project meets the requirements of near-zero energy buildings or the Technical Standards for Near-zero Energy Buildings		

Average design energy level review

Item	Numerical value	Standard requirements	Whether the requirements are met
Building comprehensive energy saving rate (%)	126.45	≥30%	Be satisfied
Conclusions of the review	The technology of this project meets the requirements of Article 3.0.1 of the General Code for Energy Conservation and Utilization of Renewable Energy in Buildings		

Review of average energy consumption index for heating and cooling

Item	Numerical value	Standard requirements	Whether the requirements are met
Heat consumption index for heating (MJ/m2·a)	52.97	<92.00	Be satisfied
Cooling power consumption index (kWh/m2·a)	0.0	-	Be satisfied
Conclusions of the review	The project technology meets the requirements of Article 2 6.2 of the General Code for Building Energy Conservation and Renewable Energy Utilization		

Carbon intensity index review

Item	Numerical value	Standard requirements	Whether the requirements are met
Reduction in carbon emission intensity(kg Co2/m2·a)	69.50	≥7	Be satisfied
Carbon emission intensity reduction rate (%)	127.50	≥40%	Be satisfied
Conclusions of the review	The technology of the project meets the requirements of Article 2 6.3 of the General Code for Building Energy Conservation and Renewable Energy Utilization		

Renewable energy produces electricity every month

■ Photovoltaic power generation ■ Wind power generation

Item	Carbon emission kgco₂	Carbon emissions per unit area kgco₂/㎡	Proportion of proportion
Carbon emissions during the use of the building	0.0	0.0	0.0%
Carbon emissions during construction	4421.50	1.87	0.65%
Carbon emissions during the building demolition phase	13264.50	5.60	1.95%
Carbon emission during the production of building materials	663731.27	280.28	97.40%
Carbon emission during the transportation of building material	28.34	0.01	0.00%
Total carbon emissions	681445.61	287.77	100%

Design the building
Base building

■ Heating energy consumption
■ Elevator energy consumption
■ Energy consumption of transmission and distribution
■ Lighting energy consumption
■ For conditioning energy consumption
■ Domestic hot water consumption

抗震救灾分析 Seismic Hazard Reduction

Seismic measure

For the masonry structure of the house, the transverse wall is often the main seismic component. If the distance between the transverse wall is too large, it is easy to cause whipping effect at the transverse wall during the earthquake, increasing the damage degree. At the same time, the direction perpendicular to the transverse wall during the earthquake will produce large shear force due to the action of gravity, and it is difficult to form sufficient resistance when the distance is too large.

Earthquake-reduction measures

1. Site height difference treatment: There is a height difference of 1.5 meters in the base, so the terrain should be flat and open as far as possible, and the upper layer should be dense, so the built part will be flattened and the part with height difference will be built outdoor platform.

2. Foundation treatment: the foundation trench should be wide and thick, and the bottom of the trough should be evenly laid with lime soil layer and ramped by layers. The foundation can be made of cement slurry brick or stone concrete, and the foundation can also be strengthened by adding piles and other technologies.

3. Building structure layout: the plane layout is simple and regular, and the wall layout is uniform and symmetrical to improve the seismic performance. The horizontal walls are denser to enhance the integrity of the house. As few holes or small holes as possible on the wall to avoid weakening the strength and integrity of the wall.

4. Roof design: The roof should be made of lightweight materials to reduce the load during earthquakes. The wall and partition wall should also be light to avoid the collapse of the heavy wall to cause disasters.

Transverse wall bearing structure

Horizontal wall bearing structure layout
Horizontal wall bearing structure scheme elevation

VS

Longitudinal wall bearing structure

Longitudinal wall bearing structure layout
Vertical wall bearing structure scheme elevation

日照分析图 Insolation Chart

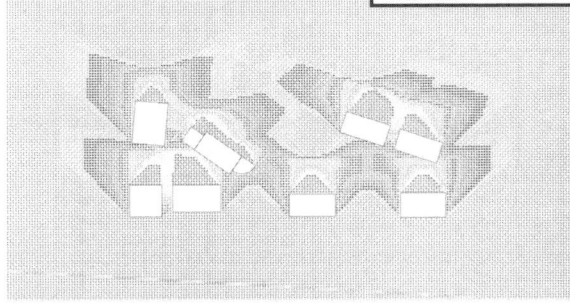

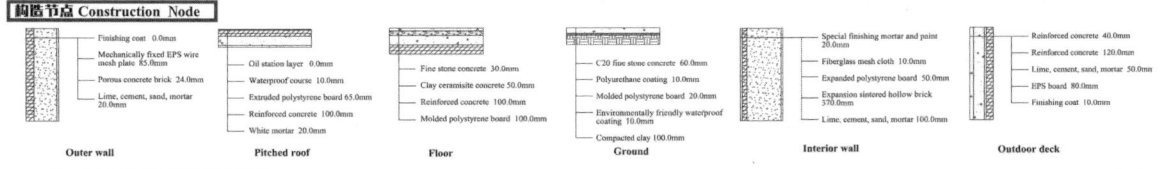

构造节点 Construction Node

Outer wall
- Finishing coat 0.0mm
- Mechanically fixed EPS wire mesh plate 85.0mm
- Porous concrete brick 24.0mm
- Lime, cement, sand, mortar 20.0mm

Pitched roof
- Oil station layer 0.0mm
- Waterproof course 10.0mm
- Extruded polystyrene board 65.0mm
- Reinforced concrete 100.0mm
- White mortar 20.0mm

Floor
- Fine stone concrete 30.0mm
- Clay ceramsite concrete 50.0mm
- Reinforced concrete 100.0mm
- Molded polystyrene board 100.0mm

Ground
- C20 fine stone concrete 60.0mm
- Polyurethane coating 10.0mm
- Molded polystyrene board 20.0mm
- Environmentally friendly waterproof coating 10.0mm
- Compacted clay 100.0mm

Interior wall
- Special finishing mortar and paint 20.0mm
- Fiberglass mesh cloth 10.0mm
- Expanded polystyrene board 50.0mm
- Expansion sintered hollow brick 370.0mm
- Lime, cement, sand, mortar 100.0mm

Outdoor deck
- Reinforced concrete 40.0mm
- Reinforced concrete 120.0mm
- Lime, cement, sand, mortar 50.0mm
- EPS board 80.0mm
- Finishing coat 10.0mm

爆炸图 Explosion Diagram

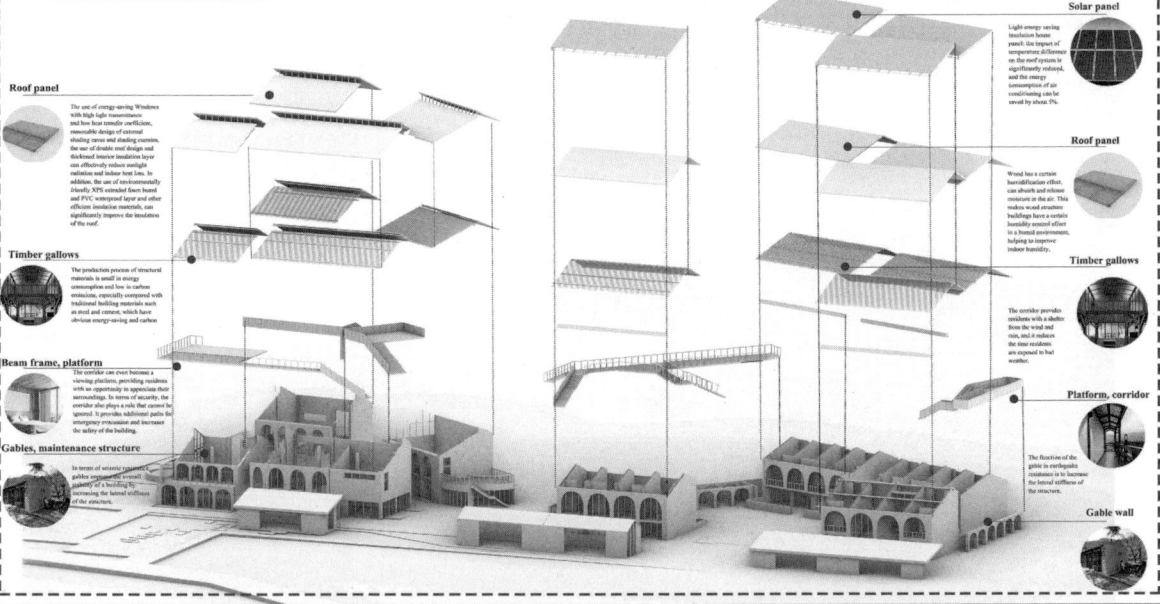

Roof panel
The use of energy-saving Windows with high light transmittance and low heat transfer coefficient, removable design of external shading curtains and shading curtains, the use of double steel design and thickened interior insulation layer can effectively reduce sunlight radiation and indoor heat loss. In addition, the use of environmentally friendly XPS extruded foam board and PVC waterproof layer and other efficient insulation materials, can significantly improve the insulation of the roof.

Timber gallows
The production process of structural materials is small in energy consumption and low in carbon emission, especially compared with traditional building materials such as steel and cement, which have obvious energy-saving and carbon...

Beam frame, platform
The corridor can even become a viewing platform, providing residents with an opportunity to appreciate their surroundings. In terms of security, the corridor also plays a role that cannot be ignored. It provides additional paths for emergency evacuation and increases the safety of the building.

Gables, maintenance structure
In terms of seismic resistance, gables significantly reduce the structural stability of a building by increasing the lateral stiffness of the structure.

Solar panel
Light energy saving insulation house panel: the impact of temperature difference on the roof system is significantly reduced, and the energy consumption of air conditioning can be saved by about 5%.

Roof panel
Wood has a certain humidification effect, can absorb and release moisture in the air. This makes wood structure buildings have a certain humidity control effect in a humid environment, helping to improve indoor humidity.

Timber gallows
The corridor provides residents with a shelter from the wind and rain, and it makes the time residents are exposed to bad weather.

Platform, corridor
The function of the gable in earthquake resistance is to increase the lateral stiffness of the structure.

Gable wall

综合奖·入围奖·张家口市怀安县二堡子村建设项目
Comprehensive Awards — Nomination — Erpuzi Village Project, Huai'an County, Zhangjiakou City

注册号：110515
Register Number：110515

项目名称：稻田里的守望者
Entry Title：The Protector in the Field

作者：李青芫、赵子意、常竞萱、
　　　张蕴泽、王阐亿、吕璐瑶
Authors：Li Qingyuan, Zhao Ziyi,
　　　Chang Jingxuan, Zhang Yunze,
　　　Wang Chanyi, and Lü Luyao

作者单位：重庆大学
Authors from：Chongqing University

指导教师：张海滨、宗德新
Tutors：Zhang Haibin, Zong Dexin

指导教师单位：重庆大学
Tutors from：Chongqing University

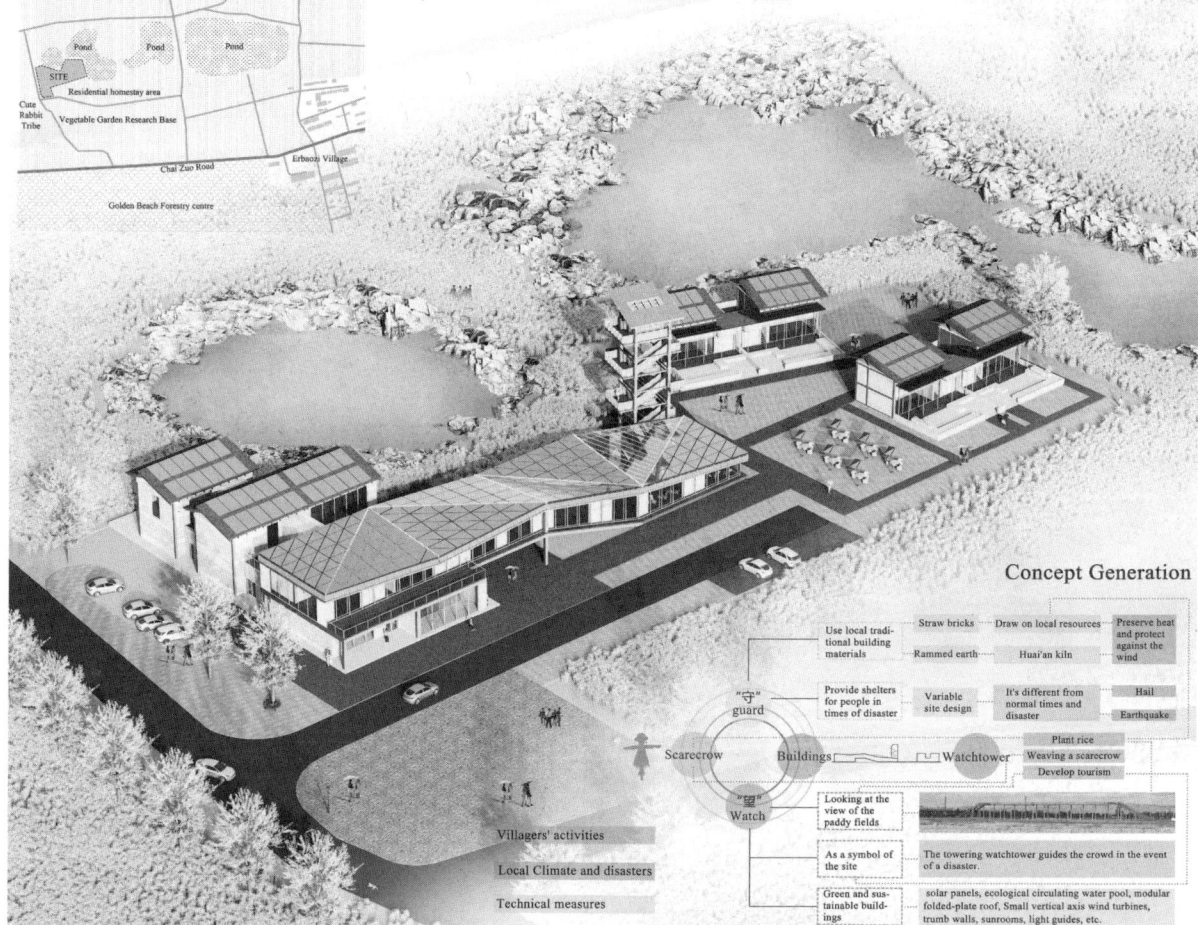

稻田里的守望者·一
The Protector in the Field·I

Design Instructions

稻草人和村民是稻田和文化的守望者，建筑是村落和居民的守望者，本设计以"守望"为灵感，建筑"守"的是当地文化和灾后安全，高塔"望"的是稻田民生，更是新技术的应用，场地采用"平灾转换模式"进行韧性设计，提供紧急避难场所。

设计结合当地气候环境，充分利用当地材料如稻草、秸秆、夯土，结合太阳能光伏板、生态循环水池、模数化折板屋顶、小型直轴风力发电机、特朗勃墙、阳光房、导光管等，进行太阳能主被动技术应用。

Scarecrows and villagers are the watchmen of rice paddies and culture, and buildings are the watchmen of villages and residents. This design is inspired by "keep watch". The building "guards" local culture and post-disaster safety, and the high tower "watches" the people's livelihood of rice fields, which is the application of new technology. The site adopts the "disaster conversion mode" for resilience design to provide an emergency shelter.

The design combines the local climate and environment, makes full use of local materials such as straw, straw, rammed soil, and combines solar panels, ecological circulating water pool, modular folded-plate roof, Small vertical axis wind turbines, trumb walls, sunrooms, light guides, etc., to carry out the application of solar main and passive technology.

Site Analysis

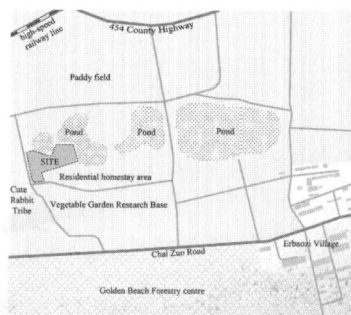

Local Climate

emperature and humidity

Huai'an County, Zhangjiakou has distinct seasons, The winter here is cold and dry, and the summer is hot.So it is necessary to pay attention to keeping warm in winter and preventing heat in summer when conducting architectural design.

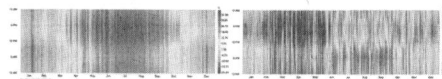

Enthalpy Diagram

Time (hr)
city: Zhangjiakou
country: CHN
time-zone: 8.0
source: Custom-544010

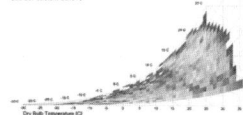

Sunlight Analysis

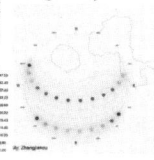

Wind Rose

The northwest wind prevails in the site in winter, so it is necessary to pay attention to the wind protection in winter.

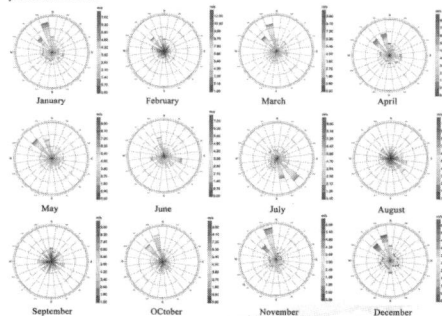

January　February　March　April

May　June　July　August

September　OCtober　November　December

Aerial View of The Site

The project site is close to the residential areas of Erbaozi Village and Beiguojiayao Village, with a straight-line distance of 1km.

high-speed railway line　Erbaozi Village　SITE

Pond　SITE　Erbaozi Village　My field

Meteorological and Geological Disasters

Hail

There are many hail disasters in Huai'an County, Zhangjiakou, so the buildings here must have the ability to resist hail.

Many countries with serious hailstorms have conducted artificial hailstorm prevention experiments. China in addition to Guangdong, Hunan and other provinces with less hail, hailstorms of varying degrees every year everywhere. Especially in the north of the mountainous and hilly areas, the terrain is complex, the weather is variable, hail, heavy damage, a great harm to agriculture.

Concept Generation

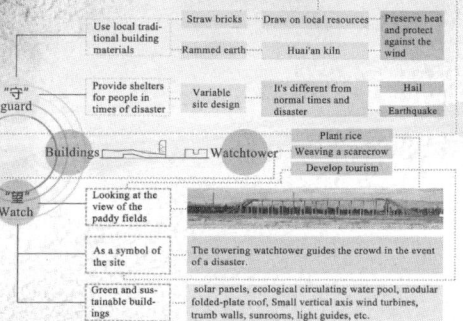

Use local traditional building materials	Straw bricks	Draw on local resources	Preserve heat and protect against the wind
	Rammed earth	Huai'an kiln	
Provide shelters for people in times of disaster	Variable site design	It's different from normal times and disaster	Hail
			Earthquake

"守" guard

Scarecrow　Buildings　Watchtower

| Plant rice |
| Weaving a scarecrow |
| Develop tourism |

"望" Watch

Looking at the view of the paddy fields	
As a symbol of the site	The towering watchtower guides the crowd in the event of a disaster.
Green and sustainable buildings	solar panels, ecological circulating water pool, modular folded-plate roof, Small vertical axis wind turbines, trumb walls, sunrooms, light guides, etc.

Villagers' activities

Local Climate and disasters

Technical measures

稻田里的守望者·二
The Protector in the Field·II

Building area: 2109 ㎡
Building footprint: 1693 ㎡
Public service space: 1425 ㎡
Public service building height 8.5m
Residential area: 676 ㎡
Residence yard area: 288 ㎡
Residence height: 7.8m
Building density: 39%
Max building storey: 2F
Floor area ratio: 0.51
Greening rate: 17.2%
Number of parking spaces:16

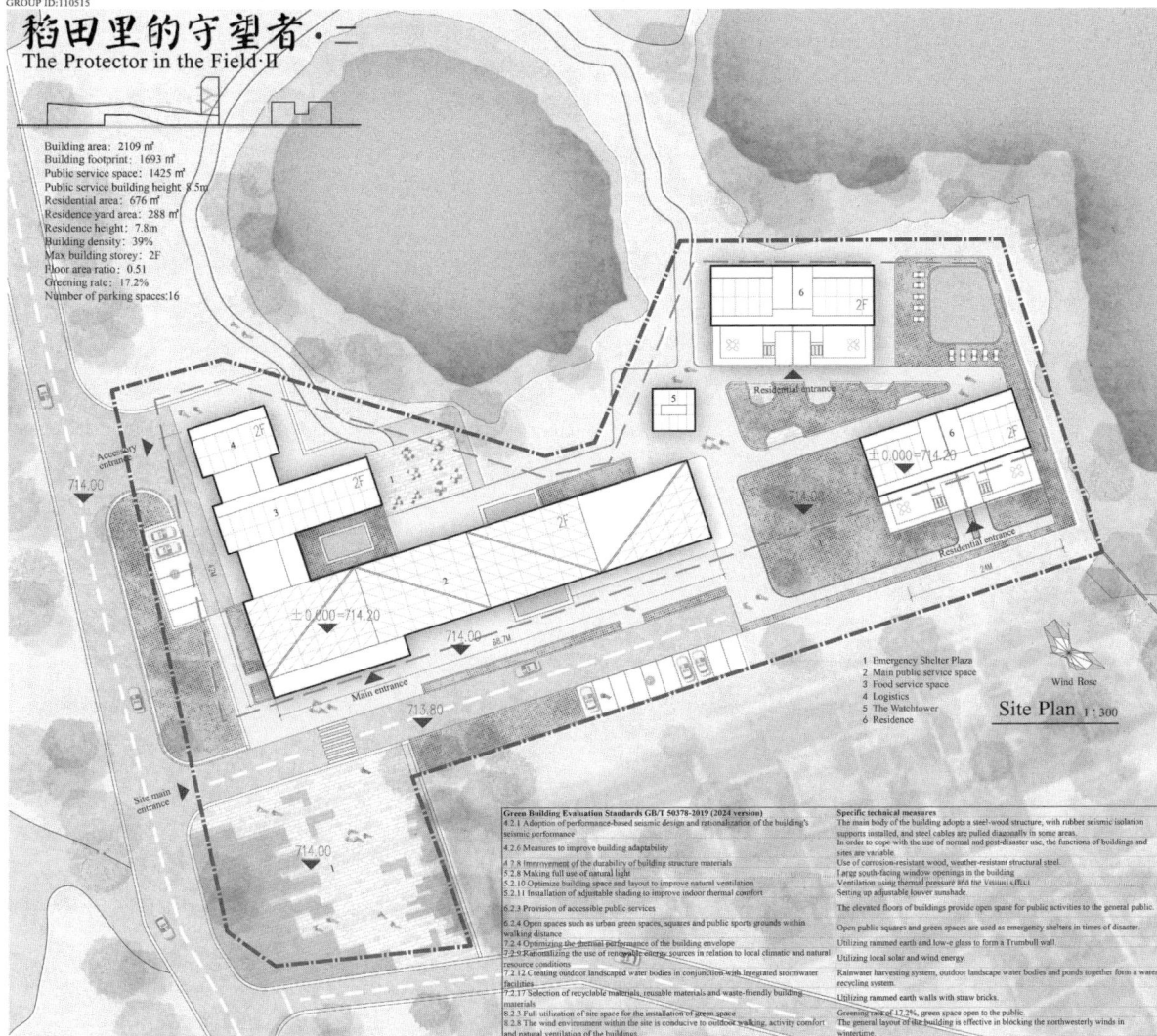

Site Plan 1 : 300

1 Emergency Shelter Plaza
2 Main public service space
3 Food service space
4 Logistics
5 The Watchtower
6 Residence

Wind Rose

Green Building Evaluation Standards GB/T 50378-2019 (2024 version)	Specific technical measures
4.2.1 Adoption of performance-based seismic design and rationalization of the building's seismic performance	The main body of the building adopts a steel-wood structure, with rubber seismic isolation supports installed, and steel cables are pulled diagonally in some areas.
4.2.6 Measures to improve building adaptability	In order to cope with the use of normal and post-disaster use, the functions of buildings and sites are variable.
4.2.8 Improvement of the durability of building structure materials	Use of corrosion-resistant wood, weather-resistant structural steel.
5.2.8 Making full use of natural light	Large south-facing window openings in the building
5.2.10 Optimize building space and layout to improve natural ventilation	Ventilation using thermal pressure and the Venturi effect
5.2.11 Installation of adjustable shading to improve indoor thermal comfort	Setting up adjustable louver sunshade
6.2.3 Provision of accessible public services	The elevated floors of buildings provide open space for public activities to the general public.
6.2.4 Open spaces such as urban green spaces, squares and public sports grounds within walking distance	Open public squares and green spaces are used as emergency shelters in times of disaster.
7.2.4 Optimizing the thermal performance of the building envelope	Utilizing rammed earth and low-e glass to form a Trumbull wall.
7.2.9 Rationalizing the use of renewable energy sources in relation to local climatic and natural resource conditions	Utilizing local solar and wind energy.
7.2.12 Creating outdoor landscaped water bodies in conjunction with integrated stormwater facilities	Rainwater harvesting system, outdoor landscape water bodies and ponds together form a water recycling system.
7.2.17 Selection of recyclable materials, reusable materials and waste-friendly building materials	Utilizing rammed earth walls with straw bricks.
8.2.3 Full utilization of site space for the installation of green space	Greening rate of 17.2%, green space open to the public.
8.2.8 The wind environment within the site is conducive to outdoor walking, activity comfort and natural ventilation of the buildings.	The general layout of the building is effective in blocking the northwesterly winds in wintertime.

Generation

The site can be roughly divided into two parts: public area and private area

Public area
Residential area

The public areas are close to the road and are well connected to the outside world.

Residential hostel

The hostel is located on the more private side and has a good view of the landscape

Generate building blocks

As a spiritual symbol, the tower guards the whole site.

Main entrance

Arrange the surrounding environment

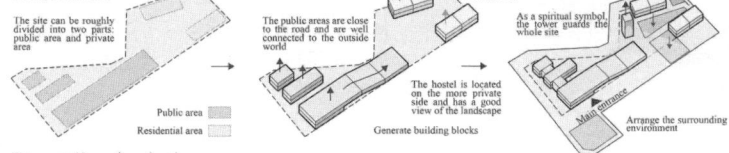

Streamline Analysis

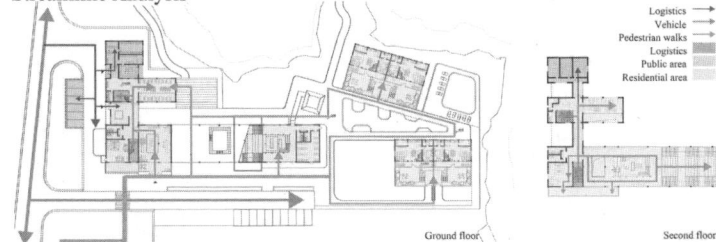

Logistics
Vehicle
Pedestrian walks
Logistics
Public area
Residential area

Ground floor

Second floor

South Elevation Plan 1:250

GROUP ID:110515

稻田里的守望者·三
The Protector in the Field·III

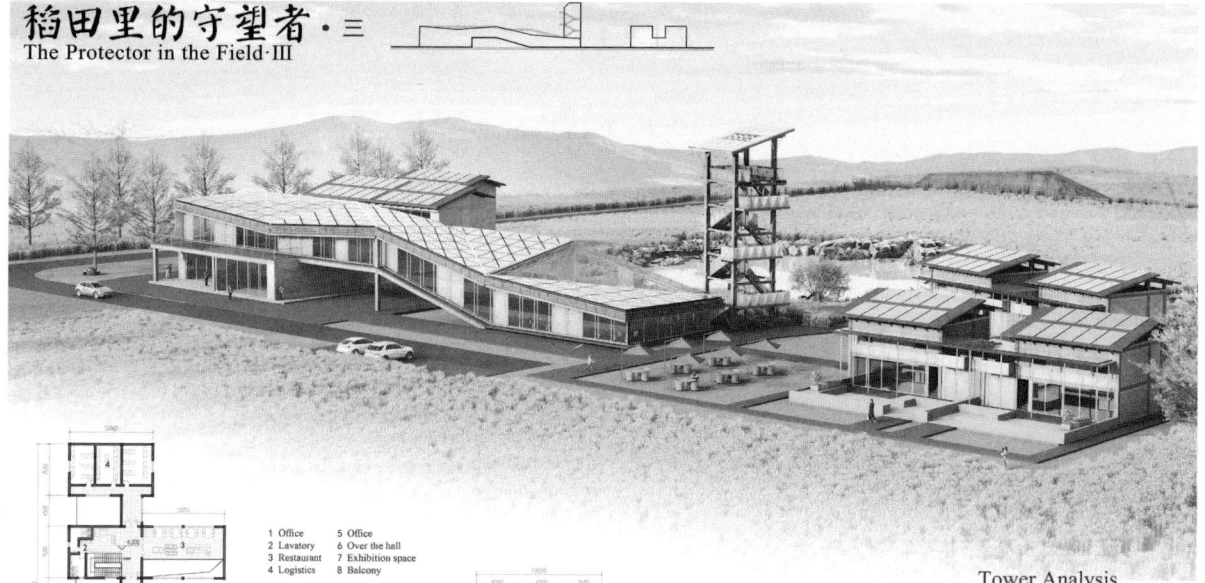

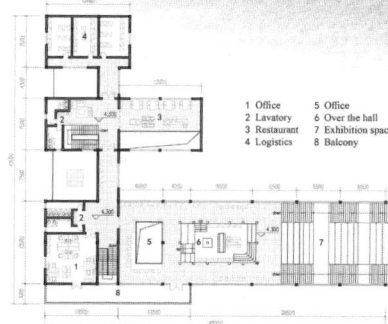

Public Facility Second Floor Plan 1:300

1 Office
2 Lavatory
3 Restaurant
4 Logistics
5 Office
6 Over the hall
7 Exhibition space
8 Balcony

Residence Ground Plan 1:200

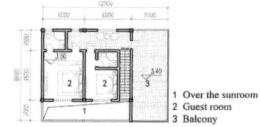

1 Yard
2 Sunroom
3 Living room
4 Restaurant
5 Kitchen
6 Master bedroom

Residence Second Floor Plan 1:200

1 Over the sunroom
2 Guest room
3 Balcony

Tower Analysis

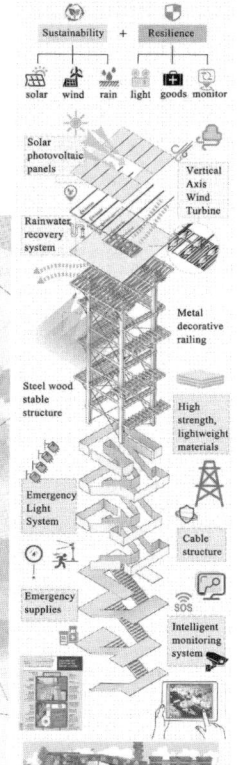

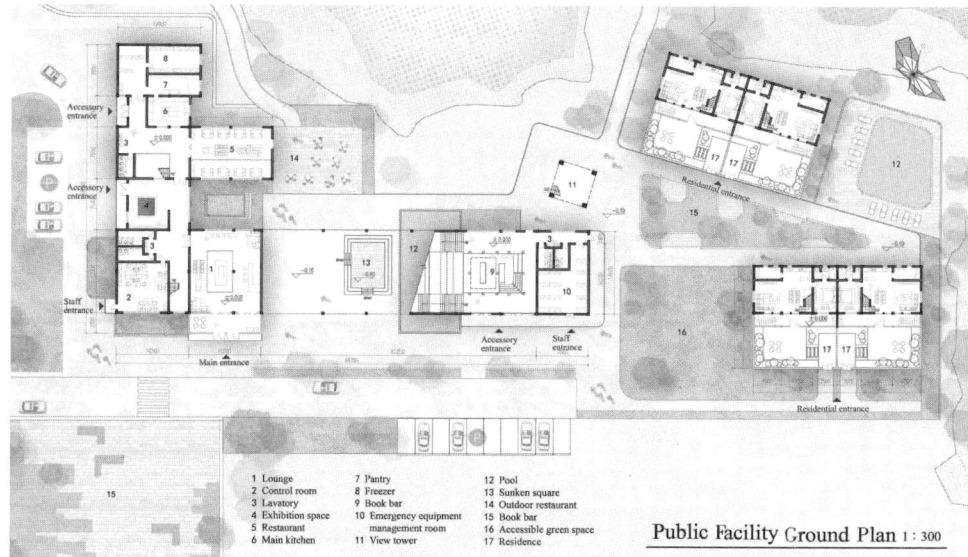

Sustainability + Resilience

solar wind rain light goods monitor

Solar photovoltaic panels

Rainwater recovery system

Steel wood stable structure

Emergency Light System

Emergency supplies

Vertical Axis Wind Turbine

Metal decorative railing

High strength, lightweight materials

Cable structure

SOS

Intelligent monitoring system

Public Facility Ground Plan 1:300

1 Lounge
2 Control room
3 Lavatory
4 Exhibition space
5 Restaurant
6 Main kitchen
7 Pantry
8 Freezer
9 Book bar
10 Emergency equipment management room
11 View tower
12 Pool
13 Sunken square
14 Outdoor restaurant
15 Book bar
16 Accessible green space
17 Residence

"The watchtower is not only a landmark building of the site, but also plays a significant role in enhancing the overall disaster resistance of the site."

Disaster Response Analysis

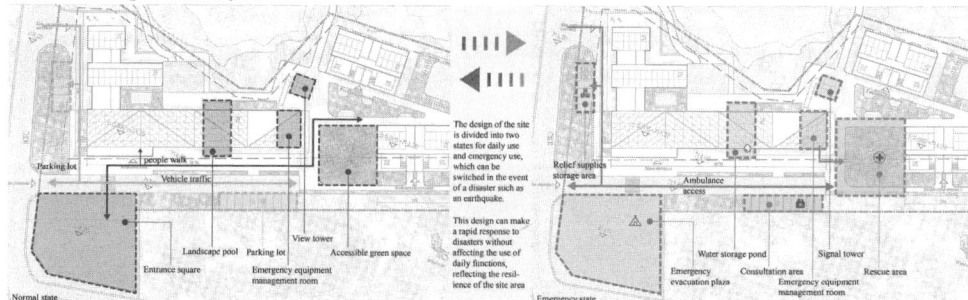

The design of the site is divided into two states for daily use and emergency use, which can be switched in the event of a disaster such as an earthquake.

This design can make a rapid response to disasters without affecting the use of daily functions, reflecting the resilience of the site area

Parking lot
people walk
Vehicle traffic

Landscape pool   Parking lot   Accessible green space
Entrance square   Emergency equipment   View tower
management room

Normal state

Relief supplies storage area
Ambulance access

Water storage pond   Signal tower
Emergency   Consultation area   Rescue area
evacuation plaza   Emergency equipment
management room

Emergency state

稻田里的守望者·四
The Protector in the Field·IV

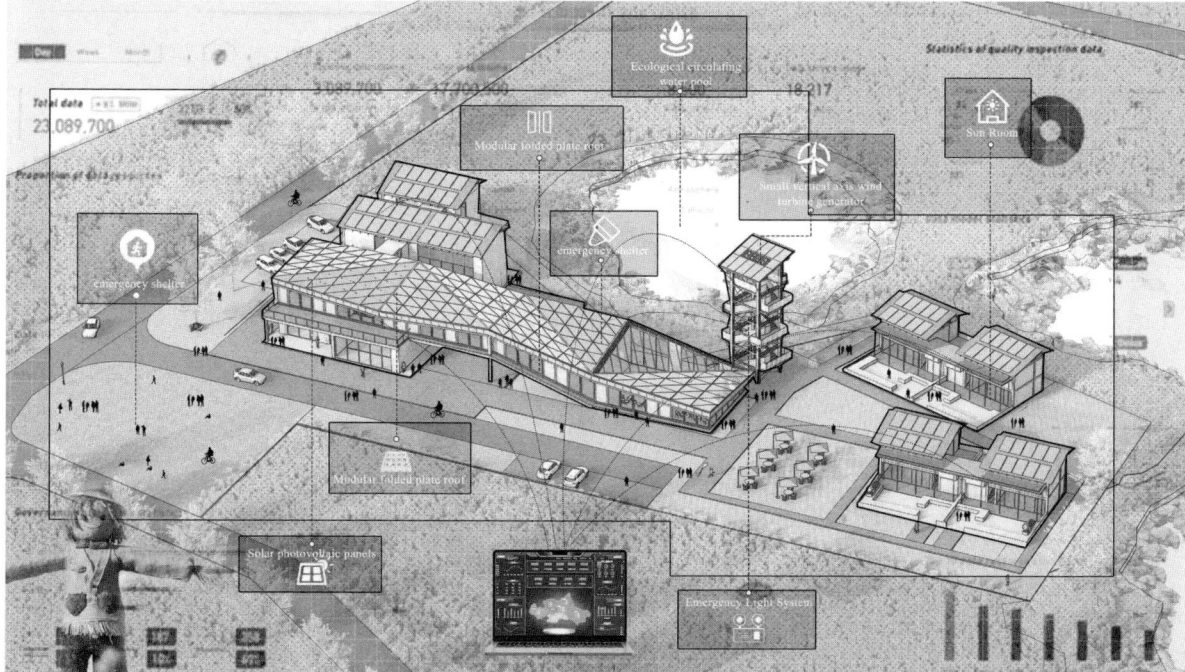

Day | Week | Month

Total data 23,089,700

Proportion of data resources

- Ecological circulating water pool
- Modular folded plate roof
- emergency shelter
- Small vertical axis wind turbine generator
- Sun Room
- emergency shelter
- Modular folded plate roof
- Solar photovoltaic panels
- Emergency Light System

Statistics of quality inspection data

Green Technology Analysis

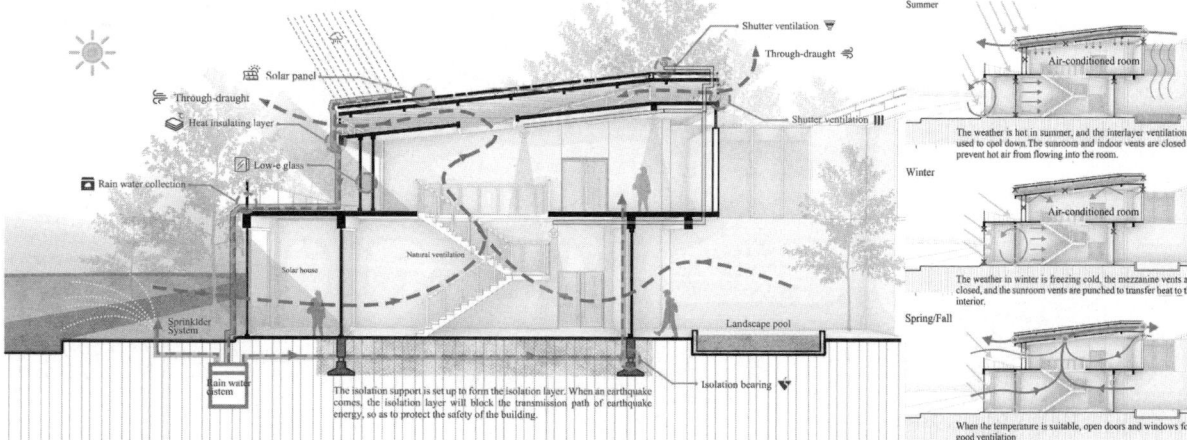

- Shutter ventilation
- Through-draught
- Solar panel
- Through-draught
- Heat insulating layer
- Low-e glass
- Rain water collection
- Shutter ventilation
- Solar house
- Natural ventilation
- Sprinkler System
- Rain water system
- Landscape pool
- Isolation bearing

The isolation support is set up to form the isolation layer. When an earthquake comes, the isolation layer will block the transmission path of earthquake energy, so as to protect the safety of the building.

Summer

Air-conditioned room

The weather is hot in summer, and the interlayer ventilation is used to cool down. The sunroom and indoor vents are closed to prevent hot air from flowing into the room.

Winter

Air-conditioned room

The weather in winter is freezing cold, the mezzanine vents are closed, and the sunroom vents are punched to transfer heat to the interior.

Spring/Fall

When the temperature is suitable, open doors and windows for good ventilation.

Residential Hotel Strategy

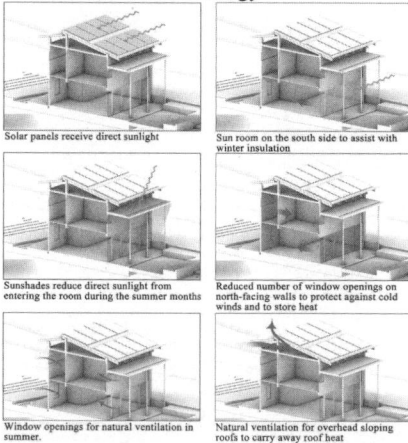

Solar panels receive direct sunlight

Sun room on the south side to assist with winter insulation

Sunshades reduce direct sunlight from entering the room during the summer months

Reduced number of window openings on north-facing walls to protect against cold winds and to store heat

Window openings for natural ventilation in summer.

Natural ventilation for overhead sloping roofs to carry away roof heat

Energy Calculation

Building power consumption

According to the standard for energy consumption of building, the civilian buildings' constraint value of energy consump-tion indicator in cold area is 2700[kW·h/(a·H)], and the public buildings' constraint value of energy consump-tion indicator in cold area is 55[kW·h/(m²·a)].

Based on four homestays and 1425 square meters of public space, the buildings' annaul electricity consumpton:

$$2700[kW·h/(a·H)]×4(H)+55[kW·h/(m²·a)×1425m²=89175kW·h$$

Solar panel power generation

The solar annual radiation in Zhangjiakou is between 1625 kW·h/(m²·a) and 1855 kW·h/(m²·a).

After 2015, if we want to get state subsidies, we must use photovoltaic modules above 255W, so we choose 260W photovoltaic modules. And the solar conversion efficiency of 60W photovoltaic modules is 16%. the roof required for 260W solar panels to generate electricity is about 1102.8 m².

$$1625 kW·h/(m²·a)×1102.8m²×16%=286728kW·h＞89175kW·h$$

conclusion

The current power generation capacity of solar panels can basically meet the electricity demand of the building.

Carbon Emissions Calculation

The energy data of this building is the sum of its power consumption and power genera-tion.

89175kW·h+286728kW·h=375903kW·h

The standard coal coefficient of electricity is 0.1229 kgce/(kW·h).

The carbon emission factor of electricity is 0.714 kg-co2/kg.

So based on Calculation formula of carbon emissions, this building's annaul carbon emissions as follows:

375903kW·h×0.1229 kgce/(kW·h)× 0.714 kg-co2/kg=32985.71kg-co2

Material Analysis

Low emissivity glass | Antiseptic Wood | Solar panel

Atmospheric corrosion resisting structural steel | Tamped earth | Straw bricks

CODE: 110577

综合奖·入围奖·张家口市
怀安县二堡子村建设项目
Comprehensive Awards —
Nomination — Erpuzi Village
Project，Huai'an County，
Zhangjiakou City

注册号：110577
Register Number：110577

项目名称：沐光乡生
Entry Title：Sunlit Countryside Thrives

作者：束家祯、陆珮瑶、韩越然、胡潘旭
Authors：Shu Jiazhen, Lu Peiyao,
　　　　Han Yueran，and Hu Panxu
作者单位：南京工业大学
Authors from：Nanjing Tech University
指导教师：罗靖
Tutor：Luo Jing
指导教师单位：南京工业大学
Tutor from：Nanjing Tech University

沐光乡生

2024台达杯国际太阳能建筑设计竞赛
Sunlit Countryside Thrives

本方案以乡村振兴和绿色乡村为目标，基于场地的环境特征，采用保温、自然通风、太阳能光伏和水循环等主被动式技术达成绿色乡村建筑的要求，并赋予建筑以自维持能力。另外，根据场地自然灾害的特点，以防灾"韧性轴"为规划概念，设置东西向主街穿过场地。基于"平灾结合"的策略，非灾时作为公共生活和乡村商业的主街；灾时充分考虑应灾需求，设置人车分流的疏散道路，并提供充足的安置和救治场所，结合太阳能建筑的就地能源自给，为灾中生活提供有效的能源保障，形成可复制性的"平灾结合"范式。方案最终让村民和来访者提升生活感受与经济创收的同时亦能践行低碳文明的生活理念。

This plan aims to promote rural revitalization and green rural areas. Based on the environmental characteristics of the site, it adopts active and passive technologies such as insulation, natural ventilation, solar photovoltaic, and water circulation to meet the requirements of green rural buildings, and endows the buildings with self-sustaining capabilities. In addition, based on the characteristics of natural disasters on the site, the planning concept of disaster prevention "resilience axis" is adopted, and an east-west main street is set up to pass through the site. Based on the strategy of "combining disaster relief", it serves as the main street for public life and rural commerce during non disaster periods; Fully consider disaster response needs during disasters, set up evacuation roads for people and vehicles, and provide sufficient resettlement and treatment facilities. Combined with the on-site energy self-sufficiency of solar buildings, provide effective energy security for life during disasters, and form a replicable "disaster relief combination" paradigm. The plan ultimately allows villagers and visitors to enhance

Aerial View

沐光乡生 2/4
2024台达杯国际太阳能建筑设计竞赛
Sunlit Countryside Thrives

Site Plan 1:600

POOL

POOL

Planning Red Line

Building Line

MAIN ENTRANCE
of BUSINESS 3

MAIN ENTRANCE
of SITE

COMMUNITY SERVICE CENTRE

2F

MAIN ENTRANCE OF
COMMUNITY SERVICE
CENTRE

RESIDENCE

SECONDARY ENTRANCE
of SITE

MENGTU TRIBE TOUR-
IST SCENIC AREA

SQUARE

RESIDENTIAL
HOMESTAY AREA

PARKING LOT

Country Road

Village Road

Village Road

Village Road

Village Road

Country Road

Country Road

Country Road

VEGETABLE GARDEN
RESEARCH BASE

Economic Indicators

Site Area	8260 ㎡
Construction Land Area	4275 ㎡
Total Floor Area	1776.4 ㎡
Total Surface Area	1080.2 ㎡
Building Density	13.1%
Floor Area Ratio	0.22
Building Storey	2F
Parking Space	54

Axis Resillence

Photovoltaic Panel

Curton Wind Power

Solar Street Lamp

Solar Seat

Window Sunlight Accumulator

Landscape Hydropower

Water Circulation System

Biogas Digester System

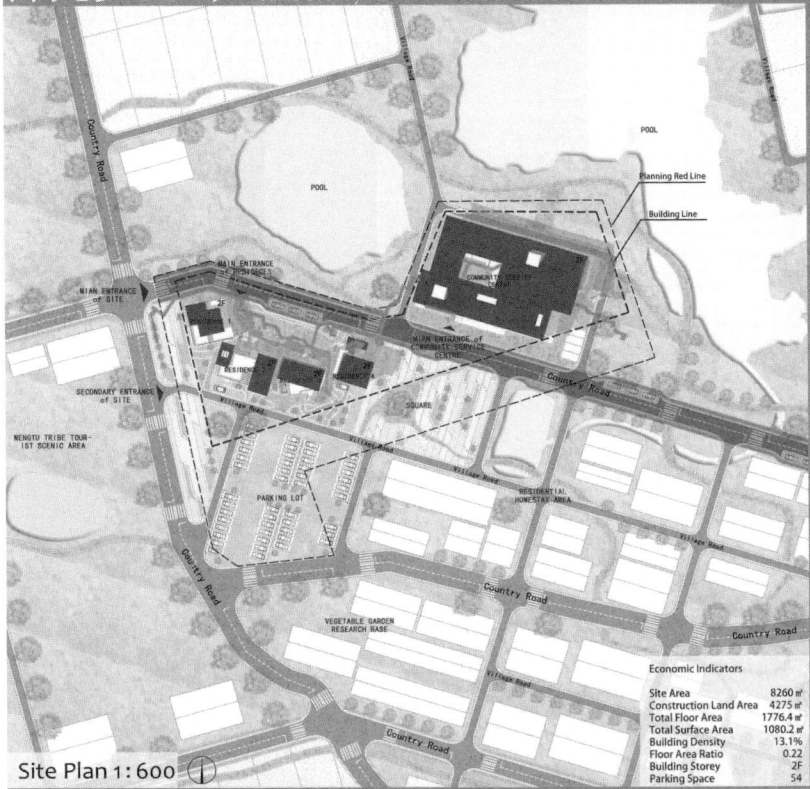

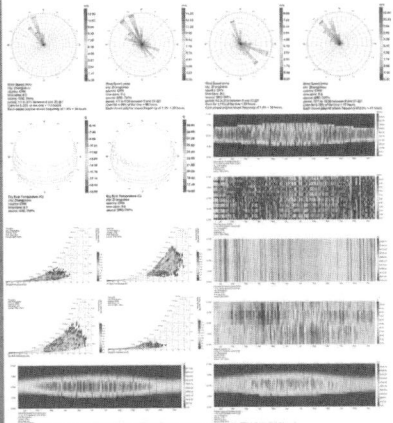

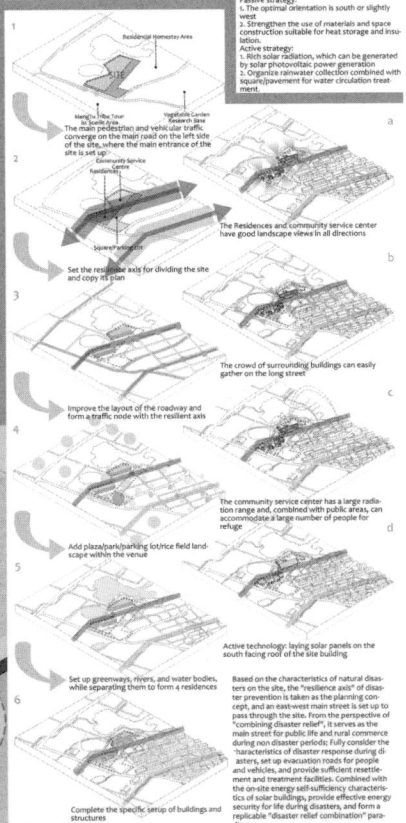

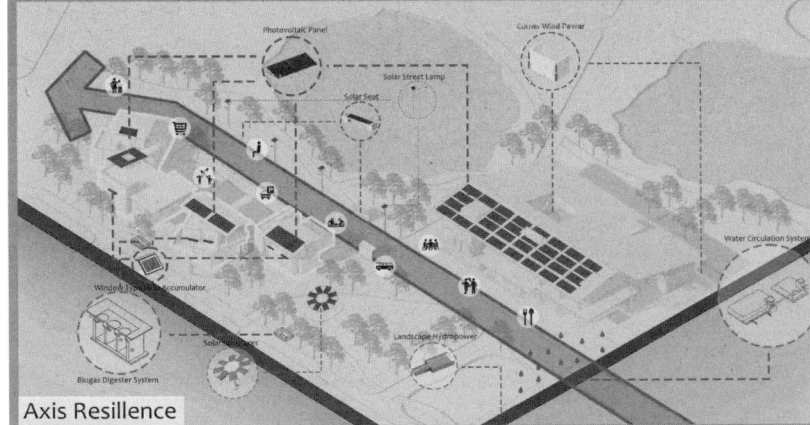

Site Generation

Passive strategy:
1. The optimal orientation is south or slightly west
2. Strengthen the use of materials and space construction suitable for heat storage and insulation.
Active strategy:
1. Rich solar radiation, which can be generated by solar photovoltaic power generation
2. Organize rainwater collection combined with square/pavement for water circulation treatment.

Residential Homestay Area

SITE

MengTu Tribe Tour-
ist Scenic Area

Vegetable Garden
Research Base

Community Service
Centre

Square/Parking

1 The main pedestrian and vehicular traffic converge on the main road on the left side of the site, where the main entrance of the site is set up

a The Residences and community service center have good landscape views in all directions

2 Set the resilience axis for dividing the site and copy its plan

b The crowd of surrounding buildings can easily gather on the long street

3 Improve the layout of the roadway and form a traffic node with the resilient axis

c The community service center has a large radiation range and, combined with public areas, can accommodate a large number of people for refuge

4 Add plaza/park/parking lot/rice field landscape within the venue

d Active technology: laying solar panels on the south facing roof of the site building

5 Set up greenways, rivers, and water bodies, while separating them to form 4 residences

6 Complete the specific setup of buildings and structures

Based on the characteristics of natural disasters on the site, the "resilience axis" of disaster prevention is taken as the planning concept, and an east-west main street is set up to pass through the site. From the perspective of "combining disaster relief", it serves as the main street for public life and rural commerce during non disaster periods; Fully consider the characteristics of disaster response during disasters, set up evacuation roads for people and vehicles, and provide sufficient resettlement and treatment facilities. Combined with the on-site energy self-sufficiency characteristics of solar buildings, provide effective energy security for life during disasters, and form a replicable "disaster relief combination" paradigm.

Entrance Perspective View

CODE: 110577

沐光乡生 3/4
2024台达杯国际太阳能建筑设计竞赛
Sunlit Countryside Thrives

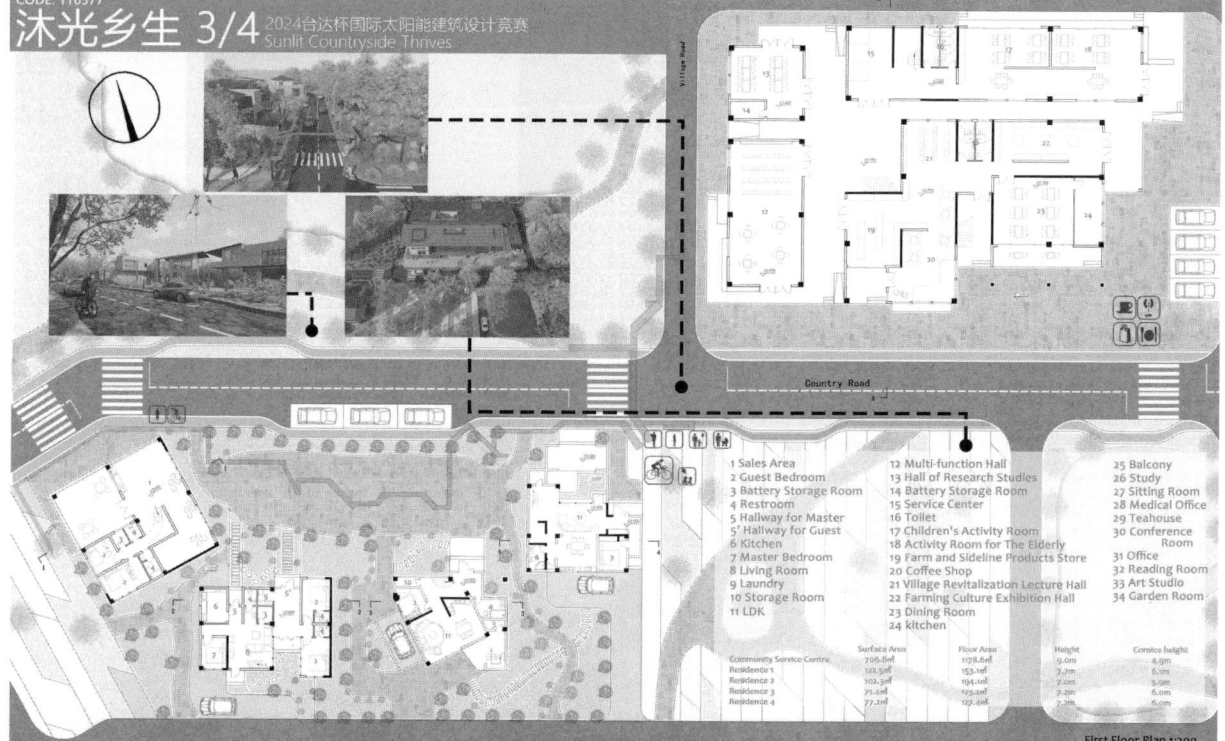

First Floor Plan 1:200

1 Sales Area
2 Guest Bedroom
3 Battery Storage Room
4 Restroom
5 Hallway for Master
5' Hallway for Guest
6 Kitchen
7 Master Bedroom
8 Living Room
9 Laundry
10 Storage Room
11 LDK
12 Multi-function Hall
13 Hall of Research Studies
14 Battery Storage Room
15 Service Center
16 Toilet
17 Children's Activity Room
18 Activity Room for The Elderly
19 Farm and Sideline Products Store
20 Coffee Shop
21 Village Revitalization Lecture Hall
22 Farming Culture Exhibition Hall
23 Dining Room
24 kitchen
25 Balcony
26 Study
27 Sitting Room
28 Medical Office
29 Teahouse
30 Conference Room
31 Office
32 Reading Room
33 Art Studio
34 Garden Room

RESIDENCE 1
Second Floor Plan 1:200
North Elevation 1:200
Section1-1 1:200

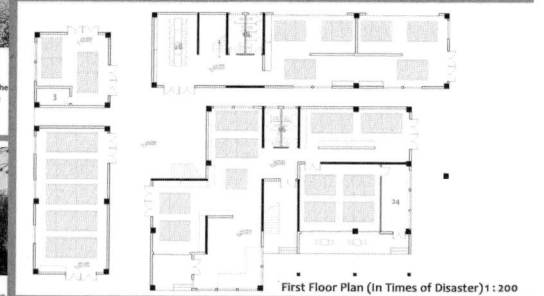

Residence 1 is a sales node that generates income for the village and provides a public space for the villagers to gather for activities. It has three different types of rooms for guests to choose from.

RESIDENCE 2
Second Floor Plan 1:200
North Elevation 1:200
Section2-2 1:200

Residence 2 functions as both a bed and breakfast and a residence, with two entrance halls separating the hosts from their guests, and a door separating the two spaces in the middle, allowing guests to choose between privacy and personal space or integration into the host's home.

RESIDENCE 3
Second Floor Plan 1:200
North Elevation 1:200
Section3-3 1:200

In the graphic design of Residence 3, the space is treated with angles and some wall settings, hoping to block the sight. Set the traffic space on the north side near the main road to provide some sound blocking. It provides a more private living space for the residents in the homestay.

RESIDENCE 4
Second Floor Plan 1:200
North Elevation 1:200
Section4-4 1:200

Residence 4 contains more open spaces, such as the front hall, open courtyard and open LDK, which enhance the interaction between people while providing a venue for activities, hoping that people can unite to promote rural revitalization.

Second Floor Plan 1:200
First Floor Plan (In Times of Disaster) 1:200
Second Floor Plan(In Times of Disaster) 1:200

190

沐光乡生 4/4 2024台达杯国际太阳能建筑设计竞赛
Sunlit Countryside Thrives

Non disaster
The resilience axis combines public places such as squares, parks, and greenways to broaden people's activity range, increase buffer space between vehicles, and create distinctive rural public landscapes.

During a disaster
Temporary relief tents can be set up on both sides of the resilience axis. The crowd can be temporarily evacuated to the long street and then moved into the community service center in batches for shelter until the disaster is over.

Carbon Emission Simulation Results

Index Class	Index Name	Annual Operating Emissions Per Unit Area (kg/·㎡)	Reference Mean (kg/·㎡)	Annual Consumption	Carbon Emission Coefficient	Total Annual Carbon Emissions
Energy Consumption	Air Conditioning	2.30	3.00	8000kWh	0.0209kg/kWh	4.00
	Heating	12.00	14.00	34000	0.0919g/GJ	13.00
	Illumination	1.30	1.50	4200kWh	0.0905kg/kWh	2.30
	Other	1.70	2.00	9900kWh	0.0009kg/kWh	2.80
	Traffic	2.30	3.00	28000/mile	0.0926g/time	5.60
	Garbage Disposal	0.90	1.30	ft	0.5kg/t	3.50
	Water Resources	1.40	1.80	850m²	0.03kg/m²	8.50
Total Carbon Emission	Total Carbon Emission					28.00
Carbon Emission Standard	Carbon Emission Standard					250.00
Condition of Reaching Standard	Condition of Reaching Standard					Reach The Standard

Self-sustaining Simulation Results

Item	Community Service Center
Building Area (㎡)	1500
Air Conditioning Energy Consumption (kWh/㎡·year)	30
Heating Energy Consumption (kWh/㎡·year)	80
Lighting Energy Consumption (kWh/㎡·year)	30
Hot Water Energy Consumption (kWh/㎡·year)	10
Energy Consumption of Other Equipment (kWh/year)	3000
Total Energy Consumption (kWh/year)	106400

Item	Community Service Center
Building Area (㎡)	1500
Rooftop Solar Photovoltaic System (kW)	80
Annual Solar Power Generation (kWh/year)	60000
Small Heat Accumulator (kWh/year)	4000
Annual Wind Energy Generation (kWh/year)	3000
Annual Hydropower Generation (kWh/year)	5000
Total Renewable Energy Production (kWh/year)	72000

Total Operating Energy Consumption (kWh/year)	Renewable Energy
129200	108200

Source	Water	Sun	methane	Wind

Business Continuity Plan

To better cope with the unknown occurrence, reduce the carbon footprint while ensuring the energy needs of residents, we built a BCP (business continuity plan) system in the site.

Not Imported BCP Construction

Import BCP Construction

Through the simulation data, it can be seen that after the introduction of the BCP system, the renewable energy output of the building is high, and the self-sustainance rate reaches a relatively considerable figure, which can support the rapid recovery of the production capacity of the building after the impact

A Large Sample of the Seismic Structure

A Large Sample of the Beam-column Connection

On the basis of the frame shear structure, structural reinforcement is carried out by making rigid connections at the joints, further enhancing the building structure's ability to resist earthquakes.

Seismic Seam
The building is divided into several units with uniform stiffness and independent deformation in the earthquake to reduce the damage degree of the building.

Floor to Floor

Lateral bracing is used to offset the load perpendicular to the direction of the pipeline in the seismic horizontal load and prevent the lateral movement of the pipeline in the earthquake.

Floor to Interior Wall

Single-tube Lateral Seismic Bracket

Leave space for furniture when designing walls, rigidly connect furniture to walls to prevent furniture from tipping over.

Built-In Furniture

The Process of Energy

Water — Precipitation — Fall to The Ground / Fall to The Roof — Infiltration / Roof Collection
Heat — Heat Preservation and Insulation / Heat Storage / Heat Accumulator — Air Source Heat Pump
Electricity — Solar Radiation — Roof Photovoltaic Panel — Solar Street Lamp / Solar Seat
Kinetic Energy of Water — Hydropower
Chemical Energy — Biogas pit
Wind Energy — Wind Power

Energy Destination — For the Residential User's Own Use — To Public Centralized Management

According to the simulation analysis of green building software, the building has good ventilation, lighting, heat storage and insulation effects, and is also exposed to abundant solar radiation

Site Sunshine Analysis Diagram
Summer Wind Speed Vector Diagram
1F Wind Speed Vector Diagram
Winter Wind Speed Vector Diagram
2F Wind Speed Vector Diagram
Summer Wind Speed Cloud Map
1F Wind Speed Cloud Map
1F Lighting Analysis diagram
Winter Wind Speed Cloud Map
2F Wind Speed Cloud Map
2F Lighting Analysis diagram

East Elevation 1:200

South Elevation 1:200

Building Material Analysis

Gray Bricks Brown Bricks Cement Wood Yellow Bricks Glass

Sewage Treatment
Winter Sun
Photovoltaic Panels
Heat Insulation Window
Atrium
Summer Sun
Regenerative Walls
Atrium Lighting
Natural Ventilation
Rain Water Collection
Plant Ponds
Footpath
Ditch
Infiltration

Sewage Treatment plants   Rainwater Collection Devices   Dehumidify   Heat Preservation
A-A Section Perspective

99-110606

岁穗年年
稻花乡里说年年

综合奖·入围奖·张家口市怀安县二堡子村建设项目
Comprehensive Awards – Nomination – Erpuzi Village Project，Huai'an County，Zhangjiakou City

"岁穗年年"设计以"阳光·乡村韧性"为核心主题，融合乡村自然景观与可持续发展理念，打造稻香旅居文化区。建筑采用单坡屋顶设计，以土窑和稻秆作为主要墙体材料呼应当地的生土与麦田。设计着重于韧性理念的应用，强调应对自然灾害的灵活性与可持续性。建筑具备平灾功能转换机制在灾害发生时可迅速转化为疏散空间以提升抗灾力。运用大阳能光伏和光热系统，日常运行中可满足村民生活能源生活需求，在灾害时则为建筑提供应能源支持。同时，建筑设计上充分利用被动式技术如墙体保温PV-Trombe墙和相变蓄热材料等，有效提升保温和节能性能。提高建筑的自我持续能力，推动低碳可持续发展。

The design of "Wheat each year" takes "sunshine, rural resilience" as the core theme, integrating rural natural landscape and the concept of sustainable development to create rice fragrance. Travelers live in the natural area. The building adopts a single-sloping roof design, with earth kilns and straw as the main wall materials, echoing the local soil and wheat fields.
The design focuses on the application of the concept of resilience, emphasizing the flexibility and sustainability of dealing with natural disasters. The building is equipped with a disaster relief function conversion mechanism, and in the event of a disaster, it can be quickly transformed into an emergency evacuation space to improve resistance. Using Dayang energy photovoltaic and solar thermal system, the daily running energy renewable energy meets the needs of life and provide energy support for buildings in the event of disaster. At the same time, make full use of passive technologies in architectural design, such as wall insulation PV-Trombe walls, etc., to effectively improve the insulation and energy-saving performance, improve the self-sustaining ability of buildings and push Low-carbon sustainable development.

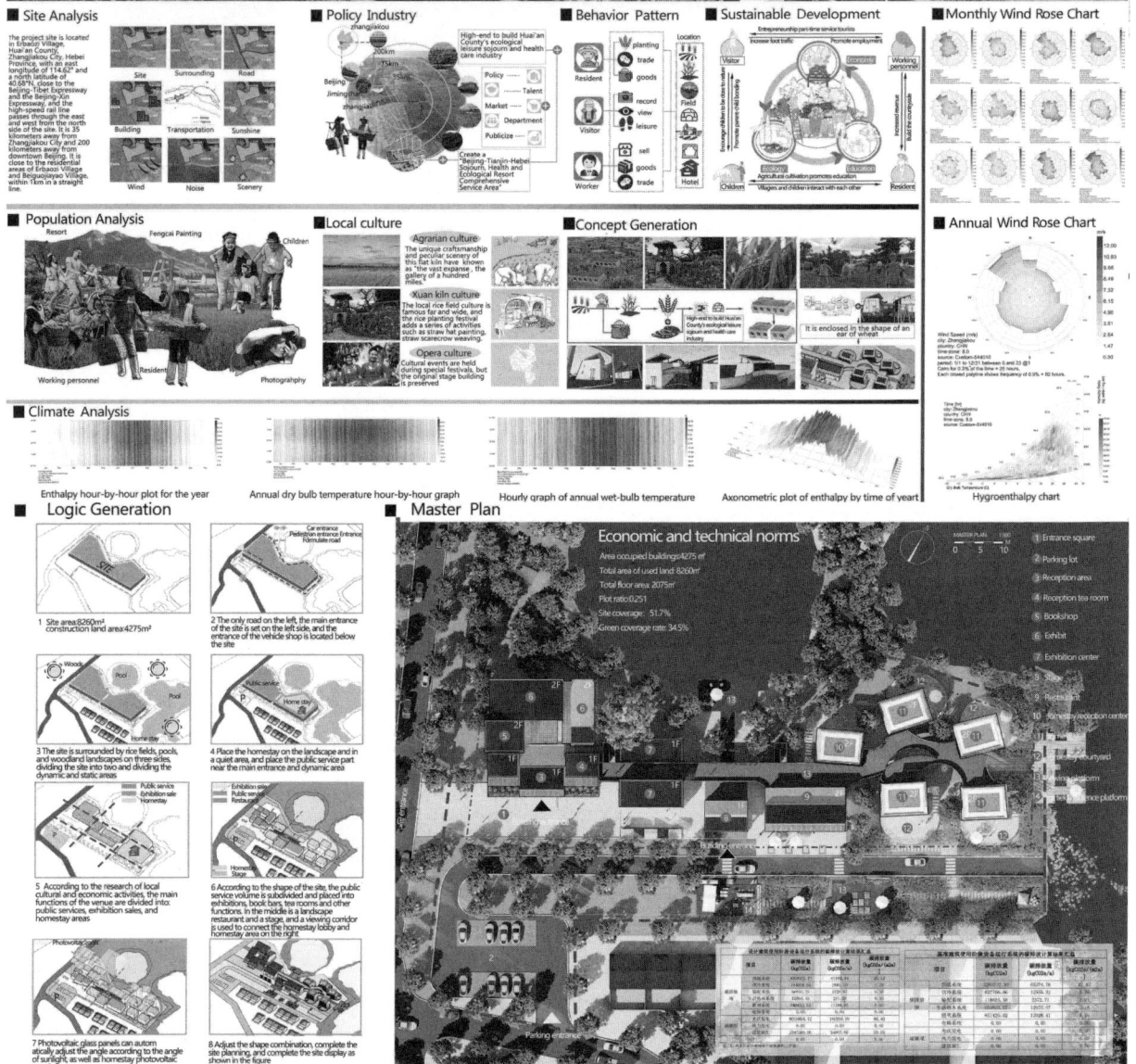

注册号：110606
Register Number：110606
项目名称：岁穗年年
Entry Title：Wheat Each Year
作者：李嘉欣、曾雅清、李思静、郝泽厚
Authors：Li Jiaxin, Zeng Yaqing, Li Sijing, and Hao Zehou
作者单位：厦门大学
Authors from：Xiamen University
指导教师：李立新
Tutor：Li Lixin
指导教师单位：厦门大学
Tutor from：Xiamen University

■ Site Analysis

The project site is located in Erbaozi Village, Huai'an County, Zhangjiakou City, Hebei Province, with an east longitude of 114.62° and a north latitude of 40.68°N, close to the Beijing-Tibet Expressway and the Beijing-Xin Expressway, and the high-speed rail line passes through the east and west from the north side of the site. It is 35 kilometers away from Zhangjiakou City and 200 kilometers away from downtown Beijing. It is close to the residential areas of Erbaozi Village and Beiguojiayao Village, within 1km in a straight line.

Site | Surrounding | Road
Building | Transportation | Sunshine
Wind | Noise | Scenery

■ Policy Industry

Create a "Beijing-Tianjin-Hebei Sojourn, Health and Ecological Resort Comprehensive Service Area"

■ Behavior Pattern

■ Sustainable Development

■ Monthly Wind Rose Chart

■ Population Analysis

■ Local culture

Agrarian culture
The unique craftsmanship and peculiar scenery of this flat kiln have known as "the vast expanse , the gallery of a hundred miles."

Xuan kiln culture
The local rice field culture is famous far and wide, and set up the rice planting field adds a series of activities such as straw hat painting, straw scarecrow weaving.

Opera culture
Cultural events are held during special festivals, but the original stage building is preserved

■ Concept Generation

It is enclosed in the shape of an ear of wheat

■ Annual Wind Rose Chart

■ Climate Analysis

Enthalpy hour-by-hour plot for the year | Annual dry bulb temperature hour-by-hour graph | Hourly graph of annual wet-bulb temperature | Axonometric plot of enthalpy by time of year

Hygroenthalpy chart

■ Logic Generation

1 Site area:8260m² construction land area:4275m²

2 The only road on the left, the main entrance of the site is set on the left side, and the entrance of the vehicle shop is located below the site

3 The site is surrounded by rice fields, pools, and woodland landscapes on three sides, dividing the site into two and dividing the dynamic and static areas

4 Place the homestay on the landscape and in a quiet area, and place the public service part near the main entrance and dynamic area

5 According to the research of local cultural and economic activities, the main functions of the venue are divided into: public services, exhibition sales, and homestay areas

6 According to the shape of the site, the public service volume is subdivided and placed into exhibitions, book bars, tea rooms and other functions. In the middle is a landscape restaurant and a stage, and a viewing corridor is used to connect the homestay lobby and homestay area on the right

7 Photovoltaic glass panels can automatically adjust the angle according to the angle of sunlight, as well as homestay photovoltaic panels

8 Adjust the shape combination, complete the site planning, and complete the site display as shown in the figure

■ Master Plan

Economic and technical norms
Area occupied buildings:4275 m²
Total area of used land: 8260m²
Total floor area: 2075m²
Plot ratio:0.251
Site coverage： 51.7%
Green coverage rate: 34.5%

1 Entrance square
2 Parking lot
3 Reception area
4 Reception tea room
5 Bookshop
6 Exhibit
7 Exhibition center
8 Restaurant
9 Restaurant
10 Homestay reception center

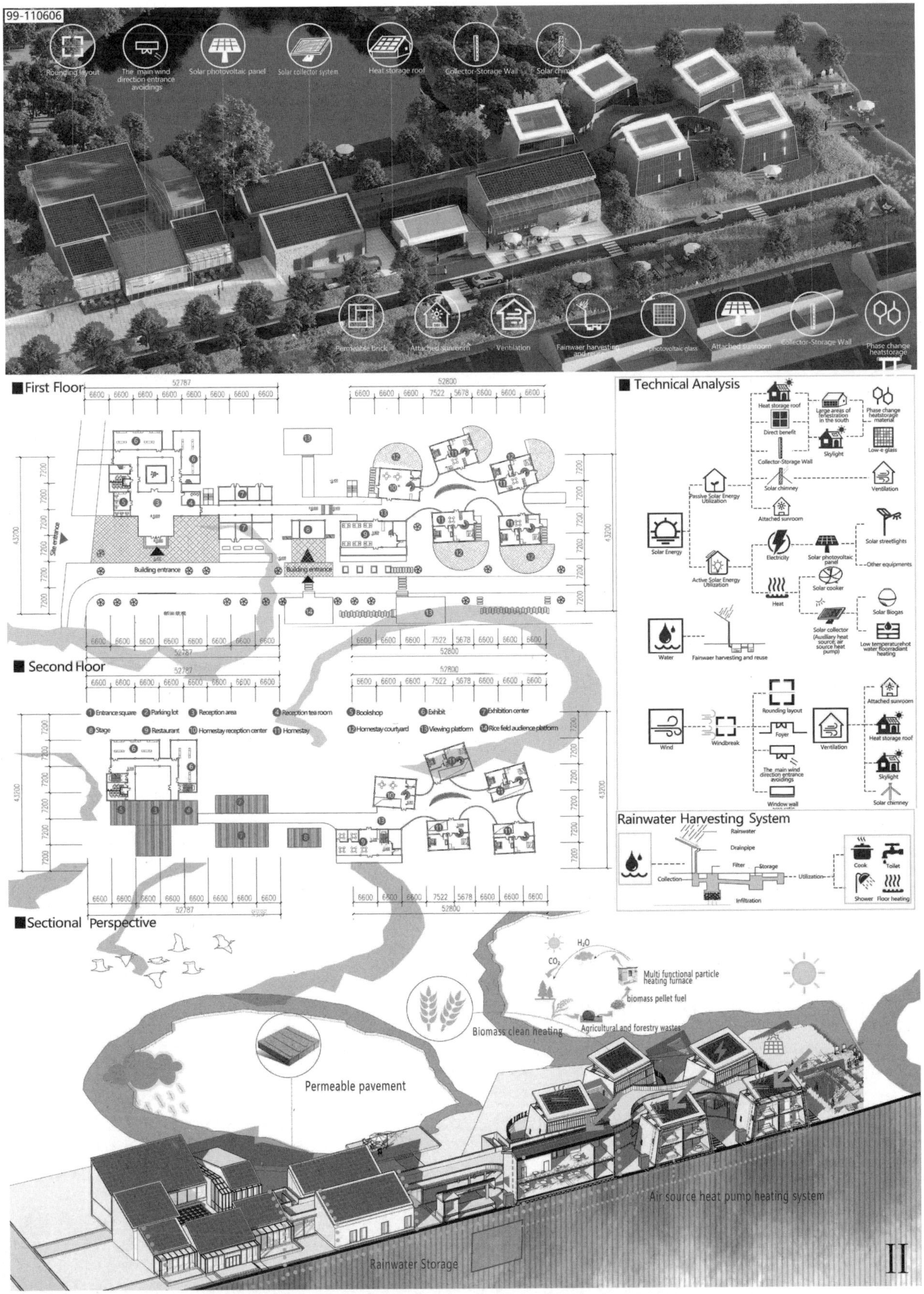
99-110606
Rounding layout
The main wind direction entrance avoidings
Solar photovoltaic panel
Solar collector system
Heat storage roof
Collector-Storage Wall
Permeable brick
Attached sunroom
Ventilation
Phase change heatstorage
First Floor
Building entrance
Site entrance
Second Floor
1 Entrance square
2 Parking lot
3 Reception area
4 Reception tea room
5 Bookshop
6 Exhibit
7 Exhibition center
8 Stage
9 Restaurant
10 Homestay reception center
11 Homestay
12 Homestay courtyard
13 Viewing platform
14 Rice field audience platform
Technical Analysis
Heat storage roof
Direct benefit
Large areas of fenestration in the south
Skylight
Phase change heatstorage material
Low-e glass
Collector-Storage Wall
Solar chimney
Ventilation
Passive Solar Energy Utilization
Attached sunroom
Solar Energy
Electricity
Solar photovoltaic panel
Solar streetlights
Other equipments
Active Solar Energy Utilization
Heat
Solar cooker
Solar Biogas
Solar collector (Auxiliary heat source: air source heat pump)
Low temperaturehot water floorradiant heating
Water
Fainwaer harvesting and reuse
Wind
Windbreak
Rounding layout
Foyer
The main wind direction entrance avoidings
Window wall
Heat storage roof
Rainwater Harvesting System
Rainwater
Drainpipe
Filter
Storage
Collection
Infiltration
Utilization
Cook
Toilet
Shower
Floor heating
Sectional Perspective
H_2O
CO_2
Multi functional particle heating furnace
biomass pellet fuel
Agricultural and forestry wastes
Biomass clean heating
Permeable pavement
Air source heat pump heating system
Rainwater Storage

99-110606

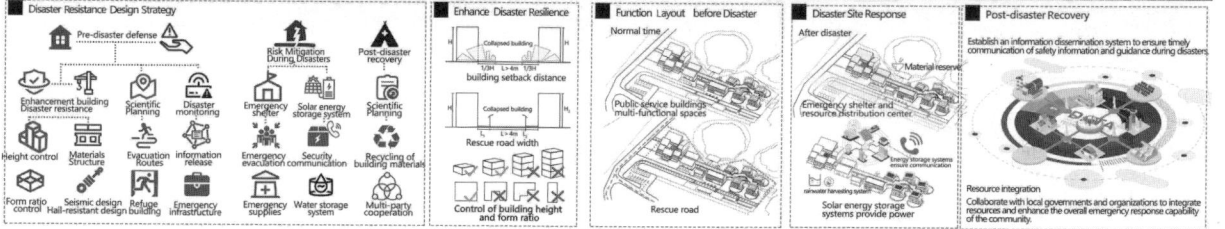

Disaster Resistance Design Strategy

Enhance Disaster Resilience

Function Layout before Disaster

Disaster Site Response

Post-disaster Recovery

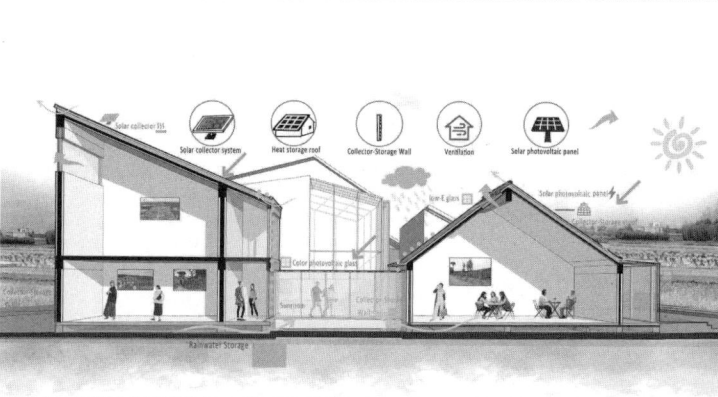

■ Analysis of Construction Strategies

■ Wall Structure

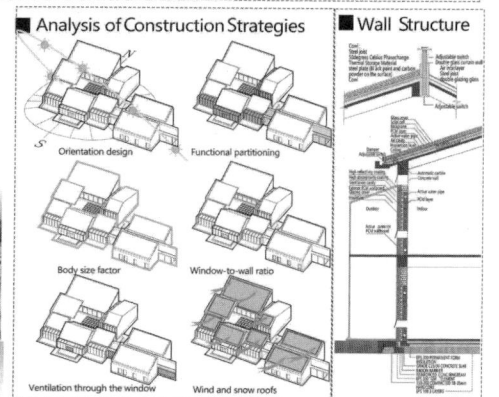

Orientation design — Functional partitioning — Body size factor — Window-to-wall ratio — Ventilation through the window — Wind and snow roofs

■ Sunroom Summer Design
■ Sunroom Winter Design
■ Braun Design

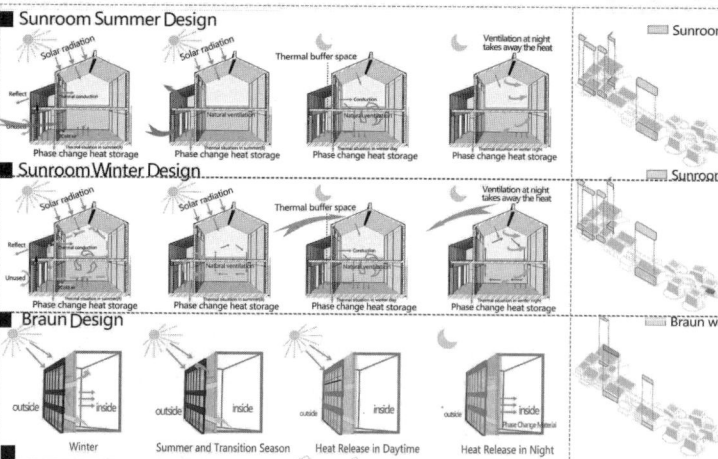

Winter — Summer and Transition Season — Heat Release in Daytime — Heat Release in Night

■ Roof Analysis
■ Integrated Wall Design

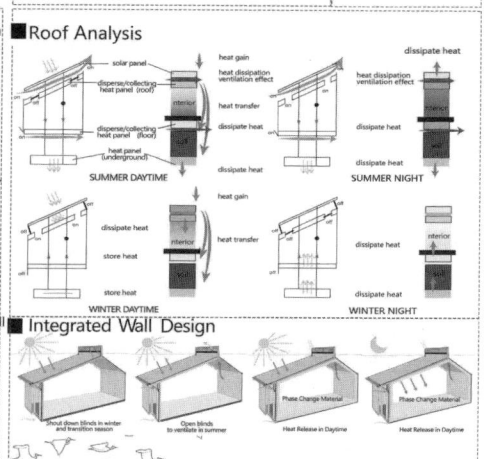

SUMMER DAYTIME — SUMMER NIGHT — WINTER DAYTIME — WINTER NIGHT

III

稻花乡里说丰年

■ Exploded View

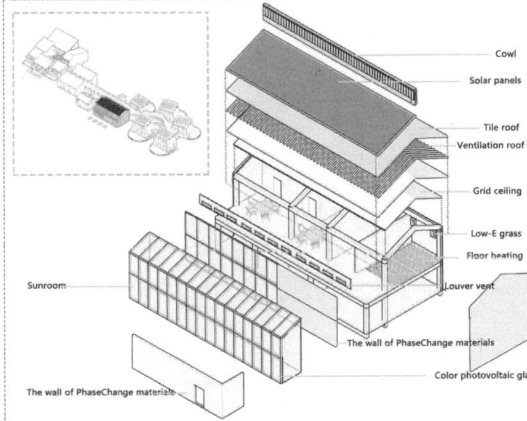

- Cowl
- Solar panels
- Tile roof
- Ventilation roof
- Grid ceiling
- Low-E grass
- Floor heating
- Louver vent
- Sunroom
- The wall of PhaseChange materials
- Color photovoltaic gla:
- The wall of PhaseChange materials

■ Technology

Solar cel construction

Solar collector

Low temperature hot water floar radiant heating

Thermal insulation construction

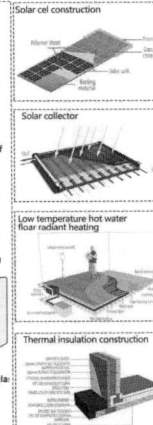

■ Concept Generation

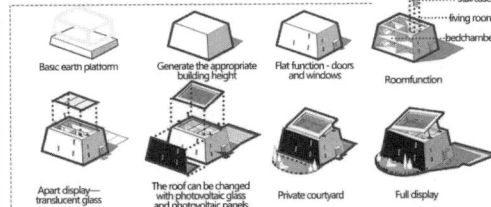

- staircase
- living room
- bedchamber

Basic earth platform | Generate the appropriate building height | Flat function - doors and windows | Roomfunction

Apart display—translucent glass | The roof can be changed with photovoltaic glass and photovoltaic panels | Private courtyard | Full display

■ Photovoltaic Technology

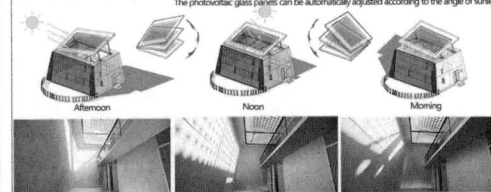

The photovoltaic glass panels can be automatically adjusted according to the angle of sunlight

Afternoon | Noon | Morning

■ Material Analysis

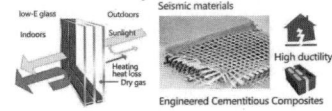

low-E glass | Outdoors | Sunlight | Indoors
Heating heat loss | Dry gas

Seismic materials
High ductility
Engineered Cementitious Composites

■ Indoor Comfort

■ Year-round Radiation Analysis

■ Heat Radiation Analysis

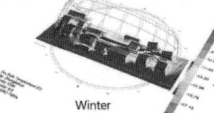

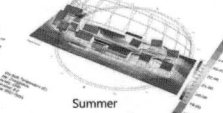

Winter

■ Heat Radiation Analysis

Summer

■ Operation Mode and Strategy

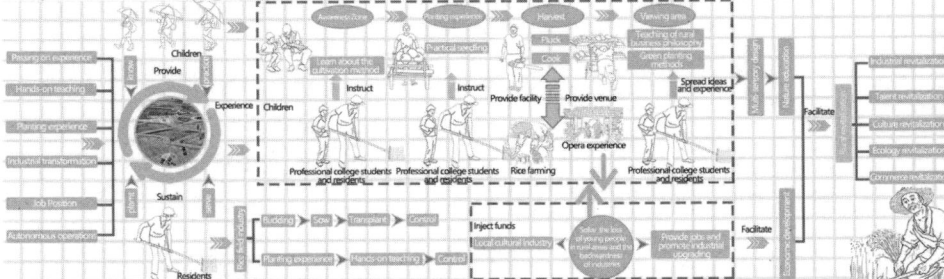

■ Activity Streamline Design

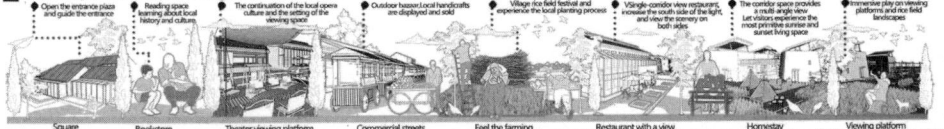

Square | Bookstore | Theater viewing platform | Commercial streets | Feel the farming | Restaurant with a view | Homestay | Viewing platform

IV

织 阳
Weave the Sunshine
Weave Sunshine, Weave Site

Economic and technical index
Floor area: 1182 m²
Plot ratio: 0.25
Building density: 16.1%

综合奖·入围奖·张家口市
怀安县二堡子村建设项目
Comprehensive Awards −
Nomination − Erpuzi Village
Project, Huai'an County,
Zhangjiakou City

注册号：110619
Register Number：110619
项目名称：织阳
Entry Title：Weave the Sunshine
作者：张鹏、李子晗、郑杭其
Authors：Zhang Peng, Li Zihan, and
　　　　Zheng Hangqi
作者单位：昆明理工大学、华侨大学
Authors from：Kunming University of
　　　　Science and Technology,
　　　　Huaqiao University
指导教师：谭良斌
Tutor：Tan Liangbin
指导教师单位：昆明理工大学
Tutor from：Kunming University of Science
　　　　and Technology

Design Specification

This design aims to improve the ability of disaster prevention and reduction in rural areas, and discusses the two themes of how to build resilient village infrastructure and improve the ability of disaster resistance and reduction in rural areas. To improve the disaster resilience and climate resilience of the site planning and design, to weave light climate design. Measure design in terms of villager activity and village space.

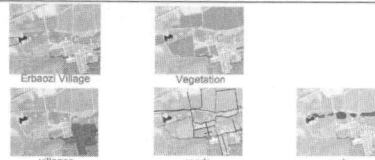

Erbaozi Village | Vegetation
villages | roads | water

Site Analysis

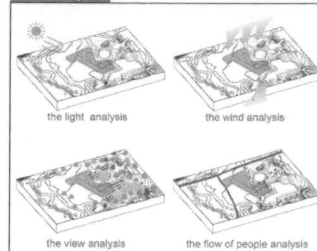

the light analysis | the wind analysis
the view analysis | the flow of people analysis

General Plan 1:500

Dwelling first
Dwelling second
Dwelling third
Building red line
site red line
Dwelling forth
Secondary entry
Residential entrance
Main entrance
Refuge square

Topic Analysis

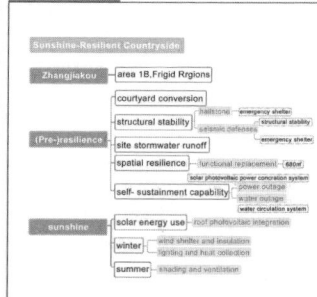

Concept Analysis

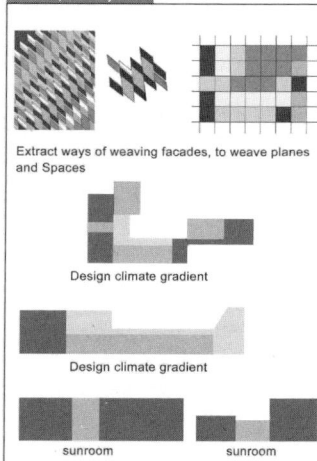

Extract ways of weaving facades, to weave planes and Spaces

Design climate gradient
Design climate gradient

sunroom | sunroom

Site Planning

According to the current planning of the site, it is planned for public buildings in the west and residential buildings on a 2-meter platform in the east.

The building is arranged according to the perennial monsoon, enclosing the building to withstand the winter wind.

Further design according to the street space.

Form Generation

1 According to the climate layout design, the northwest side layout service space

2. vOptimize the ventilation and publicity of the central space

3.Direct sunlight to the south

4. Promote ventilation in the south direct sunlight room

5.Improve central ventilation

6.Set the core observation tower according to the site planning

7.Enclose the courtyard space

8.Designed on the roof, you can see the view

9. Design integral slope roof

Form Generation

The residential courtyard is enclosed in combination with the monsoon, and an additional sunroom service space is set in the middle.

Improve lighting and ventilation in the north and south Spaces

Design roof and courtyard with drainage

Third floor plan 1：300

second floor plan 1：300

Ground floor plan 1：300

Functional Partition

Streamline Analysis

1	门厅	1 Foyer	11	供电室	111 Power Supply Room	1	客厅	6	客房
2	展厅	2 Exhibition halls	12	庭院	12 courtyards	2	餐厅	7	卧室
3	舞台	3 stages	13	办公室	13 Office	3	厨房	8	庭院
4	商店	4 Stores	14	阳台	14 Balconies	4	卫生间	9	卧室
5	简餐	5 Light Meals	15	会议室	15 Conference Room	5	阳光间	10	平台
6	备餐	6 Prepare Meals	16	门厅上空	16 Over the foyer	1 Living room		6 Rooms	
7	体息区	7 Rest Area	17	庭院上空	17 Over the Courtyard	2 Restaurants		7 Bedrooms	
8	活动室	8 Activity Room	18	室外平台	18 Outdoor deck	3 Kitchen		8 Courtyard	
9	茶室	9 Tea Room	19	多功能厅上空	19 Multi-function Hall overhead	4 Bathroom		9 Bedrooms	
10	储藏	10 Storage	20	观景台	20 Observation deck	5 Sunroom		10 Platforms	

Residential family Structure

Residence number	family structure	Room Layout	Number of rooms and size
1	Family of 3: Middle-aged 2 persons Child 1 person	2 guest bedrooms on the first floor; 2master bedrooms on the second floor.	4; 154 ㎡
2	Family of 4: Middle-aged couple 2 persons 2 children	1 guest room on the first floor; 3master bedrooms on the second floor	4; 145 ㎡
3	Family of 4: Elderly 1 person Middle-aged 2 persons Child 1 person	1 senior citizen's room and 2 guest rooms on the first floor; 2master bedrooms on the second floor	5; 146 ㎡
4	Family of 5: Elderly 2 persons Middle-aged 2 people Child 1 person	1 senior citizen's room and 2 guest rooms on the first floor;2 master bedrooms on the second floor	5; 148 ㎡

■ South Elevation Plan 1：300

■ East Elevation Plan 1：300

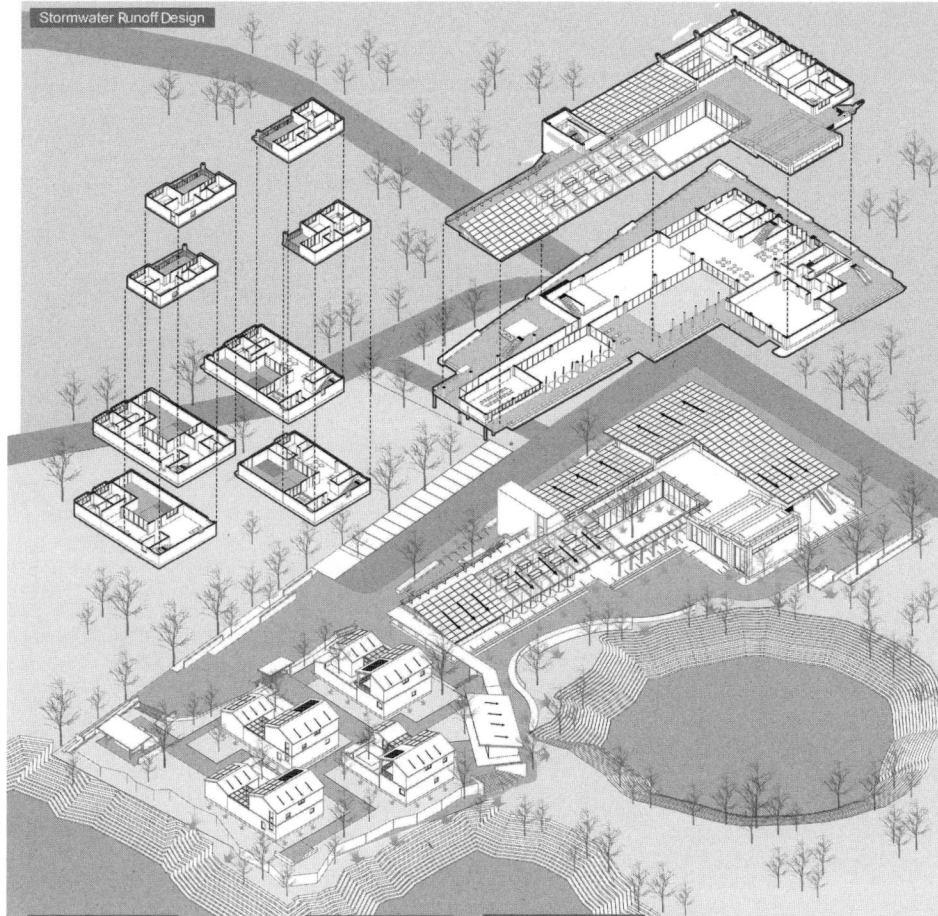

Stormwater Runoff Design

Structural Analysis

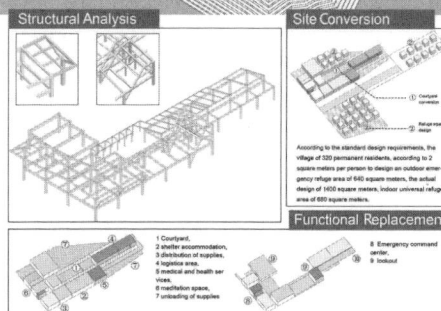

Site Conversion

According to the standard design requirements, the village of 320 permanent residents, according to 2 square meters per person to design an outdoor emergency refuge area of 640 square meters, the actual design of 1400 square meters, indoor universal refuge area of 660 square meters.

Functional Replacement

1 Courtyard,
2 shelter accommodation,
3 distribution of supplies,
4 logistics area,
5 medical and health services,
6 meditation space,
7 unloading of supplies
8 Emergency command center,
9 lookout

Solar Energy Utilization

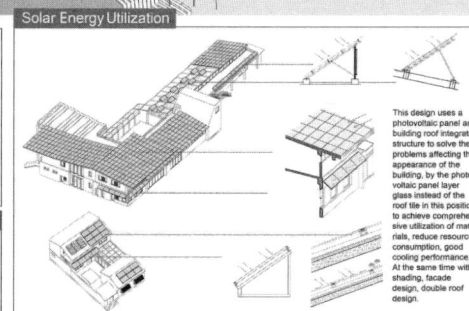

This design uses a photovoltaic panel and building roof integrated structure to solve the problems affecting the appearance of the building, by the photovoltaic panel layer glass instead of the roof tile in this position, to achieve comprehensive utilization of materials, reduce resource consumption, good cooling performance. At the same time with shading, facade design, double roof design.

Passive Technique

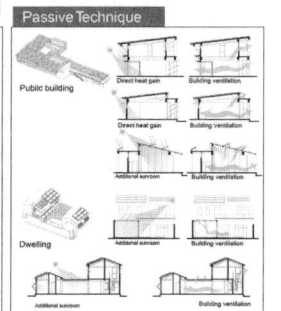

Public building

Dwelling

Direct heat gain
Building ventilation
Direct heat gain
Building ventilation
Additional sunroom
Building ventilation
Additional sunroom
Building ventilation
Additional sunroom
Building ventilation

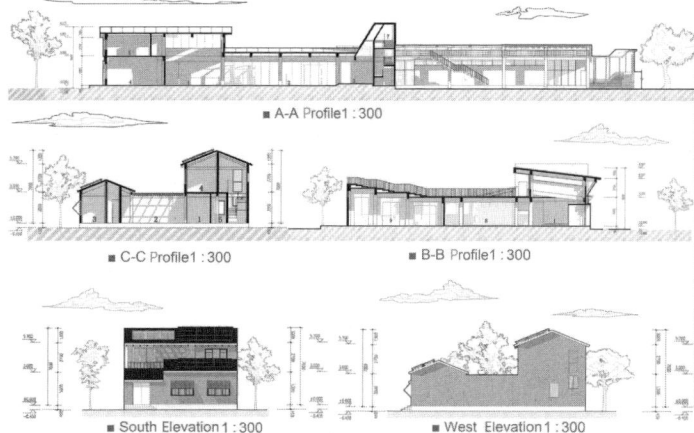

■ A-A Profile1 : 300

■ C-C Profile1 : 300

■ B-B Profile1 : 300

■ South Elevation 1 : 300

■ West Elevation 1 : 300

Thermal Radiation and Temperature Simulation

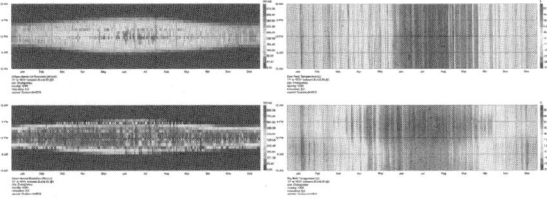

Annual Enthalpy and Humidity Chart

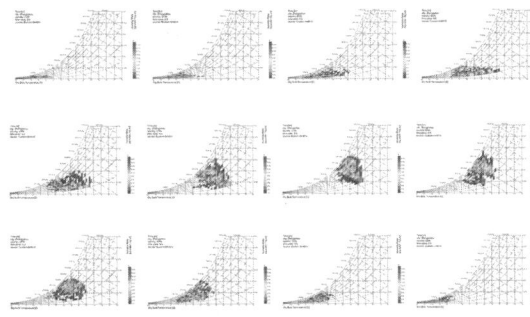

Year-round Wind Rose

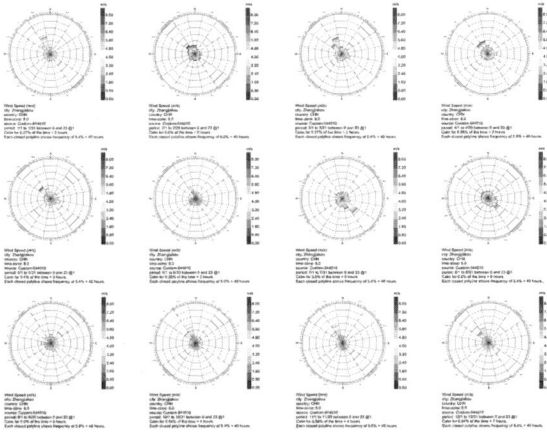

Annual thermal radiation

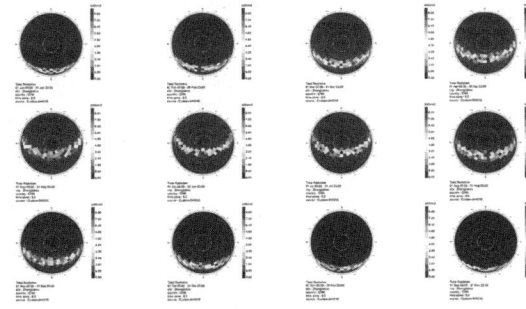

Public Building Illumination Analysis

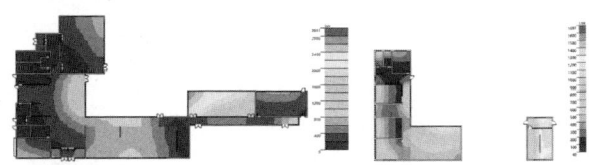

Residential Illumination Analysis

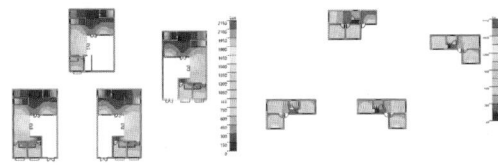

Carbon Emissions from Public Building Operations

Carbon emissions from building operations

Electricity	Category	Electricity consumption (kW·h/m²)	Carbon emission factors (kgCO₂/kW·h)	Carbon emissions (TCO₂)
Cooling (EC)	Central cold source	0.00	0.581	0.000
	Cooling water pump	0.00		
	Frozen water pump	0.00		
	Cooling Tower	0.00		
	Multi-unit air conditioning	0.00		
	Total cooling	0.00		
Heating (Eh)	Central heat source	0.00	0.581	0.000
	Heating pump	0.00		
	Heat source side pump	0.00		
	Multi-unit heat pump	0.00		
	Total heating	0.00		
Air conditioning Wind turbine (EF)	New ventilation	178.19	0.581	188.922
	Fan coil unit	33.83		
	Multi-online indoor computer	0.00		
	Full air system	0.00		
	Total fans	212.02		
	Lighting	588.45	0.581	524.353
	Socket equipment		0.581	-
Others	Elevators	0.00	0.581	4026.128
	Exhaust fan	2379.90		
	Domestic hot water (minus solar energy)	2138.41		
	Total	4518.31		

Fossil fuels	Category	Heat consumption (kW·h/m²)	Carbon emission factor (TCO₂/TJ)	Carbon emissions (TCO₂)
None 无	Heating: : heat source boiler	0.000	0	0.000
	Heating: municipal heating	0.00	0	0.000
None	Domestic hot water (minus solar energy)	0.00	0	0.000

Gas	Cooking	- (m²/m²)	55.54	
Other	Category	Consumption (kg)		Carbon emissions (TCO₂)
Refrigerant	Cooling	0		0.000
Renewable Energy (ER)	Category	Power supply (kW·h/m²)	Carbon emission factor (kgCO₂/kW·h)	Carbon reduction (TCO₂)
	Renewable Photovoltaics (EP)	4216.50	0.581	3757.196
	Wind power (EW)	0.00		
	Total carbon emissions from building operations			982.205

Index of Unit area

Category	Annual carbon emissions (kgCO₂/m²·a)	Carbon emissions (kgCO₂/m²)
Building materials production	2.33	116.35
Transportation of construction materials	0.12	5.89
Construction	0.01	0.60
Building demolition	0.22	11.10
Construction	12.81	640.42
Carbon Sink	0	0
Total	15.49	774.36

Total carbon emissions

Category	Annual carbon emissions(TCO₂/a)	Carbon emissions (TCO₂)
Building materials production	3.569	178.447
Transportation of construction materials	0.181	9.041
Construction	0.019	0.920
Building demolition	0.341	17.027
Construction	19.644	982.205
Carbon Sink	0	0
Total	23.754	1187.640

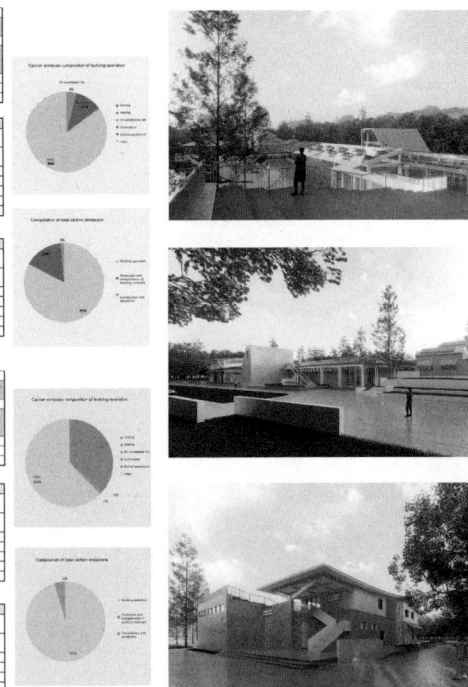

Carbon Emissions from Residential Operations

Carbon emissions from building operations

Electricity	Category	Electricity consumption (kW·h/m²)	Carbon emission factors (kgCO₂/kW·h)	Carbon emissions (TCO₂)
Cooling (EC)	Central cold source	0.00	0.581	0.000
	Cooling water pump	0.00		
	Frozen water pump	0.00		
	Cooling Tower	0.00		
	Multi-unit air conditioning	0.00		
	Total cooling	0.00		
Heating (Eh)	Central heat source	0.00	0.581	518.732
	Heating pump	5372.16		
	Heat source side pump	0.00		
	Multi-unit heat pump	0.00		
	Total heating	5372.16		
Air conditioning Wind turbine (EF)	New ventilation	138.96	0.581	43.445
	Fan coil unit	310.97		
	Multi-online indoor computer	0.00		
	Full air system	0.00		
	Total fans	449.93		
	Lighting	560.54	0.581	54.125
	Socket equipment		0.581	
Others	Elevators	0.00	0.581	2120.714
	Exhaust fan	21962.81		
	Domestic hot water (minus solar energy)	0.00		
	Total	21962.81		

Fossil fuels	Category	Heat consumption (kW·h/m²)	Carbon emission factor (TCO₂/TJ)	Carbon emissions (TCO₂)
Bituminous coal II	Heating: : heat source boiler	12573.002	89	669.478
无	Heating: municipal heating	0.00	0	0.000
None	Domestic hot water (minus solar energy)	0.00	0	0.000

Gas	Cooking	- (m²/m²)	55.54	
Other	Category	Consumption (kg)		Carbon emissions (TCO₂)
Refrigerant	Cooling	0		0.000
Renewable Energy (ER)	Category	Power supply (kW·h/m²)	Carbon emission factor (kgCO₂/kW·h)	Carbon reduction (TCO₂)
	Renewable Photovoltaics (EP)	1421.78	0.581	137.286
	Wind power (EW)	0.00		
	Total carbon emissions from building operations			3269.130

Index of Unit area

Category	Annual carbon emissions (kgCO₂/m²·a)	Carbon emissions (kgCO₂/m²)
Building materials production	18.17	908.30
Transportation of construction materials	0.63	31.53
Construction	0.12	5.57
Building demolition	0.22	11.13
Building operation	393.41	19670.45
Carbon Sink	0	0
Total	412.55	20626.98

Total carbon emissions

Category	Annual carbon emissions (TCO₂/a)	Carbon emissions (TCO₂)
Building materials production	3.019	150.949
Transportation of construction materials	0.105	5.230
Construction	0.019	0.920
Building demolition	0.037	1.845
Building operation	65.383	3269.130
Carbon Sink	0	0
Total	68.563	3428.074

综合奖·入围奖·张家口市
怀安县二堡子村建设项目
Comprehensive Awards −
Nomination − Erpuzi Village
Project，Huai'an County，
Zhangjiakou City

注册号：110693
Register Number：110693

项目名称：日影·韧境——乡村会客厅
设计

Entry Title：Sun Shadow—Resilient Realm：
Rustic Parlour Design

作者：谢欣欣、刘文洁、冯一帆、陈通

Authors：Xie Xinxin, Liu Wenjie,
Feng Yifan, and Chen Tong

作者单位：昆明理工大学
Authors from：Kunming University of
Science and Technology

指导教师：谭良斌
Tutor：Tan Liangbin

指导教师单位：昆明理工大学
Tutor from：Kunming University of Science
and Technology

110693

日影·韧境
—乡村会客厅设计

Rustic Parlour Design, Erpuzi Village, Hebei

Economic and technical Indicators

Items	Value
Project land area（㎡）	8260
Construction land area（㎡）	4275
Total building area（㎡）	1848.13
Building footprint area（㎡）	1934.98
Building density	23.4%
Floor area ratio (FAR)	0.22
Greening rate	30%

3m*3m Variable Module

tableware   bar counter   plantation unit   vertical greening

trapeze   trampoline   yoga studio   awning

Residential Space Sequence

Original Residence   3m*3m Variable Activity Space   Transitional Gray Space   Private Courtyard

First floor Plan

Village Cultural Activity Center
1. Lobby
2. Exhibition hall
3. Leisure space
4. Entertainment space
5. Reading room
6. Cafe
7. Restaurant
8. Convenience store
9. Drugstore
10. Consulting room
11. Meeting room
12. Office
13. Toilet
14. Storage
15. Kitchen
16. Parking

Homestay Group
1. Living room
2. Dining room
3. Resting room
4. Kitchen
5. Toilet
6. Courtyard
7. Waterscape

Second Floor

First Floor

0 5 10 15 20 25m

N

Section View

A-A Section View   B-B Section View

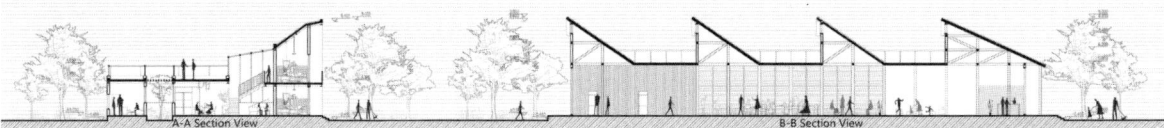

Sectional Perspective View

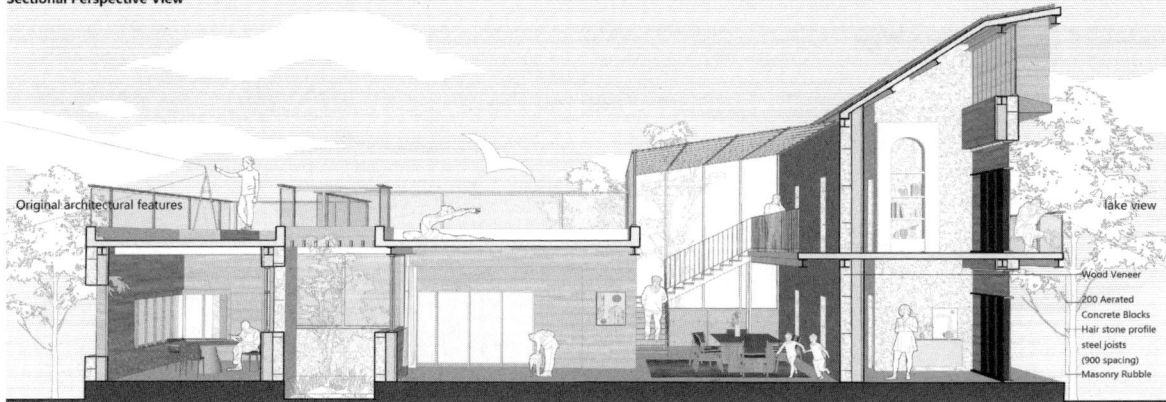

Original architectural features

lake view

Wood Veneer
200 Aerated
Concrete Blocks
Hair stone profile
steel joists
(900 spacing)
Masonry Rubble

日影·韧境
——乡村会厅设计
Rustic Parlour Design, Erpuzi Village, Hebei

Climate Analysis

The design project is located in Erdaowa Village, Zhangjiakou, Hebei Province, at the southern edge of China's cold regions. It experiences long, cold winters and short, cool summers, necessitating specific requirements for winter heating and summer shading and ventilation. Additionally, the design area is adjacent to a small lake, creating a relatively unique microclimate.

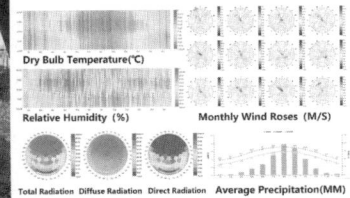

Dry Bulb Temperature(℃)

Relative Humidity (%)　　Monthly Wind Roses (M/S)

Total Radiation　Diffuse Radiation　Direct Radiation　Average Precipitation(MM)

According to the visual data analysis of meteorological information, the local climate requires enhanced ventilation in summer to lower indoor temperatures. In winter, it is essential to maximize the use of solar heat to improve thermal performance and reduce indoor heat loss.

Simulation Validation

From "resilience" to "elastic adjustment", the design starts from regulating "elastic spaces," shifting green architecture from a "measures-oriented" approach to an "outcomes-oriented" one, and from a "technology-oriented" to a "design-oriented" focus. Compared to traditional fixed spaces, this design features adaptable thermal storage and shading spaces that respond to winter and summer conditions, as well as day and night variations, while fostering localized creativity and experiences within the space.

The design incorporates 3m x 3m steel columns arranged horizontally within functional volumes, enabling rapid conversion of functions during disasters and achieving quick construction goals. The public area and guesthouse roofs are single-slope to optimally position solar photovoltaic panels, ensuring sustainable energy supply for cooling in summer and heating in winter. Additionally, ventilation is enhanced through the opening and closing of skylights and surrounding windows, promoting thermal and wind pressure ventilation. In cold seasons, the sunroom can serve as a "greenhouse," providing heat retention and insulation.

Contrast Model

In the design, we adjust different indoor natural light environments through the "virtual and real" and "open and closed" elements of the envelope interface, establishing comparative experimental groups for simulation and validation.

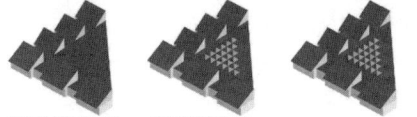

#1 Model Without Skylight　　#2 Model With Skylight

Solar Photovoltaic Power Generation(AC)

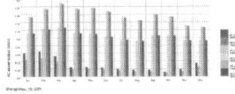

The AC power output peaks from June to August, with the highest in August at 3.24 kWh. The lowest outputs are in December and January, both at 0 kWh, likely due to shorter daylight hours and weaker sunlight in winter.

Solar Photovoltaic Power Generation(DC)

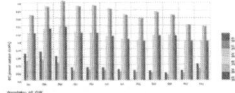

The DC power output follows a similar trend, with the highest in August at 3.6 kWh and the lowest in December at 0 kWh.

Solar Radiation In Commer Area

Site

Radiation Analysis
1 JAN 1: 00 - 31 DEC 24: 00

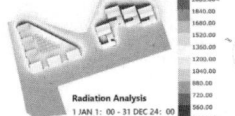

Construction

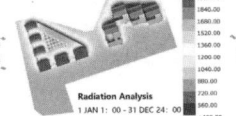

Radiation Analysis
1 JAN 1: 00 - 31 DEC 24: 00

Solar Photovoltaic　　Natural Lighting

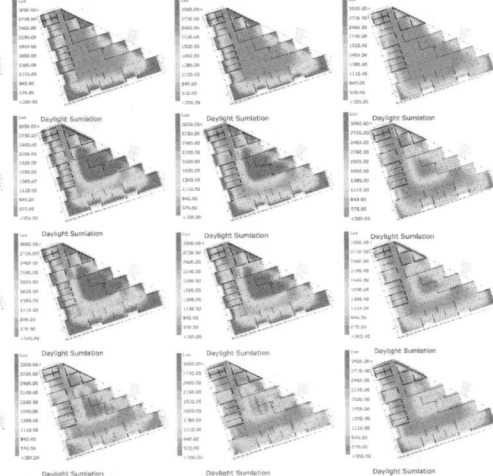

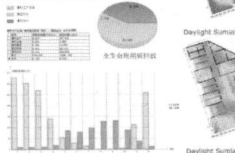

Daylight Sumation

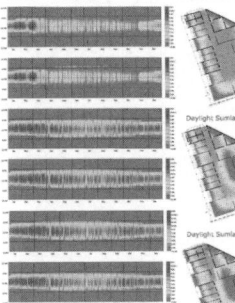

NATURAL LIGHTING

Active&Passive Energy Saving Strategy System

In this design, we selected appropriate technologies and strategies based on the local climate and environmental performance requirements, creating a green building with a sense of place. The design philosophy has shifted from "energy-consuming buildings" to "zero-energy buildings" and even "positive energy buildings." Upholding the principle of "passive first, active optimization," we integrated active technologies with passive strategies, which primarily encompass three aspects: form, skin, and structure. The active technology solutions include solar energy utilization, rainwater recycling, geothermal heat pumps, and efficient lighting systems.

Solar Radiation In B&B Area

Solar Radiation (Without Shading)
22 Jun 12: 00

Solar Radiation (Without Shading)
22 Jun 12: 00

Solar Radiation (With Shading)
22 Jun 12: 00

Solar Radiation (With Shading)
22 Jun 12: 00

On the summer solstice at noon, simulations were conducted for indoor and outdoor solar radiation in two control groups. The shading group effectively blocked the intense radiation from higher solar angles, creating a cooler and more comfortable environment. According to the simulation results, the unshaded sunroom had an internal solar radiation level of 429 kWh/m², while the addition of shading reduced internal radiation by 53.4%.

Passive Strategy Analysis

- Comfort Time 9.36%
- Evaporative Cooling 16.93%
- Mass + Night Vent 14.82%
- Occupant Use of Fans 15.10%
- Capture Internal Heat 33.57%
- Passive Solar Heating 40.03%

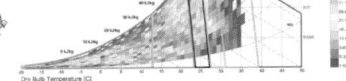

The following passive strategies are applicable for moderating the building environment in Erdaowa Village: capturing internal heat and passive solar heating can increase comfort time by 33% and 40%, respectively.

Passive Design Strategy Of Climate

In summer, the upper sunshade of the residential sunroom is opened and the external windows are opened to create a shaded space and at the same time to promote indoor and outdoor ventilation; the eaves of 1.5-3m in the public area and the design of the furniture on the facade can effectively shade the sun. In winter, the residential sunroom can capture heat from solar radiation over a large area; the public area is fitted with glass along the outermost steel columns to form a sunroom.

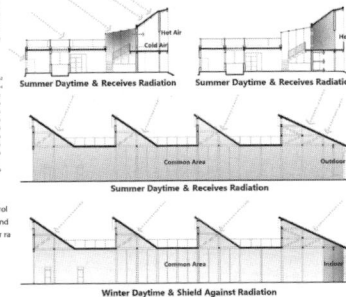

Summer Daytime & Receives Radiation　　Summer Daytime & Receives Radiation

Summer Daytime & Receives Radiation

Winter Daytime & Shield Against Radiation

Wind Simulation

Solar Radiation (With Shading)

Solar Radiation (With Shading)

By utilizing tall open spaces in multiple areas of a building to promote natural ventilation, the distribution of indoor windspeeds is optimized, energy consumptions reduced, and indoor air quality is improved, resulting in a more comfortable and healthy living environment.

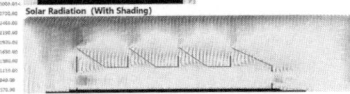

Solar Radiation (With Shading)

Analysis of Active Energy Saving Strategies

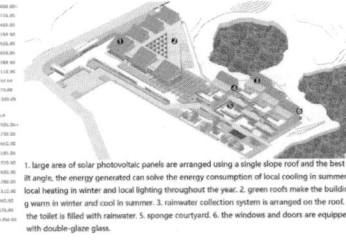

1. large area of solar photovoltaic panels are arranged using a single slope roof and the best tilt angle, the energy generated can solve the energy consumption of local cooling in summer, local heating in winter and local lighting throughout the year. 2. green roofs make the building warm in winter and cool in summer. 3. rainwater collection system is arranged on the roof. 4. the toilet is filled with rainwater. 5. sponge courtyard. 6. the windows and doors are equipped with double-glaze glass.

110693

日影·韧境
——乡村会客厅设计
Rustic Parlour Design, Erpuzi Village, Hebei

The Optimal Orientation

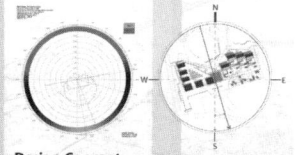

Design Concept

设计通过对当地气候及传统民居的分析，确定建筑的整体朝向和布局。针对公共大空间和居住小空间不同使用需求设置不同的气候缓冲空间策略，结合地形形成丰富、富有韵律感的空间序列。

在节能策略上，借助LadyBug、HoneyBee等模拟工具确定太阳朝向利用最佳朝向；通过计算确定太阳能光伏板功效最佳的倾角范围。在被动式策略上，民宿设计沿袭河北地区传统民居设置两进院落，并结合交通空间设置阳光间、蓄热体，借助太阳辐射解决冬季热舒适问题。在公共空间设计中，使用中庭和天窗引入更多自然光。外立面结合檐下空间设置可移动格栅形成可变的气候缓冲空间，适应夏夏不同气候条件的使用需求。

在材料和结构上，民宿采用钢框架承重，混凝土砌块做围护，毛石做外表皮，易于获取且抗震性能优越，灾害后易于回收再利用；村民活动中心采用钢框架结构，内部分隔用可移动轻质板材，方便快速搭建，也便于灾时根据使用需求进行功能置换。

Energy-saving Strategies

LadyBug and HoneyBee are used to determine the optimal direction. Determine the optimal angle range for the effectiveness of solar photovoltaic panels through calculation. As for passive strategy, the design of homestays follows the traditional two courtyard layout of residential buildings in Hebei region and combines with transportation space to set up sunroom. Skylights and courtyards are used to introduce natural light, movable grilles are setup on the exterior facade with eaves to create a variable climate buffering space adapting to Winter and Summer.

Materials and Structure

the homestay adopts a steel frame for load-bearing, concrete blocks for enclosure, and rubble for masonry External skin; The village activity center adopts a steel frame structure, and the internal partitions are made of movable lightweight panels, which facilitate functional replacement according to usage needs during disasters.

General Layout

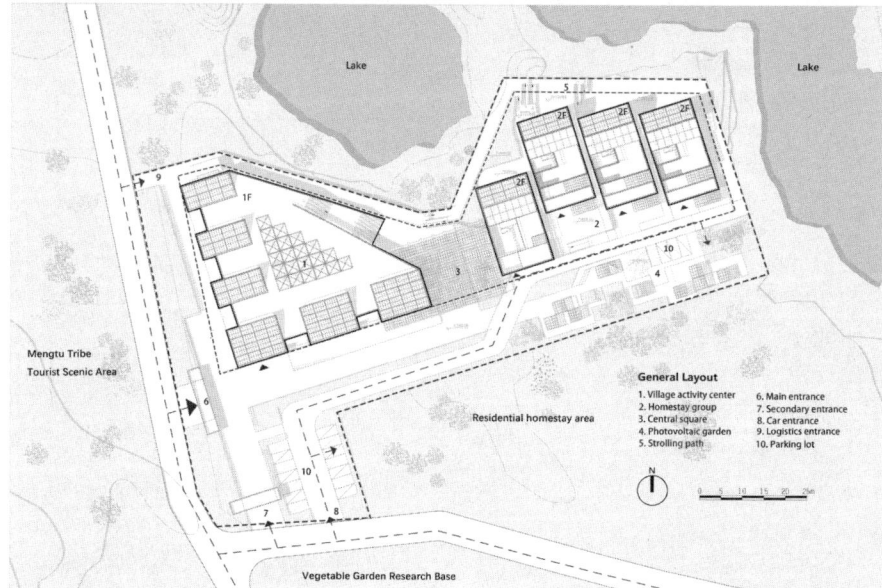

Mengtu Tribe
Tourist Scenic Area

Lake

Lake

Residential homestay area

Vegetable Garden Research Base

General Layout

1. Village activity center
2. Homestay group
3. Central square
4. Photovoltaic garden
5. Strolling path
6. Main entrance
7. Secondary entrance
8. Car entrance
9. Logistics entrance
10. Parking lot

Photovoltaic Panel Analysis

The optimal angle for photovoltaic panels

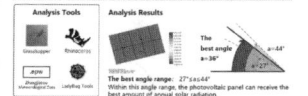

The best angle range: 27°≤a≤44°
Within this angle range, the photovoltaic panel can receive the best amount of annual solar radiation.

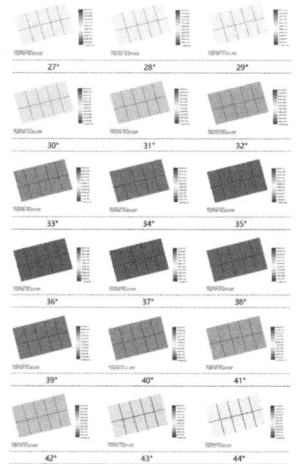

27° 28° 29°
30° 31° 32°
33° 34° 35°
36° 37° 38°
39° 40° 41°
42° 43° 44°

South Facade

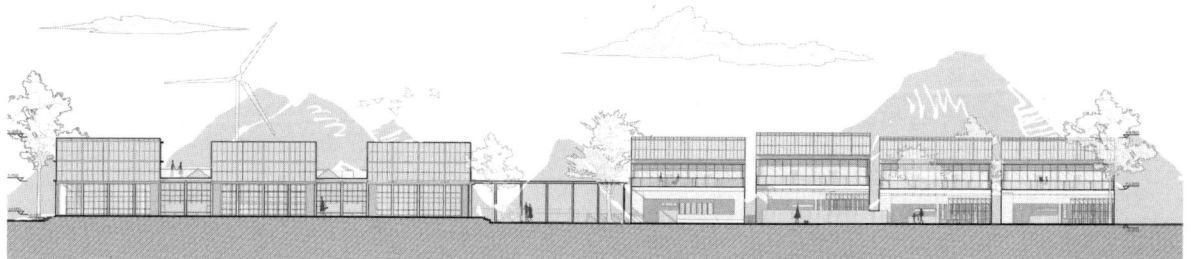

日影·韧境
——乡村会客厅设计

Rustic Parlour Design, Erpuzi Village, Hebei

Internal Perspective

Structural and Material Analysis

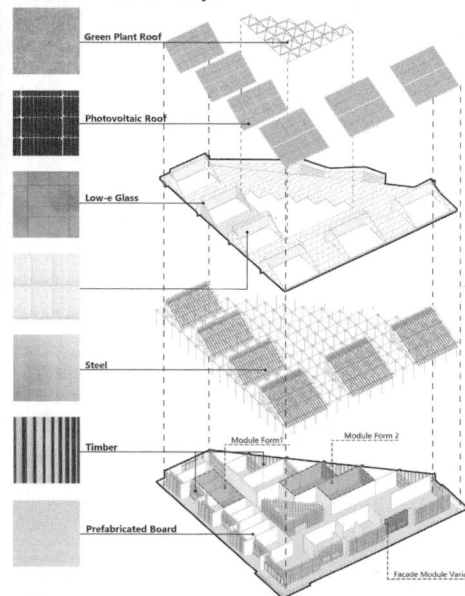

Green Plant Roof

Photovoltaic Roof

Low-e Glass

Steel

Timber

Module Form1　Module Form 2

Prefabricated Board

Facade Module Variability

Standard Unit Module

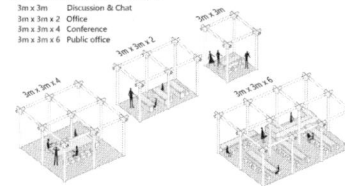

3m x 3m　Discussion & Chat
3m x 3m x 2　Office
3m x 3m x 4　Conference
3m x 3m x 6　Public office

3m x 3m x 4　　3m x 3m x 2　　2m x 3m

3m x 3m x 6

Functional Variability

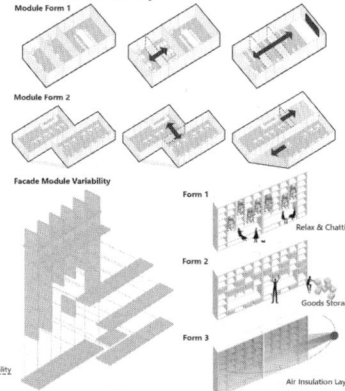

Module Form 1

Module Form 2

Facade Module Variability

Form 1　Relax & Chatting

Form 2　Goods Storage

Form 3

Air Insulation Layer

Elevation Variability

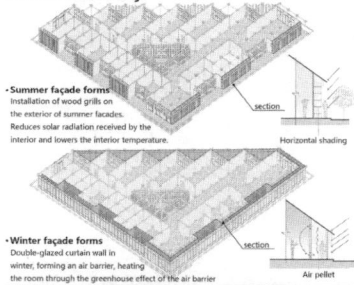

- Summer façade forms
Installation of wood grills on
the exterior of summer facades.
Reduces solar radiation received by the
interior and lowers the interior temperature.

section
Horizontal shading

- Winter façade forms
Double-glazed curtain wall in
winter, forming an air barrier, heating
the room through the greenhouse effect of the air barrier

section
Air pellet

稻梦空间
作品编号：110701

Rice Dream Space

综合奖·入围奖·张家口市
怀安县二堡子村建设项目
Comprehensive Awards -
Nomination - Erpuzi Village
Project，Huai'an County，
Zhangjiakou City

注册号：110701
Register Number：110701
项目名称：稻梦空间
Entry Title：Rice Dream Space
作者：夏陈浩、郑一涵、蓝文涛
Authors：Xia Chenhao，Zheng Yihan，and
Lan Wentao
作者单位：南京工业大学
Authors from：Nanjing Tech University
指导教师：罗佳宁
Tutor：Luo Jianing
指导教师单位：南京工业大学
Tutor from：Nanjing Tech University

Design Description

方案以"稻梦空间"为设计立意，以稻梦隐喻村民对未来的展望。

本方案以可持续发展为设计原则，整体采用模块化的设计，通过不同手法营造出丰富的空间层次，通过在建筑中植入平台、坡道、廊道等形式创造多样的空间感，将建筑部分架空，既通风防潮，又尊重原有场地现状。

另外，建筑遵循可更新的原则，后期可以根据需求对建筑进行扩展，实现了建筑与环境共同成长。

The scheme is designed with "Rice Dream Space" as the design concept, and uses Rice Dream as a metaphor for the villagers' outlook on the future.

This scheme takes sustainable development as the design principle, adopts modular design as a whole, creates a rich spatial level through different techniques, and creates a diverse sense of space by implanting platforms, ramps, corridors and other forms in the building. At the same time, the building is partially elevated, which is both ventilated and moisture-proof, and respects the original site status.

In addition, the building follows the principle of renewal, and the building can be expanded according to needs in the later stage, realizing the common growth of the building and the environment.

Site Analysis

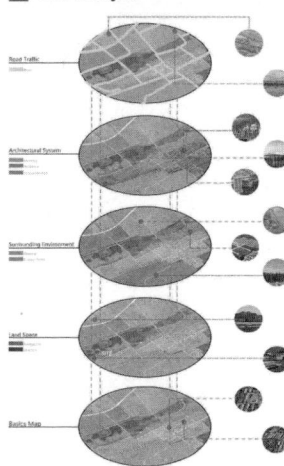

Climate Analysis

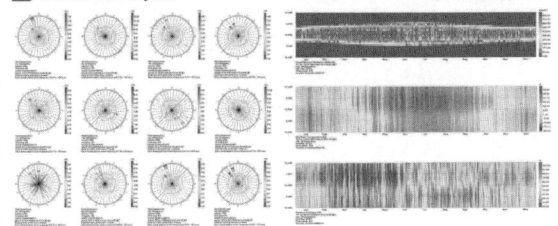

Crowd Analysis

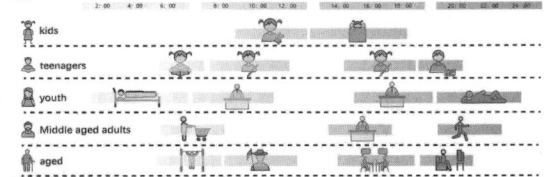

- kids
- teenagers
- youth
- Middle aged adults
- aged

Current Situation of the Site

Design Strategy

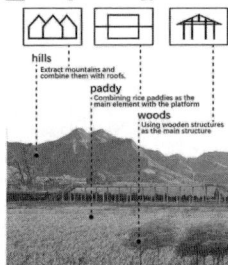

hills
Extract mountains and combine them with roofs.
paddy
Combining rice paddies as the main element with the platform
woods
Using wooden structures as the main structure

Intention extraction

Streamline design — tourist / villagers

Yanghe River is located on the north side of the village
The tourist streamline is like a river, consisting of a main stream and multiple tributaries

The original pond is located on the north side of the site
The villagers' streamline lines flow freely like a pond

Rice Dream Space

Green technology design

Layered design — tourist / villagers

Tourists will be one of the main users of the building in the future
Villagers are one of the main users of the original population

symbolized — paddy / soil

private 2F / public 1F

separate

implant

play lodging rest
recreation rally service

BIPV
Solar Energy technology
BAPV
Active energy-saving technology
Passive energy-saving technology

Design Principles

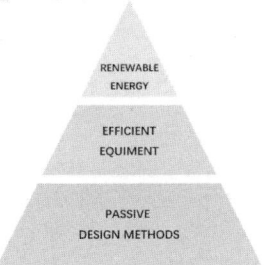

RENEWABLE ENERGY

EFFICIENT EQUIMENT

PASSIVE DESIGN METHODS

hills

paddy

woods

dream

稻梦空间

作品编号：110701

Rice Dream Space

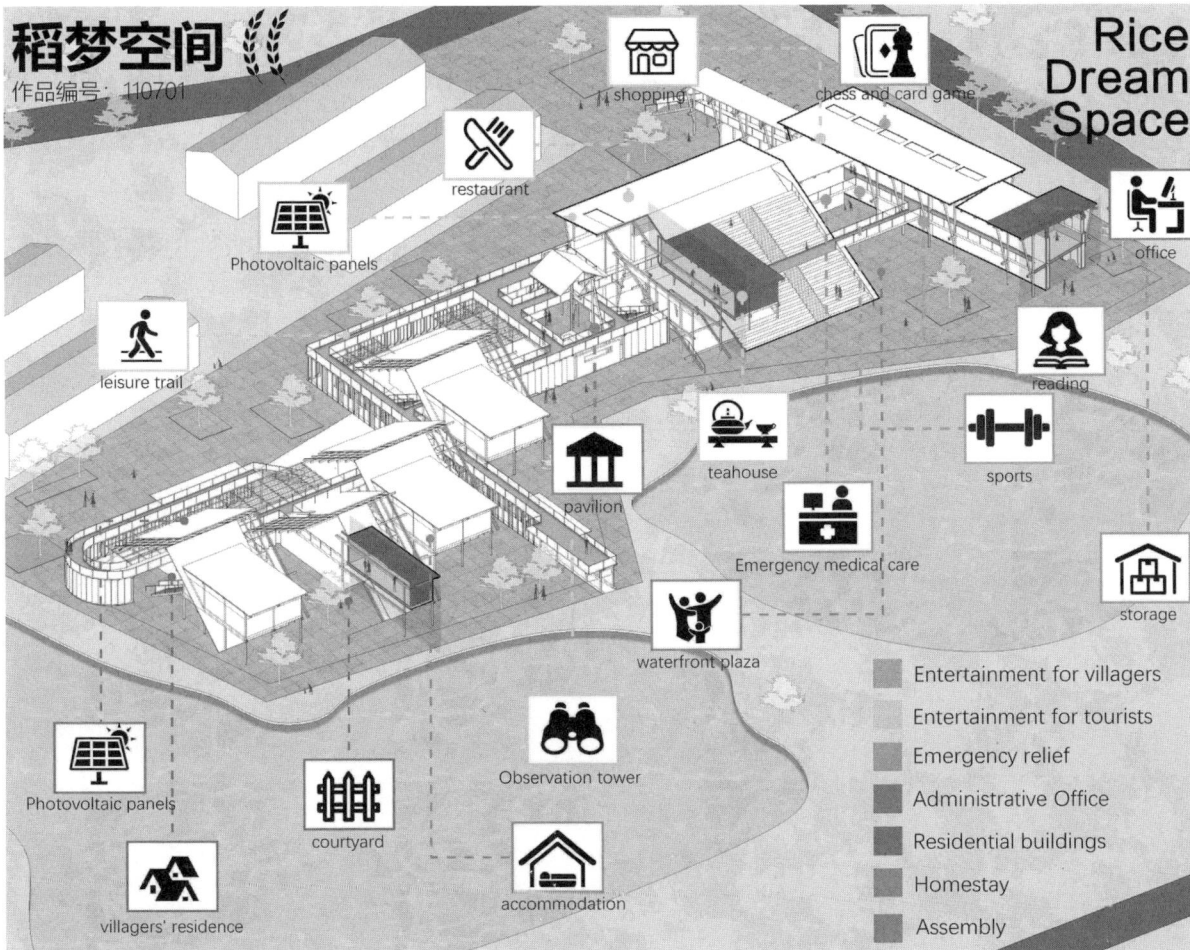

- shopping
- chess and card game
- restaurant
- Photovoltaic panels
- leisure trail
- office
- reading
- teahouse
- sports
- pavilion
- Emergency medical care
- storage
- waterfront plaza
- Photovoltaic panels
- Observation tower
- courtyard
- accommodation
- villagers' residence

- Entertainment for villagers
- Entertainment for tourists
- Emergency relief
- Administrative Office
- Residential buildings
- Homestay
- Assembly

Volume Analysis

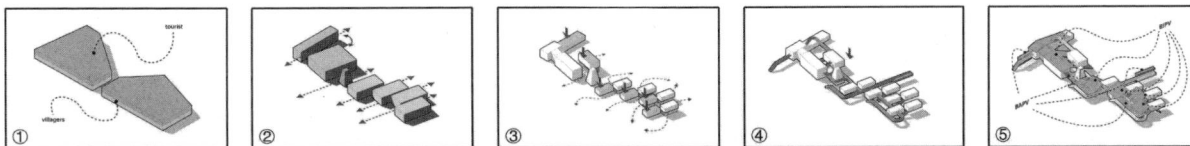

① ② ③ ④ ⑤

Streamline Analysis

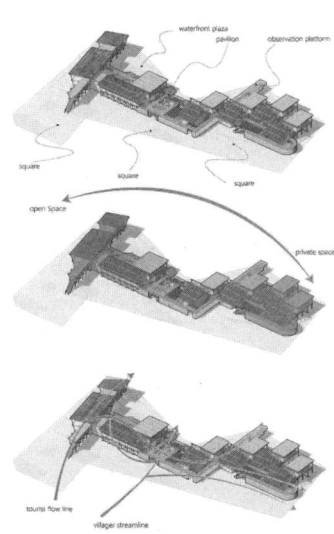

waterfront plaza
pavilion
observation platform
square
square
square
open space
private space
tourist flow line
villager streamline

General Plan

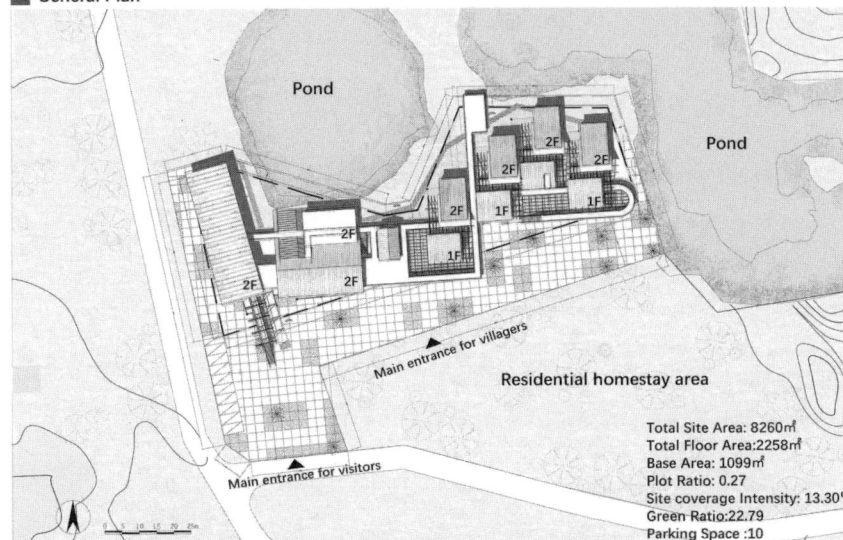

Pond

Pond

2F 2F 2F

2F 1F 1F

2F

2F 1F

2F 2F

1F

Main entrance for villagers

Residential homestay area

Main entrance for visitors

Total Site Area: 8260㎡
Total Floor Area:2258㎡
Base Area: 1099㎡
Plot Ratio: 0.27
Site coverage Intensity: 13.30%
Green Ratio:22.79
Parking Space :10

稻梦空间

作品编号：110701

Rice Dream Space

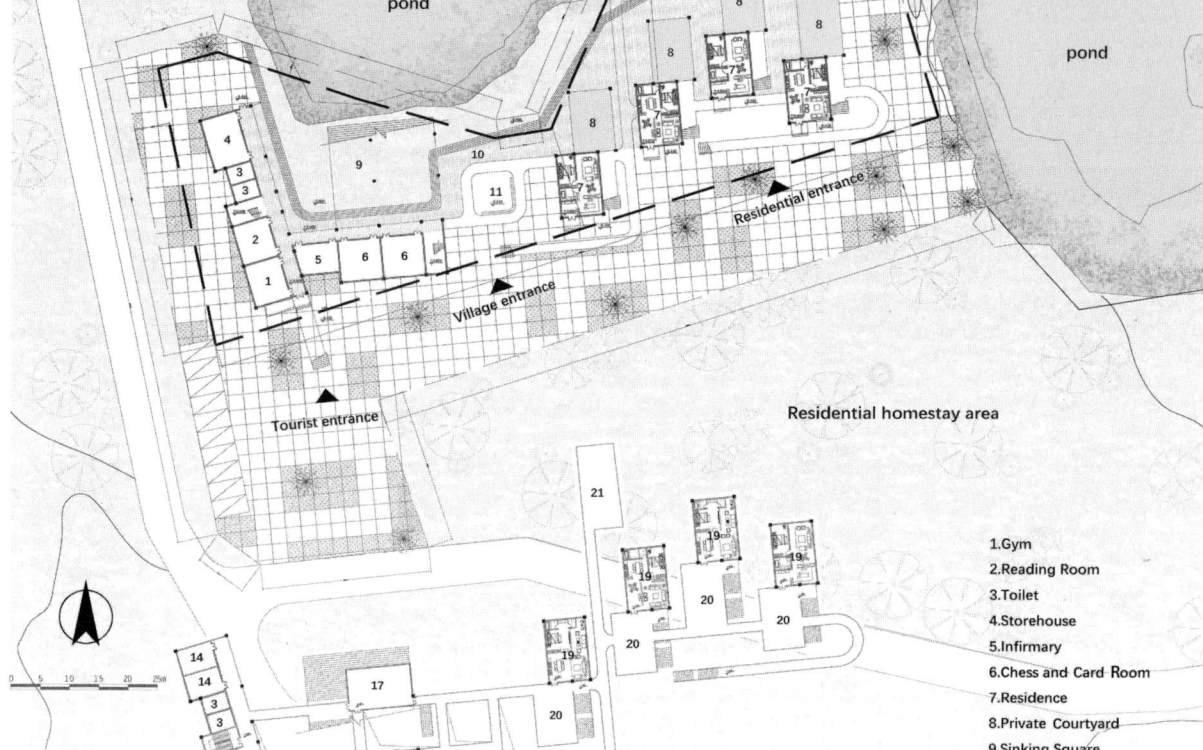

■ First Floor Plan

pond

pond

Tourist entrance

Village entrance

Residential entrance

Residential homestay area

1.Gym
2.Reading Room
3.Toilet
4.Storehouse
5.Infirmary
6.Chess and Card Room
7.Residence
8.Private Courtyard
9.Sinking Square
10.Corridor
11.Gazebo
12.Information Desk
13.Souvenir Shop
14.Office
15.Kitchen
16.Dining Room
17.Tea Room
18.Entertainment Room
19.Homestay
20.Outdoor Platform
21.Observation Tower

■ Second Floor Plan

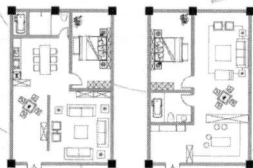

■ Residential Floor Plan

Type 1   Type 2

■ Homestay Floor Plan

Type 1   Type 2   Type 3

■ South Facade

■ Section 1-1

稻梦空间

作品编号：110701

Rice Dream Space

Thermal Insulation Analysis

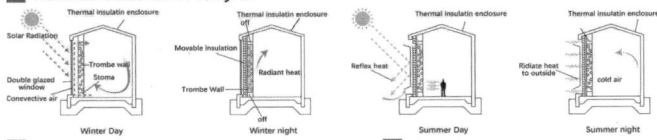

Solar Radiation
Thermal insulat enclosure
Double glazed window
Convective air
Trombe wall
Stoma

Winter Day

Thermal insulatin enclosure off
Movable insulation
Trombe Wall
Radiant heat
off

Winter night

Thermal insulatin enclosure
Reflex heat

Summer Day

Thermal insulatin enclosure
Ridiate heat to outside
cold air

Summer night

Structural and Material Analysis

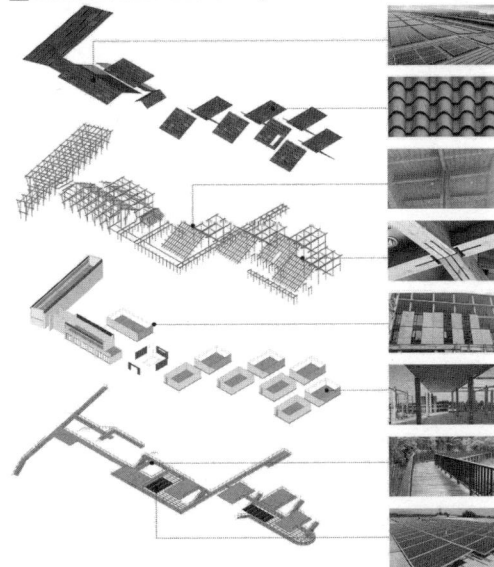

Green Technology

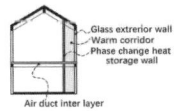

Solar photovoltaic panel
Phase change heat storage plate
Glass extrerior wall
Warm corridor
Phase change heat storage wall
Air duct inter layer

Solar photovoltaic panel
Phase change heat storage plate
Phase change heat storage wall
Glass extrerior wall
Air duct inter layer

Solar photovoltaic panel
Phase change heat storage plate
Warm corridor
Phase change heat storage wall
Glass extrerior wall
Air duct inter layer
Water heat storage
Water heat storage

Heat Analysis

Deformable Design

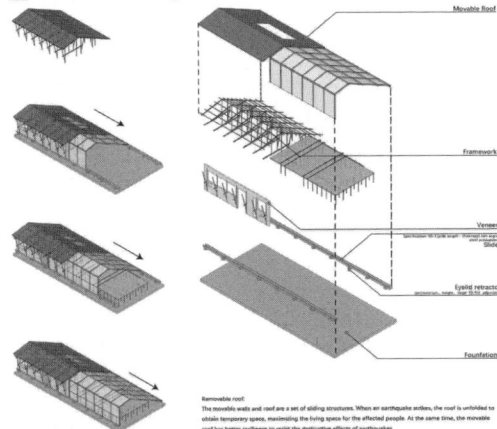

Movable Roof
Framework
Veneer
Eyelid retractor
Foundation

Removable roof:
The movable walls and roof are a set of sliding structures. When an earthquake strikes, the roof is unfolded to obtain temporary space, maximising the living space for the affected people. At the same time, the movable roof has better resilience to resist the destructive effects of earthquakes.

Sunshine Hours

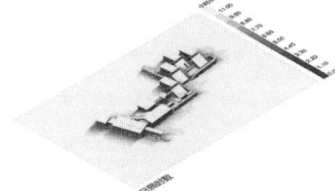

Outdoor Thermal Radiation

Shadow Masking

Daily Radiation

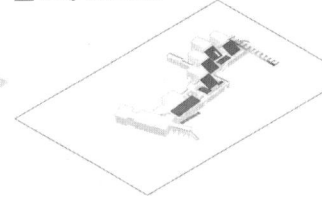

综合奖·入围奖·张家口市怀安县二堡子村建设项目
Comprehensive Awards – Nomination – Erpuzi Village Project，Huai'an County，Zhangjiakou City

注册号：110742
Register Number：110742
项目名称：光语·稻舍
Entry Title：Solar Link：Paddy's Symphony
作者：陈鑫源、陈卓玮、张雯宣、
　　　温佳坤、石斐斐、耿颜悦
Authors：Chen Xinyuan，Chen Zhuowei，
　　　　Zhang Wenxuan，Wen Jiakun，
　　　　Shi Feifei，and Geng Yanyue
作者单位：湖南大学
Authors from：Hunan University
指导教师：蒋甦琦、陈毅兴
Tutors：Jiang Suqi，Chen Yixing
指导教师单位：湖南大学
Tutors from：Hunan University

作品编号：110742

光语·稻舍
Solar Link：Paddy's Symphony

光语稻舍，一处融合住宿与社交功能的建筑，以光趣为轴，为旅居者和村民提供温馨的空间。同时为旅游业注入活力，以节能为核心，采用光伏板和地源热泵技术。实现绿色能源的自给自足。本土材料的巧妙使用，如方土墙和稻草板，不仅展现了地域特色，也彰显了建筑的可持续性和环保性。当建筑使命结束，材料将循环利用，回归自然。
"Ricelight Lodge: A blend of lodging and social hubs, energizing tourism.Sustainable and eco-designed with solar and geothermal power, local materials.Post-use, materials will be recycled, respecting nature's cycle."

■Site Location

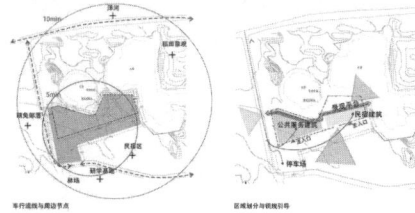

■Climatic Simulation

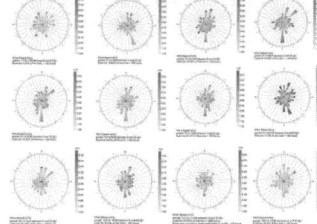

■SWOT Analysis

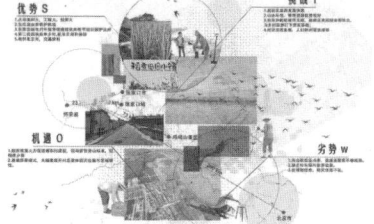

■Design Objectives

■Concept Generating

■Full lifecycle Strategy

■Scenery View

水塘

稻田

林场

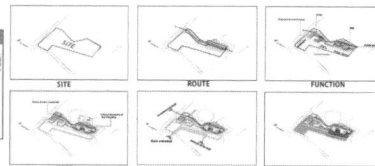

■Technology Uses

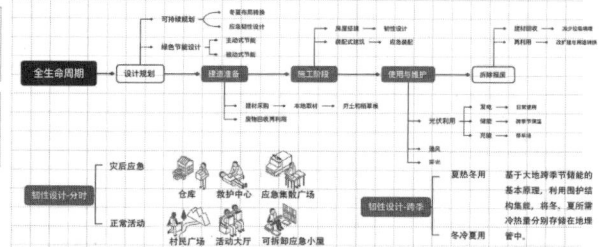

■Carbon Emission Calculations

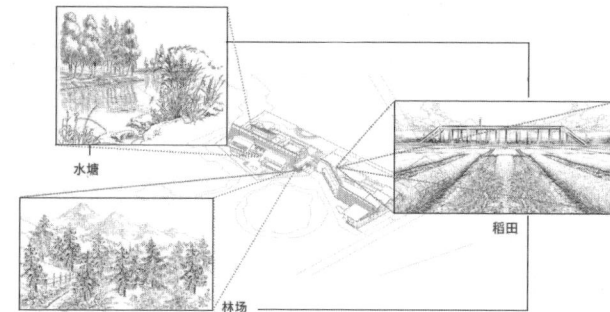

■Elevation

作品编号: 110742

光语・稻舍
Solar Link: Paddy's Symphony

■ **Total Plan**

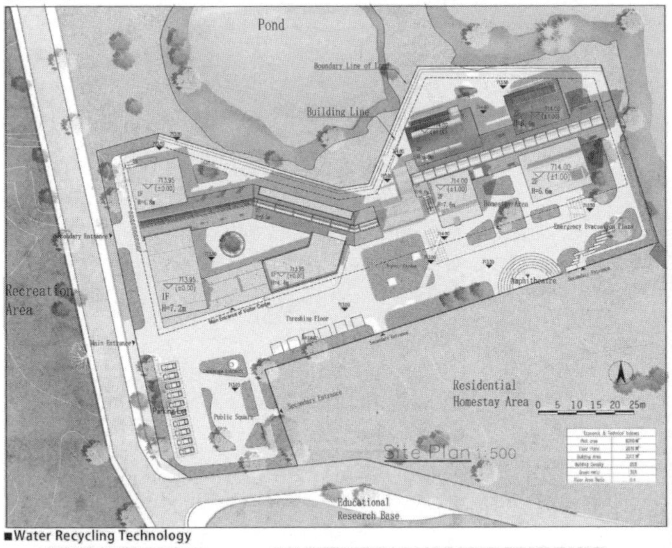

Pond

Boundary Line of Land

Building Line

Recreation Area

Residential Homestay Area

0 5 10 15 20 25m

Site Plan 1:500

Educational Research Base

■ **Exterior Renderings**

■ **Water Recycling Technology**

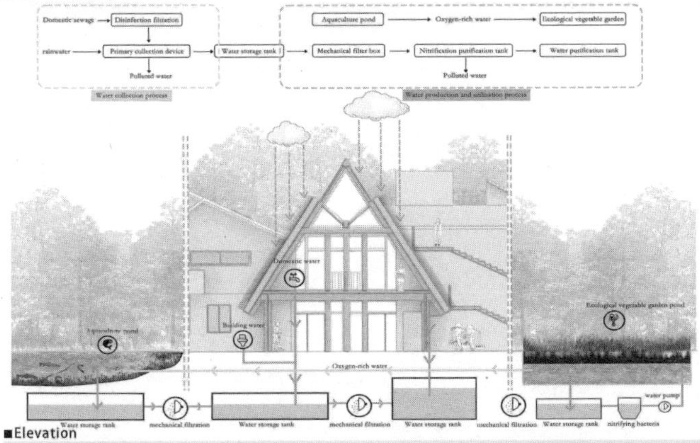

■ **Elevation**

作品编号：110742

光语·稻舍
Solar Link: Paddy's Symphony

■ Floor Plan of the First Floor 1：300

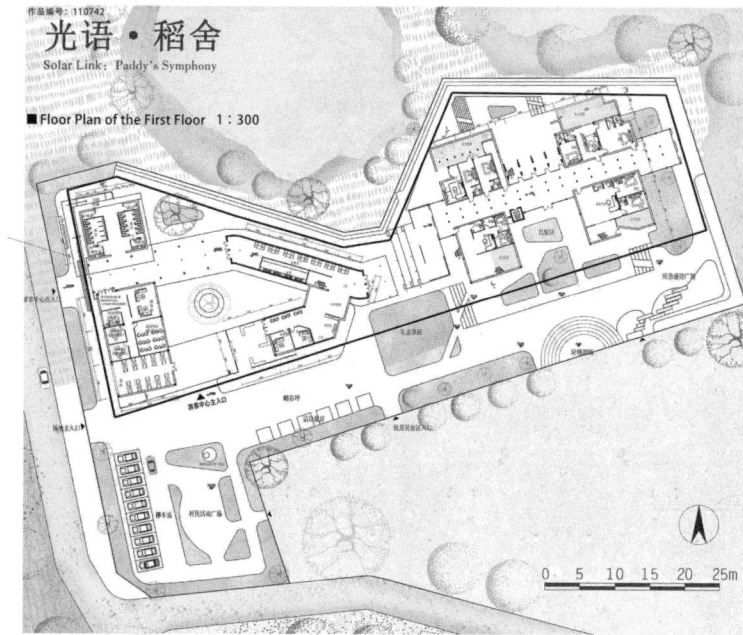

0 5 10 15 20 25m

■ Floor Plan of the Second Floor 1：300

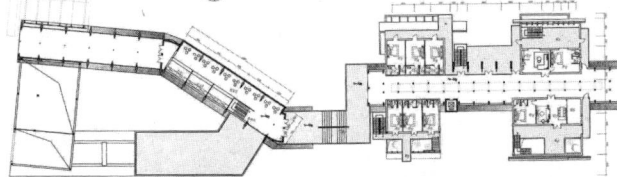

■ Node Construction

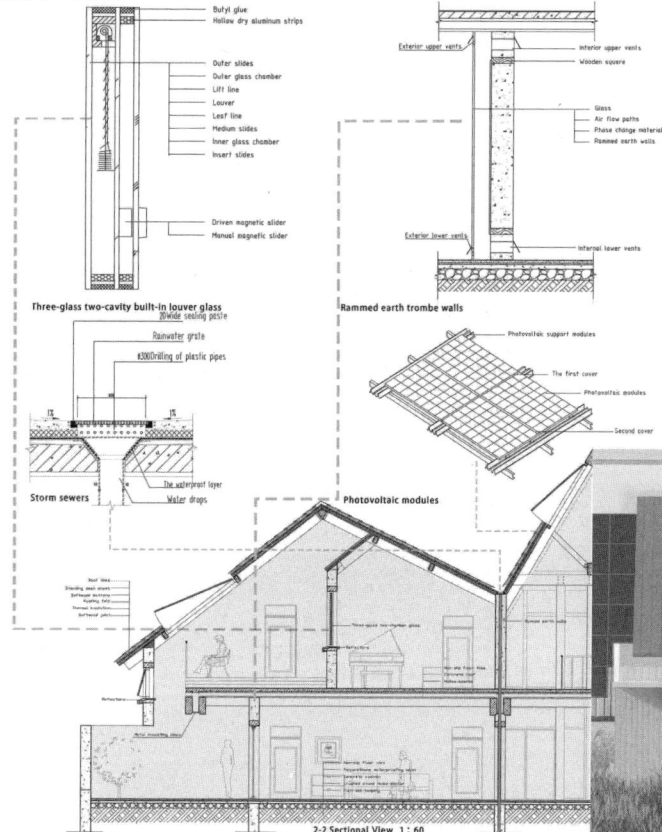

Butyl glue
Hollow dry aluminum strips

Outer slides
Outer glass chamber
Lift line
Louver
Leaf line
Medium slides
Inner glass chamber
Insert slides

Driven magnetic slider
Manual magnetic slider

Three-glass two-cavity built-in louver glass

Exterior upper vents
Interior upper vents
Wooden square

Glass
Air flow paths
Phase change material
Rammed earth walls

Exterior lower vents
Internal lower vents

Rammed earth trombe walls

20 wide sealing paste
Rainwater grate
#300 Drilling of plastic pipes

The waterproof layer

Storm sewers   Water drops

Photovoltaic support modules
The first cover
Photovoltaic modules
Second cover

Photovoltaic modules

2-2 Sectional View 1：60

■ Exterior Renderings

■ Cross-sectional Side View

光语・稻舍
Solar Link：Paddy's Symphony

■ Ventilation Analysis

Outdoor in summer

Outdoor in winter

Second floor of the homestay, top right corner

Ground floor of the homestay, top right corner

Residential hostel,Top left corner

Residential hostel,Lower right corner

Residential hostel,Lower right corner

Residential hostel,top right corner

■ Material Analysis

silicon solar cell · ultra-white glass
pv ribbon · eva
bus line · tpt
pv ribbon
bus line

10mm steel perforation ventilation
tempered glass to take off
opening with cover to open
100mm Mixed rammed earth walls

rice straw
iron frame
wood fibre board

Local crowd flow

Disaster emergency streamline

visitor streamline

Continuous slope tops Passive ventilation

Flat-slope combined Vertical space

The two-way cross skylight at the top is lit and ventilated to optimize the indoor light atmosphere and wind environment

Translucent roof Plenty of natural light

Photovoltaic panel roofing Energy saving and emission reduction

■ Exploded View of the Structure

sloping roof allows for easy ventilation

Pure steel frame Structural optimization

Herringbone steel and wood skeleton array constitutes the "spine of the building"

Double-storey corridors Rich streamlines

Translucent roof Lighting, with the overhang space

The patio is implanted to receive rain and light

Steel-timber construction with roof angles

Stable Frame pillars earthquake-resistant

Cantilevered structure Create landscapes

Herringbone steel and wood skeleton array, central double column to enhance stability, downward to open up the light and wind channel

Separate indoor stands Create a view of the landscape

Translucent roof lighting with north-facing windows

Age-friendly design Fecal-urine diversity
Ecological toilets

Detachable partition wall Tooth design
Emergency warehouses & health centers

Streamline combinations Spatially oriented
Visitor center restaurant

Separate courtyard Comfortable
guesthouse

Transition space Isolate public spaces
guesthouse

Solar car shed
Solar shutter
Solar lighting

1.Solar Parking Lot
2.Solar panel corridor
3.Solar panel corridor
4.Sports square
5.skylight
6.Three glass and two cavity windows with built-in shutters
7.Staircase theatre
8.A patio for lighting and collecting rainwater

■ Daylighting Analysis

summer

winter

summer

■ Warehouse and Multi-purpose Room （also a medical center）Time-sharing

■ Timesharing of Outdoor Multi-function Kiosks

Assemble steps

Construction details

■ Elevation

■ Photovoltaic Power Generation Calculations

Zhangjiakou small hotel, the construction area is 3313 ㎡,16 rooms, air conditioning area is about 1100 ㎡. Suppose that the photovoltaic installation area in the south (southeast-southwest), the inclination Angle is not more than 30) is 694 ㎡, and the photovoltaic installation area in other directions is 376 ㎡.Use a ground source heat pump in winter
When using the system, the annual power consumption per unit area is 137 kWh / ㎡, per unit; kWh / ㎡, then:
Annual total power load = air conditioning area annual power consumption per unit area =1100 × 137 = 150700 kWh
Total installed capacity of photovoltaic system calculation
Photovoltaic area: south (southeast-southwest), inclination no more than 30) of photovoltaic installation area is 694 ㎡, and other directions of photovoltaic installation area is 226 ㎡. PV installed capacity per unit area: When the pv panels are south or tilted, the installed capacity per unit pv area is 120 W/㎡, 100 W / ㎡, and the total installed capacity is 120.88kW. The detailed calculation is as follows:
Southbound installation:
Total installed capacity = PV area × unit PV area PV installed capacity = 694×120/1000=83.28kW
Other directions:
Total installed capacity = PV area × unit PV area PV installed capacity =376×100 / 1000=37.6 kW
Full-year power generation calculation:
When the best installation Angle of Zhangjiakou photovoltaic, the annual effective utilization hours is 1375 h. Assuming non-optimal installation angle, 5% for south and 10% to other orientation, the total generation of 155300kWh is calculated as follows:
Southbound installation:
Annual power generation = annual effective utilization hours of total installed capacity =83.28 × 137.5×95% =108800 kWh
Other directions: Annual power generation = annual effective utilization hours of total installed capacity =37.6× 1375×90% = 46500kWh
The average electricity price of Zhangjiakou is about 0.832 yuan / kWh, and the average annual income of photovoltaic power generation is about 98150 yuanAt present, the average price of photovoltaic cost is 3000 yuan / kW, and the total investment is 168,6400 yuan. The detailed calculation is as follows:
System cost = total photovoltaic installed capacity unit photovoltaic capacity cost = 120.88×3000 = 362640 yuan
The initial investment of photovoltaic is 362.6400 yuan, the annual power generation income is about 98150yuan, and the simple payback period is 3.7 years.

■ Interior Renderings

作品编号：110759

稻花香　01
——二堡子村建筑设计

综合奖・入围奖・张家口市
怀安县二堡子村建设项目
Comprehensive Awards －
Nomination － Erpuzi Village
Project，Huai'an County，
Zhangjiakou City

注册号：110759
Register Number：110759
项目名称：稻花香——二堡子村建筑设计
Entry Title：Fragrance of Rice Blossom：
　　　　　　Solar Building Technology
　　　　　　Design
作者：陈相如、严方希、康天硕、
　　　刘宇杰、徐祺煜
Authors：Chen Xiangru, Yan Fangxi,
　　　　　Kang Tianshuo, Liu Yujie,
　　　　　and Xu Qiyu
作者单位：长安大学
Authors from：Chang'an University
指导教师：夏博
Tutor：Xia Bo
指导教师单位：长安大学
Tutor from：Chang'an University

Total Land Area：8260 ㎡
Total Floor Area：2153 ㎡
Floor Area Ratio：0.26
Building Density：21.3%
Greening rate：41.7%

■ Master Plan

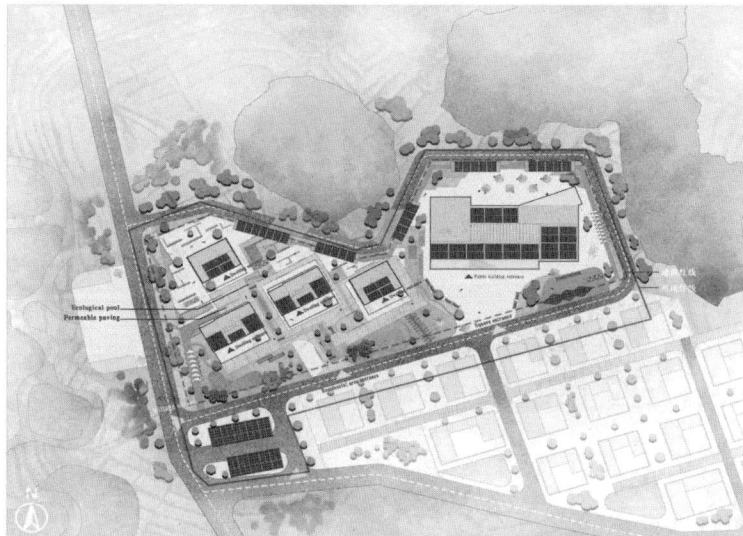

■ Description and Total Carbon Emissions

The project designs four residential buildings and one public service building in Erbaozi Village, Huai'an County, Zhangjiakou City, which has the function of a homestay. The project aims to improve the living quality of local residents through green technology under normal conditions, provide rural tourism supporting facilities, and enable villages to resist risks in special risk scenarios such as earthquakes, and have the ability to sustain themselves in extreme situations such as water and electricity outages.
This project uses large-area solar photovoltaic panels, which are installed on the sloping roof at a suitable angle, so that it can absorb solar energy with the highest efficiency and convert it into electrical energy for indoor use. At the same time, different forms of sun rooms are designed in each building to absorb solar energy and provide heat for the interior, thereby reducing carbon emissions. At the same time, there are Trombe walls in the interior of each building. Trombe walls provide heat to the interior by absorbing and storing solar energy, and acting as a heat source at night.
Public buildings are designed with the ability to respond to sudden disasters such as earthquakes. The design is a large market space with no physical partition wall on the first floor, and the column grid is 6m×6m. In the event of special circumstances, a small space with a column grid as a unit can be formed by quickly installing lightweight partitions, which can be reasonably centered into wards, temporary residences and other functions to meet the needs of emergency evacuation.

class	Annual carbon emissions (tCO₂ / a)	Carbon emission (tCO₂)
Construction material production	19.745	987.254
Transportation of building materials	1.445	72.237
Building construction	1.115	55.763
Building demolition	2.231	111.525
Building operation	22.302	1115.117
carbon sink	-0.138	-6.923
amount to	46.700	2334.973

■ Crowd Analysis

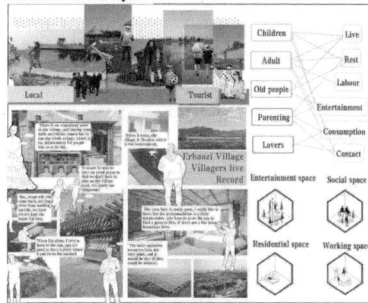

■ Logic Generation

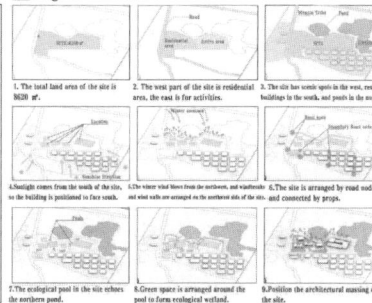

1. The total land area of the site is 8620 ㎡.
2. The west part of the site is residential area, the east is for activities.
3. The site has scenic spots in the west, residential buildings in the south, and ponds in the north.
4. Sunlight comes from the south of the site, so the building is positioned in face south.
5. The winter wind blows from the northwest, and windbreaks and wind walls are arranged on the northwest side of the site.
6. The site is arranged by road nodes and connected by props.
7. The ecological pool in the site echoes the northern pond.
8. Green space is arranged around the pool to form ecological wetland.
9. Position the architectural massing on the site.

■ Climate Analysis

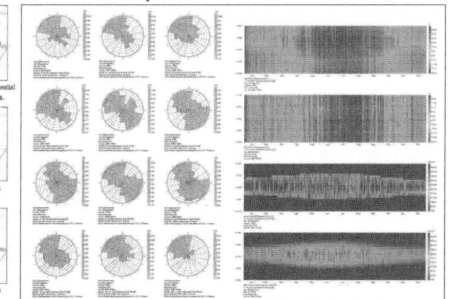

Public Building Plan

作品编号：110759

稻花香 02
——二堡子村建筑设计

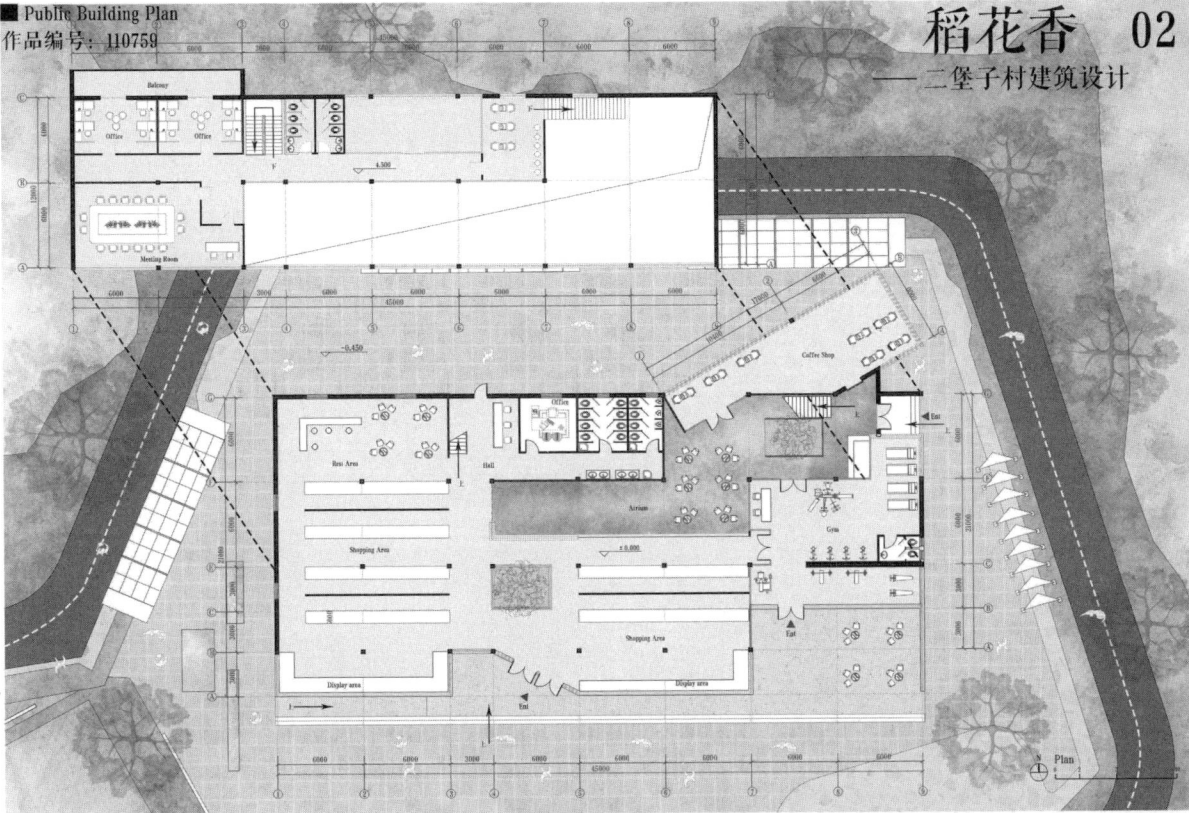

Balcony
Office Office
Meeting Room
Coffee Shop
Office
Rest Area
Hall
Atrium
Gym
Shopping Area
Display area   Shopping Area   Display area

Plan

Exploded View

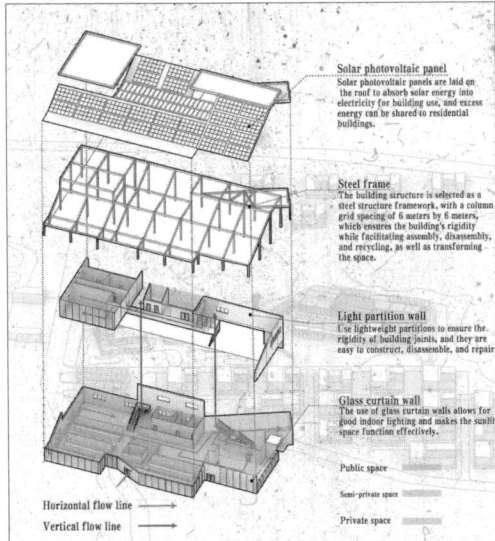

Solar photovoltaic panel
Solar photovoltaic panels are laid on the roof to absorb solar energy into electricity for building use, and excess energy can be shared to residential buildings.

Steel frame
The building structure is selected as a steel structure framework, with a column grid spacing of 6 meters by 6 meters, which ensures the building's rigidity while facilitating assembly, disassembly, and recycling, as well as transforming the space.

Light partition wall
Use lightweight partitions to ensure the rigidity of building joints, and they are easy to construct, disassemble, and repair.

Glass curtain wall
The use of glass curtain walls allows for good indoor lighting and makes the sunlit space function effectively.

Public space
Semi-private space
Private space

Horizontal flow line →
Vertical flow line →

Renderings

Usual-Emergency Conversion

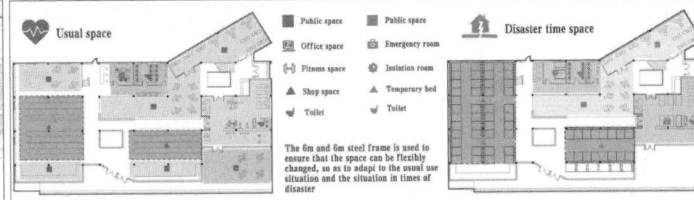

Usual space

Public space   Public space   Disaster time space
Office space   Emergency room
Fitness space   Isolation room
Shop space   Temporary bed
Toilet   Toilet

The 6m and 6m steel frame is used to ensure that the space can be flexibly changed, so as to adapt to the usual use situation and the situation in times of disaster.

Site Water System Organization

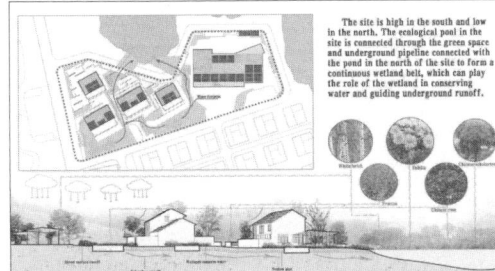

The site is high in the south and low in the north. The ecological pool in the site is connected through the green space and underground pipeline connected with the pond in the north of the site to form a continuous wetland belt, which can play the role of the wetland in conserving water and guiding underground runoff.

Green Buildingsville Analysis

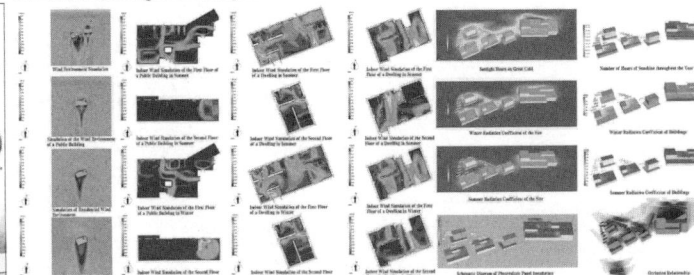

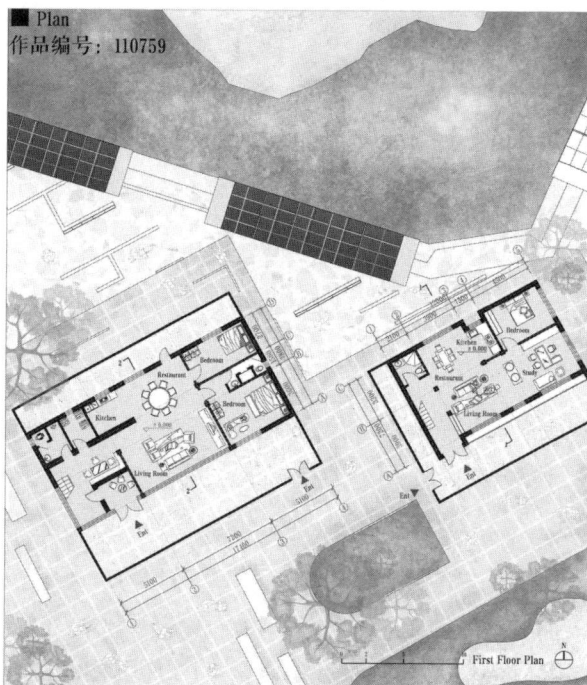

■ Plan
作品编号：110759

稻花香 03
——二堡子村建筑设计

First Floor Plan

Second Floor Plan

■ Renderings

■ Exploded View

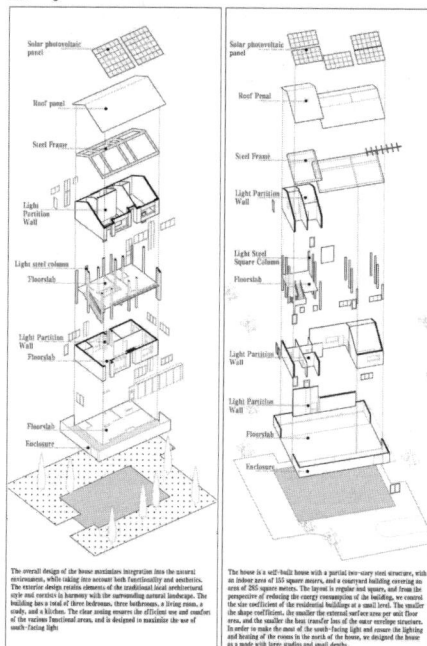

Solar photovoltaic panel
Roof Penal
Steel Frame
Light Partition Wall
Light steel column
Floorslab
Light Partition Wall
Floorslab
Floorslab
Enclosure

Solar photovoltaic panel
Roof Penal
Steel Frame
Light Partition Wall
Light Steel Square Column
Floorslab
Light Partition Wall
Light Partition Wall
Floorslab
Enclosure

■ Sectional View

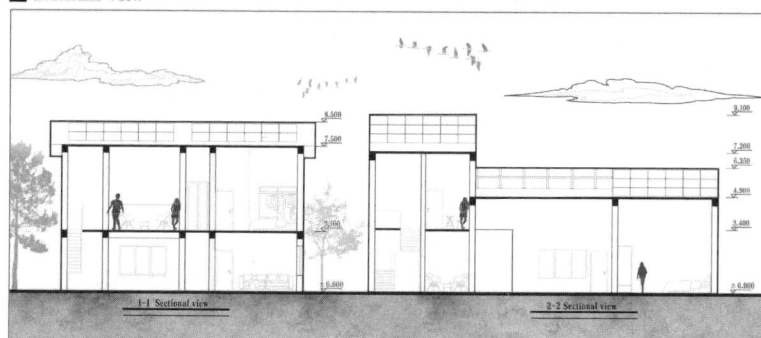

1-1 Sectional view

2-2 Sectional view

The overall design of the house maximizes integration into the natural environment, while taking into account both functionality and aesthetics. The exterior design retains elements of the traditional local architectural style and coexists in harmony with the surrounding natural landscape. The building has a total of three bedrooms, three bedrooms, a living room, a study, and a kitchen. The clear zoning ensures the efficient use and comfort of the various functional areas, and is designed to maximize the use of south-facing light.

The house is a self-built house with a partial two-story steel structure, with an indoor area of 155 square meters, and a courtyard building covering an area of 285 square meters. The layout is regular and square, and from the perspective of reducing the energy consumption of the building, we control the size coefficient of the residential buildings at a small level. The smaller the shape coefficient, the smaller the external surface area per unit floor area, and the smaller the heat transfer loss of the outer envelope structure. In order to make the most of the south-facing light and ensure the lighting and heating of the rooms in the north of the house, we designed the house as a room with large studies and small depths.

稻花香 04
——二堡子村建筑设计

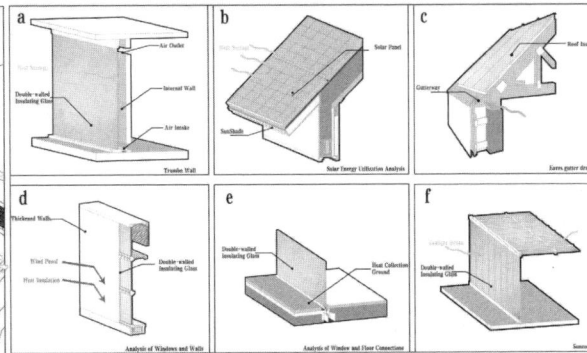

a	b	c

d	e	f

■ Passive Solar Technology

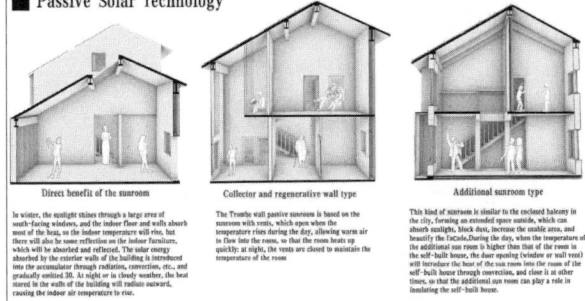

Direct benefit of the sunroom

In winter, the sunlight shines through a large area of south-facing windows, and the indoor floor and walls absorb most of the heat, so the indoor temperature will rise, but there will also be some reflection on the indoor furniture, which will be absorbed and reflected. The solar energy absorbed by the exterior walls of the building is introduced into the accumulator through radiation, convection, etc., and gradually emitted 3D. At night or in cloudy weather, the heat stored in the walls of the building will radiate outward, causing the indoor air temperature to rise.

Collector and regenerative wall type

The Trombe wall passive sunroom is based on the sunroom with vents, which open when the temperature rises during the day, allowing warm air to flow into the room, so that the room heats up quickly; at night, the vents are closed to maintain the temperature of the room.

Additional sunroom type

This kind of sunroom is similar to the enclosed balcony in the city, forming an extended space outside, which can absorb sunlight, block dust, increase the usable area, and beautify the façade. During the day, when the temperature of the additional sun room is higher than that of the room in the self-built house, the door opening (window or wall vent) will introduce the heat of the sun room into the room of the self-built house through convection, and close it at other times, so that the additional sun room can play a role in insulating the self-built house.

■ Functional Analysis

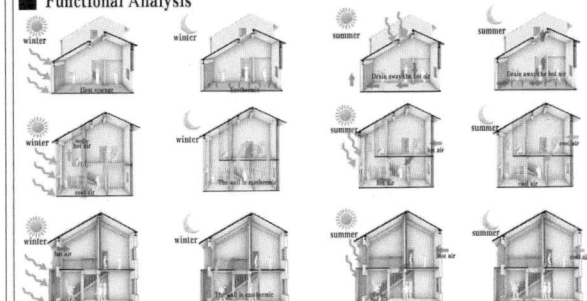

■ Section Perspective

编号：110767

▲ Land Monument

新时代建筑更加注重建筑的宜居性，本次设计重点提高农村的防灾减灾能力，贯彻落实《关于加强农村基础设施建设提高防灾减灾能力的指导意见》精神，围绕如何构建具有韧性的村镇基础设施、提升农村地区的抗灾减灾能力两个主题展开，着重关注村镇的规划布局、建筑结构、公共设施、生态环境、绿色能源技术、低碳可持续发展等方面，以期打造出安全、可持续、富有活力的农村社区。为乡村基础设施建设提供新思路，为乡村振兴注入新活力。

本项目基于二堡子村现有情况和资源，提出提升居民福祉，优化旅游体验的设计理念，融合稻香田园文化，构建田园体验、稻田艺术、文化展示、记忆民宿、休闲观光和为一体的发展模式，多方产业联动推动乡村经济的增长。同时引入绿色技术，从生态植被到建筑本身，建设可持续生态和建筑，落实低碳原则。紧跟时代发展，使用各项监测系统和云平台打造智慧生活，建设以人为本的乡村空间。

In the new era, architecture increasingly emphasizes livability. This design focuses on enhancing disaster prevention and mitigation capacities in rural areas, in line with the principles outlined in the "Guiding Opinions on Strengthening Infrastructure Construction in Rural Areas to Improve Disaster Prevention and Mitigation Capacities." The design revolves around two key themes: the development of resilient infrastructure in villages and towns and the enhancement of disaster resilience in rural areas. It concentrates on various aspects, including urban planning, structural integrity, public facilities, ecological environments, green energy technologies, and low-carbon sustainable development, with the aim of creating safe, sustainable, and vibrant rural communities. This initiative seeks to provide innovative approaches to rural infrastructure development and to inject new vitality into rural revitalization.

Based on the existing conditions and resources of Erbuzi Village, this project proposes a design philosophy that enhances resident well-being and optimizes the tourism experience, integrating the culture of aromatic rice fields. It aims to establish a development model that encompasses pastoral experiences, rice paddy art, cultural exhibitions, memory-themed accommodations, and leisure tourism, thereby promoting rural economic growth through multi-sector collaboration. Concurrently, the project will incorporate green technologies, constructing sustainable ecosystems and architecture, from ecological vegetation to the built environment, while implementing low-carbon principles. Staying attuned to contemporary developments, the project employs various monitoring systems and cloud platforms to foster smart living, creating a human-centered rural space.

综合奖·入围奖·张家口市
怀安县二堡子村建设项目
Comprehensive Awards − Nomination − Erpuzi Village Project, Huai'an County, Zhangjiakou City

注册号：110767
Register Number：110767

项目名称：土地纪念碑——基于低碳和乡村韧性的民居及公共服务建筑设计

Entry Title：Land Monument: A Public Service & Residential Building Design Based on Low Carbon and Rural Resilience Principles

作者：王学润、任广宁
Authors：Wang Xuerun, Ren Guangning

作者单位：厦门大学
Authors from：Xiamen University

指导教师：李芝也
Tutor：Li Zhiye

指导教师单位：厦门大学
Tutor from：Xiamen University

■ Location Analysis

The project is located in Erbuzi Village, Huaian County, Zhangjiakou City, Hebei Province, 114.62° E, 40.68° N, near the Beijing-Tibet Expressway and Beijing-New Expressway, and the high-speed rail line runs east-west through the north side of the site. It is 35 km from Zhangjiakou City and 200 km from Beijing city center. It is close to the residential areas of Erbaozi Village and Beiguo Jiayao Village, and the linear distance is within 1km.

■ Historical Analysis

At the end of the Jin Dynasty and the beginning of the Yuan Dynasty, there was a "Huihe River Battle" that shocked the whole world. The village is located on the south bank of the Yanghe River. At the beginning of the Ming Dynasty, the second run was formed because of the army reclamation. According to the Annals of Huaian County, during the reign of Hongwu in the Ming Dynasty (1368-1398 AD), the village population gradually increased and became the 200 households of Wanquan Zuowei, known as 200 households. The fort was built in the Jiajing years of the Ming Dynasty, and was called the second fort during the Republic of China. The village has been growing rice for more than 600 years. In 2022, Erbaozi Village introduced a technology company, and all 1022 acres of rice fields were transferred to the company to grow organic rice.

■ Climatic Condition

The site is located in the mid-latitudes. It belongs to the East Asian continental monsoon climate. It is a subtropical semi-arid area with a mild temperature. The average annual temperature is 8.3°C, the highest temperature is 37.2°C, and the lowest temperature is -22.6°C. The annual sunshine hours are 2,800 to 3,100, and the total solar radiation is 1,500 to 1,700 kilowatt-hours per square meter, making it an area with abundant solar energy. The annual precipitation is 407.4 millimeters. The seismic intensity designation of the site is 8, and there have been no flood disasters in recent years. There is a risk of hail.

1 Evacuation square	9 Viewing pavilion
2 Commercial leisure and entertainment	10 Apron
3 Logistics	11 Parking lot
4 Viewing platform	12 Green walkway
5 Exhibition hall and service center	13 Playground
6 Leisure tree pond	14 Cute rabbit tribal tourist attraction
7 Conference hall	15 Residential area
8 B&B	16 Garden research base

■ General plan

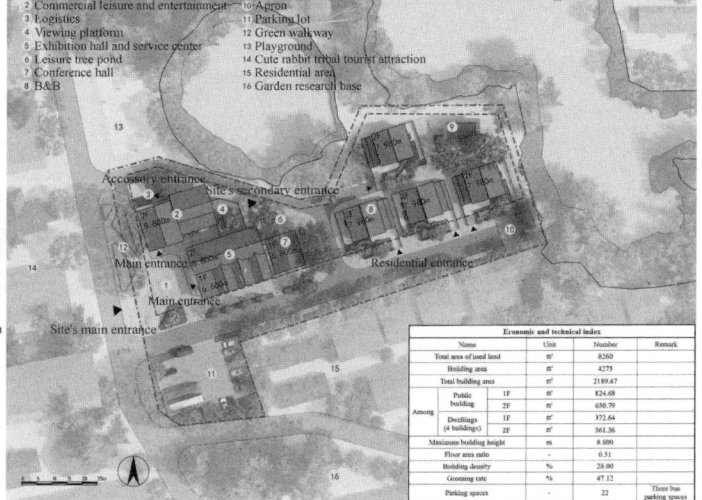

■ Site Status

Aerial view of the village | The existing rice fragrant landscape bridge | From the north to the south look at Erbaozi village and the far mountain

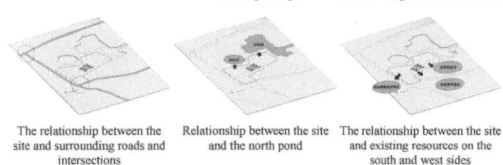

The relationship between the site and surrounding roads and intersections | Relationship between the site and the north pond | The relationship between the site and existing resources on the south and west sides

Economic and technical index			
Name	Unit	Number	Remark
Total area of used land	m²	8260	
Building area	m²	4275	
Total building area	m²	2189.47	
Among · Public building · 1F	m²	824.68	
Among · Public building · 2F	m²	630.79	
Among · Dwellings (4 buildings) · 1F	m²	372.64	
Among · Dwellings (4 buildings) · 2F	m²	361.36	
Minimum building height	m	9.890	
Floor area ratio	-	0.51	
Building density	%	26.00	
Greening rate	%	47.12	
Parking spaces	-	22	Three bus parking spaces

Land Monument II

Design Strategy

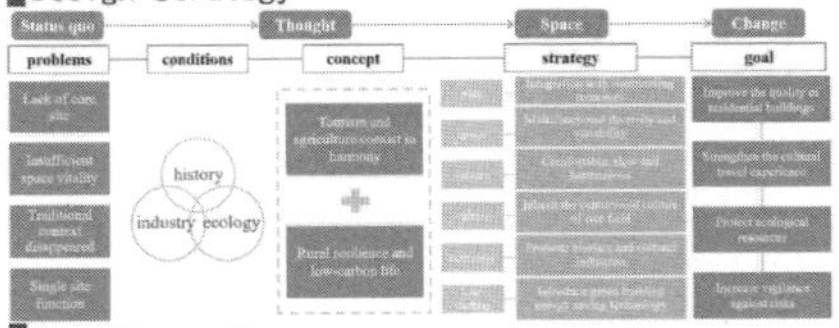

Form Generation

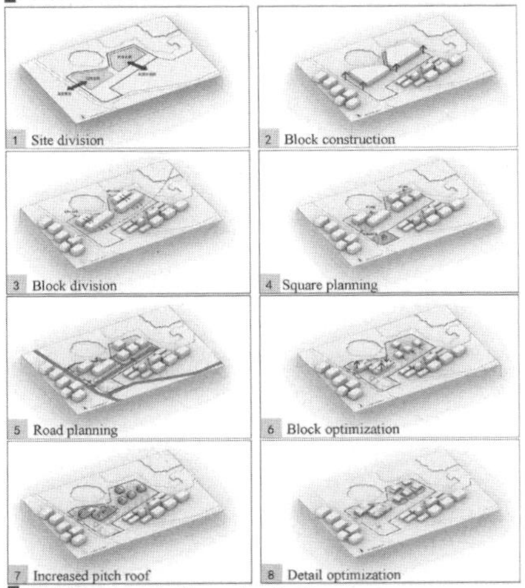

1 Site division
2 Block construction
3 Block division
4 Square planning
5 Road planning
6 Block optimization
7 Increased pitch roof
8 Detail optimization

Architectural Characteristic

Architectural intention

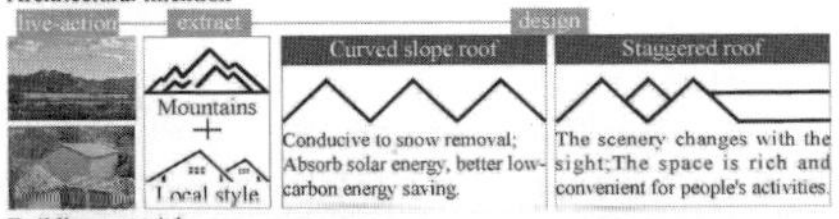

live-action — extract — design

Mountains + Local style

Curved slope root
Conducive to snow removal; Absorb solar energy, better low-carbon energy saving.

Staggered roof
The scenery changes with the sight; The space is rich and convenient for people's activities.

Building material

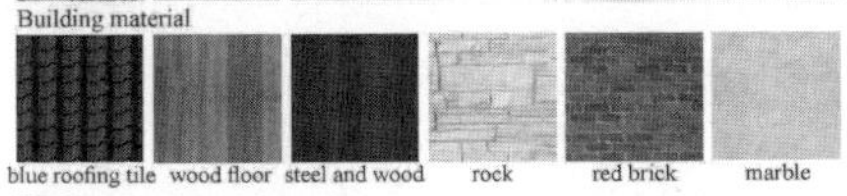

blue roofing tile | wood floor | steel and wood | rock | red brick | marble

Partial Rendering

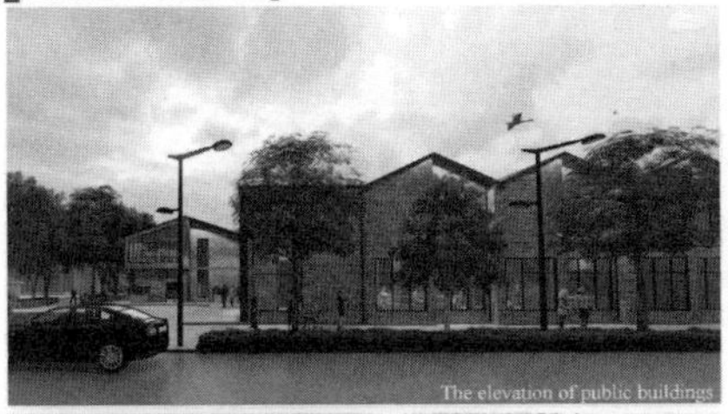

The elevation of public buildings

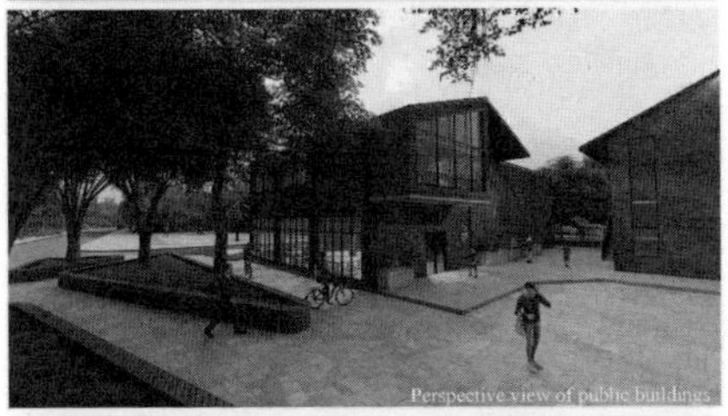

Perspective view of public buildings

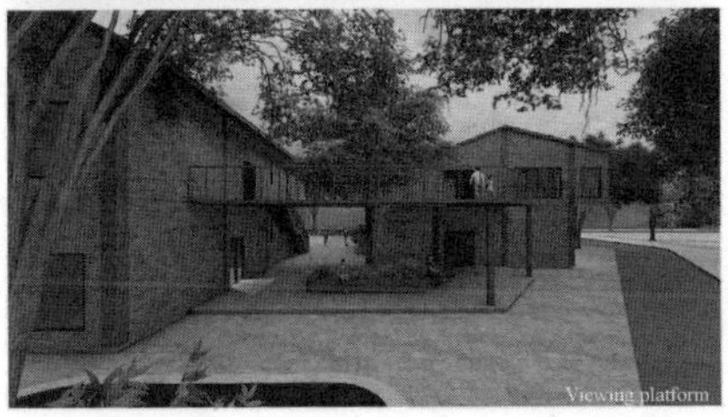

Viewing platform

Public Building plan

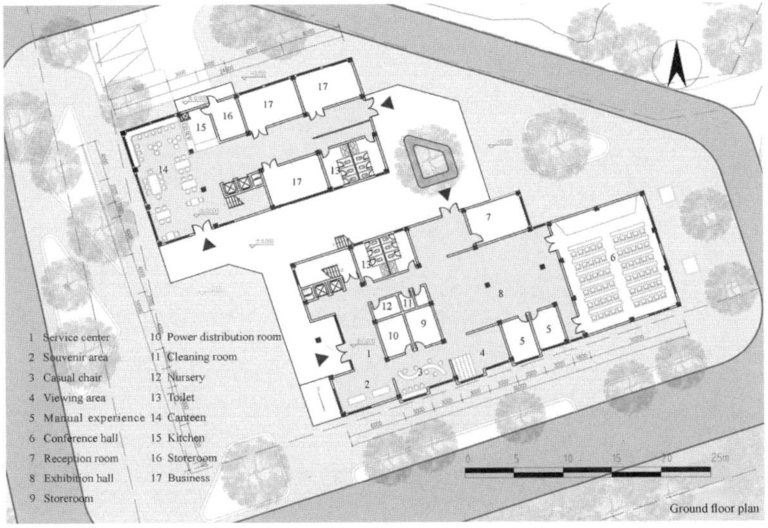

1 Service center
2 Souvenir area
3 Casual chair
4 Viewing area
5 Manual experience
6 Conference hall
7 Reception room
8 Exhibition hall
9 Storeroom
10 Power distribution room
11 Cleaning room
12 Nursery
13 Toilet
14 Canteen
15 Kitchen
16 Storeroom
17 Business

Ground floor plan

1 Over the hall
2 Reception room
3 Manager's office
4 Office
5 Meeting room
6 Break room
7 Viewing platform
8 Canteen
9 VIP room
10 Storeroom
11 Tea room
12 Café
13 Rec room
14 Toilet

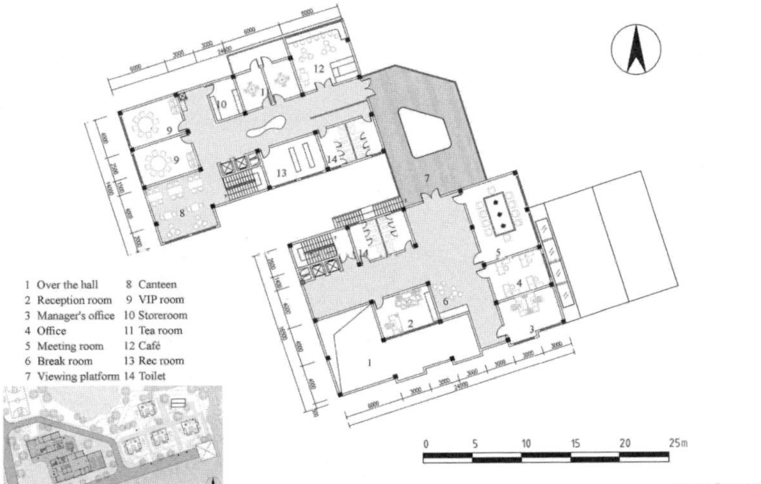

Second floor plan

Residential Plan

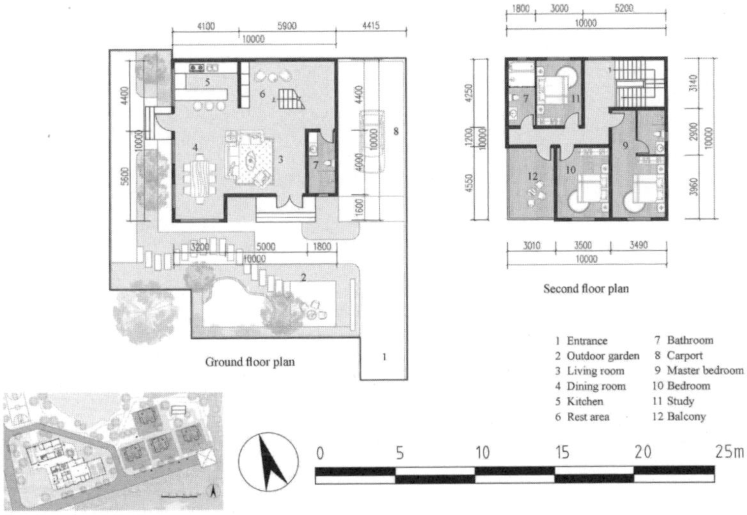

Ground floor plan

Second floor plan

1 Entrance
2 Outdoor garden
3 Living room
4 Dining room
5 Kitchen
6 Rest area
7 Bathroom
8 Carport
9 Master bedroom
10 Bedroom
11 Study
12 Balcony

编号：110767

Land Monument Ⅲ

Exploded Views

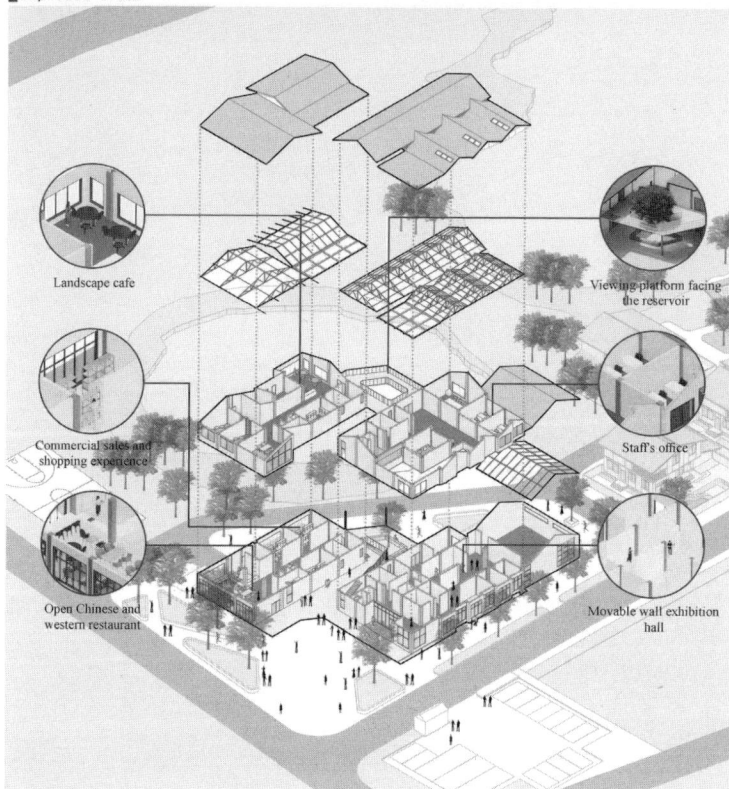

Landscape cafe

Viewing platform facing the reservoir

Commercial sales and shopping experience

Staff's office

Open Chinese and western restaurant

Movable wall exhibition hall

Population Analysis

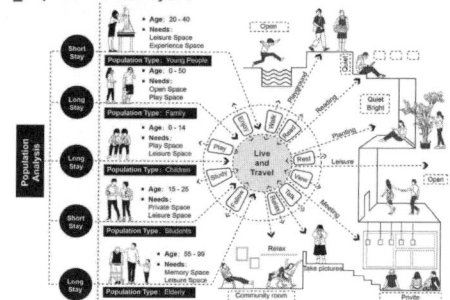

Short Stay	Age: 20 - 40 — Needs: Leisure Space Experience Space — Population Type: Young People
Long Stay	Age: 0 - 50 — Needs: Open Space Play Space — Population Type: Family
Long Stay	Age: 0 - 14 — Needs: Play Space Leisure Space — Population Type: Children
Short Stay	Age: 15 - 25 — Needs: Private Space Leisure Space — Population Type: Students
Long Stay	Age: 55 - 99 — Needs: Memory Space Leisure Space — Population Type: Elderly

Live and Travel

Diversification of Space Use

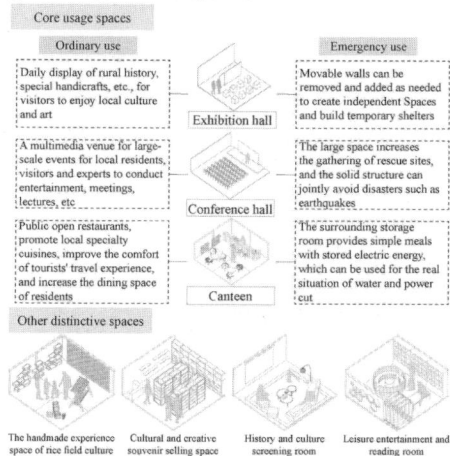

Core usage spaces

Ordinary use		Emergency use
Daily display of rural history, special handicrafts, etc., for visitors to enjoy local culture and art	Exhibition hall	Movable walls can be removed and added as needed to create independent Spaces and build temporary shelters
A multimedia venue for large-scale events for local residents, visitors and experts to conduct entertainment, meetings, lectures, etc	Conference hall	The large space increases the gathering of rescue sites, and the solid structure can jointly avoid disasters such as earthquakes
Public open restaurants, promote local specialty cuisines, improve the comfort of tourists' travel experience, and increase the dining space of residents	Canteen	The surrounding storage room provides simple meals with stored electric energy, which can be used for the real situation of water and power cut

Other distinctive spaces

The handmade experience space of rice field culture

Cultural and creative souvenir selling space

History and culture screening room

Leisure entertainment and reading room

South Elevation Plan

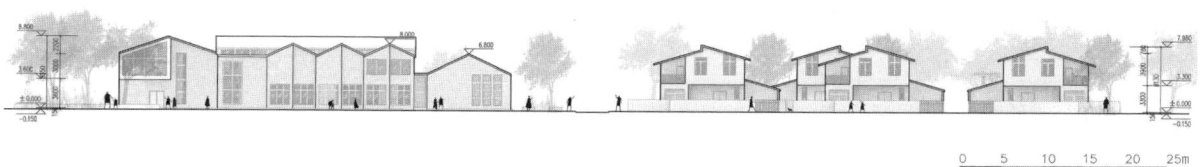

0 5 10 15 20 25m

Land Monument IV

Analysis of Energy Saving in Profile and Perspective

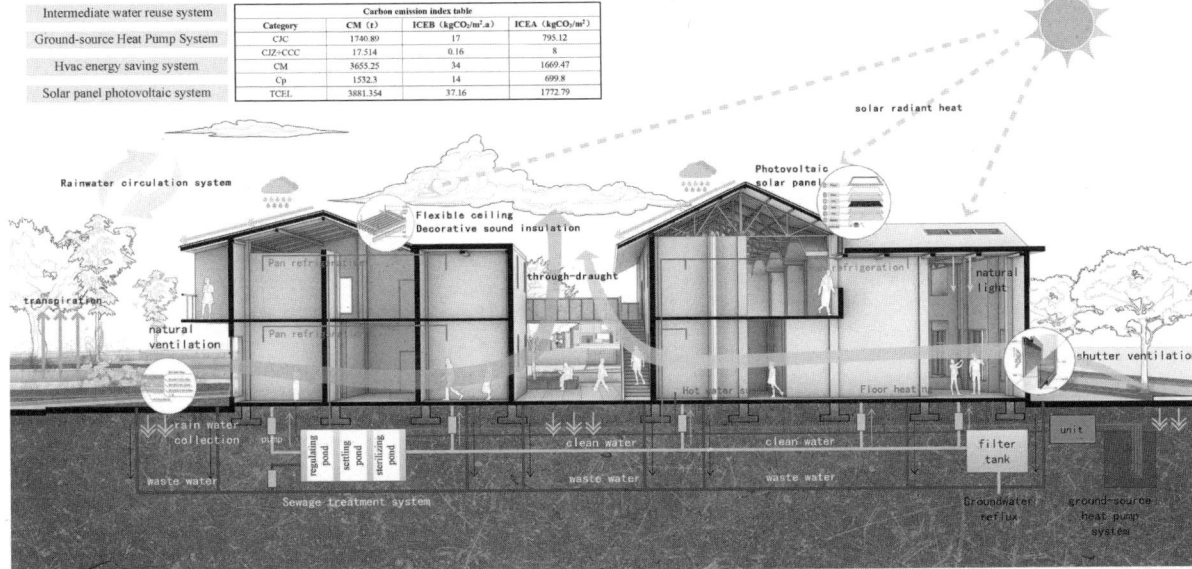

- Intermediate water reuse system
- Ground-source Heat Pump System
- Hvac energy saving system
- Solar panel photovoltaic system

Carbon emission index table			
Category	CM（t）	ICEB（kgCO₂/m².a)	ICEA（kgCO₂/m²）
CJC	1740.89	17	795.12
CJZ=CCC	17.514	0.16	8
CM	3655.25	34	1669.47
Cp	1532.3	14	699.8
TCEL	3881.354	37.16	1772.79

(Carbon emission index table, ICEB in $kgCO_2/m^2 \cdot a$, ICEA in $kgCO_2/m^2$)

Labels in section: solar radiant heat; Rainwater circulation system; Photovoltaic solar panel; Flexible ceiling Decorative sound insulation; through-draught; refrigeration; natural light; transpiration; natural ventilation; Pan refrigeration; shutter ventilation; Hot water; Floor heating; rain water collection; pump; regulating pond; settling pond; sterilizing pond; clean water; clean water; filter tank; unit; waste water; waste water; waste water; Sewage treatment system; Groundwater reflux; ground-source heat pump system

Energy-saving Strategy

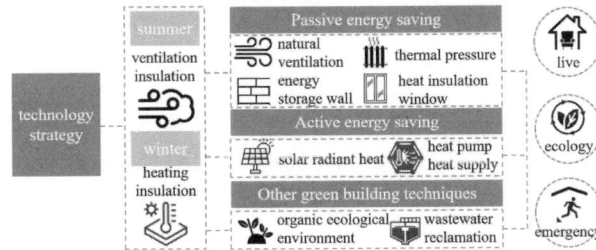

technology strategy

summer — ventilation / insulation
winter — heating / insulation

Passive energy saving
- natural ventilation
- thermal pressure
- energy storage wall
- heat insulation window

Active energy saving
- solar radiant heat
- heat pump heat supply

Other green building techniques
- organic ecological environment
- wastewater reclamation

live / ecology / emergency

Shade Analysis

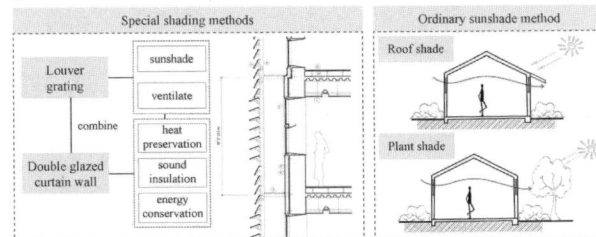

Special shading methods
- Louver grating — sunshade / ventilate
- combine
- Double glazed curtain wall — heat preservation / sound insulation / energy conservation

Ordinary sunshade method
- Roof shade
- Plant shade

Structural Technique

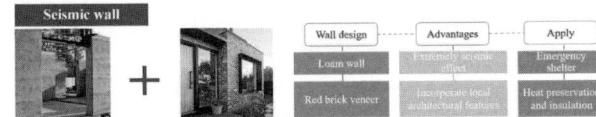

Seismic wall

Wall design	Advantages	Apply
Login wall	Extremely seismic effect	Emergency shelter
Red brick veneer	Incorporate local architectural features	Heat preservation and insulation

The building is built with the traditional material rammed earth, which has been supported by mature technology, greatly improving the technical limit of the traditional rammed earth, making it possible to use it in the seismic fortification area of eight degrees, and can be used as the structural system of the building (rather than the maintenance system).

Aseismic Structure

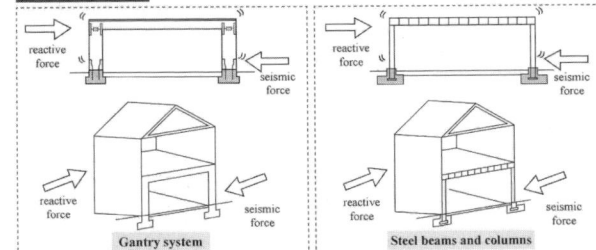

reactive force / seismic force

Gantry system | **Steel beams and columns**

Outdoor Ventilation Simulation

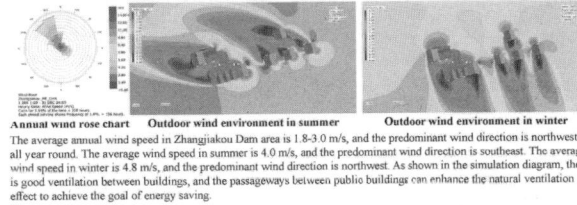

Annual wind rose chart | Outdoor wind environment in summer | Outdoor wind environment in winter

The average annual wind speed in Zhangjiakou Dam area is 1.8-3.0 m/s, and the predominant wind direction is northwest all year round. The average wind speed in summer is 4.0 m/s, and the predominant wind direction is southeast. The average wind speed in winter is 4.8 m/s, and the predominant wind direction is northwest. As shown in the simulation diagram, there is good ventilation between buildings, and the passageways between public buildings can enhance the natural ventilation effect to achieve the goal of energy saving.

Sunshine Analysis

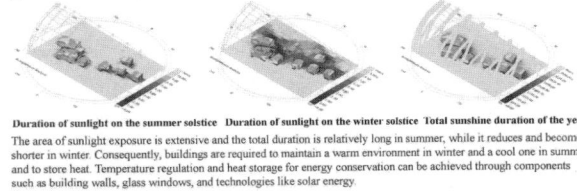

Duration of sunlight on the summer solstice　Duration of sunlight on the winter solstice　Total sunshine duration of the year

The area of sunlight exposure is extensive and the total duration is relatively long in summer, while it reduces and becomes shorter in winter. Consequently, buildings are required to maintain a warm environment in winter and a cool one in summer, and to store heat. Temperature regulation and heat storage for energy conservation can be achieved through components such as building walls, glass windows, and technologies like solar energy.

Smart System

Using the distributed hydrogen production modular brick developed by the team as the building skin, the multi-functional brick is composed of photovoltaic materials, heat storage materials, water purification structure, to create a sustainable building, and can choose different modules according to geography, climate and use needs for free combination and customization.

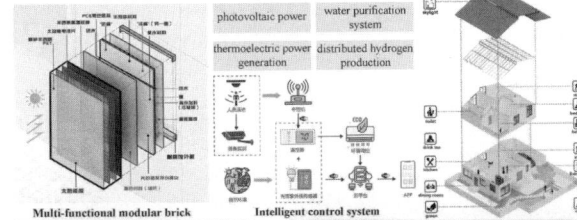

photovoltaic power / water purification system / thermoelectric power generation / distributed hydrogen production

Multi-functional modular brick　　**Intelligent control system**

Perspective view of folk house　　Recreation square

110000

洄·廊·庭 I

Upstream — Corridor — Hall

综合奖·入围奖·保定市清苑区李八庄村建设项目
Comprehensive Awards — Nomination — Libazhuang Village Project，Qingyuan District，Baoding City

注册号：110000
Register Number：110000
项目名称：洄·廊·庭
Entry Title：Upstream — Corridor — Hall
作者：吴雨航、金日、任嘉瑶、明来灵、李恩泽
Authors：Wu Yuhang，Jin Ri，Ren Jiayao，Ming Lailing，and Li Enze
作者单位：西南民族大学
Authors from：Southwest Minzu University
指导教师：熊健吾、张埝
Tutors：Xiong Jianwu，Zhang Yin
指导教师单位：西南民族大学
Tutors from：Southwest Minzu University

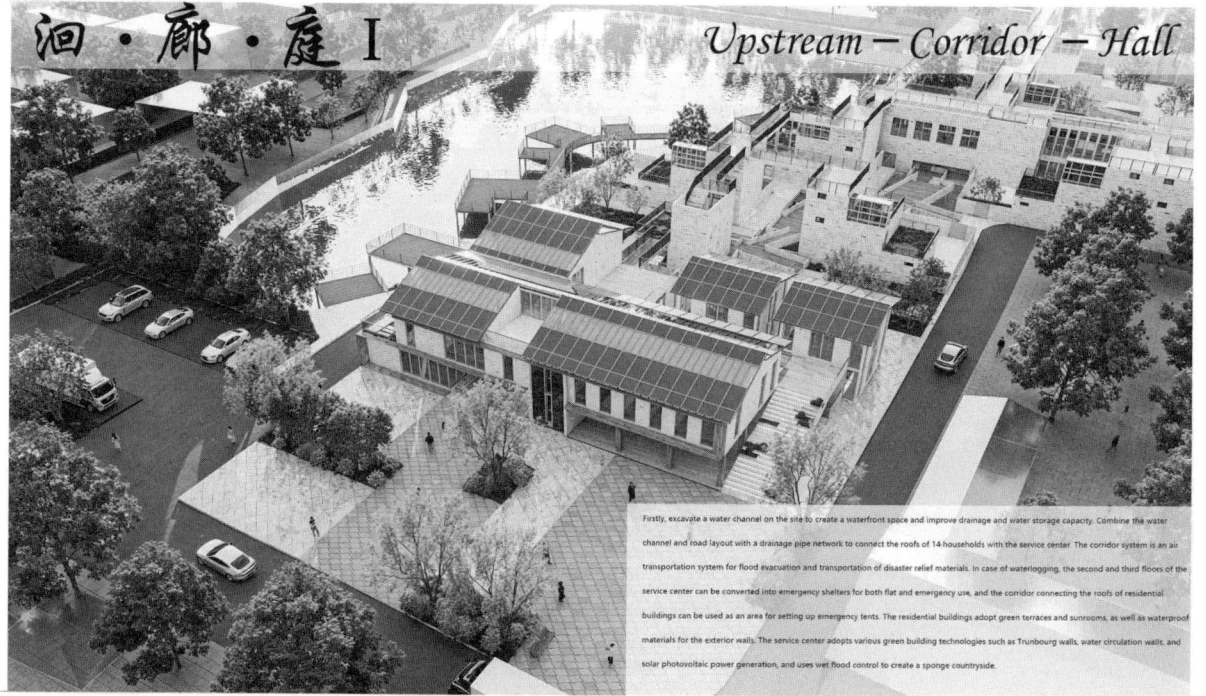

Firstly, excavate a water channel on the site to create a waterfront space and improve drainage and water storage capacity. Combine the water channel and road layout with a drainage pipe network to connect the roofs of 14 households with the service center. The corridor system is an air transportation system for flood evacuation and transportation of disaster relief materials. In case of waterlogging, the second and third floors of the service center can be converted into emergency shelters for both flat and emergency use, and the corridor connecting the roofs of residential buildings can be used as an area for setting up emergency tents. The residential buildings adopt green terraces and sunrooms, as well as waterproof materials for the exterior walls. The service center adopts various green building technologies such as Trunbourg walls, water circulation walls, and solar photovoltaic power generation, and uses wet flood control to create a sponge countryside.

Design Description

首先在场地开挖水渠，打造滨水空间，提高排水蓄水能力，结合水渠和道路布置排水管网，将14户民居屋顶与服务中心串联，连廊系统为抗灾物资运输的空中交通体系，内涝时，服务中心二、三层平急两用可转变为紧急避难场所，连接串联民居屋顶可变为急救帐篷搭建区，民居采用绿化露台及阳光房，外墙防水材料等，服务中心采用特朗勃墙、水循环墙、太阳能光伏发电等多种绿色建筑技术，运用湿式防洪，打造海绵乡村。

Site Analysis

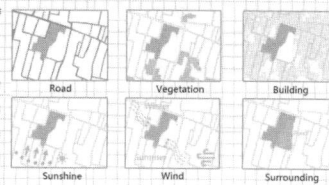

Road Vegetation Building

Sunshine Wind Surrounding

Environmental Climate Selection

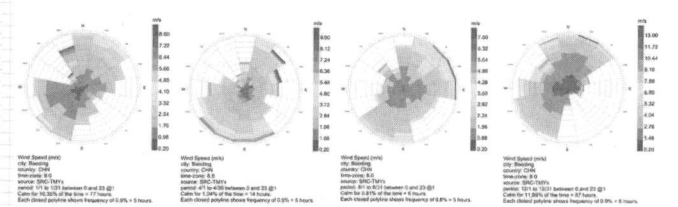

Diagram of Design Process

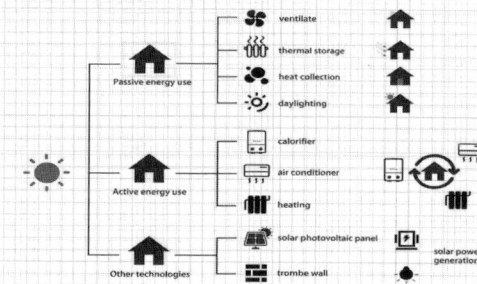

- ventilate
- thermal storage
- heat collection
- daylighting

Passive energy use

- calorifier
- air conditioner
- heating

Active energy use

- solar photovoltaic panel
- trombe wall

Other technologies

solar power generation

Pipe Network Design

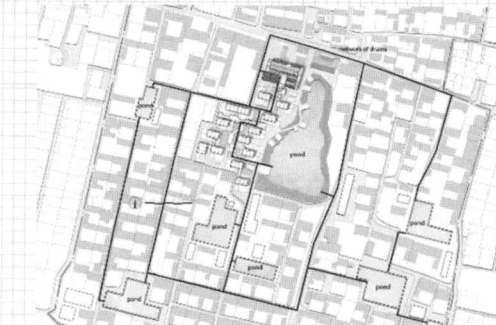

Site Plan 1：1500

Total land area: 10,200 ㎡

Rural service center total construction area: 1,130㎡

Number of households: 14 households

Total construction area: 2,250 ㎡

Residential area: 1,265 ㎡

Rural service center area: 774㎡

Floor area ratio: 0.33%

Green space rate: 22%

Building density: 0.2

Parking space: 13 vehicles

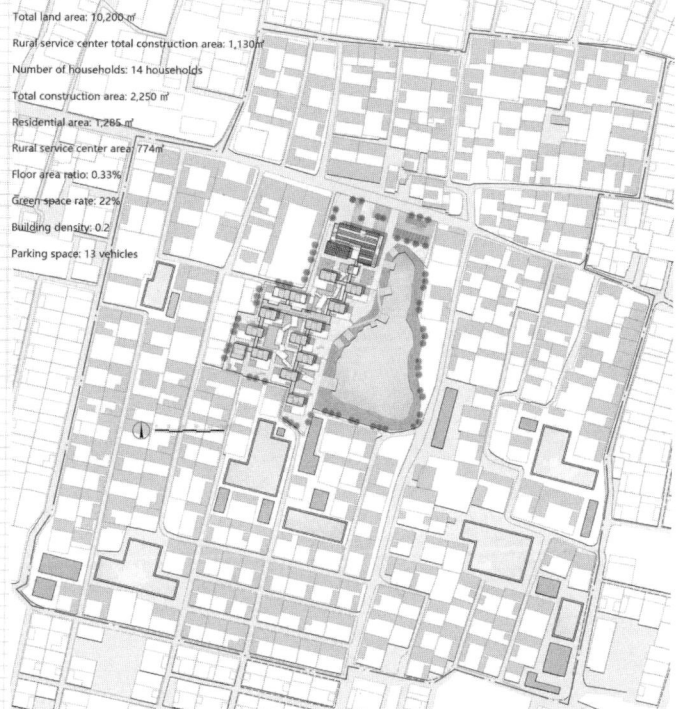

泗·廊·庭 II

Upstream — Corridor — Hall

■ First Floor Plan 1:150

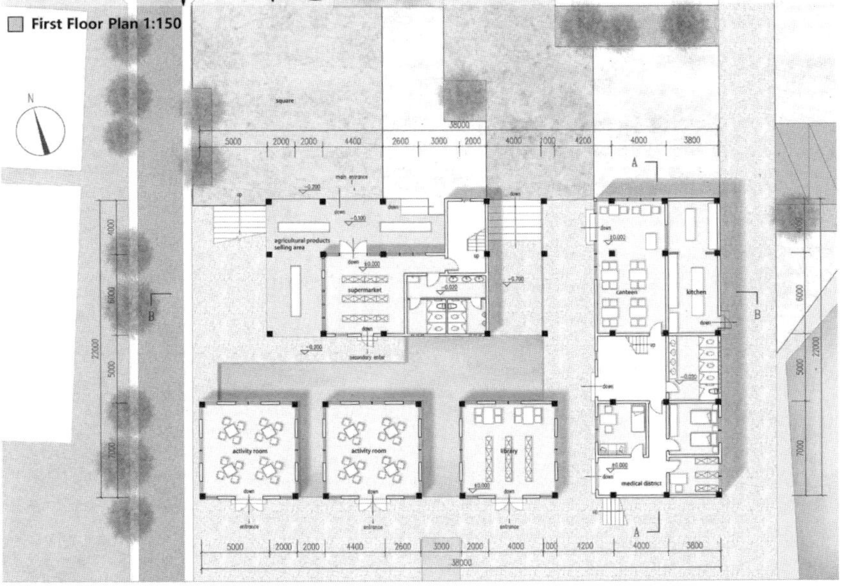

square

main entrance

agricultural products selling area

supermarket

secondary entrance

canteen

kitchen

activity room

activity room

library

medical district

entrance

entrance

entrance

■ Emergency Floor Plan

- comprehensive medical service area
- inpatient ward
- permanent refuge area

■ Function Streamline

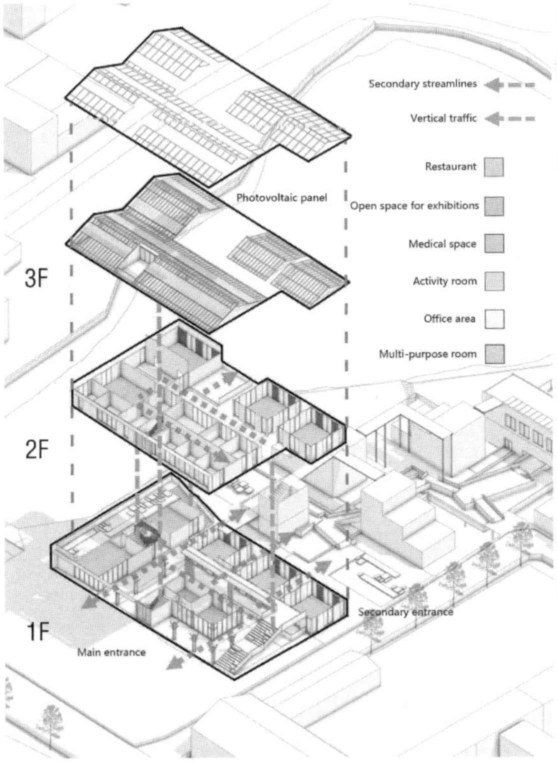

3F

2F

1F

Main entrance

Secondary entrance

Photovoltaic panel

- Secondary streamlines
- Vertical traffic
- Restaurant
- Open space for exhibitions
- Medical space
- Activity room
- Office area
- Multi-purpose room

■ Second Floor Plan 1:150

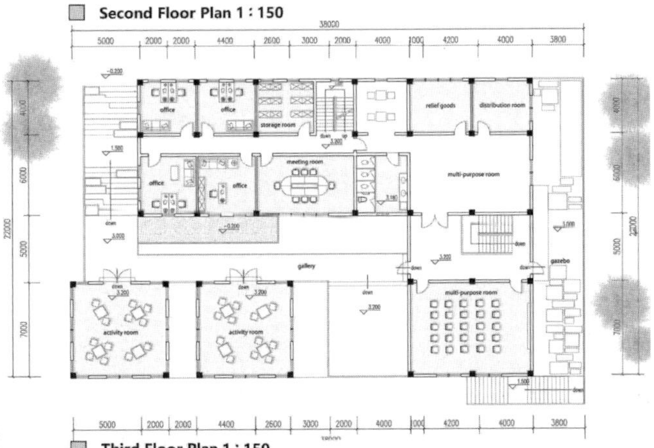

office

office

storage room

office

office

meeting room

multi-purpose room

retail goods

distribution room

activity room

activity room

gallery

multi-purpose room

gazebo

■ Third Floor Plan 1:150

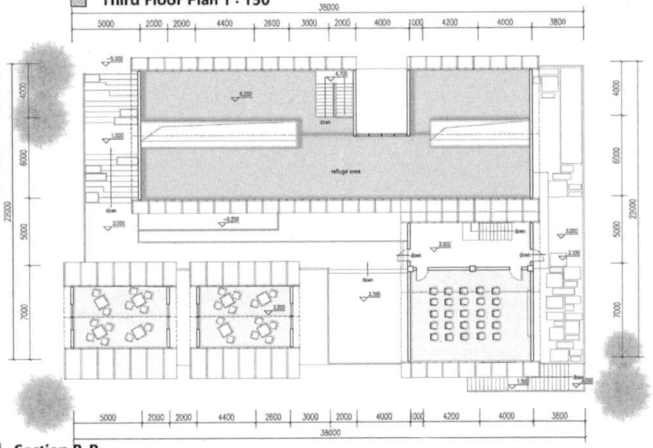

refuge area

■ Section A-A

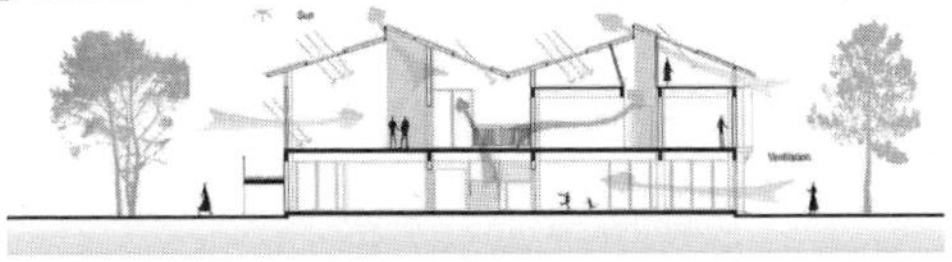

■ Section B-B

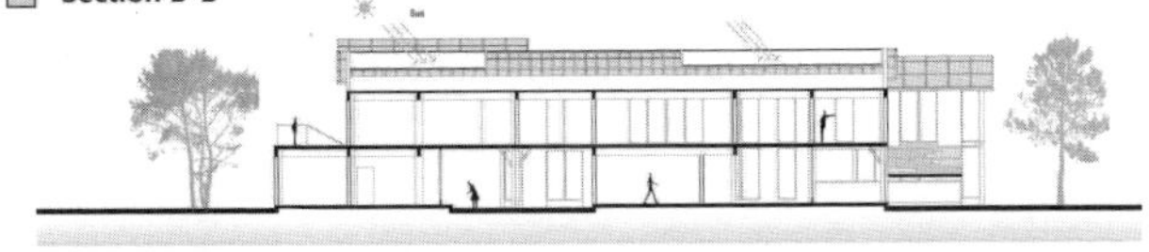

110000

泗 · 廊 · 庭 III

Upstream – Corridor – Hall

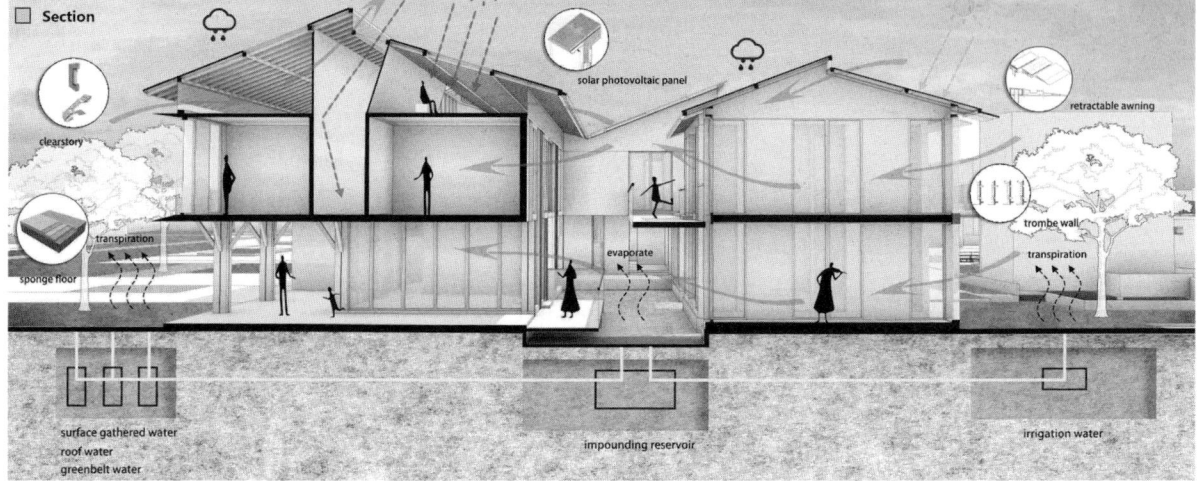

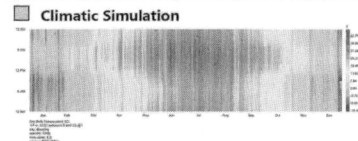

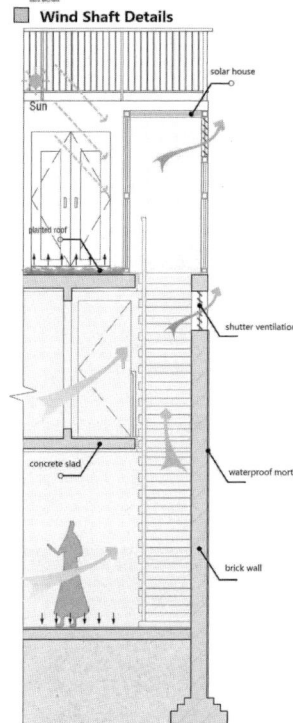

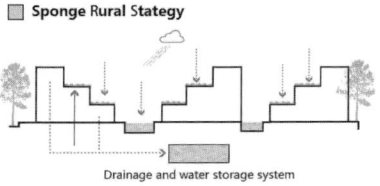

Section

clearstory

transpiration

sponge floor

solar photovoltaic panel

retractable awning

trombe wall

transpiration

evaporate

surface gathered water
roof water
greenbelt water

impounding reservoir

irrigation water

Climatic Simulation

Sunshine Simulation

city: Baoding

Wind Shaft Details

solar house

Sun

plantet roof

concrete slad

shutter ventilation

waterproof mortar

brick wall

Wind Shaft Details

storage room

bathroom

kitchen

living room

dining room

up

±0.000

reservoir

-0.250

First floor 1:120

down

bathroom

bedroom

bedroom

up

2.800

terrace

2.800

Second floor 1:120

down

bathroom

bedroom

5.400

terrace

5.300

Third floor 1:120

up

bathroom

-6.000

kitchen

reservoir

-0.250

First floor 1:120

bathroom

2.800

2.900

bedroom

bedroom

terrace

Second floor 1:120

bathroom

down

5.400

bedroom

5.200

terrace

Third floor 1:120

Sponge Rural Stategy

Drainage and water storage system

洄·廊·庭 IV

Upstream — Corridor — Hall

Solar Technology&Strategy

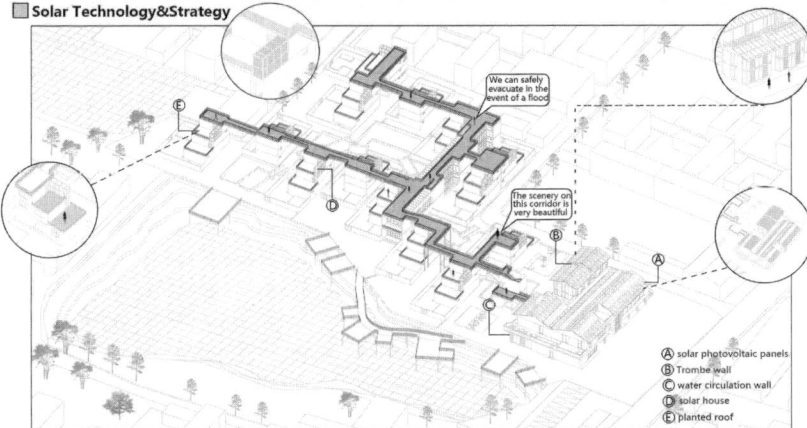

We can safely evacuate in the event of a flood

The scenery on this corridor is very beautiful

Ⓐ solar photovoltaic panels
Ⓑ Trombe wall
Ⓒ water circulation wall
Ⓓ solar house
Ⓔ planted roof

Energy Calculation

Electricity consumption of rural activity centers

According to the civilian electricity consumption index

Hourly electricity supply for buildings and public buildings

Consumption: 32W/㎡(1 hour)

Calculated based on an area of 950 square meters and an average electricity

consumption of 8 hours, the daily electricity consumption of the rural activity center :

950 ㎡ x 232W/㎡ x 8 h=243200 W/h=243.2 km·h

Solar panel power generation

Considering the 30% loss of solar panel electricity generated 4.2 times a day, the

effective sunshine duration of the 1KW module is 6 h.The area required

for 1kw solar panel power generation (roof) Approximately 6.5 ㎡ of electricity

432 ㎡ ÷ 6.8 ㎡ × 4.2 kW/h=267 kW/h>243.2 km·h

Conclusion

Current power generation capacity

Solar panels can basically meet the electricity demand of rural activity centers

Carbon emission

CO_2 emissio= (3+1.99) ×1230=6137.7 kgCO₂=6.1377 tCO₂

Node Technology Sample

Ⓑ Trombe wall

Sun
telescopic sun visor
fill up layer
double facade building
Trombe wall

Ⓒ Water circulation wall

exterior wall
irrigation line
water collection
Photovoltaic panels
water pump

air duct   functional block
cooling water institute
can open the building surface
Trombe wall   wind guide block

Ⓔ Planted roof

green plant layer
matrix layer
filter
drainage layer
building roof
drain-pipe

Planted roof
Shrub
Planted roof

Water Footprint

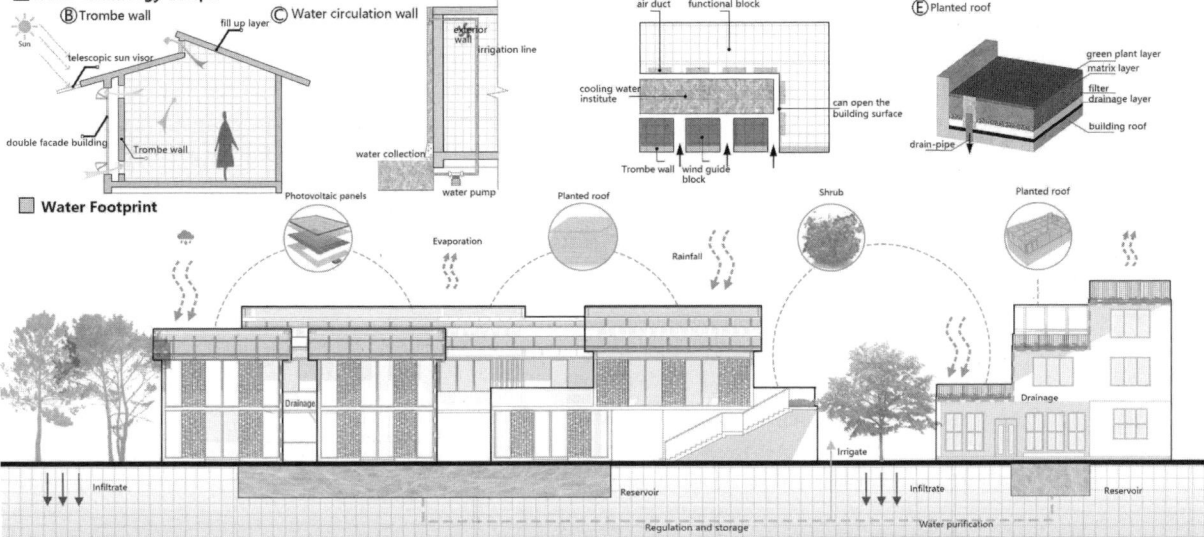

Evaporation   Rainfall

Drainage   Drainage

Infiltrate   Reservoir   Infiltrate   Reservoir

Irrigate

Regulation and storage   Water purification

综合奖 · 入围奖 · 保定市清苑区李八庄村建设项目
Comprehensive Awards − Nomination − Libazhuang Village Project，Qingyuan District，Baoding City

注册号：110195
Register Number：110195

项目名称：栖乡对野 决汩合流——三生融合下的乡村水灾防御与生态可持续设计

Entry Title：Where Village Meets Wild，Torrents Unite：Water Defense and Ecological Sustainability under the Tri−Integration Concept

作者：王方圆、东野瑞欣、周子龙、刘璇、逄雪菲、封嘉仪

Authors：Wang Fangyuan，Dongye Ruixin，Zhou Zilong，Liu Xuan，Pang Xuefei，and Feng Jiayi

作者单位：山东建筑大学
Authors from：Shandong Jianzhu University

指导教师：侯世荣
Tutor：Hou Shirong

指导教师单位：山东建筑大学
Tutor from：Shandong Jianzhu University

编号：100-110195

三生融合下的水灾防御与生态可持续设计
Water Defense and Ecological Sustainability under the Tri-Integration Concept

栖乡对野 决汩合流
Where Village Meets Wild，Torrents Unite

Design Specification

本方案从平灾结合的视角出发，旨在探索乡村空间韧性重构的技术框架和实现路径，我们保留当地传统民居的特征，对基本户型、院落布局和组团构成进行深入探讨，并且基于防洪要求对房屋构造、坑塘利用、排水线路和能源自供应能力等方面提供合理策略，打造"三生融合"的绿色韧性民居。

This scheme from the perspective of the combination, aims to explore the rural space toughness reconstruction of technical framework and implementation path, we keep the characteristics of local traditional dwellings, the basic family, courtyard layout, and group composition, and based on the housing structure of flood control requirements, pit utilization, drainage lines and energy supply ability to provide reasonable strategy, make "junior fusion" green toughness dwellings.

logic Generation

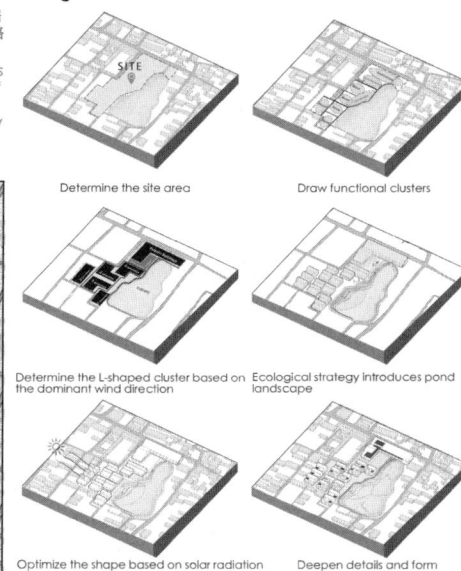

Determine the site area

Draw functional clusters

Determine the L-shaped cluster based on the dominant wind direction

Ecological strategy introduces pond landscape

Optimize the shape based on solar radiation

Deepen details and form

Site Plan 1：1000

Climate Analysis

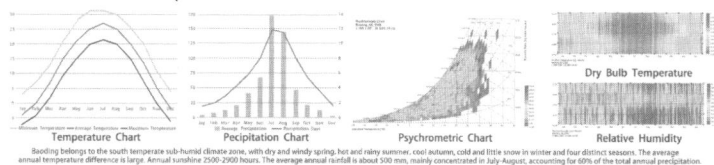

Temperature Chart

Pecipitation Chart

Psychrometric Chart

Dry Bulb Temperature

Relative Humidity

Baoding belongs to the south temperate sub-humid climate zone, with dry and windy spring, hot and rainy summer, cool autumn, cold and little snow in winter and four distinct seasons. The average annual temperature difference is large. Annual sunshine 2500-2900 hours. The average annual rainfall is about 500 mm, mainly concentrated in July-August, accounting for 60% of the total annual precipitation.

Wind Analysis

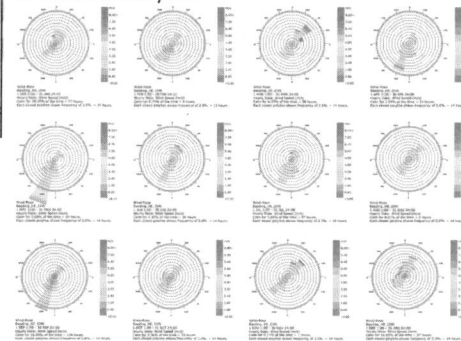

三生融合下的水灾防御与生态可持续设计
Water Defense and Ecological Sustainability
under the Tri-Integration Concept

栖乡对野 决汩合流
Where Village Meets Wild, Torrents Unite

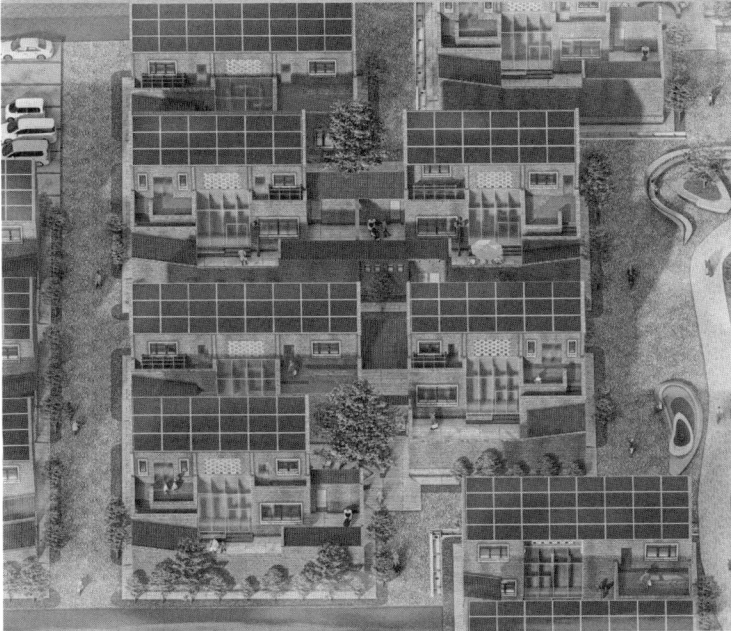

◨ First Plan

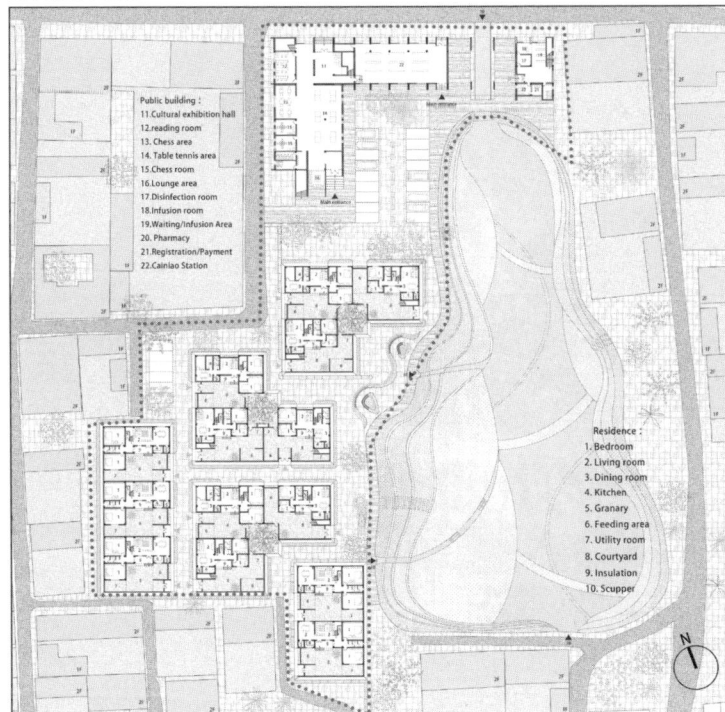

Public building :
11.Cultural exhibition hall
12.reading room
13. Chess area
14. Table tennis area
15.Chess room
16.Lounge area
17.Disinfection room
18.Infusion room
19.Waiting/Infusion Area
20. Pharmacy
21.Registration/Payment
22.Cainiao Station

Residence :
1. Bedroom
2. Living room
3. Dining room
4. Kitchen
5. Granary
6. Feeding area
7. Utility room
8. Courtyard
9. Insulation
10. Scupper

◨ Neighborhood Relationship

L-shaped cluster
Shared & Private

Outdoor Rest Seat
Elevated foundation
Drying clothes rack
Shared Courtyard

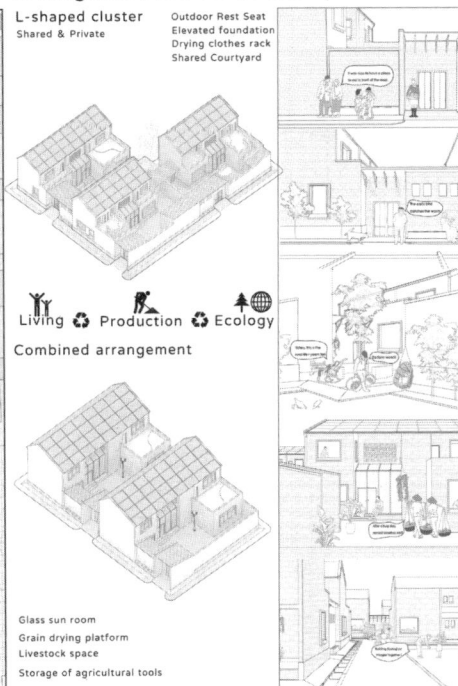

Living ♻ Production ♻ Ecology ♻

Combined arrangement

Glass sun room
Grain drying platform
Livestock space
Storage of agricultural tools

◨ Biogas Disester Analysis

◨ Trombe Wall and Sunroom

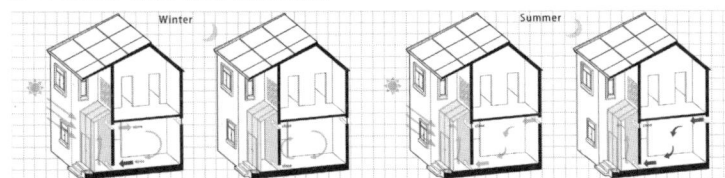

Winter   Summer

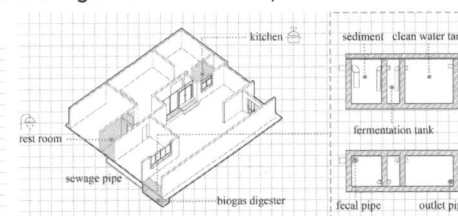

kitchen   sediment   clean water tank

rest room

fermentation tank

sewage pipe

biogas digester   fecal pipe   outlet pipe

编号: 100-110195

三生融合下的水灾防御与生态可持续设计
Water Defense and Ecological Sustainability
under the Tri-Integration Concept

栖乡对野 决汩合流
Where Village Meets Wild, Torrents Unite

Second Plan

5.Chess room
6.Borrowing area
7.Reading area
8.Reception area
9.Meeting rooms
10. Office
11.Clinic

1. Bedroom
2. Living room
3. Drying platform
4. Balcony

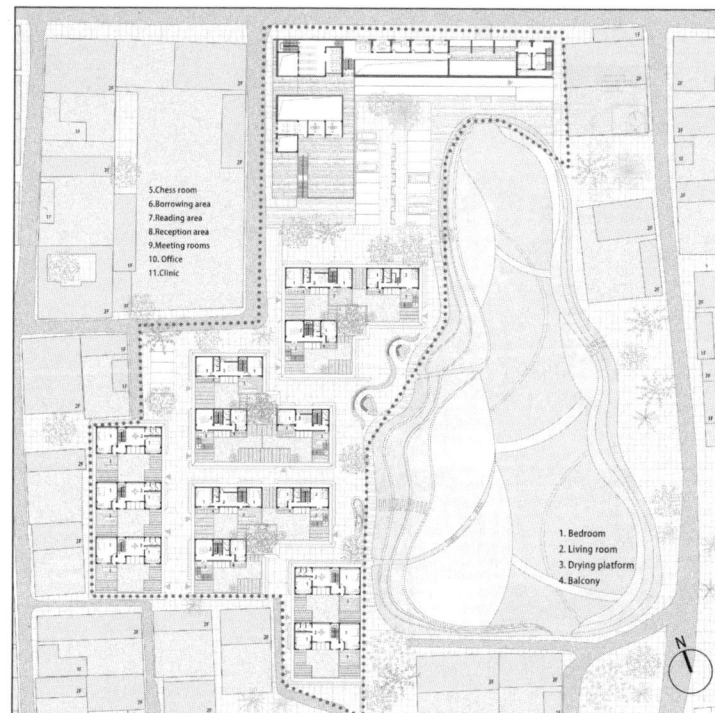

Biogas Disester Analysis

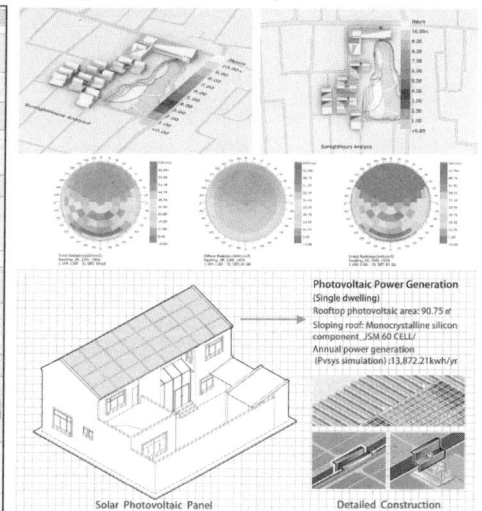

Photovoltaic Power Generation
(Single dwelling)
Rooftop photovoltaic area: 90.75 ㎡
Sloping roof: Monocrystalline silicon
component _JSM.60 CELL/
Annual power generation
(Pvsys simulation) :13,872.21kwh/yr

Solar Photovoltaic Panel   Detailed Construction

Horizontal-axis Wind Turbine

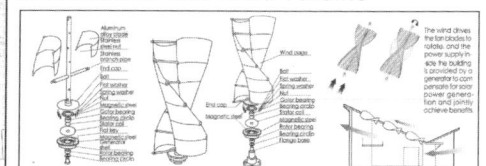

Drainage Planning

Take advantage of the situation.

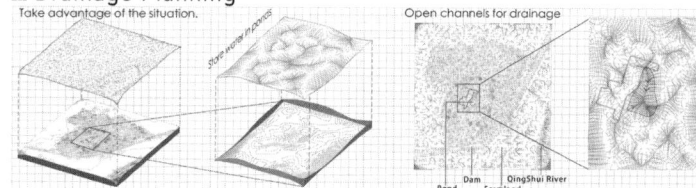

Open channels for drainage

Pond   Dam   QingShui River
Farmland

Flood Control Tecnology

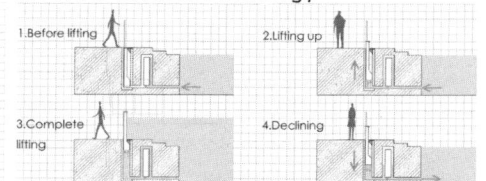

1.Before lifting   2.Lifting up

3.Complete   4.Declining
lifting

三生融合下的水灾防御与生态可持续设计
Water Defense and Ecological Sustainability
under the Tri-Integration Concept

栖乡对野 决泪合流
Where Village Meets Wild， Torrents Unite

| Sunny Courtyard 阳光庭院 | Ecological Floating Islands 生态浮岛 | Emergency Response 应急救灾 | Fast Drainage 快速排水 |

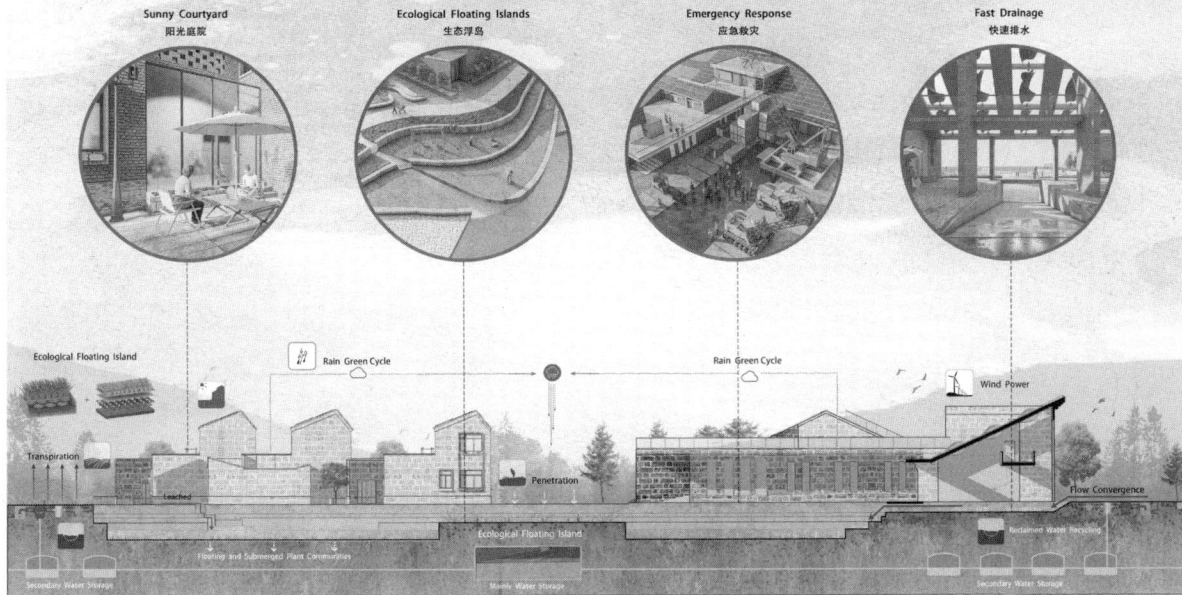

Sectional View

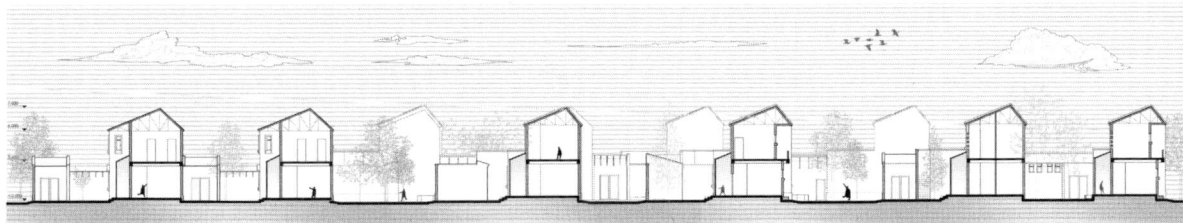

Catalogue of Green and Lowcarbon Technologies

Economic and technical indicators

The number	Project	Unit	Value
1	project land area	m²	10200
2	construction land area	m²	8840
3	total floor area	m²	5330
(1)	residential area	m²	2130
(2)	public building area	m²	1250
(3)	the yard area	m²	1350
(4)	parking area	m²	600
4	volume rate		0.6
5	building density	%	16
6	green space rate	%	22.6

Building productivity system composition

The number	capacity system	Unit	Value
1	Photovoltaic power generation	kW·h/yr	360000
2	Wind power generation	kW·h/yr	3500
3	biomass power generation	kW·h/yr	57000
4	total power generation	kW·h/yr	420500

Solar energy utilization
- photovoltaic panel
- solar house
- solar street lamp

Wind power utilization Sunshade component
- vertical axis wind power generation
- horizontal grille
- vertical grille

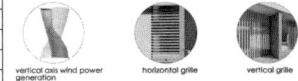

Heat insulation | **Ventilation** | **Flood control**
- trombe wall
- hollowed out brick
- foold control barrier

Crops reuse
- corn stalks
- wheat stalk
- straw straw

Construct of the Nodes

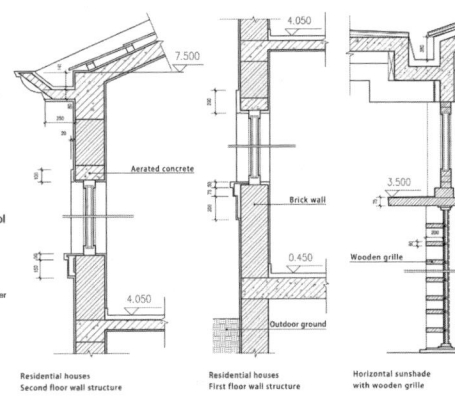

Residential houses
Second floor wall structure

Residential houses
First floor wall structure

Horizontal sunshade
with wooden grille

Elevation of Public Buildings

Sectional View of a Dwelling

ID:100-110292

泽乡守望 I
Guardian of the Marshland

综合奖·入围奖·保定市清苑区李八庄村建设项目
Comprehensive Awards – Nomination – Libazhuang Village Project, Qingyuan District, Baoding City

注册号：110292
Register Number：110292

项目名称：泽乡守望
Entry Title：Guardian of the Marshland

作者：郭梦锦、庞思怡、文一博、徐恩夏
Authors：Guo Mengjin, Pang Siyi, Wen Yibo, and Xu Enxia

作者单位：东北大学
Authors from：Northeastern University

指导教师：刘哲铭
Tutor：Liu Zheming

指导教师单位：东北大学
Tutor from：Northeastern University

Design Specification

在针对李家村雨季洪涝问题的设计中，我们巧妙地调整了传统建筑布局，引入了集水功能的共享雨水花园，旨在优化采光的同时促进邻里交往，雨水通过花园汇集，有效引导至坑塘，增强防洪能力。此外，我们将坑塘沿岸改造成台地，提升其在日常生活及灾害期间的使用价值，实现乡村空间的多功能性和可持续性。

In response to the rainy season flooding problem in Lijia Village, we cleverly adjusted the traditional building layout to introduce a shared rain garden with water collection function, aiming to optimize lighting and promote neighborhood interaction. Rainwater is collected through the garden and effectively directed to the pit to enhance the flood control capacity. In addition, we have transformed the shore of the pit into a platform to enhance its use value in daily life and during disasters, to achieve the versatility and sustainability of the rural space.

Climatic Analysis

Qingyuan District is located in the middle of Hebei Province. It has a temperate monsoon climate. It is cold and snowy in winter and hot and dry in summer. The average annual temperature in the region is about 12°C, the annual sunshine duration is about 2500 to 2900 hours, the frost-free period is about 165 to 210 days, and the average annual rainfall is about 451 mm, mainly concentrated in July to August, accounting for most of the total annual precipitation

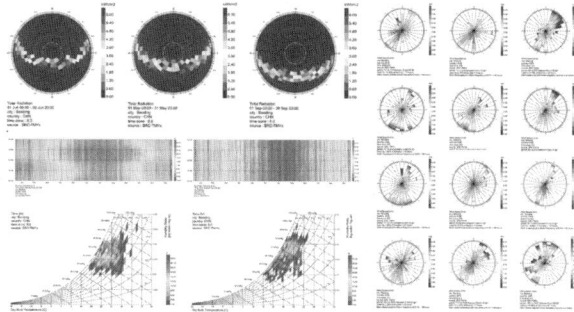

General plan 1:750

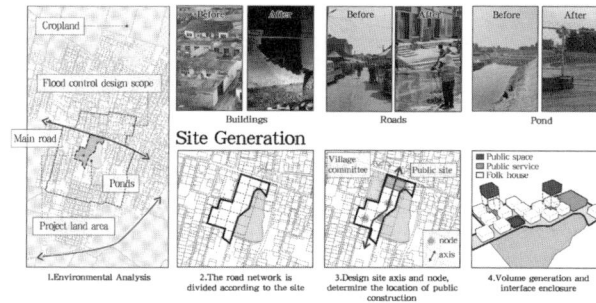

Site Status
Site Generation
Diagram of Design Process

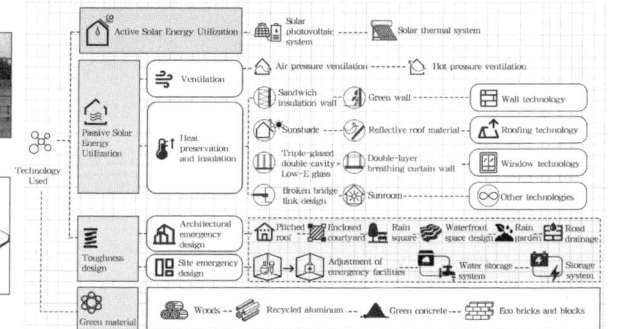

泽乡守望 Ⅱ
Guardian of the Marshland

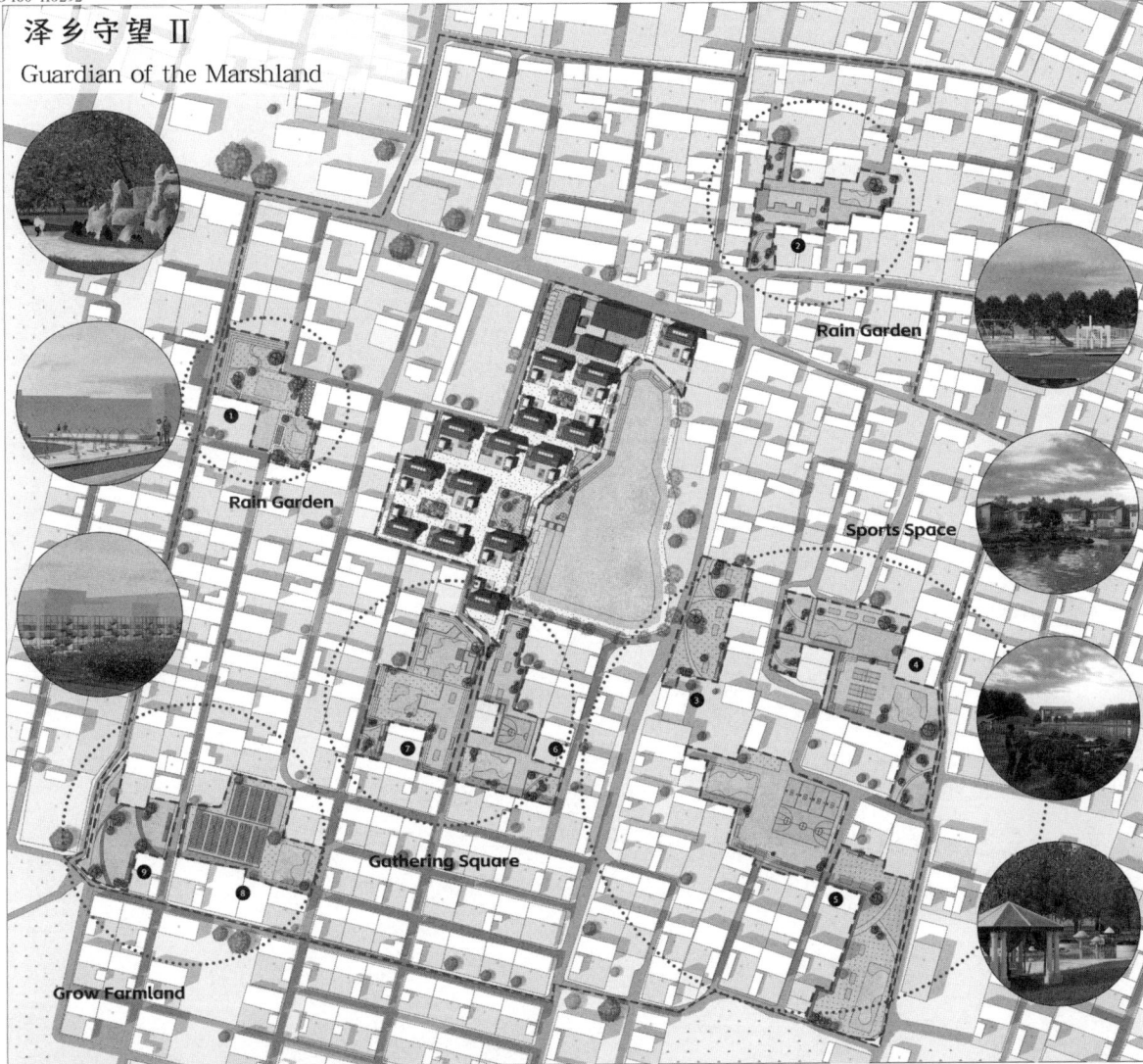

Rain Garden

Rain Garden

Sports Space

Gathering Square

Grow Farmland

Public Space Design

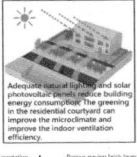

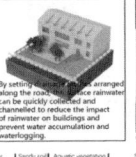

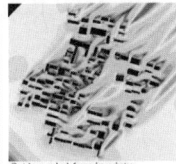

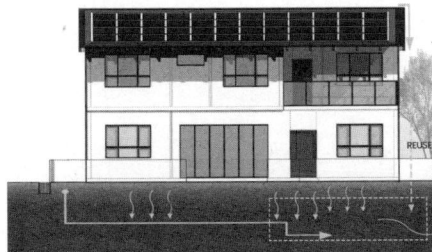

By collecting and storing rainwater, it blocks stormwater runoff and reduces the pressure on the drainage system. The reservoir can also purify water and promote groundwater recharge.

The rainwater collection system and infiltration structure are used to slowly penetrate the rainwater collected during the rainy season into the ground to filter and purify the rainwater.

Grass ditches are set up on both sides of the road to trap the rainwater flowing down the slope due to the difference in terrain height and lead to the diversion channel when it rains.

By reducing runoff and runoff peaks through infiltration and storage, stormwater is temporarily stored to prevent rapid discharge leading to waterlogging, reducing velocity, and preventing clogging of outfalls.

Through a rainwater collection system, the houses use sloping roofs to direct rainwater, and road drains are set around them to ensure that rainwater runoff reaches the reservoir or pit for further.

Adequate natural lighting and solar photovoltaic panels reduce building energy consumption. The greening in the residential courtyard can improve the microclimate and improve the indoor ventilation efficiency.

By collecting and storing rainwater, it blocks stormwater runoff and reduces the pressure on the drainage system. The reservoir can also purify water and promote groundwater recharge.

By setting rainwater pipes arranged along the road, the surface rainwater can be quickly collected and channelled to reduce the impact of rainwater on buildings and prevent water accumulation and waterlogging.

Outdoor wind farm in winter

Outdoor wind farm in summer

Road drainage ditch | The surface vegetation | Porous paving brick layer | Riparian shrub layer | Wetland layer | Sandy soil | Aquatic vegetation | Sandy soil | Pits and ponds

Rainwater collection trench

Collect rainwater and plant grass in a ditch

Wetland catchment area

Rain garden

REUSE

When waterlogging STORAGE⊠REUSE
Water storage and purification (Rainwater collection and purification System & Platform ecological wetland system)
Using the natural slope of the current slope protection, the traditional horizontal layout is changed to the ladder-shaped longitudinal layout to extend the filtering path. On the one hand, it can deal with the height difference of pits and ponds on the other hand, it can increase the contact area and duration between constructed wetlands and rainwater, purify surface runoff, and maximize the wetland area through precipitation and filtration of water sources, so as to achieve higher purification efficiency. On the basis of not affecting the flood drainage and storage function of the pit, it also provides more potential landscape display opportunities.

STORAGE

泽乡守望 III
Guardian of the Marshland

Folk House Plan

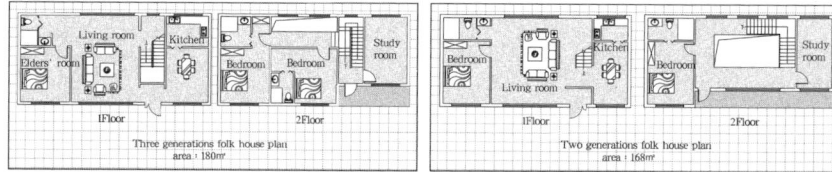

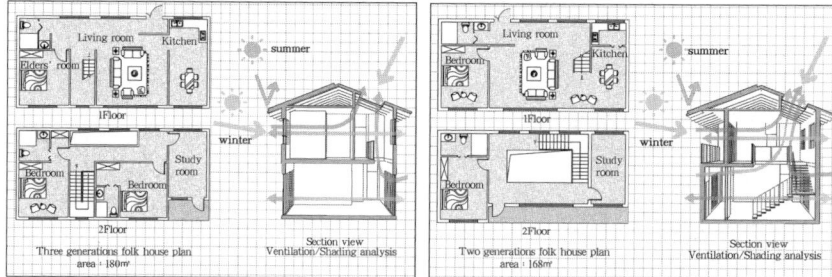

Three generations folk house plan
area：180m²

Living room　Kitchen
Elders' room
Bedroom　Bedroom　Study room
1Floor　　2Floor

Two generations folk house plan
area：168m²

Bedroom　Kitchen
Living room　Bedroom　Study room
1Floor　　2Floor

Three generations folk house plan
area：180m²

Living room　Kitchen
Elders' room
summer
Bedroom　Study room
Bedroom
winter
1Floor　　2Floor
Section view
Ventilation/Shading analysis

Two generations folk house plan
area：168m²

Living room　Kitchen
Bedroom
summer
Bedroom　Study room
winter
1Floor　　2Floor
Section view
Ventilation/Shading analysis

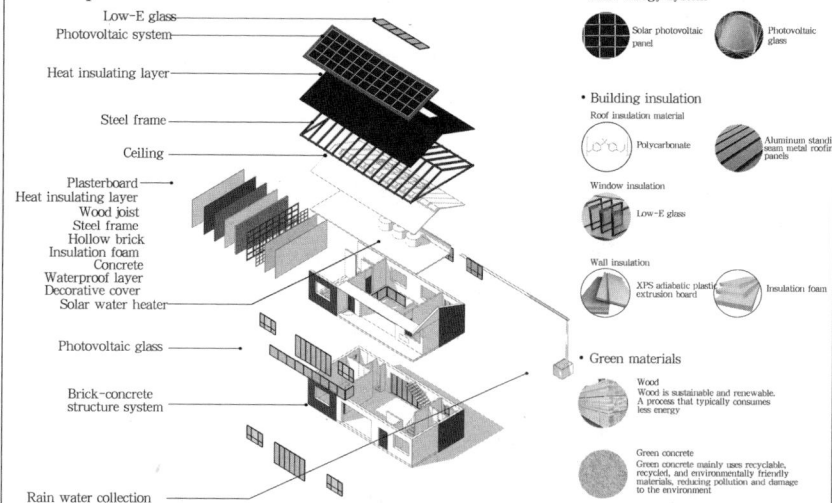

Activity Requirement Analysis

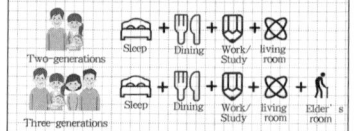

Two-generations　Sleep + Dining + Work/Study + living room

Three-generations　Sleep + Dining + Work/Study + living room + Elder's room

Functional Analysis

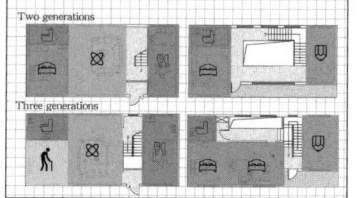

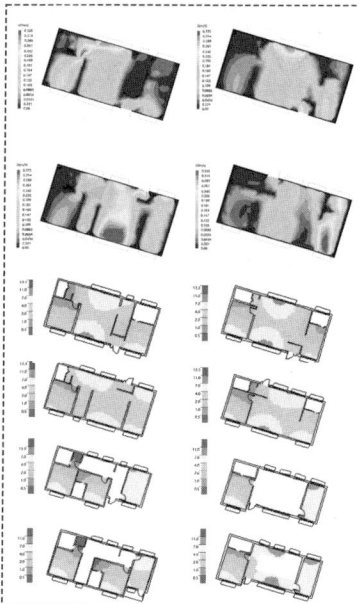

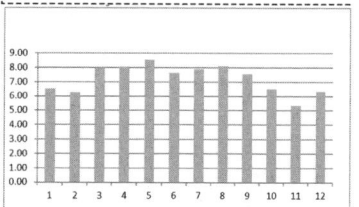

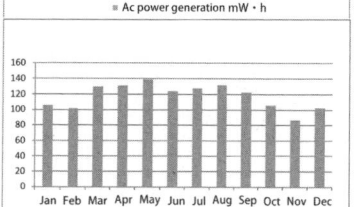

Two generations

Three generations

Rainwater Collection

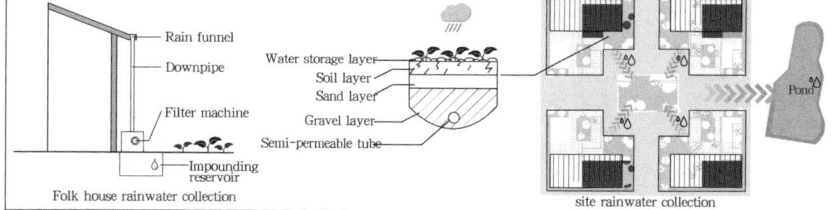

Rain funnel
Downpipe
Filter machine
Impounding reservoir
Folk house rainwater collection

Water storage layer
Soil layer
Sand layer
Gravel layer
Semi-permeable tube

Rainwater flows into the pond
Pond
site rainwater collection

Future Development

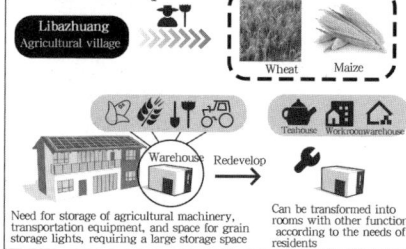

Libazhuang Agricultural village

Wheat　Maize

Warehouse　Redevelop

Teahouse　Workroom warehouse

Need for storage of agricultural machinery, transportation equipment, and space for grain storage lights, requiring a large storage space

Can be transformed into rooms with other functions according to the needs of residents

Neigh Borhood Analysis

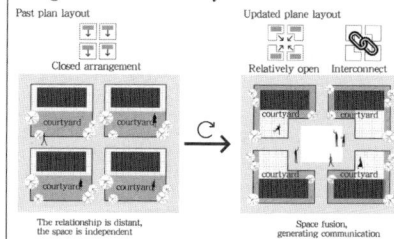

Past plan layout　　Updated plane layout

Closed arrangement　　Relatively open　Interconnect

courtyard　courtyard
courtyard　courtyard

courtyard　courtyard
courtyard　courtyard

The relationship is distant, the space is independent

Space fusion, generating communication

Technique

Low-E glass
Photovoltaic system
Heat insulating layer
Steel frame
Ceiling
Plasterboard
Heat insulating layer
Wood joist
Steel frame
Hollow brick
Insulation foam
Concrete
Waterproof layer
Decorative cover
Solar water heater
Photovoltaic glass
Brick-concrete structure system
Rain water collection

• Solar energy system

Solar photovoltaic panel　Photovoltaic glass

• Building insulation

Roof insulation material
Polycarbonate　　Aluminum standing seam metal roofing panels

Window insulation
Low-E glass

Wall insulation
XPS adiabatic plastic extrusion board　Insulation foam

• Green materials

Wood
Wood is sustainable and renewable. A process that typically consumes less energy.

Green concrete
Green concrete mainly uses recyclable, recycled, and environmentally friendly materials, reducing pollution and damage to the environment

Ac power generation mW・h

Total solar radiation kW・h/m²

Size / area	Toward the Angle	dip	quantity
1.50(1.50X1.00)	due south	34	662
Photovoltaic system information			
Component type	Copper indium gallium selenium	peak power	100Wp
Number of components	662	Total installed capacity	66.2kW
Installation mode of components	Fixed integration	Component area	993 m²
Inverter efficiency	0.96	Inverter power	6.79kW
Line loss efficiency	0.01	Material surface contamination efficiency	0.01
coefficient of correction	0.01	Integrated efficiency of the system	93.1%

Building operation	11.584		579.219
carbon sink	-10.766		-538.300
amount to	0.818		40.919
Renewable Energy Energy (Er)	photovoltaic (Ep)	6380.01	2519.804
	wind-force (Ew)	0.00	0.583	0.000
Total carbon emissions from building operations			579.219

泽乡守望 IV
Guardian of the Marshland

Wind Environment Simulation

Optical Environment Simulation

IF

IF

2F

Villagers Supporting Services Building Plan

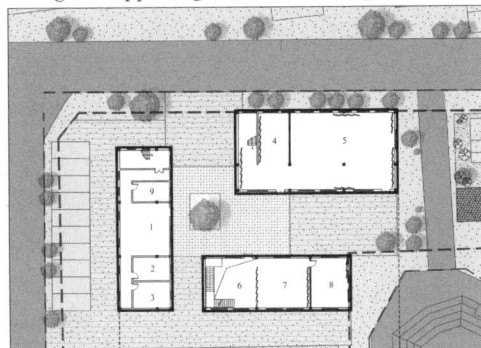

Ground Floor Plan 1:300

Second Floor Plan 1:300

1 Service hall
2 Office
3 Conference room
4 Fitness equipment
5 Table tennis room
6 Children's entertainment
7 Community talk
8 Reading room
9 Rest room
10 Chess and card room
11 Villager cinema

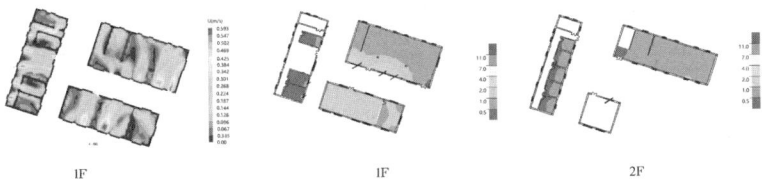

Pv system parameters

The generating capacity of the building photovoltaic system should be calculated and determined according to the local solar energy resources, the design of the photovoltaic system, the layout of the photovoltaic square array and the environmental conditions and other factors. According to the "Code for Design of Photovoltaic Power Stations" GB 50797 and other standards, the power generation value of the photovoltaic system can be obtained. Where Ep —— power generation (kWh);

Total solar radiation at HA —— level (kWh / m2);

Irradiance (constant) under Es —— standard conditions with a value of

$1 kW/m^2$;

P —— installed capacity (kWp);

AC power generation mW · h

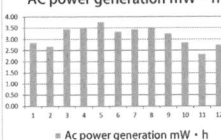

■ Ac power generation mW · h

Total solar radiation kW · h/m²

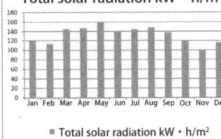

■ Total solar radiation kW · h/m²

Profile Map 1:300

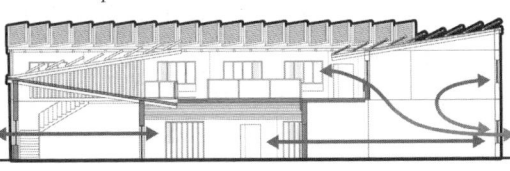

East Elevation 1:300

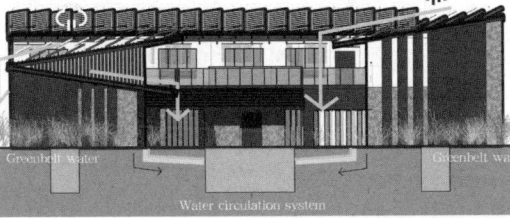

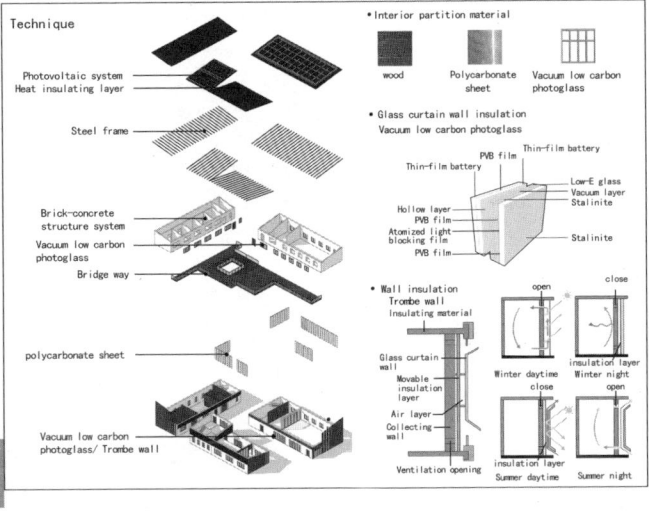

Roof water

Greenbelt Water

Greenbelt Water

Water circulation system

Open space / **Private space**

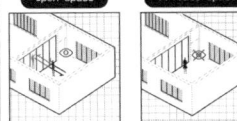

Vacuum low carbon photoglass
The electricity generated by photovoltaic glass can directly control the internal atomization effect

Space and function changes
(Polycarbonate sheet)

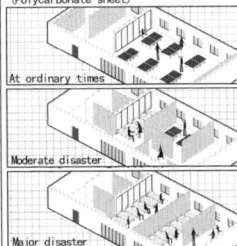

At ordinary times

Moderate disaster

Major disaster

General Disaster 1:300

Severe Disaster Situation 1:300

1 Materials sending and receiving
2 Drug storage
3 Command center
4 Information center
5 Villager activity
6 Material storage
7 Temporary medical care
8 Temporary accommodation
9 Rest room
10 Medical care
11 Emergency isolation

Level one functional shift

Technique

Photovoltaic system
Heat insulating layer

Steel frame

Brick-concrete structure system

Vacuum low carbon photoglass

Bridge way

polycarbonate sheet

Vacuum low carbon photoglass/ Trombe wall

• Interior partition material

wood
Polycarbonate sheet
Vacuum low carbon photoglass

• Glass curtain wall insulation
Vacuum low carbon photoglass

Thin-film battery
PVB film
Thin-film battery
Low-E glass
Vacuum layer
Stalinite
Hollow layer
PVB film
Atomized light blocking film
PVB film
Stalinite

• Wall insulation
Trombe wall
Insulating material

open
close

Glass curtain wall
Movable insulation layer
Air layer
Collecting wall

insulation layer
Winter daytime close
Winter night open

Ventilation opening

insulation layer
Summer daytime
Summer night

综合奖·入围奖·保定市清苑区李八庄村建设项目
Comprehensive Awards – Nomination – Libazhuang Village Project, Qingyuan District，Baoding City

注册号：110487

Register Number：110487

项目名称：溯溪而下，逆洪而上

Entry Title：Trace the River Downstream and Upstream Against the Tide

作者：邓振寰、张媛浩、许伊扬、宋雨鑫、陈宇琴、韩超辉、薛睿宜

Authors：Deng Zhenhuan, Zhang Yuanhao, Xu Yiyang, Song Yuxin, Chen Yuqin, Han Chaohui, and Xue Ruiyi

作者单位：天津大学

Authors from：Tianjin University

指导教师：原野

Tutor：Yuan Ye

指导教师单位：天津大学

Tutor from：Tianjin University

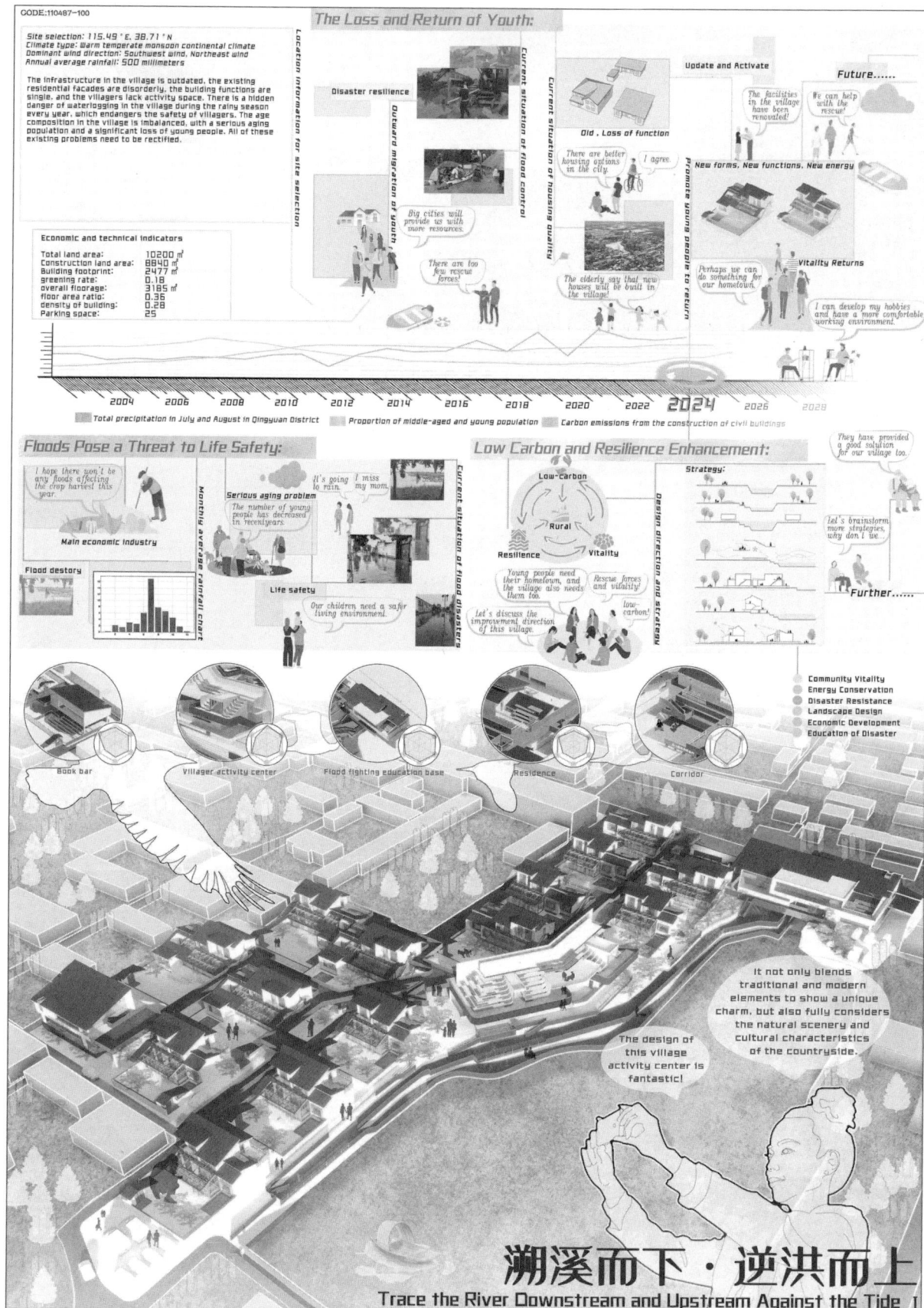

GODE:110487-100

Location information for site selection

Site selection: 115.49 ' E, 38.71 ' N
Climate type: Warm temperate monsoon continental climate
Dominant wind direction: Southwest wind, Northeast wind
Annual average rainfall: 500 millimeters

The infrastructure in the village is outdated, the existing residential facades are disorderly, the building functions are single, and the villagers lack activity space. There is a hidden danger of waterlogging in the village during the rainy season every year, which endangers the safety of villagers. The age composition in the village is imbalanced, with a serious aging population and a significant loss of young people. All of these existing problems need to be rectified.

Economic and technical indicators

Total land area	10200 ㎡
Construction land area	8840 ㎡
Building footprint	2477 ㎡
greening rate	0.18
overall floorage	3185 ㎡
floor area ratio	0.36
density of building	0.28
Parking space	25

The Loss and Return of Youth:

Disaster resilience
Outward migration of youth
Current situation of flood control
Current situation of housing quality
Promote young people to return

Update and Activate
Old. Loss of function
The facilities in the village have been renovated!
New forms, New functions, New energy
Vitality Returns
Future......
We can help with the rescue!
Big cities will provide us with more resources.
There are better housing options in the city.
I agree.
There are too few rescue forces!
The elderly say that new houses will be built in the village!
Perhaps we can do something for our hometown.
I can develop my hobbies and have a more comfortable working environment.

2004 2006 2008 2010 2012 2014 2016 2018 2020 2022 2024 2026 2028

Total precipitation in July and August in Qingyuan District   Proportion of middle-aged and young population   Carbon emissions from the construction of civil buildings

Floods Pose a Threat to Life Safety:
Monthly average rainfall chart
I hope there won't be any floods affecting the crop harvest this year.
Serious aging problem
It's going to rain.
I miss my mom.
Main economic industry
Flood destory
Life safety
Our children need a safer living environment.
Current situation of flood disasters

Low Carbon and Resilience Enhancement:
Low-carbon
Rural
Resilience   Vitality
Young people need their hometown, and the village also needs them too.
Rescue forces and vitality!
low carbon!
Let's discuss the improvement direction of this village.
Strategy:
Design direction and strategy
They have provided a good solution for our village too.
Let's brainstorm more strategies, why don't we.
Further......

Community Vitality
Energy Conservation
Disaster Resistance
Landscape Design
Economic Development
Education of Disaster

Book bar   Villager activity center   Flood fighting education base   Residence   Corridor

The design of this village activity center is fantastic!

It not only blends traditional and modern elements to show a unique charm, but also fully considers the natural scenery and cultural characteristics of the countryside.

溯溪而下·逆洪而上
Trace the River Downstream and Upstream Against the Tide I

溯溪而下·逆洪而上
Trace the River Downstream and Upstream Against the Tide Ⅱ

■ Body Gener Ation

STEP 1 — Status of the site
STEP 2 — Addition of elderly room
STEP 3 — Building elevation
STEP 4 — Increase the sun room
STEP 5 — Slope roof design
STEP 6 — Increase courtyard and roof

■ Growable Dwelling Logic

STEP 1 — Status of the site
STEP 2 — Trenching next to the building
STEP 3 — Schematic of water storage tank lifting
STEP 4 — Slope roof design

■ Master Plan 1:1000

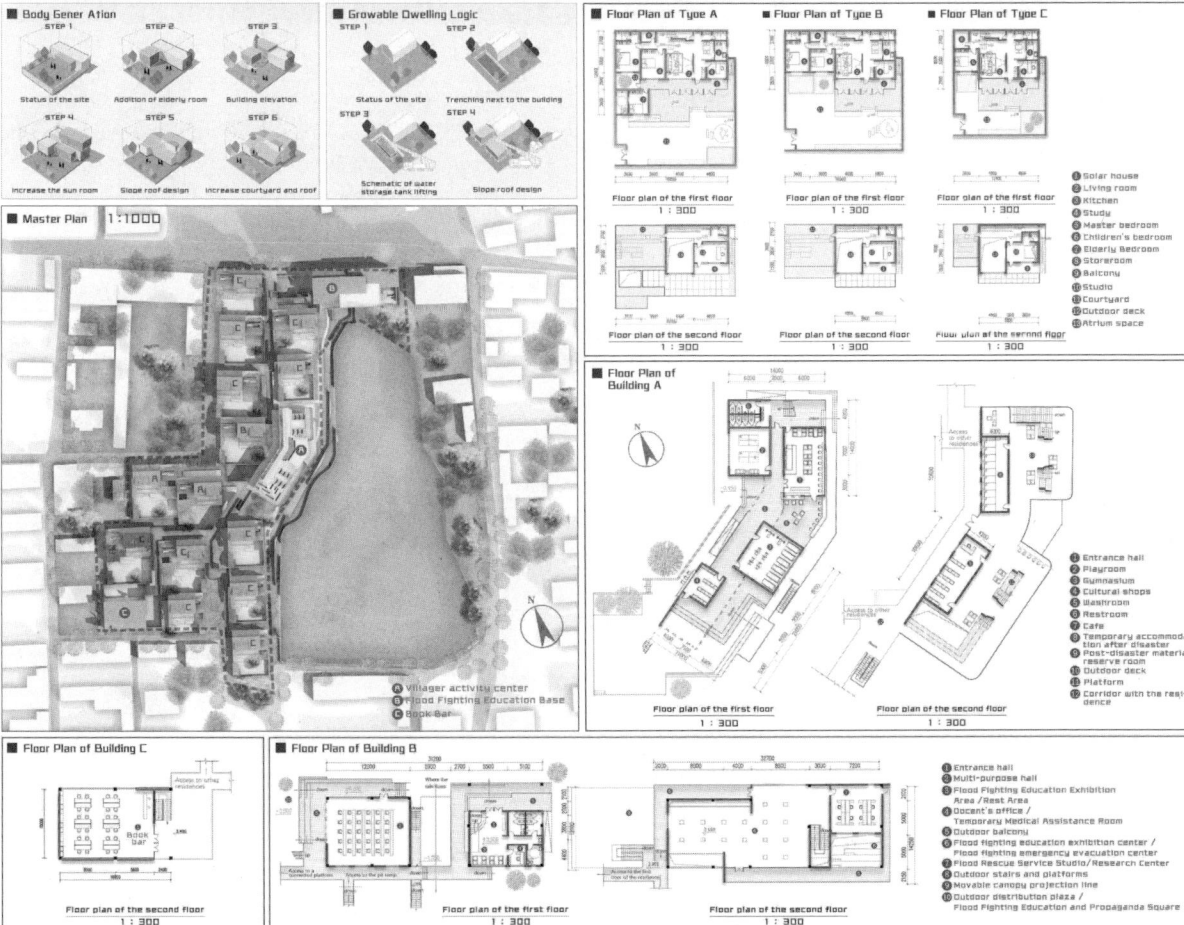

A Villager activity center
B Flood Fighting Education Base
C Book Bar

■ Floor Plan of Type A

Floor plan of the first floor 1:300
Floor plan of the second floor 1:300

■ Floor Plan of Type B

Floor plan of the first floor 1:300
Floor plan of the second floor 1:300

■ Floor Plan of Type C

Floor plan of the first floor 1:300
Floor plan of the second floor 1:300

1 Solar house
2 Living room
3 Kitchen
4 Study
5 Master bedroom
6 Children's bedroom
7 Elderly bedroom
8 Storeroom
9 Balcony
10 Studio
11 Courtyard
12 Outdoor deck
13 Atrium space

■ Floor Plan of Building A

Floor plan of the first floor 1:300
Floor plan of the second floor 1:300

1 Entrance hall
2 Playroom
3 Gymnasium
4 Cultural shops
5 Washroom
6 Restroom
7 Cafe
8 Temporary accommodation after disaster
9 Post-disaster material reserve room
10 Outdoor deck
11 Platform
12 Corridor with the residence

■ Floor Plan of Building C

Floor plan of the second floor 1:300

■ Floor Plan of Building B

Floor plan of the first floor 1:300
Floor plan of the second floor 1:300

1 Entrance hall
2 Multi-purpose hall
3 Flood Fighting Education Exhibition Area /Rest Area
4 Docent's office / Temporary Medical Assistance Room
5 Outdoor balcony
6 Flood fighting education exhibition center / Flood fighting emergency evacuation center
7 Flood Rescue Service Studio/Research Center
8 Outdoor stairs and platforms
9 Movable canopy projection line
10 Outdoor distribution plaza / Flood Fighting Education and Propaganda Square

■ East Elevation View of the Site

233

CODE:110487-100

溯溪而下 · 逆洪而上
Trace the River Downstream and Upstream Against the Tide III

Planting water storage roofs

rainfall

evaporation

evaporation

rainfall

rainfall

permeate

Ecological retention soil

evaporation

Water storage tank

Residential pond

Shui Zhen

Permeable pavement infiltration

Water storage pond

Water storage tank

Shui Zhen

Permeable pavement infiltration

Residential pond

Water storage pond

The space for collecting rainwater changes with the seasons. During the dry season, it becomes a garden for residents. During the wet season, rainwater collected through roof greening, ecological retention areas, and permeable paving is gathered here, becoming a pond filled with rainwater

residential district

Activity Sponge Square Area

Main water storage area

■ Water Canal System Diagram

■ Design Strategy

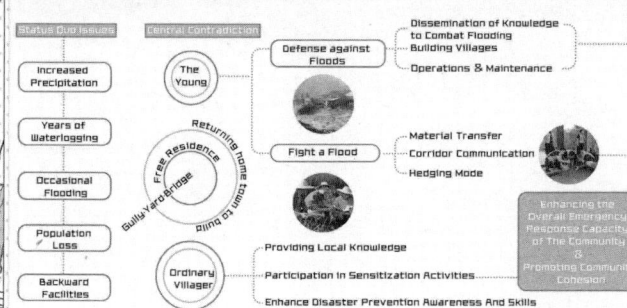

Status Quo Issues

Central Contradiction

- Increased Precipitation
- Years of Waterlogging
- Occasional Flooding
- Population Loss
- Backward Facilities

The Young

Free Residence

Ordinary Villager

Defense against Floods

Fight a Flood

Dissemination of Knowledge to Combat Flooding
Building Villages
Operations & Maintenance

Material Transfer
Corridor Communication
Hedging Mode

Enhancing the Overall Emergency Response Capacity of The Community & Promoting Community Cohesion

Providing Local Knowledge
Participation in Sensitization Activities
Enhance Disaster Prevention Awareness And Skills

■ Submerged Area

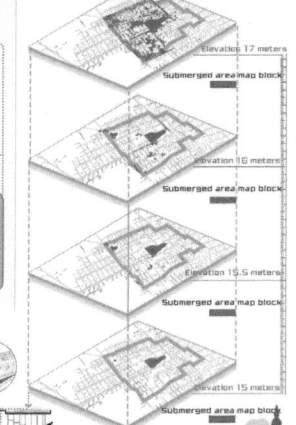

Elevation 17 meters
Submerged area map block

Elevation 16 meters
Submerged area map block

Elevation 15.5 meters
Submerged area map block

Elevation 15 meters
Submerged area map block

Main rainwater collection

Main rainwater collection area

Sponge green space distribution

Green Sponge Square
Water flow direction

Rainwater analysis

Rainwater runoff line
Rainwater collection point

Slope direction

Southeast
South direction
Southwest
west direction
Northwest
North direction
Northeast
East direction

Elevation

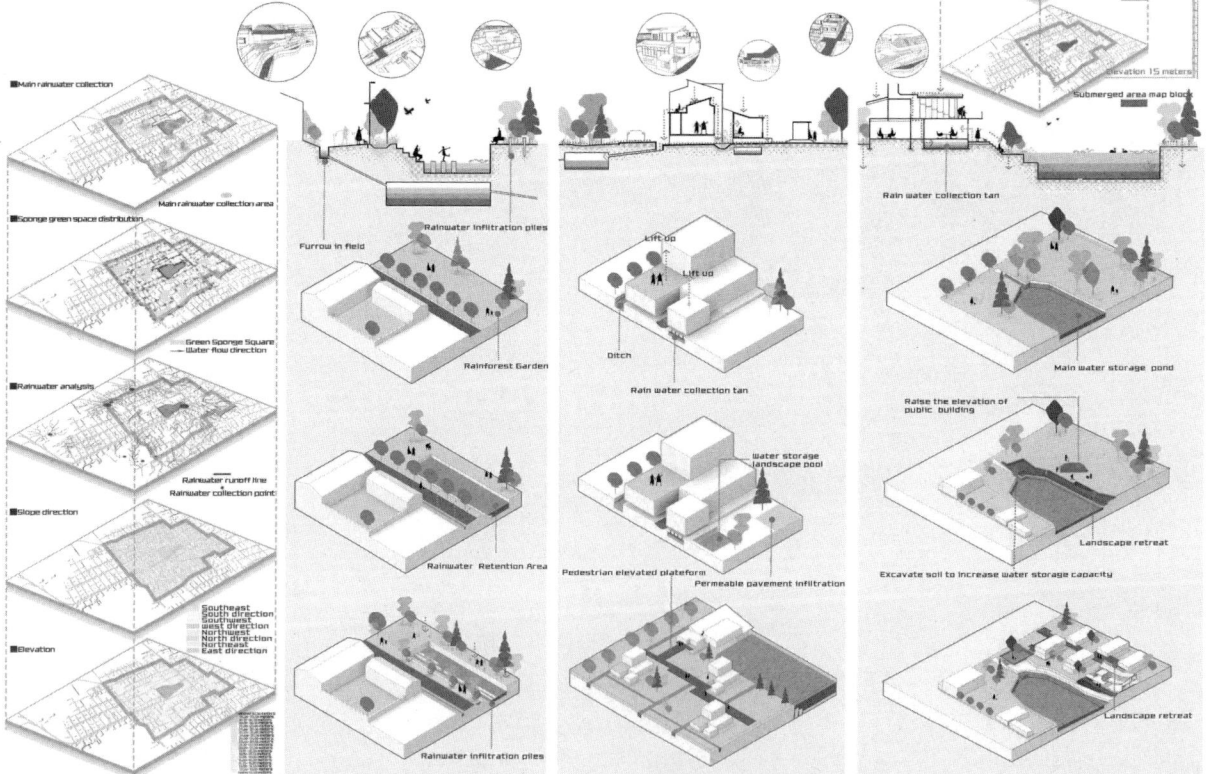

Furrow in field

Rainwater infiltration piles

Rainforest Garden

Rainwater Retention Area

Rainwater infiltration piles

Lift up

Lift up

Ditch

Rain water collection tan

Water storage landscape pool

Pedestrian elevated plateform

Permeable pavement infiltration

Rain water collection tan

Main water storage pond

Raise the elevation of public building

Landscape retreat

Excavate soil to increase water storage capacity

Landscape retreat

CODE:110487-100

溯溪而下 · 逆洪而上
Trace the River Downstream and Upstream Against the Tide IV

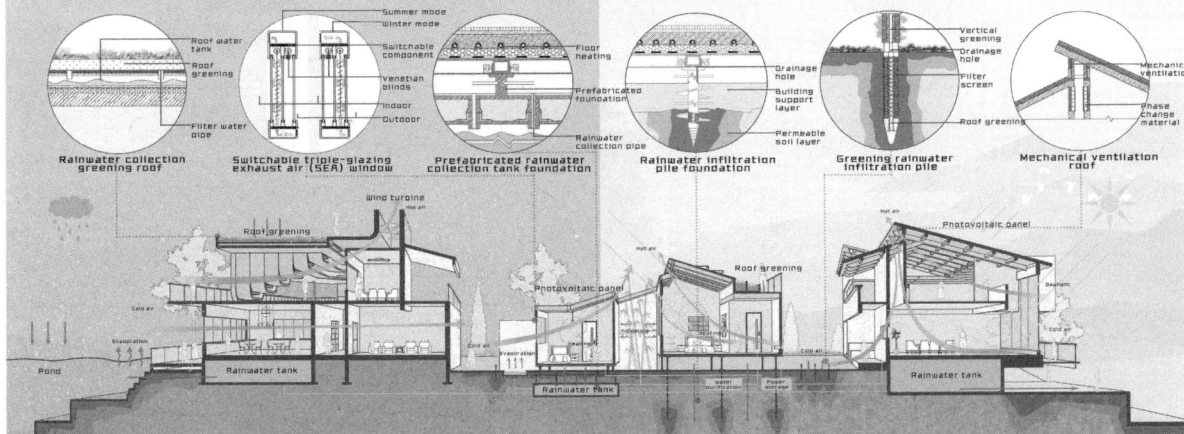

Rainwater collection greening roof | Switchable triple-glazing exhaust air (SEA) window | Prefabricated rainwater collection tank foundation | Rainwater infiltration pile foundation | Greening rainwater infiltration pile | Mechanical ventilation roof

■ Concept of Technology

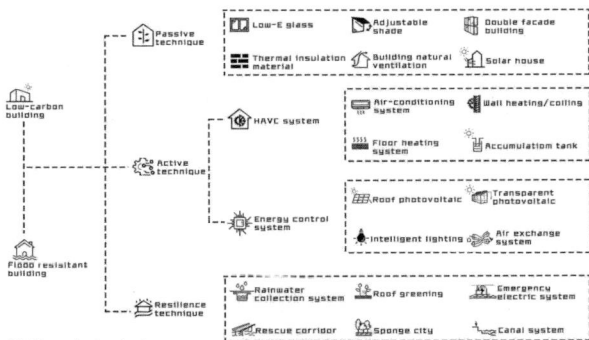

- Passive technique
- Active technique
- Energy control system
- Resilience technique

Low-carbon building

Flood resistant building

- Low-E glass
- Adjustable shade
- Double facade building
- Thermal insulation material
- Building natural ventilation
- Solar house
- HAVC system
- Air-conditioning system
- Wall heating/coiling
- Floor heating system
- Accumulation tank
- Roof photovoltaic
- Transparent photovoltaic
- Intelligent lighting
- Air exchange system
- Rainwater collection system
- Roof greening
- Emergency electric system
- Rescue corridor
- Sponge city
- Canal system

■ Builoing Fuction Conversion and Havc System

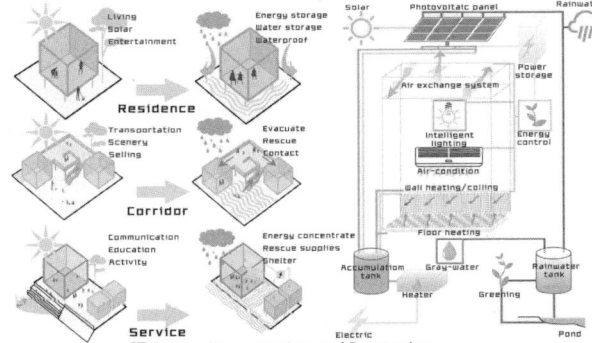

Residence — Living, Solar, Entertainment / Energy storage, Water storage, Waterproof

Corridor — Transportation, Scenery, Selling / Evacuate, Rescue, Contact

Service — Communication, Education, Activity / Energy concentrate, Rescue supplies, shelter

■ Climatic Analysis

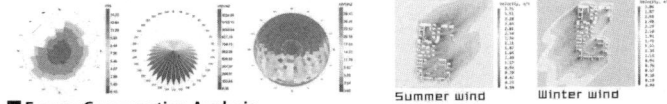

Summer wind | Winter wind

■ Energy Consumption Analysis

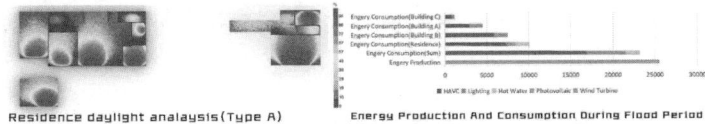

Energy Consumption(Building C)
Energy Consumption(Building A)
Energy Consumption(Building B)
Energy Consumption(Residence)
Energy Consumption(Sum)
Energy Production

HAVC ■ Lighting ■ Hot Water ■ Photovoltaic ■ Wind Turbine

Residence daylight analaysis(Type A)

Energy Production And Consumption During Flood Period

■ Energy Consumption and Prooution

■ Residence(Type A)

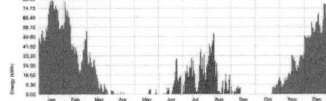

BASE BUILDING ANNUAL HVAC ENERGY CONSUMPTION

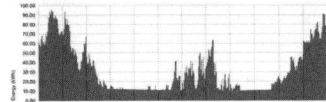

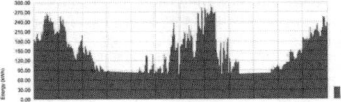

BASE BUILDING ANNUAL TOTAL ENERGY CONSUMPTION

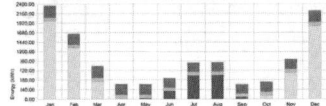

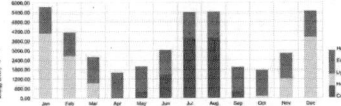

MONTHLY BUILDING ENERGY CONSUMPTION COMPOSITION

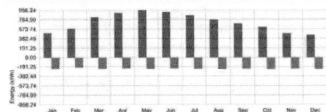

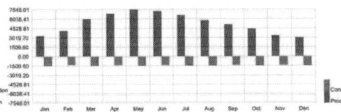

NEW BUILDING ANNUAL ELECTRICITY CONSUMPTION AND PRODUCTION

■ Service building(Building B)

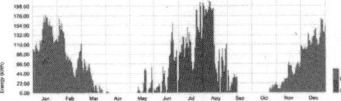

■ Enery Consumption and Prooution

Review of carbon emission intensity indicators (Type A)			
Project	Value	Standard Requirements	Result
Reduction In Carbon Emission intensity	50.4	≥7	YES
Carbon Emission Intensity Reduction Rate	57%	≥40%	YES

CARBON SEQUESTRATION				
	Residence	Building A	Open Ground	Sum
Green Area (m²)	624	108	836	1568
Carbon sequestration per unit area (kg/m²)	6.421	6.421	3.462	/
Carbon sequestration (kg)	4006.704	693.468	2894.232	7594.404

ENERGY CONSUMPITION COMPARISON						
	EHENGY TYPE	HAVC	LIGHTING	ELECTRIC EQUIPMEMT	HOT WATER	SUM TOTAL
Desicn Building (TYPE A)	Total Energy Consumption (kw·h)	7758	3472	2330	2971	16531
	Carbon emissions (kgCO₂e/a)	6128.82	2742.88	1840.7	2347.09	13059.49
Old Residential Building	Total Energy Consumption (kw·h)	9258	3472	2330	2971	18031
	Carbon emissions (kgCO₂e/a)	7313.82	2742.88	1840.7	2347.09	14244.49

TOTAL ENERGY CONSUMPTION AND PRODUCTION					
	Residence	Building A	Building B	Building C	Sum
Total Energy Production (kw·h)	130148	8400	43024	31193	212765
Total Energy Consumption (kw·h)	196718	17672	39921	8360	262671
Carbon emissions reduction (kgCO₂e/a)	102816.9	6636	33988.96	24642.47	168084.4
Carbon emissions (kgCO₂e/a)	155407.2	13960.88	31537.59	6604.4	207510.1
Photovoltaic area (m²)	1440				

PMV of Type A			
Room	Elderly Bedroom	Master bedroom	Bedroom
Percentage	0.66	0.63	0.8
Room	Study	Living room	Studio
Percentage	0.79	0.81	0.61

综合奖·入围奖·保定市清苑区李八庄村建设项目
Comprehensive Awards — Nomination — Libazhuang Village Project，Qingyuan District，Baoding City

注册号：110501
Register Number：110501
项目名称：方院·曲塘
Entry Title：Square Courtyard — Winding Pond
作者：蔡锶琦、郑泽科、袁鑫桐、殷韶璟、王依然、吴树祺、毛明俊
Authors：Cai Siqi, Zheng Zeke, Yuan Xintong, Yin Shaojing, Wang Yiran, Wu Shuqi, and Mao Mingjun
作者单位：广东工业大学
Authors from：Guangdong University of Technology
指导教师：吉慧、王平、邓寄豫、董泽豪
Tutors：Ji Hui, Wang Ping, Deng Jiyu, and Dong Zehao
指导教师单位：广东工业大学
Tutors from：Guangdong University of Technology

保定李八庄村韧性绿色设计
方院·曲塘
Square Courtyard — Winding Pond

设计说明(Design Description)：
项目位于华北平原的保定市李八庄村，包括14栋农宅和1处村民中心，为应对内涝灾害方案采用了海绵城市的方法，以场地东南侧低洼的大水塘为天然储水池，利用村中地势和规划的道路系统，开敞场地形成排蓄功能优良的生态防涝体系。青砖灰瓦的院落式建筑设计较好地适应了寒冷气候和传承了地域文化，主被动式相结合的绿色建筑技术和装配式模块设计方法，打造了健康舒适和节能降碳的人居环境。村民中心平时为提供集会、休闲活动的公共场所，灾时转变为紧急避难场所，充分体现了平急两用的设计理念。

The project is located in Libazhuang Village, Baoding City, North China Plain, including 14 farmhouses and 1 village center, in response to the waterlogging disaster, the sponge city approach is adopted, using the low-lying large pond in the southeast of the site as a natural water storage tank, using the topography of the village, the planned road system, and the open site to form an ecological flood prevention system with excellent drainage and storage functions. The courtyard building design of blue bricks and gray tiles better adapts to the cold climate and inherits the regional culture, and the combination of active and passive green building technology and prefabricated module design method creates a healthy, comfortable, energy-saving and carbon-reducing living environment. The village center is usually a public place for gatherings and leisure activities, and it is transformed into an emergency shelter during disasters, which fully embodies the design concept of dual-use in peacetime and emergency.

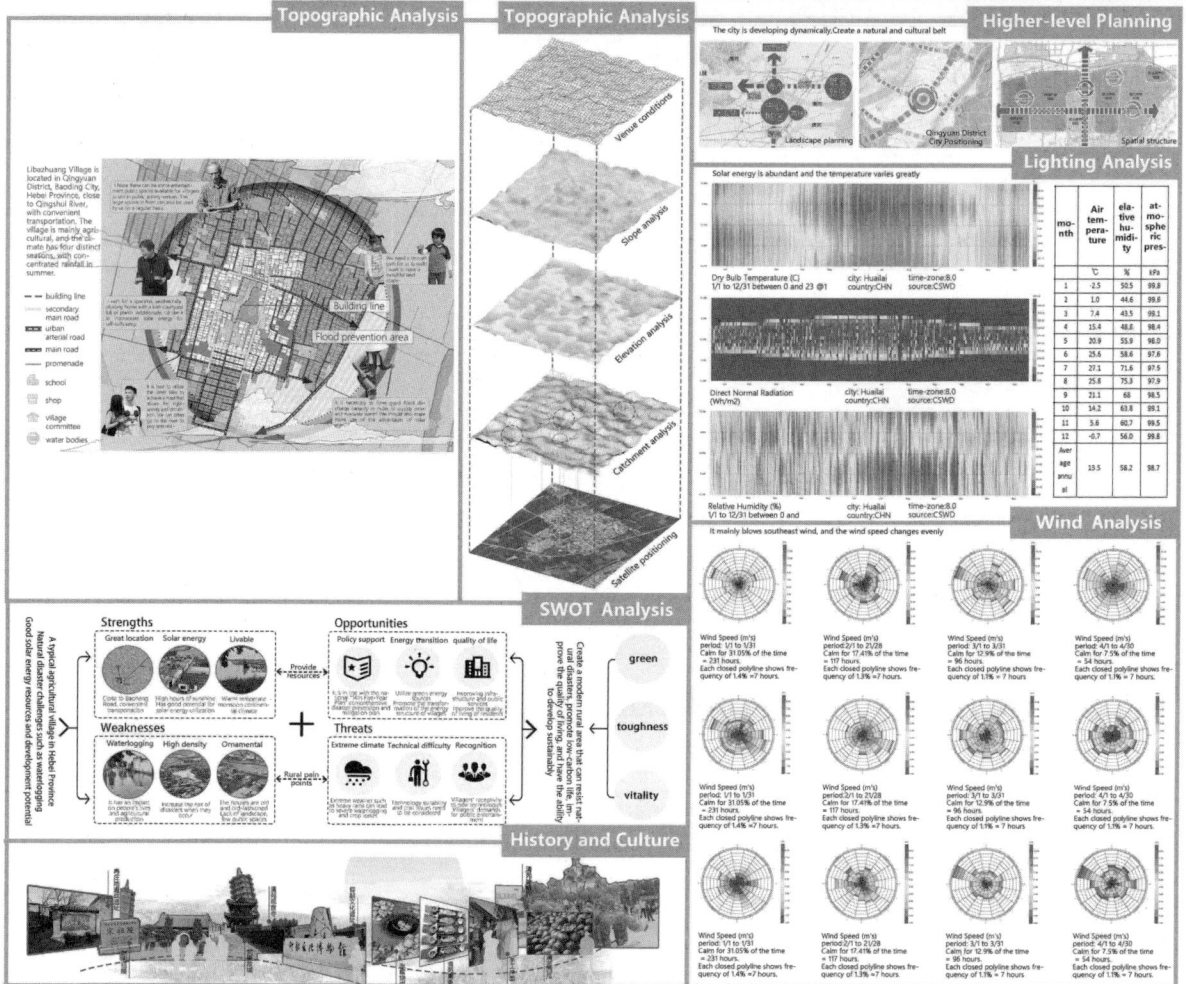

Topographic Analysis | Topographic Analysis | Higher-level Planning

Lighting Analysis

Solar energy is abundant and the temperature varies greatly

	month	Air temperature	relative humidity	atmospheric pressure
		℃	%	kPa
	1	-2.5	50.5	99.8
	2	1.0	44.6	99.8
	3	7.4	43.5	99.1
	4	15.4	48.8	98.4
	5	20.9	55.9	98.0
	6	25.6	58.6	97.6
	7	27.1	71.6	97.5
	8	25.8	75.3	97.9
	9	21.1	68	98.5
	10	14.2	63.8	99.1
	11	5.8	60.7	99.5
	12	-0.7	56.0	99.8
	Average annual	13.5	58.2	98.7

Wind Analysis

It mainly blows southeast wind, and the wind speed changes evenly

SWOT Analysis

Strengths
Great location | Solar energy | Livable

Opportunities
Policy support | Energy transition | quality of life

green
toughness
vitality

Weaknesses
Waterlogging | High density | Ornamental

Threats
Extreme climate | Technical difficulty | Recognition

History and Culture

Design the Texture

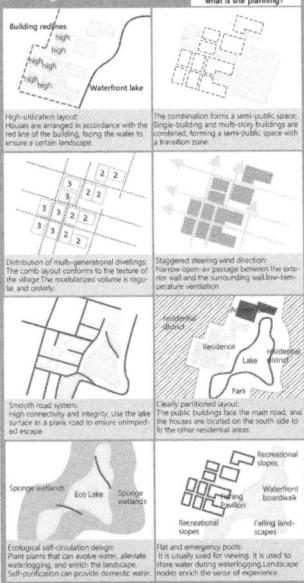

what is site planning?

Building redlines
high
high
high
high
Waterfront lake

High-utilization layout:
Houses are arranged in accordance with the red line of the building, facing the water to ensure a certain landscape.

The combination forms a semi-public space: Single-building and multi-story buildings are combined, forming a semi-public space with a transition zone.

Distribution of multi-generational dwellings:
The comb layout conforms to the texture of the village. The modularized volume is regular and orderly.

Staggered steering wind direction:
Narrow open-air passage between the exterior wall and the surrounding wall low-temperature ventilation

Residential district
Residence
Lake
Residential district
Park

Smooth road system:
High connectivity and integrity. Use the lake surface as a plank road to ensure unimpeded escape.

Clearly partitioned layout:
The public buildings face the main road, and the houses are located on the south side to the other residential areas.

Sponge wetlands
Eco Lake
Sponge wetlands

Recreational slopes
Waterfront boardwalk
Recreational slopes
Falling landscapes

Ecological self-circulation design:
Plant plants that can evolve water, alleviate waterlogging, and enrich the landscape. Self-purification can provide domestic water.

Flat and emergency pools:
It is usually used for viewing. It is used to store water during waterlogging. Landscape nodes enrich the sense of experience.

方院 · 曲塘
Square Courtyard — Winding Pond

Explosion Analysis

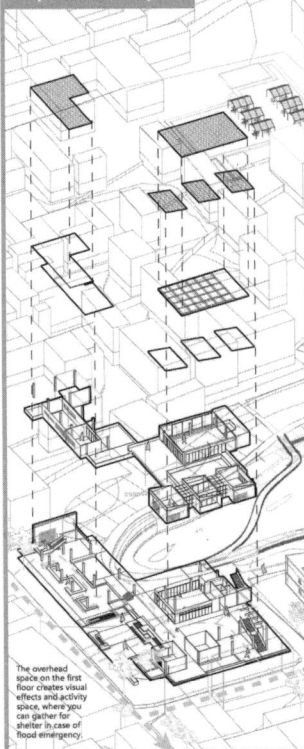

The overhead space on the first floor creates visual effects and activity space, where you can gather for shelter in case of flood emergency.

Daily Floor Plan

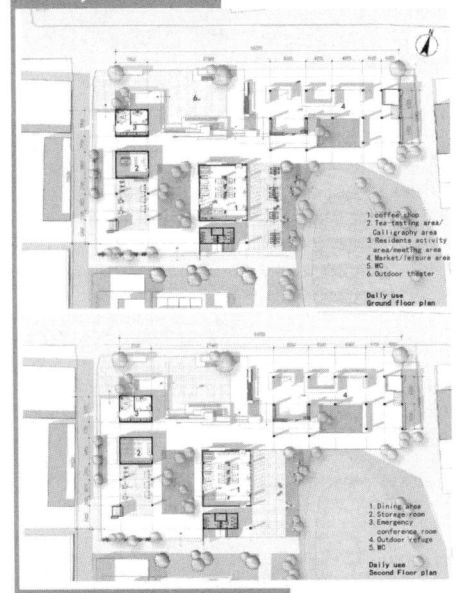

1. coffee shop
2. Tea-tasting area/ Calligraphy area
3. Residents activity area/meeting area
4. Market/leisure area
5. WC
6. Outdoor theater

Daily use
Ground floor plan

1. Dining area
2. Storage room
3. Emergency conference room
4. Outdoor refuge
5. WC

Daily use
Second floor plan

Emergency Floor Plan

1. indoor stadium
2. Pool room
3. playroom
4. Chess and card room
5. Self-service tea room
6. Conference room
7. Viewing deck

Emergency use
Second Floor plan

1. Emergency accommodation area
2. Medical treatment room
3. Rest room
4. Medical observation room
5. Self-service tea room
6. Viewing deck

Emergency use
Second Floor plan

Elevations of Public Buildings

8.504
7.450
4.300
3.300
±0.000
-0.250

North elevation of the residents' activity center 1 : 200

Site Sectioning

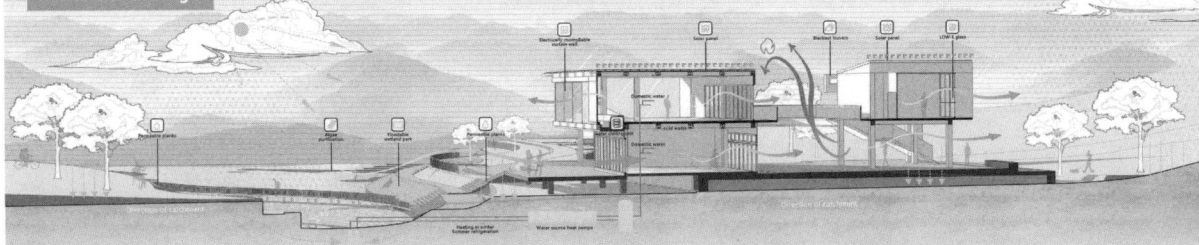

110501

Renderings

Node design

Economic and technical indicators
Building density：0.34
The total area of the building：5202 ㎡
Floor space：3664 ㎡
Green area：4658 ㎡
Green space rate：53.1%
Floor area ratio：0.59

Water-storing plants

Reservoir (Emergency)
Skateboard Pool (Daily)

Garden Square

Architectural design red lines

Reservoir (Emergency)
Landscaped fishing pond (Daily)

Sports and fitness plaza

parking lot

Spice Garden

parking lot

方 院 · 曲 塘
Square Courtyard —Winding Pond

N

2nd Generation of the Same House

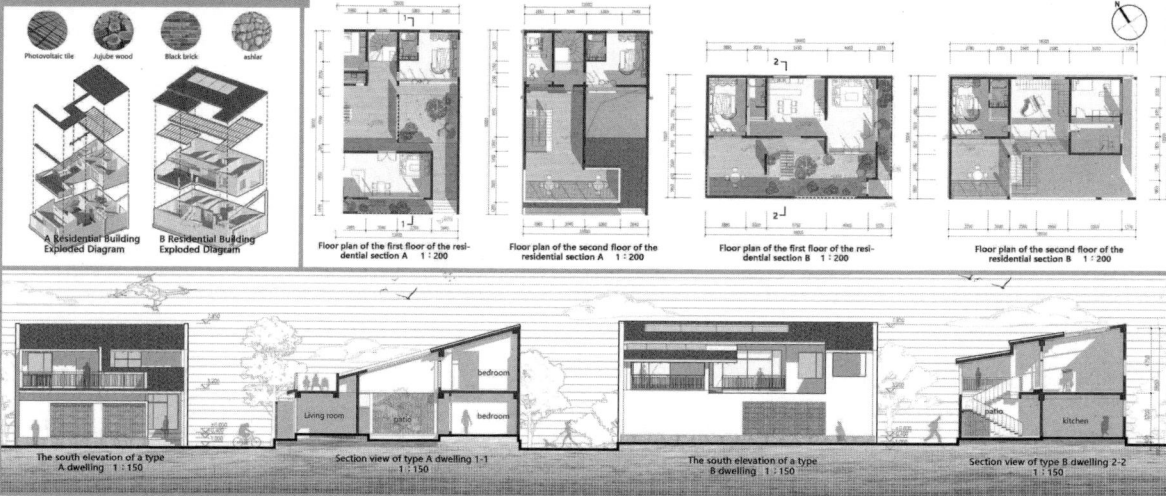

Photovoltaic tile   Jujube wood   Black brick   ashlar

A Residential Building
Exploded Diagram

B Residential Building
Exploded Diagram

Floor plan of the first floor of the residential section A   1：200

Floor plan of the second floor of the
residential section A   1：200

Floor plan of the first floor of the residential section B   1：200

Floor plan of the second floor of the
residential section B   1：200

bedroom

bedroom

Living room   patio   bedroom

patio   kitchen

The south elevation of a type
A dwelling   1：150

Section view of type A dwelling 1-1
1：150

The south elevation of a type
B dwelling   1：150

Section view of type B dwelling 2-2
1：150

3nd Generation of the Same House

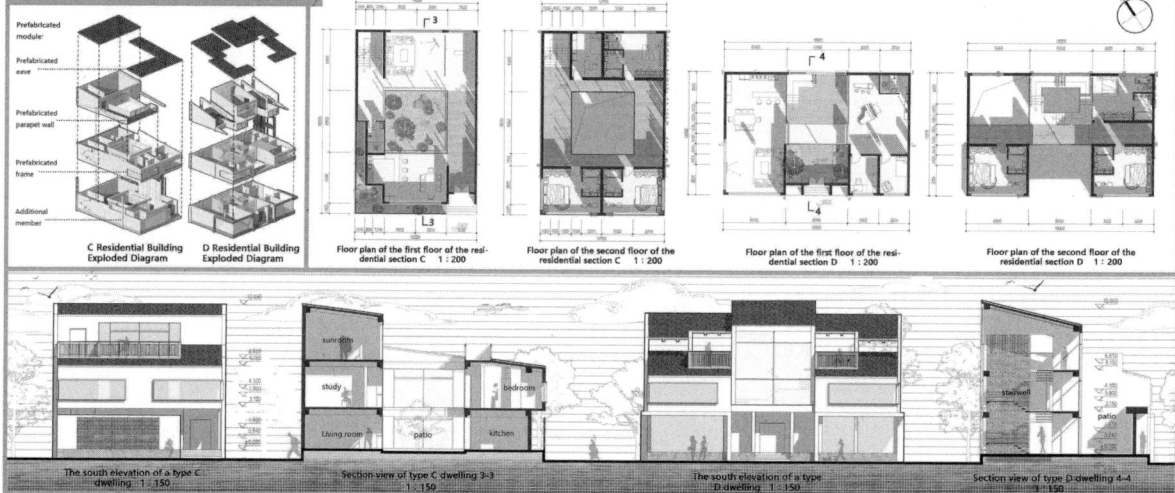

Prefabricated
module

Prefabricated
eave

Prefabricated
parapet wall

Prefabricated
frame

Additional
member

C Residential Building
Exploded Diagram

D Residential Building
Exploded Diagram

Floor plan of the first floor of the residential section C   1：200

Floor plan of the second floor of the
residential section C   1：200

Floor plan of the first floor of the residential section D   1：200

Floor plan of the second floor of the
residential section D   1：200

sunroom

study

bedroom

Living room   patio   kitchen

stairwell

patio

The south elevation of a type C
dwelling   1：150

Section view of type C dwelling 3-3
1：150

The south elevation of a type
D dwelling   1：150

Section view of type D dwelling 4-4
1：150

方院·曲塘
Square Courtyard – Winding Pond

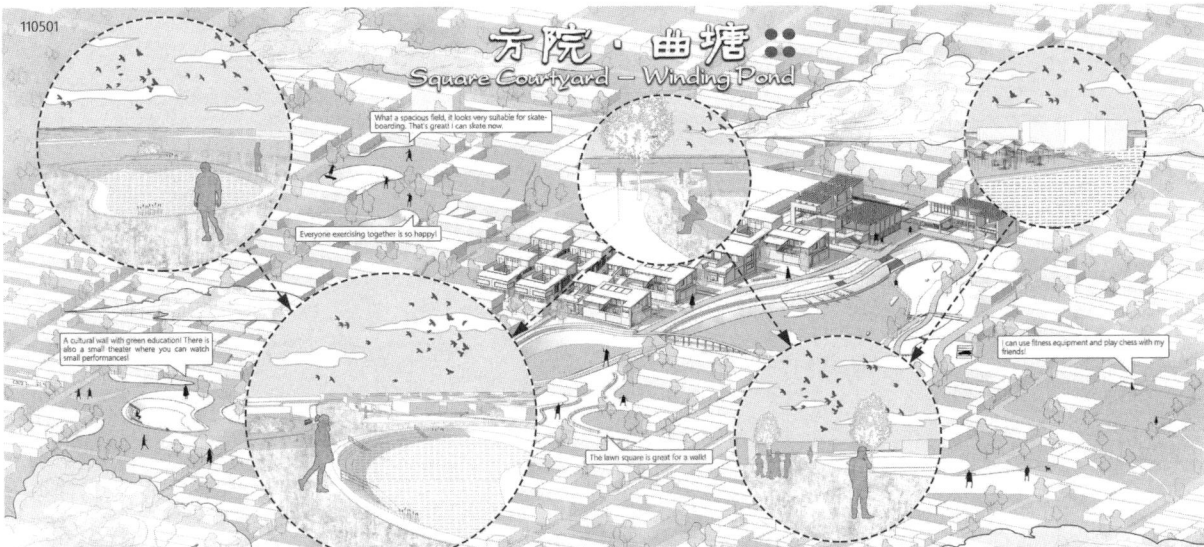

What a spacious field, it looks very suitable for skateboarding. That's great! I can skate now.

Everyone exercising together is so happy!

A cultural wall with green education! There is also a small theater where you can watch small performances!

I can use fitness equipment and play chess with my friends!

The lawn square is great for a walk!

Device Analysis

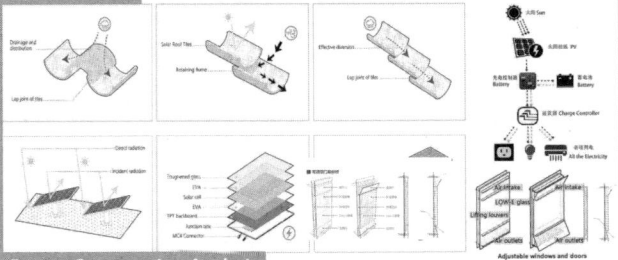

Green Technology

Ventilation
Daylighting
Roofgarden
Heat Insulation
Solar energy
Rain harvesting
Biomass energy
Floor panel heating

Passive solar energy utilization

Active solar energy utilization

Resilience Nodes

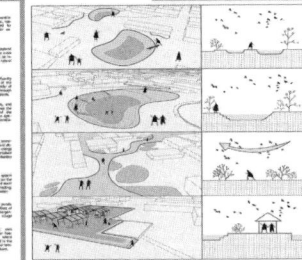

Energy Consumption Analysis

Before modification

Monthly energy consumption curve

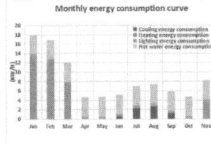

After transformation

Monthly energy consumption curve

Site wind environment analysis

The section plane is 1.2m wind velocity diagram

The YZ axial profile is the wind speed in the west of the site

The YZ axial profile is the wind speed in the middle of the site

The YZ axial profile is the wind speed map in the east of the

The section plane is 1.5m wind velocity diagram

The XZ axis profile is the wind speed diagram on the west side

The XZ axis profile is the wind speed diagram on the middle

The section plane is 1.8m wind velocity diagram

The XZ axis profile is the wind speed diagram on the east side

Daylighting performance analysis

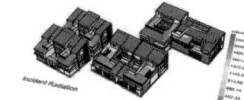

Second generation lighting illumination diagram

Three generations of daylighting diagram

Second generation lighting illumination diagram

Three generations of daylighting diagram

Solar radiation analysis

Incident Radiation

Second generation building,Three generations of building

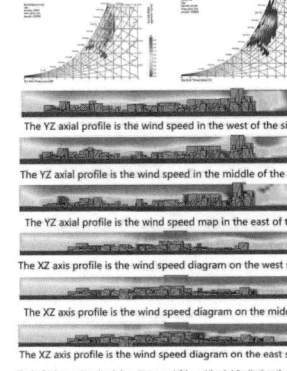

Solar radiation from public buildings

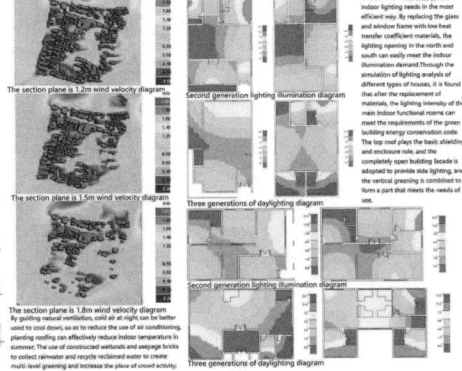

Incident Radiation

Plants are Matched with the Elevation of the Site

field | villiage | pond | villiage | field | river

CODE: 100-110564

综合奖 · 入围奖 · 保定市清苑区李八庄村建设项目
Comprehensive Awards — Nomination — Libazhuang Village Project, Qingyuan District, Baoding City

注册号：110564
Register Number：110564

项目名称：韧性家园 · 太阳绿居
Entry Title：Resilient Haven — Solar Oasis

作者：阿博莱 · 阿勒玛斯、孙君荟、巴彦 · 塞尔江、麦迪娜 · 马合木提、张疆慧、宋志辉

Authors：Abolai.Alemas, Sun Junhui, Bayan.Saierjiang, Maidina. Mahemuti, Zhang Jianghui, and Song Zhihui

作者单位：新疆大学
Authors from：Xinjiang University

指导教师：塞尔江 · 哈力克
Tutor：Saierjiang. Halike

指导教师单位：新疆大学
Tutor from：Xinjiang University

1 韧性家园·太阳绿居
Resilient Haven—Solar Oasis

Solar Building Design, Libazhuang Village, Qingyuan District, Baoding
河北省保定市清苑区李八庄村太阳能建筑设计

Location

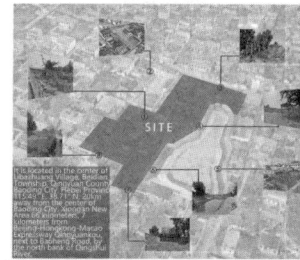

History

Effect drawing

设计说明：项目地位于河北省保定市清苑区北店乡李八庄村中心，通过规划设计对抗内涝范围内道路、空地进行韧性设计，提出针对正常降雨情况的预防措施，及应对洪涝灾害的紧急响应策略。引入海绵乡村概念，提升当地居民居住品质，采用装配式模块化的设计，高效、灵活，同时提高建筑的可持续性。屋顶使用与传统瓦片高度相似的光伏瓦片，高效地将太阳能转化为电能，满足建筑能源需求，降低碳排放。阳光房，太阳地坑，雨水收集等设计增加室内舒适度、节约能源。以及建筑在应急端情况下具备一定自维持能力。

Design Description: The project is located in the center of Libazhuang Village, Baoding City, Hebei Province. By planning and designing roads and open spaces within the scope of anti-waterlogging, resilience design is carried out, and preventive measures against normal rainfall and emergency response strategies for flood disasters are proposed. The sponge village concept is introduced to improve the living quality of local residents, and the prefabricated modular design is efficient and flexible, while improving the sustainability of the building. The roof uses photovoltaic tiles, which are highly similar to traditional tiles, to efficiently convert solar energy into electricity, meet the building's energy needs, and reduce carbon emissions. Sun room, solar kang, rainwater collection and other designs increase indoor comfort and save energy, and the building has a certain self-sustaining ability under extreme emergency conditions.

Current situation Identification → **Problem Analysis** → **Concept** → **Goals and Strategies**

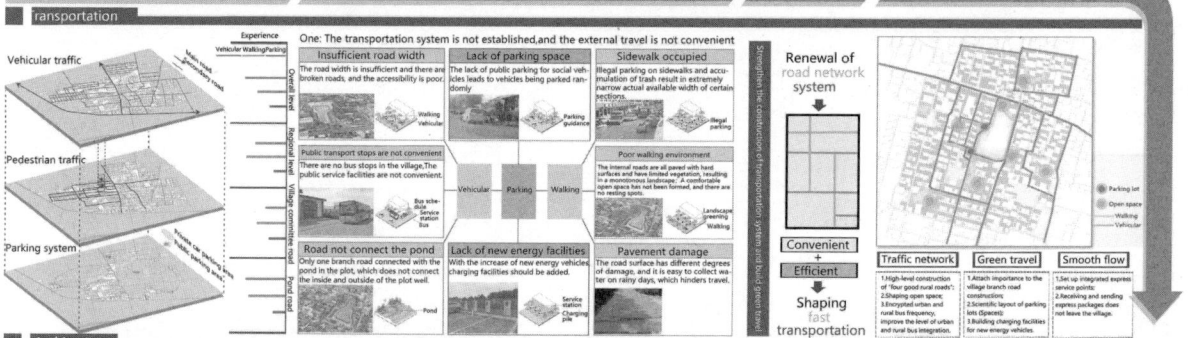

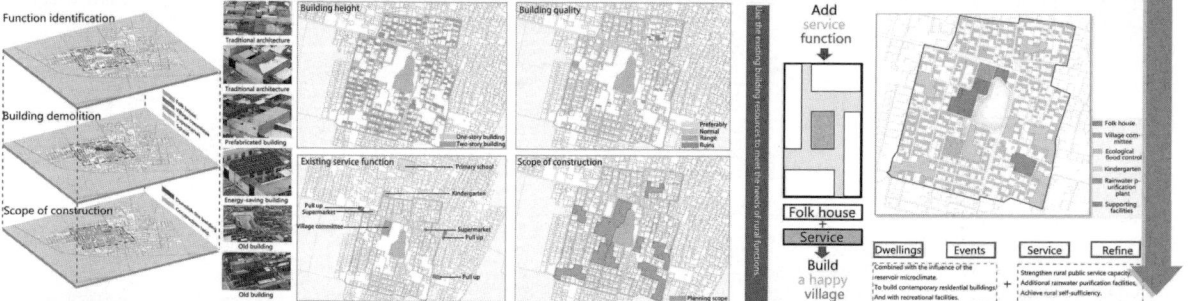

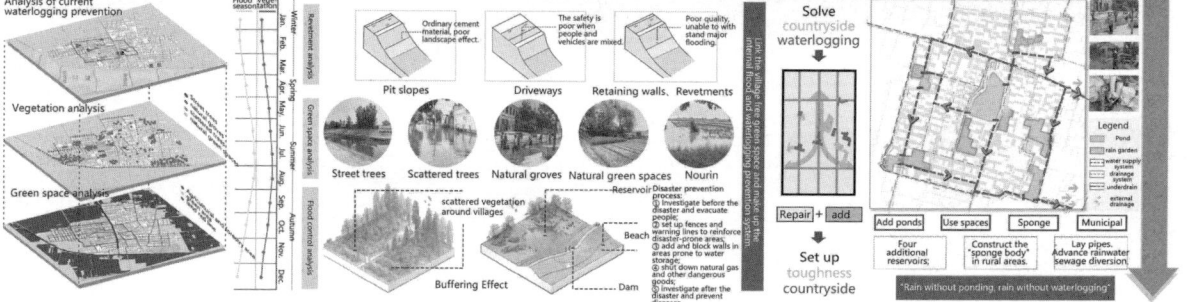

2 韧性家园·太阳绿居
Resilient Haven−Solar Oasis

Solar Building Design, Libazhuang Village, Qingyuan District, Baoding
河北省保定市清苑区李八庄村太阳能建筑设计

Theme Interpretation

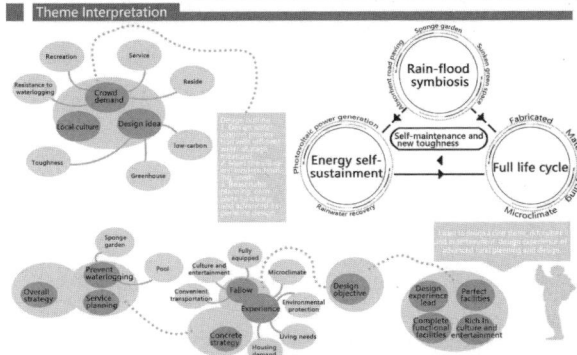

Climate Analysis

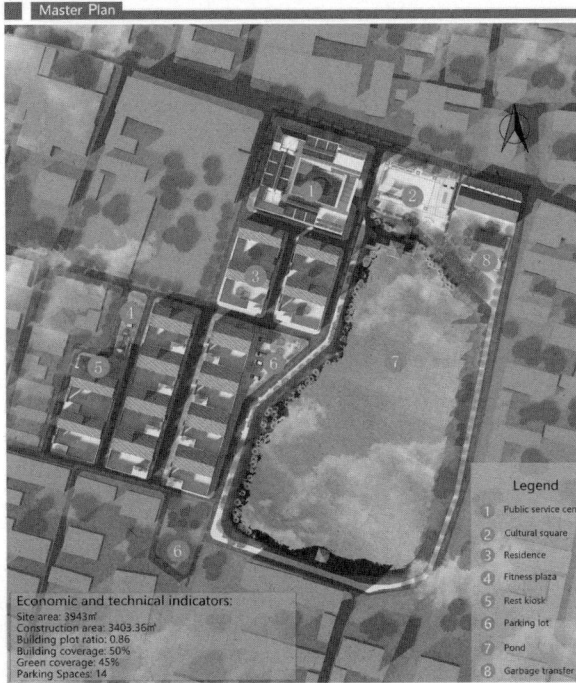

Waterlogging Prevention area

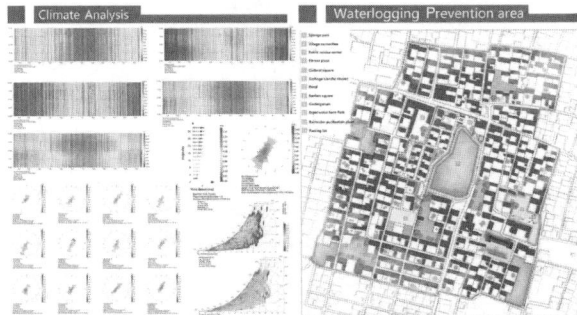

Master Plan

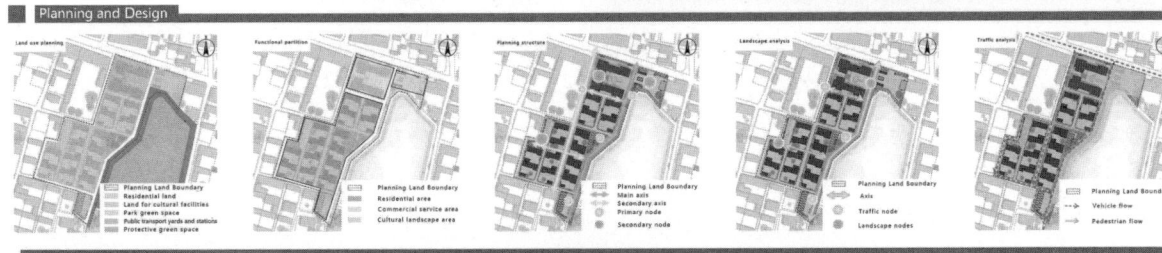

Economic and technical indicators:
Site area: 3943㎡
Construction area: 3403.36㎡
Building plot ratio: 0.86
Building coverage: 50%
Green coverage: 45%
Parking Spaces: 14

Legend
1 Public service center
2 Cultural square
3 Residence
4 Fitness plaza
5 Rest kiosk
6 Parking lot
7 Pond
8 Garbage transfer

Planning and Design

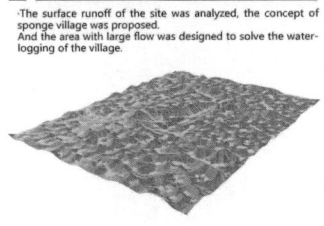

Land use planning | Functional partition | Planning structure | Landscape analysis | Traffic analysis

Sponge Country · Rain−Flood Symbiosis

Stormwater Runoff Analysis

·The surface runoff of the site was analyzed, the concept of sponge village was proposed.
And the area with large flow was designed to solve the waterlogging of the village.

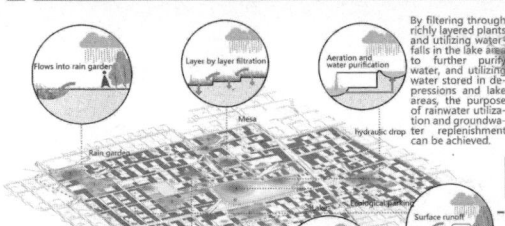

Pipeline planning diagram

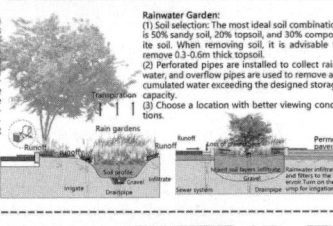

Water supply pipeline | Drainage pipeline planning

Power and telecommunications pipelines | Heating gas pipeline

Sponge Village

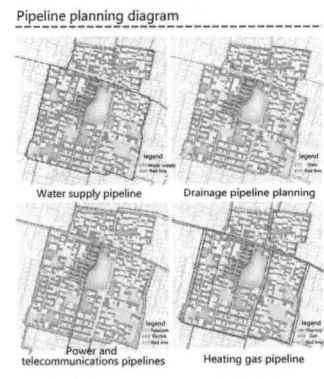

By filtering through richly layered plants and utilizing water falls in the lake areas to further purify water, and utilizing water stored in depressions and lake areas, the purpose of rainwater utilization and groundwater replenishment can be achieved.

The site reduces surface runoff through applications such as planting grass bricks and permeable paving.

Plant Configuration

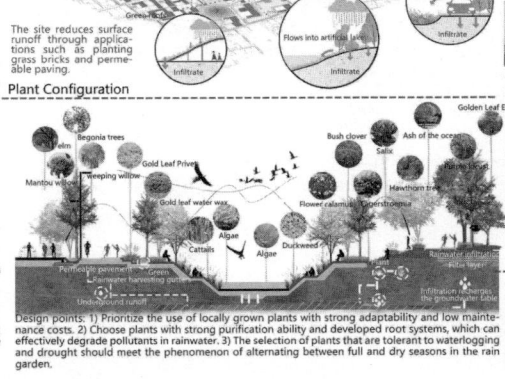

Design points: 1) Prioritize the use of locally grown plants with strong adaptability and low maintenance costs. 2) Choose plants with strong purification ability and developed root systems, which can effectively degrade pollutants in rainwater. 3) The selection of plants that are tolerant to waterlogging and drought should meet the phenomenon of alternating between full and dry seasons in the rain garden.

Rainwater Garden

Rainwater Garden:
(1) Soil selection: The most ideal soil combination is 50% sandy soil, 20% topsoil, and 30% composite soil. When removing soil, it is advisable to remove 0.3-0.6m thick topsoil.
(2) Perforated pipes are installed to collect rainwater, and overflow pipes are used to remove accumulated water exceeding the designed storage capacity.
(3) Choose a location with better viewing conditions.

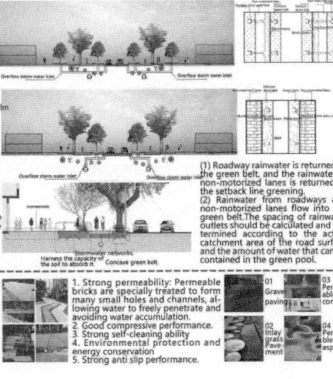

(1) Roadway rainwater is returned to the green belt, and the rainwater of non-motorized lanes is returned to the setback line greening.
(2) Rainwater from roadways and non-motorized lanes flow into the green belt. The spacing of rainwater outlets should be calculated and determined according to the actual catchment area of the road surface and the amount of water that can be contained in the green pool.

1. Strong permeability: Permeable bricks are specially treated to form many small holes and channels, allowing water to freely penetrate and avoiding water accumulation.
2. Good compressive performance.
3. Strong self-cleaning ability.
4. Environmental protection and energy conservation.
5. Strong anti slip performance.

01 Gravel paving
02 Inlay Pavement
03 Permeable concrete
04 Permeable asphalt

CODE: 100-110564

3 韧性家园·太阳绿居
Resilient Haven—Solar Oasis

Solar Building Design, Libazhuang Village, Qingyuan District, Baoding
河北省保定市清苑区李八庄村太阳能建筑设计

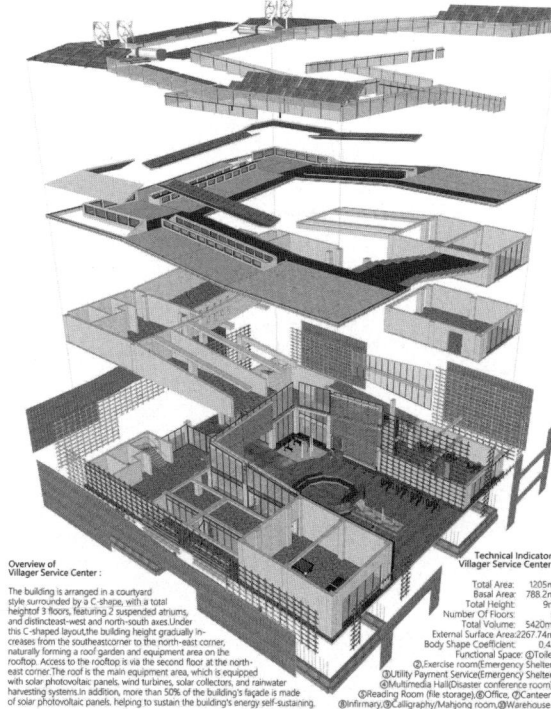

Passive Technical

Solar Thermal:
The solar collector system is designed into the design, with two 6×6m solar panels on the roof to provide a source of heat for the entire building and provide hot water bathing services for residents in times of disaster.

Ground Coupled Heat Exchanger:
Hebei Province is the most abundant region in eastern China, so it is only natural that the ground source heat pump system is designed in this building to provide heating to the building in winter. In summer, the building is air-conditioned and cooled.

Natural Daylight:
The building is distinguished by high side windows and high windows, which are symmetrically placed in the corners and middle of the building in combination with the floor plan, function and ventilation needs, providing comfortable daylight conditions.

Axis Distribution:
two axes are reinforced in this building, one is the east-west axis that directly passes through the "street building" into the ecological wetland, and the other is the north-south axis perpendicular to the main road and shuttles through the various functional areas.

Green Roof:
In addition to the greening of the surrounding blocks, the roof is also set up with planting roofs along the landscape axis of the roof, which can not only serve as a green belt, but also serve as a platform for villagers to display excellent varieties of crops.

Rain Water Harvest:
Since the roof is a platform that descends in a semi-circular manner, the rainwater collection from the roof naturally converges at the low ends of the roof according to the drop. It is also delivered directly to the reclaimed water treatment system inside the building

Sunlight Analysis

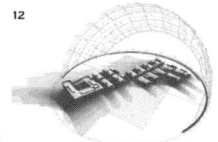
12
December Shadow Trajectory

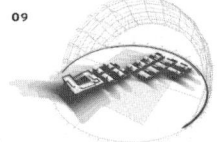
09
September Shadow Trajectory

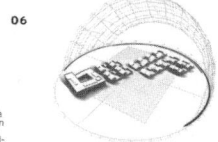
06
June Shadow Trajectory

Overview of Villager Service Center :

The building is arranged in a courtyard style surrounded by a C-shape, with a total height of 3 floors, featuring 2 suspended atriums, and distinct east-west and north-south axis. Under this C-shaped layout, the building height gradually increases from the southeast corner to the north-east corner, naturally forming a roof garden and equipment area on the rooftop. Access to the rooftop is via the second floor at the north-east corner. The roof is the main equipment area, which is equipped with solar photovoltaic panels, wind turbines, solar collectors, and rainwater harvesting systems. In addition, more than 50% of the building's facade is made of solar photovoltaic panels, helping to sustain the building's energy self-sustaining.

Technical Indicators Villager Service Center :

Total Area: 1205㎡
Basal Area: 788.2㎡
Total Height: 9m
Number Of Floors: 3
Total Volume: 5420㎡
External Surface Area:2267.74㎡
Body Shape Coefficient: 0.42
Functional Space: ①Toilet
②Exercise room(Emergency Shelter)
③Utility Payment Service(Emergency Shelter)
④Multimedia Hall(Disaster conference room).
⑤Reading Room (file storage).⑥Office, ⑦Canteen,
⑧Infirmary,⑨Calligraphy/Mahjong room,⑩Warehouse.

Wind Environment

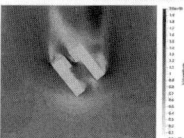

Microclimate Analysis:
The architectural design of the building harmoniously conforms to the dominant local wind patterns, featuring a graceful C-shaped silhouette that effortlessly channels summer breezes into the inviting courtyard. This courtyard, filled with refreshing cool air, ensures a continuous circulation of invigorating airflow throughout the northern and southern wings of the structure. At the heart of the C-shape lies an overhead multimedia lecture hall on the ground level, strategically positioned where the air duct narrows, accelerating the summer winds within the courtyard for an enhanced cooling effect. Furthermore, an ecological wetland gracefully connects the courtyard to the flood control pit, imparting a welcome humidity to the courtyard during the arid summer months.

Energy Self-Sustaning

Solar House
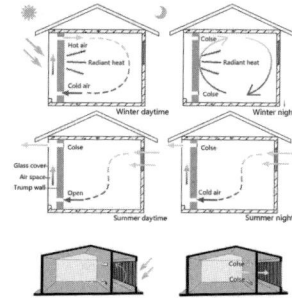

Winter daytime / Winter night / Summer daytime / Summer night

Solar Bed
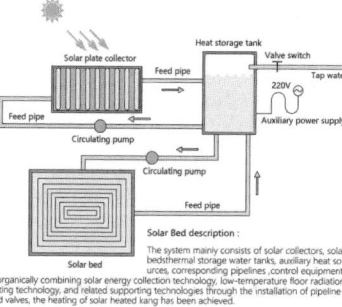

Solar Bed description :
The system mainly consists of solar collectors, solar bedsthermal storage water tanks, auxiliary heat sources, corresponding pipelines ,control equipment. By organically combining solar energy collection technology, low-temperature floor radiation heating technology, and related supporting technologies through the installation of pipeline-sand valves, the heating of solar heated kang has been achieved.

Construction Method
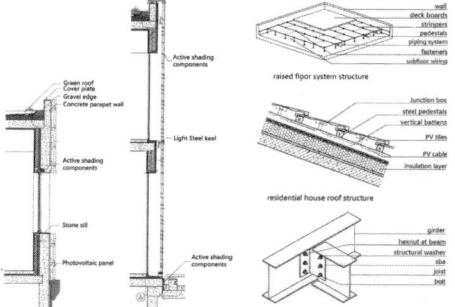

raised fipor system structure
residential house roof structure
prefabricated steel beam structure

Technology
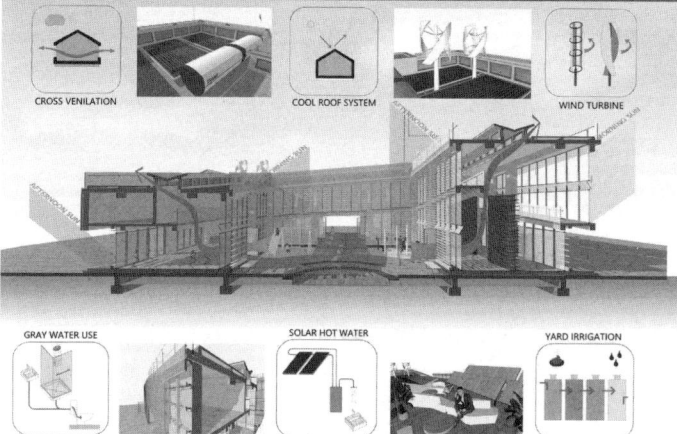

CROSS VENILATION
COOL ROOF SYSTEM
WIND TURBINE
GRAY WATER USE
SOLAR HOT WATER
YARD IRRIGATION

Rain Collection
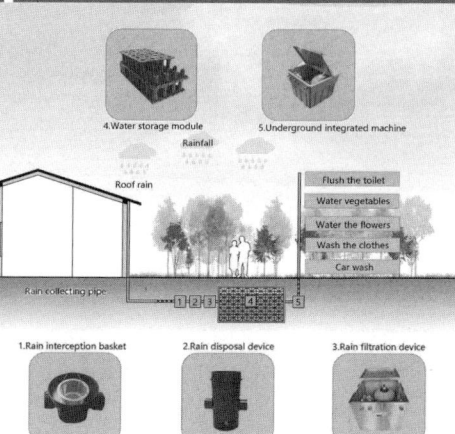

4.Water storage module
5.Underground integrated machine

Rainfall
Roof rain
flush the toilet
Water vegetables
Water the flowers
Wash the clothes
Car wash
Rain collecting pipe

1.Rain interception basket
2.Rain disposal device
3.Rain filtration device

4 韧性家园·太阳绿居
Resilient Haven—Solar Oasis

Solar Building Design, Libazhuang Village, Qingyuan District, Baoding
河北省保定市清苑区李八庄村太阳能建筑设计

Prefabricated Rebirth · Waste Utilization

Carbon emission

Wind & Light Simulation

Winter indoor airflow (center) Summer indoor airflow(center) Daylight autonomy(center)

Epidermal analysis

Method description:
Photovoltaic panels and active shading elements enhance the energy efficiency of the house by increasing the air layer and reducing the amount of sunlight incidence, respectively.

Winter indoor airflowF1(small set) Summer indoor airflowF1(small set) Daylight autonomyF1(small set)

Waste Utilization

Recycled material utilization:
Large number of houses in the village use color steel plate roofs, and the main cash crops in the village are produced by straw materials.
The construction of carbon reduction measures starts from these two local materials.

Interior surface
Straw-based planboards
Light steel keel
Straw-based planboards
External insulation
Profiled steel plate

Winter indoor airflowF1(large set) Summer indoor airflowF1(large set) Daylight autonomyF1(large set)

Winter indoor airflowF2(large set) Summer indoor airflowF2(large set) Daylight autonomyF2(large set)

Module Presentation

Modules

module A
area:9.68㎡

module B
area:14.52㎡

module C
area:19.36㎡

module D
area:10.89㎡

module E
area:7.26㎡

module F
area:4.84㎡

Modules in Residential Houses

Large set residential house

Small set residential house

■ private space   ■ public space   ■ garage

Modules Possibilities

Bedroom Modules   Kitchen Modules

Living room Modules   Staircase Modules   Bathroom Modules

Villager Service Center

First floor plan 1:200   Second floor plan 1:200   Roof plan 1:200

West elevation   South elevation   East elevation

Residential House

Small set plan 1:200   First floor plan of large set 1:200

Second floor plan of large set 1:200

2-2profile 1:200   1-1profile 1:200

Small set of front elevation   Small set of side elevation   Large set of front elevation   Large set of side elevation

作品编号：110682

单项奖·设计创意奖·张家口市怀安县二堡子村建设项目
Individual Award – Design Creativity Prize – Erpuzi Village Project，Huai'an County，Zhangjiakou City

注册号：110682
Register Number：110682
项目名称：稻香满园
Entry Title：Harvest Beneath Greenhouse
作者：董馨月、占瑶、王奕丹、许安堃、范宇飞、金宇帆
Authors：Dong Xinyue, Zhan Yao,
Wang Yidan, Xu Ankun,
Fan Yufei，and Jin Yufan
作者单位：华南理工大学
Authors from：South China University of Technology
指导教师：肖毅强、殷实
Tutors：Xiao Yiqiang, Yinshi
指导教师单位：华南理工大学
Tutors from：South China University of Technology

I 稻香满园
Harvest Beneath Greenhouse

Site Plan Analysis

Analysis of Traffic and Lanscape

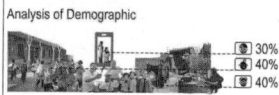

Disaster
we need：
earthquake, extremely cold, hail...

Analysis of Demographic
30%
40%
40%

Conception Analysis

Conceptual diagram

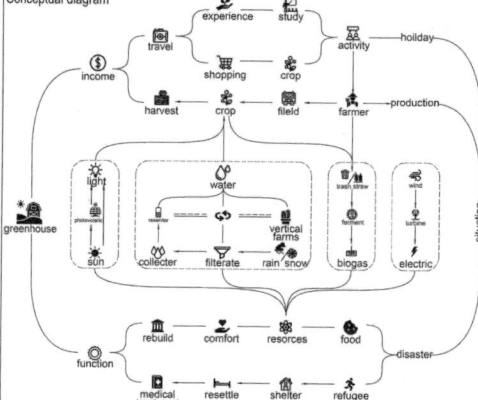

Revolution of Ruam

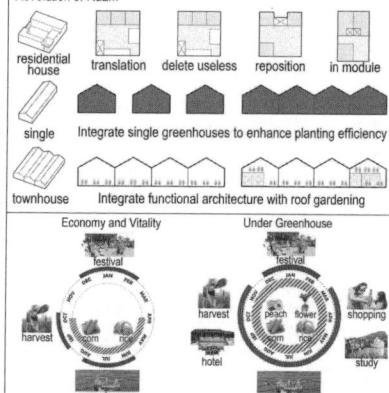

Economy and Vitality

Under Greenhouse

Design Specification

基于河北地区的气候特征，同时结合建筑节能抗灾乡村韧性建筑的必然选择与要求，选取"平灾结合，乡村韧性"为设计出发点，讨论主被动技术的平衡问题。

方案为乡村提供一种综合日常生产与生活双重性的新建筑范式，灾时在满足应急要求的同时，实现能源、资源和生产资料的自供给，设计可变家具灵活应对灾时需求。方案利用模块化可复制可量产钢木结构、自然通风、大棚阳光房等被动式技术，以及太阳能发电、风力发电、雨雪水收集净化、跨季节储能等主动式技术的整体系统，达到绿色低碳韧性建筑的要求。

Based on the climatic characteristics of Hebei Province, and considering the necessary choices and requirements for energy-efficient and disaster-resilient rural architecture, this design takes "integrating disaster preparedness and rural resilience" as its starting point and discusses the balance between active and passive technologies.

The proposal offers a new architectural paradigm that integrates daily production and living activities for rural areas. During disasters, it meets emergency needs while ensuring self-sufficiency in energy, resources, and production materials, featuring adaptable furniture to flexibly respond to disaster demands.

The plan utilizes a comprehensive system of modular, replicable, and mass-producible steel-wood structures, passive technologies like natural ventilation and sunrooms, as well as active technologies including solar power generation, wind power generation, rain and snowwater collection and purification, and inter-seasonal energy storage, in order to meet the requirements of green, low-carbon, and resilient architecture.

Project		Amount(kgCO₂)	Amount per unit area(kgCO₂/㎡)
Carbon emission	HVAC	-22146.0	28.32
	Lighting	-31004.4	11.07
	Life System	-14394.4	18.4
	Equipment	-43184.7	2159.2
Carbon reduction	Solar energy	137230.8	42486
Summation		26501.3	R/E	19.16

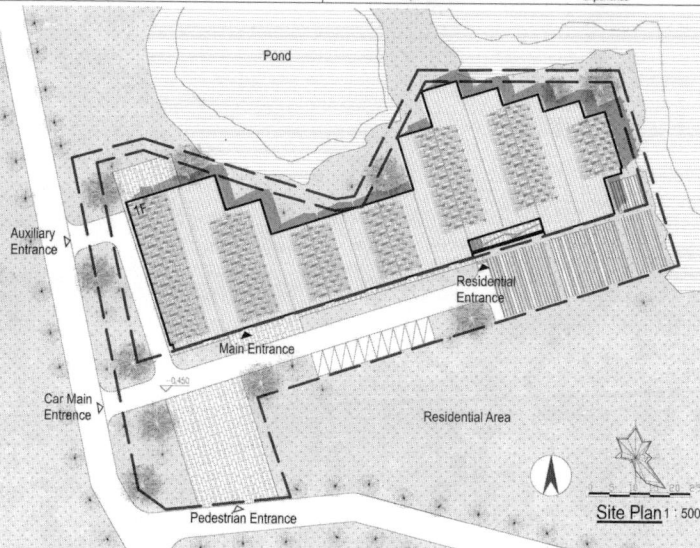

Pond
Auxiliary Entrance
1F
Residential Entrance
Main Entrance
Car Main Entrance
-0.450
Residential Area
Pedestrian Entrance
Site Plan 1：500

II 稻香满园

Harvest Beneath Greenhouse

Economic & Technical Indexes

Site area : 8260 m²
Greenhouse floor area : 3298 m²
Public usable area : 1125 m²
Homestay usable area : 588 m²
Cultivate usable area : 1786 m²
Building height : 8.2m
Floor area ratio : 0.38
Greening rate : 0.51

◉ Generation of Shape

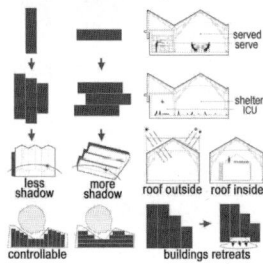

served
serve

shelter
ICU

less shadow
more shadow

roof outside
roof inside

controllable
buildings retreats

Generation of Planting Space   Variability

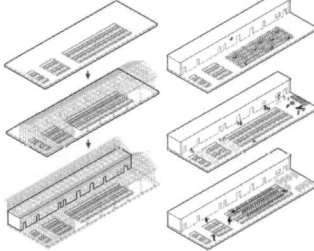

emergency
medical care
temporary bed
food supply

tourist season
market
theater restaurant
leisure space
......

production season
farm
auto irrigation
manual planting
hydroponic rack

◉ Plants List

FLOWERS & CROPS

SUNFLOWER TOMATO RICE PEONY RADISH

CORN COTTON RAPE CABBAGE JUJUBE

● VISUAL (FLOWER)
● TASTE (EDIBLE)
● HYDROPONICS
● STRONG ADAPTABILITY
● FAST GROWTH
● CARBON DIOXIDE ABSORPTION

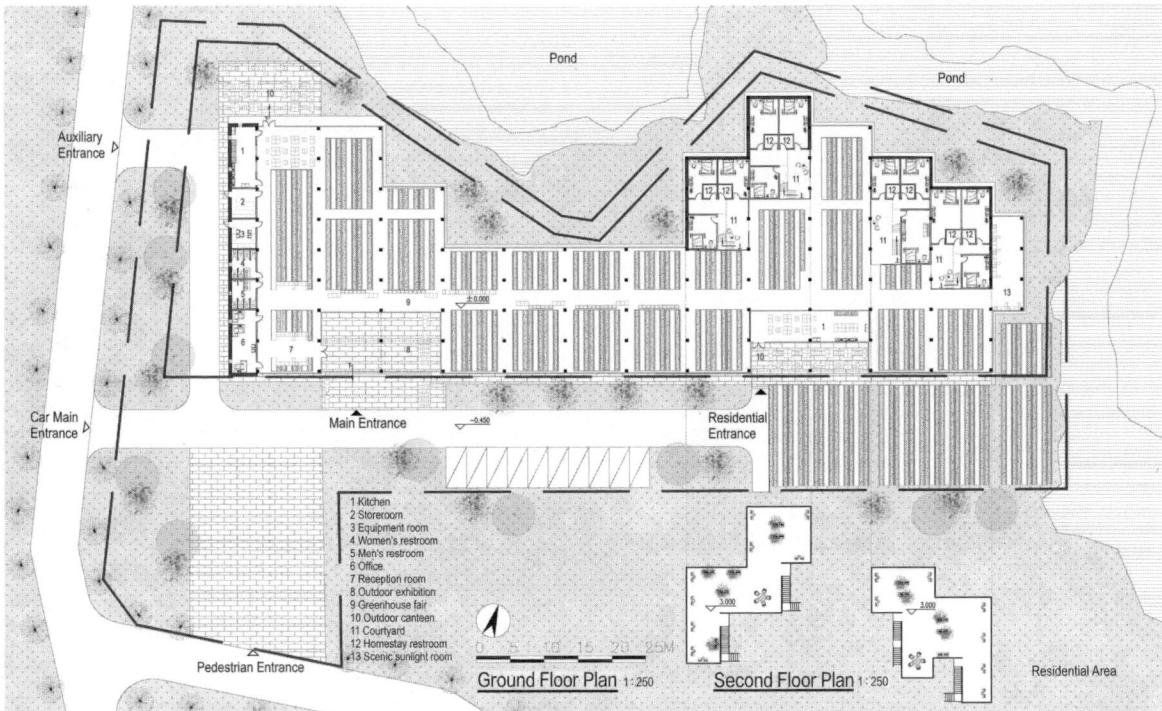

FOOD SUPPLY

SPRING
SUMMER
AUTUMN
WINTER

Greenhouse cultivation ensures a balanced growth and harvest ratio of crops in different seasons.

Pond

Pond

Auxiliary Entrance

Car Main Entrance

Main Entrance   −0.450

Residential Entrance

±0.000

1 Kitchen
2 Storeroom
3 Equipment room
4 Women's restroom
5 Men's restroom
6 Office
7 Reception room
8 Outdoor exhibition
9 Greenhouse fair
10 Outdoor canteen
11 Courtyard
12 Homestay restroom
13 Scenic sunlight room

Pedestrian Entrance

3.000

3.000

0 5 10 15 20 25M

Residential Area

Ground Floor Plan 1:250   **Second Floor Plan** 1:250

作品编号：110682

Ⅲ 稻香满园 *Harvest Beneath Greenhouse*

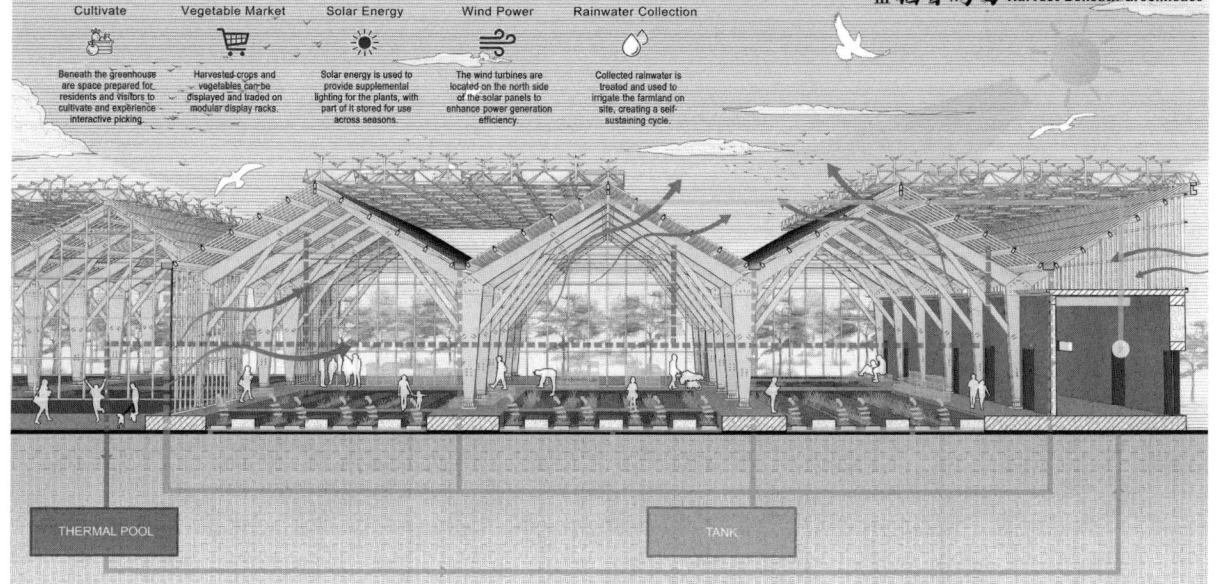

Cultivate	Vegetable Market	Solar Energy	Wind Power	Rainwater Collection
Beneath the greenhouse are space prepared for, residents and visitors to cultivate and experience interactive picking.	Harvested crops and vegetables can be displayed and traded on modular display racks.	Solar energy is used to provide supplemental lighting for the plants, with part of it stored for use across seasons.	The wind turbines are located on the north side of the solar panels to enhance power generation efficiency.	Collected rainwater is treated and used to irrigate the farmland on site, creating a self-sustaining cycle.

THERMAL POOL

TANK

◉ Activities & Energy Sources

◆ USERS

VILLAGERS　VISITORS　STUDENTS

◆ DAILY USE

FARM　CANTEEN　HOMESTAY

◆ EMERGENCY

SHELTER　AUTARKIC

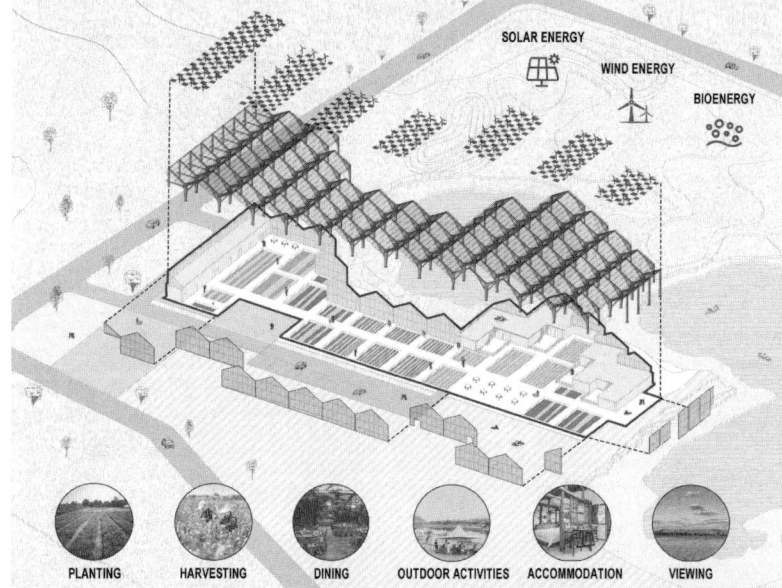

SOLAR ENERGY

WIND ENERGY

BIOENERGY

PLANTING　HARVESTING　DINING　OUTDOOR ACTIVITIES　ACCOMMODATION　VIEWING

◉ Summer Ventilation

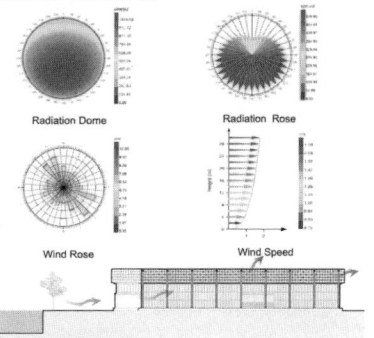

Radiation Dome　Radiation Rose

Wind Rose　Wind Speed

◉ Winter Ventilation

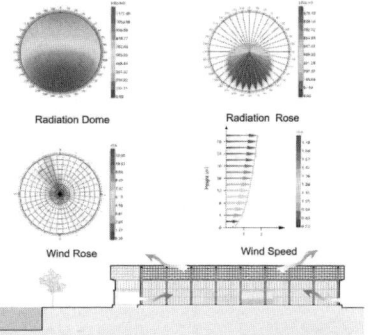

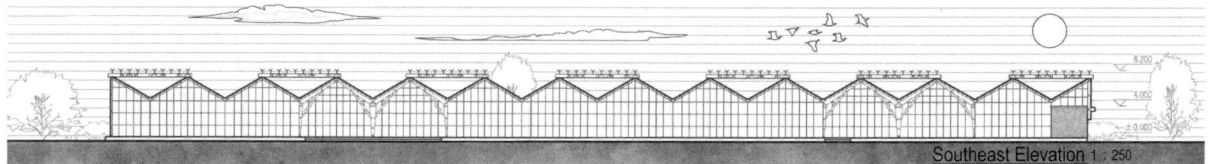

Radiation Dome　Radiation Rose

Wind Rose　Wind Speed

Southeast Elevation 1 : 250

作品编号：110682

IV 稻香满园 **Harvest Beneath Greenhouse**

Green House	Refuge	Solar Energy	Energy Storage	Snow Collection
The greenhouse provides disaster victims with a suitable temperature and sufficient supplies.	The modular furniture inside the greenhouse can be quickly converted into emergency facilities during disasters.	Solar energy maintains nighttime temperatures while providing supplemental lighting for the plants.	The cross-seasonal water storage tanks effectively address the shortage of solar energy, providing continuous heating.	Snowmelt is collected and purified for use as residential water and for irrigation.

THERMAL POOL

TANK

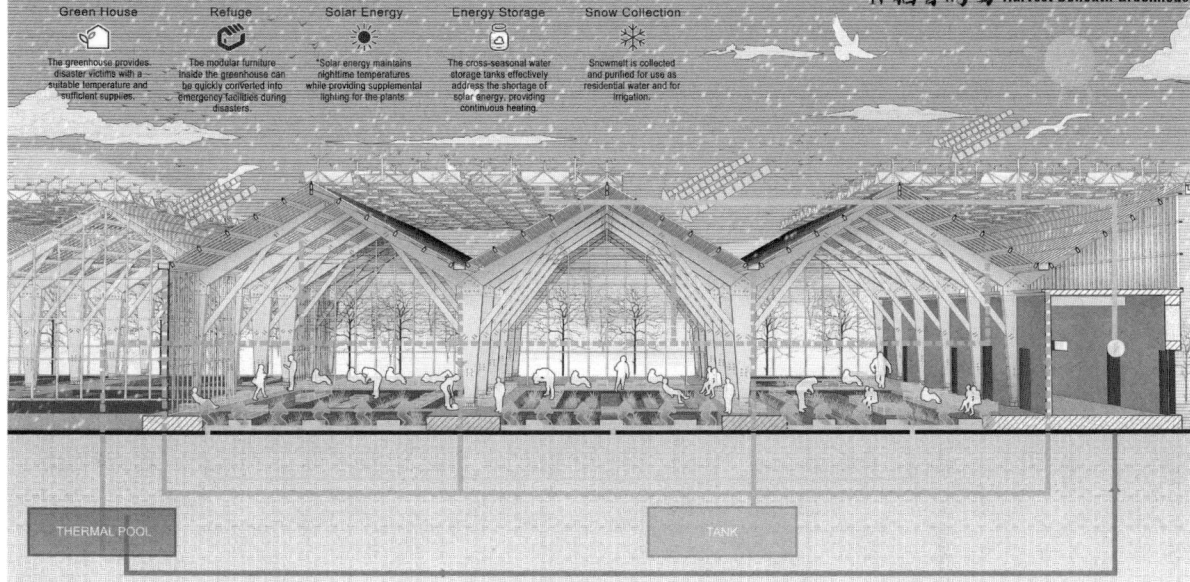

Possibilities for Modular Furniture

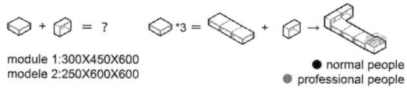

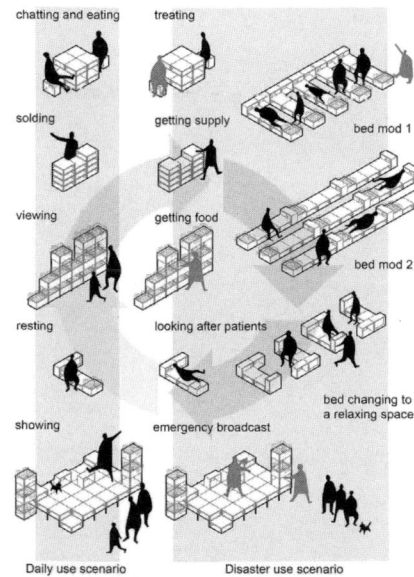

module 1：300X450X600
modele 2：250X600X600

● normal people
● professional people

chatting and eating
treating
solding
getting supply
bed mod 1
viewing
getting food
bed mod 2
resting
looking after patients
showing
emergency broadcast
bed changing to a relaxing space

Daily use scenario
Disaster use scenario

Energy Storage

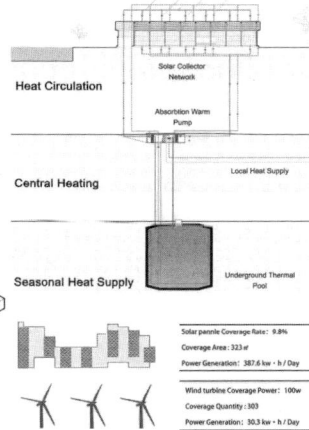

Heat Circulation

Solar Collector Network

Absorbtion Warm Pump

Central Heating

Local Heat Supply

Seasonal Heat Supply

Underground Thermal Pool

Solar pannle Coverage Rate：9.8%
Coverage Area：323 sf
Power Generation：387.6 kw·h / Day

Wind turbine Coverage Power：100w
Coverage Quantity：303
Power Generation：30.3 kw·h / Day

Structure & Material Analysis

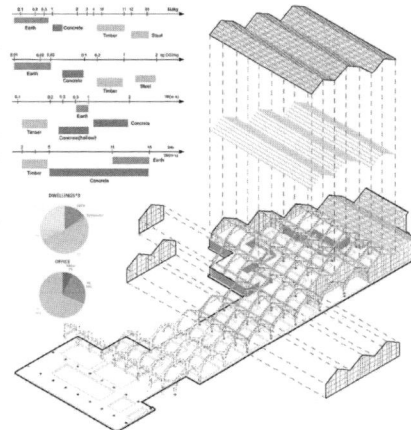

Energy Consumption Analysis

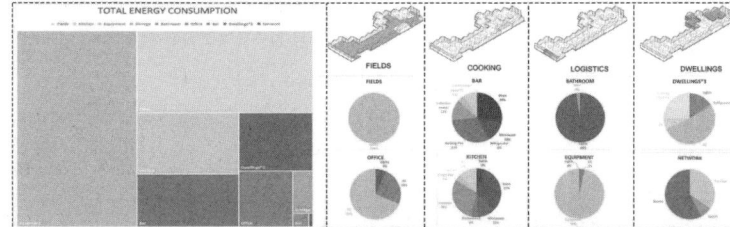

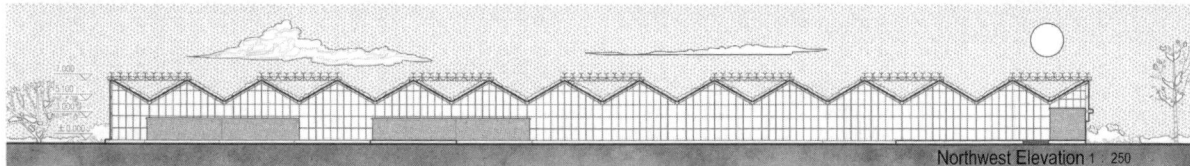

TOTAL ENERGY CONSUMPTION

FIELDS
COOKING
LOGISTICS
DWELLINGS

Northwest Elevation 1：250

247

单项奖·技术专项奖·张家口市怀安县二堡子村建设项目

Individual Awards – Technical Excellence Prize – Erpuzi Village Project, Huai'an County, Zhangjiakou City

注册号：110330
Register Number：110330

项目名称：折廊·叠院
Entry Title：Fold Verandah and Courtyard

作者：刘佳怡、孙雯昕、吴劲越
Authors：Liu Jiayi, Sun Wenxin,
          and Wu Jinyue

作者单位：河南工业大学
Authors from：Henan University of Technology

指导教师：李坤明
Tutor：Li Kunming

指导教师单位：河南工业大学
Tutor from：Henan University of Technology

折廊·叠院·1
Fold Verandah and Courtyard 1

Base unit | Fold into verandah | The verandah encloses the courtyard | The courtyard is transformed into a refuge space

▶ Location Analysis

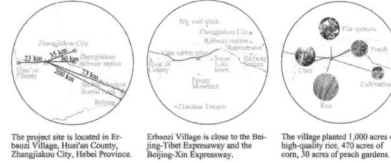

The project site is located in Erbaozi Village, Huai'an County, Zhangjiakou City, Hebei Province.

Erbaozi Village is close to the Beijing-Tibet Expressway and the Beijing-Xin Expressway.

The village planted 1,000 acres of high-quality rice, 470 acres of corn, 30 acres of peach gardens.

▶ Site features

Coupled courtyard | Huai'an kiln | Traditional roof

Straw device | Straw Pastoral Town | Agricultural technology

▶ Climate Analysis

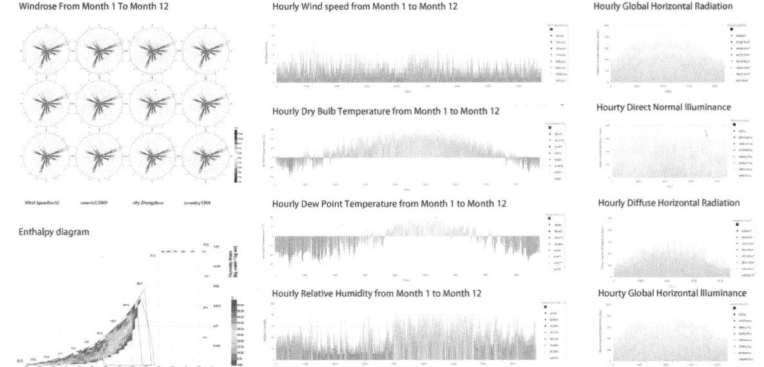

Windrose From Month 1 To Month 12

Hourly Wind speed from Month 1 to Month 12

Hourly Global Horizontal Radiation

Hourly Dry Bulb Temperature from Month 1 to Month 12

Hourly Direct Normal Illuminance

Enthalpy diagram

Hourly Dew Point Temperature from Month 1 to Month 12

Hourly Diffuse Horizontal Radiation

Hourly Relative Humidity from Month 1 to Month 12

Hourly Global Horizontal Illuminance

▶ Design Description

方案以"折廊叠院"为主题，赋予建筑、场地丰富的韧性与可变性。以被动式与韧性为出发点，采用装配式夯土结构，结合周边民居形态，传承地缘，并以艺术赋能，韧性方面，场地考虑面对紧急情况设定实用框架装置，便于组装、折卸、紧凑灵活；建筑预设功能可变。通过梁道效应加强通风，以可变结构兼顾冬季保暖，利用庭院、阳光间、折叠窗、廊等实现建筑节能。主动技术方面，设计太阳能光伏与光热系统，并通过多处水处理系统实现低冲击开发。

The scheme centers around the theme of "Folding Corridor and Layered Garden," endowing the architecture and site with rich resilience and adaptability. With passive and resilient strategies as the starting point, the design adopts a prefabricated rammed earth structure, integrating with the surrounding residential forms to carry forward the local heritage while empowering it with art. In terms of resilience, the site is equipped with practical framing devices for emergency situations, facilitating easy assembly, disassembly, and compact flexibility. The building is pre-designed with adaptable functions. Through the chimney effect, ventilation is enhanced, while the variable structure accommodates winter insulation. The use of gardens, sunrooms, folding windows, and corridors contributes to the building's energy efficiency. As for active technologies, solar photovoltaic and solar thermal systems are installed, and multiple water treatment systems are implemented to achieve low-impact development.

▶ Site Analysis

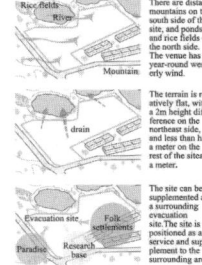

Rice fields | River
Mountain
drain
Evacuation site | Folk settlements
Paradise | Research base

There are distant mountains on the south side of the site, and ponds and rice fields on the north side. The venue has a year-round westerly wind.

The terrain is relatively flat, with a 2m height difference on the northeast side, and less than half a meter on the rest of the sitealf a meter.

The site can be supplemented as a surrounding evacuation site. The site is positioned as a service and supplement to the surrounding area.

▶ Diagram of Design Process

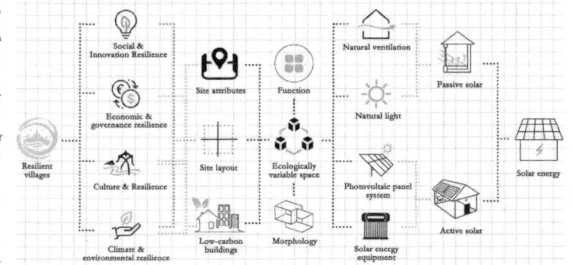

Social & Innovation Resilience | Natural ventilation | Passive solar
Economic & governance resilience | Site attributes | Function | Natural light
Resilient villages | Culture & Resilience | Site layout | Ecologically variable space | Photovoltaic panel system | Solar energy
Climate & environmental resilience | Low-carbon buildings | Morphology | Solar energy equipment | Active solar

▶ Site Generation

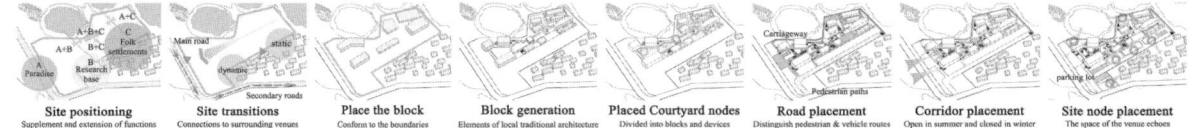

Site positioning — Supplement and extension of functions

Site transitions — Connections to surrounding venues

Place the block — Conform to the boundaries

Block generation — Elements of local traditional architecture

Placed Courtyard nodes — Divided into blocks and devices

Road placement — Distinguish pedestrian & vehicle routes

Corridor placement — Open in summer and closed in winter

Site node placement — The space of the venue echoes

▶ West Elevation 1：250

Artwork Number: 110330

▶ Site-plan 1 : 450

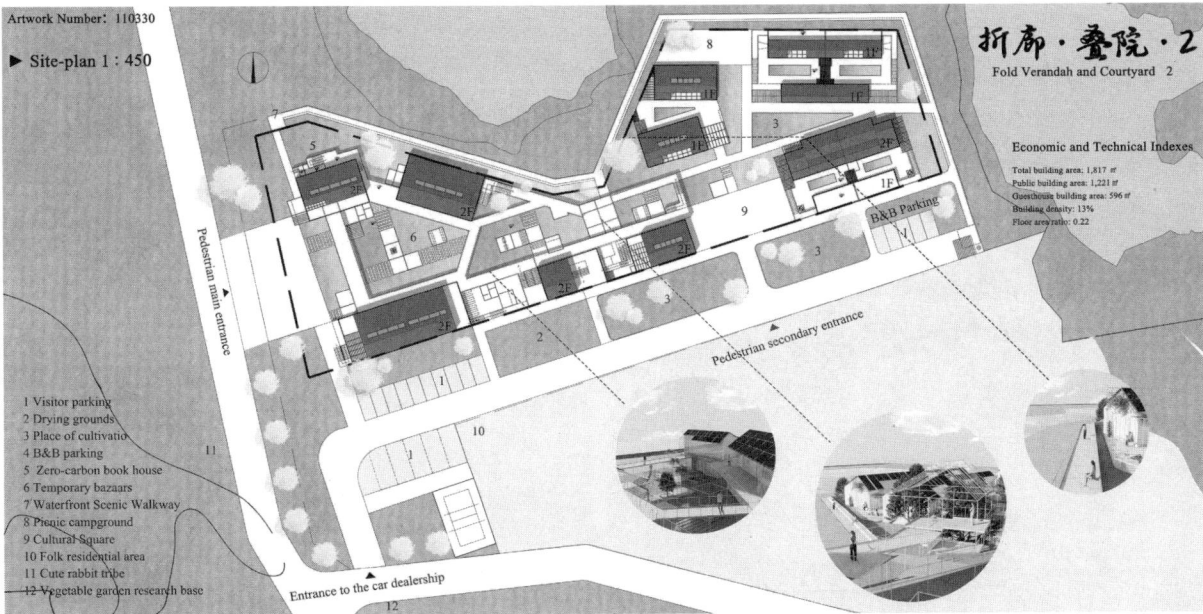

折廊·叠院·2
Fold Verandah and Courtyard 2

Economic and Technical Indexes

Total building area: 1,817 m²
Public building area: 1,221 m²
Guesthouse building area: 596 m²
Building density: 13%
Floor area ratio: 0.22

Pedestrian main entrance

Pedestrian secondary entrance

B&B Parking

1 Visitor parking
2 Drying grounds
3 Place of cultivation
4 B&B parking
5 Zero-carbon book house
6 Temporary bazaars
7 Waterfront Scenic Walkway
8 Picnic campground
9 Cultural Square
10 Folk residential area
11 Cute rabbit tribe
12 Vegetable garden research base

Entrance to the car dealership

▶ Technology and Function Analysis

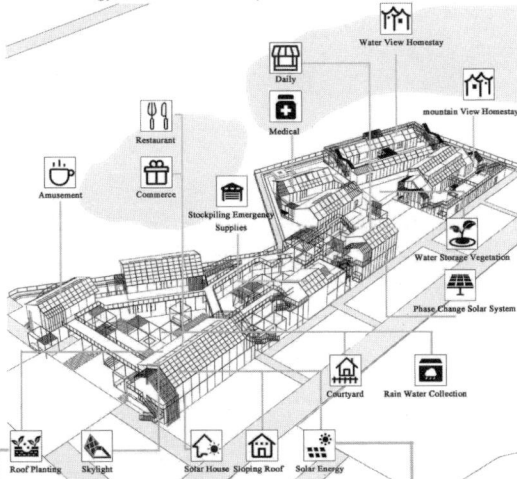

Water View Homestay
Daily
Medical
mountain View Homestay
Restaurant
Amusement
Commerce
Stockpiling Emergency Supplies
Water Storage Vegetation
Phase Change Solar System
Courtyard
Rain Water Collection
Roof Planting
Skylight
Solar House Sloping Roof
Solar Energy

▶ Design Structural Detail Analysis

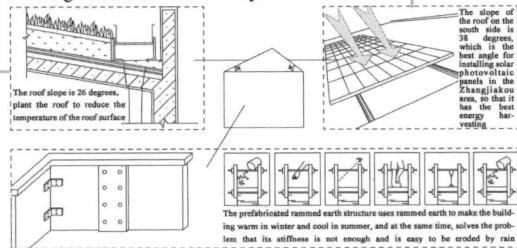

The roof slope is 26 degrees, plant the roof to reduce the temperature of the roof surface.

The slope of the roof on the south side is 38 degrees, which is the best angle for installing solar photovoltaic panels in the Zhangjiakou area, so that it has the best energy harvesting.

The prefabricated rammed earth structure uses rammed earth to make the building warm in winter and cool in summer, and at the same time, solves the problem that its stiffness is not enough and is easy to be eroded by rain.

▶ Variable Design Analysis

☐ Diversify the basic units to form units with different functions

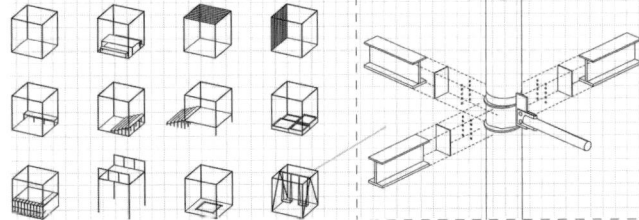

☐ The deformation elements are combined with each other

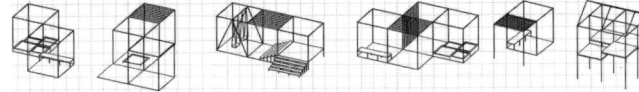

☐ The building uses prefabricated structures and variable walls to achieve resilience changes in the interior space

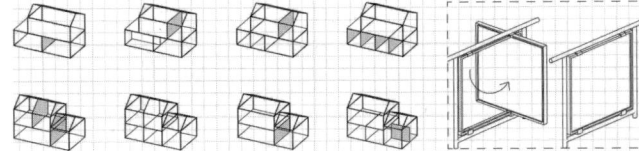

☐ The existing frame structure, combined with lightweight walls or roofs, can be used to transform the function of the space in a given situation

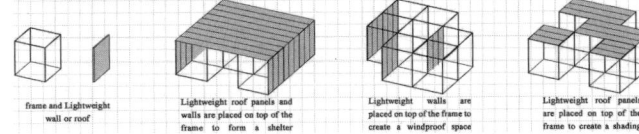

frame and Lightweight wall or roof

Lightweight roof panels and walls are placed on top of the frame to form a shelter

Lightweight walls are placed on top of the frame to create a windproof space

Lightweight roof panels are placed on top of the frame to create a shading

▶ 1-1 Section

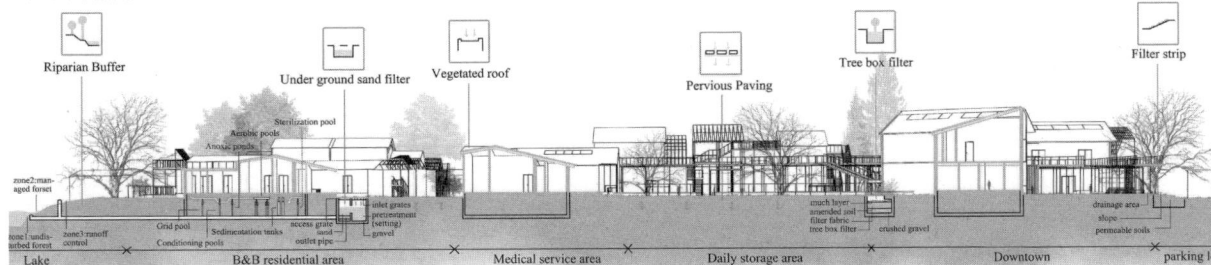

Riparian Buffer
Under ground sand filter
Vegetated roof
Pervious Paving
Tree box filter
Filter strip

Lake | B&B residential area | Medical service area | Daily storage area | Downtown | parking lot

Artwork Number: 110330

折廊 · 叠院 · 子
Fold Verandah and Courtyard 3

▶ Plan 1 : 300

1 Solar house
2 Children's reading area
3 Screening room
4 Tourist service area
5 Cultural and creative stores
6 Water bar
7 Storage of emergency supplies
8 Guesthouse
9 Postal Agency Center
10 Express Station
11 Farming Culture Exhibition
12 Lounge
13 Hall
14 Dispensing room
15 Consultation room
16 Kitchen
17 Confirmatorium
18 Restaurant
19 Master Hall
20 Master bedroom
21 Living room
22 Guest room
23 Place of cultivation
24 Terrace with a view
25 Office
26 Conference room
27 Archives
28 Reading area for adults
29 Handicraft room
30 Tearoom
31 Drying grounds
32 Parking lot
33 Zero-carbon book house
34 Waterfront Scenic Walkway
35 Landscape nodes

▶ Structural Analysis

Partition wall

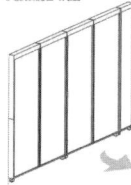

Shading elements

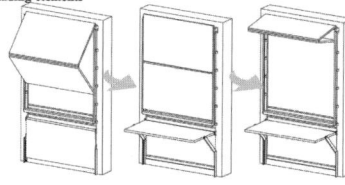

Rammed earth wall section

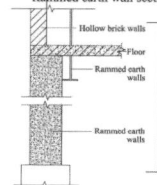

Hollow brick walls
Floor
Rammed earth walls
Galvanized steel mesh Φ5
Main body steel columns
Rammed earth walls
Structural columns
Galvanized rectangular pipe 80*80*3
Φ6@600

Seismic design

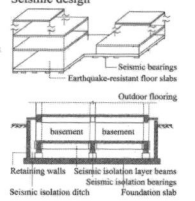

Seismic bearings
Earthquake-resistant floor slabs
Outdoor flooring
basement   basement
Retaining walls   Seismic isolation layer beams
Seismic isolation bearings
Seismic isolation ditch   Foundation slab

▶ Resilience Analysis

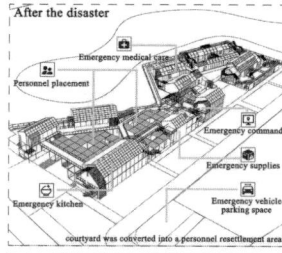

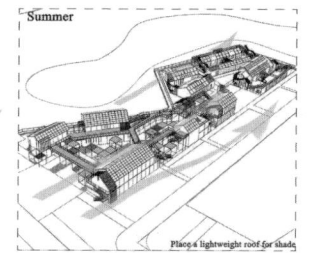

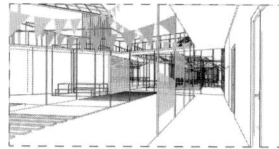

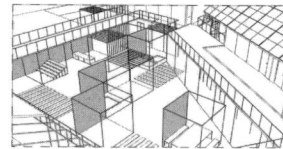

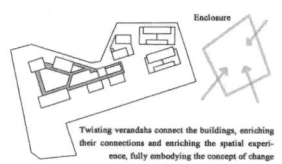

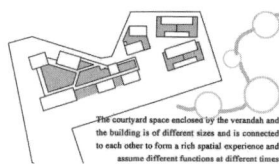

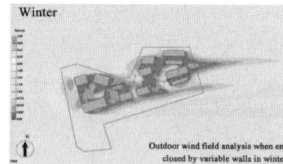

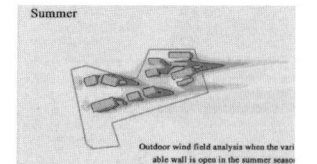

After the disaster

Emergency medical care
Personnel placement
Emergency command
Emergency supplies
Emergency kitchen
Emergency vehicle parking space

courtyard was converted into a personnel resettlement area

Peak tourist season

Experience the vegetable patch
Waterfront boardwalk
Featured bazaars
Country kitchen
Agricultural Museum
Rice processing experience
Art Exhibition   Specialty supermarket

The courtyard is converted into a recreational area for tourists

Winter

Place a vertical baffle to protect against the wind

Summer

Place a lightweight roof for shade

▶ Resilience Analysis

Enclosure

Twisting verandahs connect the buildings, enriching their connections and enriching the spatial experience, fully embodying the concept of change

The courtyard space enclosed by the verandah and the building is of different sizes and is connected to each other to form a rich experience and assume different functions at different times

Winter

Outdoor wind field analysis when enclosed by variable walls in winter

Summer

Outdoor wind field analysis when the variable wall is open in the summer season

▶ Partial Perspective

Artwork Number: 110330

折廊·叠院·4

Fold Verandah and Courtyard 4

▶ Homestay Design Analysis

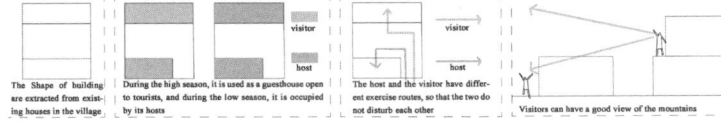

visitor
host

The Shape of building are extracted from existing houses in the village

During the high season, it is used as a guesthouse open to tourists, and during the low season, it is occupied by its hosts

The host and the visitor have different exercise routes, so that the two do not disturb each other

Visitors can have a good view of the mountains

▶ Folding window Analysis

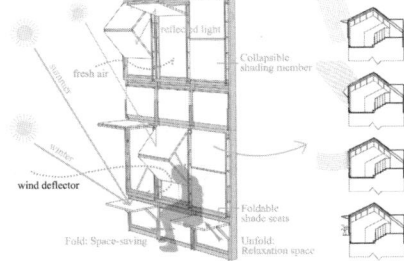

reflected light
Collapsible shading number
fresh air
summer
winter
wind deflector
Foldable shade seats
Fold: Space-saving
Unfold: Relaxation space

▶ Natural Lighting

▶ Courtyard Ventilation

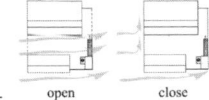

open   close

▶ Solar Radiation Heat in Winter

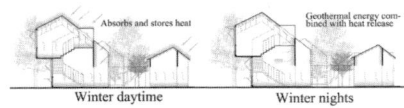

Absorbs and stores heat
Geothermal energy combined with heat release

Winter daytime   Winter nights

▶ Ventilation Analysis

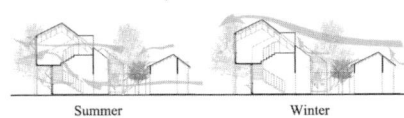

Summer   Winter

▶ Sunroom Analysis

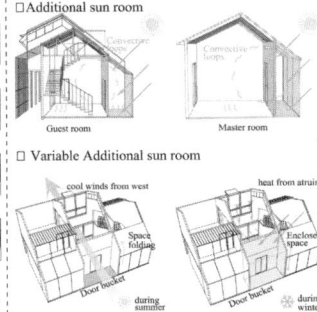

Additional sun room

Guest room   Master room

Variable Additional sun room

cool winds from west
heat from atrium
Space folding
Enclosed space
Door bucket during summer
Door bucket during winter

Solar radiation

No sunlight(in summer)   No sunlight(in winter)   With sunlight(in winter)

Summer daytime
Additional sun room
Open the window
outside   indoors

Winter daytime
outside   indoors

Summer nights
Rolled up the sunshades
Open the window
Additional sun room
outside   indoors

Highly reflective sunshade insulation curtains
Insulating glass windows
Dark surface heavy walls
Insulating glass windows can be opened
Additional sun room
Insulating glass windows
Dark surface ground

▶ Carbon Emission Simulation Calculation

Electricity	Category	Design building carbon emissions kgCO₂/(m²·a)	Refer to the carbon emissions of building kgCO₂/(m²·a)
Cooling(Ee)		2.45	5.80
Heating(Eh)		13.97	12.77
Air conditioner fan(Ef)		0.00	0.00
Illuminating		4.09	4.80
Fossil fuel	Category	Design building carbon emissions	Refer to the carbon emissions of building
Bituminous coal II	Supply heating:Heat source boilers	0.00	0.00
Nothing	Domestic hot water		(Fuel:Gas)
Renewable	Category	Design building carbon emissions	Refer to the carbon emissions of building
Renewable energy(Er)	Photovoltaic(Ep)	9.28	-
	Wind power(Ew)	0.00	-
Total carbon emissions		11.23	23.37
Relative reference to the proportion of carbon reduction in buildings(%)		51.95(Target value:40)	
Relative to the reduction of carbon emission intensity of buildings kg-CO₂/(m²·a)		12.14(Target value:7)	

▶ Simulation of Energy Consumption

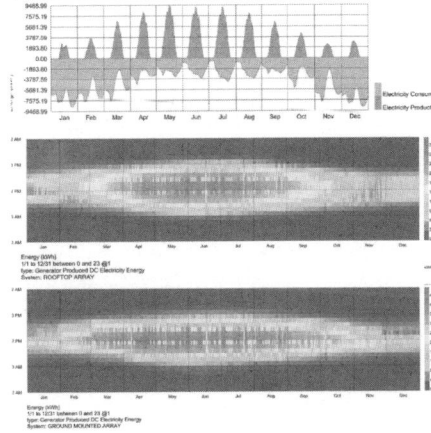

9465.96
7575.19
5681.39
3787.59
1893.80
0.00
-1893.80
-3787.59
-5681.39
-7575.19
-9465.99

Jan Feb Mar Apr May Jun Jul Aug Sep Oct Nov Dec

Electricity Consump
Electricity Production

Energy (kWh)
1/1 to 12/31 between 0 and 23 @1
type: Generator Produced DC Electricity Energy
System: ROOFTOP ARRAY

Energy (kWh)
1/1 to 12/31 between 0 and 23 @1
type: Generator Produced DC Electricity Energy
System: GROUND MOUNTED ARRAY

▶ Sectional Perspective

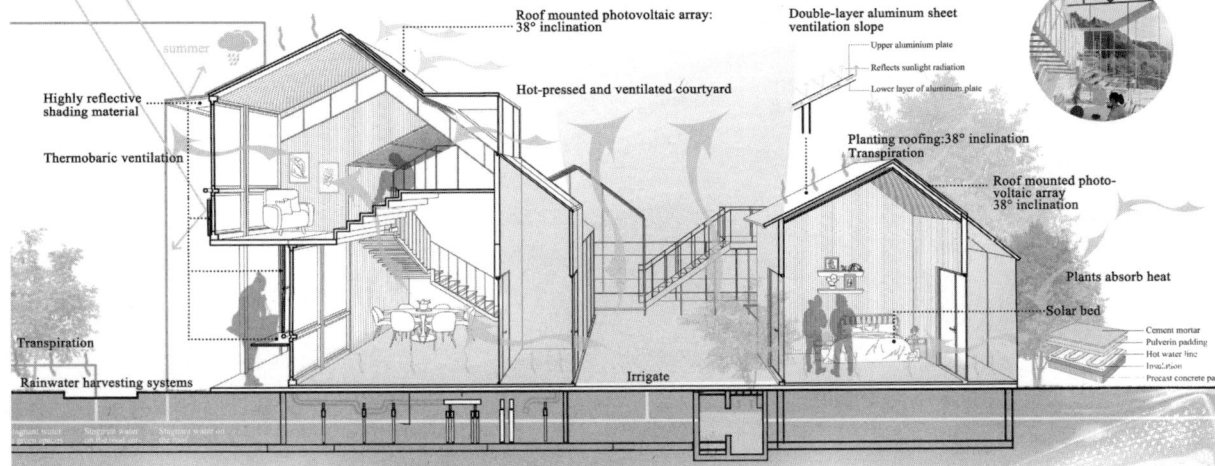

summer

Roof mounted photovoltaic array: 38° inclination

Double-layer aluminum sheet ventilation slope
Upper aluminium plate
Reflects sunlight radiation
Lower layer of aluminum plate

Hot-pressed and ventilated courtyard

Highly reflective shading material

Thermobaric ventilation

Planting roofing:38° inclination
Transpiration

Roof mounted photovoltaic array 38° inclination

Plants absorb heat

Solar bed
Cement mortar
Pulverin padding
Hot water line
Insulation
Precast concrete panel

Transpiration

Rainwater harvesting systems

Irrigate

单项奖·技术专项奖·张家口市怀安县二堡子村建设项目
Individual Awards — Technical Excellence Prize — Erpuzi Village Project, Huai'an County, Zhangjiakou City

注册号：110752
Register Number：110752

项目名称：系土而生
Entry Title：Born from the Soil

作者：张晓雨、王思雅、张宁舟、杨飞飞
Authors：Zhang Xiaoyu, Wang Siya, Zhang Ningzhou, and Yang Feifei

作者单位：合肥工业大学
Authors from：Hefei University of Technology

指导教师：王旭、杨洋
Tutors：Wang Xu, Yang Yang

指导教师单位：合肥工业大学
Tutors from：Hefei University of Technology

作品编号：110752

系土而生
Born from the Soil 1

Local Traditional House

In Huai'an County, there is a unique type of traditional dwelling - the swan kiln. Although timber is scarce in the area, there are abundant loess resources in the area, which provides local building materials that are easy to obtain for the construction of swan kiln dwellings. Swan kiln dwellings are built with arches and swan kilns, which are warm in winter and cool in summer, and are environmentally friendly. Therefore, rammed earth is used as the building material and arches are used as the reference for the building form. Energy saving and environmental protection at the same time make the building more intimate and bring the occupants closer to nature.

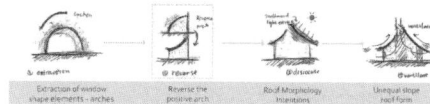

Extraction of window shape elements - arches | Reverse the positive arch | Roof Morphology Intentions | Unique slope roof form

Surrounding Resource

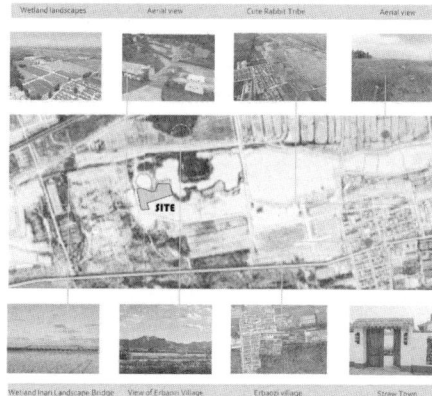

Wetland landscapes | Aerial view | Cute Rabbit Tribe | Aerial view

SITE

Wetland Inari Landscape Bridge | View of Erbaozi Village | Erbaozi village | Straw Town

Site Analysis

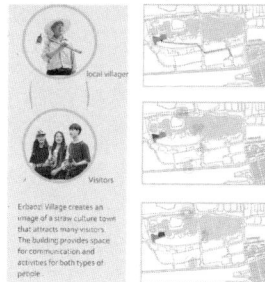

local villager

Visitors

Erbaozi Village creates an image of a straw culture town that attracts many visitors. The building provides space for communication and activities for both types of people.

Local villagers can come to this country living room via multiple paths with easy access.

The east side of the site is planned to be an ancillary building facility, providing a place for recreation and leisure on a daily basis, and a place for escape and disaster prevention in case of emergency.

The west side of the plot is planned to be four residential buildings with independent courtyards and a B&B function to serve local villagers and visiting tourists.

Thermal Comfort Polygon

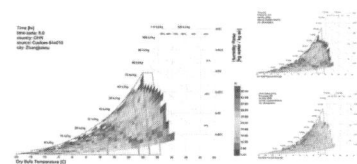

Analysis of Local Meteorological Data

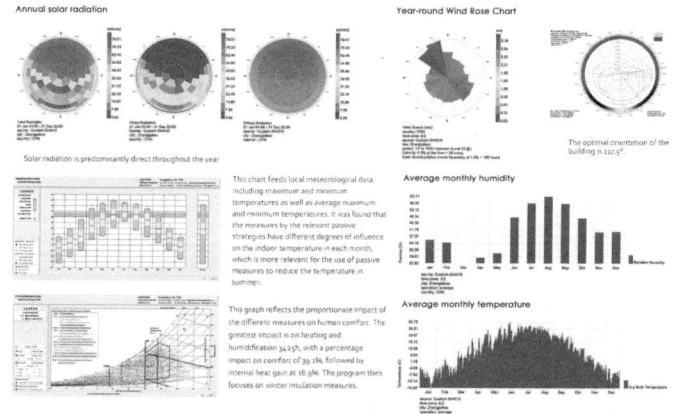

Annual solar radiation

Year-round Wind Rose Chart

The optimal orientation of the building is 117.5°.

Solar radiation is predominantly direct throughout the year

This chart feeds local meteorological data including maximum and minimum temperatures as well as average maximum and minimum temperatures. It was found that the measures by the relevant passive strategies have different degrees of influence on the indoor temperature in each month, which is more relevant for the use of passive measures to reduce the temperature in summer.

This graph reflects the proportionate impact of the different measures on human comfort. The greatest impact is on heating and humidification 34.25%, with a percentage impact on comfort of 39.1%, followed by internal heat gain at 16.9%. The program then focuses on winter insulation measures.

Average monthly humidity

Average monthly temperature

Design Statement

本设计向当地传统民居——暖窑学习，采用夯土材料搭建，实现冬季保暖、夏季防凉。以反拱的弧状拔为设计特色，展现古典之韵味。将被动式绿色技术融入到设计之中，包括南侧与中庭阳光间集热蓄热、地道通风等，增加夏季通风降温以及冬季保温供暖，节省能源。同时，采用与太阳高度角呈一定角度的太阳能光伏面板，更高效地进行太阳能发电，助力能源共享。从建筑的可持续发展角度，为当地村民提供娱乐活动场所，提升旅游业发展。同时还提供了避震逃难场所，使场所功能多元化，构建了韧性乡村。

Learning from the local traditional dwelling, the Swan Kiln, this design is constructed with rammed earth materials to realize warmth in winter and shade in summer. The design features a curved wall with a reverse arch to show the classical flavor. Passive green technologies are incorporated into the design, including heat collection and storage between the south side and the atrium sunlight, and underground ventilation, to increase ventilation and cooling in summer as well as heat preservation and heating in winter, saving energy. At the same time, solar photovoltaic panels at an angle to the sun's altitude are used to generate solar power more efficiently and help share energy. From the perspective of sustainable development of the building, it provides local villagers with places for recreational activities and enhances tourism development. It also provides a place to escape from earthquakes, diversifying the functions of the place and building a resilient village.

作品编号：110752

系土而生 Born from the Soil 2

ponds

public square

pension area

Public service facilities

Roads

Bed and Breakfast Residential Area

Parking lots

Moe Rabbit Tribe
Tourist Attractions

Vegetable Garden Research Base

0 5 10 15 20 25m

Building foot print: 4257 m²
Total area:
1181.5 (public) +593 (private) =1774.5 m²
Building coverage:27.42%
Greening rate:29.73%

First Floor Plane 1 : 300

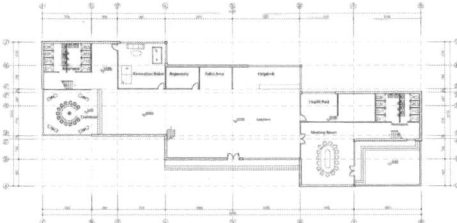

Second Floor Plan 1 : 300

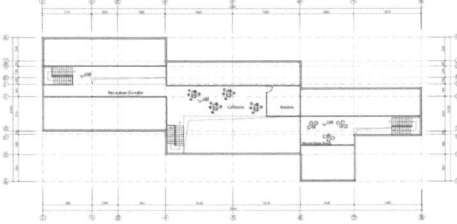

First Floor Plane 1 : 300　Second Floor Plane 1 : 300

East Elevation 1 : 200

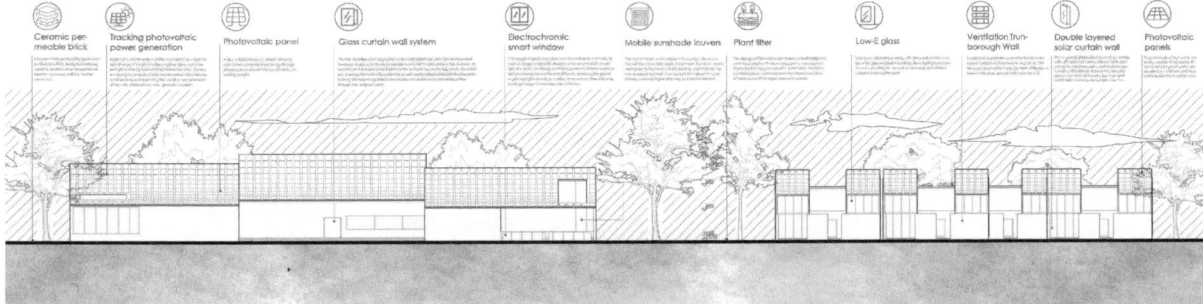

Ceramic permeable brick | Tracking photovoltaic power generation | Photovoltaic panel | Glass curtain wall system | Electrochromic smart window | Mobile sunshade louvers | Plant filter | Low-E glass | Ventilation Trunbourough Wall | Double layered solar curtain wall | Photovoltaic panels

Design Strategy

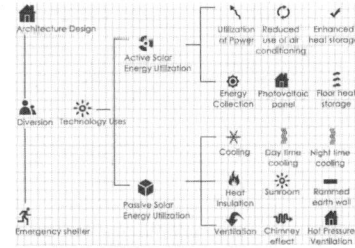

Architecture Design

Active Solar Energy Utilization — Utilization of Power / Reduced use of air conditioning / Enhanced heat storage

Energy Collection / Photovoltaic panel / Floor heat storage

Diversion Technology Uses

Passive Solar Energy Utilization — Cooling / Day-time cooling / Night time cooling

Heat Insulation / Sunroom / Rammed earth wall

Ventilation / Chimney effect / Hot Pressure Ventilation

Emergency shelter

Green Performance Data

Electrical power	Form	Carbon footprint (kgCO₂/m²·a)	Reference carbon emissions (kgCO₂/m²·a)
	Cooling(Ec)	5.49	13.86
	Heating(Eh)	0.89	0.66
	Air conditioning fans(Ef)	2.27	—
	Lighting	1.50	2.51
	Socket equipment		
Others(Eo)	Domestic hot water	0.00 (Deduction for solar energy)	0.00
	Total other	0.00	
Fossil fuels	Form	Carbon footprint (kgCO₂/m²·a)	Reference carbon emissions (kgCO₂/m²·a)
None	Heating: heat source boiler	0.00	
None	Heating: Municipal heat	0.00	
None	Domestic hot water (after deduction of solar energy)	0.00	0.00
Natural gas	Cooking	0.00	
Renewable energy	Form	Carbon footprint (kgCO₂/m²·a)	Reference carbon emissions (kgCO₂/m²·a)
Renewable energy (Er)	Solar water heating	0.00	—
	Photovoltaic (Ep)	0.00	—
	Wind power(Ew)	0.00	—
Total carbon emissions		10.13	17.33
Percentage of carbon reduction relative to reference buildings (%)		41.55(Target rate: 40)	
Carbon intensity reductions relative to reference buildings (kgCO₂/m²·a)		7.23 (Target rate: 7)	

The carbon emission intensity of the building operation of this project has been reduced by 41.55% on the basis of the energy-saving design standard implemented in 2016, and the carbon-emission intensity has been reduced by 7.23(kgCO₂/m²·a), and the emission indexes of the building operation satisfy the requirements of Clause 2.0.3 of the "General Specification for Energy Saving and Renewable Energy Utilization in Buildings" GB55015 - 2021.

Building Daylighting Analysis

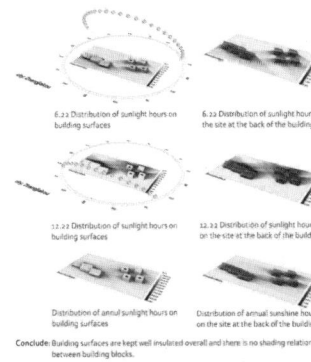

6.22 Distribution of sunlight hours on building surfaces

6.22 Distribution of sunlight hours on the site at the back of the building

12.22 Distribution of sunlight hours on building surfaces

12.22 Distribution of sunlight hours on the site at the back of the building

Distribution of annual sunlight hours on building surfaces

Distribution of annual sunshine hours on the site at the back of the building

Conclude: Building surfaces are kept well insulated overall and there is no shading relationship between building blocks.

Outdoor Wind Environment Simulation

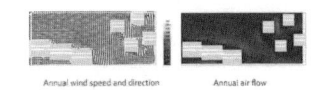

Annual wind speed and direction

Annual air flow

Indoor Sunlight Distribution

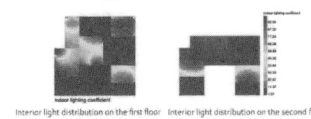

Interior light distribution on the first floor

Interior light distribution on the second floor

作品编号：110752

系土而生　Born from the Soil 3

Rammed Earth Construction Technique

Solar Radiation Analysis

Solar Trajectory and Roof Noon Radiation on Vernal Equinoxes

Summer solstice solar trajectory and roof noonday radiation

Autumnal Equinox Sun Trajectory and Roof Noon Radiation

Winter Solstice Sun Trajectory and Roof Noon Radiation

Year-round Solar Trajectory

Schematic Diagram of Solar Panel Structure

End clamp connection

Mid clamp connection

Main crail and colum connection

The solar panels used on the roofs of public buildings and residential homes are rotatable, using a size of 55*30*10cm, and can be rotated to the optimum angle according to the height of the sun to maximize the use of solar energy and help the house to get the most power.

Explosive Analysis

Interior display of public buildings

Leisure diving bar

Recreational areas

Recreation room

Teahouse

Conference room

Plant Analysis

Water retaining vegetation

Schematic diagram of soil layer

Streamline Analysis

Analysis of flow lines on the second floor of a public building

First floor flow analysis of public buildings.

Profile 1：200

作品编号：.110752

系土而生

Born from the Soil 4

Thermal Pressure Ventilation
In summer, the higher atrium forms a chimney, and the shape of the reversal arch facilitates the rise of airflow, which rises from a low level and exits through the chimney for ventilation.

Sunroom
A warm porch sunroom is provided on the south side, and heat-storage glass is used in the window openings on the public wall to guarantee the heat-storage effect and minimize the loss of temperature at night.

Wind pressure ventilation
In summer, the airflow blows from the windward side to the building. At this time, the window is opened, and the airflow is discharged from the leeward side, forming natural ventilation driven by wind pressure.

Explosive Analysis

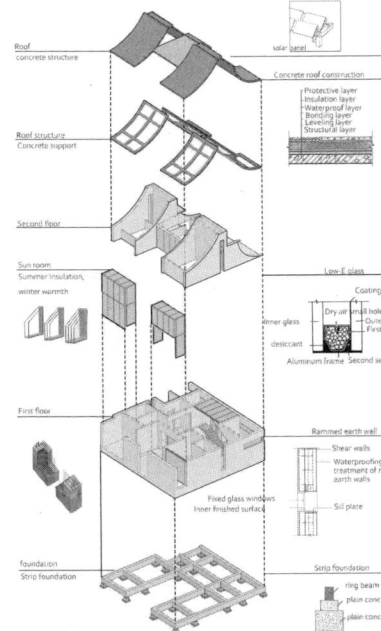

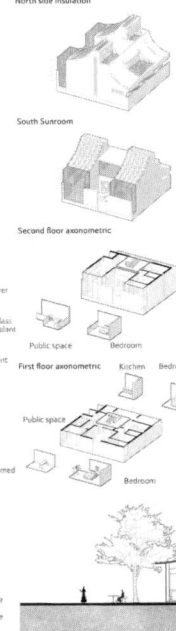

Roof
concrete structure

solar panel

Concrete roof construction

Protective layer
Insulation layer
Waterproof layer
Bonding layer
Leveling layer
Structural layer

Roof structure
Concrete support

Second floor

Sun room
Summer insulation,
winter warmth

Low-E glass

Coating layer

Dry air small hole
Inner glass
Outer glass
First sealant

desiccant

Aluminum frame Second sealant

First floor

Rammed earth wall

Shear walls

Waterproofing treatment of rammed earth walls

Fixed glass windows
Inner finished surface

Soil place

foundation
Strip foundation

Strip foundation

ring beam
plain concrete
plain concrete

Explosive Analysis

North side insulation

South Sunroom

Second floor axonometric

Public space   Bedroom

First floor axonometric   Kitchen   Bedroom

Public space

Bedroom

Passive Technical Analysis

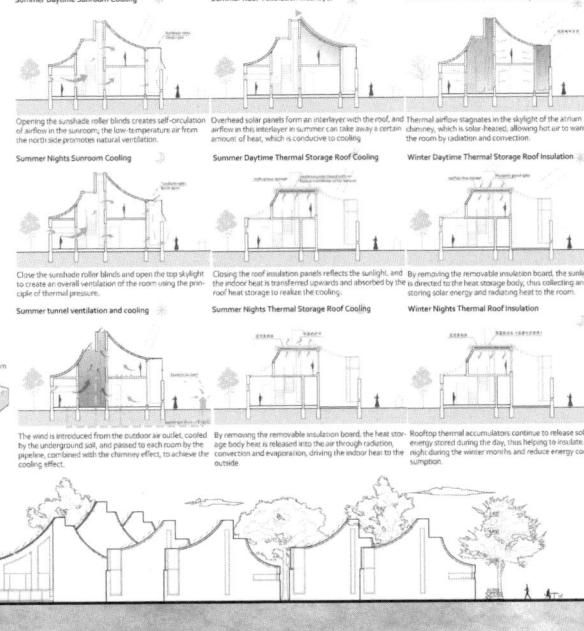

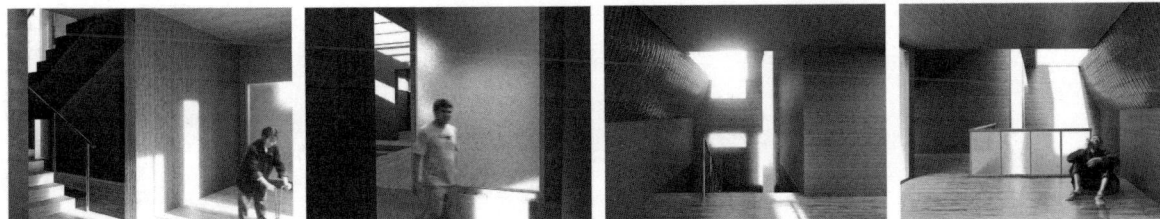

Summer Daytime Sunroom Cooling

Opening the sunshade roller blinds creates self-circulation of airflow in the sunroom; the low-temperature air from the north side promotes natural ventilation.

Summer Roof Ventilation Interlayer

Overhead solar panels form an interlayer with the roof, and airflow in this interlayer in summer can take away a certain amount of heat, which is conducive to cooling.

Winter Sunroom and Atrium Chimney Insulation

Thermal airflow stagnates in the skylight of the atrium chimney, which is solar-heated, allowing hot air to warm the room by radiation and convection.

Summer Nights Sunroom Cooling

Close the sunshade roller blinds and open the top skylight to create an overall ventilation of the room using the principle of thermal pressure.

Summer Daytime Thermal Storage Roof Cooling

Closing the roof insulation panels reflects the sunlight, and the indoor heat is transferred upwards and absorbed by the roof heat storage to realize the cooling.

Winter Daytime Thermal Storage Roof Insulation

By removing the removable insulation board, the sunlight is directed to the heat storage body, thus collecting and storing solar energy and reducing heat to the room.

Summer tunnel ventilation and cooling

The wind is introduced from the outdoor air outlet, cooled by the underground soil, and passed to each room by the pipeline, combined with the chimney effect, to achieve the cooling effect.

Summer Nights Thermal Storage Roof Cooling

By removing the removable insulation board, the heat storage body heat is released into the air through radiation, convection and evaporation, driving the indoor heat to the outside.

Winter Nights Thermal Roof Insulation

Rooftop thermal accumulators continue to release solar energy stored during the day, thus helping to insulate at night during the winter months and reduce energy consumption.

有效作品参赛团队名单
Name List of All Participants Submitting Valid Works

作品编号	作者	单位名称	指导老师	单位名称
110000	吴雨航，金日，任嘉瑶，明来灵，李恩泽	西南民族大学	熊健吾、张埕	西南民族大学
110003	董春朝、吴梦雨、朱嘉宇、张吉森、房彦娜	西北工业大学	刘煜、黄姗、杜昱民、宋戈、邵腾	西北工业大学
110004	周琪云、张焕悦	西安建筑科技大学	王芳、陈敬	西安建筑科技大学
110005	李嘉杭、车佩玉、韦彦滢	西安建筑科技大学	陈敬、王芳	西安建筑科技大学
110018	覃子倩、吴强、陈美华	苏州科技大学	刘科	苏州科技大学
110030	吴疏、朱书慧、严卓洋、龚友敏、李京、李梦楚	山东建筑大学	侯世荣	山东建筑大学
110031	殷梓芸、崔璨、贾龙庆、陈雄宇、杜钰玮、李啸跃	山东建筑大学	侯世荣、何文晶、李晓东	山东建筑大学
110046	唐奇、赵小琦、陈曦	河北水利电力学院	靳瑞峰、李黎、杨海娇	河北水利电力学院
110048	沈鑫、柴振国、李云龙、赵一凡、赵雅雯	内蒙古工业大学	伊若勒泰、许国强	内蒙古工业大学
110051	刘菊影、胡瑞鹏、黄新鹏、李乐乐	合肥工业大学	王旭	合肥工业大学
110053	吴新杰、李剑锋、周健鑫	福州大学	吴木生	福州大学
110055	李思奕、吴韵蕾、辛芃	福州大学	郑嫒、黄斯	福州大学
110066	李蔚、贺雨欣、王震	中国石油大学（华东）	李佐龙、戴昀	中国石油大学（华东）
110067	闫阳灿、李培硕、何昊泽、武子君、徐婧华、陈子凡	河北大学建筑工程学院	徐畅、李崴、李纪伟、贾慧献、黄志鹏	河北大学建筑工程学院
110068	秦正、李光美	南京工业大学	薛春霖	南京工业大学
110069	浮英媛、胡安达、袁林、陈钺、方皓宇	昆明理工大学	谭良斌	昆明理工大学
110071	杨睿涵、白慕瑶、李鑫鑫	北京交通大学	张文	北京交通大学
110072	陈诗霖、石宁	北京交通大学	张文	北京交通大学
110077	蔡安琪、邢依明、王晨曦、林俊杰	福州大学	崔育新	福州大学
110084	杨文祺、邝微	加州大学伯克利分校	—	—
110086	何彬斌、陆强、树成杰	苏州科技大学	刘长春、金雨蒙	苏州科技大学

作品编号	作者	单位名称	指导老师	单位名称
110088	莫宇豪、秦斯玲	桂林理工大学	朱文霜	桂林理工大学
110092	张蕾、高原、韩旭、刘梦飞	苏州科技大学	金雨蒙、刘长春	苏州科技大学
110093	白炳骏、蔡严宇、沈梓程、赵文清	南京工业大学	董凌	南京工业大学
110099	解雅琪、韩欣怡	北京交通大学	张文	北京交通大学
110119	吴泽同、王鑫平、傅梦婷	南京工业大学	徐琦	南京工业大学
110131	丁鑫玉、郝欣玉	长安大学	夏博	长安大学
110132	徐裴源、郁岩、蒋领、郑淳艺	合肥工业大学	黄杰、张东凯	合肥工业大学
110136	李华钊、李子乔、李少飞、赖俊昊、谈浩然	广州大学	万丰登、赵阳、李丽	广州大学
110143	吴婕、穆王宁、郑达煌、黄俊杰	广州大学	赵阳、李丽	广州大学
110160	舒心、陶禹竹、崔一然、何若珩	沈阳建筑大学	付瑶、张龙巍	沈阳建筑大学
110163	刘丽曼、聂倩宇	北京交通大学	张文	北京交通大学
110169	陈巧莺、黄斯楷、邓荣锟	福州大学	邱文明	福州大学
110170	段有斌、刘志敏、唐诗意、李成才、陈浩、王思凡	昆明理工大学	陆莹、毛志睿	昆明理工大学
110176	田爽晴、孙宇皓	山东建筑大学	薛一冰	山东建筑大学
110177	苏畅、曾喜露	中国矿业大学	韩大庆	中国矿业大学
110179	陈锦怡、吴梦雪、洪翊杰	福州大学	吴木生、杨元传	福州大学
110182	吴凯翔、徐云凡、王德旌、梅嘉懿、李晗	苏州科技大学	刘长春、金雨蒙	苏州科技大学
110189	庄严、宿树旺、骆昕、赵致博	内蒙古科技大学	马成俊	内蒙古科技大学
110195	王方圆、东野瑞欣、周子龙、刘璇、逄雪菲、封嘉仪	山东建筑大学	侯世荣	山东建筑大学
110204	李俊龙、叶可承、周高成、毕传福	合肥城市学院	张苗苗、常佳佳、高婷婷、王坤	合肥城市学院
110218	谈琳、赵佳雯、李雅倩、赵甜歌	兰州交通大学、燕山大学、河北大学	侯秋凤	兰州交通大学
110225	田畔、冯元琳、伍余林翰、王琳东	合肥工业大学	黄杰、王薇、张东凯	合肥工业大学

作品编号	作者	单位名称	指导老师	单位名称
110227	耿银、水勇、刘梦雅	郑州西亚斯学院	王冀豫	郑州西亚斯学院
110229	张淑娴、张云帆、王祎铭	厦门大学	石峰、贾令堃	厦门大学
110233	罗丽峰、徐嘉琳、余悦、武玉洁、毛穿有、齐修远	苏州大学	韩冬辰、孙磊磊、吴国栋	苏州大学
110238	欧阳宛婷、汤诗雨、郭静、王添	合肥工业大学	王旭、杨洋	合肥工业大学
110250	马学苗、陈环玉	山东建筑大学	薛一冰	山东建筑大学
110253	赵楠、郭佳音、李珑玲	西安建筑科技大学	李帆、周志菲	西安建筑科技大学
110257	娜荷芽、朱鑫宇、贺佳、刘曦远	内蒙古工业大学	王崴、甘宇田	内蒙古工业大学
110270	吴琳歆、徐凯、黄新航、黄可杰	福州大学	邱文明	福州大学
110271	张洋	福州大学	郑媛	福州大学
110275	郝煜、张毅磊、王喆、王怡然、彭俊轩	河北工程大学	杨文斌	河北工程大学
110287	李一聪、张明明	山东建筑大学	薛一冰	山东建筑大学
110292	郭梦锦、庞思怡、文一博、徐恩夏	东北大学	刘哲铭	东北大学
110307	李兆熙、梅歌、段阳	河北水利电力学院	李维珊	河北水利电力学院
110313	邓富跃、杨诗莹、陈宛灵、雷婷、郭正强、杨真广	昆明理工大学	陆莹、毛志睿	昆明理工大学
110314	侯天楠、林恺雨、张丞豪、赵城城	昆明理工大学、嘉兴大学、桂林理工大学	陆莹、毛志睿	昆明理工大学
110324	胡梓钺、徐静如	中国矿业大学	朱冬冬	中国矿业大学
110328	王玺凌、黄皓晖、唐雪晴、侯思琪	重庆大学	曾旭东、黄海静	重庆大学
110330	刘佳怡、孙雯昕、吴劲越	河南工业大学	李坤明	河南工业大学
110335	张菱桐、张奕瑾、汤语婷、骆博奥	福州大学	王炜	福州大学
110340	程微	重庆财经职业学院	尹淇淋	重庆财经职业学院
110376	刘冰霞、陈煜远	厦门大学	张乐敏	厦门大学

作品编号	作者	单位名称	指导老师	单位名称
110380	刘洪杰、李杭、李林峰、王盼	重庆交通大学	刘亚南	重庆交通大学
110382	张嘉玥、白珺	内蒙古工业大学	贺龙	内蒙古工业大学
110383	邹旭、包益槟、原建、邱伊雯	南京工业大学	罗靖	南京工业大学
110388	邬秋烨、余文翰、黄旭天	米兰理工大学	徐亮	香港中文大学
110390	王喆、许馨、关项心、林甄欣、尹思源	厦门大学	贾令堃、石峰	厦门大学
110394	高子雄、冯广北	内蒙古工业大学	伊若勒泰	内蒙古工业大学
110399	叶诗情、郑远婧	福州大学	—	—
110405	牛镜然、孙靖萱、史芳伊、宋佳怡	沈阳建筑大学	鞠叶辛	沈阳建筑大学
110411	杨洋、彭影、刘洋源、赵志鹏	去边建筑设计（广州）有限公司	杨洋	去边建筑设计（广州）有限公司
110426	黄奕、洪燕琳、梁宁晖、李雨萱、梁馨文	广州大学	庞玥、李丽	广州大学
110428	韩晓雪	西安建筑科技大学	何泉	西安建筑科技大学
110429	曹世旭、何秋实、王戊晨、巩冠锐	山东建筑大学	郑斐	山东建筑大学
110435	邓戴子阳、王磷萱、林仪铭、倪诗睿、莫旭阳	南京工业大学	倪震宇	南京工业大学
110443	李昂、张光睿、李欣雨、石桢	北京建筑大学	俞天琦	北京建筑大学
110452	郑立红、董畅、孙弈先、于萌、石万亮、赵庆、冉帆、赵星云、李炳云	天津生态城绿色建筑研究院有限公司、CLOUD 畅设计工作室	—	—
110462	涂鸿乾、崔迪	内蒙古工业大学	伊若勒泰	内蒙古工业大学
110465	刘学、厉相栋、李吉星	山东建筑大学	魏瑞涵、侯世荣	山东建筑大学
110466	张曦、曲凌寒、王美霖	昆明理工大学	高蕾、赵虎	昆明理工大学
110480	谭淏蓝、徐境奕	中国矿业大学	林岩	中国矿业大学
110484	于子正、周珈屹	沈阳建筑大学	赵钧	沈阳建筑大学
110487	邓振寰、张媛浩、许伊扬、宋雨鑫、陈宇琴、韩超辉、薛睿宜	天津大学	原野	天津大学

作品编号	作者	单位名称	指导老师	单位名称
110492	王芮佳、刘一子、张茜晨、陈新武	西安建筑科技大学	成辉	西安建筑科技大学
110495	任广宁、王学润	厦门大学	李芝也	厦门大学
110498	汪鑫宇、高紫嫣、秦锦美	武汉轻工大学	覃文柯	武汉轻工大学
110500	罗逸文、李佳玉、许青芽、黄楠、郭纪壮	长安大学	夏博	长安大学
110501	蔡锶琦、郑泽科、袁鑫桐、殷韶璟、王依然、吴树祺、毛明俊	广东工业大学	吉慧、王平、邓寄豫、董泽豪	广东工业大学
110505	刘怡然、张涵	内蒙古工业大学	贺龙	内蒙古工业大学
110508	刘桐昊、刘昱辰、李怀宇、刘云娜、张博宣	重庆大学	黄海静、聂诗东	重庆大学
110509	赵彤山、崔雨洋、田宇琪	东南大学	张彧	东南大学
110511	王亿林、罗雨成、林涵晨、黄颖、朱静、廖陈璐	昆明理工大学	陆莹、毛志睿	昆明理工大学
110513	阿侯伊木、李澳、夏青、王涛	重庆大学	张海滨、王琦、许景峰	重庆大学
110514	严章鑫、邢茹、李秋予	北京建筑大学	刘博、李勤	北京建筑大学
110515	李青芫、赵子意、常竞萱、张蕴泽、王阐亿、吕璐瑶	重庆大学	张海滨、宗德新	重庆大学
110519	樊玉雪、肖芮、陈光奥、徐淑超	潍坊科技学院	周超、武海泉	潍坊科技学院
110524	郭裕旸、王业飞、刘雨森、周一婷、徐杨阳	重庆大学	黄海静	重庆大学
110526	徐子依、康砼曦	重庆大学、成都晟墅建筑设计咨询有限公司	—	—
110532	徐睿雅、秦尉然、陶子俊、杨鑫宇	南京工业大学	吕明扬	南京工业大学
110536	王楚盈、田佳禾、贺莹洁、何诗宇、陈太正	山东建筑大学	陈林	山东建筑大学
110537	汪郑政、崔佳琳、焦佳喜	新疆大学	樊辉、袁萍	新疆大学
110540	王彦坤、闫文青、马华灿	山东建筑大学	宋晋、李晓东	山东建筑大学
110544	阳涛	广西民族大学相思湖学院	隋宏达、刘彦铭	广西民族大学相思湖学院
110547	崔珂、吴静、张梦龙、蒋贤阳、田泽轩	西南民族大学、重庆大学	熊健吾、张埕	西南民族大学

作品编号	作者	单位名称	指导老师	单位名称
110564	阿博莱·阿勒玛斯、孙君荟、巴彦·塞尔江、麦迪娜·马合木提、张疆慧、宋志辉	新疆大学	塞尔江·哈力克	新疆大学
110565	刘慧、程倩、杨宝亚	沈阳建筑大学	鞠叶辛	沈阳建筑大学
110567	雷欣桐、黄晴莎、翟俊伟	广州大学	万丰登、李丽	广州大学
110568	戴含真、黄玥、李可欣、王雅静	厦门大学	石峰	厦门大学
110577	束家祯、陆珮瑶、韩越然、胡潘旭	南京工业大学	罗靖	南京工业大学
110582	曾雅清、李嘉欣、李思静、郝泽厚	厦门大学、北方工程设计研究院有限公司	李立新	厦门大学
110586	丁瑜、赵婉灵、刘怡、黄靖媛、曾兰岚、余紫阳	昆明理工大学	赵虎	昆明理工大学
110589	张理尧、雷雅琪、周家婷、张睿、刘灿、陈艺琳	昆明理工大学	叶涧枫、高蕾	昆明理工大学
110590	姜紫璇、郁岩、周思宇、刘治论	合肥工业大学	王旭、杨洋	合肥工业大学
110591	施若瑜、马玉婷、王琪霞、罗瑜、郑清文、耿如锦	昆明理工大学	陆莹、毛志睿	昆明理工大学
110606	李嘉欣、曾雅清、李思静、郝泽厚	厦门大学	李立新	厦门大学
110609	高蕊婧、陈琳、刘琼、高楚玉、冯雨洁	长安大学	夏博	长安大学
110613	余宗盛、陈思涵、李静怡、陈雅宁	南京工业大学	刘静萍、徐琦	南京工业大学
110614	胡玉梅、高静、胡松艳、夏中天	武汉轻工大学	覃文柯	武汉轻工大学
110616	周紫育、石芮舟、林乐妍、朱钰宁、薛景匀	华侨大学	吴正旺、冉茂宇	华侨大学
110618	刘馨婷、廖雄杰、王禹翔	重庆大学	何宝杰	重庆大学
110619	张鹏、李子晗、郑杭其	昆明理工大学、华侨大学	谭良斌	昆明理工大学
110623	席语萌、张紫欣	西安建筑科技大学	王芳、陈敬	西安建筑科技大学
110627	蔡轶冰、苟歆芸、刘旭敏、李斌	中国矿业大学	林岩、郭斯凡	中国矿业大学
110628	张浩鸣、杨俊、周琴、温家诚	广州大学	赵阳、李丽	广州大学
110632	白苏日吐、李子涵	天津大学、内蒙古工业大学	朱丽	天津大学

作品编号	作者	单位名称	指导老师	单位名称
110646	李欣颖、刘好	四川农业大学、大连民族大学	孙嘉男、李响	北京优优星球教育科技有限公司
110647	杨渊钞、王思琦、康加明、王子玉	华侨大学	黄鹭红、薛佳薇	华侨大学
110649	张天晨、贺亚泉	苏州科技大学	刘长春、金雨蒙	苏州科技大学
110653	杜嘉怡、单心怡、雷佳衡、余柯颖、杨楠、蒲瑞瑞	昆明理工大学	—	—
110654	丁怡文	吉林省吉规城市建筑设计有限责任公司	—	—
110655	王伟鹏、马琳浩、杨鑫	兰州理工大学	闫幼锋	兰州理工大学
110658	张柏源、林晨啸、刘培胜、周子璇、刘文宇	山东建筑大学	郑斐、刘文、王月涛、王津红	山东建筑大学
110659	徐雪健、李哲嘉、李晓涵、靳琬頔、衡优朵、麻峰瑞、石礼贤	天津大学建筑学院	杨崴、王崎、张睿	天津大学建筑学院
110660	朱子琪、杨周圣	中国矿业大学	郭斯凡、林岩	中国矿业大学
110666	刘志波、刘洋、刘晓莹、侯玉淑、张畅、朱怡萱、高绮绮、张佳旭、柳泽	中建八局发展建设有限公司设计研究院	王剑涛	中建八局发展建设有限公司设计研究院
110674	钱明星、姜雪峰	江苏博研工程设计咨询有限公司	钱明星	江苏博研工程设计咨询有限公司
110681	杨绚、张涵旗、盛菁宇	厦门大学	石峰、洪晓强、向立群、王波、贾令堃、张喆涵	厦门大学
110682	董馨月、占瑶、王奕丹、许安堃、范宇飞、金宇帆	华南理工大学	肖毅强、殷实	华南理工大学
110685	李露昕、文亚、朱娜	西安建筑科技大学	陈敬、王芳	西安建筑科技大学
110690	舒情、李子晴、何柳莹、谭扬深	厦门大学	石峰	厦门大学
110692	唐浩鸿、赵明哲、孔淋	重庆大学	张海滨	重庆大学
110693	谢欣欣、刘文洁、冯一帆、陈通	昆明理工大学	谭良斌	昆明理工大学
110698	田如俊、谢佳利、黄恒瑞、龙我在、侯竹筠	重庆大学	张海滨	重庆大学

作品编号	作者	单位名称	指导老师	单位名称
110699	隋月超	川百建工集团有限责任公司林芝分公司	小橘子	川百建工集团有限责任公司林芝分公司
110701	夏陈浩、郑一涵、蓝文涛	南京工业大学	罗佳宁	南京工业大学
110704	周逸凡、周美言、谢文彬、李宇馨、李瑞立	长安大学	夏博	长安大学
110705	梅静波、张琳悦、李倩茹、卜天一	南京工业大学	罗靖、刘峰	南京工业大学
110706	梅雨蝶、陈默、桂琦、张怡帆、冯于泽、史浩均	南京工业大学	舒欣	南京工业大学
110708	王毅斐、田二伟、张荣峥	山东建筑大学	薛一冰、何文晶	山东建筑大学
110711	王玙、李庭杉、林欣、连菁菁	福州大学	郑媛、黄斯	福州大学
110713	安笑、刘小菁、刘真真、彭浩源	山东建筑大学	高雪莹	山东建筑大学
110722	马佳星、刘文婧、武昊楠	西南民族大学	熊健吾、张埂	西南民族大学
110732	史雨禾、李梦娟	中国矿业大学	林岩	中国矿业大学
110734	庞志祥、张春杰、武雨晨、黄向文、刘纪源	大连理工大学	李国鹏	大连理工大学
110737	张冉晨、史国崇、刘英丽、房俞含、张彦泽、焦飞鹏	大连理工大学	李国鹏	大连理工大学
110738	张浩克、杨航、王鹏举	西安建筑科技大学	王琰、陈敬	西安建筑科技大学
110742	陈鑫源、陈卓玮、张雯宣、温佳坤、石斐斐、耿颜悦	湖南大学	蒋甦琦、陈毅兴	湖南大学
110744	卢丙坤、孙速龙、李明远、姚心宇、王秉仁、张凯翔、檀楚宁、刘莹	天津大学	郭娟利、李伟	天津大学
110745	李庆斌、康雲霄、杨雅洁、郭其欣	山东建筑大学	何文晶	山东建筑大学
110746	郑好	中南林业科技大学	何玮	中南林业科技大学
110752	张晓雨、王思雅、张宁舟、杨飞飞	合肥工业大学	王旭、杨洋	合肥工业大学
110759	陈相如、严方希、康天硕、刘宇杰、徐祺煜	长安大学	夏博	长安大学
110765	卢韵莹、郭赛飞	中国矿业大学	邵泽彪、马全明	中国矿业大学
110767	王学润、任广宁	厦门大学	李芝也	厦门大学

2024台达杯国际太阳能建筑设计竞赛办法
Brief for International Solar Building Design Competition 2024

竞赛宗旨：

本竞赛旨在提高农村的防灾减灾能力，贯彻落实《"十四五"国家综合防灾减灾规划》精神，围绕如何构建具有韧性的村镇基础设施、提升农村地区的防灾减灾能力展开，重点关注村镇的规划布局、建筑结构、公共设施、生态环境、绿色能源技术、低碳可持续发展等方面，以期打造出安全、可持续、富有活力的农村社区，为乡村基础设施建设提供新思路，为乡村振兴注入新活力。

竞赛主题：阳光·乡村韧性
竞赛题目：

      张家口市怀安县二堡子村建设项目

      保定市清苑区李八庄村建设项目

主办单位：国际太阳能学会

      中国建设科技集团股份有限公司中央研究院

      中国建筑设计研究院有限公司

承办单位：国家住宅与居住环境工程技术研究中心

支持单位：保定市荣清国控集团有限公司

      际源控股（北京）有限公司

冠名单位：台达集团

技术支持：北京天正软件股份有限公司

宣发支持：中国建筑学会学生分会

      《世界建筑》

评委会专家：Deo Prasad：澳大利亚科技与工程院院士、澳大利亚勋章获得者、澳大利亚新南威尔士大学教授。

      杨经文：马来西亚汉沙杨建筑师事务所创始人、2016年梁思成建筑奖获得者。

      Peter Luscuere：荷兰代尔伏特理工大学建筑系教授。

      崔愷：中国工程院院士、全国工程勘察设计大师、中国建筑设计研究

The Purpose of Competition：

Competition 2024 aims to enhance disaster prevention and mitigation capacities in rural areas of China. It is an implementation of the 14th Five-Year Plan for National Comprehensive Disaster Prevention and Mitigation. We hope to stimulate discussions on two themes: 1) how to construct resilient infrastructure in villages and towns, and 2) how to enhance disaster resistance and mitigation capacities in rural areas. With a focus on rural development, which covers planning layout, building structure, public facilities, ecological environment, green energy technologies, low-carbon practices, and sustainable development among other aspects, our goal is to create safe, sustainable and vibrant rural communities by exploring innovative approaches for rural infrastructure construction and injecting fresh vitality into rural revitalization projects.

Competition Theme: SUNSHINE & RURAL RESILIENCE

Competition Tasks:

Construction Project of Erpuzi Village, Huai'an County, Zhangjiakou City
Construction Project of Libazhuang Village, Qingyuan District, Baoding City

Hosts:

International Solar Energy Society（ISES）
Central Research Institute of China Construction Technology Group Co., Ltd.（CCTC）
China Architecture Design & Research Group Co., Ltd.（CADG）

Organizer:

National Engineering Research Center for Human Settlements of China

Supported by:

Baoding Rongqing State Control Group Co., Ltd.
Jiyuan Holdings（Beijing）Co., Ltd.

院有限公司总建筑师。

鉾岩崇：株式会社佐藤综合计画代表取缔役社长。

张利：全国工程勘察设计大师、清华大学建筑学院院长。

钱锋：全国工程勘察设计大师、同济大学建筑与城市规划学院教授、博士生导师。

仲继寿：中国建筑设计研究院有限公司总工程师、中国建筑学会主动式建筑专业委员会主任委员。

黄秋平：华东建筑设计研究院有限公司总院总建筑师。

冯雅：中国建筑西南设计研究院顾问总工程师，中国建筑学会建筑物理分会建筑热工与节能专业委员会副主任，重庆大学建筑城规学院教授、博士生导师。

张宏：东南大学建筑学院教授，建筑技术与科学研究所所长。

杨明：华东建筑设计研究院有限公司首席总建筑师、教授级高级工程师。

袁烽：同济大学建筑与城市规划学院教授、博士生导师、副院长。

宋晔皓：清华大学建筑学院长聘教授、博士生导师、副系主任，清华大学建筑与技术研究所所长，清华大学建筑设计研究院副总建筑师。

任军：天友集团首席建筑师。

刘恒：中国建筑设计研究院有限公司副总建筑师、绿色建筑设计研究院院长、教授级高级建筑师。

组委会成员：由主办单位、承办单位及冠名单位相关人员组成。办事机构设在国家住宅与居住环境工程技术研究中心。

Sponsor:

Delta Electronics

Technical Support:

Beijing Tangent Software Co., Ltd.

Media and Publicity Support:

Student Affairs Committee of the Architectural Society of China
World Architecture

Experts of the Jury Panel:

Mr Deo Prasad: Academician of Australian Academy of Technological Sciences and Engineering, Winner of the Order of Australia, and Professor of University of New South Wales, Sydney, Australia.

Mr King Mun Yeang: Founder of T. R. Hamzah & Yeang Sdn. Bhd. of Malaysia, and Winner of Liang Sicheng Architecture Award 2016.

Mr Peter Luscuere: Porfessor of the Department of Architecture, Delft University of Technology, Netherlands.

Mr Cui Kai: Academician of China Academy of Engineering, Master of National Engineering Survey and Design of China, and Chief Architect of China Architecture Design and Research Group Co., Ltd. (CADG).

Mr Takashi Hokoiwa: Chairman of AXS Satow Inc., Janpan.

Mr Zhang Li: Master of National Engineering Survey and Design of China, Dean of School of Architecture, Tsinghua University.

Mr Qian Feng: Master of National Engineering Survey and Design of China, Professor and Doctoral Supervisor of College of Architecture and Urban Planning of Tongji University.

Mr Zhong Jishou: Chief Engineer of China Architecture Design and Research Group Co., Ltd. (CADG); Chairman of Committee of Active House of the Architectural Society of China (ASC).

Mr Huang Qiuping: Chief Architect of East China Architectural Design & Research Institute Co., Ltd. (ECADI).

Mr Feng Ya: Chief Consulting Engineer of China Southwest Architectural Design

设计任务书及专业术语等附件：

附件 1：张家口市怀安县二堡子村建设项目

附件 2：保定市清苑区李八庄村建设项目

附件 3：专业术语

奖项设置及奖励形式：

综合奖：

一等奖作品：两个项目分别评审出 1 个一等奖作品，共计 2 个，颁发奖杯、证书及人民币 100000 元奖金（税前）；

二等奖作品：两个项目共评审出 4 个，颁发奖杯、证书及人民币 20000 元奖金（税前）；

三等奖作品：两个项目共评审出 6 个，颁发奖杯、证书及人民币 5000 元奖金（税前）；

优秀奖作品：两个项目共评审出 20 个，颁发奖杯、证书及人民币 1000 元奖金（税前）。

入围奖作品：两个项目共评审出 30 个，颁发证书。

技术专项奖：名额不限，颁发证书。

设计创意奖：名额不限，颁发证书。

and Research Institute Co., Ltd.; Deputy Director of Special Committee of Building Thermal and Energy Efficiency of the Architectural Society of China (ASC); Professor and Doctoral Supervisor of School of Architecture and Urban Planning of Chongqing University.

Mr Zhang Hong: Professor of School of Architecture of Southeast University, Director of Institute of Building Technology and Science.

Mr Yang Ming: Chief Architect and Professor-level Senior Engineer of East China Architectural Design and Research Institute Co., Ltd. (ECADI).

Mr Yuan Feng: Professor, Doctoral Supervisor and Deputy Dean of College of Architecture and Urban Planning of Tongji University.

Mr Song Yehao: Tenured Professor, Doctoral Supervisor and Deputy Dean of School of Architecture of Tsinghua University; Director of Architecture and Technology Institute of Tsinghua University; Deputy Chief Architect of Architectural Design and Research Institute of Tsinghua University (THAD).

Mr Ren Jun: Chief Architect of Tenio Group.

Mr Heng Liu: Deputy Chief Architect of China Architecture Design and Research Group Co., Ltd. (CADG); Director of Green Architecture Design and Research Institute of CADG, Professor-level Senior Architect.

Members of the Organizing Committee:

The Organizing Committee, composed of selected members from the Hosts, Organizers, and Title Sponsor, assumes responsibility for the day-to-day operations of the Competition. Its office is situated at the National Engineering Research Center for Human Settlements of China.

Appendix for the Design Specifications and Technical Terms:

Appendix 1：Construction Project of Erpuzi Village, Huai'an County, Zhangjiakou City

Appendix 2：Construction Project of Libazhuang Village, Qingyuan District, Baoding City

Appendix 3：Technical Terms

Awards Setting and Reward Format:

Comprehensive Awards：

First Prize：The two competition tasks will each yield one First Prize. The two winners will be awarded First Prize with trophies, certificates and a cash prize of RMB 100,000 (pre-tax) for each team.

Second Prize：The Second Prize will be awarded to a maximum of four teams, each receiving trophies, certificates, and a cash prize of RMB 20,000 (pre-tax)．

Third Prize：The Third Prize will be awarded to a maximum of six teams, each receiving trophies, certificates, and a cash prize of RMB 5,000 (pre-tax).

Honorable Mention：A total of 20 teams will be awarded with certificates and a

参赛要求：

1. 欢迎建筑设计院、高等院校、研究机构、研发生产企业等单位，组织专业人员组成竞赛小组参加竞赛。

2. 请参赛者访问 www.isbdc.cn，按照规定步骤填写注册表，提交后会得到唯一的注册号，即作品编号，一个作品对应一个注册号。提交作品时把注册号标注在每个作品的左上角，字高 6mm。注册时间 2024 年 3 月 25 日至 2024 年 8 月 15 日。

3. 参赛者同意组委会公开刊登、出版、展览、应用其作品。

4. 被编入获奖作品集的作者，应配合组委会，按照出版要求对作品进行相应调整。

注意事项：

1. 参赛作品电子文件须在 2024 年 9 月 15 日前提交组委会，请参赛者访问 www.isbdc.cn，并上传文件，不接受其他递交方式。

2. 作品中不能出现任何与作者信息有关的标记内容，否则将视其为无效作品。

3. 组委会将及时在网上公布入选结果及评比情况，将获奖作品整理出版，并对获奖者予以表彰和奖励。

4. 获奖作品集首次出版后 30 日内，组委会向获奖作品的创作团队赠样书 2 册。

5. 竞赛活动消息发布、竞赛问题解答均可登录竞赛网站查询。

cash prize of RMB 1,000 (pre-tax) for each team.

Nomination：A total of 30 teams will receive certificates.

Technical Excellence Prize：Unlimited number, certificates issued.

Design Creativity Prize: Unlimited number, certificates issued.

Participants Requirements:

1. Professionals from architecture institutions, colleges, research organizations, manufacturers and corporations are sincerely invited to assemble specialized teams and participate in the Competition.

2. The online registration form should be completed by visiting <http：//www.isbdc.cn>. Upon successful submission, an automated Register Number will be issued to the applicant, serving as the exclusive and valid code for each submitted work. When submitting the final work, it is important to clearly mark the Register Number at the upper-left corner with a fixed height of 6mm. The registration period is valid from March 25th, 2024 to August 15th, 2024.

3. The Participant（s）(also referred to as "Author（s）", "Applicant（s）" or "Competition Team（s）") acknowledge and authorize the Organizing Committee of the Competition to publish, print, exhibit and utilize/practice their competition works.

4. The authors of the entries hereby acknowledge and agree to have their works compiled in the book of Awarded Works Collection for Competition 2024（referred to as "Collection Book"）for further editing and/or adjustment, in accordance with the publishing requirements, under the guidance of the Organizing Committee.

Important Notes:

1. All competition entries must be electronically submitted via www.isbdc.cn by September 15th, 2024. No alternative methods of submission will be accepted.

2. No disclosure of the authors' information is permitted in the submitted files, and any violation of this rule will result in disqualification from the Competition.

3. The Organizing Committee is responsible for publishing the evaluation process and results online in a timely manner, as well as issuing awards and publishing the Collection Book.

4. The awarded teams will each receive two copies of the Collection Book from the Organizing Committee within 30 days since its initial publication.

5. All competition-related news and Q & A can be accessed on the website for inquiries.

所有权及版权声明：

参赛者提交作品之前，请详细阅读以下条款，充分理解并表示同意。

依据中国有关法律法规，凡主动提交作品的"参赛者"或"作者"，主办方认为其已经对所提交的作品版权归属作如下不可撤销声明：

1. 原创声明

参赛作品是参赛者原创作品，未侵犯任何他人的任何专利、著作权、商标权及其他知识产权；该作品未在报刊、网站及其他媒体公开发表，未申请专利或进行版权登记，未参加过其他比赛，未以任何形式进入商业渠道。参赛者保证参赛作品终身不以同一作品形式参加其他的设计比赛或转让给他方。否则，主办单位将取消其参赛、入围与获奖资格，收回奖金、奖品及并保留追究法律责任的权利。

2. 参赛作品知识产权归属

为了更广泛地推广竞赛成果，所有参赛作品除作者署名权以外的全部著作权归竞赛承办单位及冠名单位所有，包括但不限于以下方式行使著作权：享有对所属竞赛作品方案进行再设计、生产、销售、展示、出版和宣传的权利；享有自行使用、授权他人使用参赛作品用于实地建设的权利。竞赛主办方对所有参赛作品拥有展示和宣传等权利。其他任何单位和个人（包括参赛者本人）未经授权不得以任何形式对作品转让、复制、转载、传播、摘编、出版、发行、许可使用等。参赛者同意竞赛承办单位及冠名单位在使用参赛作品时将对其作者予以署名，同时对作品将按出版或建设的要求作技术性处理。参赛作品均不退还。

3. 参赛者应对所提交作品的著作权承担责任，凡由于参赛作品而引发的著作权属纠纷均应由作者本人负责。

Declaration on Ownership and Copyright:

Prior to submitting the competition works, all participants are required to carefully read and comprehend the following clauses, and fully accept the stipulated terms.

In accordance with relevant national laws and regulations, the competition Hosts hereby affirm that all "participants" or "authors", who voluntarily participate in the competition, irrevocably declare the ownership and copyright of their submitted works as follows:

1. Declaration on Originality

Each entry submitted by the participants is hereby declared to be original and free from any infringement upon third-party patents, copyrights, trademarks or other forms of intellectual property rights. The entry has not been published through any media channels, including but not limited to newspapers, periodicals and websites. Furthermore, it has not been subjected to patent or copyright registrations, nor involved in any other competitions or commercial market practices. All participants assure that their works will neither be submitted to any other competition in the same form nor transferred to third parties at any time. In the event of a breach of this declaration, the competition Hosts reserve the right to disqualify participants, revoke any awarded prizes, and pursue legal action for recourse.

2. Ownership of Intellectual Property Rights

The participants hereby agree to assign all copyrights of their entries to the Competition Organizer and Title Sponsor, while retaining their rights of authorship, with the vision of promoting the entries and the competition. The Organizer and Sponsor are entitled to exercise all benefits derived from these copyrights, including but not limited to redesigning, producing, selling, displaying, publishing and promoting the works; as well as applying them in field construction projects for their own use or by authorizing third-party use. The competition Hosts are granted rights to display and disseminate the submitted works. Without prior authorization, no organization or individual may transfer, copy, reprint, disseminate, extract, edit, publish, or license the works including the authors themselves. Participants agree that when utilized by the Organizer and Sponsor their names will be associated with their works and necessary revisions or adjustments will be made in accordance with technical requirements for publication and field implementation. All entries will not be returned after submission.

3. Participants shall assume full responsibility for the originality of their entries, hereby acknowledging that any disputes arising from copyright ownership shall be exclusively their own responsibility.

声明：

1. 参与本次竞赛的活动各方（包括参赛者、评委和组委），即表明已接受上述要求。

2. 本次竞赛的参赛者，须接受评委会的评审决定作为最终竞赛结果。

3. 组委会对竞赛活动具有最终的解释权。

4. 为维护参赛者的合法权益，主办方特提请参赛者对本办法的全部条款、特别是"所有权及版权"声明部分予以充分注意。

国际太阳能建筑设计竞赛组委会

网　　址：www.isbdc.cn

组委会联系地址：北京市西城区车公庄大街 19 号（100044）

国家住宅与居住环境工程技术研究中心

联系人：鞠晓磊、张星儿、郑晶茹

联系电话：86-010-88377501、86-010-88377372

电子邮箱：isbdc2021@126.com　QQ 交流群：49266054、237927709

微信公众号：国际太阳能建筑设计竞赛

Announcement:

1. The Declaration mentioned above is hereby unanimously agreed upon, without any reservations, by all parties involved in the competition, including participants, jury members, and the Organizing Committee.

2. The participants hereby accept the decision of the Jury Panel as an ultimate judgement in its entirety.

3. The Organizing Committee reserves the exclusive right of final interpretation for this Competition.

4. To safeguard the legitimate rights and interests of the participants, it is strongly encouraged by the Organizing Committee that all participants thoroughly review the entire provisions outlined in the Guide, with particular attention to the section on "Declaration on Ownership and Copyrights".

Organizing Committee of Intentional Solar Building Design Competition
Website: www.isbdc.cn
Address: No. 19 Che Gong Zhuang Avenue, Xi Cheng District, Beijing（Postcode 100044）
National Engineering Research Center for Human Settlements of China
Liaisons: Ju Xiaolei, Zhang Xing'er, Zheng Jingru
Telephone：86-010-88377501、86-010-88377372
Email：isbdc2021@126.com　QQ（Chat group）：49266054, 237927709
Official WeChat Account: International Solar Building Construction Competition

附件1：张家口市怀安县二堡子村建设项目

Appendix 1：Construction Project of Erpuzi Village, Huai'an County, Zhangjiakou City

1. 项目背景

项目地位于河北省张家口市怀安县二堡子村内，东经114.62°，北纬40.68°，临近京藏高速、京新高速，高铁线路由场地北侧东西向穿过。距张家口市35km，距北京市中心200km。项目与二堡子村、北郭家窑村村民居住区紧邻，直线距离1km内。

二堡子村域内有成片稻田，北临大洋河、南依金沙滩林场、东西有大片林地、农田。场地周围50km范围内已有萌兔部落、天鹅湖小镇、大井沟、昭化寺、草原度假村、温泉度假村等旅游景点。全村共164户、402人。现有常住人口132户、327人。

全村种植优质水稻约666.67hm²，玉米种植约313.33hm²，水蜜桃采摘园2hm²。全村有经济林7.52hm²，主要品种为杏扁。水产养殖与家禽养殖较少。近几年开始发展乡村旅游业，以农业景观为主，提升集体经济为目标，增加经济增长点的同时保护周围生态景观环境。

2. 项目要求

本项目旨在常态下提升当地居民的居住品质，提供乡村旅游配套设施的同时，让乡村在地震等特殊风险情境下有抵抗风险能力，在断水断电等极端情况下

1. Project Background

The Project is located in Erpuzi Village, Huai'an County, Zhangjiakou City, Hebei Province, 114.62° east longitude, and 40.68° north latitude. Close to the Beijing-Tibet Expressway and Beijing-Xinjiang Expressway, with a high-speed rail line running east-west to the north of the site, the Project is 35km from Zhangjiakou City and 200km from Beijing. It is adjacent to the residential areas of Erpuzi Village and North Guojiayao Village, each with a linear distance of less than 1km.

There are stretches of paddy fields in the Erpuzi village adjacent to Dayang River in the north, the Golden Beach Forest Farm in the south, and vast forests and farms in the east and west. Scattered within a 50km radius of the site are Bunny Tribe, Swan Lake Town, Dajinggou Gully, Zhaohua Temple, grassland resorts, hot spring resorts and other tourist attractions. The village has a registered population of 402 people from 164 households, while the actual residents are 327 people from 132 households.

The village cultivates 666.67hm² of high-quality paddy, 313.33hm² of corns and 2hm² of peach orchards. The village also has 7.52hm² of economic forests, primarily consisting of apricot trees. However, there are fewer fish breeding or poultry raising business in the area. In recent years, rural tourism businesses have begun to develop with a focus on agricultural landscaping. The village aims to improve its collective economy performance by increasing economic growth while protecting the ecological landscapes and environment.

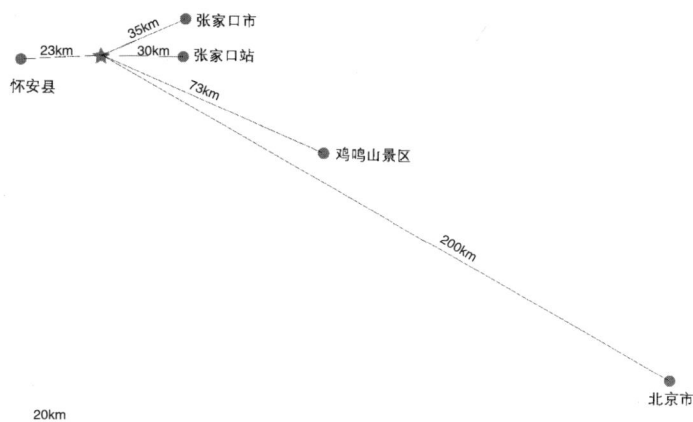

图1 项目所在区位图
Figure 1　Geographic Location of the Project

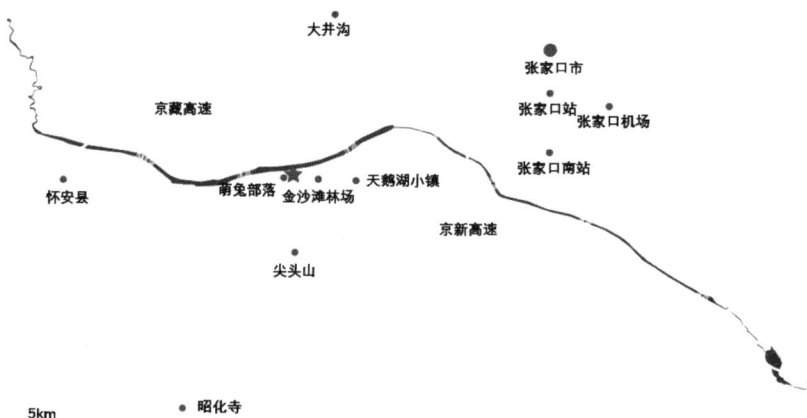

图2 项目所在地周边资源配置图
Figure 2　Peripheral Resource Configuration Diagram of the Project

有自维持能力。计划在二堡子村西北侧建设民居及公共服务建筑。民居要考虑兼具民宿功能，公共服务建筑平时服务村民，提供文化娱乐、商业服务，受灾断水断电时可作为应急避难所。

3. 气候条件

场地地处中纬度。属东亚大陆性季风气候。中温带亚干旱区。年平均气温8.3℃，最高气温37.2℃，最低气温−22.6℃。地域日照时数2800~3100h，太阳总辐射为1500~1700kW·h（m²·d），为太阳能较丰富的地区。年平均降水量407.4mm。场地抗震设防烈度为8级，近几年没有遇到过洪涝灾害，但存在冰雹灾害风险。

4. 基础设施

项目场地周边交通较为便利，给水排水、供电等市政配套设施可由场地周边市政道路接入。

2. Project Requirements

The Project aims to achieve three goals: i. improving the living quality of local residents under normal conditions; ii. Providing facilities to support rural tourism; and withstanding risks under extreme situations such as earthquakes, while enabling self-sustaining capabilities during water and power disruptions. It is planned to construct residential buildings and public service buildings in the northwest side of Erpuzi Village. The public service buildings should serve the villagers with cultural entertainment and commercial services in peacetime, and transition into emergency shelters during disasters when water and power supplies are disrupted.

3. Climate Conditions

The site is located within a mid-latitude region and belongs to the middle-temperate sub-arid zone, which is characterized by an East Asian continental monsoon climate. The average annual temperature is 8.3℃, with the highest recorded at 37.2℃ and the lowest at −22.6℃. Regional sunshine hours range from 2,800 to 3,100 hours, while the total solar radiation varies from 1,500~1,700kW·h(m²·d), proving it to be an area abundant in solar energy. The annual precipitation in the area is 407.4 mm. The

表1

月份	空气温度（℃）	相对湿度（%）	水平面日太阳辐射[kW·h（m²·d）]	大气压力（kPa）	风速（m/s）	土地温度（℃）	月供暖度日数（℃·d）	月供冷度日数（℃·d）	日照时数（h）
1 月	−7.8	37.5	2.30	89.0	2.4	−14.6	800	0	229
2 月	−4.1	33.4	3.16	88.9	2.4	−9.2	619	0	237
3 月	2.7	31.0	4.44	88.7	2.5	0.4	474	0	302
4 月	11.4	30.9	5.18	88.3	2.6	11.0	198	42	327
5 月	18.2	39.0	5.99	88.2	2.4	19.0	0	254	309
6 月	22.3	50.2	6.18	87.9	2.1	23.0	0	369	267
7 月	24.0	63.1	5.15	87.9	1.8	22.9	0	434	207
8 月	22.5	64.3	4.44	88.3	1.7	20.8	0	388	218
9 月	17.3	56.0	4.25	88.7	1.9	15.8	21	219	224
10 月	10.2	45.5	3.31	89.0	2.0	7.9	242	6	258
11 月	1.0	39.2	2.39	89.1	2.2	−2.5	510	0	200
12 月	−5.9	39.2	1.95	89.1	2.2	−11.2	741	0	232
年平均/总数据	9.4	44.2	4.06	88.6	2.2	7.0	3605（总）	1712（总）	3008.7（总）

Table 1

Month	Air Temperature (℃)	Relative Humidity (%)	Solar Radiation on Horizontal Day [kW·h(m²·d)]	Barometric Pressure (kPa)	Velocity (m/s)	Land Temperature (℃)	Monthly Heating Temperatures and Days (℃·d)	Monthly Cooling Temperatures and Days (℃·d)	Sunshine Duration (h)
January	−7.8	37.5	2.30	89.0	2.4	−14.6	800	0	229
February	−4.1	33.4	3.16	88.9	2.4	−9.2	619	0	237
March	2.7	31.0	4.44	88.7	2.5	0.4	474	0	302
April	11.4	30.9	5.18	88.3	2.6	11.0	198	42	327
May	18.2	39.0	5.99	88.2	2.4	19.0	0	254	309
June	22.3	50.2	6.18	87.9	2.1	23.0	0	369	267
July	24.0	63.1	5.15	87.9	1.8	22.9	0	434	207
August	22.5	64.3	4.44	88.3	1.7	20.8	0	388	218
September	17.3	56.0	4.25	88.7	1.9	15.8	21	219	224
October	10.2	45.5	3.31	89.0	2.0	7.9	242	6	258
November	1.0	39.2	2.39	89.1	2.2	−2.5	510	0	200
December	−5.9	39.2	1.95	89.1	2.2	−11.2	741	0	232
Yearly Average/Total Data	9.4	44.2	4.06	88.6	2.2	7.0	3,605 (Total)	1,712 (Total)	3,008.7 (Total)

5. 竞赛场地

项目用地面积 8260m²，建设用地面积 4275m²，地形较为平坦，除东北侧有 2m 高差外，其余场地高差不到 0.5m。场地在乡村会客厅规划区内，区域内主要以民居民宿、文旅体验场所为主，北侧为水池，西侧为已建成萌兔部落旅游景点，南侧为菜园研学基地，东南侧为民居和民宿。

6. 设计要求

1）在给定的竞赛用地范围内设计具有民宿功能的民居不少于 4 套，总建筑面积约 600m²，限高 9m。要求设置独立院落，每户不大于 300m²，3 个及以上独立卧室，2 个及以上卫生间。

2）村民及旅游配套服务建筑，建筑面积约 1200m²，限高 9m，包括平时的文化娱乐、商业服务，受灾断水断电时可供紧急卫生服务、救援物资储藏、周围村民应急避难所等功能。

3）需考虑区域内与周围功能相关的场地设计并设置集中停车位不少于 10 个的停车场。

4）规划与建筑单体设计要考虑应对地震、冰雹等自然灾害的能力，注重韧性设计，运用创新性建造等技术，具备功能可变潜力。

5）从民居建设、运行及全过程考虑低碳减排的方法和实施。

6）考虑项目的可实施性，技术的经济性和普适性。

seismic fortification intensity (SFI) is set at level 8. Although there have been no flood disasters in recent years, hail disasters pose a risk in this area.

4. Infrastructure Facilities

The project site benefits from convenient traffic conditions. Municipal roads surrounding the site provide easy access to essential facilities such as water supply, drainage, and power supply.

5. Site Conditions

The Project covers a land area of 8, 260 square meters, with a construction area of 4, 275 square meters. The terrain is relatively flat, except for the northeast side which has a height difference of 2m, while the rest of the site has a height difference less than 0.5m. The site is located within an area planned as "Village Hall", which mainly consists of residential accommodation, cultural experience and tourism. Around the site, there is a pool located on the north side, an already-built tourist attraction called Bunny Tribe on the west side, a Vegetable Garden for Research and Education on the south side, and residential houses with homestay services on the southeast side.

6. Design Specifications

I. The Project shall accommodate a minimum of four sets of residential houses offering homestay services, with a total construction area of approximately 600 square meters and a height restriction of 9m. Each household must have an independent courtyard not exceeding 300 square meters in size, alongside at least three separate bedrooms and no less than two bathrooms.

II. The Villagers and Tourism Service Building, with a construction area of

图 3 现有稻香景观桥
Figure 3　Fragrant Paddy Bridge for Scenic View

图 4　站在农场由北往南看二堡子村和远山
Figure 4　A View from the Farm Facing South and Overlooking Erpuzi Village and Remote Mountains

图 5　竞赛场地及邻近池塘航拍图
Figure 5　Aerial Picture of Project Site and Nearby Pool

图 6　竞赛场地周围池塘、稻田、二堡子村航拍图
Figure 6　Aerial Picture of Project Site and Nearby Pool, Paddy Fields and Erpuzi Village

图 7　二堡子村航拍图
Figure 7　Aerial Picture of Erpuzi Village

7. 评比办法

1）由组委会审查参赛资格，并确定入围作品。

2）由评委会评选出竞赛获奖作品。

8. 评比标准

1）参赛作品须符合本竞赛"作品要求"的内容。

2）作品应具有原创性，鼓励创新。

3）作品应满足使用功能、绿色低碳、安全健康的要求，建筑技术与太阳能利用技术具有适配性。

4）作品应充分体现太阳能利用技术对降低建筑使用能耗的作用，在经济、技术层面具有可实施性。

approximately 1, 200 square meters and height restriction at 9m, shall provide cultural entertainment and commercial services during normal times. In times of disasters, the Building shall transition into an emergency shelter to offer surrounding villagers essential resources such as water supply, power backup, sanitation services, and rescue materials storage.

III. The site design should consider the functional relationship with surrounding areas, including a centralized parking lot with at least10 parking spots.

IV. Planning and architectural design should consider the ability to cope with natural disasters such as earthquakes and hails with a focus on resilience design. Innovative construction technologies are encouraged and the design should demonstrate its potential for functional variability.

V. The design should incorporate possible methods and implementations to achieve low-carbon emission and carbon reduction throughout the entire life cycle of the residential buildings, from construction to operation.

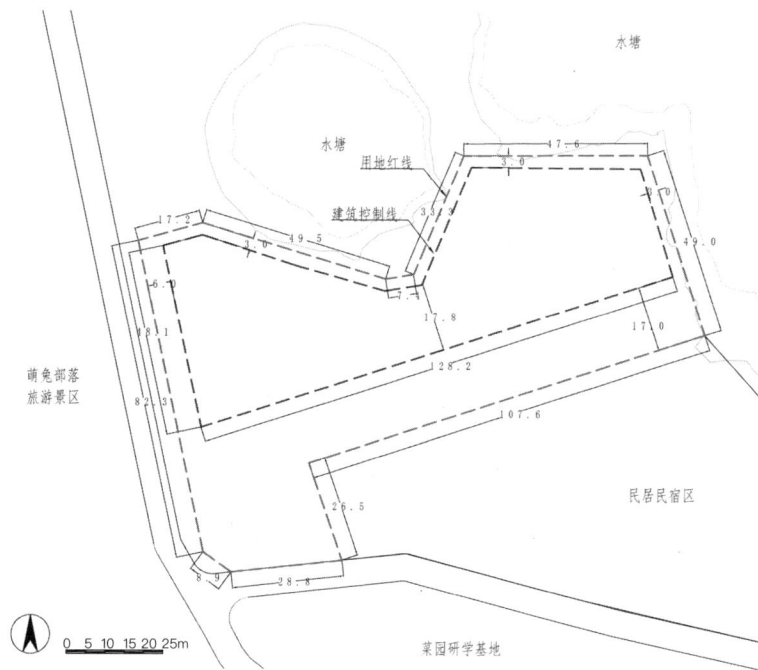

图 8 项目用地平面图
Figure 8 Plain Graph of the Project

5）作品应充分考虑区域及建筑本体韧性设计，考虑极端灾害响应，缓解区域社会、经济、环境压力。

表2

评比指标	指标说明	分值
规划与建筑设计	规划布局、建筑空间组合、功能流线组织、空间灵活性、建筑艺术	40
太阳能主、被动技术	利用建筑设计与建筑构造实现建筑隔热与通风节能降碳	25
	太阳能光伏、光热等主动太阳能技术的利用实现建筑能源自维持	
区域及建筑韧性设计	场地应急设计、建筑抗震设计、应急物资储备	15
采用的其他技术	建造与运行过程中的绿色、低碳、节能等技术	10
可操作性	作品的可实施性，技术的经济性和普适性	10

9. 作品要求

1）设计深度达到方案设计深度要求，主要技术应有相关的技术图纸和指标。作品图面、文字表达清楚，数据准确。

VI. The implementation feasibility of the Project, the cost-efficiency and general applicability of the adopted technologies should be taken into consideration.

7.Appraisal Approach

I. The Organizing Committee will review the entry qualification and determine the shortlisted works.

II. The Jury Panel is in charge of determining the winning works.

8.Appraisal Indicators

I. Entries must adhere to the requirements listed in the "Design Specifications" section.

II. Entries should be original, and innovative designs are encouraged.

III. Entries should meet the requirements for functionality, an eco-friendly orientation, low-carbon emissions, safety and health. The solar energy technologies adopted should be compatible with construction techniques.

IV. Entries should emphasize the role of solar technology in reducing energy consumption of buildings, while being economically and technically feasible.

V. Entries should incorporate resilience design into regional planning and architectural design, taking into consideration extreme disaster response to mitigate pressures on local society, economy, and environment.

Table 2

Appraisal Indicators	Explanations on the Indicators	Score
Planning and Architectural Design	Layout, architectural space combination, functions and circulation organization, spatial flexibility, architectural artistic quality	40
Active and Passive Solar Technologies	To achieve heat insulation, ventilation, energy-saving, and carbon reduction through architectural design and building structure	25
	To achieve energy self-sufficiency through the implementation of active solar energy technologies such as solar photovoltaic and solar thermal	
Regional and Building Resilience Design	Site design for emergencies, seismic design for buildings, emergency materials storage	15
Other Technologies Applicable	Green, low-carbon and energy-saving technologies applied in the entire process of architectural construction and operation	10
Practicability	The implementation possibility of the works, the cost-efficiency and general applicability of the technologies	10

9. Works Requirements

I. The design depth should meet the requirements for schematic design. The main technologies adopted should provide relevant technical drawings and indicators. The drawings and texts of the works are clearly presented with accurate data.

2）需提交方案设计说明，应包括方案构思、太阳能技术、低碳技术、韧性设计与设计创新（限 200 字以内），技术经济指标表和建筑碳排放指标表。

3）提交作品需进行竞赛项目的总平面图设计（含场地及环境设计）。

4）充分表达建筑与室内外环境关系的建筑典型平面图、立面图、剖面图，比例不小于 1 ：300。

5）能表现出技术与建筑结合的重点部位、局部详图，比例自定，相关的技术图、碳排放数据等分析图。

6）绘制场地、建筑、局部等设计效果的表现图。

7）区域及建筑韧性设计分析图。

8）提交作品需进行建筑的运行能耗及场地可再生能源产量模拟及计算，并能够实现建筑能源自维持。

10. 文字要求

1）"建筑方案设计说明"采用中英双语，其他为英文（建议使用附件 3 中提供的专业术语）。

2）排版要求：A1 展版（594mm×841mm）区域内，统一采用竖向构图，作品张数应为 2~4 张。

3）中文字体大小于 6mm，英文字体不小于 4mm。

4）文件分辨率 300dpi，格式为 JPG 或 PDF 文件。

5）提交参赛者信息表，格式为 JPG 或 PDF 文件。

6）上传方式：参赛者通过竞赛网页上传功能能将作品提交竞赛组委会，入围作品由组委会统一编辑板眉、出图、制作展板。

II. Along with the submitted files, the Schematic Design Description should include design concepts or ideas, solar energy technology, low-carbon technology, resilience consideration and innovation (with a word limit of 200). Additionally, a technical and economic index table and a carbon emission index table are required.

III. The submitted works should include a general plan which encompasses site design and environmental design.

IV. Typical building plans, elevations and sections that fully express the relationship between the architecture and the indoor/outdoor environment must be included at a scale of no less than 1 ： 300.

V. Detailed partial drawings for key positions and parts that demonstrate the integration of architecture and technologies should be included in the submission (no scale restriction). Relevant technical drawings, carbon emission data and other analysis charts are required.

VI. Renderings depict design effects, including sites, buildings, and partials.

VII. Analysis diagrams of regional and building resilience should be included.

VIII. The submitted works should simulate and calculate energy consumption and renewable energy output of buildings, and achieve self-sufficiency in energy consumption.

10. Texts Requirements

I. The Schematic Design Description should be prepared in both Chinese and English, while all other submitted documents in English only (please refer to Appendix 3：Technical Terms).

II. Layout Requirements: A1 format of exhibition panel (594mm×841mm), vertical composition for image layout, and 2~4 pictures of the works.

III. Font size：6mm of maximum size for Chinese characters, and 4mm for that of the English wording.

IV. Resolution ratio：300 dpi, in JPG or PDF format.

V. The Applicant Information Table should be submitted in JPG or PDF format.

VI. Uploading approach: all the entries should be uploaded to the Organizing Committee via the official website of the Competition. The Organizing Committee will compile headers, print, and produce the shortlisted works for display boards.

附件2：保定市清苑区李八庄村建设项目

Appendix 2：Construction Project of Libazhuang Village, Qingyuan District, Baoding City

1. 项目背景

项目地位于河北省保定市清苑县北店乡李八庄村中心，东经115.49°，北纬38.71°，距保定市中心20km，距雄安新区66km，距京港澳高速清苑出口7km，紧邻保衡路，依清水河北岸。

李八庄村域面积575hm²，以平原为主，其中村庄面积约279.67hm²，耕地面积约295.33hm²，典型农业村庄，主要作物为玉米、小麦。全村共1099户，总人口3500人。村民以务农、务工为主，吊车、铲车等建筑机械出租业务较多。

由于受村落规模较大且建筑用地集中、场地平坦、距离河道较近等因素影响，近几十年汛期多次出现农田被淹和村子内涝问题。2023年持续特大暴雨造成村内严重内涝，河滩内农作物基本绝收。村内低洼地区积水，多户村民房屋进水。房屋倒塌2间，灌溉水泵被淹28个，受灾农作物玉米（约133.33hm²）、辣椒（约6hm²）、香菜、花生、豆角、大葱等不同程度减产。最严重时村内坑塘水满，坑塘附近道路被完全淹没，道路积水500mm。各类受灾直接经济损失约590万元。

2. 项目要求

本项目旨在常态下提升当地居民居住品质的同时，让乡村在特殊风险情境下有抵抗风险能力，在断水断电等极端情况下有自维持能力。对李八庄村中心防内涝坑塘周围进行防洪涝设计，并在坑塘旁重建民居、增建公共服务建筑。公共服务建筑平时服务村民，提供文化娱乐、民生服务，受灾断水断电时可作为应急避难所。

3. 气候条件

李八庄村属暖温带季风型大陆性气候，季风气候显著，四季分明。全年平均水平面总辐照量1339.6kW·h/m²，最大值出现在5月，最小值出现在12月，年日照时长2500~2900h。最冷的1月平均气温-3℃，最热的7月平均气温27℃；极端最高气温43.3℃，极端最低气温-22℃。年均降雨量500mm左右，主要集中在7~8月，占年总降水量的60%左右，最大冻土深度59cm。

1. Project Background

The Project is located in the center of Libazhuang Village, Beidian Township, Qingyuan County, Baoding City, Hebei Province of P. R. China, at a longitude of 115.49° East and a latitude of 38.71° North. 20 kilometers away from Baoding City center, 66 kilometers away from Xiong'an New Area, 7 kilometers away from Qingyuan Exit of Beijing-Hong Kong-Macao Expressway, it is close to Baoheng Road, along the north bank of Qingshui River.

The total area of Libazhuang village is 575hm², mainly flatland. Among them, the village area covers 279.67hm² and the cultivated land area spans 295.33hm². The main crops grown in the village are corns and wheat. The total number of households in the village is 1,099 with a population of 3,500 people. Villagers are primarily engaged in farming and working activities. However, there are also many rental businesses for construction machinery such as cranes and forklifts.

Affected by the large span of lands, the concentration of buildings, flat topography and proximity to the river, farmland flooding and village waterlogging have occurred frequently during the flood season in recent decades. In 2023, continuous heavy rain caused severe waterlogging in the village, and the crops in the riverbank basically failed to harvest. Water accumulated in low-lying areas of the village, leading to flooding in many villagers' houses. Two houses collapsed and 28 irrigation pumps were submerged. Crops such as corns (more than 133.33hm²), chili peppers (more than 6hm²), cilantro, peanuts, beans, and scallions were reduced to varying degrees. In the most severe case, the village pit was completely filled with water and nearby roads were flooded with depths reaching up to 500mm. The total direct economic loss amounted to approximately 5.9 million yuan.

2. Project Requirements

The purpose of this project is to improve the living quality of local residents under normal conditions, while having the ability to resist risks under special circumstances and sustain itself in extreme situations such as water and power supply damages. The design should include flood and waterlogging control around the pit in the center of Libazhuang Village, and the folk houses should be rebuilt together with public service buildings beside the pit. The public service buildings should serve villagers in peacetime by providing cultural entertainment and livelihood services, and transition into emergency shelters during water or power disruptions caused by disasters.

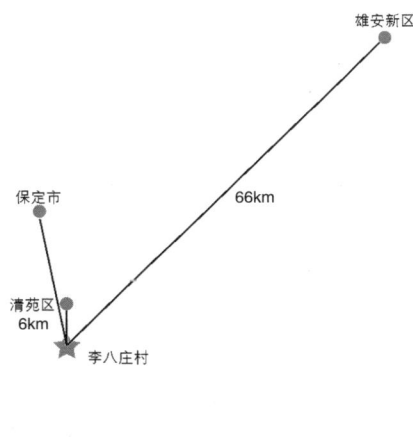

图 1 项目所在区位图
Figure 1　Geographic Location of the Project

3. Climate Conditions

Libazhuang village is located in a warm temperate monsoon continental zone characterized by significant monsoon features and four distinct seasons. The annual average total horizontal irradiation is 1, 339.6kW · h/m², with the maximum appears in May and the minimum in December. The annual sunshine period ranges from 2, 500 to 2, 900 hours. In the coldest month of January, the average temperature is −3℃ , while in the hottest month of July it reaches 27℃ . Extreme maximum temperature recorded is 43.3℃ , and the minimum is −22℃ . The average annual rainfall amounts to about 500mm and mainly occurs during July to August, accounting for approximately 60% of the total annual precipitation. Additionally, the maximum depth of frozen earth measures at 59cm.

4. Infrastructure Facilities

The transportation around the project site is convenient. The entire village has transitioned from coal to gas as its primary energy source. Basic water supply and

表1

月份	空气温度 （℃）	相对湿度 （%）	水平面日太阳辐射 [kW·h（m²·d）]	大气压力 （kPa）	风速 （m/s）	土地温度 （℃）	月供暖度日数 （℃·d）	月供冷度日数 （℃·d）	日照时数 （h）
1 月	−2.5	50.5	2.02	99.8	1.6	−3.5	636	0	200.9
2 月	1.0	44.6	3.00	99.6	1.8	1.0	476	0	192.3
3 月	7.4	43.5	3.85	99.1	2.3	9.1	329	0	120.6
4 月	15.4	48.8	5.34	98.4	2.5	19.2	78	162	207.8
5 月	20.9	55.9	5.27	98.0	2.4	26.2	0	338	262.0
6 月	25.6	58.6	5.45	97.6	2.2	29.5	0	468	191.9
7 月	27.1	71.6	4.15	97.5	1.9	28.1	0	530	151.7
8 月	25.8	75.3	4.11	97.9	1.6	26.2	0	490	201.6
9 月	21.1	68	3.87	98.5	1.7	23.1	0	333	135.0
10 月	14.2	63.8	3.09	99.1	1.6	16.6	118	130	178.6
11 月	5.6	60.7	2.26	99.5	1.6	6.3	372	0	161.6
12 月	−0.7	56.0	1.59	99.8	1.5	−1.1	580	0	200.6
年平均 / 总数据	13.5	58.2	3.67	98.7	1.9	15.1	2588（总）	2451（总）	2204.6（总）

Table 1

Month	Air Tempe-rature (℃)	Relative Humidity (%)	Solar Radiation on Horizontal Day [kW·h (m²·d)]	Barometric Pressure (kPa)	Velocity (m/s)	Land Temperature (℃)	Monthly Heating Temperatures and Days (℃·d)	Monthly Cooling Temperatures and Days (℃·d)	Sunshine Duration (h)
January	−2.5	50.5	2.02	99.8	1.6	−3.5	636	0	200.9
February	1.0	44.6	3.00	99.6	1.8	1.0	476	0	192.3
March	7.4	43.5	3.85	99.1	2.3	9.1	329	0	120.6
April	15.4	48.8	5.34	98.4	2.5	19.2	78	162	207.8
May	20.9	55.9	5.27	98.0	2.4	26.2	0	338	262.0
June	25.6	58.6	5.45	97.6	2.2	29.5	0	468	191.9
July	27.1	71.6	4.15	97.5	1.9	28.1	0	530	151.7
August	25.8	75.3	4.11	97.9	1.6	26.2	0	490	201.6
September	21.1	68	3.87	98.5	1.7	23.1	0	333	135.0
October	14.2	63.8	3.09	99.1	1.6	16.6	118	130	178.6
November	5.6	60.7	2.26	99.5	1.6	6.3	372	0	161.6
December	−0.7	56.0	1.59	99.8	1.5	−1.1	580	0	200.6
Yearly Average / Total Date	13.5	58.2	3.67	98.7	1.9	15.1	2, 588 (Total)	2, 451 (Total)	2, 204.6 (Total)

图2　李八庄村 2023 年 7 月底大雨受灾情况
Figure 2　The Heavy Rain Damage of Libazhuang Village in July，2023

4. 基础设施

项目场址周边交通较为便利，整村进行了气代煤改造，有基本给水、电力设施，但没有下水道管网。坑塘可缓解雨季内涝问题，村子堤坝以外与清水河间河滩地，可缓冲汛期清水河水位上涨带来的灾害，平时是村民耕地。

5. 竞赛场地

项目用地面积 10200m²，建设用地面积 8840m²，地形较为平坦，水塘部分等高线详见地形图。场地位于李八庄村正中心，东侧为防内涝坑溏；西北侧为村委会，西南侧为村民民居。用地红线范围内现状为 14 户荒废院落及空闲杂院。

power facilities are readily available, but there is no sewer network. The pit alleviates waterlogging issues during rainy seasons. The mudflats between the village dike and the Qingshui River serve as both a buffer against river flooding disasters and cultivated land for villagers in peacetime.

5. Site Conditions

The Project covers a land area of 10, 200 square meters, with a construction area of 8, 840 square meters. The terrain is flat and the contours of the pit are detailed in the topographic map. Situated in the center of Libazhuang Village, the Project is located alongside a waterlogging-control pit on its east side, the Village Committee on the northwest side, and villagers′ residential areas on the southwest side. Within the land property line there are currently 14 abandoned courtyards and vacant yards.

6. 设计要求

1）基于现有基础设施，对抗内涝设计范围内道路、空地进行韧性设计，提出针对正常降雨情况的预防措施及应对洪涝灾害的紧急响应策略。考虑用地红线内交通组织及与坑溏关系处理。

2）在给定的竞赛用地范围内设计民居不少于14套，户型不少于2种（考虑两代、三代同居情况），总建筑面积约2000m²，限高9m。要求设置独立院落，每户不大于300m²，需考虑村民现有生活习惯并预留足够发展可能。

3）村民配套服务建筑，建筑面积约1200m²，限高9m，包括平时的文化娱乐、民生服务，受灾断水断电时，房间功能可调整为紧急卫生服务、村民应急生活等，建筑具备一定自维持能力。

4）规划与建筑单体设计要考虑韧性发展，适当运用海绵城市等技术，并考虑功能可变潜力。

5）从民居建设、运行及全过程考虑低碳减排的方法和实施。

6）考虑项目的可实施性，技术的经济性和普适性。

7. 评比办法

1）由组委会审查参赛资格，并确定入围作品。

2）由评委会评选出竞赛获奖作品。

图3 竞赛场地范围
Figure 3 Site Scope of the Project

图4 场地航拍图
Figure 4 Aerial Picture of the Site

6. Design Specifications

I. Resilience design is required for roads and open spaces within the design scope for waterlogging control based on the existing infrastructure. It should also propose preventive measures against normal rainfall as well as emergency response strategies for flood disasters. Additionally, consideration should be given to traffic organization within the property line and its relationship with the pit.

II. At least 14 sets of residential buildings must be designed within the construction area, with a minimum of 2 types of floor plans (considering the cohabitation of two or three generations). The total construction area is approximately 2,000 square meters with a height restriction of 9m. Independent courtyards shall be constructed, with a maximum size of 300 square meters per household. It is also necessary to consider the existing living habits of villagers and allow flexibility for future development.

III. The design of a supporting service building for villagers. The construction area is approximately 1,200 square meters with a height restriction of 9m. It shall facilitate daily cultural entertainment and livelihood services for the villager. In the event of water and power disruption during disasters, the building shall transition its function to support emergency sanitation and basic living services. Furthermore, it should be designed with self-sustaining capabilities.

IV. Planning and architectural design should present the concept of resilience development with appropriate application of technical strategies like sponge city and potential for functional variability.

V. The design should incorporate possible methods and implementations to achieve low-carbon emission and carbon reduction throughout the entire life cycle of residential buildings, from construction to operation.

VI. The implementation feasibility of the Project, the cost-efficiency and general application of the adopted technologies should be taken into consideration.

7. Appraisal Approach

I. The Organizing Committee will review the entry qualification and determine the shortlisted works.

II. The Jury Panel is in charge of determining the winning works.

8. Appraisal Indicators

I. Entries must adhere to the requirements listed in the "Design Specifications" section.

图5 场地三维扫描模型
Figure 5 3D Scanning Model of the Site

图 6 抗内涝设计范围
Figure 6   Design Scope for Waterlogging Control

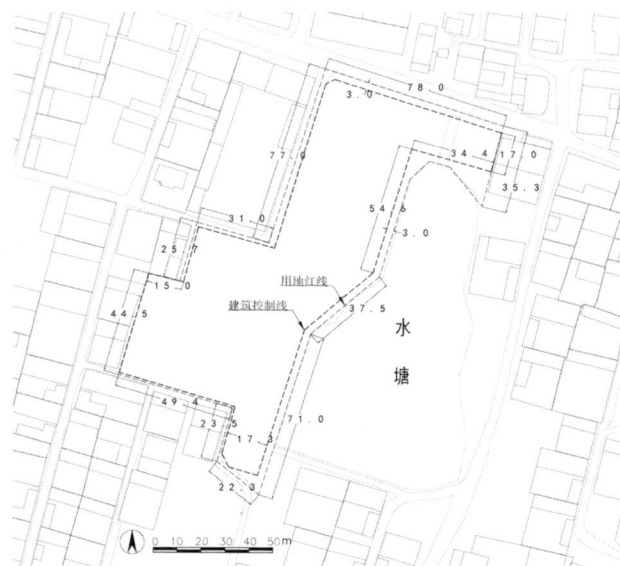

图 7   项目用地平面图
Figure 7   Project Site Plan

8. 评比标准

1）参赛作品须符合本竞赛"作品要求"的内容。

2）作品应具有原创性，鼓励创新。

3）作品应满足使用功能、绿色低碳、安全健康的要求，建筑技术与太阳能利用技术具有适配性。

4）作品应充分体现太阳能利用技术对降低建筑使用能耗的作用，在经济、技术层面具有可实施性。

5）作品应充分考虑区域及建筑本体韧性设计，考虑极端灾害响应，缓解区域社会、经济、环境压力。

表2

评比指标	指标说明	分值
规划与建筑设计	规划布局、建筑空间组合、功能流线组织、空间灵活性、建筑艺术	40
太阳能主、被动技术	利用建筑设计与建筑构造实现建筑隔热与通风节能降碳	25
	太阳能光伏、光热等主动太阳能技术的利用实现建筑能源自维持	
区域及建筑韧性设计	场地应急设计、区域及建筑防内涝设计	15
采用的其他技术	建造与运行过程中的绿色、低碳、节能等技术	10
可操作性	作品的可实施性，技术的经济性和普适性	10

II. Entries should be original, and innovative designs are encouraged.

III. Entries should meet the requirements for functionality, an eco-friendly orientation, low-carbon emissions, safety and health. The solar energy technologies adopted should be compatible with construction techniques.

IV. Entries should emphasize the role of solar technology in reducing energy consumption of buildings, while being economically and technically feasible.

V. Entries should incorporate resilience design into regional planning and architectural design, taking into consideration extreme disaster response to mitigate pressures on local society, economy, and environment.

Table 2

Appraisal Indicators	Explanations on the Indicators	Scores
Planning and Architectural Design	Layout, architectural space combination, functions and circulation organization, spatial flexibility, architectural artistic quality	40
Active and Passive Solar Technologies	To achieve heat insulation, ventilation, energy-saving, and carbon reduction through architectural design and building structure	25
	To achieve energy self-sufficiency through the implementation of active solar energy technologies such as solar photovoltaic and solar thermal	
Regional and Building Resilience Design	Site design for emergency response, regional and building waterlogging prevention	15
Other Technologies Applicable	Green, low-carbon and energy-saving technologies incorporated in the entire process of architectural construction and operation	10
Practicability	The implementation possibility of the works, the cost-efficiency and general applicability of the technologies	10

9. 作品要求

1）设计深度达到方案设计深度要求，主要技术应有相关的技术图纸和指标。作品图面、文字表达清楚，数据准确。

2）需提交方案设计说明，应包括方案构思、太阳能技术、低碳技术、韧性设计与设计创新（限200字以内），技术经济指标表和建筑碳排放指标表。

3）提交作品需进行竞赛项目的总平面图设计（含场地及环境设计）。

4）充分表达建筑与室内外环境关系的建筑典型平面图、立面图、剖面图，比例不小于1∶300。

5）能表现出技术与建筑结合的重点部位、局部详图，比例自定，相关的技术图、碳排放数据等分析图。

6）绘制场地、建筑、局部等设计效果的表现图。

7）区域及建筑韧性设计分析图。

8）提交作品需进行建筑的运行能耗及场地可再生能源产量模拟及计算，并能够实现建筑能源自维持。

10. 文字要求

1）"建筑方案设计说明"采用中英双语，其他为英文（建议使用附件3中提供的专业术语）。

2）排版要求：A1展版（594mm×841mm）区域内，统一采用竖向构图，作品张数应为2~4张。

3）中文字体大小于6mm，英文字体不小于4mm。

4）文件分辨率300dpi，格式为JPG或PDF文件。

5）提交参赛者信息表，格式为JPG或PDF文件。

6）上传方式：参赛者通过竞赛网页上传功能将作品递交竞赛组委会，入围作品由组委会统一编辑板眉、出图、制作展板。

9. Works Requirements

I. The design depth should meet the requirements for schematic design. The main technologies adopted should provide relevant technical drawings and indicators. The drawings and texts of the works are clearly presented with accurate data.

II. Along with the submitted files, the Schematic Design Description should include design concepts or ideas, solar energy technology, low-carbon technology, resilience consideration and innovation (with a word limit of 200). Additionally, a technical and economic index table and a carbon emission index table are required.

III. The submitted works should include a general plan which encompasses site design and environmental design.

IV. Typical building plans, elevations and sections that fully express the relationship between the architecture and the indoor/outdoor environment must be included at a scale of no less than 1∶300.

V. Detailed partial drawings for key positions and parts that demonstrate the integration of architecture and technologies should be included in the submission (no scale restriction). Relevant technical drawings, carbon emission data and other analysis charts are required.

VI. Renderings depict design effects, including sites, buildings, and partials.

VII. Analysis diagrams of regional and building resilience should be included.

VIII. The submitted works should simulate and calculate energy consumption and renewable energy output of buildings, and achieve self-sufficiency in energy consumption.

10. Texts Requirements

I. The Schematic Design Description should be prepared in both Chinese and English, while all other submitted documents in English only (please refer to Appendix 3: Technical Terms).

II. Layout Requirements: A1 format of display board (594mm×841mm), vertical composition for image layout, and 2~4 pictures of the works.

III. Font size: 6mm of maximum size for Chinese characters, and 4mm for that of the English wording.

IV. Resolution ratio: 300 dpi, in JPG or PDF format.

V. The Applicant Information Table should be submitted in JPG or PDF format.

VI. Uploading approach: all the entries should be uploaded to the Organizing Committee via the official website of the Competition. The Organizing Committee will compile headers, print, and produce the shortlisted works for display boards.

附件3：专业术语
Appendix 3：Technical Terms

百叶通风	— shutter ventilation	光伏发电系统	— photovoltaic system
保温	— thermal insulation	光伏幕墙	— PV facade
被动太阳能利用	— passive solar energy utilization	回流系统	— drainback system
敞开系统	— open system	回收年限	— payback time
除湿系统	— dehumidification system	集热器瞬时效率	— instantaneous collector efficiency
储热器	— thermal storage	集热器阵列	— collector array
储水量	— water storage capacity	集中供暖	— central heating
穿堂风	— through-draught	间接系统	— indirect system
窗墙面积比	— area ratio of window to wall	建筑节能率	— building energy saving rate
次入口	— secondary entrance	建筑密度	— building density
导热系数	— thermal conductivity	建筑面积	— building area
低能耗	— lower energy consumption	建筑物耗热量指标	— index of building heat loss
低温热水地板辐射供暖	— low temperature hot water floor radiant heating	节能措施	— energy saving method
地板辐射供暖	— floor panel heating	节能量	— quantity of energy saving
地面层	— ground layer	紧凑式太阳热水器	— close-coupled solar water heater
额定工作压力	— nominal working pressure	经济分析	— economic analysis
防潮层	— wetproof layer	卷帘外遮阳系统	— roller shutter sun shading system
防冻	— freeze protection	空气集热器	— air collector
防水层	— waterproof layer	空气质量检测	— air quality test（AQT）
分户热计量	— household-based heat metering	立体绿化	— tridimensional virescence
分离式系统	— remote storage system	绿地率	— greening rate
风速分布	— wind speed distribution	毛细管辐射	— capillary radiation
封闭系统	— closed system	木工修理室	— repairing room for woodworker
辅助热源	— auxiliary thermal source	耐用指标	— permanent index
辅助入口	— accessory entrance	能量储存和回收系统	— energy storage & heat recovery system
隔热层	— heat insulating layer	平屋面	— plane roof
隔热窗户	— heat insulation window	坡屋面	— sloping roof
跟踪集热器	— tracking collector	强制循环系统	— forced circulation system

热泵供暖	— heat pump heat supply	填充层	— fill up layer
热量计量装置	— heat metering device	通风模拟	— ventilation simulation
热稳定性	— thermal stability	外窗隔热系统	— external windows insulation system
热效率曲线	— thermal efficiency curve	温差控制器	— differential temperature controller
热压	— thermal pressure	屋顶植被	— roof planting
人工湿地效应	— artificial marsh effect	屋面隔热系统	— roof insulation system
日照标准	— insolation standard	相变材料	— phase change material（PCM）
容积率	— floor area ratio	相变太阳能系统	— phase change solar system
三联供	— triple co-generation	相变蓄热	— phase change thermal storage
设计使用年限	— design working life	蓄热特性	— thermal storage characteristic
使用面积	— usable area	雨水收集	— rain water collection
室内舒适度	— indoor comfort level	运动场地	— schoolyard
双层幕墙	— double facade building	遮阳系数	— sunshading coefficient
太阳方位角	— solar azimuth	直接系统	— direct system
太阳房	— solar house	值班室	— duty room
太阳辐射热	— solar radiant heat	智能建筑控制系统	— building intelligent control system
太阳辐射热吸收系数	— absorptance for solar radiation	中庭采光	— atrium lighting
太阳高度角	— solar altitude	主入口	— main entrance
太阳能保证率	— solar fraction	贮热水箱	— heat storage tank
太阳能带辅助热源系统	— solar plus supplementary system	准备室	— preparation room
太阳能电池	— solar cell	准稳态	— quasi-steady state
太阳能集热器	— solar collector	自然通风	— natural ventilation
太阳能驱动吸附式制冷	— solar driven desiccant evaporative cooling	自然循环系统	— natural circulation system
太阳能驱动吸收式制冷	— solar driven absorption cooling	自行车棚	— bike parking
太阳能热水器	— solar water heating		
太阳能烟囱	— solar chimney		
太阳能预热系统	— solar preheat system		
太阳墙	— solar wall